Electricity

$$F = \frac{kq_1q_2}{r^2}$$

$$R = \frac{\rho L}{A}$$

$$I = \frac{V}{R}$$

Series Circuits

(a) $I = I_1 = I_2 = I_3 = \cdots$

(b) $R = R_1 + R_2 + R_3 + \cdots$

(c) $E = V_1 + V_2 + V_3 + \cdots$

Parallel Circuits

(a) $I = I_1 + I_2 + I_3 + \cdots$

(b) $\frac{1}{R} = \frac{1}{R_1} + \frac{1}{R_2} + \frac{1}{R_3} + \cdots$

(c) $E = V_1 = V_2 = V_3 = \cdots$

Cells in Series

(a) $I = I_1 = I_2 = I_3 = \cdots$

(b) $r = r_1 + r_2 + r_3 + \cdots$

(c) $E = E_1 + E_2 + E_3 + \cdots$

Cells in Parallel

(a) $I = I_1 + I_2 + I_3 + \cdots$

(b) $r = \dfrac{r \text{ of one cell}}{\text{number of like cells}}$

(c) $E = E_1 = E_2 = E_3 = \cdots$

$V = E - Ir$

$P = VI = I^2R = \dfrac{V^2}{R}$

Magnetism

$$B = \frac{\mu_0 I}{2\pi R}$$

$$B = \pi_0 In$$

Transformers

$$\frac{V_P}{V_S} = \frac{N_P}{N_S}$$

$$\frac{I_S}{I_P} = \frac{N_P}{N_S}$$

ac Circuits

$$X_L = 2\pi f L$$

$$I = \frac{E}{X_L}$$

$$I = \frac{E}{Z}$$

$$Z = \sqrt{R^2 + X_L^2}$$

$$\tan \phi = \frac{X_L}{R}$$

$$X_C = \frac{1}{2\pi f C}$$

$$Z = \sqrt{R^2 + X_C^2}$$

$$\tan \phi = \frac{X_C}{R}$$

$$Z = \sqrt{R^2 + (X_L - X_C)^2}$$

$$\tan \phi = \frac{X_L - X_C}{R}$$

$$f = \frac{1}{2\pi\sqrt{LC}}$$

Light

$$c = \lambda f$$

$$E = hf$$

$$E = \frac{I}{4\pi r^2}$$

$$\frac{1}{f} = \frac{1}{s_o} + \frac{1}{s_i}$$

$$M = \frac{h_i}{h_o} = \frac{-s_i}{s_o}$$

$$n = \frac{\sin i}{\sin r} = \frac{\text{speed of light in vacuum}}{\text{speed of light in substance}}$$

$$\sin i_c = \frac{1}{n}$$

Modern Physics

$$E = -\frac{kZ^2}{n^2}$$

$$E = \Delta mc^2$$

$$Q = (M_p - M_d - m_\alpha)c^2$$

$$N = N_0 e^{-\lambda t}$$

$$T_{1/2} = \frac{0.693}{\lambda}$$

$$A = \lambda N = \lambda N_0 e^{-\lambda t} = A_0 e^{-\lambda t}$$

APPLIED PHYSICS

ELEVENTH EDITION

DALE EWEN

Parkland Community College
Champaign, Illinois

NEILL SCHURTER

P. ERIK GUNDERSEN

Pascack Valley Regional High School District
Montvale, New Jersey

PEARSON

Boston   Columbus   Indianapolis   New York   San Francisco   Amsterdam
Cape Town   Dubai   London   Madrid   Milan   Munich   Paris   Montréal   Toronto
Delhi   Mexico City   São Paulo   Sydney   Hong Kong   Seoul   Singapore   Taipei   Tokyo

Editor-in-Chief: Andrew Gilfillan
Product Manager: Anthony Webster
Program Manager: Holly Shufeldt
Project Manager: Rex Davidson
Editorial Assistant: Nancy Kesterson
Team Lead Project Manager: Bryan Pirrmann
*Team Lead Program Manage*r: Laura Weaver
Director of Marketing: David Gesell
Senior Product Marketing Manager: Darcy Betts

Field Marketing Manager: Thomas Hayward
Procurement Specialist: Deidra M. Skahill
Creative Director: Andrea Nix
Art Director: Diane Y. Ernsberger
Cover Designer: Cenveo
Full-Service Project Management: Sivakumar Krishnamoorthy/Integra Software Services, Pvt, Ltd.
Printer/Binder: LSC Communications

Library of Congress Cataloging-in-Publication Data

Ewen, Dale
   Applied physics.—Eleventh edition/Dale Ewen, Parkland Community College Champaign, Illinois, Neill Schurter, P. Erik Gundersen, Pascack Valley Regional High School District Montvale, New Jersey.
      pages cm
   Includes index.
   ISBN 978-0-13-415938-6 (alk. paper)—ISBN 0-13-415938-1 (alk. paper)   1. Physics—Textbooks.   I. Schurter, Neill.   II. Gundersen, P. Erik.   III. Title.
   QC23.2.E88 2016
    530—dc23
                                                            2015021209

ISBN 10:   0-13-415938-1
ISBN 13: 978-0-13-415938-6

CONTENTS

New to This Edition

- All art has been reviewed, evaluated, and redrawn in four-color as appropriate.
- The new feature "Physics on the Job" highlights real jobs where people use physics in their career. Understanding the connection between physics and various careers provides additional relevance for the physics content highlighted in the respective chapter.
- The new feature "New Technologies" provides up-to-date examples of how physics concepts in the chapter are being utilized in cutting-edge research and products.
- A major effort was made on color photo selection to enhance student learning.
- Some examples and problems were added or changed based on recommendations by reviewers.
- Section A.6 Law of Sines and Cosines has been moved back into the text from online for those students who have completed a trigonometry course and need a review.

Applied Physics, 11th edition, provides comprehensive and practical coverage of physics for students needing an applied physics approach or considering a vocational–technical career. It emphasizes physical concepts as applied to industrial–technical fields and uses common applications to improve the physics and mathematics competence of the student. This edition has been carefully reviewed, and special efforts have been taken to emphasize the clarity and accuracy of presentation.

This text is divided into five major areas: mechanics, matter and heat, wave motion and sound, electricity and magnetism, and light and modern physics.

Key Features

- Real-world applications are used to motivate students.
- Topic coverage is clear and to the point.
- A unique problem-solving format is consistently used throughout the text. This textbook teaches students to use a proven effective problem-solving methodology. The consistent use of this method trains students to make a sketch, identify the data elements, select the appropriate equation, solve for the unknown quantity, and substitute the data in the working equation. An icon that outlines the method is placed in the margin of most problem sets as a reminder to students. See Section 2.3 for the detailed presentation of the problem-solving method.
- Detailed, well-illustrated examples in the problem-solving format support student understanding of skills and concepts. Worked examples are consistently displayed in the problem-solving format and used to illustrate and clarify basic concepts and problems. Since many students learn by example, a large number of examples are provided.
- Problems and questions assist student learning, with extensive problem sets at the end of most sections that provide students with ample opportunity for practice.
- A four-color format with numerous drawings, diagrams, and photographs is used to illustrate the application of physics in the real world and improve student interest and comprehension.
- Try This Activity features provide students with opportunities to experiment with physics concepts. Activities involve a demonstration or mini-activity that can be performed by students on their own to experience a physics concept, allowing for more active versus passive learning.

Examples of Key Features

Chapter Openers

The chapter openers contain the learning objectives that list the major goals of the chapter.

Physics Connections

These features apply physics to familiar real-world situations and events. These brief readings help students bridge the gap between what is taught in the chapter and real-world technical applications.

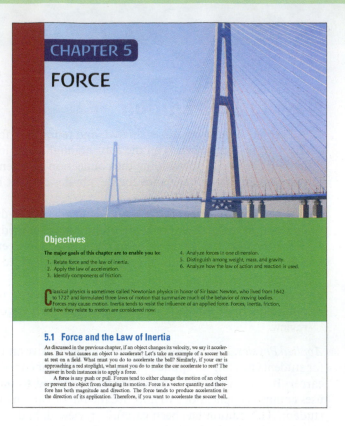

CHAPTER 5

FORCE

Objectives

The major goals of this chapter are to enable you to:

1. Relate force and the law of inertia.
2. Apply the law of acceleration.
3. Identify components of friction.
4. Analyze forces in one dimension.
5. Distinguish among weight, mass, and gravity.
6. Analyze how the law of action and reaction is used.

Classical physics is sometimes called Newtonian physics in honor of Sir Isaac Newton, who lived from 1642 to 1727 and formulated three laws of motion that summarize much of the behavior of moving bodies. Forces may cause motion. Inertia tends to resist the influence of an applied force. Forces, inertia, friction, and how they relate to motion are considered now.

5.1 Force and the Law of Inertia

As discussed in the previous chapter, if an object changes its velocity, we say it accelerates. But what causes an object to accelerate? Let's take an example of a soccer ball at rest on a field. What must you do to accelerate the ball? Similarly, if your car is approaching a red stoplight, what must you do to make the car accelerate to rest? The answer in both instances is to apply a force.

A **force** is any push or pull. Forces tend to either change the motion of an object or prevent the object from changing its motion. Force is a vector quantity and therefore has both magnitude and direction. The force tends to produce acceleration in the direction of its application. Therefore, if you want to accelerate the soccer ball,

---

PHYSICS CONNECTIONS

Global Positioning Satellites

Navigators continually struggle to find better tools to help them determine their location. The first explorers used the sun and stars to help them steer a straight course, but this method of navigation only worked under clear skies. Magnetic compasses were developed yet could only be used to determine longitude, not latitude. Finally, the mechanical clock, in conjunction with the compass, provided navigators with the most accurate method of determining location. Today, most navigators use a handheld device that functions in concert with a series of 24 orbiting satellites. This network, the Global Positioning System (GPS), can determine your position and altitude anywhere on earth.

The GPS pinpoints your location by sending out radio signals to locate any 4 of the 24 orbiting GPS satellites. Once the satellites are found, the GPS measures the length of time it takes for a radio signal to reach the handheld receiver. When the time is determined for each of four satellites, the distance is calculated, and the longitude, latitude, and altitude are displayed on the screen [Fig. 3.39(a)].

The GPS was first developed solely for military use. Eventually, the GPS was made available for civilian businesses. Shipping, airline, farming, surveying, and geological companies made use of the technology. Today, GPS receivers are affordable and are used by the general public [Fig. 3.39(b)]. More sophisticated receivers not only locate a position, but can also guide the navigator to a predetermined location. Several automobile manufacturers have included GPS receivers as an option in their cars. Such receivers come complete with voice commands, such as "Turn left at the next traffic light," as part of their option packages.

Try This Activity

These activities provide students with opportunities to experiment with physics concepts. Activities involve a demonstration or mini-activity that can be performed by students on their own to experience a physics concept, allowing for active versus passive learning.

TRY THIS ACTIVITY

Air Hockey Physics

Physics laboratory experiments often use air tracks or low-friction carts to study the conservation of momentum throughout collisions. Air hockey tables are typically more accessible for students and provide the low-friction surface that is needed to study the conservation of momentum. If you have access to an air hockey table, try the following collisions:

For elastic collisions, make sure that the pucks strike one another "head on" by aiming one puck directly toward the center of a stationary puck (Fig. 6.6). Make observations about the velocity of both pucks before and after the collision.

To observe an elastic collision for objects with different masses, place two pucks on top of one another with some double-sided tape between them so they remain secure to one another. Repeat the previous experiment by aiming a single puck at the doubled pucks and note the changes in velocity for both sets of pucks before and after the collision.

For inelastic collisions, wrap some double-sided tape around the sides of each of two pucks so they can stick together when they collide. Again, note the velocity of both pucks before and after the collision.

Figure 6.6 For collisions in one dimension, aim the puck directly at the center of the stationary puck, as indicated by the blue velocity vector in the photo.

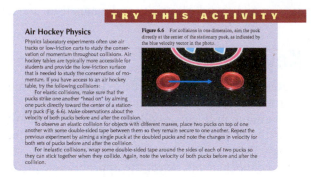

NEW TECHNOLOGIES

Tesla High-Performance Electric Automobile

Tesla is a unique automobile as it is considered to be the first-of-its-kind, fully electric and high-performance electric vehicle. Prior to the Tesla, electric cars were hybrid vehicles that relied on a combination of electric power supplemented by gasoline. Until now, pure electric cars were not considered high-performance model vehicles.

The brakes in a Tesla not only slow down the vehicle, but also utilize the energy transfer between brakes and brake pads to create additional energy for the vehicle. This regenerative braking system shifts the kinetic energy typically lost in braking and instead transfers it back to electrical energy for the automobile.

The Tesla Model S has designed a massive 1,000-lb bank of high-density, rechargeable lithium-ion batteries, similar to those used in laptops or other mobile devices. Tesla suggests charging the vehicle overnight in a home garage by plugging the vehicle into a standard 110 V or 220 V circuit. Charging speeds vary between 29 mi of driving range per charging hour at 110 V to 58 mi of driving range per charging hour using a 220 V circuit. Most impressively, Tesla supercharge stations, located at Tesla dealerships and other locations around the nation, can charge up to 340 mi of range per charging hour and can replenish half the battery in 30 minutes (Fig. 17.77)!

Figure 17.77 (a) Tesla Model S electrical vehicle

Figure 17.77 (b) Tesla charging station

New Technologies

Provides up-to-date examples of how physics concepts in the chapter are being utilized in cutting-edge research and products.

Physics on the Job

Highlights real jobs where people use physics in their career.

Unique Problem-Solving Method

This textbook teaches students to use a proven effective problem-solving methodology. The consistent use of this method trains students to make a sketch, identify the data elements, select the appropriate equation, solve for the unknown quantity, and substitute the data in the working equation. An icon that outlines the method is placed in the margin of most problem sets as a reminder to students. See Section 2.3 for the detailed presentation of the problem-solving method.

Applied Concepts

Application-based questions at the end of each chapter develop problem-solving skills in real-life physics applications.

(*Continued from page ix.*)

- Physics Connections features apply physics to familiar real-world situations and events. These brief readings help students bridge the gap between what is taught in the chapter and real-world technical applications.
- Applied Concepts features provide application-based questions at the end of chapters that develop problem-solving skills in real-life physics applications.
- There is comprehensive discussion and consistent use of the results of working with measurements and significant digits.
- Biographical sketches of important scientists appear in most chapters.
- Answers to odd-numbered problems within the chapters and all chapter review questions and problems are given in Appendix F.
- A comprehensive glossary is given as a one-stop reference in Appendix E.
- Basic scientific calculator instructions are presented in Appendix D.
- A basic math review provides students with a refresher of the mathematics needed for this course in Appendix A.

Ancillaries

- Online Instructor's Manual
- Online PowerPoints
- Online Test Generator

Download Instructor Resources from the Instructor Resource Center

To access supplementary materials online, instructors need to request an instructor access code. Go to **www.pearsonhighered.com/irc** to register for an instructor access code. Within 48 hours of registering, you will receive a confirming e-mail, including an instructor access code. Once you have received your code, locate our text in the online catalog and click on the Instructor Resources button on the left side of the catalog product page. Select a supplement, and a login page will appear. Once you have logged in, you can access instructor material for all Pearson Education textbooks. If you have any difficulties accessing the site or downloading a supplement, please contact Customer Service at **247pearsoned.custhelp.com**.

To the Faculty

This text is written at a language level and at a mathematics level that is cognizant of and beneficial to *most* students in programs that do not require a high level of mathematics. The authors have assumed that the student has successfully completed one year of high school algebra or its equivalent. Simple equations and formulas are reviewed and any mathematics beyond this level is developed in the text or in an appendix. For example, right-triangle trigonometry is developed in Appendix A.5 for those who have not studied it previously or who need a review. The manner in which the mathematics is used in the text displays the need for mathematics in technology. For the better-prepared student, the mathematics sections may be omitted with no loss in continuity. This text is designed so that faculty have flexibility in selecting the topics, as well as the order of topics, that meet the needs of their students and programs of study.

Sections are short, and each deals with only one concept. The need for the investigation of a physical principle is developed before undertaking its study, and many diagrams are used to aid students in visualizing the concept. Many examples and problems are given to help students develop and check their mastery of one concept before moving to another.

This text is designed to be used in a vocational–technical program in a community college, a technical institute, or a high school for students who plan to pursue a technical career or in a general physics course where an applied physics approach is preferred. The topics were chosen with the assistance of technicians and management in several industries and faculty consultants. Suggestions from users and reviewers of the previous edition were used extensively in this edition.

A general introduction to physics is presented in Chapter 0. Chapter 1 introduces students to basic units of measurement. For students who lack a metric background or who need a review, an extensive discussion of the metric system is given in Chapter 1, where it is shown how the results of measurements are approximate numbers, which are then used consistently throughout the text. Those who need to review some mathematical skills are referred to the appendices as necessary. Chapter 2 introduces students to a problem-solving method that is consistently used in the rest of the text. Vectors are developed in Chapter 3, followed by a comprehensive study of motion, force, work and energy, rotational energy, simple machines, and universal gravitation and satellite motion.

The treatment of matter includes a discussion of the three states of matter, density, fluids, pressure, and Pascal's principle. The treatment of heat includes temperature, specific heat, thermal expansion, change of state, and ideal gas laws.

The section on wave motion and sound deals with basic wave characteristics, the nature and speed of sound, the Doppler effect, and resonance.

The section on electricity and magnetism begins with a brief discussion of static electricity, followed by an extensive treatment of dc circuits and sources, Ohm's law, and series and parallel circuits. The chapter on magnetism, generators, and motors is largely descriptive, but it allows for a more in-depth study if desired. Then ac circuits and transformers are treated extensively.

The chapter on light briefly discusses the wave and particle nature of light, but deals primarily with illumination. The chapter on reflection and refraction treats the images formed by mirrors and lenses. A brief introduction to color includes diffraction, interference, and polarization of light.

The section on modern physics provides an introduction to the structure and properties of the atomic nucleus, radioactive decay, nuclear reactions, and radioactivity, followed by a very brief introduction to relativity.

An Online Instructor's Manual that includes Complete Solutions, Transparency Masters, and a Test Item File is available at no charge to instructors using this text.

To the Student: Why Study Physics?

Physics is useful. Architects, mechanics, builders, carpenters, electricians, plumbers, and engineers are only some of the people who use physics every day in their jobs or professions. In fact, every person uses physics principles every hour of every day. The movement of an arm can be described using principles of the lever. All building trades, as well as the entire electronics industry, also use physics.

Physics is often defined as the study of matter, energy, and their transformations. The physicist uses scientific methods to observe, measure, and predict physical events and behaviors. However, gathered data left in someone's notebook in a laboratory are of little use to society.

Physics provides a universal means of describing and communicating about physical phenomena in the language of mathematics. Mechanics is the base on which almost all other areas of physics are built. Motion, force, work, electricity, and light are topics confronted daily in industry and technology. The basic laws of conservation of energy are needed to understand heat, sound, wave motion, electricity, and electromagnetic radiation.

Physics is always changing as new frontiers are being established in the study of the nature of matter. The topics studied in this course, however, will probably not greatly change with new research and will remain a classical foundation for work in many, many fields. We begin our study with the rules of the road—measurement, followed by a systematic problem-solving method. The end result should be a firm base on which to build a career in almost any field.

ACKNOWLEDGMENTS

The authors thank the many faculty and students who have used the previous editions, and especially those who have offered suggestions. If anyone wishes to correspond with us regarding suggestions, criticisms, questions, or errors, please contact Dale Ewen through Pearson or through the web address **www.pearsonhighered.com**.

We thank the following reviewers:

Patrick Huth, Community College of Allegheny College; Michael Sampogna, Pima County Community College; Greg Tully, Morrison Institute of Technology; and Anthony Zable, Portland Community College.

We extend our sincere and special thanks to our Pearson colleagues: Holly Shufeldt, Program Manager; Rex Davidson, Project Manager; Tony Webster, Product Manager; and Andrew Gilfillan, Editor-in-Chief and to Sivakumar Krishnamoorthy of Integra Software Services.

Finally, we are especially grateful to our families for their support and encouragement.

Dale Ewen
Neill Schurter
P. Erik Gundersen

CHAPTER 0

AN INTRODUCTION TO PHYSICS

Objectives

The major goals of this chapter are to enable you to:

1. Determine what physics governs and controls.
2. Conclude that physics is a building block of all the sciences.
3. Identify areas in your life that will be impacted by studying physics.
4. Differentiate between laws and theories.
5. Provide reasons why problem-solving techniques are vital in the study of physics.

Physics plays an important role in all aspects of our lives. Before enrolling in a physics course, you may have taken physics for granted. In this chapter we will introduce physics to you and help you appreciate the impact that physics will have on your life and career.

0.1 Why Study Physics?

What do flying birds, automobiles, blue skies, and cellular phones have in common? They all involve physics. **Physics** is the branch of science that describes the motion and energy of all matter throughout the universe. Birds, for example, use the difference in air pressures above and below their wings to keep themselves aloft. Automobiles use the principles of mechanics and thermodynamics to transfer stored chemical energy in gasoline to moving energy in rotating tires. The sky appears blue when sunlight strikes and scatters off nitrogen and oxygen molecules in our atmosphere. Finally, cellular phones use electronic components and the principles of electromagnetic waves to transfer energy and information from one cellular phone to another (Fig. 0.1).

Figure 0.1 Physics is involved in all aspects of cellular phone technology. It controls everything from the electrical circuits in the phone to the transmission of radio waves between phones.

Physics is often considered to be the most fundamental of all the sciences. In order to study biology, chemistry, or any other natural science, one should have a firm understanding of the principles of physics. For example, **biology,** the branch of science that studies living organisms, uses the physics principles of fluid movement to understand how the blood flows through the heart, arteries, and veins. **Chemistry,** the branch of science that studies the composition, structure, properties, and reactions of matter, relies on the physics of subatomic particles to understand why chemical reactions take place. **Geology,** the branch of science that studies the origin, history, and structure of the earth, uses the physics of mechanical waves and energy transfer to determine the magnitude and location of earthquakes. Finally, **astronomy,** the field of science that studies everything that takes place outside the earth's atmosphere, relies on the laws of gravity and theory of relativity to describe the workings of the universe.

Students often wonder, "Why should I study physics? What is it going to do for me?" The answer is that physics plays an important role in everyday life and in the careers of many people. Choosing the right bat, golf club, or ski can be made easier with a bit of physics knowledge (Fig. 0.2). While on the job, architects, engineers, electricians, medical technicians, surveyors, and others use the principles of physics every day. When understood, physics can help us solve difficult physical problems and be better decision makers to determine the best design, tool, or process when working on a specific task.

A **physicist** is a person who is an expert in or who studies physics. It is a physicist's job to seek an understanding of how the physical universe behaves. Albert Einstein, perhaps one of the most famous physicists of all time, once said, "I am like a child, I always ask the simplest questions" (Fig. 0.3). Such **theoretical physicists** often spend their professional lives researching previous theories and mathematical models to form new theories in physics. **Experimental physicists,** however, focus on performing experiments to develop and confirm physical theories.

It is generally accepted that physics evolved from ancient Greek philosophers, including Plato (c. 428–347 BC) and Aristotle (384–322 BC). Aristotle believed that there were two types of motion: Natural motion occurred because objects wanted to seek their "natural" resting place (smoke rising or rocks falling), whereas violent motion occurred when objects were unnaturally pulled or dragged from place to place (person dragging a crate). Although he was not correct in his analysis, it was the beginning of observing and documenting physical phenomena. Plato, Aristotle, and others like them can be considered theoretical physicists.

It was not until the days of Archimedes (287–212 BC) that experiments were conducted to document and prove physical theories (Fig. 0.4). Since then, a vast number of physicists have built and improved upon the knowledge base developed by those before them. It is now your turn to use the physics that you will learn to help you understand, improve, and make advances in our technological world. You will see that physics has use!

TRY THIS ACTIVITY

Physics All Around Us

Look around and find something that may have to do with physics. Although you may not yet have studied many physics principles, you should know that physics governs things that move and transfer energy. Be as general as you need to be in your observations. The point is for you to see that physics plays a role in almost everything.

Figure 0.2 A baseball player's understanding of physics can help improve all aspects of his game, including pitching, batting, and fielding.

Courtesy of fotolia ©Paul Yates

Figure 0.3 Albert Einstein (1879–1955) is often considered one of the most influential scientists of the 20th century. His work on relativity, as made famous through the equation $E = mc^2$, and the photoelectric effect changed the way the world viewed physics.

Courtesy of National Archives and Records Administration

Figure 0.4 Archimedes is best known for observing that water was displaced when he stepped into his bath. He proceeded to conduct experiments where he measured the amount of water that overflowed when objects were placed into a tub full of water. He established a principle that states that an object immersed in a fluid will experience a buoyant force equal to the weight of the displaced fluid. Archimedes is also recognized for his work with simple machines like the screw, lever, and pulley. Legend has it that Archimedes said, "Give me a firm spot on which to stand and I will move the earth" (referring to the use of a lever).

Courtesy of fotolia ©Erica Guilane-Nachez

Figure 0.5 Thomas A. Edison

Courtesy of National Park Service, Edison National Historic Site

0.2   Physics and Its Role in Technology

Although often discussed as though they are the same thing, science and technology are quite different. **Science** is a system of knowledge that is concerned with establishing accurate conclusions about the behavior of everything in the universe. It is a field in which **hypotheses** (scientifically based predictions) are made, information is gathered, and experiments are performed to determine how something in our natural world works or behaves. **Technology,** on the other hand, is a field that uses scientific knowledge to develop material products or processes that satisfy human needs and desires. Technology and science rely closely on one another to make further advances in their respective fields.

Thomas A. Edison (1847–1931) used scientific information and the discoveries of other scientists to create over 1000 inventions (Fig. 0.5). Edison's development of the first practical lighting system was made possible by applying the science of electricity and the science of materials and then putting that knowledge to use to satisfy his technological need.

The following illustrate how science has played a role in improving technology.

Robotics: Due to advances in electronics, materials, and machines, robots commonly perform a variety of tasks from assembling cars on a production line to exploring the surface of Mars. NASA's Curiosity rover is used to travel remotely across the Martian surface collecting samples and relaying images and data back to scientists on earth (Fig. 0.6).

Bridges: Work in materials science and structural engineering has paved the way for advances in bridge design and construction. The New Clark Bridge in Alton, Illinois, is just one example of a cable-stayed bridge that has used

Figure 0.6 NASA's Curiosity Rover traveling toward the Mojave 2 region of the Martian surface in January 2015.

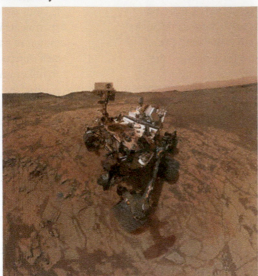

Courtesy of NASA

Figure 0.7 The New Clark Bridge, a cable-stayed bridge in Alton, Illinois.

Figure 0.8 A magnet levitates above a superconductor.

Courtesy of Brookhaven National Laboratory

scientific breakthroughs in materials science and physics to increase the structural integrity of the bridge and cut costs (Fig. 0.7).

Superconductors: Superconductors allow electric current to travel with virtually no resistance through materials. Materials such as aluminum, lead, and niobium are cooled by liquid helium to bring the temperature down to the low critical point. At the low critical point temperatures, the materials achieve zero electric resistance (Fig. 0.8). Scientific research is under way to develop superconducting materials that can operate closer to room temperature; this would bring about tremendous improvements in energy efficiency.

Active noise cancellation: Audiologists will tell you that noise increases stress levels. Acoustic and electrical engineers are now able to produce inverted noise patterns that cancel out disturbing noise (Fig. 0.9). Helicopter pilots, factory workers, and business travelers are using this technology to reduce stressful noise levels in their environment.

Liquid crystal displays: With advances in optics and electronics, physicists and chemists have created more advanced liquid crystal displays (LCDs), which are used as screens on laptop computers, cell phones, watches, and televisions (Fig. 0.10).

Magnetic levitation: The speed limitations of traditional trains have created a need for super-fast, magnetic levitation (maglev) trains (Fig. 0.11). Such a vehicle is levitated off a monorail by virtue of the magnetic repulsion between the train and the rail. Electromagnets are used to propel the train forward as it glides above the rail. Such improvements greatly reduce frictional resistance and allow trains to travel at twice the speed of conventional trains—up to 250 miles per hour.

Gyroscope: A gyroscope is a heavy wheel that uses rotational inertia to prevent tilting and is used to steady compasses, ships, airplanes, and rockets (Fig. 0.12). Advances in gyroscopes and electronic sensors have made it possible to create gyrostabilizers for ships. Such devices send signals to the ship's computer specifying how its fins should be positioned to prevent significant rolling motions.

Figure 0.9   Active noise cancellation technology for the consumer can be found in small, noise-reduction headsets.

Figure 0.10   Advances in LCD panels allow small electronic devices to have color screens.

Alternative fuels: The ever-increasing demand for energy, coupled with the earth's finite supply of fossil fuels, has led scientists and engineers to discover and develop additional sources of energy. Alternative sources include solar energy, wind energy (Fig. 0.13), hydropower, geothermal energy, hydrogen fuel cells, nuclear power, biofuel, and wave/tidal energy. Technological advances with regard to the efficiency of the various energy sources will lead to higher cost benefits and more widespread use of such alternative energy sources.

0.3   Physics and Its Connection to Other Fields and Sciences

Ancient Greeks like Plato and Aristotle did not specialize in physics. In fact, it was not until the 1800s that physics was considered a science. Prior to the 1800s, Plato, Aristotle, Copernicus, and Galileo were considered natural philosophers, not physicists. Today,

Figure 0.11   Maglev train technology has high-speed trains competing with airplane service.

Courtesy of U.S. Department of Transportation

Figure 0.12   Gyroscope

Courtesy of fotolia ©Gramper

Figure 0.13 Wind Turbines on a Texas countryside.

Courtesy of NASA

virtually every physicist specializes in a subdivision of physics. There is simply too much information to allow someone to study every type of physics.

The following is a listing of the 18 subdivisions of physics:

Mechanics: Study of forces, motion, and energy.
Thermodynamics: Study of heat energy transfer.
Cryogenics: Study of matter at extremely low temperatures.
Plasma physics: Study of electrically charged, ionized gas.
Solid state physics: Study of the physical properties of solid materials, also known as condensed matter physics.
Geophysics: Study of the interaction of forces and energy found within the earth; closely related to geology.
Astrophysics: Study of the interaction of forces and energy between interstellar objects; closely related to astronomy.
Acoustics: Study of the creation and transmission of sound under various conditions.
Optics: Study of the behavior of light in a variety of conditions.
Electromagnetism: Study of the relationship between electricity and magnetism.
Fluid dynamics: Study of how liquids and gases move from one location to another.
Mathematical physics: Study of the mathematics of physics and its related fields.
Statistical mechanics: Study of the development of statistical models that simulate the effects of systems composed of many particles.
High-energy physics: Study of new fundamental, subatomic particles using high-energy machines that send known subatomic particles colliding into one another; simulation of what the universe was like close to the time of the "big bang."
Atomic physics: Study of the structure of the atom based on the knowledge gained in the field of high-energy physics.
Molecular physics: Study of the structure of molecules based on the knowledge gained in atomic physics.
Nuclear physics: Study of nuclear interactions.
Quantum physics: Study of small particles and their energy.

0.4 Theories, Laws, and Problem Solving

Physics is constantly being refined. Although the major principles of physics do not change drastically over time, newer theories requiring a tremendous amount of experimentation can modify our understanding of physics. A **theory** is a scientific conclusion

PHYSICS CONNECTIONS

Physics, Technology, and Sports

Physics plays a major role in sports. From the padding in a baseball glove to the stance of a wrestler, a good working knowledge of physics helps athletes and sports equipment companies achieve greater successes. Ski companies employ engineers who focus solely on the physics and engineering of improving a skier's time down the mountain. At the Winter Olympics adjustments made to the length, shape, and composition of the skis play an important role in the success of the skiers. Such variables determine the amount of pressure the skier places on the snow and the friction of the ski. There are tradeoffs as well. Whereas a wider ski front increases its turning abilities, it also creates large vibrations that can slow down the skier. The use of titanium and various fibers and adhesives decreases those vibrations and results in lighter, stiffer skis. The application of physics has taken the once-simple wooden ski and has created a complex, high-performance device (Fig. 0.14).

Figure 0.14   Each ski design is created for use based on scientific knowledge.

Courtesy of fotolia ©Lulu Berlu

that attempts to explain natural occurrences. Typically it has been tested in the laboratory but has not been proven with absolute certainty. A **principle** is a step closer to a law in physics. Principles have been experimentally proven in the laboratory, have stood the test of various conditions, and continue to hold true. Laws are the final degree of scientific certainty. **Laws** are often defined using formulas. For example, Newton's second law of motion, $F = ma$, has been proven to be true and is considered a law of physics.

The **scientific method** is an orderly procedure used by scientists in collecting, organizing, and analyzing new information that refutes or supports a scientific hypothesis. The constant use of the scientific method and the development of theories, principles, and laws is similar to the problem-solving method discussed in detail in Chapter 2. The **problem-solving method** is an orderly procedure that aids in understanding questions and solving problems. Nonscientists use the problem-solving method more often than the scientific method. The problem-solving method is helpful when a problem arises in this text, in class, or on the job. An individual or a team must develop the skills needed to collect data, analyze a problem, and work toward finding its solution in a logical and orderly fashion. In order to find solutions to problems, tools are needed to make the job easier. In the next chapter, we will familiarize ourselves with two important tools of physics: measurement and mathematics.

Glossary

Astronomy The branch of science that studies everything that takes place outside the earth's atmosphere. (p. 2)

Biology The branch of science that studies living organisms. (p. 2)

Chemistry The branch of science that studies the composition, structure, properties, and reactions of matter. (p. 2)

Experimental Physicist A physicist who performs experiments to develop and confirm physical theories. (p. 2)

Geology The branch of science that studies the origin, history, and structure of the earth. (p. 2)

Hypothesis A scientifically based prediction that needs testing to verify its validity. (p. 3)

Law The highest level of certainty for an explanation of physical occurrences. A law is often accompanied by a formula. (p. 7)

Physics The branch of science that describes the motion and energy of all matter throughout the universe. (p. 1)

Physicist A person who is an expert in or who studies physics. (p. 2)

Principle A rule or fundamental assumption that has been proven in the laboratory. (p. 7)

Problem-Solving Method An orderly procedure that aids in understanding questions and solving problems. (p. 7)

Science A system of knowledge that is concerned with establishing accurate conclusions about the behavior of everything in the universe. (p. 3)

Scientific Method An orderly procedure used by scientists in collecting, organizing, and analyzing new information that refutes or supports a scientific hypothesis. (p. 7)

Technology The field that uses scientific knowledge to develop material products or processes that satisfy human needs and desires. (p. 3)

Theoretical Physicist A physicist who predominantly uses previous theories and mathematical models to form new theories in physics. (p. 2)

Theory A scientifically accepted principle that attempts to explain natural occurrences. (p. 7)

Review Questions

1. Physics is a field of study that governs
   (a) how the planets orbit the sun.
   (b) the rate at which blood flows through a person's veins.
   (c) how quickly a helium balloon will rise into the air.
   (d) all of the above.

2. Who among the following is an example of a theoretical physicist?
   (a) Archimedes, who measured the volume of water that was displaced after placing objects in a tub of water.
   (b) Albert Einstein, who performed various thought experiments in his mind to arrive at his theories of relativity.
   (c) Marie Curie, who, along with her husband, was credited with discovering radioactivity through a series of laboratory experiments.
   (d) Benjamin Franklin, who determined, through various laboratory experiments, that electricity is the flow of microscopic charged particles.

3. Why are Isaac Newton's conclusions on motion considered laws of physics?
   (a) Newton himself declared them laws.
   (b) Newton performed various thought experiments on motion.
   (c) The formulas accompanying Newton's laws have proved correct in experiments for years.
   (d) Newton's reputation alone made his scientific conclusions laws.

4. Which of the following is not considered a branch of physics?
   (a) Thermodynamics
   (b) Astronomy
   (c) Geophysics
   (d) Atomic physics

5.   Analyzing the braking distance of a sports car would most likely utilize which field of physics?
      (a)   Molecular physics
      (b)   Quantum physics
      (c)   Fluid dynamics
      (d)   Mechanics

6.   Who is considered to be the first true physicist and what did he do to deserve this recognition in scientific history?

7.   Explain the difference between science and technology. Are the two fields related?

8.   Provide two examples of scientific knowledge and a technological development that relies on that scientific knowledge.

9.   What is the difference between the scientific method and the problem-solving method?

10.   Why is it important to study physics? Provide a few examples of what an understanding of the physical world can do for you today and in your future.

Objectives

The major goals of this chapter are to enable you to:

1. Explain the need for standardization of measurement.
2. Use the metric system of measurement.
3. Convert measurements from one system to another.
4. Solve problems involving length, area, and volume.
5. Distinguish between mass and weight.
6. Use significant digits to determine the accuracy of measurements.
7. Differentiate between accuracy and precision.
8. Solve problems with measurements and consistently express the results with the correct significant digits.
9. Use a systematic approach to solving physics problems.
10. Analyze problems using the problem-solving method.

A good mechanic needs not only the right tools for the job but also to be proficient in using those tools. The same applies to physics. Analyzing the problem, choosing the correct formula, and manipulating the equation will help you become a good physics student. In this chapter, we discuss mathematical techniques, significant digits, accuracy, precision, and the problem-solving method. These will be your basic tools for physics.

1.1 Standards of Measure

When two people work together on the same job, they should both use the same standards of measure. If not, the result can be a problem (Fig. 1.1).

Figure 1.1   The trouble with inconsistent systems of measurement

Standards of measure are sets of units of measurement for length, weight, and other quantities defined in a way that is useful to a large number of people. Throughout history, there have been many standards by which measurements have been made:

- *Chain:* A measuring instrument of 100 links used in surveying. One chain has a length of 66 feet.
- *Rod:* A length determined by having each of 16 men put one foot behind the foot of the man before him in a straight line [Fig. 1.2(a)]. The rod is now standardized as $16\frac{1}{2}$ feet.
- *Yard:* The distance from the tip of the king's nose to the fingertips of his outstretched hand [Fig. 1.2(b)].
- *Foot:* The rod divided by 16; it was also common to use the length of one's own foot as the unit foot.
- *Inch:* The length of three barley corns, round and dry, taken from the center of the ear, and laid end to end [Fig. 1.2(c)].

The U.S. system of measure, which is derived from and sometimes called the English system, is a combination of makeshift units of Anglo-Saxon, Roman, and French-Norman weights and measures.

After the standards based on parts of the human body and on other gimmicks, basic standards were accepted by world governments. They also agreed to construct and distribute accurate standard copies of all the standard units. During the 1790s, a decimal system based on our number system, the metric system, was developed in France. Its acceptance was gained mostly because it was easy to use and easy to remember. Many nations began adopting it as their official system of measurement. By 1900, most of Europe and South America were metric. In 1866,

Figure 1.2   Definitions of some old units

(a) A rod was earlier defined as the length of the feet of 16 men in a line as shown. This distance divided by 16 equals one foot.

(b) The "old" yard

(c) At one time, three barley corns were used to define one inch.

TRY THIS ACTIVITY

Stepping Off

When a short distance needs to be measured and a tape measure is not available, some people measure the approximate length by using the length of their own foot as a unit. Measure the distance between two points approximately 15 to 25 ft apart by placing one foot in front of the other and counting the steps. Then, measure the same distance with a tape measure. How close to the standard foot is the length of your own foot? How much error did you generate?

For longer distances, some measure the approximate length by pacing off the distance using one stride as approximately 1 yd or 3 ft. Measure the distance between two points approximately 50 to 75 ft apart by pacing off the distance and counting the strides. Then, measure the same distance with a tape measure. How close to the standard yard is the length of your own stride? How much error did you generate?

metric measurements for official use were legalized in the United States. In 1893, the Secretary of the Treasury, by administrative order, declared the new metric standards to be the nation's "fundamental standards" of mass and length. Thus, indirectly, the United States *officially* became a metric nation. Even today, the U.S. units are officially defined in terms of the standard metric units.

Throughout U.S. history, several attempts have been made to convert the nation to the metric system. By the 1970s, the United States found itself to be the only nonmetric industrialized country left in the world. However, government inaction to implement the metric system resulted in the United States regularly using a greater variety of confusing units than any other country. Industry and business, however, found their international markets drying up because metric products were preferred. Now many segments of American industry and business

PHYSICS CONNECTIONS

Lost in Space

Even professional engineers and scientists sometimes forget to include units or make mistakes when converting from one system of measurement to another. NASA's Mars Climate Orbiter (Fig. 1.3) was lost in space on September 23, 1999, after engineers from Lockheed Martin used U.S. measurements when calculating rocket thrusts and did not convert those measurements to the metric units used by NASA engineers. Such a simple mistake, common to students on physics exams, cost Lockheed Martin and NASA a $125 million space probe and a great deal of embarrassment.

According to officials, NASA assumed that the twice-daily rocket thrust calculations were based on metric units, whereas Lockheed Martin had neither converted the numbers to metric nor labeled them in U.S. units. As a result, the Mars Climate Orbiter, which was to orbit Mars, was inadvertently sent off course and lost. The orbiter was intended to collect atmospheric and surface data while also serving as a communications link for the Mars Lander.

Figure 1.3 NASA's Mars Climate Orbiter—artist's conception

Courtesy of NASA/Jet Propulsion Laboratory

have independently gone metric because world trade is geared toward the metric system of measurement. The inherent simplicity of the metric system of measurement and standardization of weights and measures has led to major cost savings in industries that have converted to it.

Most major U.S. industries, such as the automotive, aviation, and farm implement industries, as well as the Department of Defense and other federal agencies have effectively converted to the metric system. In this text approximately 70% of our examples and exercises are metric in the sections where both U.S. and metric systems are still commonly used. In some industries, you—the student and worker—will need to know and use both systems.

1.2 Introduction to the Metric System

The modern metric system is identified in all languages by the abbreviation **SI** (for **Système International d'Unités**—the international system of units of measurement written in French). The SI metric system has seven *basic units* [Table 1.1(a)]. All other SI units are called *derived units*; that is, they can be defined in terms of these seven basic units (see Appendix C, Table 19). For example, the newton (N) is defined as 1 kg m/s^2 (kilogram metre per second per second). Many derived units will be presented and discussed in this text. Some derived SI units are given in Table 1.1(b). **Gabriel Mouton** is often credited with originating the metric system.

Because the metric system is a decimal or base-10 system, it is very similar to our decimal number system and any decimal money system. It is an easy system to use because calculations are based on the number 10 and its multiples. Special prefixes are used to name these multiples and submultiples, which may be used with most all SI units. Because the same prefixes are used repeatedly, the memorization of many conversions has been significantly reduced. Table 1.2 shows these prefixes and the corresponding symbols.

Gabriel Mouton (1618–1694),

a French vicar who spent much of his time studying mathematics and astronomy, is credited by many for originating the metric system. French scientists in the late 18th century are credited with replacing the chaotic collection of systems then in use with the metric system.

Table 1.1   SI Units of Measure

(a) Basic Unit	SI Abbreviation	Used for Measuring
metre*	m	Length
kilogram	kg	Mass
second	s	Time
ampere	A	Electric current
kelvin	K	Temperature
candela	cd	Light intensity
mole	mol	Molecular substance

(b) Derived Unit	SI Abbreviation	Used for Measuring
litre*	L or ℓ	Volume
cubic metre	m^3	Volume
square metre	m^2	Area
newton	N	Force
metre per second	m/s	Speed
joule	J	Energy
watt	W	Power

*At present, there is some difference of opinion in the United States on the spelling of metre and litre. We have chosen the "re" spellings for two reasons. First, this is the internationally accepted spelling for all English-speaking countries. Second, the word "meter" already has many different meanings—parking meter, electric meter, odometer, and so on. Many feel that the metric units of length and volume should be distinctive and readily recognizable—thus the spellings "metre" and "litre."

Table 1.2   Prefixes for SI Units

Multiple or Submultiple^a Decimal Form	Power of 10	Prefix^b	Prefix Symbol	Pronunciation	Meaning
1,000,000,000,000	10^{12}	tera	T	tĕr′ă	One trillion times
1,000,000,000	10^{9}	giga	G	jĭg′ă	One billion times
1,000,000	10^{6}	mega	M	mĕg′ă	One million times
1,000	10^{3}	kilo	k	kĭl′ō	One thousand times
100	10^{2}	hecto	h	hĕk′tō	One hundred times
10	10^{1}	deka	da	dĕk′ă	Ten times
0.1	10^{-1}	deci	d	dĕs′ĭ	One tenth of
0.01	10^{-2}	centi	c	sĕnt′ĭ	One hundredth of
0.001	10^{-3}	milli	m	mĭl′ĭ	One thousandth of
0.000001	10^{-6}	micro	μ	mī′krō	One millionth of
0.000000001	10^{-9}	nano	n	năn′ō	One billionth of
0.000000000001	10^{-12}	pico	p	pē′kō	One trillionth of

^aFactor by which the unit is multiplied.
^bThe same prefixes are used with all SI metric units.

EXAMPLE 1

Write the SI abbreviation for 36 centimetres.

The symbol for the prefix *centi* is c.
The symbol for the unit *metre* is m.

Thus, 36 cm is the SI abbreviation for 36 centimetres.

EXAMPLE 2

Write the SI metric unit for the abbreviation 45 kg.

The prefix for k is *kilo*; the unit for g is *gram*.

Thus, 45 kilograms is the SI metric unit for 45 kg.

PROBLEMS 1.2

Give the metric prefix for each value.

1. 1000
2. 0.01
3. 100
4. 0.1
5. 0.001
6. 10
7. 1,000,000
8. 0.000001

Give the metric symbol, or abbreviation, for each prefix.

9. hecto
10. kilo
11. milli
12. deci
13. mega
14. deka
15. centi
16. micro

Write the abbreviation for each quantity.

17. 135 millimetres
18. 83 dekagrams
19. 28 kilolitres
20. 52 centimetres
21. 49 centigrams
22. 85 milligrams
23. 75 hectometres
24. 15 decilitres

Write the SI unit for each abbreviation.

25. 24 m
26. 185 L
27. 59 g
28. 125 kg
29. 27 mm
30. 25 dL
31. 45 dam
32. 27 mg
33. 26 Mm
34. 275 μg
35. The basic metric unit of length is _____.
36. The basic unit of mass is _____.
37. Two common metric units of volume are _____ and _____.

38. The basic unit for electric current is _____.
39. The basic metric unit for time is _____.
40. The common metric unit for power is _____.

1.3 Scientific Notation

Scientists and technicians often need to use very large or very small numbers. For example, the thickness of an oil film on water is about 0.0000001 m. **Scientific notation** is a useful method of expressing such very small (or very large) numbers. Expressed this way, the thickness of the film is 1×10^{-7} m, or 10^{-7} m. For example:

$$0.1 = 1 \times 10^{-1} \quad \text{or} \quad 10^{-1}$$
$$10{,}000 = 1 \times 10^{4} \quad \text{or} \quad 10^{4}$$
$$0.001 = 1 \times 10^{-3} \quad \text{or} \quad 10^{-3}$$

A number in scientific notation is written as a product of a number between 1 and 10 and a power of 10. General form: $M \times 10^{n}$, where

M = a number between 1 and 10
n = the exponent or power of 10

The following numbers are written in scientific notation:

EXAMPLE 1

(a) $325 = 3.25 \times 10^{2}$
(b) $100{,}000 = 1 \times 10^{5}$ or 10^{5}

To write any decimal number in scientific notation:

1. Reading from left to right, place a decimal point after the first nonzero digit.
2. Place a caret (∧) at the position of the *original* decimal point.
3. If the newly added decimal point is to the *left* of the caret, the exponent of 10 is the number of places from the caret to the decimal point.
    Example: $83{,}662 = 8.3662_{\wedge} \times 10^{\underline{4}}$
4. If the newly added decimal point is to the *right* of the caret, the exponent of 10 is the negative of the number of places from the caret to the decimal point.
    Example: $0.00683 = {}_{\wedge}006.83 \times 10^{\underline{-3}}$
5. If the decimal point and the caret coincide, the exponent of 10 is zero.
    Example: $5.12 = 5.12 \times 10^{0}$

A number greater than 10 is expressed in scientific notation as a product of a decimal between 1 and 10 and a *positive* power of 10.

Write each number greater than 10 in scientific notation.

EXAMPLE 2

(a) $2580 = 2.58 \times 10^{3}$
(b) $54{,}600 = 5.46 \times 10^{4}$
(c) $42{,}000{,}000 = 4.2 \times 10^{7}$
(d) $715.8 = 7.158 \times 10^{2}$
(e) $34.775 = 3.4775 \times 10^{1}$

A number between 0 and 1 is expressed in scientific notation as a product of a decimal between 1 and 10 and a *negative* power of 10.

EXAMPLE 3

Write each positive number less than 1 in scientific notation.

(a) $0.0815 = 8.15 \times 10^{-2}$
(b) $0.00065 = 6.5 \times 10^{-4}$
(c) $0.73 = 7.3 \times 10^{-1}$
(d) $0.0000008 = 8 \times 10^{-7}$

A number between 1 and 10 is expressed in scientific notation as a product of a decimal between 1 and 10 and the *zero* power of 10.

EXAMPLE 4

Write each number between 1 and 10 in scientific notation.

(a) $7.33 = 7.33 \times 10^{0}$
(b) $1.06 = 1.06 \times 10^{0}$

To change a number from scientific notation to decimal form:

1.   Multiply the decimal part by the power of 10 by moving the decimal point *to the right* the same number of decimal places as indicated by the power of 10 if it is *positive*.
2.   Multiply the decimal part by the power of 10 by moving the decimal point *to the left* the same number of decimal places as indicated by the power of 10 if it is *negative*.
3.   Supply zeros as needed.

EXAMPLE 5

Write 7.62×10^{2} in decimal form.

$$7.62 \times 10^{2} = 762 \qquad \text{Move the decimal point two places to the right.}$$

EXAMPLE 6

Write 6.15×10^{-4} in decimal form.

$$6.15 \times 10^{-4} = 0.000615 \qquad \text{Move the decimal point four places to the left and insert three zeros.}$$

EXAMPLE 7

Write each number in decimal form.

(a) $3.75 \times 10^{2} = 375$
(b) $1.09 \times 10^{5} = 109,000$
(c) $2.88 \times 10^{-2} = 0.0288$
(d) $9.4 \times 10^{-6} = 0.0000094$
(e) $6.7 \times 10^{0} = 6.7$

Since calculators used in science and technology accept numbers entered in scientific notation and give some results in scientific notation, it is essential that you fully understand this topic before going to the next section. See Appendix D, Section D.1, for using a calculator with numbers in scientific notation.

PHYSICS CONNECTIONS

Powers of Ten

The importance of scientific notation is illustrated by a famous short documentary film called *Powers of Ten*. The film begins with a 1-metre-wide overhead view of a man relaxing on a blanket while on a picnic. The camera then zooms out to a shot that covers a width of 10 metres (10^1 m) with the picnicking man still in the center of the image but now surrounded by green grass. Throughout the course of another 10 seconds, the camera continues to zoom out to 100 metres (10^2 m) wide. As the camera continues to zoom out, the viewer begins to grasp how enormous the changes in distance are when a number is raised by a power of 10. Eventually, the trip outward stops at 10^{24} m, and then the picture zooms back to 10^0 m before proceeding to enter the skin of the man by moving in by negative powers of 10. Although the majority of the images are artist's renditions of what scientists imagine these objects to look like, *Powers of Ten* is a powerful film that conceptualizes how large our universe is and how small atoms and subatomic particles are within our bodies. The series of photographs demonstrating the powers of 10 can be found at http://www.powersof10.com.

Some of the views used in *Powers of Ten* include:

10^0 m or 1 metre: a view of a man on a blanket while on a picnic
10^7 m or 10 thousand kilometres: a view of the entire earth
10^{21} m or 100 thousand light-years: a view of our galaxy, the Milky Way
10^{-1} m or 10 centimetres: a view of a small section of skin on the man's hand
10^{-5} m or 10 microns: a view of a white blood cell inside a capillary in the man's hand
10^{-14} m or 10 fermis: the nucleus of a typical carbon-12 atom

PROBLEMS 1.3

Write each number in scientific notation.

1.	326	2.	798	3.	2650
4.	14,500	5.	826.4	6.	24.97
7.	0.00413	8.	0.00053	9.	6.43
10.	482,300	11.	0.000065	12.	0.00224
13.	540,000	14.	1,400,000	15.	0.0000075
16.	0.0000009	17.	0.00000005	18.	3,500,000,000
19.	732,000,000,000,000,000	20.	0.000000000000000618		

Write each number in decimal form.

21.	8.62×10^4	22.	8.67×10^2	23.	6.31×10^{-4}
24.	5.41×10^3	25.	7.68×10^{-1}	26.	9.94×10^1
27.	7.77×10^8	28.	4.19×10^{-6}	29.	6.93×10^1
30.	3.78×10^{-2}	31.	9.61×10^4	32.	7.33×10^3
33.	1.4×10^0	34.	9.6×10^{-5}	35.	8.4×10^{-6}
36.	9×10^8	37.	7×10^{11}	38.	4.05×10^0
39.	7.2×10^{-7}	40.	8×10^{-9}	41.	4.5×10^{12}
42.	1.5×10^{11}	43.	5.5×10^{-11}	44.	8.72×10^{-10}

1.4 Length

In most sections that introduce units of measure, we present the units in subsections as follows: metric units, U.S. units, and conversions between metric and U.S. units.

Metric Length

The basic SI unit of length is the **metre** (m) (Fig. 1.4). The first standard metre was chosen in the 1790s to be one ten-millionth of the distance from the earth's equator to either pole. Modern measurements of the earth's circumference show that the first length is off by about 0.02% from this initial standard. The current definition adopted in 1983 is based on the speed of light in a vacuum and reads, "The metre is the length of path traveled by light in a vacuum during a time interval of 1/299,792,458 of a second." Long distances are measured in kilometres (km) (Fig. 1.5). We use the centimetre (cm) to measure short distances, such as the length of this book or the width of a board [Fig. 1.6(a)]. The millimetre (mm) is used to measure very small lengths, such as the thickness of this book or the depth of a tire tread [Fig. 1.6(b)]. A metric ruler is shown in Fig. 1.6(c).

Figure 1.4   One metre

(a) The height of a door   (b) The length of a person's long
knob is about 1 m.   pace is also about 1 m.

(c) The length of 1 m is a little more than 1 yd.

Figure 1.5   The length of five city blocks is about 1 km.

A **conversion factor** is an expression used to change from one unit or set of units to another. We know that we can multiply any number or quantity by 1 without changing the value of the original quantity. We also know that any fraction equals 1 when its numerator and denominator are equal. For example, $\frac{5}{5} = 1$, $\frac{12 \text{ m}}{12 \text{ m}} = 1$, and $\frac{6.5 \text{ kg}}{6.5 \text{ kg}} = 1$. In addition, since 1 m = 100 cm, $\frac{1 \text{ m}}{100 \text{ cm}} = 1$. Similarly, $\frac{100 \text{ cm}}{1 \text{ m}} = 1$, because the numerator equals the denominator. We call such names for 1 *conversion factors*. The information necessary for forming a conversion factor is usually found in tables. As in the case 1 m = 100 cm, there are two conversion factors for each set of data:

$$\frac{1 \text{ m}}{100 \text{ cm}} \quad \text{and} \quad \frac{100 \text{ cm}}{1 \text{ m}}$$

Figure 1.6   Small metric length units

1 cm

1 mm

(a) The width of your small fingernail is about 1 cm.   (b) The thickness of a dime is about 1 mm.

0   1   2   3   4   5   6   7   8   9   10

(c) The large numbered divisions are centimetres shown actual size.
Each centimetre is divided into 10 equal parts, called millimetres.

CONVERSION FACTORS

Choose a conversion factor in which the old units are in the numerator of the original expression and in the denominator of the conversion factor, or the old units are in the denominator of the original expression and in the numerator of the conversion factor. That is, we want the old units to cancel each other.

EXAMPLE 1

Change 215 cm to metres.

As we saw before, the two possible conversion factors are

$$\frac{1 \text{ m}}{100 \text{ cm}} \quad \text{and} \quad \frac{100 \text{ cm}}{1 \text{ m}}$$

We choose the conversion factor with centimetres in the *denominator* so that the cm units cancel each other.

$$215 \text{ cm} \times \frac{1 \text{ m}}{100 \text{ cm}} = 2.15 \text{ m}$$

Note: Conversions *within* the metric system involve only moving the decimal point.

· · · · · · · · · · · · · · · ·

EXAMPLE 2

Change 4 m to centimetres.

$$4 \text{ m} \times \frac{100 \text{ cm}}{1 \text{ m}} = 400 \text{ cm}$$

· · · · · · · · · · · · · · · ·

EXAMPLE 3

Change 39.5 mm to centimetres.

Choose the conversion factor with millimetres in the denominator so that the mm units cancel each other.

$$39.5 \text{ mm} \times \frac{1 \text{ cm}}{10 \text{ mm}} = 3.95 \text{ cm}$$

· · · · · · · · · · · · · · · ·

EXAMPLE 4

Change 0.05 km to centimetres.
First, change to metres and then to centimetres.

$$0.05 \text{ km} \times \frac{1000 \text{ m}}{1 \text{ km}} = 50 \text{ m}$$

$$50 \text{ m} \times \frac{100 \text{ cm}}{1 \text{ m}} = 5000 \text{ cm}$$

Or,

$$0.05 \text{ km} \times \frac{1000 \text{ m}}{1 \text{ km}} \times \frac{100 \text{ cm}}{1 \text{ m}} = 5000 \text{ cm}$$

U.S. Length

The basic units of the U.S. system are the foot, the pound, and the second. The foot is the basic unit of length and may be divided into 12 equal parts or inches. Common U.S. length conversions include

1 foot (ft) = 12 inches (in.)

1 yard (yd) = 3 ft

1 mile (mi) = 5280 ft

See Table 1 of Appendix C for U.S. weights and measures.
We also use a conversion factor to change from one U.S. length unit to another.

EXAMPLE 5

Change 84 in. to feet.
Choose the conversion factor with inches in the denominator and feet in the numerator.

$$84 \text{ in.} \times \frac{1 \text{ ft}}{12 \text{ in.}} = 7 \text{ ft}$$

Metric–U.S. Conversions

To change from a U.S. unit to a metric unit or from a metric unit to a U.S. unit, again use a conversion factor, such as 1 in. = 2.54 cm.

EXAMPLE 6

Express 10 inches in centimetres.

$$1 \text{ in.} = 2.54 \text{ cm} \quad \text{so} \quad 10 \text{ in.} \times \frac{2.54 \text{ cm}}{1 \text{ in.}} = 25.4 \text{ cm}$$

The conversion factors you will need are given in Appendix C. The following examples show you how to use these tables.

EXAMPLE 7

Change 15 miles to kilometres.
From Table 2 in Appendix C, we find 1 mile listed in the left-hand column. Moving over to the fourth column, under the heading "km," we see that 1 mile (mi) = 1.61 km. Then we have

$$15 \text{ mi} \times \frac{1.61 \text{ km}}{1 \text{ mi}} = 24.15 \text{ km}$$

Change 220 centimetres to inches.

EXAMPLE 8

Find 1 centimetre in the left-hand column and move to the fifth column under the heading "in." We find that 1 centimetre = 0.394 in. Then

$$220 \text{ cm} \times \frac{0.394 \text{ in.}}{1 \text{ cm}} = 86.68 \text{ in.}$$

Change 3 yards to centimetres.

EXAMPLE 9

Since there is no direct conversion from yards to centimetres in the tables, we must first change yards to inches and then inches to centimetres:

$$3 \text{ yd} \times \frac{36 \text{ in.}}{1 \text{ yd}} \times \frac{2.54 \text{ cm}}{1 \text{ in.}} = 274.32 \text{ cm}$$

A more comprehensive discussion of dimensional analysis and conversion factors is presented in Appendix B.

PROBLEMS 1.4

Which unit is longer?

1. 1 metre or 1 centimetre
2. 1 metre or 1 millimetre
3. 1 metre or 1 kilometre
4. 1 centimetre or 1 millimetre
5. 1 centimetre or 1 kilometre
6. 1 millimetre or 1 kilometre

Which metric unit (km, m, cm, or mm) would you use to measure the following?

7. Length of a wrench
8. Thickness of a saw blade
9. Height of a barn
10. Width of a table
11. Thickness of a hypodermic needle
12. Distance around an automobile racing track
13. Distance between New York and Miami
14. Length of a hurdle race
15. Thread size on a pipe
16. Width of a house lot

Fill in each blank with the most reasonable metric unit (km, m, cm, or mm).

17. Your car is about 6 _____ long.
18. Your pencil is about 20 _____ long.
19. The distance between New York and San Francisco is about 4200 _____.
20. Your pencil is about 7 _____ thick.
21. The ceiling in my bedroom is about 240 _____ high.
22. The length of a football field is about 90 _____.
23. A jet plane usually cruises at an altitude of 9 _____.
24. The zipper on my jacket is about 70 _____ long.
25. Juan drives 9 _____ to school each day.
26. Jacob, our basketball center, is 203 _____ tall.
27. The width of your hand is about 80 _____.
28. A newborn baby is usually about 45 _____ long.

Fill in each blank.

29. 1 km = _____ m
30. 1 mm = _____ m
31. 1 m = _____ cm
32. 1 m = _____ hm
33. 1 dm = _____ m
34. 1 m = _____ mm
35. 1 hm = _____ m
36. 1 cm = _____ m
37. 1 cm = _____ mm
38. Change 250 m to cm.
39. Change 250 m to km.

40. Change 546 mm to cm.   41. Change 178 km to m.
42. Change 35 dm to dam.   43. Change 830 cm to m.
44. Change 75 hm to km.   45. Change 375 cm to mm.
46. Change 7.5 mm to μm.
47. State your height in centimetres and in metres.
48. Change 43,296 ft   (a)   to miles.   (b)   to yards.
49. Change 6.25 mi   (a)   to yards.   (b)   to feet.
50. Change 5.94 m to feet.   51. Change 7.1 cm to inches.
52. Change 1.2 in. to centimetres.

Figure 1.7   Actual sizes of 1 cm^2 and 1 in^2

One square centimetre (cm^2)

One square inch (in^2)

1.5 Area and Volume

Area

The **area** of a figure is the number of square units that it contains. To measure a surface area of an object, you must first decide on a standard unit of area. Standard units of area are based on the square and are called *square inches, square centimetres, square miles*, or some other square unit of measure. An area of 1 square centimetre (cm^2) is the amount of area found within a square 1 cm on each side. An area of 1 square inch (in^2) is the amount of area found within a square of 1 in. on each side (Fig. 1.7).

In general, when multiplying measurements of like units, multiply the numbers and then multiply the units as follows:

$$2\text{ cm} \times 4\text{ cm} = (2 \times 4)\,(\text{cm} \times \text{cm}) = 8\text{ cm}^2$$
$$3\text{ in.} \times 5\text{ in.} = (3 \times 5)\,(\text{in.} \times \text{in.}) = 15\text{ in}^2$$
$$1.4\text{ m} \times 6.7\text{ m} = (1.4 \times 6.7)\,(\text{m} \times \text{m}) = 9.38\text{ m}^2$$

Figure 1.8

One square metre (m^2)

Metric Area

The basic unit of area in the metric system is the *square metre* (m^2), the area in a square whose sides are 1 m long (Fig. 1.8). The square centimetre (cm^2) and the square millimetre (mm^2) are smaller units of area. Larger units of area are the square kilometre (km^2) and the hectare (ha).

EXAMPLE 1

Find the area of a rectangle 5 m long and 3 m wide.

Each square in Fig. 1.9 represents 1 m^2. By simply counting the number of squares (square metres), we find that the area of the rectangle is 15 m^2. We can also find the area of the rectangle by using the formula

$$A = lw = (5\text{ m})\,(3\text{ m}) = 15\text{ m}^2 \qquad (\textbf{Note:}\quad \text{m} \times \text{m} = \text{m}^2)$$

Figure 1.9

Find the area of the metal plate shown in Fig. 1.10.

EXAMPLE 2

Figure 1.10

To find the area of the metal plate, find the area of each of the two rectangles and then find the difference of their areas. The large rectangle is 10 cm long and 8 cm wide. The small rectangle is 6 cm long and 4 cm wide. Thus,

$$\text{area of large rectangle:} \quad A = lw = (10 \text{ cm})(8 \text{ cm}) = 80 \text{ cm}^2$$
$$\text{area of small rectangle:} \quad A = lw = (6 \text{ cm})(4 \text{ cm}) = \underline{24 \text{ cm}^2}$$
$$\text{area of metal plate:} \qquad\qquad\qquad\qquad\qquad\quad 56 \text{ cm}^2$$

The surface that would be seen by cutting a geometric solid with a thin plate parallel to one side of the solid represents the *cross-sectional area* of the solid.

Find the smallest cross-sectional area of the box shown in Fig. 1.11(a).

EXAMPLE 3

Figure 1.11

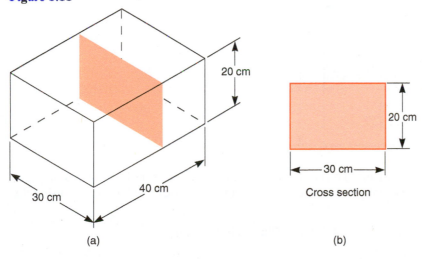

(a)

20 cm

20 cm

30 cm

Cross section

(b)

The indicated cross section of this box is a rectangle 30 cm long and 20 cm wide [Fig. 1.11(b)]. Thus,

$$A = lw = (30 \text{ cm}) \, (20 \text{ cm}) = 600 \text{ cm}^2$$

The area of this rectangle is 600 cm², which represents the cross-sectional area of the box.

The formulas for finding the areas of other plane figures are found on the inside back cover.

The *hectare* is the fundamental SI unit for land area. An area of 1 hectare equals the area of a square 100 m on a side (Fig. 1.12). The hectare is used because it is more convenient to say and use than square hectometre. The metric prefixes are *not* used with the hectare unit. That is, instead of saying the prefix "kilo" with "hectare," we say "1000 hectares."

To convert area or square units, use a conversion factor. That is, the correct conversion factor will be in fractional form and equal to 1, with the numerator expressed in the units you wish to convert to and the denominator expressed in the units given. The conversion table for area is provided as Table 3 of Appendix C.

The conversion of area units will be shown using a method of squaring the linear or length conversion factor that you are most likely to remember. An alternate method emphasizing direct use of the conversion tables will also be shown.

Figure 1.12 One hectare

1 hectare (ha) =
10,000 m² =
1 hm²

100 m

100 m

EXAMPLE 4

Change 258 cm² to m².

$$258 \text{ cm}^2 \times \left(\frac{1 \text{ m}}{100 \text{ cm}}\right)^2 = 258 \text{ cm}^2 \times \frac{1 \text{ m}^2}{100^2 \text{ cm}^2} = 0.0258 \text{ m}^2$$

Note: The intermediate step is usually not shown.

Alternate Method:

$$258 \text{ cm}^2 \times \frac{1 \text{ m}^2}{10,000 \text{ cm}^2} = 0.0258 \text{ m}^2$$

. . . .

U.S. Area

EXAMPLE 5

Find the area of a rectangle that is 6 in. long and 4 in. wide (Fig. 1.13).

Figure 1.13

6 in.

4 in.

Each square is 1 in². To find the area of the rectangle, simply count the number of squares in the rectangle. Therefore, you find that the area = 24 in², or, by using the formula,

$$A = lw = (6 \text{ in.}) (4 \text{ in.}) = 24 \text{ in}^2$$

. . . .

EXAMPLE 6

Change 324 in² to yd².

$$324 \text{ in}^2 \times \left(\frac{1 \text{ yd}}{36 \text{ in.}}\right)^2 = 0.25 \text{ yd}^2$$

Alternate Method:

$$324 \text{ in}^2 \times \frac{1 \text{ yd}^2}{1296 \text{ in}^2} = \frac{324}{1296} \text{ yd}^2 = 0.25 \text{ yd}^2$$

Metric–U.S. Area Conversions

Change 25 cm² to in².

EXAMPLE 7

$$25 \text{ cm}^2 \times \left(\frac{1 \text{ in.}}{2.54 \text{ cm}}\right)^2 = 3.875 \text{ in}^2$$

Alternate Method:

$$25 \text{ cm}^2 \times \frac{0.155 \text{ in}^2}{1 \text{ cm}^2} = 3.875 \text{ in}^2$$

Conversion factors found in tables are usually rounded. There are many rounding procedures in general use. We will use one of the simplest methods, stated as follows:

ROUNDING NUMBERS

To round a number to a particular place value:

1.   If the digit in the next place to the right is less than 5, drop that digit and all other following digits. Replace any whole number places dropped with zeros.
2.   If the digit in the next place to the right is 5 or greater, add 1 to the digit in the place to which you are rounding. Drop all other following digits. Replace any whole number places dropped with zeros.

Change 28.5 m² to in².

EXAMPLE 8

$$28.5 \text{ m}^2 \times \left(\frac{39.4 \text{ in.}}{1 \text{ m}}\right)^2 = 44{,}242.26 \text{ in}^2$$

Figure 1.14

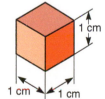

One cubic centimetre (cm³)

Alternate Method:

$$28.5 \text{ m}^2 \times \frac{1550 \text{ in}^2}{1 \text{ m}^2} = 44{,}175 \text{ in}^2$$

Note: The choice of rounded conversion factors will often lead to results that differ slightly. When checking your answers, you must allow for such rounding differences.

To convert between metric and U.S. land area units, use the relationship

1 hectare = 2.47 acres

Volume

The **volume** of a figure is the number of cubic units that it contains. Standard units of volume are based on the cube and are called *cubic centimetres, cubic inches, cubic yards,* or some other cubic unit of measure. A volume of 1 cubic centimetre (cm³) is the same as the amount of volume contained in a cube 1 cm on each side. One cubic inch (in³) is the volume contained in a cube 1 in. on each side (Fig. 1.14).

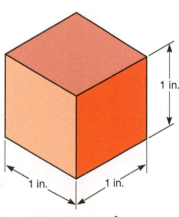

One cubic inch (in³)

Note: When multiplying measurements of like units, multiply the numbers and then multiply the units as follows:

$$3 \text{ in.} \times 5 \text{ in.} \times 4 \text{ in.} = (3 \times 5 \times 4) \, (\text{in.} \times \text{in.} \times \text{in.}) = 60 \text{ in}^3$$

$$2 \text{ cm} \times 4 \text{ cm} \times 1 \text{ cm} = (2 \times 4 \times 1) \, (\text{cm} \times \text{cm} \times \text{cm}) = 8 \text{ cm}^3$$

$$1.5 \text{ ft} \times 8.7 \text{ ft} \times 6 \text{ ft} = (1.5 \times 8.7 \times 6) \, (\text{ft} \times \text{ft} \times \text{ft}) = 78.3 \text{ ft}^3$$

Metric Volume

Find the volume of a rectangular prism 6 cm long, 4 cm wide, and 5 cm high.

Each cube shown in Fig. 1.15 is 1 cm³. To find the volume of the rectangular solid, count the number of cubes in the bottom layer of the rectangular solid and then multiply that number by the number of layers that the solid can hold. Therefore, there are 5 layers of 24 cubes, which is 120 cubes or 120 cubic centimetres.

Or, by formula, $V = Bh$, where B is the area of the base and h is the height. However, the area of the base is found by lw, where l is the length and w is the width of the rectangle. Therefore, the volume of a rectangular solid can be found by the formula

$$V = lwh = (6 \text{ cm})(4 \text{ cm})(5 \text{ cm}) = 120 \text{ cm}^3$$

Note: cm $\times$ cm $\times$ cm = cm³.

EXAMPLE 9

Figure 1.15

A common unit of volume in the metric system is the *litre* (L) (Fig. 1.16). The litre is commonly used for liquid volumes.

Figure 1.16　One litre of milk is a little more than 1 quart of milk.

(a)　　　　(b)

The cubic metre (m³) is used to measure large volumes. The cubic metre is the volume in a cube 1 m on an edge. For example, the usual teacher's desk could be boxed into 2 cubic metres side by side.

The relationship between the litre and the cubic centimetre deserves special mention. The litre is defined as the volume in 1 cubic decimetre (dm³). That is, 1 litre of liquid fills a cube 1 dm (10 cm) on an edge (Fig. 1.17). The volume of this cube can be found by using the formula

$$V = lwh = (10 \text{ cm}) (10 \text{ cm}) (10 \text{ cm}) = 1000 \text{ cm}^3$$

That is,

$$1 \text{ L} = 1000 \text{ cm}^3$$

Figure 1.17   One litre contains 1000 cm³.

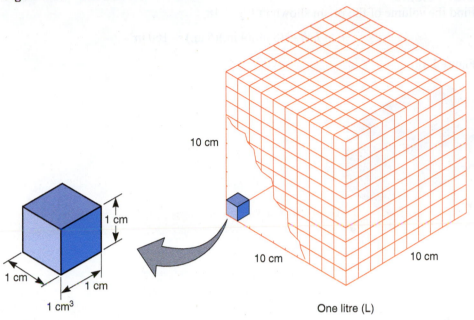

10 cm

1 cm

1 cm

1 cm

1 cm³

10 cm

10 cm

One litre (L)

Then

$$\frac{1}{1000} \text{L} = 1 \text{ cm}^3$$

But

$$\frac{1}{1000} \text{L} = 1 \text{ mL}$$

Therefore,

$$1 \text{ mL} = 1 \text{ cm}^3$$

Milk, soda, and gasoline are usually sold by the litre in countries using the metric system. Liquid medicine, vanilla extract, and lighter fluid are usually sold by the millilitre. Many metric cooking recipes are given in millilitres. Very large quantities of oil are sold by the kilolitre (1000 L).

Change 0.75 L to millilitres.

EXAMPLE 10

$$0.75 \text{ L} \times \frac{1000 \text{ mL}}{1 \text{ L}} = 750 \text{ mL}$$

Similarly, the conversion of volume cubic units will be shown using a method of cubing the linear, or length, conversion factor that you are most likely to remember. An alternate method emphasizing direct use of the conversion tables will also be shown.

Change 0.65 cm³ to cubic millimetres.

EXAMPLE 11

$$0.65 \text{ cm}^3 \times \left(\frac{10 \text{ mm}}{1 \text{ cm}}\right)^3 = 0.65 \text{ cm}^3 \times \frac{10^3 \text{ mm}^3}{1^3 \text{ cm}^3} = 650 \text{ mm}^3$$

Note: The intermediate step is usually not shown.

Alternate Method:

$$0.65 \text{ cm}^3 \times \frac{1000 \text{ mm}^3}{1 \text{ cm}^3} = 650 \text{ mm}^3$$

EXAMPLE 12

U.S. Volume
Find the volume of the prism shown in Fig. 1.18.

$$V = lwh = (8 \text{ in.})(4 \text{ in.})(5 \text{ in.}) = 160 \text{ in}^3$$

Figure 1.18

5 in.

8 in.

4 in.

EXAMPLE 13

Change 24 ft^3 to in^3.

$$24 \text{ ft}^3 \times \left(\frac{12 \text{ in.}}{1 \text{ ft}} \right)^3 = 41{,}472 \text{ in}^3$$

Alternate Method:

$$24 \text{ ft}^3 \times \frac{1728 \text{ in}^3}{1 \text{ ft}^3} = 41{,}472 \text{ in}^3$$

EXAMPLE 14

Metric–U.S. Volume Conversions
Change 56 in^3 to cm^3.

$$56 \text{ in}^3 \times \left(\frac{2.54 \text{ cm}}{1 \text{ in.}} \right)^3 = 917.68 \text{ cm}^3$$

Alternate Method:

$$56 \text{ in}^3 \times \frac{16.4 \text{ cm}^3}{1 \text{ in}^3} = 918.4 \text{ cm}^3$$

EXAMPLE 15

Change 28 m^3 to ft^3.

$$28 \text{ m}^3 \times \left(\frac{3.28 \text{ ft}}{1 \text{ m}} \right)^3 = 988.1 \text{ ft}^3$$

Alternate Method:

$$28 \text{ m}^3 \times \frac{35.3 \text{ ft}^3}{1 \text{ m}^3} = 988.4 \text{ ft}^3$$

Surface Area
The **lateral** (side) **surface area** of any geometric solid is the area of all the lateral faces. The **total surface area** of any geometric solid is the lateral surface area plus the area of the bases.

Find the lateral surface area of the prism shown in Fig. 1.19.

EXAMPLE 16

area of lateral face 1 = (6 in.)(5 in.) = 30 in²
area of lateral face 2 = (5 in.)(4 in.) = 20 in²
area of lateral face 3 = (6 in.)(5 in.) = 30 in²
area of lateral face 4 = (5 in.)(4 in.) = 20 in²
lateral surface area = 100 in²

· · · · · · · · · · · · · · · · · ·

Find the total surface area of the prism shown in Fig. 1.19.

EXAMPLE 17

total surface area = lateral surface area + area of the bases
area of base = (6 in.)(4 in.) = 24 in²
area of both bases = 2(24 in²) = 48 in²
total surface area = 100 in² + 48 in² = 148 in²

Figure 1.19

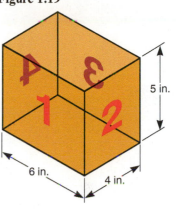

· · · · · · · · · · · · · · · · · ·

Area formulas, volume formulas, and lateral surface area formulas are provided on the inside back cover.

PROBLEMS 1.5

Find the area of each figure.

1.
5 cm
8 cm

2.
15 cm
28 cm

3.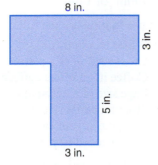
8 in.
3 in.
5 in.
3 in.

4.
6 cm   3 cm
7 cm
12 cm
15 cm

5. Find the cross-sectional area of the I-beam.

1 in.
1 in.
6 in.
1 in.
8 in.

6. Find the largest cross-sectional area of the figure.

35 cm
15 cm
15 cm
12 cm
40 cm

Find the volume in each figure.

7.

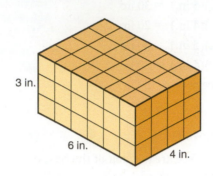

3 in.

6 in.   4 in.

8.

4 cm

7 cm   3 cm

7 cm   4 cm

3 cm

9.

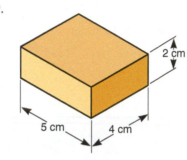

2 cm

5 cm   4 cm

10.

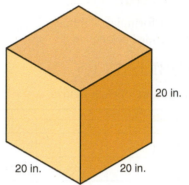

20 in.

20 in.   20 in.

Which unit is larger?

11.   1 litre or 1 centilitre
12.   1 millilitre or 1 kilolitre
13.   1 cubic millimetre or 1 cubic centimetre
14.   1 cm^3 or 1 m^3
15.   1 square kilometre or 1 hectare
16.   1 mm^2 or 1 dm^2

Which metric unit (m^3, L, mL, m^2, cm^2, ha) would you use to measure the following?

17.   Oil in your car's crankcase
18.   Water in a bathtub
19.   Floor space in a house
20.   Cross section of a piston
21.   Storage space in a miniwarehouse
22.   Coffee in an office coffeepot
23.   Size of a field of corn
24.   Page size of a newspaper
25.   A dose of cough syrup
26.   Size of a cattle ranch
27.   Cargo space in a truck
28.   Gasoline in your car's gas tank
29.   Piston displacement of an engine
30.   Paint needed to paint a house

Fill in the blank with the most reasonable metric unit (m^3, L, mL, m^2, cm^2, ha).

31.   Go to the store and buy 4 _____ of root beer for the party.
32.   I drank 200 _____ of orange juice for breakfast.
33.   Craig bought a 30-_____ tarpaulin for his truck.
34.   The cross section of a log is 3200 _____.
35.   A farmer's gasoline storage tank holds 4000 _____.
36.   Our city water tower holds 500 _____ of water.
37.   Brian planted 60 _____ of soybeans this year.
38.   David needs some copper tubing with a cross section of 3 _____.
39.   Paula ordered 15 _____ of concrete for her new driveway.
40.   Barbara heats 420 _____ of living space in her house.
41.   Joyce's house has 210 _____ of floor space.
42.   Kurt mows 5 _____ of grass each week.
43.   Amy is told by her doctor to drink 2 _____ of water each day.
44.   My favorite coffee cup holds 225 _____ of coffee.

Fill in each blank.

45. $1 \text{ L} = \underline{\hspace{1cm}} \text{ mL}$
46. $1 \text{ kL} = \underline{\hspace{1cm}} \text{ L}$
47. $1 \text{ L} = \underline{\hspace{1cm}} \text{ daL}$
48. $1 \text{ L} = \underline{\hspace{1cm}} \text{ kL}$
49. $1 \text{ L} = \underline{\hspace{1cm}} \text{ hL}$
50. $1 \text{ mL} = \underline{\hspace{1cm}} \text{ L}$
51. $1 \text{ mL} = \underline{\hspace{1cm}} \text{ cm}^3$
52. $1 \text{ L} = \underline{\hspace{1cm}} \text{ cm}^3$
53. $1 \text{ m}^3 = \underline{\hspace{1cm}} \text{ cm}^3$
54. $1 \text{ cm}^3 = \underline{\hspace{1cm}} \text{ mL}$
55. $1 \text{ cm}^3 = \underline{\hspace{1cm}} \text{ L}$
56. $1 \text{ dm}^3 = \underline{\hspace{1cm}} \text{ L}$
57. $1 \text{ m}^2 = \underline{\hspace{1cm}} \text{ cm}^2$
58. $1 \text{ km}^2 = \underline{\hspace{1cm}} \text{ m}^2$
59. $1 \text{ cm}^2 = \underline{\hspace{1cm}} \text{ mm}^2$
60. $1 \text{ ha} = \underline{\hspace{1cm}} \text{ m}^2$
61. $1 \text{ km}^2 = \underline{\hspace{1cm}} \text{ ha}$
62. $1 \text{ ha} = \underline{\hspace{1cm}} \text{ km}^2$
63. Change 7500 mL to L.
64. Change 0.85 L to mL.
65. Change 275 cm^3 to mm^3.
66. Change 5 m^3 to cm^3.
67. Change 275 cm^3 to mL.
68. Change 520 mm^3 to cm^3.
69. Change 1 m^3 to L.
70. Change 125 cm^3 to L.
71. Change 7.5 L to cm^3.
72. Change 450 L to m^3.
73. Change 5000 mm^2 to cm^2.
74. Change 1.75 km^2 to m^2.
75. Change 5 m^2 to cm^2.
76. Change 250 cm^2 to mm^2.
77. Change $4 \times 10^8 \text{ m}^2$ to km^2.
78. Change $5 \times 10^7 \text{ cm}^2$ to m^2.
79. Change 5 yd^2 to ft^2.
80. How many m^2 are in 225 ft^2?
81. Change 15 ft^2 to cm^2.
82. Change 108 in^2 to ft^2.
83. How many in^2 are in 51 cm^2?
84. How many ft^2 are in a rectangle 15 m long and 12 m wide?
85. How many in^2 are in a square 11 yd on a side?
86. How many m^2 are in a doorway whose area is 20 ft^2?
87. Change 19 yd^3 to ft^3.
88. How many in^3 are in 29 cm^3?
89. How many yd^3 are in 23 m^3?
90. How many cm^3 are in 88 in^3?
91. Change 8 ft^3 to in^3.
92. How many in^3 are in 12 m^3?
93. The volume of a casting is 38 in^3. What is its volume in cm^3?
94. How many castings of 14 cm^3 can be made from a 12-ft^3 block of steel?
95. (a) Find the lateral surface area and (b) find the total surface area of the figure in Problem 9.
96. (a) Find the lateral surface area and (b) find the total surface area of the figure in Problem 10.
97. How many mL of water would the figure in Problem 9 hold?
98. How many mL of water would the figure in Problem 8 hold?

1.6 Other Units

Mass and Weight

The **mass** of an object is the quantity of material making up the object. One unit of mass in the metric system is the *gram* (g) (Fig. 1.20). The gram is defined as the mass of 1 cubic centimetre (cm^3) of water at its maximum density. Since the gram is so small, the **kilogram** (kg) is the basic unit of mass in the metric system. One kilogram is defined as the mass of 1 cubic decimetre (dm^3) of water at its maximum density. The standard kilogram is a special platinum–iridium cylinder at the International Bureau of Weights and Measures near Paris, France. Since $1 \text{ dm}^3 = 1 \text{ L}$, 1 litre of water has a mass of 1 kilogram.

For very, very small masses, such as medicine dosages, we use the *milligram* (mg). One grain of salt has a mass of about 1 mg. The *metric ton* (1000 kg) is used to measure the mass of very large quantities, such as the coal on a barge, a trainload of grain, or a shipload of ore.

Figure 1.20

(a) A common paper clip has a mass of about 1 g.   (b) Three aspirin have a mass of about 1 g.

EXAMPLE 1

Change 74 kg to grams.

Choose the conversion factor with kilograms in the denominator so that the kg units cancel each other.

$$74 \, \text{kg} \times \frac{1000 \, \text{g}}{1 \, \text{kg}} = 74{,}000 \, \text{g}$$

EXAMPLE 2

Change 600 mg to grams.

$$600 \, \text{mg} \times \frac{1 \, \text{g}}{1000 \, \text{mg}} = 0.6 \, \text{g}$$

Figure 1.21   Spring balance

The **weight** of an object is a measure of the gravitational force or pull acting on it. The weight unit in the metric system is the *newton* (N). A small apple weighs about one newton.

The *pound* (lb), a unit of force, is one of the basic U.S. system units. It is defined as the pull of the earth on a cylinder of a platinum–iridium alloy that is stored in a vault at the U.S. Bureau of Standards. The *ounce* (oz) is another common unit of weight in the U.S. system. The relationship between pounds and ounces is

$$1 \, \text{lb} = 16 \, \text{oz}$$

The following relationships can be used for conversion between systems of units:

$$1 \, \text{N} = 0.225 \, \text{lb} \qquad \text{or} \qquad 1 \, \text{lb} = 4.45 \, \text{N}$$

The mass of an object remains constant, but its weight changes according to its distance from the earth or another planet. Mass and weight and their units of measure are discussed in more detail in Section 5.3.

A **spring balance** (Fig. 1.21) is an instrument containing a spring with a pointer attached to it. The spring stretches in proportion to the weight of the item being weighed. The weight is shown on a calibrated scale read directly in pounds or newtons. The common bathroom scale uses this principle to measure weight.

Figure 1.22   Platform balance

A **platform balance** (Fig. 1.22) consists of two platforms connected by a horizontal rod that balances on a knife edge. This device compares the pull of gravity on objects that are on the two platforms. The platforms are at the same height only when the unknown mass of the object on the left is equal to the known mass placed on the right. It is also possible to use one platform and a mass that slides along a calibrated scale. Variations of this basic design are found in some meat market and truck scales.

Courtesy of Dorling Kindersley Media Library

EXAMPLE 3

The weight of the intake valve of an auto engine is 0.18 lb. What is its weight in ounces and in newtons?

To find the weight in ounces, we simply use a conversion factor as follows:

$$0.18 \, \text{lb} \times \frac{16 \, \text{oz}}{1 \, \text{lb}} = 2.88 \, \text{oz}$$

To find the weight in newtons, we again use a conversion factor:

$$0.18 \, \text{lb} \times \frac{4.45 \, \text{N}}{1 \, \text{lb}} = 0.801 \, \text{N}$$

Time

Airlines and other transportation systems run on time schedules that would be meaningless if we did not have a common unit for time measurement. All the common units for time measurement are the same in both systems. These units are based on the motion of the earth (Fig. 1.23). The year is the amount of time required for one complete revolution of the earth about the sun. The month is the amount of time for one complete revolution of the moon about the earth. The day is the amount of time for one rotation of the earth about its axis.

Figure 1.23   Time units are based on the motion of the earth.

(a) One year is the amount of time it takes for the earth to revolve about the sun.

(b) One day is the amount of time it takes for the earth to rotate about its axis.

The basic time unit is the **second** (s). For many years, the second was defined as $\frac{1}{86,400}$ of a mean solar day. The standard second adopted in 1967 is defined more precisely in terms of the frequency of radiation emitted by cesium atoms when they pass between two particular states; this is the time required for 9,192,631,770 periods of this radiation. The second is not always convenient to use, so other units are necessary. The *minute* (min) is 60 seconds, the *hour* (h) is 60 minutes, and the *day* is 24 hours. The *year* is 365 days in length except for every fourth year, when it is 366 days long. This difference is necessary to keep the seasons at the same time each year, since one revolution of the earth about the sun takes approximately $365\frac{1}{4}$ days.

The Julian calendar introduced by Julius Caesar in 46 BC provided for an ordinary year of 365 days and a leap year of 366 days every fourth year. Astronomers have found that the length of a year has varied from 365.24253 days in 5000 BC to 365.24219 days in 2000 AD. The Gregorian calendar now used in most countries of the world was introduced by Pope Gregory XIII in 1582 to correct the Julian calendar discrepancies. The Gregorian calendar provides for an ordinary year of 365 days and a leap year of 366 days in years divisible by four except in century years not divisible by 400. Thus, years 1600 and 2000 are leap years but years 1700, 1800, 1900, and 2100 are not. In 1582, 10 days were omitted from the calendar to adjust for the accumulated difference of the Julian calendar since 46 BC. By decree, the day following October 4, 1582, became October 15, 1582. Now, small fractions-of-a-second adjustments are made in the calendar annually by international agreement to compensate for the variation of the earth's orbit around the sun.

Common devices for time measurement are the electric clock, the mechanical watch, and the quartz crystal watch. The accuracy of an electric clock depends on how accurately the 60-Hz (hertz = cycles per second) line voltage is controlled. In the United States this is controlled very accurately. Most mechanical watches have a balance wheel that oscillates near a given frequency, usually 18,000 to 36,000 vibrations per hour, and drives the hands of the watch (Fig. 1.24). The quartz crystal in a watch is excited by a small power cell and vibrates 32,768 times per second. The accuracy of the watch depends on how well the frequency of oscillation is controlled.

Figure 1.24   Oscillation of balance wheel

EXAMPLE 4

Change 2 h 15 min to seconds.
    First,

$$2 \, \cancel{h} \times \frac{60 \text{ min}}{1 \, \cancel{h}} = 120 \text{ min}$$

Then

$$2 \text{ h } 15 \text{ min} = 120 \text{ min} + 15 \text{ min} = 135 \text{ min}$$

and

$$135 \, \cancel{\text{min}} \times \frac{60 \text{ s}}{1 \, \cancel{\text{min}}} = 8100 \text{ s}$$

Very short periods of time are measured in parts of a second, given with the appropriate metric prefix. Such units are commonly used in electronics.

EXAMPLE 5

What is the meaning of each unit?

(a)   1 ms = 1 millisecond = 10^{-3} s and means one one-thousandth of a second.
(b)   1 μs = 1 microsecond = 10^{-6} s and means one one-millionth of a second.
(c)   1 ns = 1 nanosecond = 10^{-9} s and means one one-billionth of a second.
(d)   1 ps = 1 picosecond = 10^{-12} s and means one one-trillionth of a second.

    Note: The Greek letter μ is pronounced "my$\overline{oo}$." However, 1 μs is stated or read as "one microsecond."

EXAMPLE 6

Change 45 ms to seconds.
    Since 1 ms = 10^{-3} s,

$$45 \, \cancel{\text{ms}} \times \frac{10^{-3} \text{ s}}{1 \, \cancel{\text{ms}}} = 45 \times 10^{-3} \text{ s} = 0.045 \text{ s}$$

EXAMPLE 7

Change 0.000000025 s to nanoseconds.
    Since 1 ns = 10^{-9} s,

$$0.000000025 \, \cancel{s} \times \frac{1 \text{ ns}}{10^{-9} \, \cancel{s}} = 25 \text{ ns}$$

PROBLEMS 1.6

Which unit is larger?

1.   1 gram or 1 centigram
2.   1 gram or 1 milligram
3.   1 gram or 1 kilogram
4.   1 centigram or 1 milligram
5.   1 centigram or 1 kilogram
6.   1 milligram or 1 kilogram

Which metric unit (kg, g, mg, or metric ton) would you use to measure the following?

7.   Your mass
8.   An aspirin
9.   A bag of lawn fertilizer
10.   A bar of hand soap
11.   A trainload of grain
12.   A sewing needle
13.   A small can of corn
14.   A channel catfish
15.   A vitamin capsule
16.   A car

Fill in each blank with the most reasonable metric unit (kg, g, mg, or metric ton).

17. A newborn's mass is about 3 _____.
18. An elevator in a local department store has a load limit of 2000 _____.
19. Margie's diet calls for 250 _____ of meat.
20. A 200-car train carries 11,000 _____ of soybeans.
21. A truckload shipment of copper pipe has a mass of 900 _____.
22. A carrot has a mass of 75 _____.
23. A candy recipe calls for 175 _____ of chocolate.
24. My father has a mass of 70 _____.
25. A pencil has a mass of 10 _____.
26. A nickel has a mass of 5 _____.
27. My favorite spaghetti recipe calls for 1 _____ of ground beef.
28. My favorite spaghetti recipe calls for 150 _____ of tomato paste.
29. Our local grain elevator shipped 10,000 _____ of wheat last year.
30. A slice of bread has a mass of about 25 _____.
31. I bought a 5-_____ bag of potatoes at the store today.
32. My grandmother takes 250-_____ capsules for her arthritis.

Fill in each blank.

33. 1 kg = _____ g
34. 1 mg = _____ g
35. 1 g = _____ cg
36. 1 g = _____ hg
37. 1 dg = _____ g
38. 1 dag = _____ g
39. 1 g = _____ mg
40. 1 cg = _____ g
41. 1 g = _____ kg
42. 1 mg = _____ μg
43. Change 575 g to mg.
44. Change 575 g to kg.
45. Change 650 mg to g.
46. Change 375 kg to g.
47. Change 30 kg to mg.
48. Change 4 metric tons to kg.
49. Change 400 μg to mg.
50. Change 30,000 kg to metric tons.
51. What is the mass of 750 mL of water?
52. What is the mass of 1 m^3 of water?
53. The weight of a car is 3500 lb. Find its weight in newtons.
54. A certain bridge is designed to support 150,000 lb. Find the maximum weight that it will support in newtons.
55. Jose weighs 200 lb. What is his weight in newtons?
56. Change 80 lb to newtons.
57. Change 2000 N to pounds.
58. Change 2000 lb to newtons.
59. Change 120 oz to pounds.
60. Change 3.5 lb to ounces.
61. Change 10 N to ounces.
62. Change 25 oz to newtons.
63. Find the metric weight of a 94-lb bag of cement.
64. What is the weight in newtons of 500 blocks if each weighs 3 lb?

Fill in each blank.

65. The basic metric unit of time is _____. Its abbreviation is _____.
66. The basic metric unit of mass is _____. Its abbreviation is _____.
67. The common metric unit of weight is _____. Its abbreviation is _____.

Which is larger?

68. 1 second or 1 millisecond
69. 1 millisecond or 1 nanosecond
70. 1 ps or 1 μs
71. 1 ms or 1 μs
72. Write the abbreviation for each unit.
    (a) 8.6 microseconds   (b) 45 nanoseconds   (c) 75 picoseconds
73. Change 345 μs to s.
74. Change 1 h 25 min to min.
75. Change 4 h 25 min 15 s to s.
76. Change 7×10^6 s to h.
77. Change 4 s to ns.
78. Change 1 h to ps.

1.7 Measurement: Significant Digits and Accuracy

Up to this time in your studies, probably all numbers and all measurements have been treated as exact numbers. An **exact number** is a number that has been determined as a result of counting, such as 24 students are enrolled in this class, or by some definition, such as 1 h = 60 min or 1 in. = 2.54 cm, a conversion definition agreed to by the world governments' bureaus of standards. Generally, the treatment of the addition, subtraction, multiplication, and division of exact numbers is the emphasis or main content of elementary mathematics.

However, nearly all data of a technical nature involve **approximate numbers**, that is, numbers determined as a result of some measurement process—some direct, as with a ruler, and some indirect, as with a surveying transit or reading an electric meter. First, realize that no measurement can be found exactly. The length of the cover of this book can be found using many instruments. The better the measuring device used, the better is the measurement.

A measurement may be expressed in terms of its accuracy or its precision. The **accuracy** of a measurement refers to the number of digits, called **significant digits**, which indicates the number of units that we are reasonably sure of having counted when making a measurement. The greater the number of significant digits given in a measurement, the better is the accuracy, and vice versa.

EXAMPLE 1

The average distance between the moon and the earth is 385,000 km. This measurement indicates measuring 385 thousands of kilometres; its accuracy is indicated by three significant digits.

EXAMPLE 2

A measurement of 0.025 cm indicates measuring 25 thousandths of a centimetre; its accuracy is indicated by two significant digits.

EXAMPLE 3

A measurement of 0.0500 s indicates measuring $50\overline{0}$ ten-thousandths of a second; its accuracy is indicated by three significant digits.

Notice that sometimes a zero is significant and sometimes it is not. To clarify this, we use the following rules for significant digits:

SIGNIFICANT DIGITS

1. All nonzero digits are significant: 156.4 m has four significant digits (this measurement indicates 1564 tenths of metres).
2. All zeros between significant digits are significant: 306.02 km has five significant digits (this measurement indicates 30,602 hundredths of kilometres).
3. In a number greater than 1, a zero that is specially tagged, such as by a bar above it, is significant: $23\overline{0},000$ km has three significant digits (this measurement indicates $23\overline{0}$ thousands of kilometres).
4. All zeros to the right of a significant digit *and* a decimal point are significant: 86.10 cm has four significant digits (this measurement indicates $861\overline{0}$ hundredths of centimetres).
5. In whole-number measurements, zeros at the right that are not tagged are *not* significant: 2500 m has two significant digits (25 hundreds of metres).
6. In measurements of less than 1, zeros at the left are *not* significant: 0.00752 m has three significant digits (752 hundred-thousandths of a metre).

When a number is written in scientific notation, the decimal part indicates the number of significant digits. For example, $20\bar{0},000$ m would be written in scientific notation as 2.00×10^5 m.

In summary:

> To find the number of significant digits:
>
> 1. All nonzero digits are significant.
> 2. Zeros are significant when they
>   (a) are between significant digits;
>   (b) follow the decimal point and a significant digit; or
>   (c) are in a whole number and a bar is placed over the zero.

Determine the accuracy (the number of significant digits) of each measurement.

EXAMPLE 4

Measurement	Accuracy (significant digits)
(a) 2642 ft	4
(b) 2005 m	4 (Both zeros are significant.)
(c) 2050 m	3 (Only the first zero is significant.)
(d) 2500 m	2 (No zero is significant.)
(e) $250\bar{0}$ m	3 (Only the first zero is significant.)
(f) $250\bar{0}$ m	4 (Both zeros are significant.)
(g) 34,000 mi	2 (No zeros are significant.)
(h) 15,670,000 lb	4 (No zeros are significant.)
(i) 203.05 km	5 (Both zeros are significant.)
(j) 0.000345 kg	3 (No zeros are significant.)
(k) 75 N	2
(l) 2.3 s	2
(m) 0.02700 g	4 (Only the right two zeros are significant.)
(n) 2.40 cm	3 (The zero is significant.)
(o) 4.050 μs	4 (All zeros are significant.)
(p) 100.050 km	6 (All zeros are significant.)
(q) 0.004 s	1 (No zeros are significant.)
(r) 2.03×10^4 m^2	3 (The zero is significant.)
(s) 1.0×10^{-3} N	2 (The zero is significant.)
(t) 5×10^6 kg	1
(u) 3.060×10^8 m^3	4 (Both zeros are significant.)

PROBLEMS 1.7

Determine the accuracy (the number of significant digits) of each measurement.

1. 536 ft
2. 307.3 mi
3. 5007 m
4. 5.00 cm
5. 0.0070 in.
6. 6.010 cm
7. $84\bar{0}0$ km
8. $30\bar{0}0$ ft
9. 187.40 m
10. $5\bar{0}0$ g
11. 0.00700 in.
12. 10.30 cm
13. 376.52 m
14. 3.05 mi
15. 4087 kg
16. 35.00 mm
17. 0.0160 in.
18. $37\bar{0}$ lb
19. $40\bar{0}0$ N
20. 5010 ft^3
21. 7 N
22. 32,000 tons
23. 70.00 m^2
24. 0.007 m
25. 2.4×10^3 kg
26. 1.20×10^{-5} ms
27. 3.00×10^{-4} kg
28. 4.0×10^6 ft
29. 5.106×10^7 kg
30. 1×10^{-9} m

1.8 Measurement: Precision

The **precision** of a measurement refers to the smallest unit with which a measurement is made, that is, the position of the last significant digit.

EXAMPLE 1

The precision of the measurement 385,000 km is 1000 km. (The position of the last significant digit is in the thousands place.)

EXAMPLE 2

The precision of the measurement 0.025 cm is 0.001 cm. (The position of the last significant digit is in the thousandths place.)

EXAMPLE 3

The precision of the measurement 0.0500 s is 0.0001 s. (The position of the last significant digit is in the ten-thousandths place.)

Unfortunately, the terms *accuracy* and *precision* have several different common meanings. Here we will use each term consistently as we have defined them. A measurement of 0.0004 cm has good precision and poor accuracy when compared with the measurement 378.0 cm.

Measurement	Precision	Accuracy
0.0004 cm	0.0001 cm	1 significant digit
378.0 cm	0.1 cm	4 significant digits

EXAMPLE 4

Determine the precision of each measurement given in Example 4 of Section 1.7.

Measurement	Precision	Accuracy (significant digits)
(a) 2642 ft	1 ft	4
(b) 2005 m	1 m	4
(c) 2050 m	10 m	3
(d) 2500 m	100 m	2
(e) $25\overline{0}0$ m	10 m	3
(f) $250\overline{0}$ m	1 m	4
(g) 34,000 mi	1000 mi	2
(h) 15,670,000 lb	10,000 lb	4
(i) 203.05 km	0.01 km	5
(j) 0.000345 kg	0.000001 kg	3
(k) 75 N	1 N	2
(l) 2.3 s	0.1 s	2
(m) 0.02700 g	0.00001 g	4
(n) 2.40 cm	0.01 cm	3
(o) 4.050 μs	0.001 μs	4
(p) 100.050 km	0.001 km	6
(q) 0.004 s	0.001 s	1
(r) 2.03×10^4 m^2	0.01×10^4 m^2 or 100 m^2	3
(s) 1.0×10^{-3} N	0.1×10^{-3} N or 0.0001 N	2
(t) 5×10^6 kg	1×10^6 kg or 1,000,000 kg	1
(u) 3.060×10^8 m^3	0.001×10^8 m^3 or 1×10^5 m^3 or 100,000 m^3	4

PHYSICS CONNECTIONS

Precision and the New Clark Bridge

Bridges are usually not built from one end to the other. Construction typically begins in the middle of the bridge or on each end and meets in the middle. In doing so, it becomes extremely important that every section is in precise alignment so that the bridge will meet at the critical connection points.

For example, the New Clark Bridge crossing the Mississippi at Alton, Illinois, spans over 4600 ft and was designed to high precision so that each member of the bridge would be no more than $\frac{1}{8}$ in. out of alignment. As frames and towers were built in the flowing river, teams of surveyors used fixed points of reference and laser beams to survey the placement of each tower and pier. Ignoring the importance of accuracy and precision would have caused serious problems.

TRY THIS ACTIVITY

Accuracy and Precision

Measure the time it takes someone to run the 100-yard dash. Use a digital stopwatch and a regular wristwatch with a second hand. How accurate is each of the measurements? How precise is each of the measurements?

PROBLEMS 1.8

Determine the precision of each measurement.

1. 536 ft
2. 307.3 mi
3. 5007 m
4. 5.00 cm
5. 0.0070 in.
6. 6.010 cm
7. $84\overline{0}0$ km
8. $30\overline{0}0$ ft
9. 187.40 m
10. $5\overline{0}0$ g
11. 0.00700 in.
12. 10.30 cm
13. 376.52 m
14. 3.05 mi
15. 4087 kg
16. 35.00 mm
17. 0.0160 in.
18. $37\overline{0}$ lb
19. $4\overline{0}00$ N
20. 5010 ft^3
21. 7 N
22. 32,000 tons
23. 70.00 m^2
24. 0.007 m
25. 2.4×10^3 kg
26. 1.20×10^{-5} ms
27. 3.00×10^{-4} kg
28. 4.0×10^6 ft
29. 5.106×10^7 kg
30. 1×10^{-9} m

In each set of the measurements, find the measurement that is (a) the most accurate and (b) the most precise.

31. 15.7 in.; 0.018 in.; 0.07 in.
32. 368 ft; 600 ft; 180 ft
33. 0.734 cm; 0.65 cm; 16.01 cm
34. 3.85 m; 8.90 m; 7.00 m
35. 0.0350 s; 0.025 s; 0.00040 s; 0.051 s
36. 125.00 g; 8.50 g; 9.000 g; 0.05 g
37. $27,0\overline{0}0$ L; 350 L; 27.6 L; 4.75 L
38. 8.4 m; 15 m; 180 m; 0.40 m
39. 500 N; 10,000 N; 500,000 N; 50 N
40. 7.5 ms; 14.2 ms; 10.5 ms; 120.0 ms

In each set of measurements, find the measurement that is (a) the least accurate and (b) the least precise.

41. 16.4 in.; 0.075 in.; 0.05 in.
42. 475 ft; 300 ft; 360 ft
43. 27.5 m; 0.65 m; 12.02 m
44. 5.7 kg; 120 kg; 0.025 kg
45. 0.0250 g; 0.015 g; 0.00005 g; 0.75 g
46. 185.0 m; 6.75 m; 5.000 m; 0.09 m
47. 45,000 N; 250 N; 16.8 N; 0.25 N; 3 N
48. 2.50 kg; 42.0 kg; $15\overline{0}$ kg; 0.500 kg
49. $200\overline{0}$ kg; $10,\overline{0}00$ kg; $40\overline{0},000$ kg; 20 kg
50. 80 ft; 250 ft; 12,550 ft; $26\overline{0}0$ ft

PHYSICS ON THE JOB

Regulatory Inspection Official & Metrologist

Consumer pricing of gasoline is usually given to the nearest tenth of a cent per gallon. Many ask how gasoline can be measured with such precision? Are you really getting precisely one gallon for the given price? A legal regulatory inspection official and metrologist is responsible for visiting commercial and industrial entities to test equipment to ensure that measurements are being made accurately. Measurement standards and methods measurements are set by the National Metrology Institute. Consumers can rest assured that standards are being met when they observe the inspection stickers on scales, pumps, and meters!

Inspection officials must have a strong working knowledge of weights and measures and unit conversion. Many inspection official positions require a minimum of a two-year postsecondary degree in Metrology and Precision Measurement Technology with coursework in physics, chemistry, and methods of modern instrumental methods (Fig. 1.25).

Figure 1.25 (a) Regulatory inspection official inspecting the pump at a gas station

(b) Inspection sticker at a gas station

Courtesy of Arizona Department of Weights and Measures

Figure 1.26 Micrometer with precision 0.01 mm

Courtesy of fotolia © sima

1.9 Calculations with Measurements

If one person measures the length of one of two parts of a shaft with a micrometer calibrated in 0.01 mm as 42.28 mm and another person measures the second part with a ruler calibrated in mm as 54 mm, would the total length be 96.28 mm? Note that the sum 96.28 mm indicates a precision of 0.01 mm. The precision of the ruler is 1 mm, which means that the measurement 54 mm with the ruler could actually be anywhere between 53.50 mm and 54.50 mm using the micrometer (which has a precision of 0.01 mm and is shown in Fig. 1.26). That is, using the ruler, any measurement between 53.50 mm and 54.50 mm can only be read as 54 mm. Of course, this means that the tenths and hundredths digits in the sum 96.28 mm are really meaningless. In other words, *the sum or difference of measurements can be no more precise than the least precise measurement*. That is,

To add or subtract measurements:

1. Make certain that all the measurements are expressed in the same unit. If they are not, convert them all to the same unit.
2. Add or subtract.
3. Round the results to the same precision as the least precise measurement.

EXAMPLE 1

Add the measurements 16.6 mi, 124 mi, 3.05 mi, and 0.837 mi.
All measurements are in the same unit, so add.

$$
\begin{array}{r}
16.6 \ \text{mi} \\
124 \quad \text{mi} \\
3.05 \ \text{mi} \\
\underline{0.837 \ \text{mi}} \\
144.487 \ \text{mi} \rightarrow 144 \ \text{mi}
\end{array}
$$

Then, round this sum to the same precision as the least precise measurement, which is 124 mi. Thus, the sum is 144 mi.

....

EXAMPLE 2

Add the measurements 1370 cm, 1575 mm, 2.374 m, and 8.63 m.
First, convert all measurements to the same unit, say, m.

$$
\begin{array}{l}
1370 \ \text{cm} = 13.7 \ \text{m} \\
1575 \ \text{mm} = 1.575 \ \text{m}
\end{array}
$$

Then add.

$$
\begin{array}{r}
13.7 \quad \text{m} \\
1.575 \ \text{m} \\
2.374 \ \text{m} \\
\underline{8.63 \quad \text{m}} \\
26.279 \ \text{m} \rightarrow 26.3 \ \text{m}
\end{array}
$$

Then, round this sum to the same precision as the least precise measurement, which is 13.7 m. Thus, the sum is 26.3 m.

....

EXAMPLE 3

Subtract the measurements 3457.8 g − 2.80 kg.
First, convert both measurements to the same unit, say, g.

$$
2.80 \ \text{kg} = 28\overline{0}0 \ \text{g}
$$

Then subtract.

$$
\begin{array}{r}
3457.8 \ \text{g} \\
\underline{28\overline{0}0 \quad \text{g}} \\
657.8 \ \text{g} \rightarrow 660 \ \text{g}
\end{array}
$$

Then, round this difference to the same precision as the least precise measurement, which is $28\overline{0}0$ g. Thus, the difference is 660 g.

....

Now suppose that you wish to find the area of the base of a rectangular building. You measure its length as 54.7 m and its width as 21.5 m. Its area is then

$$
\begin{aligned}
A &= lw \\
A &= (54.7 \ \text{m})(21.5 \ \text{m}) \\
&= 1176.05 \ \text{m}^2
\end{aligned}
$$

Note that the result contains six significant digits, whereas each of the original measurements contains only three significant digits. To rectify this inconsistency, we say that the product or quotient of measurements can be no more accurate than the least accurate measurement. That is,

To multiply or divide measurements:

1. Multiply or divide the measurements as given.
2. Round the result to the same number of significant digits as the measurement with the least number of significant digits.

Using the preceding rules, we find that the area of the base of the rectangular building is 1180 m^2.

Note: We assume throughout that you are using a calculator to do all calculations.

EXAMPLE 4

Multiply the measurements (124 ft)(187 ft).

$$(124 \text{ ft})(187 \text{ ft}) = 23{,}188 \text{ ft}^2$$

Round this product to three significant digits, which is the accuracy of the least accurate measurement (and also the accuracy of each measurement in the example). That is,

$$(124 \text{ ft})(187 \text{ ft}) = 23{,}200 \text{ ft}^2$$

EXAMPLE 5

Multiply the measurements (2.75 m)(1.25 m)(0.75 m).

$$(2.75 \text{ m})(1.25 \text{ m})(0.75 \text{ m}) = 2.578125 \text{ m}^3$$

Round this product to two significant digits, which is the accuracy of the least accurate measurement (0.75 m). That is,

$$(2.75 \text{ m})(1.25 \text{ m})(0.75 \text{ m}) = 2.6 \text{ m}^3$$

EXAMPLE 6

Divide the measurements 144,000 ft^3 ÷ 108 ft.

$$144{,}000 \text{ ft}^3 \div 108 \text{ ft} = 1333.333\ldots \text{ ft}^2$$

Round this quotient to three significant digits, which is the accuracy of the least accurate measurement (the accuracy of both measurements in this example). That is,

$$144{,}000 \text{ ft}^3 \div 108 \text{ ft} = 1330 \text{ ft}^2$$

EXAMPLE 7

Find the value of $\dfrac{(68 \text{ ft})(10{,}\overline{0}00 \text{ lb})}{95.6 \text{ s}}$.

$$\frac{(68 \text{ ft})(10{,}\overline{0}00 \text{ lb})}{95.6 \text{ s}} = 7112.9707\ldots \frac{\text{ft lb}}{\text{s}}$$

Round this result to two significant digits, which is the accuracy of the least accurate measurement (68 ft). That is,

$$\frac{(68 \text{ ft})(10{,}\overline{0}00 \text{ lb})}{95.6 \text{ s}} = 7100 \text{ ft lb/s}$$

Find the value of $\dfrac{(58.0 \text{ kg})(2.40 \text{ m/s})^2}{5.40 \text{ m}}$.

EXAMPLE 8

$$\frac{(58.0 \text{ kg})(2.40 \text{ m/s})^2}{5.40 \text{ m}} = 61.8666 \ldots \frac{\text{kg m}}{\text{s}^2}$$

Carefully simplify the units:

$$\frac{(\text{kg})(\text{m/s})^2}{\text{m}} = \frac{(\text{kg})(\text{m}^2/\text{s}^2)}{\text{m}} = \frac{\text{kg m}}{\text{s}^2}$$

Round this result to three significant digits, which is the accuracy of the least accurate measurement (the accuracy of all measurements in this example). That is,

$$\frac{(58.0 \text{ kg})(2.40 \text{ m/s})^2}{5.40 \text{ m}} = 61.9 \text{ kg m/s}^2$$

$\cdots\cdots\cdots\cdots\cdots$

Note: To multiply or divide measurements, the units do not need to be the same. The units must be the same to add or subtract measurements. Also, the units are multiplied and/or divided in the same manner as the corresponding numbers.

Any power or root of a measurement should be rounded to the same accuracy as the given measurement.

COMBINATIONS OF OPERATIONS WITH MEASUREMENTS

For combinations of additions, subtractions, multiplications, divisions, and powers involving measurements, follow the usual order of operations used in mathematics:

1. Perform all operations inside parentheses first.
2. Evaluate all powers.
3. Perform any multiplications or divisions, in order, from left to right; then express each product or quotient using its correct accuracy.
4. Perform any additions or subtractions, in order, from left to right; then express the final result using the correct precision.

Find the value of $(4.00 \text{ m})(12.65 \text{ m}) + (24.6 \text{ m})^2 + \dfrac{235.0 \text{ m}^3}{16.00 \text{ m}}$.

EXAMPLE 9

$$(4.00 \text{ m})(12.65 \text{ m}) + (24.6 \text{ m})^2 + \frac{235.0 \text{ m}^3}{16.00 \text{ m}} =$$

$$50.6 \text{ m}^2 + 605 \text{ m}^2 + 14.69 \text{ m}^2 = 67\overline{0} \text{ m}^2$$

$\cdots\cdots\cdots\cdots\cdots$

Obviously, such calculations with measurements should be done with a calculator. When no calculator is available, you may round the original measurements or any intermediate results to one more digit than the required accuracy or precision as required in the final result.

If both exact numbers and approximate numbers (measurements) occur in the same calculation, only the approximate numbers are used to determine the accuracy or precision of the result.

The procedures for operations with measurements shown here are based on methods followed and presented by the American Society for Testing and Materials. There are even more sophisticated methods for dealing with the calculations of measurements. The method one uses, and indeed whether one should even follow any given procedure, depends on the number of measurements and the sophistication needed for a particular situation.

In this book, we generally follow the customary practice of expressing measurements in terms of three significant digits, which is the accuracy used in most engineering and design work.

PROBLEMS 1.9

Use the rules for addition of measurements to add each set of measurements.

1.	3847 ft	2.	8,560 m	3.	42.8 cm	4.	0.456 g
	5800 ft		84,000 m		16.48 cm		0.93 g
	4520 ft		18,476 m		1.497 cm		0.402 g
			12,500 m		12.8 cm		0.079 g
					9.69 cm		0.964 g

5.   39,000 N; 19,600 N; 8470 N; 2500 N
6.   6800 ft; 2760 ft; 4$\overline{0}$00 ft; 20$\overline{0}$0 ft
7.   467 m; 970 cm; 12$\overline{0}$0 cm; 1352 cm; 30$\overline{0}$ m
8.   36.8 m; 147.5 cm; 1.967 m; 125.0 m; 98.3 cm
9.   12 s; 1.004 s; 0.040 s; 3.9 s; 0.87 s
10.   160,000 N; 84,200 N; 4300 N; 239,000 N; 17,450 N

Use the rules for subtraction of measurements to subtract each second measurement from the first.

11.	2876 kg	12.	14.73 m	13.	45.585 g	14.	34,500 kg
	2400 kg		9.378 m		4.6 g		9,5$\overline{0}$0 kg

15.   4200 km − 975 km
16.   64.73 g − 9.4936 g
17.   1,600,000 kg − 685,000 kg
18.   170 mm − 10.2 cm
19.   3.00 m − 26$\overline{0}$ cm
20.   1.40 ms − 0.708 ms

Use the rules for multiplication of measurements to multiply each set of measurements.

21.   (125 m)(39 m)
22.   (470 ft)(120$\overline{0}$ ft)
23.   (1637 km)(857 km)
24.   (9100 m)(6$\overline{0}$0 m)
25.   (18.70 m)(39.45 m)
26.   (565 cm)(180 cm)
27.   (14.5 cm)(18.7 cm)(20.5 cm)
28.   (0.046 m)(0.0317 m)(0.0437 m)
29.   (45$\overline{0}$ in.)(315 in.)(205 in.)
30.   (18.7 kg)(217 m)

Use the rules for division of measurements to divide.

31. $360 \text{ ft}^3 \div 12 \text{ ft}^2$
32. $125 \text{ m}^2 \div 3.0 \text{ m}$
33. $275 \text{ cm}^2 \div 90.0 \text{ cm}$
34. $185 \text{ mi} \div 4.5 \text{ h}$
35. $\dfrac{347 \text{ km}}{4.6 \text{ h}}$
36. $\dfrac{2700 \text{ m}^3}{9\overline{0}0 \text{ m}^2}$
37. $\dfrac{8800 \text{ mi}}{8.5 \text{ h}}$
38. $\dfrac{4960 \text{ ft}}{2.95 \text{ s}}$

Use the rules for multiplication and division of measurements to find the value of each of the following.

39. $\dfrac{(18 \text{ ft})(290 \text{ lb})}{4.6 \text{ s}}$
40. $\dfrac{(18.5 \text{ kg})(4.65 \text{ m})}{19.5 \text{ s}}$
41. $\dfrac{4500 \text{ mi}}{12.3 \text{ h}}$
42. $\dfrac{48.9 \text{ kg}}{(1.5 \text{ m})(3.25 \text{ m})}$
43. $\dfrac{(48.7 \text{ m})(68.5 \text{ m})(18.4 \text{ m})}{(35.5 \text{ m})(40.0 \text{ m})}$
44. $\frac{1}{2}(270 \text{ kg})(16.4 \text{ m/s})^2$
45. $\dfrac{(85.7 \text{ kg})(25.7 \text{ m/s})^2}{12.5 \text{ m}}$
46. $\dfrac{(45.2 \text{ kg})(13.7 \text{ m})}{(2.65 \text{ s})^2}$

47. $\frac{4}{3}\pi(13.5\text{ m})^3$

48. $\dfrac{140\text{ g}}{(3.4\text{ cm})(2.8\text{ cm})(5.6\text{ cm})}$

49. $(213\text{ m})(65.3\text{ m}) - (175\text{ m})(44.5\text{ m})$

50. $(4.5\text{ ft})(7.2\text{ ft})(12.4\text{ ft}) + (5.42\text{ ft})^3$

51. $\dfrac{(125\text{ ft})(295\text{ ft})}{44.7\text{ ft}} + \dfrac{(215\text{ ft})^3}{(68.8\text{ ft})(12.4\text{ ft})} + \dfrac{(454\text{ ft})^3}{(75.5\text{ ft})^2}$

52. $(12.5\text{ m})(46.75\text{ m}) + \dfrac{(6.76\text{ m})^3}{4910\text{ m}} + \dfrac{(41.5\text{ m})(21\text{ m})(28.8\text{ m})}{31.7\text{ m}}$

Glossary

Accuracy The number of digits, called significant digits, in a measurement, which indicates the number of units that we are reasonably sure of having counted. The greater the number of significant digits, the better is the accuracy. (p. 36)

Approximate Number A number that has been determined by some measurement or estimation process. (p. 36)

Area The number of square units contained in a figure. (p. 22)

Conversion Factor An expression used to convert from one set of units to another. Often expressed as a fraction whose numerator and denominator are equal to each other although in different units. (p. 18)

Exact Number A number that has been determined as a result of counting, such as 21 students enrolled in a class, or by some definition, such as 1 h = 60 min. (p. 36)

Kilogram The basic metric unit of mass. (p. 31)

Lateral Surface Area The area of all the lateral (side) faces of a geometric solid. (p. 28)

Mass A measure of the quantity of material making up an object. (p. 31)

Metre The basic metric unit of length. (p. 17)

Platform Balance An instrument consisting of two platforms connected by a horizontal rod that balances on a knife edge. The pull of gravity on objects placed on the two platforms is compared. (p. 32)

Precision Refers to the smallest unit with which a measurement is made, that is, the position of the last significant digit. (p. 38)

Scientific Notation A form in which a number can be written as a product of a number between 1 and 10 and a power of 10. The general form is $M \times 10^n$, where M is a number between 1 and 10 and n is the exponent or power of 10. (p. 15)

Second The basic unit of time. (p. 33)

SI (Système International d'Unités) The international modern metric system of units of measurement. (p. 13)

Significant Digits The number of digits in a measurement, which indicates the number of units we are reasonably sure of having counted. (p. 36)

Spring Balance An instrument containing a spring, which stretches in proportion to the force applied to it, and a pointer attached to the spring with a calibrated scale read directly in given units. (p. 32)

Standards of Measure A set of units of measurement for length, weight, and other quantities defined in such a way as to be useful to a large number of people. (p. 11)

Total Surface Area The total area of all the surfaces of a geometric solid, that is, the lateral surface area plus the area of the bases. (p. 28)

Volume The number of cubic units contained in a figure. (p. 25)

Weight A measure of the gravitational force or pull acting on an object. (p. 32)

Review Questions

1. What are the basic metric units for length, mass, and time?
   - (a) Foot, pound, hour
   - (b) Newton, litre, second
   - (c) Metre, kilogram, second
   - (d) Mile, ton, day
2. When a value is multiplied or divided by 1, the value is
   - (a) increased.
   - (b) unchanged.
   - (c) decreased.
   - (d) none of the above.
3. The lateral surface area of a solid is
   - (a) always equal to total surface area.
   - (b) never equal to total surface area.
   - (c) usually equal to total surface area.
   - (d) rarely equal to total surface area.
4. Accuracy is
   - (a) the same as precision.
   - (b) the smallest unit with which a measurement is made.
   - (c) the number of significant digits.
   - (d) all of the above.
5. When multiplying or dividing two or more measurements, the units
   - (a) must be the same.   (b) must be different.   (c) can be different.
6. Cite three examples of problems that would arise in the construction of a home by workers using different systems of measurement.
7. Why is the metric system preferred worldwide to the U.S. system of measurement?
8. List a very large and a very small measurement that could be usefully written in scientific notation.
9. When using conversion factors, can units be treated like other algebraic quantities?
10. What is the meaning of cross-sectional area?
11. Can a brick have more than one cross-sectional area?
12. What is the fundamental metric unit for land area?
13. Which is larger, a litre or a quart?
14. List three things that might conveniently be measured in millilitres.
15. How do weight and mass differ?
16. What is the basic metric unit of weight?
17. A microsecond is one-____ of a second.
18. Why must we concern ourselves with significant digits?
19. Can the sum or difference of two measurements ever be more precise than the least precise measurement?
20. When rounding the product or quotient of two measurements, is it necessary to consider significant digits?

Review Problems

Give the metric prefix for each value:

1. 1000

2. 0.001

Give the metric symbol, or abbreviation, for each prefix:

3. micro

4. mega

Write the abbreviation for each quantity:

5. 45 milligrams

6. 138 centimetres

Which is larger?

7. 1 L or 1 mL   8.   1 kg or 1 mg   9.   1 L or 1 m^3

Fill in each blank (round to three significant digits when necessary):

10.   250 m = ____ km   11.   850 mL = ____ L
12.   5.4 kg = ____ g   13.   0.55 s = ____ μs
14.   25 kg = ____ g   15.   75 μs = ____ ns
16.   275 cm^2 = ____ mm^2 17.   350 cm^2 = ____ m^2
18.   0.15 m^3 = ____ cm^3 19.   500 cm^3 = ____ mL
20.   150 lb = ____ kg   21.   36 ft = ____ m
22.   250 cm = ____ in.   23.   150 in^2 = ____ cm^2
24.   24 yd^2 = ____ ft^2 25.   6 m^3 = ____ ft^3
26.   16 lb = ____ N   27.   15,600 s = ____ h ____ min

Determine the accuracy (the number of significant digits) in each measurement:

28.   5.08 kg   29.   20,570 lb   30.   0.060 cm   31.   2.00 × 10^{-4} s

Determine the precision of each measurement:

32.   30.6 ft   33.   0.0500 s   34.   18,000 mi   35.   4 × 10^5 N

For each set of measurements, find the measurement that is

(a)   the most accurate.   (b)   the least accurate.
(c)   the most precise.   (d)   the least precise.

36.   12.00 m; 0.150 m; 2600 m; 0.008 m   37.   208 L; 18,050 L; 21.5 L; 0.75 L

Use the rules of measurements to add the following measurements:

38.   0.0250 s; 0.075 s; 0.00080 s; 0.024 s
39.   2100 N; 36,800 N; 24,000 N; 14.5 N; 470 N

Use the rules for multiplication and division of measurements to find the value of each of the following:

40.   (450 cm)(18.5 cm)(215 cm)   41. $\dfrac{1480 \text{ m}^3}{9.6 \text{ m}}$ 42. $\dfrac{(25.0 \text{ kg}) (1.20 \text{ m/s})^2}{3.70 \text{ m}}$

43.   Find the area of a rectangle 4.50 m long and 2.20 m wide.
44.   Find the volume of a rectangular box 9.0 cm long, 6.0 cm wide, and 13 cm high.

CHAPTER 2

PROBLEM SOLVING

Objectives

The major goals of this chapter are to enable you to:

1. Use formulas in problem solving.
2. Use a systematic approach to solving technical problems.
3. Analyze technical problems using a problem-solving method.

A formula is an equation, usually expressed in letters, called *variables*, and numbers. Much technical work includes the substitution of measured data into known formulas or relationships to find solutions to problems. A systematic approach to solving problems is a valuable tool.

The problem-solving method presented will assist you in processing data, analyzing the problems present, and finding the solution in an orderly manner.

2.1 Formulas

A **formula** is an equation, usually expressed in letters (called *variables*) and numbers. A **variable** is a symbol, usually a letter, used to represent some unknown number or quantity.

The formula $s = vt$ states that the distance traveled, s, equals the product of the velocity, v, and the time, t.

EXAMPLE 1

The formula $I = \dfrac{Q}{t}$ states that the current, I, equals the quotient of the charge, Q, and the time, t.

EXAMPLE 2

To solve a formula for a given letter means to express the given letter or variable in terms of all the remaining letters. That is, by using the equation-solving principles, rewrite the formula so that the given letter appears on one side of the equation by itself and all the other letters appear on the other side.

Solve $s = vt$ for v.

EXAMPLE 3

$$s = vt$$

$$\frac{s}{t} = \frac{vt}{t} \qquad \text{Divide both sides by } t.$$

$$\frac{s}{t} = v$$

Solve $I = Q/t$

EXAMPLE 4

(a) for Q. (b) for t.

(a)
$$I = \frac{Q}{t}$$

$$(I)t = \left(\frac{Q}{t}\right)t \quad \text{Multiply both sides by } t.$$

$$It = Q$$

(b) Starting with $It = Q$, we obtain

$$\frac{It}{I} = \frac{Q}{I} \qquad \text{Divide both sides by } I.$$

$$t = \frac{Q}{I}$$

Solve $V = E - Ir$ for r.

EXAMPLE 5

Method 1:

$$V = E - Ir$$
$$V - E = E - Ir - E \qquad \text{Subtract } E \text{ from both sides.}$$
$$V - E = -Ir$$
$$\frac{V - E}{-I} = \frac{-Ir}{-I} \qquad \text{Divide both sides by } -I.$$
$$\frac{V - E}{-I} = r$$

Method 2:

$$V = E - Ir$$

$$V + Ir = E - Ir + Ir \qquad \text{Add } Ir \text{ to both sides.}$$

$$V + Ir = E$$

$$V + Ir - V = E - V \qquad \text{Subtract } V \text{ from both sides.}$$

$$Ir = E - V$$

$$\frac{Ir}{I} = \frac{E - V}{I} \qquad \text{Divide both sides by } I.$$

$$r = \frac{E - V}{I}$$

Note that the two results are equivalent. Take the first result and multiply both numerator and denominator by -1. That is,

$$\frac{V - E}{-I} = \left(\frac{V - E}{-I}\right)\left(\frac{-1}{-1}\right) = \frac{-V + E}{I} = \frac{E - V}{I}$$

which is the same as the second result.

· · · · · · · · · · · · · · · · · ·

We often use the same quantity in more than one way in a formula. For example, we may wish to use a certain measurement of a quantity, such as velocity, at a given time, say, at $t = 0$ s, then use the velocity at a later time, say, at $t = 6$ s. To write these desired values of the velocity is rather awkward. We simplify this written statement by using *subscripts* (small letters or numbers printed a half-space below the printed line and to the right of the variable) to shorten what we must write.

For the example given, v at time $t = 0$ s will be written as v_i (initial velocity); v at time $t = 6$ s will be written as v_f (final velocity). Mathematically, v_i and v_f are two different quantities, which in most cases are unequal. The sum of v_i and v_f is written as $v_i + v_f$. The product of v_i and v_f is written as $v_i v_f$. The subscript notation is used only to distinguish the general quantity, v, velocity, from the measure of that quantity at certain specified times.

EXAMPLE 6

Solve the formula $x = x_i + v_i t + \frac{1}{2}at^2$ for v_i.

Method 1:

$$x = x_i + v_i t + \frac{1}{2}at^2$$

$$x - v_i t = x_i + v_i t + \frac{1}{2}at^2 - v_i t \qquad \text{Subtract } v_i t \text{ from both sides.}$$

$$x - v_i t = x_i + \frac{1}{2}at^2$$

$$x - v_i t - x = x_i + \frac{1}{2}at^2 - x \qquad \text{Subtract } x \text{ from both sides.}$$

$$-v_i t = x_i + \frac{1}{2}at^2 - x$$

$$\frac{-v_i t}{-t} = \frac{x_i + \frac{1}{2}at^2 - x}{-t} \qquad \text{Divide both sides by } -t.$$

$$v_i = \frac{x_i + \frac{1}{2}at^2 - x}{-t}$$

Method 2:

$$x = x_i + v_i t + \tfrac{1}{2}at^2$$

$$x - x_i - \tfrac{1}{2}at^2 = x_i + v_i t + \tfrac{1}{2}at^2 - x_i - \tfrac{1}{2}at^2 \qquad \text{Subtract } x_i \text{ and } \tfrac{1}{2}at^2 \text{ from both sides.}$$

$$x - x_i - \tfrac{1}{2}at^2 = v_i t$$

$$\frac{x - x_i - \tfrac{1}{2}at^2}{t} = \frac{v_i t}{t} \qquad\qquad \text{Divided both sides by } t.$$

$$\frac{x - x_i - \tfrac{1}{2}at^2}{t} = v_i$$

Solve the formula $v_{avg} = \tfrac{1}{2}(v_f + v_i)$ for v_f (*avg* is used here as a subscript meaning "average").

EXAMPLE 7

$$v_{avg} = \tfrac{1}{2}(v_f + v_i)$$

$$2v_{avg} = v_f + v_i \qquad \text{Multiply both sides by 2.}$$

$$2v_{avg} - v_i = v_f \qquad \text{Subtract } v_i \text{ from both sides.}$$

Solve $A = \dfrac{\pi d^2}{4}$ for d, where d is a diameter.

EXAMPLE 8

$$A = \frac{\pi d^2}{4}$$

$$4A = \left(\frac{\pi d^2}{4}\right)(4) \qquad \text{Multiply both sides by 4.}$$

$$4A = \pi d^2$$

$$\frac{4A}{\pi} = \frac{\pi d^2}{\pi} \qquad \text{Divide both sides by } \pi.$$

$$\frac{4A}{\pi} = d^2$$

$$\pm\sqrt{\frac{4A}{\pi}} = d \qquad \text{Take the square root of both sides.}$$

In this case, a negative diameter has no physical meaning, so the result is

$$d = \sqrt{\frac{4A}{\pi}}$$

PROBLEMS 2.1

Solve each formula for the quantity given.

1. $v = \dfrac{s}{t}$ for s
2. $a = \dfrac{v}{t}$ for v
3. $w = mg$ for m
4. $F = ma$ for a
5. $E = IR$ for R
6. $V = lwh$ for w
7. $E_P = mgh$ for g
8. $E_P = mgh$ for h
9. $v^2 = 2gh$ for h

10. $X_L = 2\pi f L$ for f 11. $P = \dfrac{W}{t}$ for W 12. $p = \dfrac{F}{A}$ for F

13. $P = \dfrac{W}{t}$ for t 14. $p = \dfrac{F}{A}$ for A 15. $E_k = \frac{1}{2}mv^2$ for m

16. $E_k = \frac{1}{2}mv^2$ for v^2 17. $W = Fs$ for s 18. $v_f = v_i + at$ for a

19. $V = E - Ir$ for I 20. $v_2 = v_1 + at$ for t 21. $R = \dfrac{\pi}{2P}$ for P

22. $R = \dfrac{kL}{d^2}$ for L 23. $F = \frac{9}{5}C + 32$ for C 24. $C = \frac{5}{9}(F - 32)$ for F

25. $X_C = \dfrac{1}{2\pi fC}$ for f 26. $R = \dfrac{\rho L}{A}$ for L

27. $R_T = R_1 + R_2 + R_3 + R_4$ for R_3 28. $Q_1 = P(Q_2 - Q_1)$ for Q_2

29. $\dfrac{I_S}{I_P} = \dfrac{N_P}{N_S}$ for I_P 30. $\dfrac{V_P}{V_S} = \dfrac{N_P}{N_S}$ for N_S

31. $v_{avg} = \frac{1}{2}(v_f + v_i)$ for v_i 32. $2a(s - s_i) = v^2 - v_i^2$ for a

33. $2a(s - s_i) = v^2 - v_i^2$ for s 34. $Ft = m(V_2 - V_1)$ for V_1

35. $Q = \dfrac{I^2 Rt}{J}$ for R 36. $x = x_i + v_i t + \frac{1}{2}at^2$ for x_i

37. $A = \pi r^2$ for r, where r is a radius   38. $V = \pi r^2 h$ for r, where r is a radius

39. $R = \dfrac{kL}{d^2}$ for d, where d is a diameter   40. $V = \frac{1}{3}\pi r^2 h$ for r, where r is a radius

41. $Q = \dfrac{I^2 Rt}{J}$ for I 42. $F = \dfrac{mv^2}{r}$ for v

2.2   Substituting Data into Formulas

An important part of problem solving is substituting the given data into the appropriate formula to find the value of the unknown quantity. Basically, there are two ways of substituting data into formulas to solve for the unknown quantity:

1.   Solve the formula for the unknown quantity and then make the substitution of the data.
2.   Substitute the data into the formula first and then solve for the unknown quantity.

When using a calculator, the first way is more useful. We will be using this way most of the time in this text.

EXAMPLE 1

Given the formula $A = bh$, $A = 120\,\text{m}^2$, and $b = 15\,\text{m}$, find h.

First, solve for h.

$$A = bh$$

$$\frac{A}{b} = \frac{bh}{b} \qquad \text{Divide both sides by } b.$$

$$\frac{A}{b} = h$$

Then substitute the data.

$$h = \frac{A}{b} = \frac{120 \text{ m}^2}{15 \text{ m}} = 8.0 \text{ m}$$

(Remember to follow the rules of measurement discussed in Chapter 1. We use them consistently throughout.)

· · · · · · · · · · · · · · · · ·

Given the formula $P = 2a + 2b$, $P = 824$ cm, and $a = 292$ cm, find b.

EXAMPLE 2

First, solve for b.

$$P = 2a + 2b$$
$$P - 2a = 2a + 2b - 2a \qquad \text{Subtract } 2a \text{ from both sides.}$$
$$P - 2a = 2b$$
$$\frac{P - 2a}{2} = \frac{2b}{2} \qquad \text{Divide both sides by 2.}$$
$$\frac{P - 2a}{2} = b \qquad \left(\text{or } b = \frac{P}{2} - a \right)$$

Then substitute the data.

$$b = \frac{P - 2a}{2} = \frac{824 \text{ cm} - 2\,(292 \text{ cm})}{2}$$
$$= \frac{824 \text{ cm} - 584 \text{ cm}}{2}$$
$$= \frac{24\bar{0} \text{ cm}}{2} = 12\bar{0} \text{ cm}$$

· · · · · · · · · · · · · · · · ·

Given the formula $A = \left(\dfrac{a + b}{2} \right)h$, $A = 15\bar{0}\,\text{m}^2$, $b = 18.0$ m, and $h = 10.0$ m, find a.

EXAMPLE 3

First, solve for a.

$$A = \left(\frac{a - b}{2} \right)h$$

$$2A = \left[\left(\frac{a - b}{2} \right)h \right](2) \qquad \text{Multiply both sides by 2.}$$

$$2A = (a + b)h$$

$$2A = ah + bh \qquad \text{Remove the parentheses.}$$

$$2A - bh = ah + bh - bh \qquad \text{Subtract } bh \text{ from both sides.}$$

$$2A - bh = ah$$

$$\frac{2A - bh}{h} = \frac{ah}{h} \qquad \text{Divide both sides by } h.$$

$$\frac{2A - bh}{h} = a \qquad \left(\text{or } a = \frac{2A}{h} - b \right)$$

Then substitute the data.

$$a = \frac{2A - bh}{h}$$

$$= \frac{2(15\bar{0} \text{ m}^2) - (18.0 \text{ m})(10.0 \text{ m})}{10.0 \text{ m}}$$

$$= \frac{30\bar{0} \text{ m}^2 - 18\bar{0} \text{ m}^2}{10.0 \text{ m}}$$

$$= \frac{12\bar{0} \text{ m}^2}{10.0 \text{ m}} = 12.0 \text{ m} \qquad \boxed{\frac{\text{m}^2}{\text{m}} = \text{m}}$$

Note: We often include a units analysis box to show how the final units are determined.

EXAMPLE 4

Given the formula $V = \frac{1}{3}\pi r^2 h$, $V = 64{,}400 \text{ mm}^3$, and $h = 48.0 \text{ mm}$, find r, where r is a radius.

First, solve for r.

$$V = \tfrac{1}{3}\pi r^2 h$$

$$3V = \left(\tfrac{1}{3}\pi r^2 h\right)(3) \qquad \text{Multiply both sides by 3.}$$

$$3V = \pi r^2 h$$

$$\frac{3V}{\pi h} = \frac{\pi r^2 h}{\pi h} \qquad \text{Divide both sides by } \pi h.$$

$$\frac{3V}{\pi h} = r^2$$

$$\pm\sqrt{\frac{3V}{\pi h}} = r \qquad \text{Take the square root of both sides.}$$

In this case, a negative radius has no physical meaning, so the result is

$$r = \sqrt{\frac{3V}{\pi h}}$$

Then substitute the data.

$$r = \sqrt{\frac{3(64{,}400 \text{ mm}^3)}{\pi(48.0 \text{ mm})}}$$

$$= 35.8 \text{ mm} \qquad \boxed{\sqrt{\frac{\text{mm}^3}{\text{mm}}} = \sqrt{\text{mm}^2} = \text{mm}}$$

PROBLEMS 2.2

For each formula, (a) solve for the indicated letter and then (b) substitute the given data to find the value of the indicated letter. Follow the rules of calculations with measurements.

Note: In Problems 14 and 16, r is a radius, and in Problem 15, b is the length of the side of a square.

Formula	Data	Find
1. $A = bh$	$b = 14.5$ cm, $h = 11.2$ cm	A
2. $V = lwh$	$l = 16.7$ m, $w = 10.5$ m, $h = 25.2$ m	V
3. $A = bh$	$A = 34.5$ cm^2, $h = 4.60$ cm	b
4. $P = 4b$	$P = 42\overline{0}$ in.	b
5. $P = a + b + c$	$P = 48.5$ cm, $a = 18.2$ cm, $b = 24.3$ cm	c
6. $C = \pi d$	$C = 495$ ft	d
7. $C = 2\pi r$	$C = 68.5$ yd	r
8. $A = \frac{1}{2}bh$	$A = 468$ m^2, $b = 36.0$ m	h
9. $P = 2(a + b)$	$P = 88.7$ km, $a = 11.2$ km	b
10. $V = \pi r^2 h$	$r = 61.0$ m, $h = 125.3$ m	V
11. $V = \pi r^2 h$	$V = 368$ m^3, $r = 4.38$ m	h
12. $A = 2\pi rh$	$A = 51\overline{0}$ cm^2, $r = 14.0$ cm	h
13. $V = Bh$	$V = 2185$ m^3, $h = 14.2$ m	B
14. $A = \pi r^2$	$A = 463.5$ m^2	r
15. $A = b^2$	$A = 465$ in^2	b
16. $V = \frac{1}{3}\pi r^2 h$	$V = 2680$ m^3, $h = 14.7$ m	r
17. $C = 2\pi r$	$r = 19.36$ m	C
18. $V = \frac{4}{3}\pi r^3$	$r = 25.65$ m	V
19. $V = \frac{1}{3}Bh$	$V = 19{,}850$ ft^3, $h = 486.5$ ft	B
20. $A = \left(\dfrac{a + b}{2}\right)h$	$A = 205.2$ m^2, $a = 16.50$ m, $b = 19.50$ m	h

2.3 Problem-Solving Method

Problem solving in technical fields is more than substituting numbers and units into formulas. You must develop skill in taking data, analyzing the problem, and finding the solution in an orderly manner. Understanding the principle involved in solving a problem is more important than blindly substituting into a formula. By following an orderly procedure for problem solving, we develop an approach to problem solving that you can use in your studies and on the job.

The following **problem-solving method** aids in understanding and solving problems and will be applied to all problems in this book where appropriate.

1. ***Read the problem carefully.*** This might appear obvious, but it is the most important step in solving a problem. As a matter of habit, you should read the problem at least twice.
    (a) The first time you should read the problem straight through from beginning to end. Do not stop to think about setting up an equation or formula. You are only trying to get a general overview of the problem during this first reading.
    (b) Read through a second time slowly and *completely*, beginning to think ahead to the following steps.
2. ***Make a sketch.*** Some problems may not lend themselves to a sketch. However, make a sketch whenever possible. Many times, seeing a sketch will show if you have forgotten important parts of the problem and may suggest the solution. This is a *very important* part of problem solving and is often overlooked.
3. ***Write all given information including units.*** This is necessary to have all essential facts in mind before looking for the solution. There are some common phrases that have understood physical meanings. For example, the term *from rest* means the initial velocity equals zero, or $v_i = 0$; the term *smooth surface* means assume that no friction is present.

4. ***Write the unknown or quantity asked for in the problem.*** Many students have difficulty solving problems because they don't know what they are looking for and solve for the wrong quantity.

5. ***Write the basic equation or formula that relates the known and unknown quantities.*** Find the basic formula or equation to use by studying what is given and what you are asked to find. Then look for a formula or equation that relates these quantities. Sometimes, you may need to use more than one equation or formula in a problem.

6. ***Find a working equation by solving the basic equation or formula for the unknown quantity.***

7. ***Substitute the data in the working equation, including the appropriate units.*** It is important that you *carry the units all the way through the problem* as a check that you have solved the problem correctly. For example, if you are asked to find the weight of an object in newtons and the units of your answer work out to be metres, you need to review your solution for the error. (When the unit analysis is not obvious, we will go through it step by step in a box nearby.)

8. ***Perform the indicated operations and work out the solution.*** Although this will be your final written step, you should always ask yourself, "Is my answer reasonable?" Here and on the job you will be dealing with practical problems. A quick estimate will often reveal an error in your calculations.

9. ***Check your answer.*** Ask yourself, "Did I answer the questions?"

To help you recall this procedure, with almost every problem set that follows, you will find Fig. 2.1 as shown here. This figure is not meant to be complete and is only an outline to assist you in remembering and following the procedure for solving problems. *You should follow this outline in solving all problems in this course.*

This problem-solving method will now be demonstrated in terms of relationships and formulas with which you are probably familiar. The formulas for finding area and volume can be found on the inside back cover.

Figure 2.1

SKETCH

12 cm^2 w

4.0 cm

DATA

$A = 12 \text{ cm}^2, l = 4.0 \text{ cm}, w = ?$

BASIC EQUATION

$A = lw$

WORKING EQUATION

$w = \dfrac{A}{l}$

SUBSTITUTION

$w = \dfrac{12 \text{ cm}^2}{4.0 \text{ cm}} = 3.0 \text{ cm}$

EXAMPLE 1

Find the volume of concrete required to fill a rectangular bridge abutment whose dimensions are $6.00 \text{ m} \times 3.00 \text{ m} \times 15.0 \text{ m}$.

Sketch:

15.0 m

6.00 m

3.00 m

Data:

$$\left.\begin{array}{l} l = 6.00 \text{ m} \\ w = 3.00 \text{ m} \\ h = 15.0 \text{ m} \end{array}\right\}$$ This is a listing of the information that is known.

$V = ?$ This identifies the unknown.

Basic Equation:

$$V = lwh$$

Working Equation: Same

Substitution:

$$V = (6.00 \text{ m})(3.00 \text{ m})(15.0 \text{ m})$$
$$= 27\overline{0} \text{ m}^3$$

Note: m × m × m = m³.

· · · · · · · · · · · · · · · ·

A rectangular holding tank 24.0 m in length and 15.0 m in width is used to store water for short periods of time in an industrial plant. If 2880 m³ of water is pumped into the tank, what is the depth of the water?

EXAMPLE 2

Sketch:

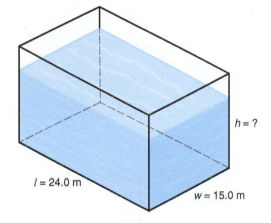

$h = ?$

$l = 24.0$ m

$w = 15.0$ m

Data:

$$V = 2880 \text{ m}^3$$
$$l = 24.0 \text{ m}$$
$$w = 15.0 \text{ m}$$
$$h = ?$$

Basic Equation:

$$V = lwh$$

Working Equation:

$$h = \frac{V}{lw}$$

Substitution:

$$h = \frac{2880 \text{ m}^3}{(24.0 \text{ m})(15.0 \text{ m})}$$

$$= 8.00 \text{ m} \qquad \boxed{\frac{\text{m}^3}{\text{m} \times \text{m}} = \text{m}}$$

· · · · · · · · · · · · · · · ·

EXAMPLE 3

A storage bin in the shape of a cylinder contains 814 m³ of storage space. If its radius is 6.00 m, find its height.

Sketch:

$r = 6.00$ m

$h = ?$

Data:

$$V = 814 \, \text{m}^3$$
$$r = 6.00 \, \text{m}$$
$$h = ?$$

Basic Equation:

$$V = \pi r^2 h$$

Working Equation:

$$h = \frac{V}{\pi r^2}$$

Substitution:

$$h = \frac{814 \, \text{m}^3}{\pi (6.00 \, \text{m})^2}$$

$$= 7.20 \, \text{m} \qquad \boxed{\frac{\text{m}^3}{\text{m}^2} = \text{m}}$$

· · · · · · · · · · · · · · · · · ·

EXAMPLE 4

A rectangular piece of sheet metal measures 45.0 cm by 75.0 cm. A 10.0-cm square is then cut from each corner. The metal is then folded to form a box-like container without a top. Find the volume of the container.

Sketch:

(a)

(b)

Data:

$$l = 55.0 \, \text{cm}$$
$$w = 25.0 \, \text{cm}$$
$$h = 10.0 \, \text{cm}$$
$$V = ?$$

Basic Equation:

$$V = lwh$$

Working Equation: Same

Substitution:

$$V = (55.0 \, \text{cm})(25.0 \, \text{cm})(10.0 \, \text{cm})$$
$$= 13,800 \, \text{cm}^3 \qquad \boxed{\text{cm} \times \text{cm} \times \text{cm} = \text{cm}^3}$$

· · · · · · · · · · · · · · · · ·

The cross-sectional area of a hole is 725 cm². Find its radius. ◄ **EXAMPLE 5**

Sketch:

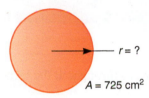

$r = ?$

$A = 725 \, \text{cm}^2$

Data:

$$A = 725 \, \text{cm}^2$$
$$r = ?$$

Basic Equation:

$$A = \pi r^2$$

Working Equation:

$$r = \sqrt{\frac{A}{\pi}}$$

Substitution:

$$r = \sqrt{\frac{725 \, \text{cm}^2}{\pi}}$$
$$= 15.2 \, \text{cm} \qquad \boxed{\sqrt{\text{cm}^2} = \text{cm}}$$

· · · · · · · · · · · ·

PHYSICS CONNECTIONS

Eratosthenes, a third-century Egyptian, used a problem-solving method to determine that the earth was not flat, but round, a fact that Columbus has been credited with discovering more than 1000 years later. Eratosthenes wondered why it was that at noon of the summer solstice, towers in Syene, Egypt (modern Aswan on the Nile), made no shadows, whereas documentation showed that towers in Alexandria, Egypt, did make distinct shadows. Eratosthenes decided to determine why towers in one city would cast shadows while towers in another city would not.

Eratosthenes sketched the problem, gathered data, and collected geometrical equations to solve this complex problem. He hired a person to pace the distance between the two cities (800 km) and used geometry to solve the problem (Fig. 2.2). After calculating the difference in the positions of the two cities to be approximately 7° out of a 360° sphere, he concluded that the earth's circumference was 40,000 km—remarkably accurate when compared to today's calculations. Eratosthenes had a problem to solve, so he, like all good scientists and problem solvers, followed several steps that included analyzing the problem, collecting data, selecting appropriate equations, and making the calculations.

Figure 2.2   Towers in Alexandria and Syene, the shadow cast by the tower in Alexandria, and the curvature of the earth with the angle and the distance between the two cities.

PROBLEMS 2.3

Use the problem-solving method to work each problem. (Here, as throughout the text, follow the rules for calculations with measurements.)

1. Find the volume of the box in Fig. 2.3.
2. Find the volume of a cylinder whose height is 7.50 in. and diameter is 4.20 in. (Fig. 2.4).
3. Find the volume of a cone whose height is 9.30 cm if the radius of the base is 5.40 cm (Fig. 2.5).

Figure 2.3

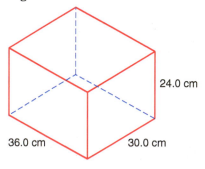

24.0 cm

36.0 cm   30.0 cm

Figure 2.4

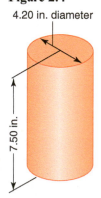

4.20 in. diameter

7.50 in.

Figure 2.5

9.30 cm

5.40 cm radius

The cylinder in an engine of a road grader as shown in Fig. 2.6 is 11.40 cm in diameter and 24.00 cm high. Use Fig. 2.6 for Problems 4 through 6.

4. Find the volume of the cylinder.
5. Find the cross-sectional area of the cylinder.
6. Find the lateral surface area of the cylinder.
7. Find the total volume of the building shown in Fig. 2.7.
8. Find the cross-sectional area of the concrete retaining wall shown in Fig. 2.8.
9. Find the volume of a rectangular storage facility 9.00 ft by 12.0 ft by 8.00 ft.

Figure 2.6

24.00 cm

11.40 cm diameter

Figure 2.7

4.70 ft

12.0 ft

10.0 ft

22.0 ft

37.0 ft

Figure 2.8

3.70 ft

19.3 ft

6.80 ft

10. Find the cross-sectional area of a piston head with a diameter of 3.25 cm.
11. Find the area of a right triangle that has legs of 4.00 cm and 6.00 cm.
12. Find the length of the hypotenuse of the right triangle in Problem 11.
13. Find the cross-sectional area of a pipe with outer diameter 3.50 cm and inner diameter 3.20 cm.
14. Find the volume of a spherical water tank with radius 8.00 m.
15. The area of a rectangular parking lot is $90\overline{0}$ m². If the length is 25.0 m, what is the width?
16. The volume of a rectangular crate is 192 ft³. If the length is 8.00 ft and the width is 4.00 ft, what is the height?
17. Find the volume of a brake cylinder whose diameter is 4.00 cm and whose length is 4.20 cm.
18. Find the volume of a tractor engine cylinder whose radius is 3.90 cm and whose length is 8.00 cm.
19. A cylindrical silo has a circumference of 29.5 m. Find its diameter.
20. If the silo in Problem 19 has a capacity of $100\overline{0}$ m³, what is its height?
21. A wheel 30.0 cm in diameter moving along level ground made 145 complete rotations. How many metres did the wheel travel?
22. The side of the silo in Problems 19 and 20 needs to be painted. If each litre of paint covers 5.0 m², how many litres of paint will be needed? (Round up to the nearest litre.)
23. You are asked to design a cylindrical water tank that holds $50\overline{0},000$ gal with radius 18.0 ft. Find its height (1 ft³ = 7.50 gal).
24. If the height of the water tank in Problem 23 were 42.0 ft, what would be its radius?
25. A ceiling is 12.0 ft by 15.0 ft. How many suspension panels 1.00 ft by 3.00 ft are needed to cover the ceiling?
26. Find the cross-sectional area of the dovetail slide shown in Fig. 2.9.

Figure 2.9

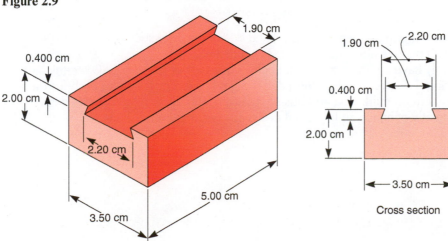

0.400 cm

2.00 cm

1.90 cm

2.20 cm

5.00 cm

3.50 cm

1.90 cm

2.20 cm

0.400 cm

2.00 cm

3.50 cm

Cross section

Figure 2.10

1.50 m

2.75 m

2.00 m

27. Find the volume of the storage bin shown in Fig. 2.10.
28. The maximum cross-sectional area of a spherical propane storage tank is 3.05 m². Will it fit into a 2.00-m-wide trailer?
29. How many cubic yards of concrete are needed to pour a patio 12.0 ft × 20.0 ft and 6.00 in. thick?
30. What length of sidewalk 4.00 in. thick and 4.00 ft wide could be poured with 2.00 yd³ of concrete?

Find the volume of each figure.

31.

8.00 in.

8.00 in.   5.00 in.

6.00 in.

32.

60.0 cm

Inside diameter: 20.0 cm
Outside diameter: 50.0 cm

PHYSICS ON THE JOB

Industrial Designer

The advent of inexpensive, three-dimensional printing technology, in conjunction with advances in computer-aided design (CAD) software, has expanded employment opportunities for industrial designers. Industrial designers develop the schematic as well as the detailed parts for products. Industrial designers perform research, develop virtual models, utilize three-dimensional printers to create prototypes, and determine materials to be used in production.

Mechanical skills are important in industrial design as measurement, manipulating formulas for design, and problem-solving skills are critical in the development of a product. The combination of artistic talent along with a background in engineering-related skills is vital for success in this career. Drafting as well as basic physics courses are necessary for such a degree.

Industrial designers typically have a minimum of a bachelor's degree, yet specific CAD technicians can often work with an associate's degree in computer-aided design. In addition, a career in industrial design can also evolve into a focus on marketing and business. As a result, pursuing a master's degree in business administration can greatly enhance an industrial designer's resume (Figure 2.11).

Figure 2.11   Industrial designer

Courtesy of fotolia © Marzky Ragsac Jr.

Glossary

Formula An equation, usually expressed in letters (called variables) and numbers. (p. 48)

Problem-Solving Method An orderly procedure that aids in understanding and solving problems. (p. 55)

Variable A symbol, usually a letter, used to represent some unknown number or quantity. (p. 48)

Review Questions

1. A formula is
   - (a) the amount of each value needed.
   - (b) a solution for problems.
   - (c) an equation usually expressed in letters and numbers.
2. Subscripts are
   - (a) the same as exponents.
   - (b) used to shorten what must be written.
   - (c) used to make a problem look hard.
3. A working equation
   - (a) is derived from the basic equation.
   - (b) is totally different from the basic equation.
   - (c) comes before the basic equation in the problem.
   - (d) none of the above.
4. Cite two examples in industry in which formulas are used.
5. How are subscripts used in measurement?
6. Why is reading the problem carefully the most important step in problem solving?
7. How can making a sketch help in problem solving?
8. What do we call the relationship between data that are given and what we are asked to find?
9. How is a working equation different from a basic equation?
10. How can analysis of the units in a problem assist in solving the problem?
11. How can making an estimate of your answer assist in the correct solution of problems?

Review Problems

1. Solve $F = ma$ for (a) m and (b) a.
2. Solve $v = \sqrt{2gh}$ for h.
3. Solve $s = \frac{1}{2}(v_f + v_i)t$ for v_f.
4. Solve $E_k = \frac{1}{2}mv^2$ for v.
5. Given $P = a + b + c$, with $P = 36$ ft, $a = 12$ ft, and $c = 6$ ft, find b.
6. Given $A = \left(\dfrac{a + b}{2}\right)h$, with $A = 21\bar{0}$ m^2, b $= 16.0$ m, and $h = 15.0$ m, find a.
7. Given $A = \pi r^2$, if $A = 15.0$ m^2, find r.
8. Given $A = \frac{1}{2}bh$, if $b = 12.2$ cm and $h = 20.0$ cm, what is A?
9. A cone has a volume of 314 cm^3 and radius of 5.00 cm. What is its height?
10. A right triangle has a side of 41.2 mm and a side of 9.80 mm. Find the length of the hypotenuse.
11. Given a cylinder with a radius of 7.20 cm and a height of 13.4 cm, find the lateral surface area.

SKETCH

| 12 cm^2 | w |

4.0 cm

DATA

$A = 12$ cm^2, $l = 4.0$ cm, $w = ?$

BASIC EQUATION

$A = lw$

WORKING EQUATION

$w = \dfrac{A}{l}$

SUBSTITUTION

$w = \dfrac{12\,\text{cm}^2}{4.0\,\text{cm}} = 3.0\,\text{cm}$

Figure 2.12

6.0 cm

9.0 cm

3.0 cm

12 cm

6.0 cm

12.   A rectangle has a perimeter of 40.0 cm. One side has a length of 14.0 cm. What is the length of an adjacent side?

13.   The formula for the volume of a cylinder is $V = \pi r^2 h$. If $V = 21\overline{0}0$ m^3 and $h = 17.0$ m, find r.

14.   The formula for the area of a triangle is $A = \frac{1}{2} bh$. If $b = 12.3$ m and $A = 88.6$ m^2, find h.

15.   Find the volume of the lead sleeve with the cored hole in Fig. 2.12.

16.   A rectangular plot of land measures 40.0 m by $12\overline{0}$ m with a parcel 10.0 m by 12.0 m out of one corner for an electrical transformer. What is the area of the remaining plot?

A P P L I E D   C O N C E P T S

Use the problem-solving method outlined in Section 2.3 to solve each problem.

1.   You run a landscaping business and know that you want to charge $50.00 to mow a person's lawn whose property is $10\overline{0}$ ft × $20\overline{0}$ ft. If the house dimensions take up a 35.0-ft × 80.0-ft area, how much are you charging per square yard?

2.   A room that measures 10.0 ft wide, 32.0 ft long, and 8.00 ft high needs a certain amount of air pumped into it per minute to keep the air quality up to regulations. If the room needs completely new air every 20.0 minutes, what is the volume of air per second that is being pumped into the room?

3.   Instead of using a solid iron beam, structural engineers and contractors use I-beams to save materials and money. How many I-beams can be molded from the same amount of iron contained in the solid iron beam as shown in Fig. 2.13?

Figure 2.13

20.0 ft

8.00 in.

(a)   8.00 in.

1.00 in.

1.00 in.

6.00 in.

1.00 in.

20.0 ft

(b)   8.00 in.

4.   A shipping specialist at a craft store needs to pack Styrofoam balls of radius 4.00 in. into a 1.40-ft × 2.80-ft × 1.40-ft rectangular cardboard container. What is the maximum number of balls that can fit in the container? Hint: Spherical balls have spaces around them when packed in rectangular containers.

5.   A crane needs to lift a spool of fine steel cable to the top of a bridge deck. The type of steel in the cable has a density of 7750 kg/m^3. The maximum lifting mass of the crane is 43,400 kg. (a) Given the dimensions of the spool in Fig. 2.14, find the volume of the spool. (b) Can the crane safely lift the spool?

Figure 2.14

1.53 m

0.915 m

CHAPTER 3

VECTORS

Objectives

The major goals of this chapter are to enable you to:

1. Distinguish between a vector and a scalar quantity.
2. Add vectors graphically.
3. Find the components of a vector.

4. Work with vectors in standard position.
5. Apply the basic concepts of right-triangle trigonometry using displacement vectors.

Some physical quantities, called *scalars*, may be described by and involve calculations with numerical quantities alone. Other physical quantities, called *vectors*, require both a numerical quantity and a direction to be completely described and often involve calculations using trigonometry. Vectors are developed in this chapter prior to their use in the following chapters.

3.1 Vectors and Scalars*

Every physical quantity can be classified as either a scalar or a vector quantity. A **scalar** is a quantity that can be completely described by a number (called its magnitude) and a unit. Examples of scalars are length, temperature, and volume. All these quantities

---

*Right-triangle trigonometry is developed in Appendix A.5, and instructions for using sin, cos, and tan keys on a scientific calculator are included in Appendix D.3 for those who have not studied this before or who need a review.

Figure 3.1 Displacement

can be expressed by a number with the appropriate units. For example, the length of a steel beam is expressed as 18 ft; the temperature at 11:00 A.M. is 15°C; the volume of a room is 300 m³.

A **vector** is a quantity that requires both *magnitude* (size) and *direction* to be completely described. Examples of vectors are force, displacement, and velocity. To completely describe a force, you must give not only its magnitude (size or amount), but also its direction.

To describe the change of position of an object, such as an airplane flying from one city to another, we use the term *displacement*. **Displacement** is the net change in position of an object, or the direct distance and direction it moves. For example, to completely describe the flight of a plane between two cities requires both the *distance* between them and the *direction from* the first city *to* the second (Fig. 3.1). *The units of displacement are length units*, such as metres, kilometres, feet, or miles.

Suppose that a friend asks you how to reach your home from school. If you replied that he should walk four blocks, you would not have given him enough information [Fig. 3.2(a)]. Obviously, you would need to tell him which direction to follow. If you had replied, "Four blocks north," your friend could then find your home [Fig. 3.2(b)]. If your friend decides to walk one block west, four blocks north, and then one block east, he will still arrive at your house. This resultant displacement is the same as if he had walked four blocks north [Fig. 3.2(c)].

Figure 3.2 Displacement involves both a distance and a direction.

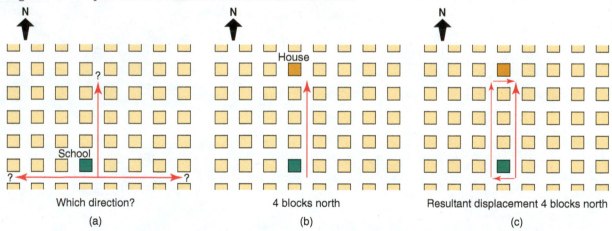

Which direction?

(a)

4 blocks north

(b)

Resultant displacement 4 blocks north

(c)

Figure 3.3

The magnitude of the displacement vector "15 miles NE" is 15 miles and its direction is northeast (Fig. 3.3).

To represent a vector in a diagram, we draw an arrow that points in the correct direction. The magnitude of the vector is indicated by the length of the arrow. We usually choose a scale, such as 1.0 cm = 25 mi, for this purpose (Fig. 3.4). Thus, a displacement of $10\overline{0}$ mi west is drawn as an arrow (pointing west) 4.0 cm long [Fig. 3.4(a)] since

$$10\overline{0} \text{ mi} \times \frac{1.0 \text{ cm}}{25 \text{ mi}} = 4.0 \text{ cm}$$

Displacements of $5\overline{0}$ mi north [Fig. 3.4(b)] and $5\overline{0}$ mi east [Fig. 3.4(c)] using the same scale are also shown.

Figure 3.4 Use a scale to draw the proper length of a given vector.

0   25
Scale
1.0 cm = 25 mi

> **EXAMPLE 1**

Using the scale 1.0 cm = $5\bar{0}$ km, draw the displacement vector 275 km at 45° north of west.

First, find the length of the vector.

$$275 \text{ km} \times \frac{1.0 \text{ cm}}{5\bar{0} \text{ km}} = 5.5 \text{ cm}$$

Then draw the vector at an angle 45° north of west (Fig. 3.5).

Figure 3.5 Displacement vector 275 km at 45° north of west drawn to scale: 1.0 cm = 50 km

> **EXAMPLE 2**

Using the scale $\frac{1}{4}$ in. = $2\bar{0}$ mi, draw the displacement vector $15\bar{0}$ mi at 22° east of south.

First, find the length of the vector.

$$15\bar{0} \text{ mi} \times \frac{\frac{1}{4} \text{ in.}}{2\bar{0} \text{ mi}} = 1\frac{7}{8} \text{ in.}$$

Then draw the vector at 22° east of south (Fig. 3.6).

Figure 3.6   Displacement vector $15\overline{0}$ mi at 22° east of south drawn to scale: ¼ in. = $2\overline{0}$ mi

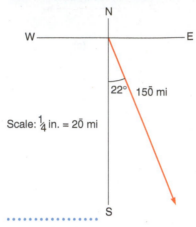

A vector may be denoted by a single letter with a small arrow above, such as $\vec{A}$, $\vec{v}$, or $\vec{R}$ [Fig. 3.7(a)]. This notation is especially useful when writing vectors on paper or on a chalkboard. In this book we use the traditional boldface type to denote vectors, such as **A**, **v**, or **R** [Fig. 3.7(b)]. The length of vector $\vec{A}$ is written $|\vec{A}|$; the length of vector **A** is written $|\mathbf{A}|$.

Figure 3.7

(a) Vector quantities $\vec{A}$, $\vec{v}$, and $\vec{R}$ usually have arrows when they are written on paper or on a chalkboard.

(b) Vector quantities **A**, **v**, and **R** usually are written in boldface type in textbooks.

TRY THIS ACTIVITY

New York Vectors

In 1811, a comprehensive plan was mapped out to create a rectangular grid of roadways on the island of Manhattan in New York City. As a result, giving directions in Manhattan can be done in a number of different ways while still achieving the same result. Assuming that the streets and avenues in the map in Fig. 3.8 are at right angles with one another and that the distance between streets is 0.05 mi and the distance between avenues in 0.20 mi, determine three different ways that someone could travel from Macy's at Herald Square to Times Square. What would be the distance traveled and the displacement for each of these paths?

Figure 3.8 A midtown Manhattan map is a good way to demonstrate the usefulness of vectors.

Courtesy of A midtown Manhattan map is a good way to demonstrate the usefulness of vectors, (c) 2009 City Maps, City Maps Inc.

PROBLEMS 3.1

Using the scale 1.0 cm = $5\overline{0}$ km, find the length of the vector that represents each displacement.

1. Displacement $10\overline{0}$ km east   length = ____ cm
2. Displacement 125 km south   length = ____ cm
3. Displacement $14\overline{0}$ km at 45° east of south   length = ____ cm
4. Displacement 260 km at $3\overline{0}$° south of west   length = ____ cm
5. Displacement 315 km at 65° north of east   length = ____ cm
6. Displacement 187 km at 17° north of west   length = ____ cm
7–12. Draw the vectors in Problems 1 through 6 using the scale indicated.

Using the scale $\frac{1}{4}$ in. = $2\overline{0}$ mi, find the length of the vector that represents each displacement.

13. Displacement $10\overline{0}$ mi west   length = ____ in.
14. Displacement $17\overline{0}$ mi north   length = ____ in.
15. Displacement $21\overline{0}$ mi at 45° south of west   length = ____ in.
16. Displacement 145 mi at $6\overline{0}$° north of east   length = ____ in.
17. Displacement 75 mi at 25° west of north   length = ____ in.
18. Displacement $16\overline{0}$ mi at 72° west of south   length = ____ in.
19–24. Draw the vectors in Problems 13 through 18 using the scale indicated.

René Descartes (1596–1650), mathematician and philosopher, was born in France. He founded analytic or coordinate geometry, often called Cartesian geometry, and made major contributions in motion and optics.

Figure 3.9 Number plane

3.2 Components of a Vector

Before we study vectors further, we need to discuss components of vectors and the number plane. The **number plane** (sometimes called the *Cartesian coordinate system,* after René Descartes) consists of a horizontal line called the *x*-axis and a vertical line called the *y*-axis intersecting at a right angle at a point called the *origin* as shown in Fig. 3.9. These two lines divide the number plane into four quadrants, which we label as quadrants I, II, III, and IV. The *x*-axis contains positive numbers to the right of the origin and negative numbers to the left of the origin. The *y*-axis contains positive numbers above the origin and negative numbers below the origin.

Graphically, a vector is represented by a directed line segment. The length of the line segment indicates the magnitude of the quantity. An arrowhead indicates the direction. If *A* and *B* are the end points of a line segment as in Fig. 3.10, the symbol **AB** denotes the *vector from A to B*. Point *A* is called the *initial point*. Point *B* is called the *terminal point* or *end point* of the vector. Vector **BA** has the same length as vector **AB** but has the opposite direction. Vectors may also be denoted by a single letter, such as **u**, **v**, or **R**.

The sum of two or more vectors is called the **resultant vector**. When two or more vectors are added, each of these vectors is called a **component** of the resultant vector. The components of vector **R** in Fig. 3.11(a) are vectors **A**, **B**, and **C**. *Note:* A vector may have more than one set of component vectors. The components of vector **R** in Fig. 3.11(b) are vectors **E** and **F**.

Figure 3.10 Vector from *A* to *B*

Figure 3.11 Components of resultant vectors

(a) Vectors **A**, **B**, and **C** are components of the resultant vector **R**.

(b) Vector **E** is a horizontal component and vector **F** is a vertical component of the resultant vector **R**.

We are often interested in the components of a vector that are perpendicular to each other and that are on or parallel to the *x*- and *y*-axes. In particular, we are interested in the type of component vectors shown in Fig. 3.11(b) (component vectors **E** and **F**). The horizontal component vector that lies on or is parallel to the *x*-axis is called the ***x*-component**. The vertical component vector that lies on or is parallel to the *y*-axis is called the ***y*-component**. Three examples are shown in Fig. 3.12.

Figure 3.12 *x*- and *y*-components of resultant vectors

(a) (b) (c)

R_x = the *x*-component of vector **R**
R_y = the *y*-component of vector **R**

The x- and y-components of vectors can also be expressed as signed numbers. The absolute value of the signed number corresponds to the magnitude (length) of the component vector. The sign of the number corresponds to the direction of the component as follows:

x-component	*y*-component
+, if right	+, if up
−, if left	−, if down

EXAMPLE 1

Find the x- and y-components of vector **R** in Fig. 3.13.

$$\mathbf{R}_x = x\text{-component of } \mathbf{R} = +4$$
$$\mathbf{R}_y = y\text{-component of } \mathbf{R} = +3$$

Figure 3.13

EXAMPLE 2

Find the x- and y-components of vector **R** in Fig. 3.14.

$$\mathbf{R}_x = x\text{-component of } \mathbf{R} = +6$$
$$\mathbf{R}_y = y\text{-component of } \mathbf{R} = -8$$

(The y-component points in a negative direction.)

Figure 3.14

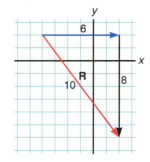

EXAMPLE 3

Find the x- and y-components of vector **R** in Fig. 3.15.

$$\mathbf{R}_x = -12$$
$$\mathbf{R}_y = -9$$

(Both x- and y-components point in a negative direction.)

Figure 3.15

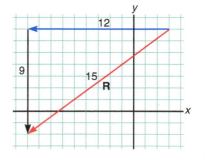

A vector may be placed in any position in the number plane as long as its magnitude and direction are not changed. The vectors in each set in Fig. 3.16 are equal because they have the same magnitude (length) and the same direction.

A vector is in **standard position** when its initial point is at the origin of the number plane. A vector in standard position is expressed in terms of its magnitude (length) and its angle θ, where θ *is measured counterclockwise from the positive x-axis to the vector*. The vectors shown in Fig. 3.17 are in standard position.

Figure 3.16 Sets of equal vectors

(a)

(b)

Figure 3.17 Vectors in standard position

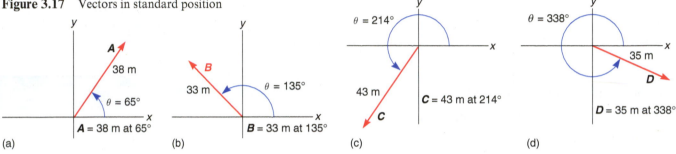

(a) $A = 38$ m at $65°$

(b) $B = 33$ m at $135°$

(c) $C = 43$ m at $214°$

(d) $D = 35$ m at $338°$

Finding the Components of a Vector

EXAMPLE 4

Figure 3.18

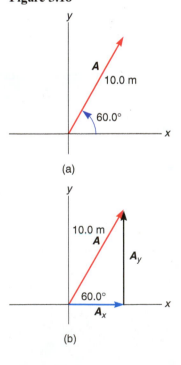

(a)

(b)

Find the *x*- and *y*-components of the vector $\mathbf{A} = 10.0$ m at $60.0°$.

First, draw the vector in standard position [Fig. 3.18(a)]. Then, draw a right triangle where the legs represent the *x*- and *y*-components [Fig. 3.18(b)]. The absolute value of the *x*-component of the vector is the length of the side adjacent to the $60.0°$ angle. Therefore, to find the *x*-component,

$$\cos 60.0° = \frac{\text{side adjacent to } 60.0°}{\text{hypotenuse}} = \frac{|\mathbf{A}_x|}{10.0 \text{ m}}$$

$$\cos 60.0° = \frac{|\mathbf{A}_x|}{10.0 \text{ m}}$$

$$(\cos 60.0°)(10.0 \text{ m}) = \left(\frac{|\mathbf{A}_x|}{10.0 \text{ m}}\right)(10.0 \text{ m}) \qquad \text{Multiply both sides by } 10.0 \text{ m.}$$

$$5.00 \text{ m} = |\mathbf{A}_x|$$

Since the *x*-component is pointing in the positive *x*-direction, $\mathbf{A}_x = +5.00$ m.

The absolute value of the *y*-component of the vector is the length of the side opposite the $60.0°$ angle. Therefore, to find the *y*-component,

$$\sin 60.0° = \frac{\text{side opposite } 60.0°}{\text{hypotenuse}} = \frac{|\mathbf{A}_y|}{10.0 \text{ m}}$$

$$\sin 60.0° = \frac{|\mathbf{A}_y|}{10.0 \text{ m}}$$

$$(\sin 60.0°)\,(10.0\ \text{m}) = \left(\frac{|\mathbf{A}_y|}{10.0\ \text{m}}\right)(10.0\ \text{m}) \qquad \text{Multiply both sides by 10.0 m.}$$

$$8.66\ \text{m} = |\mathbf{A}_y|$$

Since the y-component is pointing in the positive y-direction, $\mathbf{A}_y = +8.66\ \text{m}$.

···················

EXAMPLE 5

Find the x- and y-components of the vector $\mathbf{B} = 13.0$ km at $220.0°$.

First, draw the vector in standard position [Fig. 3.19(a)]. Then, complete a right triangle with the x- and y-components being the two legs [Fig. 3.19(b)].

We will let angle α (Greek letter alpha) be the acute angle (an angle whose measure is less than $90°$) between the vector in standard position and the x-axis.

Find angle α as follows:

$$180° + \alpha = 220.0°$$
$$\alpha = 40.0°$$

The absolute value of the x-component is the length of the side adjacent to angle α. Therefore, to find the x-component,

$$\cos \alpha = \frac{\text{side adjacent to } \alpha}{\text{hypotenuse}}$$

$$\cos 40.0° = \frac{|\mathbf{B}_x|}{13.0\ \text{km}}$$

$$(\cos 40.0°)\,(13.0\ \text{km}) = \left(\frac{|\mathbf{B}_x|}{13.0\ \text{km}}\right)(13.0\ \text{km}) \qquad \text{Multiply both sides by 13.0 km.}$$

$$9.96\ \text{km} = |\mathbf{B}_x|$$

Since the x-component is pointing in the negative x-direction, $\mathbf{B}_x = -9.96$ km.

The absolute value of the y-component of the vector is the length of the side opposite angle α. Therefore, to find the y-component,

$$\sin \alpha = \frac{\text{side opposite } \alpha}{\text{hypotenuse}}$$

$$\sin 40.0° = \frac{|\mathbf{B}_y|}{13.0\ \text{km}}$$

$$(\sin 40.0°)\,(13.0\ \text{km}) = \left(\frac{|\mathbf{B}_y|}{13.0\ \text{km}}\right)(13.0\ \text{km}) \qquad \text{Multiply both sides by 13.0 km.}$$

$$8.36\ \text{km} = |\mathbf{B}_y|$$

Since the y-component is pointing in the negative y-direction, $\mathbf{B}_y = -8.36$ km.

···················

Figure 3.19

(a)

(b)

In general, find the x- and y-components of a vector as follows. First, draw any vector $\mathbf{A}$ in standard position; then, draw its x- and y-components as shown in Fig. 3.20. Use the right triangle to find the x-component as follows:

$$\cos \alpha = \frac{\text{side adjacent to } \alpha}{\text{hypotenuse}}$$

$$\cos \alpha = \left(\frac{|\mathbf{A}_x|}{|\mathbf{A}|}\right)$$

Figure 3.20

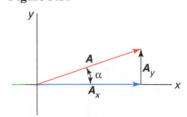

Vector $\mathbf{A}$ in standard position with its horizontal component $\mathbf{A}_x$ and its vertical component $\mathbf{A}_y$

$$|\mathbf{A}|(\cos \alpha) = \left(\frac{|\mathbf{A}_x|}{|\mathbf{A}|}\right)|\mathbf{A}| \qquad \text{Multiply both sides by } |\mathbf{A}|.$$

$$|\mathbf{A}|(\cos \alpha) = |\mathbf{A}_x|$$

Similarly, we use the right triangle to find the *y*-component as follows:

$$\sin \alpha = \frac{\text{side opposite to } \alpha}{\text{hypotenuse}}$$

$$\sin \alpha = \left(\frac{|\mathbf{A}_y|}{|\mathbf{A}|}\right)$$

$$|\mathbf{A}|(\sin \alpha) = \left(\frac{|\mathbf{A}_y|}{|\mathbf{A}|}\right)|\mathbf{A}| \qquad \text{Multiply both sides by } |\mathbf{A}|.$$

$$|\mathbf{A}|(\sin \alpha) = |\mathbf{A}_y|$$

The signs of the *x*- and *y*-components are determined by the quadrants in which the vector in standard position lies.

In general:

To find the *x*- and *y*-components of a vector **A** given in standard position:

1. Complete the right triangle with the legs being the *x*- and *y*-components of the vector.
2. Find the lengths of the legs of the right triangle as follows:

$$|\mathbf{A}_x| = |\mathbf{A}|(\cos \alpha)$$

$$|\mathbf{A}_y| = |\mathbf{A}|(\sin \alpha)$$

where angle α is the acute angle between vector **A** in standard position and the *x*-axis.

3. Determine the signs of the *x*- and *y*-components.

EXAMPLE 6

Figure 3.21

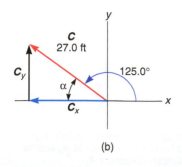

(a)

(b)

Find the *x*- and *y*-components of the vector **C** = 27.0 ft at 125.0°.

First, draw the vector in standard position [Fig. 3.21(a)]. Then, complete a right triangle with the *x*- and *y*-components being the two legs [Fig. 3.21(b)]. Find angle α as follows:

$$\alpha + 125.0° = 180°$$

$$\alpha = 55.0°$$

Next, find the *x*-component as follows:

$$|\mathbf{C}_x| = |\mathbf{C}|(\cos \alpha)$$

$$|\mathbf{C}_x| = (27.0 \text{ ft})(\cos 55.0°)$$

$$= 15.5 \text{ ft}$$

Since the *x*-component is pointing in the negative *x*-direction,

$$\mathbf{C}_x = -15.5 \text{ ft}$$

Then, find the *y*-component as follows:

$$|\mathbf{C}_y| = |\mathbf{C}|(\sin \alpha)$$

$$|\mathbf{C}_y| = (27.0 \text{ ft})(\sin 55.0°)$$

$$= 22.1 \text{ ft}$$

Since the *y*-component is pointing in the positive *y*-direction,

$$\mathbf{C}_y = +22.1 \text{ ft}$$

PROBLEMS 3.2

Find the *x*- and *y*-components of each vector in the following diagram. (Express them as signed numbers and then graph them as vectors.)

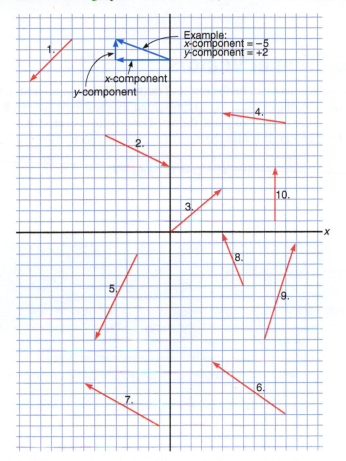

Example:
x-component = −5
y-component = +2

x-component
y-component

Make a sketch of each vector in standard position. Use the scale 1.0 cm = $1\overline{0}$ m.

11. $\mathbf{A} = 2\overline{0}$ m at 25°   12. $\mathbf{B} = 25$ m at 125°   13. $\mathbf{C} = 25$ m at 245°

14. $\mathbf{D} = 2\overline{0}$ m at 345°   15. $\mathbf{E} = 15$ m at 105°   16. $\mathbf{F} = 35$ m at 291°

17. $\mathbf{G} = 3\overline{0}$ m at 405°   18. $\mathbf{H} = 25$ m at 525°

Find the *x*- and *y*-components of each vector.

19.
13.4 m
42.0°

20.
284.5°
275 mi

21.
248.0°
48.6 km

22.
9780 m
111.4°

23.
351.6°
98.5 km

24.
48.5 ft
165.0°

Figure 3.22

The resultant vector **R** is the graphic sum of the component sets of vectors **A**, **B**, **C**, and **D**, and **E** and **F**. That is, **A** + **B** + **C** + **D** = **R** and **E** + **F** = **R**.

Find the *x*- and *y*-components of each vector given in standard position.

25. **A** = 38.9 m at 10.5°   26. **B** = 478 ft at 195.0°   27. **C** = 9.60 km at 310.0°

28. **D** = 5430 mi at 153.7°   29. **E** = 29.5 m at 101.5°   30. **F** = 154 mi at 273.2°

3.3 Addition of Vectors

Any given displacement can be the result of many different combinations of displacements. In Fig. 3.22, the displacement represented by the arrow labeled **R** for resultant is the result of either of the two paths shown. The resultant vector, **R**, is the sum of the vectors **A**, **B**, **C**, and **D**. It is also the sum of vectors **E** and **F**. That is,

$$\mathbf{A} + \mathbf{B} + \mathbf{C} + \mathbf{D} = \mathbf{R} \quad \text{and} \quad \mathbf{E} + \mathbf{F} = \mathbf{R}$$

To solve a vector addition problem graphically such as displacement:

1. Choose a suitable scale and calculate the length of each vector.
2. Draw the north–south reference line or other suitable reference line. Graph paper should be used.
3. Using a ruler and protractor, draw the first vector and then draw the other vectors so that the initial end of each vector is placed at the terminal end of the previous vector.
4. Draw the resultant vector from the initial end of the first vector to the terminal end of the last vector.
5. Measure the length of the resultant and use the scale to find the magnitude of the vector. Use a protractor to measure the angle of the resultant.

Note: We find it useful and standard practice to label the coordinate axes for solving many vector problems involving a geographic orientation as shown in Fig. 3.23.

EXAMPLE 1

Figure 3.23

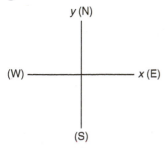

Standard practice for labeling the coordinate axes involving many vector problems with a geographic orientation: Label the vertical *y*-axis *up for N (north)* and *down for S (south)* and the horizontal *x*-axis *right for E (east)* and *left for W (west)*.

Find the resultant displacement of an airplane that flies $2\overline{0}$ mi due east, then $3\overline{0}$ mi due north, and then $1\overline{0}$ mi at 60° west of south.

We choose a scale of 1.0 cm = 5.0 mi so that the vectors are large enough to be accurate and small enough to fit on the paper. (Here each block represents 0.5 cm.) The length of the first vector is

$$|\mathbf{A}| = 2\overline{0}\ \text{mi} \times \frac{1.0\ \text{cm}}{5.0\ \text{mi}} = 4.0\ \text{cm}$$

The length of the second vector is

$$|\mathbf{B}| = 3\overline{0}\ \text{mi} \times \frac{1.0\ \text{cm}}{5.0\ \text{mi}} = 6.0\ \text{cm}$$

The length of the third vector is

$$|\mathbf{C}| = 1\overline{0}\ \text{mi} \times \frac{1.0\ \text{cm}}{5.0\ \text{mi}} = 2.0\ \text{cm}$$

Draw the north–south reference line, and draw the first vector as shown in Fig. 3.24(a). The second and third vectors are then drawn as shown in Fig. 3.24(b) and 3.24(c).

Figure 3.24

1. Draw the north–south reference line and the first vector: 2̄0 mi due east.

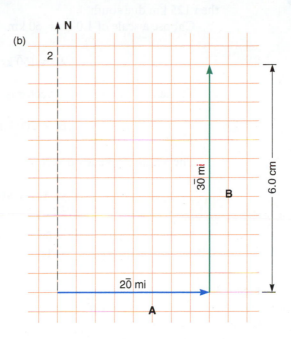

2. Draw the second vector: 3̄0 mi due north.

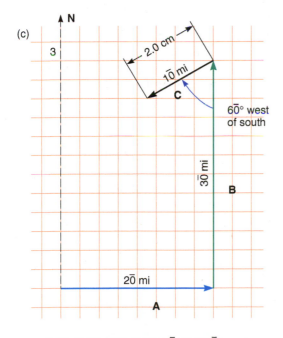

3. Draw the third vector: 1̄0 mi at 6̄0° west of south.

4. Draw the resultant vector, which is 28 mi at 24° east of north.

Scale: 1.0 cm = 5.0 mi

Using a ruler, we find that the length of the resultant vector measures 5.5 cm [Fig. 3.24(d)]. Since 1.0 cm = 5.0 mi, this represents a displacement with magnitude

$$|\mathbf{R}| = 5.5 \text{ cm} \times \frac{5.0 \text{ mi}}{1.0 \text{ cm}} = 28 \text{ mi}$$

The angle between vector **R** and north measures 24°, so the resultant is 28 mi at 24° east of north.

EXAMPLE 2

Find the resultant of the displacements $15\overline{0}$ km due west, then $20\overline{0}$ km due east, and then 125 km due south.

Choose a scale of 1.0 cm = $5\overline{0}$ km. The length of the first vector is

$$|\mathbf{A}| = 15\overline{0} \text{ km} \times \frac{1.0 \text{ cm}}{5\overline{0} \text{ km}} = 3.0 \text{ cm}$$

The length of the second vector is

$$|\mathbf{B}| = 20\overline{0} \text{ km} \times \frac{1.0 \text{ cm}}{5\overline{0} \text{ km}} = 4.0 \text{ cm}$$

The length of the third vector is

$$|\mathbf{C}| = 125 \text{ km} \times \frac{1.0 \text{ cm}}{5\overline{0} \text{ km}} = 2.5 \text{ cm}$$

Draw the north–south reference line, and draw the first vector as shown in Fig. 3.25(a). Then, draw the second and third vectors as shown in Figs. 3.25(b) and 3.25(c).

Figure 3.25

1. Draw the north–south reference line and the first vector: $15\overline{0}$ km due west.

2. Draw the vector: $20\overline{0}$ km due east.

3. Draw the vector: 125 mi due south.

4. The length of the resultant is 2.6 cm, which represents $13\overline{0}$ km at 22° east of south.

Scale: 1.0 cm = $5\overline{0}$ km

The length of the resultant vector measures 2.6 cm in Fig. 3.25(d). Since $1.0 \text{ cm} = 5\overline{0} \text{ km}$,

$$|\mathbf{R}| = 2.6 \text{ cm} \times \frac{5\overline{0} \text{ km}}{1.0 \text{ cm}} = 130 \text{ km}$$

The angle between vector $\mathbf{R}$ and south measures 22°, so the resultant vector is 130 km at 22° east of south.

••••••••••••••••••

Expressing the x- and y-components as signed numbers, we find the resultant vector of several vectors as follows:

1. Find the x-component of each vector and then find the sum of these x-components. This sum is the x-component of the resultant vector.
2. Find the y-component of each vector and then find the sum of these y-components. This sum is the y-component of the resultant vector.

Given vectors $\mathbf{A}$ and $\mathbf{B}$ in Fig. 3.26, graph and find the x- and y-components of the resultant vector $\mathbf{R}$.

EXAMPLE 3

Graph resultant vector $\mathbf{R}$ by connecting the initial point of vector $\mathbf{A}$ to the end point of vector $\mathbf{B}$ [Fig. 3.27(a)]. The resultant vector $\mathbf{R}$ is shown in Fig. 3.27(b).

Find the x-component of $\mathbf{R}$ by finding and adding the x-components of $\mathbf{A}$ and $\mathbf{B}$.

$$\mathbf{A}_x = +3$$
$$\mathbf{B}_x = \underline{+2}$$
$$\mathbf{R}_x = +5$$

Figure 3.26

Figure 3.27

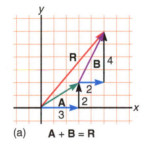

(a) $\mathbf{A} + \mathbf{B} = \mathbf{R}$

(b)

Find the y-component of $\mathbf{R}$ by finding and adding the y-components of $\mathbf{A}$ and $\mathbf{B}$.

$$\mathbf{A}_y = +2$$
$$\mathbf{B}_y = \underline{+4}$$
$$\mathbf{R}_y = +6$$

••••••••••••••••••

Given vectors $\mathbf{A}$, $\mathbf{B}$, and $\mathbf{C}$ in Fig. 3.28, graph and find the x- and y-components of the resultant vector $\mathbf{R}$.

EXAMPLE 4

Graph resultant vector $\mathbf{R}$ by connecting the initial point of vector $\mathbf{A}$ to the end point of vector $\mathbf{C}$ [Fig. 3.29(a)]. The resultant vector $\mathbf{R}$ is shown in Fig. 3.29(b).

Figure 3.28

Figure 3.29

(a)   **A + B + C = R**

(b)

Find the *x*-component of **R** by finding and adding the *x*-components of **A**, **B**, and **C** as shown below. Find the *y*-component of **R** by finding and adding the *y*-components of **A**, **B**, and **C**.

Vector	*x*-component	*y*-component
A	+7	+2
B	−2	+1
C	−1	+4
R	+4	+7

EXAMPLE 5

Given vectors **A**, **B**, **C**, and **D** in Fig. 3.30, graph and find the *x*- and *y*-components of the resultant vector **R**.

Graph resultant vector **R** by connecting the initial point of vector **A** to the end point of vector **D** [Fig. 3.31(a)]. The resultant vector **R** is shown in Fig. 3.31(b).

Figure 3.30

Figure 3.31

(a)   **A + B + C + D = R**

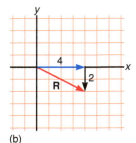

(b)

Find the *x*-component of **R** by finding and adding the *x*-components of **A**, **B**, **C**, and **D** as shown below. Find the *y*-component of **R** by finding and adding the *y*-components of **A**, **B**, **C**, and **D**.

Vector	*x*-component	*y*-component
A	+2	+3
B	+5	−4
C	−2	−3
D	−1	+2
R	+4	−2

Two vectors are equal when they have the same magnitude and the same direction [Fig. 3.32(a)]. *Two vectors are opposites or negatives* of each other when they have the same magnitude but opposite directions [Fig. 3.32(b)].

Figure 3.32

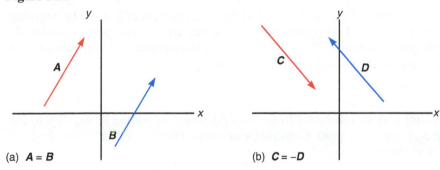

(a) **A = B**   (b) **C = –D**

To add two or more vectors in any position graphically, construct the first vector with its initial point at the origin and parallel to its given position. Then, construct the second vector with its initial point on the end point of the first vector and parallel to its given position. Then, construct the third vector with its initial point on the end point of the second vector and parallel to its given position. Continue this process until all vectors have been so constructed. The resultant vector is the vector joining the initial point of the first vector (origin) to the end point of the last vector. (The order of adding or constructing the given vectors does not matter.)

Given vectors **A**, **B**, and **C** in Fig. 3.33(a), graph and find the *x*- and *y*-components of the resultant vector **R**.

EXAMPLE 6

Figure 3.33

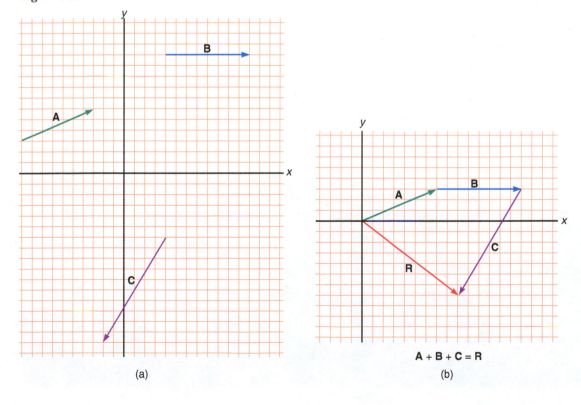

(a)   (b)

A + B + C = R

Construct vector **A** with its initial point at the origin and parallel to its given position as in Fig. 3.33(b). Next, construct vector **B** with its initial point on the end point of vector **A** and parallel to its given position. Then, construct vector **C** with its initial point on the end point of vector **B** and parallel to its given position. The resultant vector **R** is the vector with its initial point at the origin and its end point at the end point of vector **C**.

From the graph in Fig. 3.33(b), we read the x-component of **R** as +9 by counting the number of squares *to the right* between the y-axis and the end point of vector **R**. We read the y-component of **R** as −7 by counting the number of squares *below* and between the x-axis and the end point of vector **R**.

· · · · · · · · · · · · · · · ·

One vector may be subtracted from a second vector by adding its negative to the first. That is, **v** − **w** = **v** + (−**w**). Construct **v** as usual. Then construct −**w** and find the resultant **R** as shown in Fig. 3.34.

Figure 3.34
R = **v** − **w** = **v** + (−**w**)

Finding a Vector from Its Components

EXAMPLE 7

Find vector **R** in standard position with $\mathbf{R}_x = +3.00$ m and $\mathbf{R}_y = +4.00$ m.

First, graph the x- and y-components (Fig. 3.35) and complete the right triangle. The hypotenuse is the resultant vector. Find angle α as follows:

Figure 3.35

$$\tan \alpha = \frac{\text{side opposite } \alpha}{\text{side adjacent to } \alpha}$$

$$\tan \alpha = \frac{4.00 \text{ m}}{3.00 \text{ m}} = 1.333$$

$$\alpha = 53.1° \qquad \text{(see Appendix D, Section D.3)}$$

Find the magnitude of **R** using the Pythagorean theorem.

$$|\mathbf{R}| = \sqrt{|\mathbf{R}_x|^2 + |\mathbf{R}_y|^2}$$

$$|\mathbf{R}| = \sqrt{(3.00 \text{ m})^2 + (4.00 \text{ m})^2}$$

$$= 5.00 \text{ m}$$

That is, **R** = 5.00 m at 53.1°.

· · · · · · · · · · · · · · · ·

In general:

To find resultant vector **R** in standard position when its x- and y-components are given:

1. Complete the right triangle with the legs being the x- and y-components of the vector.
2. Find the acute angle α of the right triangle whose vertex is at the origin by using tan α.
3. Find angle θ in standard position as follows:

$$\theta = \alpha \qquad \text{(θ in first quadrant)}$$
$$\theta = 180° - \alpha \qquad \text{(θ in second quadrant)}$$
$$\theta = 180° + \alpha \qquad \text{(θ in third quadrant)}$$
$$\theta = 360° - \alpha \qquad \text{(θ in fourth quadrant)}$$

The Greek letter θ (theta) is often used to represent the measure of an angle.
4. Find the magnitude of the vector using the Pythagorean theorem:

$$\mathbf{R} = \sqrt{|\mathbf{R}_x|^2 + |\mathbf{R}_y|^2}$$

Find vector **R** in standard position whose *x*-component is +7.00 mi and *y*-component is –5.00 mi.

EXAMPLE 8

First, graph the *x*- and *y*-components (Fig. 3.36) and complete the right triangle. The hypotenuse is the resultant vector. Find angle α as follows:

Figure 3.36

$$\tan \alpha = \frac{\text{side opposite } \alpha}{\text{side adjacent to } \alpha}$$

$$\tan \alpha = \frac{5.00 \text{ mi}}{7.00 \text{ mi}} = 0.7143$$

$$\alpha = 35.5°$$

Then

$$\theta = 360° - \alpha \qquad \textbf{R} \text{ is in the fourth quadrant.}$$

$$= 360° - 35.5°$$

$$= 324.5°$$

Find the magnitude of **R** using the Pythagorean theorem.

$$|\textbf{R}| = \sqrt{|\textbf{R}_x|^2 + |\textbf{R}_y|^2}$$

$$|\textbf{R}| = \sqrt{|7.00 \text{ mi}|^2 + |-5.00 \text{ mi}|^2}$$

$$= 8.60 \text{ mi}$$

That is, **R** = 8.60 mi at 324.5°.

. . . .

Find vector **R** in standard position with $\textbf{R}_x = -115$ km and $\textbf{R}_y = +175$ km.

EXAMPLE 9

First, graph the *x*- and *y*-components (Fig. 3.37) and complete the right triangle. The hypotenuse is the resultant vector. Find angle α as follows:

Figure 3.37

$$\tan \alpha = \frac{\text{side opposite } \alpha}{\text{side adjacent to } \alpha}$$

$$\tan \alpha = \frac{175 \text{ km}}{115 \text{ km}} = 1.522$$

$$\alpha = 56.7°$$

Then

$$\theta = 180° - \alpha \qquad \textbf{R} \text{ is in the second quadrant.}$$

$$= 180° - 56.7°$$

$$= 123.3°$$

Find the magnitude of **R** using the Pythagorean theorem.

$$|\textbf{R}| = \sqrt{|\textbf{R}_x|^2 + |\textbf{R}_y|^2}$$

$$|\textbf{R}| = \sqrt{|-115 \text{ km}|^2 + |175 \text{ km}|^2}$$

$$= 209 \text{ km}$$

That is, **R** = 209 km at 123.3°.

. . . .

To find the resultant vector **R** of any set of vectors, such as **R** = **A** + **B** + **C**, using right-triangle trigonometry:

1. Find the *x*-component of each vector and add: $R_x = A_x + B_x + C_x$.
2. Find the *y*-component of each vector and add: $R_y = A_y + B_y + C_y$.
3. Find the magnitude of the resultant vector **R** using the Pythagorean theorem: $|\mathbf{R}| = \sqrt{|R_x|^2 + |R_y|^2}$.
4. Find the direction of the resultant vector **R** using right-triangle trigonometry by (a) first finding the acute angle α between the resultant vector and the *x*-axis and then finding angle θ in standard position or (b) expressing the direction of the resultant vector using some other reference.

EXAMPLE 10

Figure 3.38

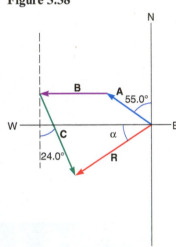

A ship travels 105 km from port on a course of 55.0° west of north to an island. Then it travels 124 km due west to a second island. Then it travels 177 km on a course of 24.0° east of south to a third island. Find the displacement from the starting point to the ending point.

First, draw a vector diagram as in Fig. 3.38. Then, find the *x*- and *y*-components of each of the three vectors as follows:

$\|A_x\| = \|\mathbf{A}\| \cos 35.0°$	The acute angle between vector **A** and the *x*-axis is $90° - 55.0° = 35.0°$.
$= (105 \text{ km})(\cos 35.0°)$	
$= -86.0 \text{ km}$	The *x*-component is in the negative *x*-direction.
$\|A_y\| = \|\mathbf{A}\| \sin 35.0°$	
$= (105 \text{ km})(\sin 35.0°)$	
$= 60.2 \text{ km}$	The *y*-component is in the positive *y*-direction.
$\|B_x\| = \|\mathbf{B}\| = -124 \text{ km}$	This *x*-component is in the negative *x*-direction.
$\|B_y\| = \|\mathbf{B}\| = 0 \text{ km}$	This *y*-component of due west is 0.
$\|C_x\| = \|\mathbf{C}\| \cos 66.0°$	The acute angle between vector **C** in standard position and the *x*-axis is $90° - 24.0° = 66.0°$.
$= (177 \text{ km})(\cos 66.0°)$	
$= 72.0 \text{ km}$	The *x*-component is in the positive *x*-direction.
$\|C_y\| = \|\mathbf{C}\| \sin 66.0°$	
$= (177 \text{ km})(\sin 66.0°)$	
$= -162 \text{ km}$	The *y*-component is in the negative *y*-direction.

Thus,

$$\mathbf{R}_x = \mathbf{A}_x + \mathbf{B}_x + \mathbf{C}_x = -86.0 \text{ km} + (-124 \text{ km}) + 72.0 \text{ km} = -138 \text{ km}$$

$$\mathbf{R}_y = \mathbf{A}_y + \mathbf{B}_y + \mathbf{C}_y = 60.2 \text{ km} + 0 + (-162 \text{ km}) = -102 \text{ km}$$

$$|\mathbf{R}| = \sqrt{|R_x|^2 + |R_y|^2}$$

$$= \sqrt{|-138 \text{ km}|^2 + |-102 \text{ km}|^2}$$

$$= 172 \text{ km}$$

$$\tan \alpha = \frac{102 \text{ km}}{138 \text{ km}} = 0.7391$$

$$\alpha = 36.5°$$

So, the displacement is 172 km at 36.5° south of west. That is, the ship stops at a port that is 172 km at 36.5° south of west from its starting point.

PROBLEMS 3.3

Use graph paper to find the resultant of each displacement pair.

1. 35 km due east, then 5̄0 km due north
2. 6̄0 km due west, then 9̄0 km due south
3. 5̄00 mi at 75° east of north, then 15̄00 mi at 2̄0° west of south
4. 2̄0 mi at 3° north of east, then 17 mi at 9° west of south
5. 67 km at 55° north of west, then 46 km at 25° south of east
6. 4.0 km at 25° west of south, then 2.0 km at 15° north of east

Use graph paper to find the resultant of each set of displacements.

7. 6̄0 km due south, then 9̄0 km at 15° north of west, and then 75 km at 45° north of east
8. 110 km at 5̄0° north of east, then 170 km at 3̄0° east of south, and then 145 km at 2̄0° north of east
9. 1700 mi due north, then 2400 mi at 1̄0° north of east, and then 2̄000 mi at 2̄0° south of west
10. 9̄0 mi at 1̄0° west of north, then 75 mi at 3̄0° west of south, and then 55 mi at 2̄0° east of south
11. 75 km at 25° north of east, then 75 km at 65° south of west, and then 75 km due south
12. 17 km due north, then 1̄0 km at 7° south of east, and then 15 km at 1̄0° west of south
13. 12 mi at 58° north of east, then 16 mi at 78° north of east, then 1̄0 mi at 45° north of east, and then 14 mi at 1̄0° north of east
14. 1̄0 km at 15° west of south, then 27 km at 35° north of east, then 31 km at 5° north of east, and then 22 km at 2̄0° west of north

Find the x- and y-components of each resultant vector **R** and graph the resultant vector **R**.

	Vector	x-component	y-component		Vector	x-component	y-component
15.	A	+2	+3	16.	A	+9	−5
	B	+7	+2		B	−4	−6
	R				R		
17.	A	−2	+13	18.	A	+10	−5
	B	−11	+1		B	−13	−9
	C	+3	−4		C	+4	+3
	R				R		
19.	A	+17	+7	20.	A	+1	+7
	B	−14	+11		B	+9	−4
	C	+7	+9		C	−4	+13
	D	−6	−15		D	−11	−4
	R				R		
21.	A	+1.5	−1.5	22.	A	+1	−1
	B	−3	−2		B	−4	−2
	C	+7.5	−3		C	+2	+4
	D	+2	+2.5		D	+5	−3
	R				E	+3	+5
					R		
23.	A	+1.5	+2.5	24.	A	−7	+15
	B	−2	−3		B	+13.5	−17.5
	C	+3.5	−7.5		C	−7.5	−20
	D	−4	+6		D	+6	+13.5
	E	−5.5	+2		E	+2.5	+2.5
	R				F	−11	+11.5
					R		

For each set of vectors, graph and find the *x*- and *y*-components of the resultant vector **R**.

25.

26.

27.

28.

29.

30.

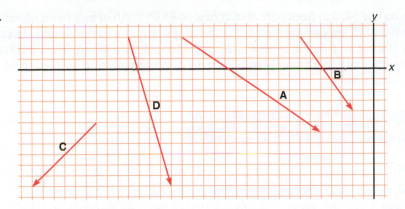

In Problems 31 through 42, find each resultant vector **R**. Give **R** in standard position.

31.

R, 6.00 m, 8.00 m

32.

15.0 ft, 45.0 ft, R

33.

18.5 mi, 18.5 mi, R

34.

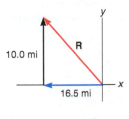

10.0 mi, R, 16.5 mi

35.

9.60 m, R, 1.40 m

36.

425 km, 365 km, R

	x-component	y-component		x-component	y-component
37.	+19.5 m	−49.6 m	38.	−158 km	+236 km
39.	+14.7 mi	+16.8 mi	40.	−3240 ft	−1890 ft
41.	−9.65 m	+4.36 m	42.	+375 km	−408 km

43. A road grader must go around a pond by traveling 10̄0 m south and then 15̄0 m east. If the road grader could go directly from the beginning point to the end point, how far would it travel?

44. An earthmover must go north 35̄0 m and then west 275 m to avoid a pipeline hazard. What distance would it travel if it could go directly to the endpoint?

45. An airplane takes off and flies 225 km on a course of 25.0° north of west and then changes direction and flies 135 km due north where it lands. Find the displacement from its starting point to its landing point.

46. A ship travels 50.0 mi on a course of 15.0° south of east and then travels 85.5 mi on a course of 60.0° west of south. Find the displacement from its starting point to its ending point.

47. A ship travels 135 km from port on a course of 25.0° south of east to an island. It then travels 122 km on a course of 35.5° west of south to a second island. Then it travels 135 km on a course of 10.4° north of west to a third island. Find the displacement from its starting point to its ending point.

48. A ship travels 145 km from port on a course of 65.0° north of east to an island. It then travels 112 km on a course of 30.5° west of north to a second island. Then it travels 182 km on a course of 10.4° west of south to a third island. Then it travels 42.5 km due south to a fourth island. Find the displacement from its starting point to its ending point.

PHYSICS CONNECTIONS

Global Positioning Satellites

Navigators continually struggle to find better tools to help them determine their location. The first explorers used the sun and stars to help them steer a straight course, but this method of navigation only worked under clear skies. Magnetic compasses were developed yet could only be used to determine longitude, not latitude. Finally, the mechanical clock, in conjunction with the compass, provided navigators with the most accurate method of determining location. Today, most navigators use a handheld device that functions in concert with a series of 24 orbiting satellites. This network, the Global Positioning System (GPS), can determine your position and altitude anywhere on earth.

The GPS pinpoints your location by sending out radio signals to locate any 4 of the 24 orbiting GPS satellites. Once the satellites are found, the GPS measures the length of time it takes for a radio signal to reach the handheld receiver. When the time is determined for each of four satellites, the distance is calculated, and the longitude, latitude, and altitude are displayed on the screen [Fig. 3.39(a)].

The GPS was first developed solely for military use. Eventually, the GPS was made available for civilian businesses. Shipping, airline, farming, surveying, and geological companies made use of the technology. Today, GPS receivers are affordable and are used by the general public [Fig. 3.39(b)]. More sophisticated receivers not only locate a position, but can also guide the navigator to a predetermined location. Several automobile manufacturers have included GPS receivers as an option in their cars. Such receivers come complete with voice commands, such as "Turn left at the next traffic light," as part of their option packages.

Figure 3.39 (a) The screen on the GPS receiver shows the position and strength of the signal between the receiver and the various satellites. At the time this photograph was taken, the receiver picked up 7 of the 12 overhead satellites, bringing the precision to within 20 ft of the actual location. (b) Global Positioning Systems have allowed for an enormous step forward in navigation. The GPS receiver shown has monitored and recorded precisely where the person has traveled and is now helping the user find his way back to camp.

(a)   (b)

Glossary

Component Vector When two or more vectors are added, each of the vectors is called a component of the resultant, or sum, vector. (p. 71)

Displacement The net change in position of an object, or the direct distance and direction it moves; a vector. (p. 66)

Number Plane A plane determined by the horizontal line called the x-axis and a vertical line called the y-axis intersecting at a right angle at a point called the origin. These two lines divide the number plane into four quadrants. The x-axis contains positive numbers to the right of the origin and negative numbers to the left of the origin. The y-axis contains positive numbers above the origin and negative numbers below the origin. (p. 70)

Resultant Vector The sum of two or more vectors. (p. 70)

Scalar A physical quantity that can be completely described by a number (called its magnitude) and a unit. (p. 65)

Standard Position A vector is in standard position when its initial point is at the origin of the number plane. The vector is expressed in terms of its length and its angle θ, where θ is measured counterclockwise from the positive x-axis to the vector. (p. 71)

Vector A physical quantity that requires both magnitude (size) and direction to be completely described. (p. 66)

x-component The horizontal component of a vector that lies along the x-axis. (p. 70)

y-component The vertical component of a vector that lies along the y-axis. (p. 70)

Formulas

3.2 To find the x- and y-components of a vector **v** given in standard position (Fig. 3.40):

Figure 3.40

1. Complete the right triangle with the legs being the x- and y-components of the vector.

2. Find the lengths of the legs of the right triangle as follows:

$$|\mathbf{A}_x| = |\mathbf{A}|(\cos \alpha)$$
$$|\mathbf{A}_y| = |\mathbf{A}|(\sin \alpha)$$

3. Determine the signs of the x- and y-components.

3.3 To find resultant vector **R** in standard position when its x- and y-components are given:

1. Complete the right triangle with the legs being the x- and y-components of the vector.

2. Find the acute angle α of the right triangle whose vertex is at the origin by using $\tan \alpha$.

3. Find angle θ in standard position as follows:

$$\theta = \alpha \qquad (\theta \text{ in first quadrant})$$

$$\theta = 180° - \alpha \qquad (\theta \text{ in second quadrant})$$

$$\theta = 180° + \alpha \qquad (\theta \text{ in third quadrant})$$

$$\theta = 360° - \alpha \qquad (\theta \text{ in fourth quadrant})$$

4. Find the magnitude of the vector using the Pythagorean theorem:

$$\mathbf{R} = \sqrt{|\mathbf{R}_x|^2 + |\mathbf{R}_y|^2}$$

To find the resultant vector $\mathbf{R}$ of any set of vectors, such as $\mathbf{R} = \mathbf{A} + \mathbf{B} + \mathbf{C}$, using right-triangle trigonometry:

1. Find the x-component of each vector and add: $\mathbf{R}_x = \mathbf{A}_x + \mathbf{B}_x + \mathbf{C}_x$.
2. Find the y-component of each vector and add: $\mathbf{R}_y = \mathbf{A}_y + \mathbf{B}_y + \mathbf{C}_y$.
3. Find the magnitude of the resultant vector $\mathbf{R}$ using the Pythagorean theorem: $|\mathbf{R}| = \sqrt{|\mathbf{R}_x|^2 + |\mathbf{R}_y|^2}$.
4. Find the direction of the resultant vector $\mathbf{R}$ using right-triangle trigonometry by (a) first finding the acute α between the resultant vector and the x-axis and then finding angle θ in standard position or (b) expressing the direction of the resultant vector using some other reference.

Review Questions

1. Displacement
   (a) can be interchanged with direction.
   (b) is a measurement of volume.
   (c) can be described only with a number.
   (d) is the net distance an object travels, showing direction and distance.
2. When adding vectors, the order in which they are added
   (a) is not important.   (b) is important.
   (c) is important only in certain cases.
3. A vector is in standard position when its initial point is
   (a) at the origin.   (b) along the x-axis.
   (c) along the y-axis.
4. Discuss number plane, origin, and axis in your own words.
5. Can every vector be described in terms of its components?
6. Can a vector have more than one set of component vectors?
7. Describe how to add two or more vectors graphically.
8. Describe how to find a resultant vector if given its x- and y-components.
9. Is a vector limited to a single position in the number plane?
10. Is the angle of a vector in standard position measured clockwise or counterclockwise?
11. What are the limits on the angle measure of a vector in standard position in the third quadrant?
12. Describe how to find the x- and y-components of a vector given in standard position.
13. Describe how to find a vector in standard position when the x- and y-components are given.

Review Problems

1. Find the x- and y-components of vector $\mathbf{R}$, which has a length of 13.0 cm at 30.0°.
2. Find the x- and y-components of vector $\mathbf{R}$, which has a length of 10.0 cm at 60.0°.
3. Find the x- and y-components of vector $\mathbf{R}$, which has a length of 20.0 cm at 30.0°.
4. Vector $\mathbf{R}$ has length 9.00 cm at 240.0°. Find its x- and y-components.
5. Vector $\mathbf{R}$ has length 9.00 cm at 40.0°. Find its x- and y-components.
6. Vector $\mathbf{R}$ has length 18.0 cm at 305.0°. Find its x- and y-components.
7. A hiker is plotting his course on a map with a scale of 1.00 cm = 3.00 km. If the hiker walks 2.50 cm north, then turns south and walks 1.50 cm, what is the actual displacement of the hiker in km?
8. A hiker is plotting his course on a map with a scale of 1.00 cm = 3.00 km. If the hiker walks 1.50 cm north, then turns south and walks 2.50 cm, what is the actual displacement of the hiker in km?

9. A co-pilot is charting her course on a map with a scale of 1.00 cm = 20.0 km. If the plane is charted to head 13.0 cm west, 9.00 cm north, and 2.00 cm east, what is the actual displacement of the plane in km?

10. A co-pilot is charting her course on a map with a scale of 1.00 cm = 20.0 km. If the plane is charted to head 25.0° north of east for 16.0 cm, north for 6.00 cm, and west for 5.00 cm, what is the actual displacement of the plane in km?

11. Vector **R** has x-component = +14.0 and y-component = +3.00. Find its length.

12. Vector **R** has x-component = −5.00 and y-component = +10.0. Find its length.

13. Vector **R** has x-component = +8.00 and y-component = −2.00. Find its length.

14. Vector **R** has x-component = −3.00 and y-component = −4.00. Find its length.

15. Vectors **A**, **B**, and **C** are given. Vector **A** has x-component = +3.00 and y-component = +4.00. Vector **B** has x-component = +5.00 and y-component = −7.00. Vector **C** has x-component = −2.00 and y-component = +1.00. Find the resultant vector **R**.

16. Vectors **A**, **B**, and **C** are given. Vector **A** has x-component = +5.00 and y-component = +7.00. Vector **B** has x-component = +9.00 and y-component = −3.00. Vector **C** has x-component = −5.00 and y-component = +5.00. Find the x- and y-components of the resultant vector **R**.

17. Vectors **A**, **B**, and **C** are given. Vector **A** has x-component = −3.00 and y-component = +4.00. Vector **B** has x-component = −5.00 and y-component = +7.00. Vector **C** has x-component = +2.00 and y-component = −1.00. Find the x- and y-components of the resultant vector **R**.

18. Vectors **A**, **B**, and **C** are given. Vector **A** has x-component = −5.00 and y-component = −7.00. Vector **B** has x-component = −9.00 and y-component = +3.00. Vector **C** has x-component = +5.00 and y-component = −5.00. Find the x- and y-components of the resultant vector **R**.

Graph and find the x- and y-components of each resultant vector **R**, where **R = A + B + C + D**.

19.

20.

21. An airplane takes off and flies 245 km on a course of 45.0° south of west and then changes direction and flies 175 km due south, where it lands. Find the displacement from its starting point to its landing point.

22. A ship travels 155 km from port on a course of 35.0° south of west to an island. It then travels 142 km on a course of 55.5° east of south to a second island. Then it travels 138 km on a course of 9.4° north of east to a third island. Then it travels 185 km due east to a fourth island. Find the displacement from its starting point to its ending point.

APPLIED CONCEPTS

1. The New Clark Bridge is an elegant cable-stayed bridge. Its design requires cables to reach from the road deck up to the tower and back down to the road deck on the other side of the tower, as shown in Fig. 3.41. In order to determine the best method for shipping the cables, the shipping company needs to know the lengths of the shortest and longest cables. Given the measurements in the diagram, determine the indicated total lengths *BEC* and *AED*, respectively.

2. Frank just learned that the $80\overline{0}$-m section of Broadway that he uses to get to work will be closed for several days. Given the information from a map of Manhattan (Fig. 3.42), what is the distance of Frank's next-shortest route?

Figure 3.41

Tower's height above the road deck = 176 ft

Figure 3.42

3. Power cables need to be suspended by the power company across a river to a new condominium development. Find the distance across the river in Fig. 3.43.

Figure 3.43

4. Bill has set his GPS to track his route. At the conclusion of his hike, the receiver indicates that he walked 3.50 mi north, 1.00 mi northeast, and 1.50 mi south. How far away is Bill from his original position?

5. With the airplane cruising at $30,\overline{0}00$ ft, the navigator indicates to the captain that the plane should continue traveling north for $50\overline{0}$ km and then turn to a heading of 45.0° east of north for $20\overline{0}$ km. What will be the resultant distance traveled?

CHAPTER 4

MOTION

Objectives

The major goals of this chapter are to enable you to:

1. Distinguish between speed and velocity.
2. Use vectors to illustrate and solve velocity problems.
3. Distinguish between velocity and acceleration.
4. Utilize vectors to illustrate and solve acceleration problems.
5. Analyze the motion of an object in free fall.
6. Solve two-dimensional motion problems.
7. Calculate the range of projectile motion.

Motion is a change of position. Velocity and acceleration describe important kinds of motion. An analysis of motion helps introduce the real nature of physics—to understand the nature and behavior of the physical world.

4.1 Speed Versus Velocity

Motion is defined as an object's change in position. One of the most frequently used descriptors of motion is speed.

Describing and measuring an object's motion is critical in many aspects of science, engineering, and everyday life. In baseball, the speed of a fastball is measured to determine the prowess of a pitcher. Medical technicians use Doppler radar devices to measure the speed of blood passing through arteries. Automotive engineers synchronize the motion of pistons, valves, and drive shafts to ensure that an engine operates efficiently. Finally, automobile drivers frequently refer to their speedometers to gauge how fast they are traveling.

Figure 4.1   A speedometer measures speed but not velocity.

Figure 4.2   A person walks along path *a* and then turns to follow path *b*. The distance traveled is the total distance of *a* and *b*. The displacement is represented by distance *c*.

Speed, as measured on a speedometer, is the distance traveled per unit of time. The speed of an automobile is represented in either miles per hour or kilometres per hour (Fig. 4.1). These units actually help define the formula for calculating speed:

$$\text{speed} = \frac{\text{distance traveled}}{\text{time to move that distance}}$$

Speed is a scalar value, for it shows only the magnitude of the position change per unit of time and does not indicate a direction. The unit for speed is a distance unit divided by a time unit, such as miles per hour (mi/h), kilometres per hour (km/h), metres per second (m/s), and feet per second (ft/s). For example, if you drive $35\overline{0}$ mi in 7.00 h, your average speed is

$$\frac{35\overline{0} \text{ mi}}{7.00 \text{ h}} = 50.0 \text{ mi/h}$$

Speed represents how fast something is moving, yet it does not indicate the direction in which it is traveling. Suppose you started driving from Chicago at a speed of 50 mi/h for 6 h. Where did you end your trip? You may have driven 50 mi/h southwest toward St. Louis, 50 mi/h northeast toward Detroit, 50 mi/h southeast toward Louisville, or 50 mi/h in a loop that brought you back to Chicago. Although speed may indicate how fast you are moving, it may not give you all the information you need to solve a problem.

Distance traveled must be distinguished from displacement. Whereas *distance* traveled may follow a path that is not straight, *displacement* is the net change of position of an object. It is represented by a straight line from the initial position to the final position and is a vector because it has both magnitude and direction (Fig. 4.2).

The **velocity** of an object is the rate of motion in a particular direction. Velocity is a vector that not only represents the speed, but also indicates the direction of motion. The relationship may be expressed by the equation

$$v_{\text{avg}} = \frac{s}{t}$$

or

$$s = v_{\text{avg}}t$$

where s = displacement
v_{avg} = average velocity
t = time

This equation is used to find either average speed (a scalar quantity) or the magnitude of the velocity (a vector quantity). Remember that if indicating velocity, the direction must be included with the speed. Therefore, a speed of 50 mi/h would be written 50 mi/h northeast, 50 mi/h up, or 50 mi/h 30° east of south as a velocity.

Figure 4.3 shows an illustration of a car traveling at a constant velocity of 10 m/s to the right. Note that its displacement is 10 m to the right during each second of travel.

Figure 4.3   The velocity, displacement, and time for a car traveling at a constant velocity of 10 m/s to the right is shown in 1-s intervals.

Find the average speed of an automobile that travels 160 km in 2.0 h.   **EXAMPLE 1**

Data:

$$s = 160 \text{ km}$$
$$t = 2.0 \text{ h}$$
$$v_{avg} = ?$$

Basic Equation:

$$s = v_{avg}t$$

Working Equation:

$$v_{avg} = \frac{s}{t}$$

Substitution:

$$v_{avg} = \frac{160 \text{ km}}{2.0 \text{ h}}$$
$$= 8\bar{0} \text{ km/h}$$

An airplane flies $35\bar{0}0$ mi in 5.00 h. Find its average speed.   **EXAMPLE 2**

Data:

$$s = 35\bar{0}0 \text{ mi}$$
$$t = 5.00 \text{ h}$$
$$v_{avg} = ?$$

Basic Equation:

$$s = v_{avg}t$$

Working Equation:

$$v_{avg} = \frac{s}{t}$$

Substitution:

$$v_{avg} = \frac{35\bar{0}0 \text{ mi}}{5.00 \text{ h}}$$
$$= 70\bar{0} \text{ mi/h}$$

Find the velocity of a plane that travels $60\bar{0}$ km due north in 3 h 15 min.   **EXAMPLE 3**

Data:

$$s = 60\bar{0} \text{ km}$$
$$t = 3 \text{ h } 15 \text{ min} = 3.25 \text{ h}$$
$$v_{avg} = ?$$

Basic Equation:

$$s = v_{avg}t$$

Working Equation:

$$v_{avg} = \frac{s}{t}$$

Substitution:

$$v_{avg} = \frac{60\overline{0}\ km}{3.25\ h}$$

$$= 185\ km/h$$

The direction is north. Thus, the velocity is 185 km/h due north.

· · · · · · · · · · · · · · · · ·

Until now, our study of velocity has assumed a fixed observation point. The frame of reference can also be important for determining relative velocity. Paddling a canoe into a headwind may produce a net velocity of zero for the paddler. Another example is the flight of an airplane in which a crosswind, or in fact wind from any direction, will affect the airplane's velocity with respect to the ground. The airplane's final velocity is calculated by taking into account the velocity of the airplane in calm air and the velocity of any wind that the airplane encounters.

To find the sum (resultant vector) of velocity vectors, use the component method as outlined in Chapter 3.

EXAMPLE 4

A plane is flying due north (at 90°) at 265 km/h and encounters a wind from the east (at 180°) at 55.0 km/h. What is the plane's new velocity with respect to the ground in standard position? Assume that the plane's new velocity is the vector sum of the plane's original velocity and the wind velocity.

First, graph the plane's old velocity as the y-component and the wind velocity as the x-component (Fig. 4.4). The resultant vector is the plane's new velocity with respect to the ground. Find angle α as follows:

Figure 4.4

$$\tan \alpha = \frac{\text{side opposite } \alpha}{\text{side adjacent to } \alpha}$$

$$\tan \alpha = \frac{265\ km/h}{55.0\ km/h} = 4.818$$

$$\alpha = 78.3°$$

then

$$\theta = 180° - 78.3° = 101.7°$$

Find the magnitude of the new velocity (ground speed) using the Pythagorean theorem:

$$|\mathbf{R}| = \sqrt{|\mathbf{R}_x|^2 + |\mathbf{R}_y|^2}$$

$$|\mathbf{R}| = \sqrt{(55.0\ km/h)^2 + (265\ km/h)^2}$$

$$= 271\ km/h$$

That is, the new velocity of the plane is 271 km/h at 101.7°.

Note: Recall that standard practice is to use the vertical *y*-axis *up for N (north)* and *down for S (south)* and the horizontal *x*-axis *right for E (east)* and *left for W (west)* as shown in *Fig.* 3.23 on page 76.

. . . .

EXAMPLE 5

A plane is flying northwest (at 135.0°) at 315 km/h and encounters a wind from 30.0° south of west (at 30.0°) at 65.0 km/h. What is the plane's new velocity with respect to the ground in standard position? Assume that the plane's new velocity is the vector sum of the plane's original velocity and the wind velocity.

First, graph the plane's old velocity and the wind velocity as vectors in standard position (Fig. 4.5). The resultant vector is the plane's new velocity with respect to the ground.

Figure 4.5

Then, find the *x*- and *y*-components of the plane's old velocity and the wind velocity using Fig. 4.6.

Figure 4.6

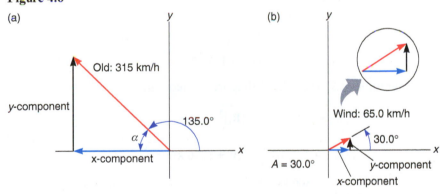

Plane: See Fig. 4.6(a). $\alpha = 180° - 135.0° = 45.0°$.

x-component	*y*-component
$\cos \alpha = \dfrac{\text{side adjacent to } \alpha}{\text{hypotenuse}}$	$\sin \alpha = \dfrac{\text{side opposite } \alpha}{\text{hypotenuse}}$
$\cos 45.0° = \dfrac{x\text{-component}}{315 \text{ km/h}}$	$\sin 45.0° = \dfrac{y\text{-component}}{315 \text{ km/h}}$
$(315 \text{ km/h})(\cos 45.0°) = x\text{-component}$	$(315 \text{ km/h})(\sin 45.0°) = y\text{-component}$
$223 \text{ km/h} = x\text{-component}$	$223 \text{ km/h} = y\text{-component}$
Thus, $x\text{-component} = -223 \text{ km/h}$	$y\text{-component} = +223 \text{ km/h}$

Wind: See Fig. 4.6(b).

x-component	y-component
$\cos \alpha = \dfrac{\text{side adjacent to } \alpha}{\text{hypotenuse}}$	$\sin \alpha = \dfrac{\text{side opposite } \alpha}{\text{hypotenuse}}$
$\cos 30.0° = \dfrac{x\text{-component}}{65.0 \text{ km/h}}$	$\sin 30.0° = \dfrac{y\text{-component}}{65.0 \text{ km/h}}$
$(65.0 \text{ km/h})(\cos 30.0°) = x\text{-component}$	$(65.0 \text{ km/h})(\sin 30.0°) = y\text{-component}$
$56.3 \text{ km/h} = x\text{-component}$	$32.5 \text{ km/h} = y\text{-component}$
Thus, x-component $= +56.3$ km/h	y-component $= +32.5$ km/h

To find **R**:

	x-component	y-component	
Plane:	− 223 km/h	+ 223 km/h	
Wind:	+ 56.3 km/h	+ 32.5 km/h	
Sum:	− 167 km/h	+ 256 km/h	(Round each component sum to its least precise component.)

Figure 4.7

Find angle α from Fig. 4.7 as follows:

$$\tan \alpha = \frac{\text{side opposite } \alpha}{\text{side adjacent to } \alpha}$$

$$\tan \alpha = \frac{256 \text{ km/h}}{167 \text{ km/h}} = 1.533$$

$$\alpha = 56.9°$$

and

$$\theta = 180° - 56.9° = 123.1°$$

Find the magnitude of **R** using the Pythagorean theorem:

$$|\mathbf{R}| = \sqrt{|\mathbf{R}_x|^2 + |\mathbf{R}_y|^2}$$

$$|\mathbf{R}| = \sqrt{(167 \text{ km/h})^2 + (256 \text{ km/h})^2}$$

$$= 306 \text{ km/h}$$

That is, the new velocity of the plane is 306 km/h at 123.1°.

····················

PHYSICS CONNECTIONS

Vectors Across Rivers

Crossing the Hudson River, which separates New York City from New Jersey, can be challenging. A working knowledge of velocity and vectors is absolutely essential, especially when attempting to cross the river in strong currents, brisk winds, driving rain, and dense fog. In addition, maneuvering between barges, cruise ships, recreational boaters, and driftwood can make the job even more difficult.

Ferry captains like Mike and John combine vectors every time they cross the river. Captain Mike said, "At times crossing the river can be quite tricky. Your ferry might be pointing directly across the river, but the current is pushing you farther down river. In order to combat this, you need to change the boat's heading so when your velocity and the current's velocity combine, you arrive at your planned destination" (Fig. 4.8).

Sometimes, when the current and wind are strong and headed in the same direction, ferryboats can appear to be heading up to 45° away from their destination, yet still travel directly across the river. In such situations, docking can be nerve-racking (Fig. 4.9).

Figure 4.8 An example of how the velocity of a boat and the velocity of the current are combined so the resultant velocity is directed toward the desired location.

Figure 4.9 Although the boat is not pointed toward the dock, the combination of the boat's velocity (green vector) plus the current's velocity (blue vector) results in a perfect docking (red vector).

The two experienced sea captains say that vectors are even more important in the open seas. Captain Mike said, "Combining your boat's velocity vector with the current and wind vectors can mean the difference between arriving at your home port or at a port 100 miles away." These days the process is easier with the use of radar and computer navigation equipment and software. Looking at the monitor, Captain Mike said, "We can see the velocity of the current and our boat's intended velocity directly on the monitor. Instead of our combining the vectors, the computer can instantaneously combine them and provide the resultant velocity" (Fig. 4.10).

Figure 4.10(a) Captain Mike combines velocity vectors every time he navigates across the Hudson River.

Figure 4.10(b) Captain John's high-speed ferry is equipped with computers that automatically combine and display the boat's resultant velocity vector.

SKETCH

$$12 \text{ cm}^2 \quad w$$

$$4.0 \text{ cm}$$

DATA

$A = 12 \text{ cm}^2, l = 4.0 \text{ cm}, w = ?$

BASIC EQUATION

$A = lw$

WORKING EQUATION

$w = \dfrac{A}{l}$

SUBSTITUTION

$w = \dfrac{12 \text{ cm}^2}{4.0 \text{ cm}} = 3.0 \text{ cm}$

PROBLEMS 4.1

Find the average speed (in the given units) of an auto that travels each distance in the given time.

1. 150 mi in 3.0 h (in mi/h)
2. 190 m in 8.5 s (in m/s)
3. 8550 m in 6 min 35 s (in m/s)
4. 45 km in 0.50 h (in km/h)
5. 785 ft in 11.5 s (in ft/s)
6. Find the average speed (in mi/h) of a racing car that turns a lap on a 1.00-mi oval track in 30.0 s.
7. While driving at $9\overline{0}$ km/h, how far can you travel in 3.5 h?
8. While driving at $9\overline{0}$ km/h, how far (in metres) do you travel in 1.0 s?
9. An automobile is traveling at 55 mi/h. Find its speed
    (a) in ft/s.   (b) in m/s.   (c) in km/h.
10. An automobile is traveling at 22.0 m/s. Find its speed
    (a) in km/h.   (b) in mi/h.   (c) in ft/s.
11. A semi-trailer truck traveling $10\overline{0}$ km/h continues for 2.75 h. How far does it go?
12. A flatbed truck travels for 3.85 h at 105 km/h. How far does it go?
13. The average speed of a garbage truck is 60.0 km/h. How long does it take for the truck to travel 265 km?
14. A highway maintenance truck has an average speed of 55.0 km/h. How far does it travel in 3.65 h?

Find the velocity for each displacement and time.

15. 160 km east in 2.0 h
16. $10\overline{0}$ km north in 3.0 h
17. $100\overline{0}$ mi south in 8.00 h
18. 31.0 mi west in 0.500 h
19. 275 km at $3\overline{0}°$ south of east in 4.50 h
20. 426 km at 45° north of west in 2.75 h
21. Milwaukee is 121 mi (air miles) due west of Grand Rapids. Maria drives 255 mi in 4.75 h from Grand Rapids to Milwaukee around Lake Michigan. Find (a) her average driving speed and (b) her average travel velocity.
22. Telluride, Colorado, is 45 air miles at 11° east of north of Durango. On a winter day, Chuck drove 120 mi from Durango to Telluride around a mountain in $4\frac{1}{4}$ h including a traffic delay. Find (a) his average driving speed and (b) his average travel velocity.

In Problems 23–30, assume that the plane's new velocity is the vector sum of the plane's original velocity and the wind velocity.

23. A plane is flying due north at 325 km/h and encounters a wind from the south at 45 km/h. What is the plane's new velocity with respect to the ground in standard position?
24. A plane is flying due west at 275 km/h and encounters a wind from the west at $8\overline{0}$ km/h. What is the plane's new velocity with respect to the ground in standard position?
25. A plane is flying due west at 235 km/h and encounters a wind from the north at 45.0 km/h. What is the plane's new velocity with respect to the ground in standard position?
26. A plane is flying due north at 185 mi/h and encounters a wind from the west at 35.0 mi/h. What is the plane's new velocity with respect to the ground in standard position?
27. A plane is flying southwest at 155 mi/h and encounters a wind from the west at 45.0 mi/h. What is the plane's new velocity with respect to the ground in standard position?
28. A plane is flying southeast at 215 km/h and encounters a wind from the north at 75.0 km/h. What is the plane's new velocity with respect to the ground in standard position?

29. A plane is flying at 25.0° north of west at $19\overline{0}$ km/h and encounters a wind from 15.0° north of east at 45.0 km/h. What is the plane's new velocity with respect to the ground in standard position?

30. A plane is flying at 36.0° south of west at $15\overline{0}$ mi/h and encounters a wind from 75.0° north of east at 55.0 mi/h. What is the plane's new velocity with respect to the ground in standard position?

4.2 Acceleration

When shopping for a new vehicle, most consumers choose to ignore its maximum speed but instead pay particular attention to how quickly the vehicle can change its speed from 0 mi/h to 60 mi/h. From the new-car-buyer's perspective, the less time the car takes to achieve a speed of 60 mi/h, the greater will be its acceleration and the easier it will be to get onto a highway or pass a slow-moving vehicle.

Acceleration *is the change in velocity per unit time.* In other words, acceleration measures how quickly velocity changes. That is,

$$\text{average acceleration} = \frac{\text{change in velocity (or speed)}}{\text{elapsed time}}$$

$$= \frac{\text{final velocity} - \text{initial velocity}}{\text{time}}$$

This relationship can be expressed by the equation

$$a = \frac{\Delta v}{t} = \frac{v_f - v_i}{t}$$

or

$$\Delta v = at$$

where $\Delta v = $ change in velocity (or speed)
$\quad a = $ acceleration
$\quad t = $ time

The Greek letter Δ (capital delta) is used to mean "change in."

A dragster starts from rest (velocity = 0 ft/s) and attains a speed of $15\overline{0}$ ft/s in 10.0 s. Find its acceleration.

EXAMPLE 1

Data:

$$\Delta v = 15\overline{0} \text{ ft/s} - 0 \text{ ft/s} = 15\overline{0} \text{ ft/s}$$

$$t = 10.0 \text{ s}$$

$$a = ?$$

Basic Equation:

$$\Delta v = at$$

Working Equation:

$$a = \frac{\Delta v}{t}$$

Substitution:

$$a = \frac{15\overline{0} \text{ ft/s}}{10.0 \text{ s}}$$

$$= 15.0 \frac{\text{ft/s}}{\text{s}} \text{ or } 15.0 \text{ feet per second per second}$$

Recall from arithmetic that to simplify fractions in the form

$$\frac{\dfrac{a}{b}}{\dfrac{c}{d}}$$

we divide by the denominator; that is, invert and multiply:

$$\frac{\dfrac{a}{b}}{\dfrac{c}{d}} = \frac{a}{b} \div \frac{c}{d} = \frac{a}{b} \cdot \frac{d}{c} = \frac{ad}{bc}$$

Use this idea to simplify the units 15.0 feet per second per second:

$$\frac{\dfrac{15.0 \text{ ft}}{\text{s}}}{\dfrac{\text{s}}{1}} = \frac{15.0 \text{ ft}}{\text{s}} \div \frac{\text{s}}{1} = \frac{15.0 \text{ ft}}{\text{s}} \cdot \frac{1}{\text{s}} = \frac{15.0 \text{ ft}}{\text{s}^2} \text{ or } 15.0 \text{ ft/s}^2$$

The units of acceleration are usually ft/s^2 or m/s^2.

· · · · · · · · · · · · · · · · · ·

When the speed of an automobile increases from rest to 5 mi/h in the first second to 10 mi/h in the next second and to 15 mi/h in the third second, its acceleration is 5 $\dfrac{\text{mi/h}}{\text{s}}$.

That is, its increase in speed is 5 mi/h during each second. In Fig. 4.11 an automobile increases in speed from 6 m/s to 9 m/s in the first second, to 12 m/s in the next second, and to 15 m/s in the third second, so its acceleration is 3 $\dfrac{\text{m/s}}{\text{s}}$, usually written 3 m/s^2. This means that the speed of the automobile increases 3 m/s during each second.

Figure 4.11   This car is speeding up with a constant acceleration. Note how the distance covered and the velocity change during each time interval.

A car accelerates from 45 km/h to 8$\overline{0}$ km/h in 3.00 s. Find its acceleration (in m/s^2).

EXAMPLE 2

Data:

$$\Delta v = 8\overline{0} \text{ km/h} - 45 \text{ km/h} = 35 \text{ km/h}$$
$$t = 3.00 \text{ s}$$
$$a = ?$$

Basic Equation:

$$\Delta v = at$$

Working Equation:

$$a = \frac{\Delta v}{t}$$

Substitution:

$$a = \frac{35 \text{ km} / \text{h}}{3.00 \text{ s}} \times \frac{1000 \text{ m}}{1 \text{ km}} \times \frac{1 \text{ h}}{3600 \text{ s}}$$

$$= 3.2 \text{ m/s}^2$$

Note the use of the conversion factors to change the units km/h/s to m/s^2.

· · · · · · · · · · · · · · · ·

A plane accelerates at 8.5 m/s^2 for 4.5 s. Find its increase in speed (in m/s).   **EXAMPLE 3**

Data:

$$a = 8.5 \text{ m/s}^2$$
$$t = 4.5 \text{ s}$$
$$\Delta v = ?$$

Basic Equation:

$$\Delta v = at$$

Working Equation: Same

Substitution:

$$\Delta v = (8.5 \text{ m/s}^2)(4.5 \text{ s})$$
$$= 38 \text{ m/s}$$

$$\boxed{\frac{\text{m}}{\text{s}^2} \times \text{s} = \frac{\text{m}}{\text{s}}}$$

· · · · · · · · · · · · · · · ·

Acceleration means more than just an increase in speed. In fact, since velocity is the speed of an object and its direction of motion, acceleration can mean speeding up, slowing down, or changing direction. The next example illustrates negative acceleration (sometimes called deceleration). **Deceleration** is an acceleration that usually indicates that an object is slowing down (Fig. 4.12). Acceleration when an object changes direction will be discussed in Chapter 9.

Figure 4.12   This car is slowing down with a constant acceleration of −10 m/s^2. Note how the distance covered and the velocity change during each unit of time interval.

Sometimes the direction of the velocity or the acceleration of an object is understood and does not need to be stated explicitly. For example, if a car is accelerating on a straight road, both the direction of the velocity and the direction of the acceleration are along the road in the direction in which the car is moving. When the car begins to slow down, the direction of the velocity is understood to be along the road in the direction pointing ahead of the car, and the direction of its acceleration (deceleration) is understood to be along the road in the direction pointing behind the car.

EXAMPLE 4

A driver steps off the gas pedal and coasts at a rate of −3.00 m/s² for 5.00 s. Find the driver's new speed if she was originally traveling at a velocity of 20.0 m/s. (The negative acceleration indicates that the acceleration is in the opposite direction of the velocity; that is, the object is slowing down.)

Data:

$$a = -3.00 \text{ m/s}^2$$
$$t = 5.00 \text{ s}$$
$$v_i = 20.0 \text{ m/s}$$
$$v_f = ?$$

Basic Equation:

$$\Delta v = at$$
$$v_f - v_i = at$$

Working Equation:

$$v_f = v_i + at$$

Substitution:

$$v_f = 20.0 \text{ m/s} + (-3.00 \text{ m/s}^2)(5.00 \text{ s})$$
$$= 5.0 \text{ m/s}$$

Figure 4.13 Motion of a high-speed train going from one station to another. When the speed increases, the acceleration is positive. When the speed is constant, the acceleration is zero. When the speed decreases, the acceleration is negative.

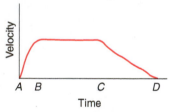

Acceleration has both magnitude and direction. Consider a high-speed train moving out of the station due east (positive direction). Its velocity increases in magnitude until it reaches its cruising speed (Fig. 4.13). Its acceleration is greatest at the start (from A to B), when its increase in speed is largest. When the train is moving at relatively constant velocity (from B to C), its acceleration is near zero. Then as the train approaches the next station (from C to D), its speed decreases as it decelerates, thus creating a negative acceleration.

SKETCH

12 cm²   w

4.0 cm

DATA

$A = 12 \text{ cm}^2, l = 4.0 \text{ cm}, w = ?$

BASIC EQUATION

$A = lw$

WORKING EQUATION

$w = \dfrac{A}{l}$

SUBSTITUTION

$w = \dfrac{12 \text{ cm}^2}{4.0 \text{ cm}} = 3.0 \text{ cm}$

PROBLEMS 4.2

An automobile changes speed as shown. Find its acceleration.

	Speed Change	Time Interval	Find a
1.	From 0 to 15 m/s	1.0 s	in m/s²
2.	From 0 to 18 m/s	3.0 s	in m/s²
3.	From 6$\overline{0}$ ft/s to 7$\overline{0}$ ft/s	1.0 s	in ft/s²
4.	From 45 m/s to 65 m/s	2.0 s	in m/s²
5.	From 25 km/h to 9$\overline{0}$ km/h	5.6 s	in m/s²
6.	From 1$\overline{0}$ mi/h to 5$\overline{0}$ mi/h	3.5 s	in ft/s²

7. A dragster starts from rest and reaches a speed of 62.5 m/s in 10.0 s. Find its acceleration (in m/s²).

8. A car accelerates from 25 mi/h to 55 mi/h in 4.5 s. Find its acceleration (in ft/s²).

9. A train accelerates from $1\bar{0}$ km/h to $11\bar{0}$ km/h in 2 min 15 s. Find its acceleration (in m/s^2).

A plane accelerates at 30.0 ft/s^2 for 3.30 s. Find its increase in speed

10. in ft/s.   11. in mi/h.

A rocket accelerates at 10.0 m/s^2 from rest for 20.0 s. Find its increase in speed

12. in m/s.   13. in km/h.

14. How long (in seconds) does it take for a rocket sled accelerating at 15.0 m/s^2 to change its speed from 20.0 m/s to 65.0 m/s?

15. What is the acceleration of a road grader that goes from rest to 10.0 km/h in 5.20 s?

16. What is the acceleration of a compactor that goes from rest to 20.0 km/h in 4.80 s?

17. How long (in seconds) does it take for a truck accelerating at 1.50 m/s^2 to go from rest to 90.0 km/h?

18. How long (in seconds) does it take for a car accelerating at 3.50 m/s^2 to go from rest to $12\bar{0}$ km/h?

19. A bulldozer accelerates from rest to 3.00 m/s in 4.20 s. What is its acceleration?

20. A pickup truck pulling a trailer accelerates at 3.25 m/s^2 for 5.80 s. If it starts from rest, what is its final velocity?

21. The speed of a delivery van increases from 2.00 m/s at 1.00 s to 16.0 m/s at 4.50 s. What is its average speed?

22. A go-cart rolls backward down a driveway. We define forward speed as positive and backward speed as negative. The cart's speed changes from −2.00 m/s to −9.00 m/s in 2.00 s. What is its acceleration?

23. A stock car is moving at 25.0 m/s when the driver applies the brakes. If it stops in 3.00 s, what is its average acceleration?

24. If the car in Problem 23 took twice as long to stop, what would its acceleration be?

25. If the car in Problem 23 was going twice as fast but was able to stop in the same time, what would its acceleration be?

26. If the car in Problem 23 was going twice the speed and stopped in twice the time, what would its acceleration be?

4.3 Uniformly Accelerated Motion and Free Fall

Uniformly accelerated motion of an object occurs when its acceleration is constant; examples are a ball rolling down a straight incline, a car increasing its speed at a constant rate, and a ball dropped from a building. The following equations apply to uniformly accelerated motion and freely falling bodies. Some of the derivations are beyond the scope of this text.

ACCELERATED MOTION

1. $v_{avg} = \dfrac{v_f + v_i}{2}$ 3. $v_f = v_i + at$ 5. $s = \frac{1}{2}(v_f + v_i)t$

2. $a = \dfrac{v_f - v_i}{t}$ 4. $s = v_i t + \frac{1}{2}at^2$ 6. $2as = v_f^2 - v_i^2$

where s = displacement v_{avg} = average velocity
        v_f = final velocity a = constant acceleration
        v_i = initial velocity t = time

Now consider some problems using these equations, applying our problem-solving method.

EXAMPLE 1

The average velocity of a rolling freight car is 2.00 m/s. How long does it take for the car to roll 15.0 m?

Data:

$$s = 15.0 \text{ m}$$
$$v_{avg} = 2.00 \text{ m/s}$$
$$t = ?$$

Basic Equation:

$$s = v_{avg}t$$

Working Equation:

$$t = \frac{s}{v_{avg}}$$

Substitution:

$$t = \frac{15.0 \text{ m}}{2.00 \text{ m/s}}$$
$$= 7.50 \text{ s}$$

$$\boxed{\frac{\text{m}}{\text{m/s}} = \text{m} \div \frac{\text{m}}{\text{s}} = \text{m} \cdot \frac{\text{s}}{\text{m}} = \text{s}}$$

EXAMPLE 2

A dragster starting from rest reaches a final velocity of 318 km/h. Find its average velocity.

Data:

$$v_i = 0$$
$$v_f = 318 \text{ km/h}$$
$$v_{avg} = ?$$

Basic Equation:

$$v_{avg} = \frac{v_f + v_i}{2}$$

Working Equation: Same

Substitution:

$$v_{avg} = \frac{318 \text{ km/h} + 0 \text{ km/h}}{2}$$
$$= 159 \text{ km/h}$$

EXAMPLE 3

A train slowing to a stop has an average acceleration of -3.00 m/s^2. [Note that a minus ($-$) acceleration is commonly called *deceleration*, meaning that the train is slowing down.] If its initial velocity is 30.0 m/s, how far does it travel in 4.00 s?

Data:

$$a = -3.00 \text{ m/s}^2$$
$$v_i = 30.0 \text{ m/s}$$
$$t = 4.00 \text{ s}$$
$$s = ?$$

Basic Equation:

$$s = v_i t + \tfrac{1}{2} a t^2$$

Working Equation: Same

Substitution:

$$s = (30.0 \text{ m/s}) (4.00 \text{ s}) + \tfrac{1}{2} (-3.00 \text{ m/s}^2) (4.00 \text{ s})^2$$
$$= 12\overline{0} \text{ m} - 24.0 \text{ m}$$
$$= 96 \text{ m}$$

· · · · · · · · · · · · · · · ·

An automobile accelerates from 67.0 km/h to 96.0 km/h in 7.80 s. What is its acceleration (in m/s²)?

EXAMPLE 4

Data:

$$v_f = 96.0 \text{ km/h}$$
$$v_i = 67.0 \text{ km/h}$$
$$t = 7.80 \text{ s}$$
$$a = ?$$

Basic Equation:

$$a = \frac{v_f - v_i}{t}$$

Working Equation: Same

Substitution:

$$a = \frac{96.0 \text{ km/h} - 67.0 \text{ km/h}}{7.80 \text{ s}}$$

$$= \frac{29.0 \text{ km/h}}{7.80 \text{ s}}$$

$$= \frac{29.0 \dfrac{\cancel{\text{km}}}{\cancel{\text{h}}} \times \dfrac{1000 \text{ m}}{1 \cancel{\text{km}}} \times \dfrac{1 \cancel{\text{h}}}{3600 \text{ s}}}{7.80 \text{ s}}$$

$$= 1.03 \text{ m/s}^2$$

· · · · · · · · · · · · · · · ·

Freely falling bodies undergo constant acceleration. In a vacuum (in the absence of air resistance) a ball and a feather fall at the same rate. In 1971, astronaut David Scott demonstrated this during a moon mission by dropping a hammer and a feather simultaneously on the surface of the moon, where there is no atmosphere. Both objects fell at the same rate and landed at the same time.

TRY THIS ACTIVITY

Free Fall in a Vacuum

Drop a piece of paper and a book at the same time and note the relative time it takes for each to hit the floor. Now place that paper on top of the book as shown in Fig. 4.14. (*Note:* The top surface area of the book must be larger than that of the paper.) What happens to the time it takes the book and paper to fall? What does this show about objects falling in a vacuum? (A vacuum is a space in which there is no air resistance present.)

Figure 4.14   Place the paper on top of the book. The book must be larger than the paper.

The acceleration of a freely falling body, also called the **acceleration due to gravity**, is denoted by the symbol g, where $a = g = 9.80 \text{ m/s}^2$ (metric system) or $a = g = 32.2 \text{ ft/s}^2$ (U.S. system) on the earth's surface.

What does $a = 9.80 \text{ m/s}^2$ mean? When a ball is dropped from a building, the speed of the ball increases by 9.80 m/s during each second. Figure 4.15 shows the total distance traveled and the speed at 1-s intervals.

We limit our discussion to uniformly accelerated motion because the mathematical tools needed to study other kinds of motion are beyond the scope of this book. We also need to assume that air resistance of a freely falling body is negligible for our calculations. However, air resistance is, in fact, an important factor in the design of machines that move through the atmosphere.

When the air resistance of a falling object equals its weight, the net force is zero and no further acceleration occurs. That is, a falling object reaches its **terminal speed**.

TRY THIS ACTIVITY

Calculating Height

Drop a ball from a window that is at least two stories high (or from some other height) and time how long it takes the ball to fall to the ground. Use the formulas for accelerated motion to find the height. Measure the actual height with a tape measure. How does your calculated height compare with the actual height? What factors may have made your calculated height different from the actual height? Remember to use the problem-solving method to help you solve this problem.

EXAMPLE 5

A rock is thrown straight down from a cliff with an initial velocity of 10.0 ft/s. Its final velocity when it strikes the water below is $31\bar{0}$ ft/s. The acceleration due to gravity is 32.2 ft/s^2. How long is the rock in flight?

Data:

$$v_i = 10.0 \text{ ft/s}$$
$$a = 32.2 \text{ ft/s}^2$$
$$v_f = 31\bar{0} \text{ ft/s}$$
$$t = ?$$

Figure 4.15 A ball falls with constant acceleration $a = g = 9.80$ m/s^2 with the speed and the distance traveled calculated at the given times. Because the ball was dropped, $v_i = 0$ and the formulas for v_f and s are shown simplified. Note how the velocity and the distance traveled increase during each successive time interval.

Time	Distance Traveled		Speed
	$s = \frac{1}{2}at^2$		$v_f = at$
$t = 0$	0 m		0 m/s
$t = 1.00$ s	$s = \frac{1}{2}(9.80$ m/s$^2)(1.00$ s$)^2$		$v_f = (9.80$ m/s$^2)(1.00$ s$)$
	$s = 4.90$ m		$v_f = 9.80$ m/s
$t = 2.00$ s	$s = \frac{1}{2}(9.80$ m/s$^2)(2.00$ s$)^2$		$v_f = (9.80$ m/s$^2)(2.00$ s$)$
	$s = 19.6$ m		$v_f = 19.6$ m/s
$t = 3.00$ s	$s = \frac{1}{2}(9.80$ m/s$^2)(3.00$ s$)^2$		$v_f = (9.80$ m/s$^2)(3.00$ s$)$
	$s = 44.1$ m		$v_f = 29.4$ m/s
$t = 4.00$ s	$s = \frac{1}{2}(9.80$ m/s$^2)(4.00$ s$)^2$		$v_f = (9.80$ m/s$^2)(4.00$ s$)$
	$s = 78.4$ m		$v_f = 39.2$ m/s

Note the importance of listing all the data as an aid to finding the basic equation.

Basic Equation:

$$v_f = v_i + at \quad \text{or} \quad a = \frac{v_f - v_i}{t} \quad \text{(two forms of the same equation)}$$

Working Equation:

$$t = \frac{v_f - v_i}{a}$$

Substitution:

$$t = \frac{31\bar{0} \text{ ft/s} - 10.0 \text{ ft/s}}{32.2 \text{ ft/s}^2}$$

$$= \frac{30\bar{0} \text{ ft/s}}{32.2 \text{ ft/s}^2}$$

$$= 9.32 \text{ s}$$

$$\boxed{\frac{\text{ft/s}}{\text{ft/s}^2} = \frac{\text{ft}}{\text{s}} \div \frac{\text{ft}}{\text{s}^2} = \frac{\cancel{\text{ft}}}{\cancel{\text{s}}} \cdot \frac{\cancel{\text{s}^2}}{\cancel{\text{ft}}} = \text{s}}$$

When any object is launched vertically upward, its upward speed is uniformly decreased by the force of gravity until it stops for an instant at its peak before falling back to the ground. As it is falling to the ground, it is uniformly accelerated by gravity the same as it would have been if dropped from its peak height. If an object is thrown vertically upward and if the initial velocity is known, the previous acceleration/gravity formulas may be used to find how high the object rises, how long it is in flight, and so on.

Note: When we consider a problem involving an object being thrown upward, we will consider an upward direction to be negative and the opposing gravity in its normal downward direction to be positive.

EXAMPLE 6

Figure 4.16

$v_i = -25.0$ m/s

A baseball is thrown vertically upward with an initial velocity of 25.0 m/s (see Fig. 4.16). (a) How high does it go? (b) How long does it take to reach its maximum height? (c) How long is it in flight?

(a) Data:

$$v_i = -25.0 \text{ m/s}$$

(v_i is negative because the initial velocity is directed opposite gravity, g.)

$$v_f = 0$$

(At the instant of the ball's maximum height, its velocity is zero.)

$$a = g = 9.80 \text{ m/s}^2$$
$$s = ?$$

Basic Equation:

$$2as = v_f^2 - v_i^2$$

Working Equation:

$$s = \frac{v_f^2 - v_i^2}{2a}$$

Substitution:

$$s = \frac{0^2 - (-25.0 \text{ m/s})^2}{2(9.80 \text{ m/s}^2)}$$

$$\frac{(\text{m/s})^2}{\text{m/s}^2} = \frac{\text{m}^2/\text{s}^2}{\text{m/s}^2} = \frac{\text{m}^2}{\text{s}^2} \div \frac{\text{m}}{\text{s}^2} = \frac{\frac{\text{m}^2}{\text{m}}}{\text{s}^2} \times \frac{\text{s}^2}{\text{m}} = \text{m}$$

$$= -31.9 \text{ m}$$

(s being negative indicates an upward displacement.)

(b) Data:

$$v_i = -25.0 \text{ m/s}$$
$$v_f = 0$$
$$a = g = 9.80 \text{ m/s}^2$$
$$t = ?$$

Basic Equation:

$$v_f = v_i + at$$

Working Equation:

$$t = \frac{v_f - v_i}{a}$$

Substitution:

$$t = \frac{0 - (-25.0 \text{ m/s})}{9.80 \text{ m/s}^2}$$

$$\boxed{\frac{\text{m/s}}{\text{m/s}^2} = \frac{\text{m}}{\text{s}} \div \frac{\text{m}}{\text{s}^2} = \frac{\text{m}}{\text{s}} \times \frac{\text{s}^2}{\text{m}} = \text{s}}$$

$$= 2.55 \text{ s}$$

(c) The ball decelerates on the way up and accelerates on the way down at the same rate because the acceleration due to gravity is constant (9.80 m/s²). Therefore, the time for the ball to reach its peak is the same as the time for it to fall to the ground. The total time in flight is 2(2.55 s) = 5.10 s.

With what speed does the ball in Example 6 hit the ground? The answer is 25.0 m/s. Can you explain why?

Earlier, we assumed no air resistance. We know that two dense and compact objects, such as a bowling ball and a marble, will fall at the same rate in air. We also know that two unlike objects, such as a marble and a feather, fall at different rates because of air resistance.

A parachute takes advantage of air resistance to slow a sky diver's descent [Fig. 4.17(a)]. What would happen if a parachute does not open [Fig. 4.17(b)]? Would the sky diver's velocity increase constantly until he or she hits the ground? As the velocity increases, the air resistance also increases. Since the gravitational pull and the air resistance are directed opposite each other, they tend to oppose or equalize each other. (Here, the velocity and acceleration are both directed downward, while the air resistance is directed upward.) This equalization occurs when the friction of the air's resistance equals the force of gravity. When this equalization occurs, the falling object stops accelerating and continues to fall at a constant velocity, called *terminal velocity*.

The terminal velocity or speed of a sky diver varies from 150 to 200 km/h, depending on the person's weight and position. A heavier person will attain a greater terminal speed than a lighter person because the larger weight results in a larger acceleration before the air resistance equals the weight. Body position also makes a difference. When a body is spread out like that of a bird gliding with outstretched wings, its surface area increases, which results in more air resistance. Terminal speed can be controlled by varying the body position. A light sky diver and a heavy sky diver can remain in close proximity to each other if the light person decreases his or her air resistance by

Figure 4.17 The sky diver in (a) has a different acceleration toward the ground than those in (b) as a result of different air resistance.

(a)

(b)

falling head or feet first while the heavy person spreads out and increases his or her air resistance. A parachute greatly increases air resistance and reduces the terminal speed to approximately 15 to 25 km/h, which is slow enough for a safe landing.

In general, the terminal velocity of an object varies with its weight and its aerodynamic features, which include the following:

1. The shape of the object. (A symmetrical object is more aerodynamic than a nonsymmetrical one.)
2. The orientation of the object as it is traveling. (A sky diver slows the fall by spreading out his or her arms and legs parallel to the ground while falling. The speed of the fall increases if the person falls head or feet first.)
3. The smoothness of the surface. (A body with a smooth surface provides less air resistance than a body with a rough surface and falls or flies faster as a result.)

TRY THIS ACTIVITY

Air Resistance and Acceleration

Safely drop several objects of varying sizes, weights, and shapes from a second- or third-floor window or balcony. Which objects reach the ground faster than others?

Gravity acts equally on all objects, but air resistance may cause some objects to accelerate less than 9.80 m/s^2. What aerodynamic features may have caused some objects to encounter more air resistance than others?

Finally, drop a coffee filter and observe the filter's descent to the ground (Fig. 4.18). Does it seem to accelerate as it falls? Why or why not?

Figure 4.18 Falling coffee filter

SKETCH

$$\boxed{12 \text{ cm}^2 \quad w}$$
4.0 cm

DATA
$A = 12 \text{ cm}^2, l = 4.0 \text{ cm}, w = ?$

BASIC EQUATION
$A = lw$

WORKING EQUATION
$w = \dfrac{A}{l}$

SUBSTITUTION
$w = \dfrac{12 \text{ cm}^2}{4.0 \text{ cm}} = 3.0 \text{ cm}$

PROBLEMS 4.3

Substitute in the given equation and find the unknown quantity.

1. Given: $v_{avg} = \dfrac{v_f + v_i}{2}$
   $v_f = 6.20$ m/s
   $v_i = 3.90$ m/s
   $v_{avg} = ?$

2. Given: $a = \dfrac{v_f - v_i}{t}$
   $a = 3.07$ m/s^2
   $v_f = 16.8$ m/s
   $t = 4.10$ s
   $v_i = ?$

3. Given: $s = v_i t + \frac{1}{2} at^2$
   $t = 3.00$ s
   $a = 6.40$ m/s^2
   $v_i = 33.0$ m/s
   $s = ?$

4. Given: $2as = v_f^2 - v_i^2$
   $a = 8.41$ m/s^2
   $s = 4.81$ m
   $v_i = 1.24$ m/s
   $v_f = ?$

5. Given: $v_f = v_i + at$
   $v_f = 10.40$ ft/s
   $v_i = 4.01$ ft/s
   $t = 3.00$ s
   $a = ?$

6. The average velocity of a mini-bike is 15.0 km/h. How long does it take for the bike to go 35.0 m?

7. A sprinter starting from rest reaches a final velocity of 18.0 mi/h. What is her average velocity?

8. A coin is dropped with no initial velocity. Its final velocity when it strikes the earth is 50.0 ft/s. The acceleration due to gravity is 32.2 ft/s^2. How long does it take to strike the earth?

9. A front endloader accelerates from rest to 1.75 m/s in 2.50 s. How far does it travel in that time?

10. A mechanic test driving a car that she has just given a tune-up accelerates from rest to 50.0 m/s in 9.80 s. How far does she travel in that time?

11. A rocket lifting off from earth has an average acceleration of 44.0 ft/s^2. Its initial velocity is zero. How far into the atmosphere does it travel during the first 5.00 s, assuming that it goes straight up?

12. The final velocity of a truck is 74.0 ft/s. If it accelerates at a rate of 2.00 ft/s^2 from an initial velocity of 5.00 ft/s, how long will it take for it to attain its final velocity?

13. A truck accelerates from 85 km/h to $12\overline{0}$ km/h in 9.2 s. Find its acceleration in m/s^2.

14. How long does it take a rock to drop 95.0 m from rest? Find the final speed of the rock.

15. An aircraft with a landing speed of 295 km/h lands on an aircraft carrier with a landing area 205 m long. Find the minimum constant deceleration required for a safe landing.

16. A ball is thrown downward from the top of a 43.0-ft building with an initial speed of 62.0 ft/s. Find its final speed as it strikes the ground.

17. A car is traveling at $7\overline{0}$ km/h. It then uniformly decelerates to a complete stop in 12 s. Find its acceleration (in m/s^2).

18. A car is traveling at $6\overline{0}$ km/h. It then accelerates at 3.6 m/s^2 to $9\overline{0}$ km/h. (a) How long does it take to reach the new speed? (b) How far does it travel while accelerating?

19. A rock is dropped from a bridge to the water below. It takes 2.40 s for the rock to hit the water. (a) Find the speed (in m/s) of the rock as it hits the water. (b) How high (in metres) is the bridge above the water?

20. A bullet is fired vertically upward from a gun and reaches a height of $70\overline{0}0$ ft. (a) Find its initial velocity. (b) How long does it take to reach its maximum height? (c) How long is it in flight?

21. A bullet is fired vertically upward from a gun with an initial velocity of $25\overline{0}$ m/s. (a) How high does it go? (b) How long does it take to reach its maximum height? (c) How long is it in flight?

22. A rock is thrown down with an initial speed of 30.0 ft/s from a bridge to the water below. It takes 3.50 s for the rock to hit the water. (a) Find the speed (in ft/s) of the rock as it hits the water. (b) How high is the bridge above the water?

23. John stands at the edge of a deck that is 25.0 m above the ground and throws a rock straight up with an initial speed of 10.0 m/s. (a) How long does it take to reach its maximum height? (b) What maximum height above the deck does it reach? (c) Assuming it misses the deck on its way down, at what speed does it hit the ground? (d) What total length of time is the rock in the air?

24. John stands at the edge of a deck that is 40.0 m above the ground and throws a rock straight up that reaches a height of 15.0 m above the deck. (a) What is the initial speed of the rock? (b) How long does it take to reach its maximum height? (c) Assuming it misses the deck on its way down, at what speed does it hit the ground? (d) What total length of time is the rock in the air?

25. John is standing on a steel beam 255.0 ft above the ground. Linda is standing 30.0 ft directly above John. (a) For John to throw a hammer up to Linda, at what initial speed must John throw the hammer for it to just reach Linda? (b) Suppose the hammer reaches the correct height, but Linda just misses catching it. How long does someone on the ground have to move out of the way from the time the hammer reaches its maximum height? (c) At what speed does the hammer hit the ground?

26. Kurt is standing on a steel beam 275.0 ft above the ground and throws a hammer straight up at an initial speed of 40.0 ft/s. At the instant he releases the hammer, he also drops a wrench from his pocket. Assume that neither the hammer nor the wrench hits anything while in flight. (a) Find the time difference between when the wrench and the hammer hit the ground. (b) Find the speed at which the wrench hits the ground. (c) Find the speed at which the hammer hits the ground. (d) How long does it take for the hammer to reach its maximum height? (e) How high above the ground is the wrench at the time the hammer reaches its maximum height?

27. One ball is dropped from a cliff. A second ball is thrown down 1.00 s later with an initial speed of 40.0 ft/s. How long after the second ball is thrown will the second ball overtake the first?

28. A car with velocity 2.00 m/s at $t = 0$ accelerates at 4.00 m/s^2 for 2.50 s. What is its velocity at $t = 2.50$ s?

29. A truck moving at 30.0 km/h accelerates at a constant rate of 3.50 m/s^2 for 6.80 s. Find its final velocity in km/h.

30. A bus accelerates from rest at a constant 5.50 m/s^2. How long will it take to reach 28.0 m/s?

31. A motorcycle slows from 22.0 m/s to 3.00 m/s with constant acceleration -2.10 m/s^2. How much time is required to slow down?

4.4 Projectile Motion

A **projectile** is a launched object that travels through the air but has no capacity to propel itself. Up to this point, we have discussed projectiles either being thrown straight up or dropped straight down. We now discuss objects being thrown at an angle. This type of motion uses the same principles and formulas as in the previous section on uniformly accelerated motion. However, two dimensions are needed to analyze projectile motion; x will represent horizontal and y will represent vertical.

A baseball pop fly to right field, a car driving off a cliff, and water flowing out of a hose are all examples of projectile motion. **Projectile motion** is the movement of a projectile as it travels through the air influenced only by its initial velocity and gravitational acceleration.

Consider the example of a ball being rolled across a table at a constant 2.00 m/s. The ball moves 2.00 m for every second that it travels. Assuming no friction, the ball would continue to roll at that same speed (Fig. 4.19). As the ball approaches and rolls off the edge of the table, the ball continues to move horizontally at 2.00 m/s until it strikes the floor. Horizontally, there is nothing causing the ball to speed up or slow down, so it continues with that same horizontal motion.

Figure 4.20 If a ball is simply dropped from the edge of a table, it will accelerate toward the ground, picking up speed and covering a greater and greater distance as time passes.

Figure 4.19　A ball rolling across a table with a constant velocity.

2.00 m/s

As the ball rolls off the table, gravity is accelerating the ball downward at a rate of $g = 9.80$ m/s^2 (Fig. 4.20). As a result, the ball increases its vertical speed by 9.80 m/s for every second it falls. Vertically, the problem can be treated as any other uniformly accelerated motion problem, as seen in the previous section.

A projectile motion problem may be separated into the horizontal frame, the nongravitational acceleration component, and into the vertical frame, the gravitational acceleration component. Using formulas in both the horizontal and vertical frames allows you to solve problems such as finding the speed the ball was traveling as it came off the table, the distance the ball landed from the edge of the table, and the time the ball was in the air (Fig. 4.21).

Figure 4.21

(a) This schematic shows the positions of a ball being dropped and a ball being projected horizontally from the same height at the same time. Note that the blue vertical component of the velocity of both balls is increasing and is the same for both balls at each instant, that the yellow horizontal component of the projected ball is constant at each instant, and that the incremental red resultant velocity of the projected ball is the resultant of each corresponding set of vertical and horizontal components, which results in the curved motion shown.

Courtesy of Richard Megna/Fundamental Photographs, NYC

(b) Here a series of times flash photos shows that a dropped red ball and a yellow ball projected horizontally from the same height at the same time fall vertically at the same rate, and so they hit the ground at the same time, as demonstrated in the schematic in part (a).

TRY THIS ACTIVITY

Quarter and the Ruler Trick

Figure 4.22

Set up two coins and a ruler as shown in Fig. 4.22. Flick the ruler so the two coins are launched off the table at the same time. The coin closer to your stationary hand will travel more slowly than the coin farther away from your stationary hand. Which coin hits the ground first? Why?

EXAMPLE 1

A ball rolls at a constant speed of 0.700 m/s as it reaches the end of a 1.30-m-high table (Fig. 4.23). How far from the edge of the table does the ball land?

Sketch:

Figure 4.23

$v_x = 0.700$ m/s

$s_y = 1.30$ m

$s_x = ?$

Data:

$$v_{iy} = 0 \text{ m/s} \qquad v_x = 0.700 \text{ m/s}$$
$$s_y = 1.30 \text{ m} \qquad s_x = ?$$

Basic Equations:

$$s_y = v_{iy}t + \tfrac{1}{2} a_y t^2 \qquad s_x = v_x t$$

Working Equations (with $v_{iy} = 0$):

$$t = \sqrt{\frac{2s_y}{a}} \qquad s_x = v_x t$$

Substitution:

$$t = \sqrt{\frac{2(1.30 \text{ m})}{9.80 \text{ m/s}^2}}$$
$$= 0.515 \text{ s}$$

$$s_x = (0.700 \text{ m/s})(0.515 \text{ s})$$
$$= 0.361 \text{ m}$$

· · · · · · · · · · · · · · · · ·

Projectiles launched at an angle need to be treated the same as a ball rolled off a table. However, unlike such a ball, which has an initial vertical velocity of zero, a projectile launched at an angle has an initial vertical velocity. The **range** is the horizontal distance that a projectile will travel before striking the ground. The range and the flight time may be found as follows.

1. Separate the original speed of the projectile into *horizontal* (x-component of the velocity) and *vertical components* (y-component of the velocity) using vectors and trigonometry (Fig. 4.24).
2. Use the equation $s_x = v_x t$ to find the range. The horizontal component of the velocity does not change because gravitational acceleration does not act in the horizontal direction (Fig. 4.25). (Gravity pulls objects down in only the vertical direction.)
3. Use the vertical component of the velocity and uniformly accelerated motion equations to determine how long the projectile will be in the air. Since gravity accelerates the projectile, use the following formula and solve for time: $v_f = v_i + at$.

Figure 4.24   The ball is kicked with a horizontal (yellow vector) and vertical (blue vector) velocity. The red vector represents the resultant of the horizontal and vertical velocities.

Figure 4.25   Velocity vectors are shown as a ball kicked into the air travels through its trajectory. Note that the yellow horizontal component remains constant at each instant while the blue vertical component decreases as the ball rises until the ball reaches its peak, at which the vertical component is zero, and then the vertical component increases as the ball falls.

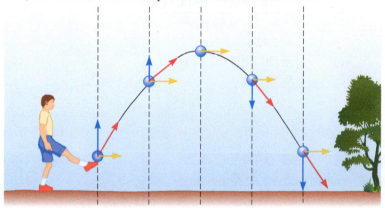

Note: Since the vertical components of v_i and v_f are equal but opposite in direction and therefore have opposite signs, the time for the projectile going up is the same as the time for it going down.

A baseball is hit and moves initially at an angle of 35.0° above the horizontal ground with a velocity of 25.0 m/s as shown in Fig. 4.26. (a) What are the vertical and horizontal components of the initial velocity of the ball? (b) How long will the ball be in the air? (c) What will be the range for this projectile?

EXAMPLE 2

Figure 4.26

Data:

$$v = 25.0 \text{ m/s}$$
$$A = 35.0°$$

Basic Equations:

$$|\mathbf{v}_x| = |\mathbf{v}| \cos A$$
$$|\mathbf{v}_y| = |\mathbf{v}| \sin A$$
$$v_{fy} = v_{iy} + a_y t$$
$$s_x = v_x t$$

Working Equations: Same except for

$$t = \frac{v_{fy} - v_{iy}}{a_y}$$

Substitutions:

(a) $|\mathbf{v}_x| = (25.0 \text{ m/s}) (\cos 35.0°)$
        $= 20.5 \text{ m/s}$

$|\mathbf{v}_y| = (25.0 \text{ m/s}) (\sin 35.0°)$
        $= 14.3 \text{ m/s}$

(b) $t = \dfrac{(14.3 \text{ m/s}) - (-14.3 \text{ m/s})}{9.80 \text{ m/s}^2}$
        $= 2.92 \text{ s}$

(c) $s_x = (20.5 \text{ m/s}) (2.92 \text{ s})$
        $= 59.9 \text{ m}$

PHYSICS CONNECTIONS

Orbiting Cannonballs?

Isaac Newton once said that if he could fire a cannonball with enough velocity, he could get it to circle the globe. Is this true? Can you really put something into orbit by launching it fast enough? In reality, there is too much air resistance and too many obstacles to allow this to happen. Theoretically, though, if we launch something with enough horizontal velocity, the earth itself would curve away before the cannonball strikes it. Figure 4.27 shows a cannonball launched with varying initial velocities. As the horizontal velocity of the cannonball increases, its range increases as well. In addition, the ball appears to fall farther due to the curvature of the earth. Finally, when the ball is launched with a large enough velocity, it completely misses the earth and achieves orbit.

To place a satellite into orbit, a rocket or space shuttle must bring the satellite to a point above the earth's atmosphere where it will not experience any air resistance. The object is then given enough horizontal velocity so that, although it is falling toward the earth, its horizontal velocity will prevent it from getting any closer to the surface. The concept of a falling object continuously missing the earth was a revolutionary concept developed by Isaac Newton over 300 years ago.

Communications satellites, the moon, and the astronauts in a space shuttle all are in a constant state of free fall but have the correct amount of horizontal velocity to prevent them from striking the earth. The speed required to keep a cannonball in orbit around the earth is approximately 17,700 mi/h. A space shuttle travels at approximately 17,400 mi/h because it orbits farther away from the earth.

Figure 4.27 Newton's diagram of a cannonball orbiting the earth

Earth

PROBLEMS 4.4

Find the horizontal range for each projectile with an initial speed of 35.0 m/s at the given angle:

1. 15.0°   2. 75.0°   3. 35.0°
4. 55.0°   5. 45.0°
6. Draw a conclusion about range and angles based on the answers to Problems 1 through 5.
7. Part of military training involves aiming and shooting a cannon. If a soldier sets the cannon at an angle of 62.0° and a launch velocity of 67.0 m/s, how far will the projectile travel horizontally?
8. A faulty fireworks rocket launches but never discharges. If the rocket launches with an initial velocity of 33.0 ft/s at an angle of 85.0°, how far away from the launch site does the rocket land?
9. An outfielder throws a baseball at a speed of 110 ft/s at an angle of 25.0° above the horizontal to home plate $29\overline{0}$ ft away. Will the ball reach the catcher on the fly or will it bounce first?
10. A bearing rolls off a 1.40-m-high workbench with an initial horizontal speed of 0.600 m/s. How far from the edge of the bench does the bearing land?
11. A mechanic's socket rolls off a 1.50-m-high bench with an initial horizontal speed of 0.800 m/s. How far from the edge of the bench does the socket hit the floor?

Glossary

Acceleration Change in velocity per unit time. (p. 101)

Acceleration due to Gravity The acceleration of a freely falling object. On the earth's surface the acceleration due to gravity is 9.80 m/s^2 (metric) or 32.2 ft/s^2 (U.S.) (p. 108)

Deceleration An acceleration that indicates an object is slowing down. (p. 103)

Motion A change of position. (p. 93)

Projectile A launched object that travels through the air but has no capacity to propel itself. (p. 114)

Projectile Motion The motion of a projectile as it travels through the air influenced only by its initial velocity and gravitational acceleration. (p. 114)

Range The horizontal distance that a projectile will travel before striking the ground. (p. 116)

Speed The distance traveled per unit of time; a scalar because it is described by a number and a unit, not a direction. (p. 94)

Terminal Speed The speed attained by a freely falling body when the air resistance equals its weight and no further acceleration occurs. (p. 108)

Velocity The rate of motion in a particular direction. The time rate of change of an object's displacement. Velocity is a vector that gives the direction of travel and the distance traveled per unit of time. (p. 94)

Formulas

4.1 $\quad s = v_{avg}t$

4.2 $\quad \Delta v = at$

4.3 $\quad v_{avg} = \dfrac{v_f + v_i}{2} \qquad\qquad s = v_i t + \frac{1}{2}at^2$

$\qquad\quad a = \dfrac{v_f - v_i}{t} \qquad\qquad s = \frac{1}{2}(v_f + v_i)t$

$\qquad\quad v_f = v_i + at \qquad\qquad 2as = v_f^2 - v_i^2$

Review Questions

1. Velocity is
   (a) the distance traveled per unit of time.
   (b) the same as speed.
   (c) direction of travel and distance traveled per unit of time.
   (d) only the direction of travel.
2. A large heavy rock and a small marble are dropped at the same time from the roof of a three-story building. Neglecting air resistance, which object will strike the ground first?
   (a) The marble.
   (b) The rock.
   (c) They both strike the ground at the same time.
3. One ball is thrown horizontally while another is dropped vertically. Which ball will strike the ground first?
   (a) Both strike the ground at the same time.
   (b) The horizontally thrown ball strikes first.
   (c) The vertically dropped ball strikes first.
4. At what launch angle with the ground does a projectile have the greatest horizontal range?
   (a) 0° (b) 45° (c) 60° (d) 90°
5. Where in a projectile's path would its speed be the least?
   (a) Just as it is launched.
   (b) Just before it lands.
   (c) It has the same speed throughout its entire motion.
   (d) At the top of its path.
6. Explain your answer to Question 2.
7. Explain your answer to Question 3.
8. Distinguish between velocity and speed.
9. Is velocity always constant?
10. Why are vectors important in measuring motion? Provide two examples where vectors are used to help measure motion.
11. Give three familiar examples of acceleration.
12. Distinguish among acceleration, deceleration, and average acceleration.
13. State the values of the acceleration due to gravity for freely falling bodies in both the metric and U.S. systems.

Review Problems

1. A boat travels at 17.0 mi/h for 1.50 h. How far does the boat travel?
2. A commercial jet flies at $55\overline{0}$ mi/h for $30\overline{0}0$ mi. For how much time does the jet fly?
3. A plane flies north at 215 km/h. A wind from the east blows at 69 km/h. What is the plane's new velocity with respect to the ground in standard position?
4. A glider flies southeast (at 320.0°) at 25.0 km/h. A wind blows at 12.0 km/h from 15.0° south of west. What is the new velocity of the glider with respect to the ground in standard position?
5. A runner starts from rest and attains a speed of 8.00 ft/s after 2.00 s. What is the runner's acceleration?
6. A race car goes from rest to 150 km/h with an acceleration of 6.0 m/s². How many seconds does it take?
7. A sailboat has an initial velocity of 10.0 km/h and accelerates to 20.0 km/h. Find its average velocity.
8. A skateboarder starts from rest and accelerates at a rate of 1.30 m/s² for 3.00 s. What is his final velocity?
9. A plane has an average velocity of $50\overline{0}$ km/h. How long does it take to travel 1.5×10^4 km?
10. A train has a final velocity of 110 km/h. It accelerated for 36 s at 0.50 m/s². What was its initial velocity?
11. A boulder is rolling down a hill at 8.00 m/s before it comes to rest 17.0 s later. What is its average velocity?
12. A truck accelerates from rest to 120 km/h in 13 s. Find its acceleration.
13. An airplane reaches a velocity of 71.0 m/s when it takes off. What must its acceleration be if the runway is 1.00 km long?
14. An airplane accelerates at 3.00 m/s² from a velocity of 21.0 m/s over a distance of 535 m. What is its final velocity?
15. A bullet is fired vertically upward and reaches a height of 2150 m. (a) Find its initial velocity. (b) How long does it take to reach its maximum height? (c) How long is it in flight?
16. A rock is thrown down with an initial speed of 10.0 m/s from a bridge to the water below. It takes 2.75 s for the rock to hit the water. (a) Find the speed of the rock as it hits the water. (b) How high is the bridge above the water?
17. A shot put is hurled at 9.43 m/s at an angle of 55.0°. Ignoring the height of the shot-putter, what is the range of the shot?
18. An archer needs to hit a bull's eye on a target at eye level 60.0 ft away. If the archer fires the arrow from eye level with a speed of 47.0 ft/s at an angle of 25.0° above the horizontal, will the arrow hit the target?

SKETCH

12 cm^2 w

4.0 cm

DATA

$A = 12 \text{ cm}^2, l = 4.0 \text{ cm}, w = ?$

BASIC EQUATION

$A = lw$

WORKING EQUATION

$w = \dfrac{A}{l}$

SUBSTITUTION

$w = \dfrac{12 \text{ cm}^2}{4.0 \text{ cm}} = 3.0 \text{ cm}$

APPLIED CONCEPTS

1. Amy walks at an average speed of 1.75 m/s toward her airport gate. When she comes within 57.5 m of her gate, she gets onto and continues to walk on a people mover at the same rate. (a) If she arrives at the gate 15.3 s after getting on the people mover, how fast was she moving relative to the ground? (b) What was the speed of the people mover?

2. A novice captain is pointing his ferryboat directly across the river at a speed of 15.7 mi/h. If he does not pay attention to the current that is headed downriver at 5.35 mi/h, what will be his resultant speed and direction? (Consider that the current and the boat's initial heading are perpendicular to each other.)

3. Anette is a civil engineer and needs to determine the length of a highway on-ramp before construction begins. If the average vehicle takes 10.8 s to go from 20.0 mi/h to 60.0 mi/h, how long should she design the separate merge lane to be so a car can reach a speed of 60.0 mi/h before merging?

4. As a movie stunt coordinator, you need to be sure a stunt will be safe before it is performed. If a stuntwoman is to run horizontally off the roof of a three-story building and land on a foam pad 7.35 m away from the base of the building, how fast should she run as she leaves the roof? (Each story is 4.00 m.)

5. As a newspaper delivery boy, Jason needs to know his projectile motion to throw a paper horizontally from a height of 1.40 m to a door that is 13.5 m away. What must the paper's velocity be for it to reach the door?

CHAPTER 5

FORCE

Objectives

The major goals of this chapter are to enable you to:

1. Relate force and the law of inertia.
2. Apply the law of acceleration.
3. Identify components of friction.

4. Analyze forces in one dimension.
5. Distinguish among weight, mass, and gravity.
6. Analyze how the law of action and reaction is used.

Classical physics is sometimes called Newtonian physics in honor of Sir Isaac Newton, who lived from 1642 to 1727 and formulated three laws of motion that summarize much of the behavior of moving bodies. Forces may cause motion. Inertia tends to resist the influence of an applied force. Forces, inertia, friction, and how they relate to motion are considered now.

5.1 Force and the Law of Inertia

As discussed in the previous chapter, if an object changes its velocity, we say it accelerates. But what causes an object to accelerate? Let's take an example of a soccer ball at rest on a field. What must you do to accelerate the ball? Similarly, if your car is approaching a red stoplight, what must you do to make the car accelerate to rest? The answer in both instances is to apply a force.

A **force** is any push or pull. Forces tend to either change the motion of an object or prevent the object from changing its motion. Force is a vector quantity and therefore has both magnitude and direction. The force tends to produce acceleration in the direction of its application. Therefore, if you want to accelerate the soccer ball, you need to kick it in the direction you want the ball to move. To get the car to slow down, you apply the brakes, which creates a force in the opposite direction of the car's motion and causes a negative acceleration.

Not all forces result in acceleration. For example, if you attempt to push a large crate, it may not move because another force, in this case a frictional force, matches your pushing force, preventing you from moving it (Fig. 5.1). This concept of concurrent forces will be covered in Chapter 7.

The units for measuring force are the newton (N) in the metric system and the pound (lb) in the U.S. system. The conversion factor is

$$4.45 \text{ N} = 1 \text{ lb}$$

Let's examine the relationship between forces and motion. There are three relationships or laws that were discovered by **Isaac Newton** during the late 17th century. The three laws are often called Newton's laws. The first law, called the **law of inertia**, is as follows:

LAW OF INERTIA: NEWTON'S FIRST LAW

A body that is in motion continues in motion with the same velocity (at constant speed and in a straight line) and a body at rest continues at rest unless an unbalanced (outside) force acts upon it.

If an automobile is stopped (at rest) on level ground, it resists being moved. That is, a person is required to exert a tremendous push to get it moving. Similarly, if an automobile is moving—even slowly—it takes a large force to stop it. This property of resisting a change in motion is called inertia. **Inertia** is the property of a body that causes it to remain at rest if it is at rest or to continue moving with a constant velocity unless an unbalanced force acts upon it. When the accelerating force of an automobile engine is no longer applied to a moving car, the car will slow down. This is not a violation of the law of inertia because there are forces being applied to the car through air resistance, friction in the bearings, and the rolling resistance of the tires [Fig. 5.2(a)]. If these forces could be removed, the auto would continue moving with a constant velocity. Anyone who has tried to stop quickly on ice knows the effect of the law of inertia when frictional forces are small [Fig. 5.2(b)].

Figure 5.1　The pushing force and the frictional force cancel each other out, resulting in zero acceleration.

Figure 5.2

(a) Air resistance and friction slow the car.

(b) Inertia makes it hard to stop a car on ice.

The engine of a train makes use of the concept of inertia in incrementally starting a train that has far too much inertia to start moving all at once. The train is set in motion one car at a time. The couplers are loosely connected, which allows each car to start moving separately from the others. Then the inertia of each moving car aids in getting the rest of the cars moving. With use of this technique trains up to 1 mi long can be set in motion.

Inertia Tricks

Place a card on top of a glass. Then place a quarter on top of the card as in Fig. 5.3(a). Quickly flick the card horizontally. The inertia of the coin tends to keep it at rest horizontally. The force that causes the coin to move is the vertical force of gravity, which pulls it straight down into the glass as in Fig. 5.3(b).

Figure 5.3   When the card is flicked, the inertia of the quarter keeps it at rest until it falls into the glass.

(a)   (b)

How could you remove a tablecloth from under a set of dishes without removing the dishes from the table? The trick to keeping the dishes on the table is to use a very smooth tablecloth without any hem and very quickly jerk the tablecloth backward in a slightly downward direction. We suggest that you practice with heavy objects on plastic plates. Remember that the more mass placed on the plates, the greater is the inertia.

Some objects more than others tend to resist changes in their motion. It is much easier to push a small automobile than to push a large truck into motion (Fig. 5.4). **Mass** *is a measure of the inertia of a body*, that is, a measure of the resistance a body has to changing its motion. The common units of mass are the kilogram (kg) in the metric system and the slug in the U.S. system. The conversion factor is 1 kg = 0.0685 slug.

Figure 5.4   A larger body (in mass) has a greater resistance to a change in its motion than does a smaller one.

5.2 Force and the Law of Acceleration

The second law of motion, called the **law of acceleration**, relates the applied force, the mass, and the acceleration of an object.

LAW OF ACCELERATION: NEWTON'S SECOND LAW
The total force acting on a body is equal to the mass of the body times its acceleration.

In equation form this law is

$$F = ma$$

where F = total force
m = mass
a = acceleration

The formula states that when a force is applied to an object, the force causes the object to accelerate. The stronger the force, the larger is the acceleration. The weaker the force, the smaller is the acceleration. In addition, when pushed with the same force, a more massive object will accelerate less and a less massive object will accelerate more.

If the mass is kept constant,

$$F \sim a \qquad \text{(The force is directly proportional to the acceleration.)}$$

If the force is kept constant,

$$m \sim 1/a \qquad \text{(The mass is inversely proportional to the acceleration.)}$$

In SI units, the mass unit is the kilogram (kg) and the acceleration unit is metre/second/second (m/s^2). The force required to accelerate 1 kg of mass at a rate of 1 m/s^2 is

$$F = ma$$
$$= (1 \text{ kg}) (1 \text{ m/s}^2)$$
$$= 1 \text{ kg m/s}^2$$

The SI force unit is the newton (N), named in honor of Isaac Newton, and is defined as

$$1 \text{ N} = 1 \text{ kg m/s}^2$$

In the U.S. system, the mass unit is the slug and the acceleration unit is foot/second/second (ft/s^2). The force required to accelerate 1 slug of mass at the rate of 1 ft/s^2 is

$$F = ma$$
$$= (1 \text{ slug}) (1 \text{ ft/s}^2)$$
$$= 1 \text{ slug ft/s}^2$$

The U.S. force unit is the pound (lb) and is defined as

$$1 \text{ lb} = 1 \text{ slug ft/s}^2$$

Note: The other metric unit of force is the dyne. One dyne is the force required to accelerate 1 g of mass at the rate of 1 cm/s^2. The dyne is not an SI unit, and its use is less common.

EXAMPLE 1

What force is necessary to produce an acceleration of 6.00 m/s^2 on a mass of 5.00 kg?

Data:

$$m = 5.00 \text{ kg}$$
$$a = 6.00 \text{ m/s}^2$$
$$F = ?$$

Basic Equation:

$$F = ma$$

Working Equation: Same

Substitution:

$$F = (5.00 \text{ kg}) (6.00 \text{ m/s}^2)$$
$$= 30.0 \text{ kg m/s}^2$$
$$= 30.0 \text{ N} \qquad (1 \text{ N} = 1 \text{ kg m/s}^2)$$

EXAMPLE 2

What force is necessary to produce an acceleration of 2.00 ft/s^2 on a mass of 3.00 slugs?

Data:

$$m = 3.00 \text{ slugs}$$
$$a = 2.00 \text{ ft/s}^2$$
$$F = ?$$

Basic Equation:

$$F = ma$$

Working Equation: Same

Substitution:

$$F = (3.00 \text{ slugs}) (2.00 \text{ ft/s}^2)$$
$$= 6.00 \text{ slug ft/s}^2$$
$$= 6.00 \text{ lb} \qquad (1 \text{ lb} = 1 \text{ slug ft/s}^2)$$

· · · · · · · · · · · · · · · ·

EXAMPLE 3

Find the acceleration produced by a force of $50\overline{0}$ N applied to a mass of 20.0 kg.

Data:

$$F = 50\overline{0} \text{ N}$$
$$m = 20.0 \text{ kg}$$
$$a = ?$$

Basic Equation:

$$F = ma$$

Working Equation:

$$a = \frac{F}{m}$$

Substitution:

$$a = \frac{50\overline{0} \text{ N}}{20.0 \text{ kg}}$$
$$= 25.0 \frac{\text{N}}{\text{kg}}$$
$$= 25.0 \frac{\cancel{\text{N}}}{\cancel{\text{kg}}} \times \frac{1 \text{ kg m/s}^2}{1 \cancel{\text{N}}}$$
$$= 25.0 \text{ m/s}^2$$

Note: We use a conversion factor to obtain acceleration units.

· · · · · · · · · · · · · · · ·

PROBLEMS 5.2

Find the total force necessary to give each mass the given acceleration.

1. $m = 15.0$ kg, $a = 2.00$ m/s^2
2. $m = 4.00$ kg, $a = 0.500$ m/s^2
3. $m = 111$ slugs, $a = 6.70$ ft/s^2
4. $m = 91.0$ kg, $a = 6.00$ m/s^2
5. $m = 28.0$ slugs, $a = 9.00$ ft/s^2
6. $m = 42.0$ kg, $a = 3.00$ m/s^2
7. $m = 59.0$ kg, $a = 3.90$ m/s^2
8. $m = 2.20$ slugs, $a = 1.53$ ft/s^2

SKETCH

$$12\ \text{cm}^2 \quad w$$

$$4.0\ \text{cm}$$

DATA

$A = 12\ \text{cm}^2, l = 4.0\ \text{cm}, w = ?$

BASIC EQUATION

$A = lw$

WORKING EQUATION

$w = \dfrac{A}{l}$

SUBSTITUTION

$w = \dfrac{12\ \text{cm}^2}{4.0\ \text{cm}} = 3.0\ \text{cm}$

Find the acceleration of each mass with the given total force.

9. $m = 19\overline{0}$ kg, $F = 76\overline{0}0$ N
10. $m = 7.00$ slugs, $F = 12.0$ lb
11. $m = 3.60$ kg, $F = 42.0$ N
12. $m = 0.790$ kg, $F = 13.0$ N
13. $m = 11\overline{0}$ kg, $F = 57.0$ N
14. $m = 84.0$ kg, $F = 33.0$ N
15. $m = 9.97$ slugs, $F = 13.9$ lb
16. $m = 21\overline{0}$ kg, $F = 41.0$ N

17. Find the total force necessary to give an automobile of mass 1750 kg an acceleration of 3.00 m/s².

18. Find the acceleration produced by a total force of 93.0 N on a mass of 6.00 kg.

19. Find the total force necessary to give an automobile of mass $12\overline{0}$ slugs an acceleration of 11.0 ft/s².

20. Find the total force necessary to give a rocket of mass $25,0\overline{0}0$ slugs an acceleration of 28.0 ft/s².

21. A forklift has a mass of 975 kg. What force must be applied to give it an acceleration of 2.50 m/s²?

22. A power wheelbarrow has a mass of 432 kg. What force must be applied to give it an acceleration of 1.75 m/s²?

23. What is the rate of deceleration of a $14\overline{0}0$-kg SUV that is going 75.0 km/h and then slows to 25.0 km/h in 8.20 s?

24. An earthmover slows from 15.0 km/h to 3.00 km/h in 2.70 s. What is its rate of deceleration?

25. Find the total force necessary to give a $14\overline{0}$-kg mass an acceleration of 41.0 m/s².

26. Find the acceleration produced by a total force of $30\overline{0}$ N on a mass of 0.750 kg.

27. Find the mass of an object with acceleration 15.0 m/s² when an unbalanced force of 90.0 N acts on it.

28. An automobile has a mass of $10\overline{0}$ slugs. The passengers it carries have a mass of 5.00 slugs each. (a) Find the acceleration of the auto and one passenger if the total force acting on them is $15\overline{0}0$ lb. (b) Find the acceleration of the auto and six passengers if the total force is again $15\overline{0}0$ lb.

29. Find the acceleration produced by a force of 6.75×10^6 N on a rocket of mass 5.27×10^5 kg.

30. An astronaut has a mass of 80.0 kg. His space suit has a mass of 15.5 kg. Find the acceleration of the astronaut during his space walk when his backpack propulsion unit applies a force to him (and his suit) of 85.0 N.

31. A discus thrower exerts a force of $14\overline{0}$ N on the discus. (a) If the discus has an acceleration of 19.0 m/s², what is the mass of the discus? (b) If the mass of the discus is doubled and the applied force remains the same, what is the new acceleration? (c) If the force is doubled and the mass is unchanged, how is the acceleration affected?

32. A scooter and rider together have a mass of 275 kg. (a) If the scooter slows with an acceleration of −4.50 m/s², what is the net force on the scooter and rider? (b) Describe the direction of the force and state the meaning of the (−) sign.

33. A pickup truck with mass of 1230 kg moving at 105 km/h stops within a distance of 53.0 m. (a) What is the direction and size of the force that acts on the truck? (b) How much more slowing force would be required to stop the truck in half the distance?

Figure 5.5 Balancing gravity in holding a brick.

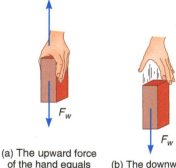

(a) The upward force of the hand equals the downward force of the weight.

(b) The downward force of the weight is now greater.

5.3 Gravity and Weight

The **weight** of an object is the amount of *gravitational pull* exerted on an object by the earth. If this force is not balanced by other forces, an acceleration is produced. When you hold a brick in your hand as in Fig. 5.5, you exert an upward force on the brick that balances the downward force (weight). If you remove your hand, the brick moves downward due to the unbalanced force. The velocity of the falling brick increases, but the acceleration (rate of change of the velocity) is constant.

The acceleration of all objects near the surface of the earth is the same if air resistance is ignored. We call this acceleration due to the gravitational pull of the earth g. Its value is 9.80 m/s² in the metric system and 32.2 ft/s² in the U.S. system.

The *weight* of an object is the force exerted by the earth (or by another large body) and gives the object an acceleration g. This force can be found using $F = ma$, where $a = g$. If we abbreviate weight by F_w, the equation for weight is

$$F_w = mg$$

where F_w = weight
   m = mass
   g = acceleration due to gravity
   g = 9.80 m/s² (earth, metric)
   g = 32.2 ft/s² (earth, U.S.)

Find the weight of 5.00 kg.

EXAMPLE 1

Data:

$$m = 5.00 \text{ kg}$$
$$g = 9.80 \text{ m/s}^2$$
$$F_w = ?$$

Basic Equation:

$$F_w = mg$$

Working Equation: Same

Substitution:

$$F_w = (5.00 \text{ kg})(9.80 \text{ m/s}^2)$$
$$= 49.0 \text{ kg m/s}^2$$
$$= 49.0 \text{ N} \qquad (1 \text{ N} = 1 \text{ kg m/s}^2)$$

Find the weight of 12.0 slugs.

EXAMPLE 2

Data:

$$m = 12.0 \text{ slugs}$$
$$g = 32.2 \text{ ft/s}^2$$
$$F_w = ?$$

Basic Equation:

$$F_w = mg$$

Working Equation: Same

Substitution:

$$F_w = (12.0 \text{ slugs})(32.2 \text{ ft/s}^2)$$
$$= 386 \text{ slug ft/s}^2$$
$$= 386 \text{ lb} \qquad (1 \text{ lb} = 1 \text{ slug ft/s}^2)$$

Figure 5.6 Since the brick is at rest, the normal force is equal in magnitude but opposite in direction to the weight of the brick.

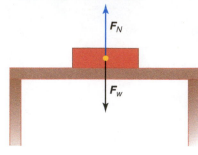

Weight Versus Normal Force

When an object is in contact with a surface, a force is exerted on that object by the surface. This force, called a **normal force**, F_N, *is perpendicular to the contact surface.*

Picture a brick at rest on a table (Fig. 5.6) Since the brick is at rest, the contact force, pointing directly up from the table, must be equal to the magnitude of the brick's weight. If additional bricks are stacked on top of the original brick, the total weight of the bricks will increase, causing the normal force from the table to increase as well. Eventually, the table will not be able to support more and more bricks with enough of a normal force. When this occurs, the table will break, causing the bricks to accelerate down toward the ground.

Normal forces are an essential component of understanding how to calculate frictional forces, as discussed in Section 5.4.

Mass Versus Weight

Note that the mass of an object remains the same, but its weight varies according to the gravitational pull. For example, an astronaut of mass 75.0 kg has a weight of

$$F_w = mg = (75.0 \text{ kg}) (9.80 \text{ m/s}^2) = 735 \text{ N}$$

on the earth. If that astronaut lands on the moon, where the acceleration due to gravity is less than it is on the earth, the astronaut will still have the same 75.0 kg mass but will weigh significantly less than 735 N. (The moon has less acceleration due to gravity than the earth in part because the moon is less massive than the earth. A full explanation will be given in Chapter 11.) The following example shows the astronaut's weight on the moon.

EXAMPLE 3

Find the 75.0-kg astronaut's weight on the moon, where $g = 1.63$ m/s^2.

Data:

$$m = 75.0 \text{ kg}$$
$$g = 1.63 \text{ m/s}^2 \text{ (on moon)}$$
$$F_w = ?$$

Basic Equation:

$$F_w = mg$$

Working Equation: Same

Substitution:

$$F_w = (75.0 \text{ kg}) (1.63 \text{ m/s}^2)$$
$$= 122 \text{ N} \quad (1 \text{ N} = 1 \text{ kg m/s}^2)$$

Note that the astronaut's *mass* on the moon remains 75.0 kg.

Mass Versus Volume

Do not confuse mass and volume. The volume of an object is the measure of the space it occupies. Volume is measured in cubic units such as cm^3, ft^3, or L. The mass of an object is the amount of inertia or the amount of material it contains. The more mass contained in an object, the greater its inertia and the more force it takes to move it or change its motion. Compare the masses of two boxes of identical size, one filled with books and one empty. The box filled with books has more mass and requires more force to move it.

As an astronaut goes from the earth to the moon, his or her weight changes but mass and volume remain the same. Weight and mass are directly proportional in a given place, as we saw earlier.

PROBLEMS 5.3

Find the weight for each mass.

1. $m = 30.0$ kg
2. $m = 60.0$ kg
3. $m = 10.0$ slugs
4. $m = 9.00$ kg

Find the mass for each weight.

5. $F_w = 17.0$ N
6. $F_w = 21.0$ lb
7. $F_w = 12,\overline{0}00$ N
8. $F_w = 25,\overline{0}00$ N
9. $F_w = 6.7 \times 10^{12}$ N
10. $F_w = 5.5 \times 10^6$ lb

11. Find the weight of an 1150-kg automobile.
12. Find the weight of an 81.5-slug automobile.
13. Find the mass of a 2750-lb automobile.
14. What is the mass of a $20,\overline{0}00$-N truck?
15. What is the mass of a $75\overline{0}0$-N trailer?
16. Find the mass of an 11,500-N automobile.
17. Find the weight of a 1350-kg automobile (a) on the earth and (b) on the moon.
18. Maria weighs 115 lb on the earth. What are her (a) mass and (b) weight on the moon?
19. John's mass is 65.0 kg on the earth. What are his (a) mass and (b) weight on the moon?
20. What is your weight in newtons and in pounds?
21. What is your mass in kilograms and in slugs?
22. What are your U.S. mass and weight on the moon?
23. What are your metric mass and weight on the moon?
24. John's mass is 65.0 kg on the earth. What are his U.S. (a) mass and (b) weight $40\overline{0}0$ mi above the surface of the earth, where $g = 7.85$ ft/s^2?
25. Maria weighs 115 lb on the earth. What are her U.S. (a) mass and (b) weight on Jupiter, where $g = 85.0$ ft/s^2?
26. John's mass is 65.0 kg on the earth. What are his metric (a) mass and (b) weight on Mars, where $g = 3.72$ m/s^2?
27. What are your metric mass and weight on Jupiter, where $g = 25.9$ m/s^2?
28. What are your metric mass and weight on Mars, where $g = 3.72$ m/s^2?
29. An automobile transmission weighs 995 N. What is its mass?
30. A power wheelbarrow weighs $21\overline{0}$ N. What is its mass?

5.4 Friction

Friction is a force that resists the relative motion of two objects in contact caused by the irregularities of the two surfaces sliding or rolling across each other, which tend to catch on each other (Fig. 5.7). In general, it is found that if two rough surfaces are polished, the frictional force between them is lessened. However, there is a point beyond which this decrease in friction is not observed. If two objects are polished such that the surfaces are very smooth, then the frictional force actually increases. (Two panes of glass provide an example.) This is related to the fact that friction can

Figure 5.7 Friction resists the motion of objects in contact with each other.

Motion

Friction

Friction opposing motion

(a)

Motion

Magnified portion

(b)

also be caused by the adhesion of molecules of one surface to the molecules of the other surface. This adhesive force is similar to the electric forces that hold atoms together in solids.

Static friction is the force that opposes the start of the relative motion of two objects in contact with each other. It is sometimes called starting friction. Once the touching objects are in motion relative to each other, the friction is generally less and is called *kinetic friction* or sliding friction.

Friction is both a necessity and a hindrance in our everyday lives. Without friction, walking, driving, swimming, and many other normal activities would not be possible. Experiments with frictional forces indicate the following general characteristics:

1. ***Friction is a force that always acts parallel to the surface in contact and opposite to the direction of motion.*** If there is no motion, friction acts in the direction opposite any force that tends to produce motion [Fig. 5.8(a)]. The resistance to motion is the frictional force.

Figure 5.8   Friction increases as the force between the surfaces increases.

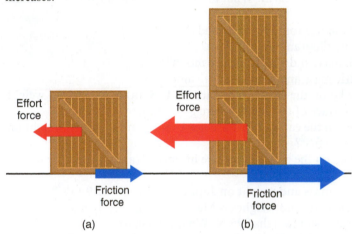

(a)   (b)

2. ***Static friction is greater than kinetic friction.*** When you push a large box across the floor, you probably notice that it takes more force to start it moving than to keep it moving. This is due to inertia. A box at rest tends to remain at rest, whereas a moving box tends to continue moving. Pushing someone on a sled is a good way to experience the difference between static and kinetic friction.
3. ***Friction increases as the force between the surfaces increases.*** It is much easier to slide a light crate than a heavy one across the floor [Fig. 5.8(b)]. *The area of contact is not relevant.* Friction depends only on the nature of the materials in contact and the force pressing them together.

The characteristics of friction can be described by the following equation:

$$F_f = \mu F_N$$

where F_f = frictional force
F_N = normal force (force perpendicular to the contact surface)
μ = coefficient of friction

The **coefficient of friction** is the ratio between the frictional force and the normal force of the object. The coefficient describes how rough or smooth the two surfaces are when they are in contact with each other. A higher coefficient of friction indicates two rough surfaces, whereas a lower number indicates two smooth surfaces.

Representative values for the coefficients of friction for some surfaces are given in Table 5.1. Values may vary with surface conditions.

Table 5.1   Coefficients of Friction (μ)

Material	Static Friction	Kinetic Friction
Steel on steel	0.58	0.20
Steel on steel (lubricated)	0.13	0.13
Glass on glass	0.95	0.40
Hardwood on hardwood	0.40	0.25
Steel on concrete		0.30
Aluminum on aluminum	1.9	
Rubber on dry concrete	2.0	1.0
Rubber on wet concrete	1.5	0.97
Aluminum on wet snow	0.4	0.02
Steel on Teflon	0.04	0.04

TRY THIS ACTIVITY

How to Find μ

Establish your own chart of coefficients of friction for various pairs of surfaces. Take any fairly massive object, perhaps a textbook, and measure its normal force by weighing it with a spring scale. (We will be doing this activity only on horizontal surfaces so the normal force is the same as the weight of the object.) To determine the coefficient of friction between the two surfaces, slowly pull horizontally on the spring scale until the object moves at a constant velocity. As the object moves, the reading on the scale gives the applied pulling force, which is also the force of friction. Record your data in a table and compare. The rougher materials should have a larger coefficient of friction than the smoother materials.

In general:

To reduce kinetic friction:

1. Use smoother surfaces.
2. Use lubrication to provide a thin film between surfaces.
3. Use Teflon to greatly reduce friction between surfaces when an oil lubricant is not desirable, such as in electric motors.
4. Substitute rolling friction for sliding friction. Rolling friction is the resistance that occurs when an object, such as a wheel or a tire, rolls on a surface. It is usually much smaller than sliding friction. Using ball bearings and roller bearings greatly reduces friction.

A force of 170 N is needed to keep a 530-N wooden box sliding on a wooden floor. What is the coefficient of kinetic friction?

EXAMPLE 1

Sketch:

Push
170 N

$F_N = 530$ N

$F_f = 170$ N $F_w = 530$ N

Data:

$$F_f = 170 \text{ N}$$
$$F_N = 530 \text{ N}$$
$$\mu = ?$$

Basic Equation:

$$F_f = \mu F_N$$

Working Equation:

$$\mu = \frac{F_f}{F_N}$$

Substitution:

$$\mu = \frac{170 \text{ N}}{530 \text{ N}}$$
$$= 0.32$$

Note that μ does not have a unit because the force units always cancel.

························

PHYSICS CONNECTIONS

Friction and Antilock Brakes

Since the invention of the automobile, engineers have worked to design brakes that reduce the stopping distance for automobiles. The most impressive advance in automobile braking technology is the antilock braking system (ABS). ABS technology is found on most new automobiles and works to prevent changing the friction between the tires and the road to sliding or kinetic friction. The ratio of the coefficients of static friction to kinetic friction of rubber on dry cement is 2.0 to 1.0.

Automobiles stop by using friction to slow down during normal braking situations. This friction takes place while the tires are in contact with the pavement. Locking the wheels or braking on ice or a film of water can quickly change static friction between the road and the tires to kinetic friction between the tires and the road. Sliding tires cause cars to go out of control and result in longer stopping distances.

Antilock braking systems use computerized sensors to monitor the rotational speed of the tires while braking. If the sensors detect a sudden decrease in the turning speed of the tires (indicating a skid), the control unit repeatedly releases and then restores the pressure to the brakes at a rate of up to 15 times per second. In essence, the ABS pulses the brakes on and off to keep the tires from sliding. Maintaining the static frictional force throughout braking keeps the car in control and reduces its stopping distance.

PROBLEM SOLVING

SKETCH

12 cm²	w

4.0 cm

DATA
$A = 12 \text{ cm}^2, l = 4.0 \text{ cm}, w = ?$

BASIC EQUATION
$A = lw$

WORKING EQUATION
$w = \dfrac{A}{l}$

SUBSTITUTION
$w = \dfrac{12 \text{ cm}^2}{4.0 \text{ cm}} = 3.0 \text{ cm}$

PROBLEMS 5.4

1. A cart on wheels weighs 2400 N. The coefficient of rolling friction between the wheels and floor is 0.16. What force is needed to keep the cart rolling uniformly?
2. A wooden crate weighs 780 lb. What force is needed to start the crate sliding on a wooden floor when the coefficient of static friction is 0.40?
3. A piano weighs 4700 N. What force is needed to start the piano rolling across the floor when the coefficient of static friction is 0.23?
4. A force of 850 N is needed to keep the piano in Problem 3 rolling uniformly. What is the coefficient of rolling friction?
5. A dog sled weighing 750 lb is pulled over level snow at a uniform speed by a dog team exerting a force of $6\overline{0}$ lb. Find the coefficient of friction.
6. A horizontal conveyor belt system has a coefficient of kinetic friction of 0.65. The motor driving the system can deliver a maximum force of 2.5×10^6 N. What maximum total weight can be placed on the conveyor system?
7. A tow truck can deliver 2500 lb of pulling force. What is the maximum-weight truck that can be pulled by the tow truck if the coefficient of rolling friction of the truck is 0.10?

8. A snowmobile is pulling a large sled across a snow-covered field. The weight of the sled is 3560 N. If the coefficient of friction of the sled is 0.12, what is the pulling force supplied by the snowmobile?

9. An automobile weighs 12,000 N and has a coefficient of static friction of 0.13. What force is required to start the auto rolling?

10. A light truck weighs 14,$\overline{0}$00 N with a coefficient of static friction of 0.140. What force is required to start the truck rolling?

11. A stake truck weighs 20,$\overline{0}$00 N with a coefficient of static friction of 0.130. What force is required to start the truck rolling?

12. If the coefficient of rolling friction of the auto in Problem 11 is 0.080, what force is required to keep it moving once it is in motion?

13. An alloy block is placed on a smooth composite table. If a force of 14.0 N is required to keep the 40.0-N block moving at a constant velocity, what is the coefficient of kinetic friction for the table and block?

14. If a 20.0-N casting is placed on the block in Problem 13, what force is necessary to keep the block and casting moving at constant velocity?

15. Rubber tires and wet blacktop have a coefficient of kinetic friction of 0.500. A pickup truck with mass 75$\overline{0}$ kg traveling 30.0 m/s skids to a stop. (a) What are the size and direction of the frictional force that the road exerts on the truck? (b) Find the acceleration of the truck. (c) How far would the truck travel before coming to rest?

16. The coefficient of friction in Problem 15 is 0.700 if the tires do not skid. (a) Would the force of friction be greater than, less than, or equal to that in Problem 15? (b) Would the truck in (a) stop in a shorter, a longer, or an equal distance compared to the truck in Problem 15?

NEW TECHNOLOGIES

Hydrophobic Metals— Reducing Friction

Teflon coatings have long been revered as the low-friction solution for metal coatings. Frying pans are a typical example of Teflon-coated surfaces where, when tilted beyond 70°, an egg slides right off the pan and onto the breakfast plate.

Professor Chunlei Guo and his team at The Institute of Optics at the University of Rochester have developed a technique that, by exposing the surface of a metal to the power of a high-intensity laser, allows some metals to have their surface structures altered to make a near-frictionless surface! Transforming an otherwise rough metal surface into a hydrophobic surface results in water, eggs, and other fluid-like substances sliding off a surface by angling it less than 10°!

Figure 5.9   Hydrophobic metal and water droplet

Hydrophobic metals are not subject to corrosion. They repel water droplets as the surface of the metal is so smooth that it does not apply enough pressure on the droplet to break its surface tension. Dust placed on a hydrophobic surface can be cleaned with just a few water droplets as the droplets absorb the dust and the droplets leave the surface. Dr. Chunlei Guo sees a future of hydrophobic surfaces where they would play a role in cleaning surfaces, keeping ice off airplane wings, and maintaining sterile medical equipment.

Research continues on hydrophobic metals as high-intensity lasers currently take an hour to create a hydrophobic surface 1 in. × 1 in. (Fig. 5.9)!

5.5 Total Forces in One Dimension

In the examples used to illustrate the law of acceleration, we discussed only total forces. We need to remember that forces are vectors and have magnitude and direction. The total force acting on an object is the resultant of the separate forces.

When forces act in the same or opposite directions (in one dimension), the total, or net, force can be found by adding the forces that act in one direction and subtracting the forces that act in the opposite direction.

It is often useful to draw the forces as vectors.

EXAMPLE 1

Two workers push in the same direction (to the right) on a crate. The force exerted by one worker is $15\overline{0}$ lb. The force exerted by the other is 175 lb. Find the net force exerted.

Sketch:

Both forces act in the same direction, so the total force is the sum of the two.

Note: The Greek letter Σ (sigma) means "sum of."

$$\Sigma \mathbf{F} = 175 \text{ lb} + 15\overline{0} \text{ lb}$$
$$= 325 \text{ lb to the right}$$

· · · · · · · · · · · · · · · ·

EXAMPLE 2

The same two workers push the crate to the right, and the motion is opposed by a static frictional force of $30\overline{0}$ lb. Find the net force.

Sketch:

The workers push in one direction and static friction pushes in the opposite direction, so we add the forces exerted by the workers and subtract the frictional force.

$$\Sigma F = 175 \text{ lb} + 15\overline{0} \text{ lb} - 30\overline{0} \text{ lb}$$
$$= 25 \text{ lb to the right}$$

. . . .

The crate in Example 2 has a mass of 5.00 slugs. What is its acceleration when the workers are pushing against the frictional force?

EXAMPLE 3

Data:

$$F = 25 \text{ lb (from Example 2)}$$
$$m = 5.00 \text{ slugs}$$
$$a = ?$$

Basic Equation:

$$F = ma$$

Working Equation:

$$a = \frac{F}{m}$$

Substitution:

$$a = \frac{25 \text{ lb}}{5.00 \text{ slugs}}$$
$$= 5.0 \frac{\text{lb}}{\text{slugs}} \times \frac{1 \text{ slug ft/s}^2}{1 \text{ lb}}$$
$$= 5.0 \text{ ft/s}^2$$

Note: We use a conversion factor to obtain acceleration units.

. . . .

Two workers push in the same direction on a large pallet. The force exerted by one worker is 645 N. The force exerted by the other worker is 755 N. The motion is opposed by a frictional force of 1175 N. Find the net force.

EXAMPLE 4

$$\Sigma F = 645 \text{ N} + 755 \text{ N} - 1175 \text{ N}$$
$$= 225 \text{ N}$$

. . . .

PROBLEMS 5.5

Find the net force including its direction when each force acts in the direction indicated.

1. 17.0 N to the left, 20.0 N to the right.
2. 265 N to the left, 85 N to the right.
3. 100.0 N to the left, 75.0 N to the right, and 10.0 N to the right.
4. $19\overline{0}$ lb to the left, 87 lb to the right, and 49 lb to the right.
5. 346 N to the right, 247 N to the left, and 103 N to the left.
6. 37 N to the right, 24 N to the left, 65 N to the right, and 85 N to the right.
7. Find the acceleration of an automobile of mass $10\overline{0}$ slugs acted upon by a driving force of $50\overline{0}$ lb that is opposed by a frictional force of $10\overline{0}$ lb (Fig. 5.10).
8. Find the acceleration of an automobile of mass $150\overline{0}$ kg acted upon by a driving force of $220\overline{0}$ N that is opposed by a frictional force of 450 N.
9. A truck of mass 13,100 kg is acted upon by a driving force of $890\overline{0}$ N. The motion is opposed by a frictional force of 2230 N. Find the acceleration.

SKETCH

| 12 cm² | w |

4.0 cm

DATA
$A = 12 \text{ cm}^2, l = 4.0 \text{ cm}, w = ?$

BASIC EQUATION
$A = lw$

WORKING EQUATION
$w = \dfrac{A}{l}$

SUBSTITUTION
$w = \dfrac{12 \text{ cm}^2}{4.0 \text{ cm}} = 3.0 \text{ cm}$

Figure 5.10

100 lb   500 lb

Figure 5.11   The ground supplies the force to accelerate the car.

Force of ground   Force of tires on
on tires (reaction)   ground (action)

$$F_1 = -F_2$$

10. A speedboat of mass 30.0 slugs has a 300-lb force applied by the propellers. The friction of the water on the hull is a force of 100 lb. Find the acceleration.
11. A truck with a mass of 14,000 kg is pushed with a force of 9200 N. If the frictional force is 2150 N, what is the truck's acceleration?
12. A trailer has a mass of 5000 kg. It is pulled by a truck with a force of 8750 N. If the frictional force is 2000 N, what is its acceleration?
13. A refrigeration unit on a job site must be slid into place. If the frictional force of 975 N opposes the motion and two workers apply a total force of 1095 N to the unit, what is the net force on the unit?
14. A light truck of 2000-kg mass has to be pushed with a force of 1975 N. If the frictional force is 550 N, what is its acceleration?

5.6   Law of Action and Reaction

When an automobile accelerates, we know that a force is being applied to it. What applies this force? You may think that the tires exert this force on the auto. This is not correct because the tires move along with the auto and there must be a force applied to them also. The ground below the tires actually supplies the force that accelerates the car (Fig. 5.11). This force is called a *reaction* to the force exerted by the tires on the ground, which is called the *action force*. The third law of motion, the **law of action and reaction**, can be stated as follows:

> **LAW OF ACTION AND REACTION: NEWTON'S THIRD LAW**
> For every force applied by object *A* to object *B* (action), there is an equal but opposite force exerted by object *B* to object *A* (reaction).

Figure 5.12   The magnitude of the reaction force always equals the magnitude of the action force.

Action on bullet: F_1

Reaction on hand:
F_2

$$F_1 = -F_2$$

When a bullet is fired from a handgun (action), the recoil felt is the reaction. These forces are shown in Fig. 5.12. Note that the action and reaction forces *never* act on the same object.

For every interaction, the forces always occur in pairs and are equal and opposite. When you sit on a chair, your weight pushes *down* on the chair; the chair pushes *up* with a force equal to your weight. If the chair pushed up with a force less than your weight, you would fall through it. If the chair pushed up with a force greater than your weight, you would be pushed up above the seat.

Most interactions depend on force. What would happen if you were standing in a small boat and tried to jump across to the nearby boat dock? You would fall into the water. Why? The force you would exert against the boat as you jumped would push it away, and the equal and opposite force the boat would exert on you would not result in much forward motion for you.

TRY THIS ACTIVITY

Balloon Rocket

Blow up a balloon and release it. What causes the balloon to accelerate? What causes it to stop accelerating? Use Newton's third law of motion to describe how the balloon accelerates forward.

   To create a straight path for the balloon, tape the balloon to a straw that is on a string that serves as a guide wire between two points or people, as shown in Fig. 5.13.

Figure 5.13   A balloon rocket

PHYSICS CONNECTIONS

Shuttle Launch

Newton's second and third laws of motion play a vital role during the launch of the Space Shuttle (see Fig. 5.14). Just as a massive gun recoils as a less-massive bullet is fired, the massive shuttle recoils or "lifts off" as the less-massive, high-velocity gaseous exhaust is ejected from the rocket engines.

In order to overcome the force of gravity that the earth applies to all objects, the Space Shuttle must carry enough fuel in its external tank and solid rocket boosters to overcome the gravitational pull of the earth. To do so, the shuttle carries more than 4.4 million pounds of rocket fuel for use during its launch. Compare this to the weight of the Space Shuttle itself, which is 170,000 pounds, and it is clear that the shuttle uses a tremendous amount of fuel to lift off.

In order to take advantage of Newton's third law and send the shuttle into space, the gaseous exhaust from the burned fuel is ejected from the shuttle's rocket boosters at speeds of nearly 6000 mi/h. As the rocket fuel is burned and its gaseous exhaust is ejected out of the rocket, the overall weight of the shuttle's fuel decreases. According to Newton's second law, if the same force is applied but the mass is reduced, the acceleration increases.

Figure 5.14 The gas particles are expelled at a high velocity, causing the shuttle to recoil or launch into space.

Courtesy of NASA/John F. Kennedy Space Center

Glossary

Coefficient of Friction The ratio between the frictional force and the normal force of an object. The number represents how rough or smooth two surfaces are when moving across one another. (p. 132)

Force A push or a pull that tends to change the motion of an object or prevent an object from changing motion. Force is a vector quantity and thus has both magnitude and direction. (p. 123)

Friction A force that resists the relative motion of two objects in contact caused by the irregularities of two surfaces sliding or rolling across each other. (p. 131)

Inertia The property of a body that causes it to remain at rest if it is at rest or to continue moving with a constant velocity unless an unbalanced force acts upon it. (p. 124)

Law of Acceleration The total force acting on a body is equal to the mass of the body times its acceleration (Newton's second law). (p. 125)

Law of Action and Reaction For every force applied by object A to object B (action), there is an equal but opposite force exerted by object B on object A (reaction) (Newton's third law). (p. 138)

Law of Inertia A body that is in motion continues in motion with the same velocity (at constant speed and in a straight line) and a body at rest continues at rest unless an unbalanced (outside) force acts upon it (Newton's first law). (p. 124)

Mass A measure of the inertia of a body. (p. 125)

Normal Force Force perpendicular to the contact surface. (p. 130)

Weight The amount of gravitational pull exerted on an object by the earth or by another large body. (p. 128)

Formulas

5.2 $F = ma$

5.3 $F_w = mg$

where $g = 9.80 \text{ m/s}^2$ (metric)
        $g = 32.2 \text{ ft/s}^2$ (U.S.)

5.4 $F_f = \mu F_N$

Review Questions

1. Force
   - (a) is a vector quantity.
   - (b) may be different from weight.
   - (c) does not always cause motion.
   - (d) all of the above.
2. The metric weight of a 10-lb bag of sugar is approximately
   - (a) 4.45 N.
   - (b) 44.5 N.
   - (c) 445 N.
   - (d) none of the above.
3. Mass and weight
   - (a) are the same.
   - (b) are different.
   - (c) do not change wherever you are.
4. According to Newton's second law, the law of acceleration,
   - (a) acceleration is equal to mass times force.
   - (b) mass is equal to mass times acceleration.
   - (c) force is equal to mass times acceleration.
   - (d) none of the above.
5. Friction
   - (a) always acts parallel to the surface of contact and opposite to the direction of motion.
   - (b) acts in the direction of motion.
   - (c) is smaller when starting than when moving.
   - (d) is an imaginary force.
6. Cite three examples of forces acting without motion being produced.
7. (a) Does a pound of feathers have more inertia than a pound of lead?
   (b) Does the pound of feathers have more mass than the pound of lead?
8. How is inertia a factor in multicar pileups?
9. Using your own words, state Newton's first law, the law of inertia.
10. Distinguish between velocity and acceleration.

11. When the same force is applied to two different masses, which will have a greater acceleration?
12. Is 3 lb heavier than 10 N?
13. Explain how life would be easier or more difficult without friction.
14. Explain how the weight of an astronaut is different on the moon than on the earth. Would the astronaut's mass be different?
15. Explain the difference between action and reaction forces.
16. State Newton's third law of motion, the law of action and reaction, in your own words.

Review Problems

1. A crate of mass 6.00 kg is moved by a force of 18.0 N. What is its acceleration?
2. An 825-N force is required to pedal a bike with an acceleration of 11.0 m/s². What is the mass of the bike and rider?
3. A block of mass 0.89 slug moves with a force of 17.0 lb. Find the block's acceleration.
4. What is the force necessary for a 2400-kg truck to accelerate at a rate of 8.0 m/s²?
5. Two movers push a piano across a frictionless surface. One pushes with 29.0 N of force and the other mover exerts 35.0 N. What is the total force?
6. A 340-N box has a frictional force of 57 N. Find the coefficient of kinetic friction.
7. A truck pulls a trailer with a frictional force of 870 N and a coefficient of friction of 0.23. What is the trailer's normal force?
8. A steel box is slid along a steel surface. It has a normal force of 57 N. What is the frictional force?
9. A rock of a mass 13.0 kg is dropped from a cliff. Find its weight.
10. A projectile has a mass of 0.37 slug. Find its weight.
11. What force is required to produce an acceleration of 4.00 m/s² on a wrecking ball with a mass of 50.0 kg?
12. Find the total force necessary to give a 28$\overline{0}$-kg motorcycle an acceleration of 3.20 m/s².
13. A force of 175 N is needed to keep a 64$\overline{0}$-N stationary engine on wooden skids from sliding on a wooden floor. What is the coefficient of static friction?
14. A crated garden tractor weighs 375 N. What force is needed to start the crate sliding on a wooden floor when the coefficient of static friction is 0.40?
15. Find the acceleration of a forklift of mass 14$\overline{0}$0 kg pushed by a force of 21$\overline{0}$0 N that is opposed by a frictional force of 425 N.
16. What is the weight of a 375-kg air compressor?
17. What is the mass of a 405-N welder?
18. What is the mass of a 12.0-N hammer?

SKETCH

12 cm² w

4.0 cm

DATA

$A = 12$ cm², $l = 4.0$ cm, $w = ?$

BASIC EQUATION

$A = lw$

WORKING EQUATION

$w = \dfrac{A}{l}$

SUBSTITUTION

$w = \dfrac{12 \text{ cm}^2}{4.0 \text{ cm}} = 3.0 \text{ cm}$

APPLIED CONCEPTS

1. Engineers at Boeing developing specs for their "next-generation" 737 aircraft needed to know the acceleration of the 737-900 during a typical take-off. (a) What acceleration would they calculate given the plane's 78,200-kg mass and its maximum engine force of 121,000 N? (b) How fast would the plane be traveling after the first $50\overline{0}$ m of runway? (c) How fast would it travel after the first $150\overline{0}$ m of runway?

2. The Apollo spacecrafts were launched toward the moon using the Saturn rocket, the most powerful rocket available. Each rocket had five engines producing a total of 33.4×10^6 N of force to launch the 2.77×10^6-kg spacecraft toward the moon. (a) Find the average acceleration of the spacecraft. (b) Calculate the altitude of the rocket 2.50 min after launch—the point when the spacecraft loses its first stage.

3. Kirsten's mass is 3.73 slugs. Being the physics fan that she is, she decides to see what her apparent weight will be during an elevator ride. Beginning at rest, the elevator accelerates upward at 4.50 ft/s^2 for 3.00 s and then continues at a constant upward velocity. Finally, as the elevator comes to a stop at the top floor, the elevator slows down (accelerates downward but continues to move upward) at a rate of -5.5 ft/s^2 (the negative sign represents the downward direction). Find Kirsten's weight while the elevator is (a) at rest, (b) speeding up, (c) moving at a constant velocity, and (d) slowing down. The next time you ride in an elevator, concentrate on when you feel heavier and when you feel lighter.

4. A motorcycle racer traveling at 145 km/h loses control in a corner of the track and slides across the concrete surface. The combined mass of the rider and bike is 243 kg. The steel of the motorcycle rubs against the concrete road surface. (a) What is the frictional force between the road and the motorcycle and rider? (b) What would be the acceleration of the motorcycle and rider during the wipeout? (c) Assuming there were no barriers to stop the motorcycle and rider, how long would it take the bike and the rider to slow to a stop?

5. The motorcycle and rider are sliding with the same acceleration as found in Problem 4. If the motorcycle and rider have been sliding for 4.55 s, what will be the force applied to the motorcycle and the rider when they strike the side barrier and come to rest in another 0.530 s?

CHAPTER 6

MOMENTUM

Objectives

The major goals of this chapter are to enable you to:

1. Use momentum and impulse in describing motion.
2. State the law of conservation of momentum and apply it to physical problems.
3. Analyze elastic and inelastic collisions of two objects.

Momentum and impulse are important concepts in describing and understanding the motion of objects and the related effects on those objects. The law of conservation of momentum is an important law of physics; it helps us analyze how two objects interact with each other when they are in contact with each other and when they collide.

6.1 Momentum and Impulse

We know that it is much more difficult to stop a large truck than a small car traveling at the same speed. The truck has more *inertia* and is more difficult to bring to a stop or to begin moving than the car. The truck has more inertia because it has a greater mass than the small car. In a head-on collision, common sense tells us that the small car will be damaged far more than the truck. However, if the car is traveling at a higher rate of speed, the damage it could cause in a collision might be greater. Momentum is the concept that describes the combined effect of inertia and motion. As an illustration, compare the effects of a tennis ball being rolled or even thrown at an array of pins to the effects of a heavy bowling ball. Clearly the bowling ball has more momentum.

Momentum, however, is not just a function of greater mass or inertia. A bullet fired from a rifle has great momentum because, although its mass and inertia are small, its high velocity gives it a large momentum.

Momentum is a measure of the amount of inertia and motion an object has or of the difficulty in bringing a moving object to rest. Since momentum is partially determined by velocity, it, like velocity, is a relative physical quantity. It must have a frame of reference. For example, a canoe being paddled upstream at the same rate at which the current is flowing downstream has a velocity of zero and zero momentum relative to the shore. Expressing these concepts in mathematical terms, we have **momentum** *equals the product of the mass times the velocity of an object.*

$$p = mv$$

where p = momentum
m = mass
v = velocity

The momentum of a train makes it impossible to stop within a short distance, and this explains why it cannot stop at a railroad crossing when the engineer sees someone stopped or stalled at it.

The units of momentum are kg m/s in the metric system and slug ft/s in the U.S. system. Momentum is a vector quantity whose direction is the same as the velocity.

Note: A constant *linear momentum p* is the momentum of an object of mass m that is moving in a straight line with velocity v. In Section 9.2 we introduce a similar quantity, *angular momentum*, to describe the momentum of an object that is rotating. In general, the term "momentum" refers to linear momentum. We will always specify angular momentum when referring to momentum relating to rotation.

EXAMPLE 1

Find the momentum of an auto with mass 1350 kg traveling 75.0 km/h.

Data:

$$m = 1350 \text{ kg}$$

$$v = 75.0 \, \frac{\text{km}}{\text{h}} \times \frac{1000 \text{ m}}{1 \text{ km}} \times \frac{1 \text{ h}}{3600 \text{ s}} = 20.8 \text{ m/s}$$

$$p = ?$$

Basic Equation:

$$p = mv$$

Working Equation: Same

Substitution:

$$p = (1350 \text{ kg}) (20.8 \text{ m/s})$$

$$= 28{,}100 \text{ kg m/s}$$

EXAMPLE 2

Find the momentum of a small truck with load weighs 6920 lb traveling 60.0 mi/h.

Data:

$$m = 6920 \text{ lb} \times \frac{1 \text{ slug}}{32.2 \text{ lb}} = 215 \text{ slugs}$$

$$v = 60.0 \text{ mi/h} = 88.0 \text{ ft/s}$$

$$p = ?$$

Basic Equation:

$$p = mv$$

Working Equation: Same

Substitution:

$$p = (215 \text{ slugs})(88.0 \text{ ft/s})$$

$$= 18,900 \text{ slugs ft/s}$$

- - - - - - - - - - - - - - - - - -

Find the velocity that an automobile weighing 3250 lb would have to have so that it has the same momentum as a loaded truck weighing 29,500 lb and traveling 50.0 mi/h.

EXAMPLE 3

Sketch:

$m_1 = 3250 \text{ lb}$ $m_2 = 29,500 \text{ lb}$

$v_1 = ?$ $v_2 = 50.0 \text{ mi/h}$

Data:

Automobile	Truck
$m_1 = 3250 \text{ lb} \times \dfrac{1 \text{ slug}}{32.2 \text{ lb}} = 101 \text{ slugs}$ $v_1 = ?$	$m_2 = 29,500 \text{ lb} \times \dfrac{1 \text{ slug}}{32.2 \text{ lb}} = 916 \text{ slugs}$ $v_2 = 50.0 \dfrac{\text{mi}}{\text{h}} \times \dfrac{1 \text{ h}}{3600 \text{ s}} \times \dfrac{5280 \text{ ft}}{1 \text{ mi}} = 73.3 \text{ ft/s}$

Basic Equations:

$$p_1 = m_1 v_1$$
$$p_2 = m_2 v_2$$

We want

$$p_1 = p_2$$

or

$$m_1 v_1 = m_2 v_2$$

Working Equation:

$$v_1 = \frac{m_2 v_2}{m_1}$$

Substitution:

$$v_1 = \frac{(916 \text{ slugs})(73.3 \text{ ft/s})}{101 \text{ slugs}}$$

$$= 665 \frac{\text{ft}}{\text{s}} \times \frac{1 \text{ mi}}{5280 \text{ ft}} \times \frac{3600 \text{ s}}{1 \text{ h}}$$

$$= 453 \text{ mi/h (Wow!)}$$

- - - - - - - - - - - - - - - - - -

When a truck accelerates or stops, its momentum changes. If the momentum of any object changes, either its mass or its velocity changes. Usually the mass (for example, the truck) does not change, so any change in momentum is a result of its velocity changing. Its velocity changes because a force acts on an object for an interval of time. A stalled truck does not move unless a force or push is applied to it. How fast and how far it moves are also determined by how long the force is applied to it. Both force and time determine the change in momentum.

Decreasing momentum can be as important as increasing momentum. Bringing a moving truck to a stop reduces its momentum to zero. Brakes are used to apply a retarding force to the truck. Common experience tells us that the longer the retarding force is applied, the more the truck is slowed. Impulse—the force applied and the length of time it is applied—determines stopping distance. Everyday examples of impulse include jumping off a deck or dropping a glass. If you do not bend your knees when you land so as to increase the time of impact, you risk breaking bones. Similarly, a glass dropped on a hardwood floor will probably shatter, whereas the same glass dropped on a pile rug may not. The rug has more "give," which means that the pile of the rug provides a longer time of contact during the impact. Both are examples of impulse.

The **impulse** on an object is the product of the force applied and the time interval during which the force acts on the object. That is,

$$\text{impulse} = Ft$$

where F = force
t = time interval during which the force acts

How are impulse and momentum related? Recall that

$$a = \frac{v_f - v_i}{t}$$

If we substitute this equation into Newton's second law of motion, we have

$$F = ma$$

$$F = m\left(\frac{v_f - v_i}{t}\right)$$

$$F = \frac{mv_f - mv_i}{t} \qquad \text{Remove parentheses.}$$

$$Ft = mv_f - mv_i \qquad \text{Multiply both sides by } t.$$

Note that mv_f is the final momentum and mv_i is the initial momentum. That is,

$$\text{impulse} = \Delta p \text{ (change in momentum)} = Ft = mv_f - mv_i$$

Note: The Greek letter Δ ("delta") is used to designate "change in."

Impulse is the measure of the change in momentum of an object in response to an exerted force over an interval of time. The amount of force and the length of time the force is applied will determine the change in momentum. A common example that illustrates this relationship is a golf club hitting a golf ball (Fig. 6.1). When a golf ball is on the tee, it has zero momentum because its velocity is zero. To give it or change its momentum (impulse), you apply a force for a given period of time. During the time that the club and ball are in contact, the force of the swinging club is transferring most of its momentum to the ball. The impulse given to the ball is the product of the *force* with which the ball is hit and the length of *time* that the club and ball are in direct contact. You can increase its momentum by increasing the *force* (by swinging the golf club faster) or increasing the *time* (by keeping the golf club in contact with the ball longer, which shows the importance of "follow-through").

Figure 6.1   When a person hits a golf ball with a golf club, the club applies a force F during the time t that the club is in contact with the ball. The impulse (change in momentum) is $Ft = mv_f - mv_i = mv_f$ because $v_i = 0$.

Courtesy of Stephen Marks/Getty Images

(a)

Courtesy of Mike Powell/Getty Images

mv_i

(b)

A 17.5-g bullet is fired at a muzzle velocity of 582 m/s from a gun with a mass of 8.00 kg and a barrel length of 75.0 cm. (a) How long is the bullet in the barrel? (b) What is the force on the bullet while it is in the barrel? (c) Find the impulse exerted on the bullet while it is in the barrel. (d) Find the bullet's momentum as it leaves the barrel.

EXAMPLE 4

Sketch:

75.0 cm

582 m/s

(a) Data:

$$s = 75.0 \text{ cm} = 0.750 \text{ m}$$

$$v_f = 582 \text{ m/s}$$

$$v_i = 0 \text{ m/s}$$

$$v_{avg} = \frac{v_f + v_i}{2} = \frac{582 \text{ m/s} + 0 \text{ m/s}}{2} = 291 \text{ m/s}$$

$$t = ?$$

Basic Equation:

$$s = v_{avg}t$$

Working Equation:

$$t = \frac{s}{v_{avg}}$$

Substitution:

$$t = \frac{0.750 \text{ m}}{291 \text{ m/s}}$$

$$= 0.00258 \text{ s}$$

Note: This is the length of time that the force is applied to the bullet.

(b) Data:

$$t = 0.00258 \text{ s}$$
$$m = 17.5 \text{ g} = 0.0175 \text{ kg}$$
$$v_f = 582 \text{ m/s}$$
$$v_i = 0 \text{ m/s}$$
$$F = ?$$

Basic Equation:

$$Ft = mv_f - mv_i$$

Working Equation:

$$F = \frac{mv_f - mv_i}{t}$$

Substitution:

$$F = \frac{(0.0175 \text{ kg}) (582 \text{ m/s}) - (0.0175 \text{ kg}) (0 \text{ m/s})}{0.00258 \text{ s}}$$
$$= 3950 \text{ kg m/s}^2$$
$$= 3950 \text{ N} \qquad (1 \text{ N} = 1 \text{ kg m/s}^2)$$

(c) Data:

$$t = 0.00258 \text{ s}$$
$$F = 3950 \text{ N}$$
$$\text{impulse} = ?$$

Basic Equation:

$$\text{impulse} = Ft$$

Working Equation: Same

Substitution:

$$\text{impulse} = (3950 \text{ N}) (0.00258 \text{ s})$$
$$= 10.2 \text{ N s}$$
$$= 10.2 (\text{kg m/s}^2) (\text{s}) \qquad (1 \text{ N} = 1 \text{ kg m/s}^2)$$
$$= 10.2 \text{ kg m/s}$$

(d) Data:

$$m = 17.5 \text{ g} = 0.0175 \text{ kg}$$
$$v = 582 \text{ m/s}$$
$$p = ?$$

Basic Equation:

$$p = mv$$

Working Equation: Same

Substitution:

$$p = (0.0175 \text{ kg}) (582 \text{ m/s})$$
$$= 10.2 \text{ kg m/s}$$

Note: The impulse equals the change in momentum.

TRY THIS ACTIVITY

Scrambled Eggs

Figure 6.2

Drop a raw egg from a height of a few feet onto a surface that can be cleaned. Observe the motion of the egg as it hits the surface and note the time the egg takes to come to rest. Drop another raw egg from the same height into a suspended bed sheet. Again observe the motion of the egg as it hits the bed sheet and note the time the egg takes to come to rest (Fig. 6.2). Explain the connection between what happened to these eggs and how airbags in automobiles work.

One of the most important laws of physics, the **law of conservation of momentum**, is the following:

LAW OF CONSERVATION OF MOMENTUM

When no outside forces are acting on a system of moving objects, the total momentum of the system remains constant.

For example, consider a 35-kg boy and a 75-kg man standing next to each other on ice skates on "frictionless" ice (Fig. 6.3). The man pushes on the boy, which gives the boy a velocity of 0.40 m/s. What happens to the man? Initially, the total momentum was zero because the initial velocity of each was zero. According to the law of conservation of momentum, the total momentum must still be zero. That is,

$$m_{boy}v_{boy} + m_{man}v_{man} = 0$$
$$(35 \text{ kg})(0.40 \text{ m/s}) + (75 \text{ kg})v_{man} = 0$$
$$v_{man} = -0.19 \text{ m/s}$$

Figure 6.3 Momentum is conserved by the lighter boy moving faster than the heavier man.

$v_{boy} = 0.40$ m/s $v_{man} = 0.19$ m/s

(a)   (b)

Wernher von Braun (1912–1977),

engineer and rocket expert, was born in Germany. He was chiefly responsible for the manufacture and launching of the first American artificial earth satellite, Explorer I, in 1958. As director of the Marshall Space Flight Center from 1960 to 1970, he developed the Saturn rocket for the Apollo 8 moon landing in 1969.

Note: The minus sign indicates that the man's velocity and the boy's velocity are in opposite directions.

Rene Descartes (1596–1650) is credited with developing three laws of nature, which are essentially laws of bodily motion and later would be incorporated into Newton's laws of motion. A consequence of his first law of motion led to the principle of conservation of momentum.

Rocket propulsion is another illustration of conservation of momentum. **Wernher von Braun** was a pioneering rocket scientist. As in the example of the skaters, the total momentum of a rocket on the launch pad is zero. When the rocket engines are fired, hot exhaust gases (actually gas molecules) are expelled downward through the rocket nozzle at tremendous speeds. As the rocket takes off, the sum of the total momentums of the rocket and the gas particles must remain zero. The total momentum of the gas particles is the sum of the products of each mass and its corresponding velocity and is directed down. The momentum of the rocket is the product of its mass and its velocity and is directed up.

When the rocket is in space, its propulsion works in the same manner. The conservation of momentum is still valid except that when the rocket engines are fired, the total momentum is a nonzero constant. This is because the rocket has velocity.

Actually, repair work is more difficult in space than it is on the earth because of the conservation of momentum and the "weightlessness" of objects in orbit. On the earth, when a hammer is swung, the person is coupled to the earth by frictional forces, so that the person's mass includes that of the earth. In space orbit, because the person is weightless, there is no friction to couple him or her to the spaceship. A person in space has roughly the same problem driving a nail as a person on the earth would have wearing a pair of "frictionless" roller skates.

Other examples of conservation of momentum include nuclear reactions and high-energy particle collisions.

A change in momentum takes force and time because

$$\text{change in momentum} = \text{impulse} = Ft$$

As we noted earlier in this section, it is more difficult to stop a large truck than a small car traveling at the same speed and impossible to stop a rapidly moving train within a short distance. These events can be explained in terms of the **impulse–momentum theorem** as follows.

If the mass of an object is constant, then a change in its velocity results in a change in its momentum. That is,

$$\Delta p = m\Delta v$$

The impulse of an object equals its change in momentum. That is,

$$F\Delta t = \Delta p$$

Then,

Impulse–Momentum Theorem

$$F\Delta t = \Delta p = m\Delta v = mv_f - mv_i$$

This theorem is a result of Newton's second law as follows:

$$F = ma = m\,\frac{\Delta v}{\Delta t}$$

Or,

$$F = m \frac{\Delta v}{\Delta t}$$

$$F\Delta t = m\Delta v \ (\textit{Multiply both sides by } \Delta t.)$$

That is,

$$\text{Impulse} = F\,\Delta t = m\,\Delta v = m(v_f - v_i) = mv_f - mv_i$$

A standard 0.145-kg baseball is thrown and reaches a batter with a velocity of 35.0 m/s when it is struck with a bat that causes it to reverse direction with a velocity of 44.0 m/s. Find the impulse and the force exerted on the baseball if the bat is in contact with the ball for 7.0 ms.

EXAMPLE 5

Sketch:

Before impact | Impulse | After impact

$v_i = -35.0$ m/s   during contact $v_f = 44.0$ m/s

(a)   (b)   (c)

Data:

$$m = 0.145 \text{ kg}$$
$$v_f = 44.0 \text{ m/s}$$
$$v_i = -35.0 \text{ m/s} \quad (v_i \text{ is in the opposite direction of } v_f, \text{ as indicated by the negative sign.})$$
$$t = 7.0 \text{ ms} = 0.0070 \text{ s}$$
$$\text{Impulse} = ?$$
$$F = ?$$

Basic Equation:

$$\text{Impulse} = F\Delta t = m(v_f - v_i) = mv_f - mv_i$$

Working Equation: Same

Substitution:

$$\text{Impulse} = (0.145 \text{ kg})(44.0 \text{ m/s}) - (0.145 \text{ kg})(-35.0 \text{ m/s})$$
$$= 11.5 \text{ kg m/s}$$

To find the force:

$$F\Delta t = \text{Impulse}$$

$$F = \frac{\text{Impulse}}{\Delta t}$$

$$= \frac{11.5 \text{ kg m/s}}{0.0070 \text{ s}}$$

$$= 1600 \text{ kg m/s}^2$$

$$= 1600 \text{ N}$$

. . . .

EXAMPLE 6

What force is required to slow a 1450-kg car traveling 115 km/h to 45.0 km/h within 3.00 s? How far does the car travel during its deceleration?

Data:

$$m = 1450 \text{ kg}$$

$$v_f = 45\frac{\text{km}}{\text{h}} \times \frac{1 \text{ h}}{3600 \text{ s}} \times \frac{1000 \text{ m}}{1 \text{ km}} = 12.5 \text{ m/s}$$

$$v_i = 115\frac{\text{km}}{\text{h}} \times \frac{1 \text{ h}}{3600 \text{ s}} \times \frac{1000 \text{ m}}{1 \text{ km}} = 31.9 \text{ m/s}$$

$$\Delta t = 3.00 \text{ s}$$

$$F = ?$$

Basic Equation:

$$F\Delta t = mv_f - mv_i$$

Working Equation:

$$F = \frac{mv_f - mv_i}{\Delta t}$$

Substitution:

$$F = \frac{(1450 \text{ kg}) (12.5 \text{ m/s}) - (1450 \text{ kg}) (31.9 \text{ m/s})}{3.00 \text{ s}}$$

$$= -9380 \text{ kg m/s}^2 = -9380 \text{ N}$$

Note: The negative sign indicates a deceleration force.

Basic Equation:

$$s = \tfrac{1}{2} (v_f + v_i)t$$

Working Equation: Same

Substitution:

$$s = \tfrac{1}{2} (12.5 \text{ m/s} + 31.9 \text{ m/s}) (3.00 \text{ s})$$

$$= 66.6 \text{ m}$$

. . . .

PHYSICS CONNECTIONS

Airbags

During an automobile front-end collision, passengers will continue to travel forward until the dashboard, seat belt, or airbag applies a force on them to stop them. Airbags are designed to provide a cushion-like effect to gradually bring passengers to rest. Airbags increase the time it takes to bring passengers to a stop and reduce the force of the impact (Fig. 6.4). Airbags used in conjunction with seat belts help prevent death and serious injury.

Airbags expand from the steering wheel or dashboard when a sudden impulse or a change in momentum of the vehicle triggers a sensor that is connected to a heating element. The heating element causes a chemical reaction with a propellant that fills the airbag with nitrogen gas within $\frac{1}{20}$ s. This short inflation time gives the airbag enough time to inflate before the passenger strikes it. Within $\frac{1}{2}$ s, the collision is completed and the airbag deflates.

Figure 6.4 An airbag increases the time it takes to bring a passenger to a stop in a collision by reducing the force of the impact applied to the passenger.

Courtesy of fotolia © benjaminnolte

Airbags are designed to strike the average seat-belted man in the midsection of the body. An airbag, which expands at a rate of 150 mi/h, can be quite dangerous if the bag strikes short individuals in the face. Injuries to women and children caused by airbags are a serious problem. Efforts are being made to automatically adjust airbag deployment to make airbags safer for all passengers. Airbags are also used for side-impact collisions.

PROBLEMS 6.1

Find the momentum of each object.

1. $m = 2.00$ kg, $v = 40.0$ m/s
2. $m = 5.00$ kg, $v = 90.0$ m/s
3. $m = 17.0$ slugs, $v = 45.0$ ft/s
4. $m = 38.0$ kg, $v = 97.0$ m/s
5. $m = 3.8 \times 10^5$ kg, $v = 2.5 \times 10^3$ m/s
6. $m = 3.84$ kg, $v = 1.6 \times 10^5$ m/s
7. $F_w = 1.50 \times 10^5$ N, $v = 4.50 \times 10^4$ m/s
8. $F_w = 3200$ lb, $v = 6\overline{0}$ mi/h (change to ft/s)
9. (a) Find the momentum of a heavy automobile weighing 5850 lb traveling 70.0 ft/s.
   (b) Find the velocity of a light automobile weighing 2580 lb so that it has the same momentum as the heavy automobile.
10. (a) Find the momentum of a bullet of mass 1.00×10^{-3} slug traveling $70\overline{0}$ ft/s.
    (b) Find the velocity of a bullet of mass 5.00×10^{-4} slug so that it has the same momentum as the bullet in part (a).
11. (a) Find the momentum of an automobile of mass 2630 kg traveling 21.0 m/s.
    (b) Find the velocity (in km/h) of a light auto of mass 1170 kg so that it has the same momentum as the auto in part (a).
12. A ball of mass 0.50 kg is thrown straight up at 6.0 m/s. (a) What is the initial momentum of the ball? (b) What is the momentum of the ball at its peak? (c) What is the momentum of the ball as it hits the ground?
13. A bullet with mass 60.0 g is fired with an initial velocity of 575 m/s from a gun with mass 4.50 kg. What is the speed of the recoil of the gun?
14. A cannon is mounted on a railroad car. The cannon shoots a 1.75-kg ball with a muzzle velocity of $30\overline{0}$ m/s. The cannon and the railroad car together have a mass of $45\overline{0}0$ kg. If the ball, cannon, and railroad car are initially at rest, what is the recoil velocity of the car and cannon?

SKETCH

12 cm²   w

4.0 cm

DATA
$A = 12$ cm², $l = 4.0$ cm, $w = ?$

BASIC EQUATION
$A = lw$

WORKING EQUATION
$w = \dfrac{A}{l}$

SUBSTITUTION
$w = \dfrac{12 \text{ cm}^2}{4.0 \text{ cm}} = 3.0$ cm

15. A 125-kg pile driver falls from a height of 10.0 m to hit a piling. (a) What is its speed as it hits the piling? (b) With what momentum does it hit the piling?

16. A person is traveling 75.0 km/h in an automobile and throws a bottle of mass 0.500 kg out the window. (a) With what momentum does the bottle hit a roadway sign? (b) With what momentum does the bottle hit an oncoming automobile traveling 85.0 km/h in the opposite direction? (c) With what momentum does the bottle hit an automobile passing and traveling 85.0 km/h in the same direction?

17. A 75.0-g bullet is fired with a muzzle velocity of $46\overline{0}$ m/s from a gun with mass 3.75 kg and barrel length of 66.0 cm. (a) How long is the bullet in the barrel? (b) What is the force on the bullet while it is in the barrel? (c) Find the impulse exerted on the bullet while it is in the barrel. (d) Find the bullet's momentum as it leaves the barrel.

18. A 40.0-grain bullet is fired at a muzzle velocity of 1125 ft/s from a gun weighing 9.90 lb and a barrel length of 21.6 in. (1 grain = 64.8 mg) (a) How long is the bullet in the barrel? (b) What is the force on the bullet while it is in the barrel? (c) Find the impulse exerted on the bullet while it is in the barrel. (d) Find the bullet's momentum as it leaves the barrel.

19. (a) What force is required to stop a 1250-kg car traveling 95.0 km/h within 4.00 s? (b) How far does the car travel during its deceleration?

20. (a) What force is required to slow a 1350-kg car traveling 90.0 km/h to 25.0 km/h within 4.00 s? (b) How far does the car travel during its deceleration? (c) How long does it take for the car to come to a complete stop at this same rate of deceleration?

21. What force is required to stop a $300\overline{0}$-kg truck going 35.0 km/h within 5.00 s?

22. What force is needed to stop a piece of heavy equipment moving 10.0 km/h in 6.00 s if its mass is $500\overline{0}$ kg?

23. A standard 5.0-oz baseball is thrown and reaches a batter with a velocity of 85 mi/h when it is struck with a bat that causes it to reverse direction with a velocity of 95 mi/h. Find (a) the impulse and (b) the force exerted on the baseball if the bat is in contact with the ball for 7.0 ms.

PHYSICS ON THE JOB

Forensic Science Technician

Forensic science technicians and criminalists are typically employed by medical examiners, criminal investigation laboratories, and police or FBI offices. They often are engaged in collecting evidence in the field and then processing the information collected in the laboratory environment.

General forensic science technicians must be familiar with techniques used to perform chemical, biological, and physical analysis of evidence. The connection to physics is most often in relation to ballistics, accident investigation, and blood splatter. A strong understanding of force, centripetal force, projectile motion, gravity, viscosity, impulse, and momentum are important concepts that must be understood to be an effective forensic science technician.

Critical thinking and problem-solving skills, in addition to a strong background in physics, chemistry, and biology, are critical qualifications for a criminalist or forensic science technician.

A bachelor's degree in biology, chemistry, or physics is preferred, but some colleges also offer criminalistics or forensic science majors. In addition to science majors, many police officers will return for additional coursework to obtain certification in the field of forensic science (Fig. 6.5).

Figure 6.5   Forensic technician

Courtesy of fotolia © Paul Fleet

6.2 Collisions in One Dimension

The collision of two objects is an excellent example that demonstrates the law of conservation of momentum. Whenever objects collide in the absence of any external forces, the total momentum of the objects *before* the collision equals the total momentum *after* the collision. That is,

total momentum _before collision_ = total momentum _after collision_

We will now study collisions in one dimension (along a straight line) by discussing the two extreme types, perfectly elastic and inelastic.

Elastic Collisions

In an **elastic collision**, two objects collide and return to their original shape without being permanently deformed. This happens when two billiard balls collide.

One ball of mass 0.600 kg traveling 9.00 m/s to the right collides with a second ball of mass 0.300 kg traveling 8.00 m/s to the left. After the collision, the heavier ball is traveling 2.33 m/s to the left. What is the velocity of the lighter ball after the collision?

EXAMPLE 1

Sketch:

(a) Before impact   (b) After impact

Data:

$$m_1 = 0.600 \text{ kg}$$
$$m_2 = 0.300 \text{ kg}$$
$$v_1 = 9.00 \text{ m/s} \qquad \text{(Right is the positive direction.)}$$
$$v_2 = -8.00 \text{ m/s} \qquad \text{(Left is the negative direction.)}$$
$$v_1' = -2.33 \text{ m/s}$$
$$v_2' = ?$$

Basic Equation:

$$m_1 v_1 + m_2 v_2 = m_1 v_1' + m_2 v_2'$$

Working Equation:

$$v_2' = \frac{m_1 v_1 + m_2 v_2 - m_1 v_1'}{m_2}$$

Substitution:

$$v_2' = \frac{(0.600 \text{ kg})(9.00 \text{ m/s}) + (0.300 \text{ kg})(-8.00 \text{ m/s}) - (0.600 \text{ kg})(-2.33 \text{ m/s})}{0.300 \text{ kg}}$$

$$= 14.66 \text{ m/s} \qquad \text{(to the right)}$$

Inelastic Collisions

In an **inelastic collision**, two objects collide and couple together. This happens when two railroad cars collide and couple together and move along the tracks.

EXAMPLE 2

A 1.75×10^4-kg railroad car traveling 8.00 m/s to the east collides and couples with a stopped 2.25×10^4-kg railroad car. What is the velocity of the joined railroad cars after the collision?

Sketch:

8.00 m/s   0 m/s $v' = ?$

1.75×10^4 kg 2.25×10^4 kg 1.75×10^4 kg 2.25×10^4 kg

(a) Before impact   (b) After impact

Data:

$$m_1 = 1.75 \times 10^4 \text{ kg}$$
$$m_2 = 2.25 \times 10^4 \text{ kg}$$
$$v_1 = 8.00 \text{ m/s}$$
$$v_2 = 0 \text{ m/s}$$
$$v' = ?$$

Basic Equation:

$$m_1 v_1 + m_2 v_2 = (m_1 + m_2)v'$$

Working Equation:

$$v' = \frac{m_1 v_1 + m_2 v_2}{m_1 + m_2}$$

Substitution:

$$v' = \frac{(1.75 \times 10^4 \text{ kg})(8.00 \text{ m/s}) + (2.25 \times 10^4 \text{ kg})(0 \text{ m/s})}{1.75 \times 10^4 \text{ kg} + 2.25 \times 10^4 \text{ kg}}$$

$$= 3.50 \text{ m/s (east)}$$

SKETCH

12 cm² | w

4.0 cm

DATA
$A = 12$ cm², $l = 4.0$ cm, $w = ?$

BASIC EQUATION
$A = lw$

WORKING EQUATION
$w = \dfrac{A}{l}$

SUBSTITUTION
$w = \dfrac{12 \text{ cm}^2}{4.0 \text{ cm}} = 3.0$ cm

PROBLEMS 6.2

Assume a frictionless surface in each of the following problems.

1. One ball of mass 0.500 kg traveling 6.00 m/s to the right collides with a ball of mass 0.200 kg initially at rest. After the collision, the heavier ball is traveling 2.57 m/s to the right. What is the velocity of the lighter ball after the collision?
2. A ball of mass 625 g traveling 4.00 m/s to the right collides with another ball of equal mass that is initially at rest. The first ball is at rest after the collision. (a) What is the velocity of the second ball after the collision? (b) What is the velocity of the second ball if the mass of the two balls of equal mass changes?

TRY THIS ACTIVITY

Air Hockey Physics

Physics laboratory experiments often use air tracks or low-friction carts to study the conservation of momentum throughout collisions. Air hockey tables are typically more accessible for students and provide the low-friction surface that is needed to study the conservation of momentum. If you have access to an air hockey table, try the following collisions:

For elastic collisions, make sure that the pucks strike one another "head on" by aiming one puck directly toward the center of a stationary puck (Fig. 6.6). Make observations about the velocity of both pucks before and after the collision.

Figure 6.6 For collisions in one dimension, aim the puck directly at the center of the stationary puck, as indicated by the blue velocity vector in the photo.

To observe an elastic collision for objects with different masses, place two pucks on top of one another with some double-sided tape between them so they remain secure to one another. Repeat the previous experiment by aiming a single puck at the doubled pucks and note the changes in velocity for both sets of pucks before and after the collision.

For inelastic collisions, wrap some double-sided tape around the sides of each of two pucks so they can stick together when they collide. Again, note the velocity of both pucks before and after the collision.

3. A 0.600-kg ball traveling 4.00 m/s to the right collides with a 1.00-kg ball traveling 5.00 m/s to the left. After the collision, the lighter ball is traveling 7.25 m/s to the left. What is the velocity of the heavier ball after the collision?

4. A 90.0-g disk traveling 3.00 m/s to the right collides with a 75.0-g disk traveling 8.00 m/s to the left. After the collision, the heavier disk is traveling 7.00 m/s to the left. What is the velocity of the lighter disk after the collision?

5. A 98.0-kg parts cart with rubber bumpers rolling 1.20 m/s to the right crashes into a similar cart of mass 125 kg moving left at 0.750 m/s. After the collision, the lighter cart is traveling 0.986 m/s to the left. What is the velocity of the heavier cart after the collision?

6. A 75.0-kg paint cart with rubber bumpers is rolling 0.965 m/s to the right and strikes a second cart of mass 85.0 kg moving 1.30 m/s to the left. After the collision, the heavier cart is traveling 0.823 m/s to the right. What is the velocity of the lighter cart after the collision?

7. A railroad car of mass 2.00×10^4 kg is traveling north 6.00 m/s and collides with a railroad car of mass 1.50×10^4 kg traveling south 4.00 m/s. Find the velocity of the railroad cars that become coupled after the collision.

8. Find the velocity of the railroad cars in Problem 7 if the lighter car is initially at rest.

9. One cart of mass 12.0 kg is moving 6.00 m/s to the right on a frictionless track and collides with a cart of mass 4.00 kg moving in the opposite direction 3.00 m/s. Find the final velocity of the carts that become stuck together after the collision.

10. One cart of mass 15.0 kg is moving 5.00 m/s to the right on a frictionless track and collides with a cart of mass 3.00 kg. The final velocity of the carts that become stuck together after the collision is 1.50 m/s to the right. Find the velocity of the second cart before the collision.

11. A 1650-kg automobile moving south 12.0 m/s collides with a 2450-kg automobile moving north on an icy road. The automobiles stick together and move 3.00 m/s

to the north after the collision. What is the speed of the heavier automobile before the collision?

12. A 16.0-g bullet is shot into a wooden block at rest with mass 4550 g on a frictionless surface. The block moves 1.20 m/s after the bullet strikes and becomes lodged in the block. Find the speed of the bullet before striking the block.

13. A 2450-kg automobile moving north 12.0 m/s collides with a 1650-kg automobile moving 8.00 m/s on an icy road. The automobiles stick together and move after the collision. Find the velocity of the automobiles after the collision if the automobiles were traveling in (a) opposite directions and (b) the same direction before the collision.

6.3　Collisions of Equal Masses in Two Dimensions

If the two colliding objects do not join together to form a single mass but ricochet off each other, the collision is elastic. *The same principles of conservation of momentum apply to elastic collisions in two dimensions; that is, the total momentum in the system must be the same before and after the collision.* The following example shows how such problems can be solved using the above principles.

EXAMPLE 1

A 2.00-kg mass A is moving in an easterly direction at a velocity of 5.00 m/s. It crashes into a stationary mass B, which also has a mass of 2.00 kg. Mass A is deflected 30.0° north of its original path, and mass B is thrust in a direction 90.0° to the right of the final path of mass A. (a) What is the momentum of mass A after the collision? (b) What is the momentum of mass B after the collision? (c) What is the velocity of mass A after the collision? (d) What is the velocity of mass B after the collision?

Sketch:

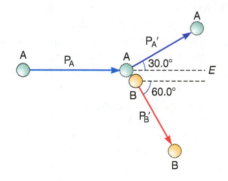

Data:

$$m_A = 2.00 \text{ kg}$$
$$v_A = 5.00 \text{ m/s}$$
$$m_B = 2.00 \text{ kg}$$
$$v_B = 0.00 \text{ m/s}$$

Basic Equation:

$$p = p'$$

Working Equation:

$$p_A + p_B = p'_A + p'_B$$

Before the collision:

Substitution:

$$p_A = m_A v_A \qquad p_B = 0$$
$$= (2.00 \text{ kg}) (5.00 \text{ m/s})$$
$$= 10.0 \text{ kg m/s}$$

Use vector addition to find p'_A and p'_B after the collision.

Since momentum must be conserved, the vector sum of p'_A and p'_B must equal p_A.

(a) Mass A

$$\cos 30.0° = \frac{p'_A}{p_A}$$
$$p'_A = p_A \cos 30.0°$$
$$= (10.0 \text{ kg m/s})(\cos 30.0°)$$
$$= 8.66 \text{ kg m/s}$$

(b) Mass B

$$\sin 30.0° = \frac{p'_B}{p_A}$$
$$p'_B = p_A \sin 30.0°$$
$$= (10.0 \text{ kg m/s})(\sin 30.0°)$$
$$= 5.00 \text{ kg m/s}$$

(c) Also, $p = mv$
so $p'_A = m_A v'_A$
$$v'_A = \frac{p'_A}{m_A}$$
$$= \frac{8.66 \text{ kg m/s}}{2.00 \text{ kg}}$$
$$= 4.33 \text{ m/s}$$

(d) Also, $p = mv$,
so $p'_B = m_B v'_B$
$$v'_B = \frac{p'_B}{m_B}$$
$$= \frac{5.00 \text{ kg m/s}}{2.00 \text{ kg}}$$
$$= 2.50 \text{ m/s}$$

The law of conservation of momentum holds for all isolated, closed systems, regardless of the directions of the masses before and after the collision.

What appears to be a special case in Example 1 when the angle of deflection between the masses is 90° is actually verifiable experimentally. When the masses are equal and one is at rest before the collision, the angle of deflection between them is always 90°. The concept of conservation of momentum applies not only in games like billiards but also in nuclear physics in collisions between identical subatomic particles.

Example 1 also shows that since momentum is a vector quantity, momentum along the x-axis p_x must be conserved, so p_{A_x} (before the collision) equals $p'_A \cos 30.0° + p'_B \cos 60.0°$ (after the collision). Similarly, momentum along the y-axis is also conserved, so p_y (zero in the example before the collision) equals $p'_A \sin 30.0° + p'_B \sin 60.0°$ (after the collision). The sum of the x-components before the collision must equal the sum of the x-components after the collision and the sum of the y-components before the collision must equal the sum of the y-components after the collision.

Collisions that demonstrate both elastic and inelastic characteristics as well as collisions in three-dimensional space will not be discussed in this text.

TRY THIS ACTIVITY

Two-Dimensional Air Hockey Physics

The advantage of using an air hockey table to model the conservation of momentum is that it can easily be used to create two-dimensional collisions (Fig. 6.7). Repeat the previous Try This Activity with elastic and inelastic collisions, but now make sure the moving puck is aimed off center from the stationary puck. Carefully note the angle between the objects after they collide (the holes in the table can be used as guidelines for measuring the angle).

Figure 6.7   For collisions in two dimensions, aim the puck off center of the stationary puck, as indicated by the blue velocity vector in the photo.

To add a level of complexity, create a two-dimensional collision with pucks of two different masses. Finally, attempt an inelastic collision at an angle. Note what happens. Get creative with the collisions, and make predictions and observe what actually happens and how it reinforces what you know about the conservation of momentum.

PHYSICS CONNECTION

Football Positions

The mass of football players has a significant role in the position they play on the football field. The function of offensive linemen is to protect the quarterback as defensive linemen attempt to cross the line of scrimmage (Fig. 6.8) and tackle the quarterback before a pass or a handoff can be made. Linemen make their way through the opposing team's line by using their momentum. Since linemen are positioned at the line of scrimmage, they don't have the opportunity to gain a lot of speed; as a result, players take advantage of their large mass to increase their momentum. According to the National Football League, the average mass of a lineman is 135 kg (approximately 295 lb).

On the other hand, the function of a running back is to carry the football as far into the defensive territory as possible while avoiding tackles from the opposing team. Thus, a running back tends not to be as massive as a lineman and therefore can change velocity, and thus momentum, more easily than can a lineman attempting to run at such high speeds. According to the National Football League, the average mass of a running back is 110 kg (approximately 240 lb).

Figure 6.8   Players on the line of scrimmage are typically more massive than those at other positions.

Courtesy of fotolia © Brocreative

PROBLEMS 6.3

1. Two motorcycles of equal mass collide at a 90° intersection. If the momentum of motorcycle A is 450 kg km/h west and the momentum of motorcycle B is 725 kg km/h south, what is the magnitude of the resulting momentum of the final mass?
2. Two pickup trucks crash at a 90° intersection. If the momentum of pickup A is 4.60×10^4 kg km/h north and the momentum of pickup B is 6.25×10^4 kg km/h east, what is the magnitude of the resulting momentum of the final mass?

3. Two vehicles collide at a 90° intersection. If the momentum of vehicle A is 6.10×10^5 kg km/h south and the momentum of vehicle B is 7.20×10^5 kg km/h east, what is the magnitude of the resulting momentum of the final mass?

4. Two vehicles of equal mass collide at a 90° intersection. If the momentum of vehicle A is 1.20×10^5 kg km/h east and the momentum of vehicle B is 8.50×10^4 kg km/h north, what is the resulting momentum of the final mass?

5. A vehicle with a mass of $10\overline{0}0$ kg is going east at a velocity of 30.0 m/s. It collides with a stationary vehicle of the same mass and is deflected 35.0° north of its original course. The second vehicle's final path is 90° to the right of the final path of the first vehicle. (a) What is the momentum of the first vehicle after the collision? (b) What is the momentum of the second vehicle after the collision? (c) What is the velocity of the first vehicle after the collision? (d) What is the velocity of the second vehicle after the collision?

6. Ball A with a mass of 0.500 kg is moving east at a velocity of 0.800 m/s. It strikes ball B, also of mass 0.500 kg, which is stationary. Ball A glances off B at an angle of 40.0° north of its original path. Ball B is pushed along a path perpendicular to the final path of ball A. (a) What is the momentum of ball A after the collision? (b) What is the momentum of ball B after the collision? (c) What is the velocity of ball A after the collision? (d) What is the velocity of ball B after the collision?

7. A vehicle with mass of $95\overline{0}$ kg is driving east with velocity 12.0 m/s. It crashes into a stationary vehicle of the same mass. Assume an elastic collision. The first vehicle is deflected at an angle of 40.0° north of its original path. The second vehicle's path is 90° to the right of the first vehicle's final path. (a) What is the momentum of the first vehicle after the crash? (b) What is the momentum of the second vehicle after the crash? (c) What is the velocity of the first vehicle after the crash? (d) What is the velocity of the second vehicle after the crash?

8. A vehicle with a mass of $80\overline{0}$ kg is traveling west with a velocity of 20.0 m/s. It collides with a vehicle of the same mass that is not moving. Assume an elastic collision. the first vehicle is deflected at 30.0° north from its original path. The second vehicle's path is 90° to the left of the first vehicle's final path. (a) What is the momentum of the first vehicle after the crash? (b) What is momentum of the second vehicle after the crash? (c) What is the velocity of the first vehicle after the crash? (d) What is the velocity of the second vehicle after the crash?

Glossary

Elastic Collision A collision in which two objects return to their original shape without being permanently deformed. (p. 155)

Impulse The product of the force exerted and the time interval during which the force acts on the object. Impulse equals the change in momentum of an object in response to the exerted force. (p. 146)

Impulse–Momentum Theorem If the mass of an object is constant, then a change in its velocity results in a change of its momentum. That is,
$F\Delta t = \Delta p = m\Delta v = mv_f - mv_i$ (p. 150)

Inelastic Collision A collision in which two objects couple together. (p. 156)

Law of Conservation of Momentum When no outside forces are acting on a system of moving objects, the total momentum of the system remains constant. (p. 149)

Momentum A measure of the amount of inertia and motion an object has or the difficulty in bringing a moving object to rest. Momentum equals the mass times the velocity of an object. (p. 144)

Formulas

6.1 $p = mv$
impulse $= Ft$
impulse = change in momentum
impulse − momentum theorem

$$F\Delta t = \Delta p = m\Delta v = mv_f - mv_i$$

6.2   total momentum$_{\text{before collision}}$ = total momentum$_{\text{after collision}}$

Review Questions

1. Momentum is
   (a) equal to speed times weight.   (b) equal to mass times velocity.
   (c) the same as force.
2. Impulse is
   (a) a force applied to an object.   (b) the initial force applied to an object.
   (c) the initial momentum applied to an object.
   (d) the change in momentum due to a force being applied to an object during a given time.
3. Why do a slow-moving loaded truck and a speeding rifle bullet each have a large momentum?
4. How are impulse and change in momentum related?
5. Why is "follow-through" important in hitting a baseball or a golf ball?
6. Describe in your own words the law of conservation of momentum.
7. Describe conservation of momentum in terms of a rocket being fired.
8. One billiard ball striking another is an example of a(n) _____ collision.
9. One moving loaded railroad car striking and coupling with a parked empty railroad car and then both moving on down the track is an example of a(n) _____ collision.
10. A father and 8-year-old son are standing on ice skates in an ice arena. The father then pushes the son on the back to give him a quick start. What do we know about the momentum of each person?

Review Problems

1. A truck with mass 1475 slugs travels 57.0 mi/h. Find its momentum.
2. A projectile with mass 27.0 kg is fired with a momentum of 5.50 kg m/s. Find its velocity.
3. A box is pushed with a force of 125 N for 2.00 min. What is the impulse?
4. What is the momentum of a bullet of mass 0.034 kg traveling at 250 m/s?
5. A 4.00-g bullet is fired from a 4.50-kg gun with a muzzle velocity of 625 m/s. What is the speed of the recoil of the gun?
6. A 150-kg pile driver falls from a height of 7.5 m to hit a piling. (a) What is its speed as it hits the piling? (b) With what momentum does it hit the piling?
7. A 15.0-g bullet is fired at a muzzle velocity of 3250 m/s from a high-powered rifle with a mass of 4.75 kg and barrel of length 75.0 cm. (a) How long is the bullet in the barrel? (b) What is the force on the bullet while it is in the barrel? (c) Find the impulse exerted on the bullet while it is in the barrel. (d) Find the bullet's momentum as it leaves the barrel.

8. What force is required to slow a 1250-kg car traveling 115 km/h to 30.0 km/h within 3.50 s? (a) How far does the car travel during its deceleration? (b) How long does it take for the car to come to a complete stop at this same rate of deceleration?

9. One ball of mass 575 g traveling 3.50 m/s to the right collides with another ball of mass 425 g that is initially at rest. After the collision, the lighter ball is traveling 4.03 m/s. What is the velocity of the heavier ball after the collision?

10. A railroad car of mass 2.25×10^4 kg is traveling east 5.50 m/s and collides with a railroad car of mass 3.00×10^4 kg traveling west 1.50 m/s. Find the velocity of the railroad cars that become coupled after the collision.

11. A 195-g ball traveling 4.50 m/s to the right collides with a 125-g ball traveling 12.0 m/s to the left. After the collision, the heavier ball is traveling 8.40 m/s to the left. What is the velocity of the lighter ball after the collision?

12. Two trucks of equal mass collide at a 90° intersection. If the momentum of truck A is 9.50×10^4 kg km/h east and the momentum of truck B is 1.05×10^5 kg km/h north, what is the resulting momentum of the final mass, assuming the trucks remain joined together following the crash?

13. Ball A, of mass 0.35 kg, has a velocity 0.75 m/s east. It strikes a stationary ball, also of mass 0.35 kg. Ball A deflects off B at an angle of 37.0° north of A's original path. Ball B moves in a line 90° right of the final path of A. (a) Find ball A's momentum after the collision. (b) Find ball B's momentum after the collision. (c) Find the velocity of A after the collision. (d) Find the velocity of B after the collision.

SKETCH

12 cm² _w_

4.0 cm

DATA

$A = 12$ cm², $l = 4.0$ cm, $w = ?$

BASIC EQUATION

$A = lw$

WORKING EQUATION

$w = \dfrac{A}{l}$

SUBSTITUTION

$w = \dfrac{12 \text{ cm}^2}{4.0 \text{ cm}} = 3.0 \text{ cm}$

A P P L I E D   C O N C E P T S

1. A coach knows it is vital that the volleyballs be fully inflated before a match. (a) Calculate the impulse on a spiked 0.123-slug volleyball when the incoming velocity of the ball is −11.5 ft/s and the outgoing velocity is 57.3 ft/s. (b) Using physics terms, explain what would happen to the outgoing velocity and the impulse on the ball if it were not fully inflated.

2. An automobile accident causes both the driver and passenger front airbags to deploy. (a) If the vehicle was traveling at a speed of 88.6 km/h and is now at rest, find the change in momentum for both the 68.4-kg adult driver and the 34.2-kg child passenger. (b) The adult took 0.564 s and the child took 0.260 s to come to rest. Find the force that the airbag exerted on each individual. Explain why airbags tend to be dangerous for children.

3. Several African tribes engage in a ritual much like bungee jumping, in which a tree vine is used instead of a bungee cord. (a) A 70.8-kg person falls from a cliff with such a vine attached to his ankles. Find the force applied to his ankles if it takes 0.355 s to change his velocity from −18.5 *m/s* to rest (the negative sign represents the downward direction). (b) Find the force applied to the person's ankles when he takes 1.98 s to change his velocity from −18.5 *m/s* to + 9.75 *m/s* using a manufactured bungee cord (remember that a bungee cord causes the jumper to bounce back upward). (c) Using physics terms, describe why it is safer to use a bungee cord than a tree vine.

4. Sally, who weighs 125 lb, knows that getting out of a 65.5-lb canoe can be a difficult experience. (a) What happens to the canoe's velocity if she attempts to step out of the canoe and onto the dock with a velocity of 3.50 ft/s? (b) If the canoe were heavier, would it be easier or harder to step out of it?

5. An automobile accident investigator needs to determine the initial westerly velocity of a Jeep ($m = 1720$ kg) that may have been speeding before colliding head-on with a Volkswagen ($m = 1510$ kg) that was moving with a velocity of 75.7 km/h east. The speed limit on this road is 90 km/h. After the collision, the Jeep and the Volkswagen stuck together and continued to travel with a velocity of 15.5 km/h west. (a) Find the initial westerly velocity of the Jeep. (b) Was the Jeep speeding?

CHAPTER 7

CONCURRENT AND PARALLEL FORCES

Objectives

The major goals of this chapter are to enable you to:

1. Find the vector sum of concurrent forces.
2. Analyze equilibrium in one dimension.
3. Analyze concurrent force situations using force diagrams.
4. Distinguish compression and tension.

5. Apply the torque equation to rotational problems.
6. Solve parallel force problems.
7. Express the conditions of equilibrium using torque concepts.
8. Use the center of gravity to solve parallel force problems.

Not all forces cause motion. A body that is static, or not moving, may be in equilibrium. Huge forces may be acting on a bridge without producing motion. We will consider concurrent forces (forces acting at the same point) in equilibrium.

When forces act nonconcurrently (that is, not at the same point), they may tend to produce rotational motion. We will consider torque (an applied force that causes a rotation), parallel force problems, equilibrium, and the concept of center of gravity.

7.1 Forces in Two Dimensions

Concurrent forces are those forces that are applied to or act at the same point, as in Fig. 7.1. When two or more forces act at the same point, the **resultant force** is the sum of the forces applied at that point. The resultant force is the single force that has the same effect as two or more forces acting together.

As we saw in Section 5.5, when forces act in the same or opposite directions (in one dimension), the total, or net, force can be found by adding the forces that act in one direction and subtracting the forces that act in the opposite direction. What is the result when the forces are acting in two dimensions? In Chapter 3, addition of two vectors was shown by connecting the end point of the first vector to the initial point of the second vector, as shown in Fig. 7.2. This is often called the *vector triangle method*.

This sum may also be obtained by constructing a parallelogram using the two vectors as adjacent sides and then constructing the opposite sides parallel, as shown in Fig. 7.3. The diagonal of the parallelogram is the resultant, or sum, of the two vectors. This is often called the *parallelogram method*.

Figure 7.1   Concurrent forces are applied to or act at the same point.

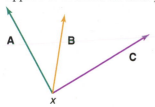

Figure 7.2   Vector triangle method of adding two vectors

Figure 7.3   Parallelogram method of adding two vectors

In Fig. 7.4, two people on opposite banks are pulling a boat up a small river. Note that the effort forces must be equal to keep the boat in the middle of the river, the direction of the equal-effort forces is along the ropes, and the direction of the resultant force **R** is upriver. What would happen if the effort forces were not equal?

Figure 7.4

Two workers move a large crate by applying two ropes at the same point. The first worker applies a force of 525 N while the second worker applies a force of 763 N at the same point at right angles, as shown in Fig. 7.5. Find the resultant force.

EXAMPLE 1

Figure 7.5

To find $\mathbf{F_R}$, find the x- and y-components of each vector and add the components as follows:

Vector	x-component	y-component
$\mathbf{F}_1$	0 N	525 N
$\mathbf{F}_2$	763 N	0 N
$\mathbf{F_R}$	763 N	525 N

Find angle α as follows:

$$\tan \alpha = \frac{\text{side opposite } \alpha}{\text{side adjacent to } \alpha} = \frac{|\mathbf{F}_{Ry}|}{|\mathbf{F}_{Rx}|}$$

$$\tan \alpha = \frac{525 \text{ N}}{763 \text{ N}} \qquad \text{(In a parallelogram, opposite sides are equal.)}$$

$$= 0.6881$$

$$\alpha = 34.5°$$

Find the magnitude of $\mathbf{F_R}$ using the Pythagorean theorem:

$$|\mathbf{F_R}| = \sqrt{|\mathbf{F}_{Rx}|^2 + |\mathbf{F}_{Ry}|^2}$$

$$|\mathbf{F_R}| = \sqrt{(763 \text{ N})^2 + (525 \text{ N})^2}$$

$$= 926 \text{ N}$$

That is, $\mathbf{F_R} = 926$ N at 34.5°.

· · · · · · · · · · · · · · · · ·

EXAMPLE 2

Two workers move a large crate by applying two ropes at the same point. The first worker applies a force of 525 N while the second worker applies a force of 763 N at the same point, as shown in Fig. 7.6. Find the resultant force.

Figure 7.6

To find $\mathbf{F_R}$, find the x- and y-components of each vector and add the components as follows:

Vector	x-component	y-component
$\mathbf{F}_1$	$\begin{aligned}&\lvert\mathbf{F}_1\rvert \cos \alpha =\\ &(525 \text{ N}) \cos 40.0° = 402 \text{ N}\end{aligned}$	$\begin{aligned}&\lvert\mathbf{F}_1\rvert \sin \alpha =\\ &(525 \text{ N}) \sin 40.0° = 337 \text{ N}\end{aligned}$
$\mathbf{F}_2$	763 N	0 N
$\mathbf{F_R}$	1165 N	337 N

Find angle α as follows:

$$\tan \alpha = \frac{|\mathbf{F}_{Ry}|}{|\mathbf{F}_{Rx}|} = \frac{337 \text{ N}}{1165 \text{ N}} = 0.2893$$

$$\alpha = 16.1°$$

Find the magnitude of $\mathbf{F}_R$ using the Pythagorean theorem:

$$|\mathbf{F}_R| = \sqrt{|\mathbf{F}_{Rx}|^2 + |\mathbf{F}_{Ry}|^2}$$

$$|\mathbf{F}_R| = \sqrt{(1165 \text{ N})^2 + (337 \text{ N})^2}$$

$$= 1210 \text{ N}$$

That is, $\mathbf{F}_R = 1210$ N at $16.1°$.

· · · · · · · · · · · · · · · · ·

EXAMPLE 3

Forces of $\mathbf{F}_1 = 375$ N, $\mathbf{F}_2 = 575$ N, and $\mathbf{F}_3 = 975$ N are applied at the same point. The angle between $\mathbf{F}_1$ and $\mathbf{F}_2$ is $60.0°$ and the angle between $\mathbf{F}_2$ and $\mathbf{F}_3$ is $80.0°$. $\mathbf{F}_2$ is between $\mathbf{F}_1$ and $\mathbf{F}_3$. Find the resultant force.

First, draw a force diagram as in Fig. 7.7. Place the point of application at the origin and one of the forces on the x-axis for ease in computing the components.

Figure 7.7

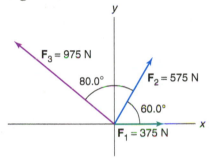

To find $\mathbf{F}_R$, find the x- and y-components of each vector and add the components as follows:

Vector	x-component	y-component
$\mathbf{F}_1$	375 N	0 N
$\mathbf{F}_2$	$\|\mathbf{F}_2\| \cos \alpha =$ $(575 \text{ N}) \cos 60.0° =$ 288 N	$\|\mathbf{F}_2\| \sin \alpha =$ $(575 \text{ N}) \sin 60.0° =$ 498 N
$\mathbf{F}_3$	$\|\mathbf{F}_3\| \cos \alpha =$ $-(975 \text{ N}) \cos 40.0° =$ -747 N	$\|\mathbf{F}_3\| \sin \alpha =$ $(975 \text{ N}) \sin 40.0° =$ 627 N
	Note for $\mathbf{F}_3$: $\alpha = 180° - 140.0° = 40.0°$	
$\mathbf{F}_R$	-84 N	1125 N

Find angle α of the resultant vector as follows:

$$\tan \alpha = \frac{|\mathbf{F}_{Ry}|}{|\mathbf{F}_{Rx}|} = \frac{1125 \text{ N}}{84 \text{ N}} = 13.39$$

$$\alpha = 85.7°$$

Note: The *x*-component of $\mathbf{F_R}$ is negative and its *y*-component is positive; this means that $\mathbf{F_R}$ is in the second quadrant. Its angle in standard position is $180° - 85.7° = 94.3°$.

Find the magnitude of $\mathbf{F_R}$ using the Pythagorean theorem:

$$|\mathbf{F_R}| = \sqrt{|\mathbf{F}_{Rx}|^2 + |\mathbf{F}_{Ry}|^2}$$

$$|\mathbf{F_R}| = \sqrt{(84\ \text{N})^2 + (1125\ \text{N})^2}$$

$$= 1130\ \text{N}$$

That is, $\mathbf{F_R} = 1130$ N at $94.3°$, or $94.3°$ from $\mathbf{F_1}$.
The resultant vector is shown in Fig. 7.8.

Figure 7.8

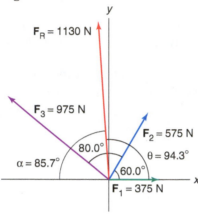

....

TRY THIS ACTIVITY

Tension Forces

An easy way to experiment with concurrent forces in cables is to suspend a 1.0-kg mass from two equal-length strings (the actual length of the strings is not important) (see Fig. 7.9). Attach a spring scale at the end of each string and observe the reading on the scales as they suspend the mass vertically. In this position, each spring scale should have the same reading, which is half the weight of the suspended mass.

Now, slowly separate the two strings from one another, gradually creating a larger and larger angle between them. Use the spring scales to observe the tension force in the strings and use a protractor to determine the angle between the two strings. What happens to the tension force as the angle between the suspending cables increases? What is the resultant upward force from the two strings?

Figure 7.9 Two spring scales suspending a mass by two cables that are separated from the vertical at equal angles

1.0 kg

PHYSICS CONNECTIONS

The Cable-Stayed Bridge

All bridges are designed and constructed according to the needs of the community, the desired aesthetics, the costs, and the geographic and geological conditions around the bridge site. One of the most popular, attractive, and cost-effective designs is the cable-stayed bridge. The physical strength and relatively low cost of the design made the cable-stayed bridge ideal for the mid-length span across the Mississippi River at Alton, Illinois (Fig. 7.10).

Cable-stayed bridges support the roadbed by attaching one end of multiple cables directly to the deck, passing them through a vertical tower, and attaching them to the deck on the opposite side of the tower. Through the use of lighter, stronger materials, engineers are able to avoid the need for the heavy and expensive steel and massive anchorages that are needed to support more traditional suspension bridges.

The combination of compression, tension, shear, and bending forces keeps the cable-stayed New Clark Bridge static. This particular cable-stayed bridge was designed to replace a deteriorating truss bridge that the community had outgrown. The New Clark Bridge meets the needs of the growing community and a busy shipping channel, is aesthetically pleasing, and is economically viable. The bridge also met the geographic and geological conditions as dictated by the Mississippi River and surrounding landscape.

Figure 7.10   The new cable-stayed design of the New Clark Bridge at Alton, Illinois

PROBLEMS 7.1

Find the sum of each set of forces acting at the same point in a straight line.

1. 355 N (right); 475 N (right); 245 N (left); 555 N (left)
2. 703 N (right); 829 N (left); 125 N (left); 484 N (left)
3. Forces of 225 N and 175 N act at the same point. (a) What is the magnitude of the maximum net force the two forces can exert together? (b) What is the magnitude of the minimum net force the two forces can exert together?
4. Three forces with magnitudes of 225 N, 175 N, and 125 N act at the same point. (a) What is the magnitude of the maximum net force the three forces can exert together? (b) What is the magnitude of the minimum net force the three forces can exert together?

Find the sum of each set of vectors. Give angles in standard position.

5.

$F_1 = 1350$ N F_R

$F_2 = 925$ N

6.

$F_2 = 805$ lb

$F_1 = 1150$ lb F_R

SKETCH

12 cm^2 w

4.0 cm

DATA

$A = 12$ cm^2, $l = 4.0$ cm, $w = ?$

BASIC EQUATION

$A = lw$

WORKING EQUATION

$w = \dfrac{A}{l}$

SUBSTITUTION

$w = \dfrac{12 \text{ cm}^2}{4.0 \text{ cm}} = 3.0$ cm

7.   If forces of $10\bar{0}0$ N acting in a northerly direction and $15\bar{0}0$ N acting in an easterly direction both act on the same point, what is the resultant force?

8.   If two forces of $10\bar{0}$ N and 50.0 N, respectively, act in a westerly direction on a point and a force of 175 N acts in a northerly direction on the same point, what is the resultant force?

Find the sum of each set of vectors. Give angles in standard position.

9.

10.

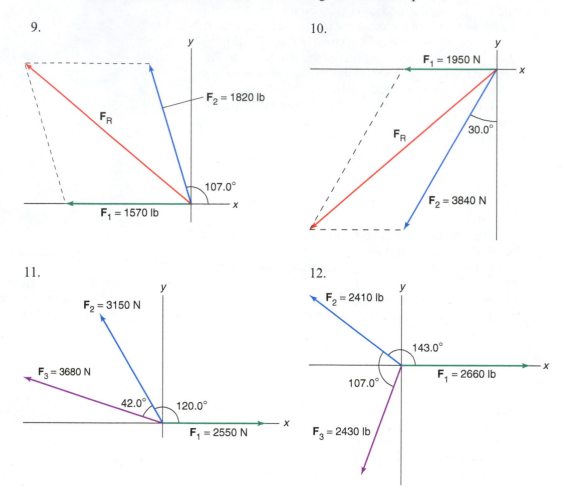

11.

12.

13.   Forces of $F_1 = 1150$ N, $F_2 = 875$ N, and $F_3 = 1450$ N are applied at the same point. The angle between F_1 and F_2 is 90.0° and the angle between F_2 and F_3 is 120.0°. F_2 is between F_1 and F_3. Find the resultant force.

14.   Four forces, each of magnitude 2750 lb, act at the same point. The angle between adjacent forces is 30.0°. Find the resultant force.

7.2   Concurrent Forces in Equilibrium

Equilibrium in One Dimension

Equilibrium *is the state of a body in which there is no change in its motion.* A body is in equilibrium when the net force acting on it is zero. That is, it is not accelerating; it is either at rest or moving at a constant velocity. The study of objects in equilibrium is called **statics.**

The forces applied to an object in one dimension act in the same direction or in opposite directions. For the net force to be zero, the forces in one direction must equal the forces in the opposite direction. We can write the equation for equilibrium in one dimension as

$$F_+ = F_-$$

where F_+ = the sum of all forces acting in one direction (call it the positive direction)
F_- = the sum of all the forces acting in the opposite (negative) direction

Note in Fig. 7.11 that the downward force (weight of the bridge) must equal the sum of the upward forces produced by the two bridge supports for the bridge to be in equilibrium.

Figure 7.11

A cable supports a large crate of weight 1250 N (Fig. 7.12). What is the upward force on the crate if it is in equilibrium?

EXAMPLE 1

Sketch:
Draw a force diagram of the crate in equilibrium, and show the forces that act on it. Note that we call the upward direction positive as indicated by the arrow.

Figure 7.12

S = Supporting force

F_w = 1250 N

Data:

$$F_w = 1250 \text{ N}$$
$$S = ?$$

Basic Equation:

$$F_+ = F_-$$

Working Equation:

$$S = F_w$$

Substitution:

$$S = 1250 \text{ N}$$

EXAMPLE 2

Four persons are having a tug-of-war with a rope. Harry and Mary are on the left; Bill and Jill are on the right. Mary pulls with a force of 105 lb, Harry pulls with a force of 255 lb, and Jill pulls with a force of 165 lb. With what force must Bill pull to produce equilibrium?

Sketch:

H: 255 lb   M: 105 lb   J: 165 lb   B: ?

Data:

$$M = 105 \text{ lb}$$
$$H = 255 \text{ lb}$$
$$J = 165 \text{ lb}$$
$$B = \text{?}$$

Basic Equation:

$$\mathbf{F}_+ = \mathbf{F}_- \quad \text{or}$$
$$M + H = J + B$$

Working Equation:

$$B = M + H - J$$

Substitution:

$$B = 105 \text{ lb} + 255 \text{ lb} - 165 \text{ lb}$$
$$= 195 \text{ lb}$$

Equilibrium in Two Dimensions

A body is in equilibrium when it is either at rest or moving at a constant speed in a straight line. Figure 7.13(a) shows the resultant force of the sum of two forces from Example 1 in Section 7.1. When two or more forces act at a point, the **equilibrant force** is the force that, when applied at that same point as the resultant force, produces equilibrium. *The equilibrant force is equal in magnitude to that of the resultant force, but it acts in the opposite direction* [see Fig. 7.13(b)]. In this case, the equilibrant force is 926 N at 214.5° (180° + 34.5°).

Figure 7.13

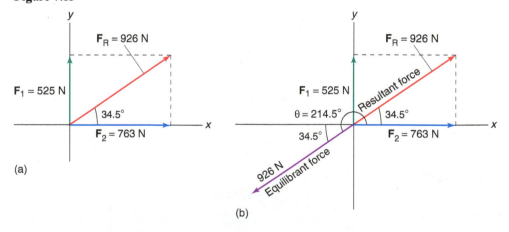

(a)

(b)

If an object is in equilibrium in two dimensions, the net force acting on it must be zero. For the net force to be zero, the sum of the *x*-components must be zero and the sum of the *y*-components must be zero. For forces **A**, **B**, and **C** with *x*-components A_x, B_x, and C_x, respectively, and with *y*-components A_y, B_y, and C_y, respectively, to be in equilibrium, both of the following conditions must hold:

CONDITIONS FOR EQUILIBRIUM
1. The sum of *x*-components = 0; that is, $A_x + B_x + C_x = 0$.
2. The sum of *y*-components = 0; that is, $A_y + B_y + C_y = 0$.

In general, to solve equilibrium problems:

1. Draw a force diagram from the point at which the unknown forces act.
2. Find the *x*- and *y*-components of each force.
3. Substitute the components in the equations

$$\text{sum of } x\text{-components} = 0$$
$$\text{sum of } y\text{-components} = 0$$

4. Solve for the unknowns. This may involve two simultaneous equations.

We may need to find the tension or compression in part of a structure, such as in a beam or a cable. **Tension** is a stretching force produced by forces pulling outward on the ends of an object [Fig. 7.14(a)]. **Compression** is a force produced by forces pushing inward on the ends of an object [Fig. 7.14(b)]. A rubber band being stretched is an example of tension [Fig. 7.15(a)]. A valve spring whose ends are pushed together is an example of compression [Fig. 7.15(b)].

Figure 7.14   Tension and compression forces

(a)

Tension

(b)

Compression

Figure 7.15

(a) Tension in a rubber band

(b) Compression in a valve spring

Find the forces **F** and **F′** necessary to produce equilibrium in the force diagram shown in Fig. 7.16.

EXAMPLE 3

1. **Figure 7.16**

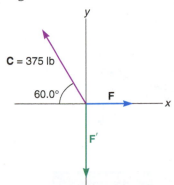

2. | *x-components* | *y-components* |
    |---|---|
    | $\mathbf{F}_x = \mathbf{F}$ | $\mathbf{F}_y = 0$ |
    | $\mathbf{F}'_x = 0$ | $\mathbf{F}'_y = -\mathbf{F}'$ |
    | $\mathbf{C}_x = -(375\text{ lb})(\cos 60.0°)$ | $\mathbf{C}_y = (375\text{ lb})(\sin 60.0°)$ |
    | $= -188\text{ lb}$ | $= 325\text{ lb}$ |

3. Sum of *x*-components = 0   Sum of *y*-components = 0

    $\mathbf{F} + 0 + (-188\text{ lb}) = 0$ $0 + (-\mathbf{F}') + 325\text{ lb} = 0$

4. $\mathbf{F} = 188\text{ lb}$ $\mathbf{F}' = 325\text{ lb}$

· · · · · · · · · · · · · · · · · ·

EXAMPLE 4

Find the forces **F** and **F**′ necessary to produce equilibrium in the force diagram shown in Fig. 7.17.

1. **Figure 7.17**

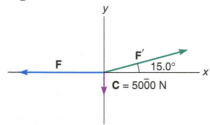

2. | *x-components* | *y-components* |
    |---|---|
    | $\mathbf{F}_x = -\mathbf{F}$ | $\mathbf{F}_y = 0$ |
    | $\mathbf{F}'_x = \mathbf{F}'\cos 15.0°$ | $\mathbf{F}'_y = \mathbf{F}'\sin 15.0°$ |
    | $\mathbf{C}_x = 0$ | $\mathbf{C}_y = -50\overline{0}0\text{ N}$ |

3.   Sum of *x*-components = 0   Sum of *y*-components = 0

    $(-\mathbf{F}) + \mathbf{F}'\cos 15.0° + 0 = 0$ $0 + \mathbf{F}'\sin 15.0° + (-50\overline{0}0\text{ N}) = 0$

4. *Note:* Solve for **F** in the right-hand equation first. Then substitute this value in the left-hand equation to solve for **F**:

$$\mathbf{F}' = \frac{50\overline{0}0\text{ N}}{\sin 15.0°}$$

$$= 19{,}300\text{ N}$$

$$\mathbf{F} = \mathbf{F}'\cos 15.0°$$
$$= (19{,}300\text{ N})(\cos 15.0°)$$
$$= 18{,}600\text{ N}$$

· · · · · · · · · · · · · · · · · ·

EXAMPLE 5

The crane shown in Fig. 7.18 is supporting a beam that weighs $60\overline{0}0$ N. Find the tension in the horizontal supporting cable and the compression in the boom. The horizontal cable is attached to the boom at point *A*. The separate vertical cable holding the beam is attached through the pulley at point *A*.

Figure 7.18

1. Draw the force diagram showing the forces acting at point A.

T = tension

R = compression

E = equilibrant force

F_w = weight

T is the force exerted at A by the horizontal supporting cable.
E is the force exerted by the boom at A.
F_w is the force (weight of the beam) pulling straight down at A.
R is the sum of forces F_w and **T**, which is equal in magnitude but opposite in direction to force **E** ($\mathbf{R} = -\mathbf{E}$).

2. *x-components*   *y-components*

$E_x = E \cos 20.0°$ $E_y = E \sin 20.0°$

$T_x = -T$ $T_y = 0$

$F_{wx} = 0$ $F_{wy} = -60\overline{0}0$ N

3. Sum of *x*-components = 0   Sum of *y*-components = 0

$E \cos 20.0° + (-T) = 0$ $E \sin 20.0° + (-60\overline{0}0 \text{ N}) = 0$

4. $T = E \cos 20.0°$

$$E = \frac{60\overline{0}0 \text{ N}}{\sin 20.0°}$$

$$= 17,500 \text{ N}$$

$T = (17,500 \text{ N})(\cos 20.0°)$

$= 16,400 \text{ N}$

················

A homeowner pushes a 40.0-lb lawn mower at a constant velocity (Fig. 7.19). The frictional force on the mower is 20.0 lb. What force must the person exert on the handle, which makes an angle of 30.0° with the ground? Also, find the normal (perpendicular to ground) force.

This is an equilibrium problem because the mower is not accelerating and the net force is zero.

EXAMPLE 6

Figure 7.19

1. Draw the force diagram.

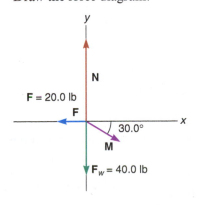

M is the force exerted on the mower by the person; this compression force is directed down along the handle.
F_w is the weight of the mower directed straight down.

N is the force exerted upward on the mower by the ground, which keeps the mower from falling through the ground.

F is the frictional force that opposes the motion.

2. *x-components*   *y-components*

    $\mathbf{N}_x = 0$ $\mathbf{N}_y = \mathbf{N}$

    $\mathbf{F}_{wx} = 0$ $\mathbf{F}_{wy} = -40.0 \text{ lb}$

    $\mathbf{F}_x = -20.0 \text{ lb}$ $\mathbf{F}_y = 0$

    $\mathbf{M}_x = \mathbf{M} \cos 30.0°$ $\mathbf{M}_y = -\mathbf{M} \sin 30.0°$

3.   Sum of *x*-components = 0   Sum of *y*-components = 0

    $0 + 0 + (-20.0 \text{ lb}) + \mathbf{M} \cos 30.0° = 0$ $\mathbf{N} + (-40.0 \text{ lb}) + 0$

    $\qquad\qquad\qquad + (-\mathbf{M} \sin 30.0°) = 0$

    $\qquad\qquad\qquad \mathbf{N} = \mathbf{M} \sin 30.0 + 40.0 \text{ lb}$

4. $\mathbf{M} = \dfrac{20.0 \text{ lb}}{\cos 30.0°}$

    $\quad = 23.1 \text{ lb}$

    $\mathbf{N} = (23.1 \text{ lb})(\sin 30.0°) + 40.0 \text{ lb}$

    $\quad = 51.6 \text{ lb}$

- - - - - - - - - - - - - - - - - -

Note: This is an example of a body in equilibrium moving at a constant speed in a straight line. This type of equilibrium is often referred to as *dynamic* equilibrium, whereas a body in equilibrium at rest is referred to as *static* equilibrium.

EXAMPLE 7

The crane shown in Fig. 7.20 is supporting a beam that weighs $60\overline{0}0$ N. Find the tension in the supporting cable and the compression in the boom.

1. Draw the force diagram showing the forces acting at point *A*.

Figure 7.20

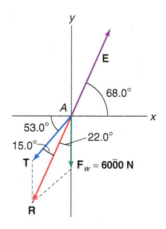

$\mathbf{F}_w$ is the weight of the beam, which pulls straight down.

T is the force exerted at *A* by the supporting cable.

E is the force exerted by the boom at *A*.

R is the sum of forces $\mathbf{F}_w$ and **T**, which is equal in magnitude but opposite in direction to force **E** ($\mathbf{R} = -\mathbf{E}$).

2. *x-components*   *y-components*

    $\mathbf{E}_x = \mathbf{E} \cos 68.0°$ $\mathbf{E}_y = \mathbf{E} \sin 68.0°$

    $\mathbf{T}_x = -\mathbf{T} \cos 53.0°$ $\mathbf{T}_y = -\mathbf{T} \sin 53.0°$

    $\mathbf{F}_{wx} = 0$ $\mathbf{F}_{wy} = -60\overline{0}0 \text{ N}$

3. Sum of x-components = 0   Sum of y-components = 0

    E cos 68.0° +   E sin 68.0° +

        (−T cos 53.0°) + 0 = 0   (−T sin 53.0°) + (−6000 N) = 0

4. *Note:* Solve the left equation for **E**. Then substitute this quantity in the right equation and solve for **T**:

$$E = \frac{T \cos 53.0°}{\cos 68.0°}$$

$$\left(\frac{T \cos 53.0°}{\cos 68.0°}\right)(\sin 68.0°) - T \sin 53.0° = 60\overline{0}0 \text{ N}$$

$$1.490T - 0.799T = 60\overline{0}0 \text{ N}$$

$$0.691T = 60\overline{0}0 \text{ N}$$

$$T = \frac{60\overline{0}0 \text{ N}}{0.691}$$

$$= 8680 \text{ N}$$

$$E = \frac{(8680 \text{ N})(\cos 53.0°)}{\cos 68.0°}$$

$$= 13{,}900 \text{ N}$$

Alternate Method: You can orient a force diagram any way you want on the x–y axes. You should orient it so that as many of the vectors as possible are on an x- or a y-axis. The result will be the same. Let's rework Example 7 as follows:

1. Draw the force diagram showing the forces acting at point A using the same notation as follows:

2. *x-components*   *y-components*

    $E_x = E$ $E_y = 0$

    $T_x = -T \cos 15.0°$ $T_y = T \sin 15.0°$

    $F_{wx} = -(60\overline{0}0 \text{ N})(\cos 22.0°)$ $F_{wy} = -(60\overline{0}0 \text{ N})(\sin 22.0°)$

3. Sum of x-components = 0   Sum of y-components = 0

    $E + (-T \cos 15.0°) +$ $0 + T \sin 15.0° +$

    $(-60\overline{0}0 \text{ N})(\cos 22.0°) = 0$ $(-60\overline{0}0 \text{ N})(\sin 22.0°) = 0$

4. Solve the right equation for **T** (since it has only one variable). Then solve the left equation for **E** and substitute this quantity:

$$T = \frac{(60\overline{0}0 \text{ N})(\sin 22.0°)}{\sin 15.0°}$$

$$= 8680 \text{ N}$$

$$E = T \cos 15.0° + (60\overline{0}0 \text{ N})(\cos 22.0°)$$

$$E = (8680 \text{ N})(\cos 15.0°) + (60\overline{0}0 \text{ N})(\cos 22.0°)$$

$$= 13{,}900 \text{ N}$$

SKETCH

12 cm² | w

4.0 cm

DATA
$A = 12 \text{ cm}^2, l = 4.0 \text{ cm}, w = ?$

BASIC EQUATION
$A = lw$

WORKING EQUATION
$w = \dfrac{A}{l}$

SUBSTITUTION
$w = \dfrac{12 \text{ cm}^2}{4.0 \text{ cm}} = 3.0 \text{ cm}$

PROBLEMS 7.2

Find the force **F** that will produce equilibrium in each force diagram.

1. 100 N, F = ?

2. F = ?, 100 lb

3. F = ?, 295 N, 184 N, 136 N, 235 N

4. F = ?, 250 N, 250 N, 250 N, 1250 N

5. F = ?, 1250 N, 1750 N, 1050 N, 2100 N, 750 N

6. 3600 lb, $F_1 = ?$ $F_2 = ?$ $F_3 = ?$ $F_4 = ?$ $F_5 = ?$
   $F_1 = F_2 = F_3 = F_4 = F_5$

7. Five persons are having a tug-of-war. Kurt and Brian are on the left; Amy, Barbara, and Joyce are on the right. Amy pulls with a force of 225 N, Barbara pulls with a force of 495 N, Joyce pulls with a force of 455 N, and Kurt pulls with a force of 605 N. With what force must Brian pull to produce equilibrium?

8. A certain wire can support 6450 lb before it breaks. Seven 820-lb weights are suspended from the wire. Can the wire support an eighth weight of 820 lb?

9. The frictional force of a loaded pallet in a warehouse is 385 lb. Can three workers, each exerting a force of 135 lb, push it to the side?

10. A bridge has a weight limit of 7.0 tons. How heavy a load can a 2.5-ton truck carry across?

11. A tractor transmission weighing 260 N and a steering gear box weighing 62.0 N are on a workbench. What upward force must the bench exert to maintain equilibrium?

12. A skid loader lifts a compressor weighing 672 N and a hose weighing 26.0 N. What upward force must the loader exert to maintain equilibrium?

Find the forces F_1 and F_2 that produce equilibrium in each force diagram.

13. F_2, F_1, 45.0°, $C = 100$ N

14. $C = 950$ N, 30.0°, F_1, F_2

15. F_1, 30.0°, $C = 500$ lb, F_2

16. 10.0°, F_2, F_1, $C = 1000$ lb

17.

$\mathbf{C} = 25\bar{0}$ lb
$60.0°$

18.

$\mathbf{C} = 40\bar{0}$ N
$20.0°$

19. A rope is attached to two buildings and supports a $50\bar{0}$-lb sign (Fig. 7.21). Find the tensions in the two ropes $\mathbf{T}_1$ and $\mathbf{T}_2$. (*Hint:* Draw the force diagram of the forces acting at the point labeled *A*.)

20. If the angle between the horizontal and the ropes in Problem 19 is changed to 10.0°, what are the tensions in the two ropes $\mathbf{T}_1$ and $\mathbf{T}_2$?

21. If the angles between the horizontal and the ropes in Problem 19 are changed to 20.0° and 30.0°, find the tension in each rope.

22. Find the tension in the horizontal supporting cable and the compression in the boom of the crane that supports an $89\bar{0}0$-N beam shown in Fig. 7.22.

Figure 7.21

$20.0°$ $20.0°$
T_1 A T_2

Building

$\mathbf{F}_w = 50\bar{0}$ lb

Figure 7.22

A

$90.0°$

$25.0°$

$\mathbf{F}_w = 8900$ N

23. Find the tension in the horizontal supporting cable and the compression in the boom of the crane that supports a $150\bar{0}$-lb beam shown in Fig. 7.23.

Figure 7.23

A

$90.0°$

$30.0°$

$\mathbf{F}_w = 8900$ N

24. An automobile of mass 1550 kg is towed at a steady speed by a cable held at an angle of 18.0° below the horizontal. A frictional force of 345 N opposes the motion of the automobile. Find the tension in the cable.

25. A vehicle that weighs 16,200 N is parked on a 20.0° hill (Fig. 7.24). What braking force is necessary to keep it from rolling? Neglect frictional forces. (*Hint:* When you draw the force diagram, tilt the *x*- and *y*-axes as shown. **B** is the braking force directed up the hill and along the *x*-axis.)

26. Find the tension in the cable and the compression in the support of the sign shown in Fig. 7.25.

Figure 7.24

Figure 7.25

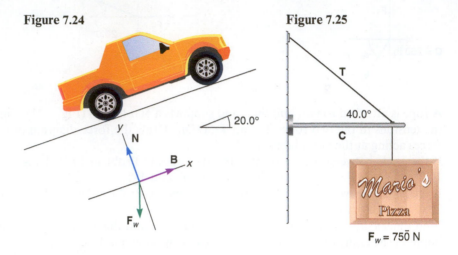

$F_w = 75\bar{0}$ N

27. The crane shown in Fig. 7.26 is supporting a load of 1850 lb. Find the tension in the supporting cable and the compression in the boom.

28. The crane shown in Fig. 7.27 is supporting a load of 11,500 N. Find the tension in the supporting cable and the compression in the boom.

Figure 7.26

Figure 7.27

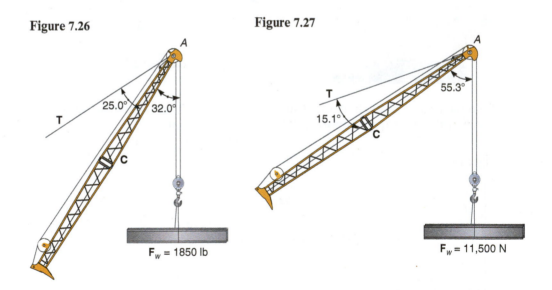

$F_w = 1850$ lb

$F_w = 11,500$ N

N E W   T E C H N O L O G I E S

The Freedom Tower

At a symbolic height of 1776 ft, the majestic Freedom Tower has risen above all other buildings on the New York Skyline to become the tallest building in the United States. Starting from a square base, the tower angularly shifts to an octagon in the mid-section of the tower and gradually transforms back into a square, shifted 45°, at the top of the building.

Just as impressive as its sleek, reflective outer skin are the engineering and technological features that have been incorporated into the structural and safety features of the tower. The steel frame of the foundation was installed in December 2006 and includes twenty-four 18-m-long steel columns that form the perimeter of the foundation. The tower contains 40,000 metric tons of structural steel and 150,000 m^3 of concrete!

Enhanced safety features include reinforced concrete elevator shafts, multiple extra-wide staircases to improve egress, and the introduction of an emergency-responder-only staircase so officials can climb the tower as others are evacuating. Staircase crossover features were introduced to link staircases in the event of a staircase blockage or congestion. Also, communications, sprinklers, and risers are enclosed in a steel-reinforced, 1-m-thick concrete shaft to protect such critical safety and communications systems from damage.

The 104-story, $4.0 billion Freedom Tower is complete and is designated as an environmentally sustaining LEED Gold facility. The tower reclaims rainwater and has its own power-generation system that incorporates on-site fuel cells for energy use in the tower (Fig. 7.28).

Figure 7.28   The Freedom Tower

7.3 Torque

A *torque* is produced when a force is applied to produce a rotation, such as, for example, when a wrench is used to turn a bolt or a claw hammer is used to pull a nail from wood. **Torque** is the tendency to produce change in rotational motion. The torque developed depends on two factors:

1.   The amount of force applied
2.   How far from the point of rotation the force is applied

Torque is expressed by the equation

$$\tau = Fr$$

where τ = torque (N m or lb ft) (τ is the lowercase Greek letter "tau.")
          F = applied force (N or lb)
          r = length of torque arm (m or ft)

Note that r, the length of the torque arm, is different from s in the equation defining work ($W = Fs$). Recall that s in the work equation is the linear distance over which the force acts.

When you use a wrench to turn a bolt, less effort is used (greater torque is produced) as the distance you place your hand from the bolt increases (Fig. 7.29). Plumbers often use a wrench with a long torque arm to loosen or tighten large bolts and fittings.

Figure 7.29 Even though the same force is used, the torque applied to the bolt increases as the distance from your hand to the bolt increases. In (b), you produce more torque by placing your hand on the end of the wrench handle. In (c), you produce even more torque by using an extender sleeve.

In all torque problems, we are concerned with motion about a point or axis of rotation, as in pedaling a bicycle (Fig. 7.30). In pedaling, we apply a force to the pedal, causing the sprocket to rotate. The torque arm is the *perpendicular* distance from the point of rotation to the applied force [Fig. 7.30(a)]. In torque problems, r is always perpendicular to the force [Fig. 7.30(a)]. Note that r is the distance from the pedal to the axle.

Figure 7.30 Torque produced in pedaling a bicycle

(a)   (b)

Figure 7.31 Maximum torque is produced only when the pedal reaches a position perpendicular to the applied force.

If the force is not exerted tangent to the circle made by the pedal (Fig. 7.31), the length of the torque arm is *not* the length of the pedal arm. The torque arm, r, is measured as the perpendicular distance to the force. Since r is therefore shorter, the product $F \cdot r$ is smaller, and the turning effect, the torque, is less in the pedal position shown in Fig. 7.30. Maximum torque is produced when the pedals are horizontal and the force applied is straight down.

Torque is a vector quantity that acts along the axis of rotation (not along the force) and points in the direction in which a right-handed screw would advance if turned by the torque as in Fig. 7.32(a). The *right-hand rule* is often used to determine the direction of the torque as follows: Grasp the axis of rotation with your right hand so that your fingers circle it in the direction that the torque tends to induce rotation. Your thumb will point in the direction of the torque vector [Fig. 7.32(b)]. Thus, the torque vector in Fig. 7.32(a) is perpendicular to and points out of the page.

Figure 7.32   Torque is a vector quantity that acts along the axis of rotation according to the right-hand rule.

(a)   (b)

A force of 10.0 lb is applied to a bicycle pedal. If the length of the pedal arm is 0.850 ft, what torque is applied to the shaft?

EXAMPLE 1

Sketch:

10.0 lb

0.850 ft

Data:

$$F = 10.0 \text{ lb}$$
$$r = 0.850 \text{ ft}$$
$$\tau = ?$$

Basic Equation:

$$\tau = Fr$$

Working Equation: Same

Substitution:

$$\tau = (10.0 \text{ lb})(0.850 \text{ ft})$$
$$= 8.50 \text{ lb ft}$$

T R Y   T H I S   A C T I V I T Y

Hammers and Screwdrivers

Torque is an essential component of most hand tools. Drive a nail into a piece of wood by holding a hammer near its head. Count the number of hits it takes to drive the nail into the wood. Then, drive a like nail into the wood by holding the hammer near the end of the handle. Count the number of hits it takes. Using physics terminology, explain which handle grip is better for driving nails into wood.

Use two screwdrivers with different diameter handles to screw two similar screws into a board. Which screwdriver is able to apply more torque to a screw? If one screwdriver applies more torque to a screw than the other, why would anyone want to use a screwdriver that cannot exert the maximum torque?

PROBLEMS 7.3

Assume that each force is applied perpendicular to the torque arm.

1. Given: $F = 16.0$ lb
   $r = 6.00$ ft
   $\tau = ?$

2. Given: $F = 10\overline{0}$ N
   $r = 0.420$ m
   $\tau = ?$

3. Given: $\tau = 60.0$ N m
   $F = 30.0$ N
   $r = ?$

4. Given: $\tau = 35.7$ lb ft
   $r = 0.0240$ ft
   $F = ?$

5. Given: $\tau = 65.4$ N m
   $r = 35.0$ cm
   $F = ?$

6. Given: $F = 63\overline{0}$ N
   $r = 74.0$ cm
   $\tau = ?$

7. If the torque on a shaft of radius 2.37 cm is 38.0 N m (Fig. 7.33), what force is applied to the shaft?

8. If a force of 56.2 lb is applied to a torque wrench 1.50 ft long (Fig. 7.34), what torque is indicated by the wrench?

Figure 7.33

Figure 7.34

F = ?

r = 2.37 cm

56.2 lb

1.50 ft

9. A motorcycle head bolt is torqued to 25.0 N m. What length shaft do we need on a wrench to exert a maximum force of 70.0 N?

10. A force of 112 N is applied to a shaft of radius 3.50 cm. What is the torque on the shaft?

11. A torque of 175 lb ft is needed to free a large rusted-on nut. The length of the wrench is 1.10 ft. What force must be applied to free it?

12. A torque wrench reads 14.5 N m. If its length is 25.0 cm, what force is being applied to the handle?

13. The torque on a shaft of radius 3.00 cm is 12.0 N m. What force is being applied to the shaft?

14. An engine bolt is torqued to 30.0 N m. If the length of the wrench is 29.0 cm, what force is applied to the wrench?

15. A mower bolt is torqued to 65.0 N m. If the length of the wrench is 30.0 cm, what force is applied to the wrench?

16. An automobile bolt is torqued to 27.0 N m. If the length of the wrench is 30.0 cm, what force is applied to the wrench?

17. A torque wrench reads 25 lb ft. (a) If its length is 1.0 ft, what force is being applied to the wrench? (b) What is the force if the length is doubled? Explain the results.

18. If 13 N m of torque is applied to a bolt with an applied force of 28 N, what is the length of the wrench?

19. If the torque required to loosen a nut on the wheel of a pickup truck is 40.0 N m, what minimum force must be applied to the end of a wrench 30.0 cm long to loosen the nut?

20. How is the required force to loosen the nut in Problem 19 affected if the length of the wrench is doubled?

21. A truck mechanic must loosen a rusted lug nut. If the torque required to loosen the nut is 60.0 N m, what force must be applied to a 35.0-cm wrench?

22. An agricultural mechanic tries to loosen a nut on a tractor wheel with a wrench that is 32.5 cm long. If the torque required to loosen the nut is 55.0 N m, what force must she apply to the wrench?

7.4 Parallel Forces

A painter stands 2.00 ft from one end of a 6.00-ft plank that is supported at each end by a scaffold [Fig. 7.35(a)]. How much of the painter's weight must each end of the scaffold support? Problems of this kind are often faced in the construction industry, particularly in the design of bridges and buildings. Using some things we learned about torques and equilibrium, we can now solve problems of this type.

Figure 7.35 Parallel forces shown by the example of a painter on a scaffold

(a) (b)

Let's look more closely at the painter problem. The force diagram [Fig. 7.35(b)] shows the forces and distances involved. The arrow pointing down represents the weight of the person, F_w. The arrows pointing up represent the forces exerted by each end of the scaffold in supporting the plank and painter. (For now, we will disregard the weight of the plank.) We have a condition of equilibrium. The plank and painter are not moving. The sum of the forces exerted by the ends of the scaffold is equal to the weight of the painter (Fig. 7.36). Since these forces are vectors and are parallel, we can show that their sum is zero. Using engineering notation, we write

$$\Sigma F = 0$$

where Σ (Greek capital letter sigma) means summation or "the sum of" and F is force, a vector quantity. So ΣF means "the sum of forces," in this case the sum of parallel forces. Thus, the **first condition of equilibrium** may be written:

Figure 7.36 In equilibrium, the sum of the forces is zero.

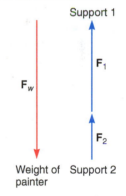

FIRST CONDITION OF EQUILIBRIUM
The sum of all parallel forces on a body in equilibrium must be zero.

If the vector sum is not zero (forces up unequal to forces down), we have an unbalanced force tending to cause motion.

Now consider this situation: One end of the scaffold remains firmly in place, supporting the man, and the other is removed. What happens to the painter? The plank, supported only on one end, falls (Fig. 7.37), and the painter has a mess to clean up!

Figure 7.37 The position of the supporting force is important!

(a)

F_1 F_w

Force diagram

(b)

EXAMPLE 1

A sign of weight $150\overline{0}$ lb is supported by two cables (Fig. 7.38). If one cable has a tension of $60\overline{0}$ lb, what is the tension in the other cable?

Sketch: Draw the force diagram.

Figure 7.38

$T_1 = 60\overline{0}$ lb

T_2

GEM

Weight $= 150\overline{0}$ lb

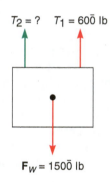

$T_2 = ?$ $T_1 = 60\overline{0}$ lb

$\mathbf{F}_w = 150\overline{0}$ lb

Data:

$$F_w = 150\overline{0} \text{ lb}$$
$$T_1 = 60\overline{0} \text{ lb}$$
$$T_2 = ?$$

Basic Equation:

$$F_+ = F_-$$

Working Equation:

$$T_1 + T_2 = F_w$$
$$T_2 = F_w - T_1$$

Substitution:

$$T_2 = 150\overline{0} \text{ lb} - 60\overline{0} \text{ lb}$$
$$= 90\overline{0} \text{ lb}$$

Not only must the forces balance each other (vector sum = 0), but they must also be positioned so that there is no rotation in the system. To avoid rotation, we can have no unbalanced torques.

Sometimes there will be a natural point of rotation, as in our painter problem. We can, however, choose any point as our center of rotation as we consider the torques present. We will soon see that one of any number of points could be selected. What is necessary, though, is that there be no rotation (no unbalanced torques).

Again, using engineering notation, we write

$$\Sigma \tau_{\text{any point}} = 0$$

where $\Sigma \tau_{\text{any point}}$ is the sum of the torques about any chosen point.

The **second condition of equilibrium** may be written:

SECOND CONDITION OF EQUILIBRIUM

The sum of the clockwise torques on a body in equilibrium must equal the sum of the counterclockwise torques about any point.

$$\Sigma \tau_{\text{clockwise (cw)}} = \Sigma \tau_{\text{counterclockwise (ccw)}}$$

EXAMPLE 2

To illustrate these principles, we will find how much weight each end of the scaffold must support if our painter weighs $15\overline{0}$ lb.

Sketch:

Data:

$$F_w = 15\overline{0} \text{ lb}$$

$$\text{plank} = 6.00 \text{ ft}$$

$$F_w \text{ is } 2.00 \text{ ft from one end}$$

Basic Equations:

1. $$\Sigma F = 0$$

sum of forces = 0

$$F_1 + F_2 - F_w = 0$$

or $F_1 + F_2 = F_w$

$$F_1 + F_2 = 15\overline{0} \text{ lb}$$

(**Note:** F_w is negative because its direction is opposite F_1 and F_2.)

2. $$\Sigma \tau_{\text{clockwise}} = \Sigma \tau_{\text{counterclockwise}}$$

Figure 7.39 Torque arm of painter about point A

Figure 7.40 Torque arm of F_2 about point A

First, select a point of rotation. Choosing an end is usually helpful in simplifying the calculations. Choose the left end (point A) where F_1 acts. What are the clockwise torques about this point?

The force due to the weight of the painter tends to cause clockwise motion. The torque arm is 2.00 ft (Fig. 7.39). Then $\tau = (15\overline{0}\ \text{lb})(2.00\ \text{ft})$. This is the only clockwise torque.

The only counterclockwise torque is F_2 times its torque arm, 6.00 ft (Fig. 7.40). $\tau = (F_2)(6.00\ \text{ft})$. There is no torque involving F_1 because its torque arm is zero. Setting $\Sigma\tau_{\text{clockwise}} = \Sigma\tau_{\text{counterclockwise}}$, we have the equation:

$$(15\overline{0}\ \text{lb})(2.00\ \text{ft}) = (F_2)(6.00\ \text{ft})$$

Note that by selecting an end as the point of rotation, we were able to have an equation with just one variable (F_2). Solving for F_2 gives the working equation:

$$F_2 = \frac{(15\overline{0}\ \text{lb})(2.00\ \text{ft})}{6.00\ \text{ft}} = 50.0\ \text{lb}$$

Since $\Sigma F = F_1 + F_2 = F_w$, substitute for F_2 and F_w to find F_1:

$$F_1 + 50.0\ \text{lb} = 15\overline{0}\ \text{lb}$$
$$F_1 = 15\overline{0}\ \text{lb} - 50.0\ \text{lb}$$
$$= 10\overline{0}\ \text{lb}$$

· · · · · · · · · · · · · · · · · ·

To solve parallel force problems:

1. Sketch the problem; position the parallel forces vertically when possible.
2. Write an equation setting the sums of the opposite forces equal to each other.
3. Choose a point of rotation. Eliminate a variable if possible (by making its torque arm zero).
4. Write the sum of all clockwise torques.
5. Write the sum of all counterclockwise torques.
6. Set $\Sigma\tau_{\text{clockwise}} = \Sigma\tau_{\text{counterclockwise}}$.
7. Solve the equation $\Sigma\tau_{\text{clockwise}} = \Sigma\tau_{\text{counterclockwise}}$ for the unknown quantity.
8. Substitute the value found in step 7 into the equation in step 2 to find the other unknown quantity.

EXAMPLE 3

A bricklayer weighing 175 lb stands on an 8.00-ft scaffold 3.00 ft from one end (Fig. 7.41). He has a pile of bricks, which weighs 40.0 lb, 3.00 ft from the other end. How much weight must each end support?

Figure 7.41

1.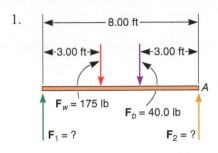

2. $\Sigma F = F_1 + F_2 = 175 \text{ lb} + 40.0 \text{ lb}$

3. Choose a point of rotation. Choose either end to eliminate one of the variables F_1 or F_2. Let us choose the right end and label it A.

4. $\Sigma \tau_{\text{clockwise}} = (F_1)(8.00 \text{ ft})$

5. $\Sigma \tau_{\text{counterclockwise}} = (40.0 \text{ lb})(3.00 \text{ ft}) + (175 \text{ lb})(5.00 \text{ ft})$

   Note that there are two counterclockwise torques.

6. $\Sigma \tau_{\text{clockwise}} = \Sigma \tau_{\text{counterclockwise}}$

   $F_1(8.00 \text{ ft}) = (40.0 \text{ lb})(3.00 \text{ ft}) + (175 \text{ lb})(5.00 \text{ ft})$

7. $F_1 = \dfrac{(40.0 \text{ lb})(3.00 \text{ ft}) + (175 \text{ lb})(5.00 \text{ ft})}{8.00 \text{ ft}}$

   $= \dfrac{12\overline{0} \text{ lb ft} + 875 \text{ lb ft}}{8.00 \text{ ft}} = \dfrac{995 \text{ lb ft}}{8.00 \text{ ft}} = 124 \text{ lb}$

8. $F_1 + F_2 = 175 \text{ lb} + 40.0 \text{ lb}$

   $124 \text{ lb} + F_2 = 215 \text{ lb} \qquad (F_1 = 124 \text{ lb})$

   $F_2 = 91 \text{ lb}$

TRY THIS ACTIVITY

The Physics of Window Washing

Window washers on tall buildings typically stand on a platform that is suspended from the top of the building by ropes or cables. Designers of these devices must make sure that the cables can safely suspend the workers while they are all working on one side of the platform. To simulate this activity, take the two strings, the scales, and the mass from the previous Try This Activity and use a metre stick or a wooden dowel as the platform.

Set up the activity as indicated in Fig. 7.42, making sure both suspending strings are the same length and that the metre stick is horizontal. Place the weight directly in the middle of the metre stick and note the readings in the scales. Then gradually shift the mass over to one side and observe what happens to the readings in the spring scales. What happens to the sum of the tension forces as the mass is shifted to one of the sides?

Figure 7.42   After suspending the mass from the middle of the platform, shift the mass toward one side and observe what happens to the tension in the strings.

1.0 kg

PROBLEMS 7.4

Find the force *F* that will produce equilibrium for each force diagram. Use the same procedure as in Example 1.

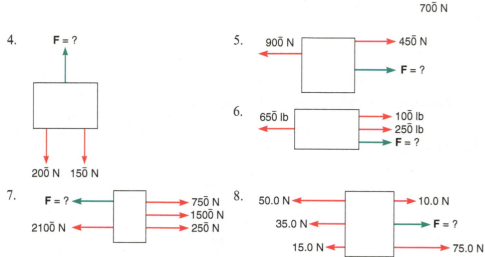

9. A 90.0-kg painter stands 3.00 m from one end of an 8.00-m scaffold. If the scaffold is supported at each end by a stepladder, how much of the weight of the painter must each ladder support?

10. A 50̄00-lb truck is 20.0 ft from one end of a 50.0-ft bridge. A 40̄00-lb car is 40.0 ft from the same end. How much weight must each end of the bridge support? (Neglect the weight of the bridge.)

11. A 24̄00-kg truck is 6.00 m from one end of a 27.0-m-long bridge. A 15̄00-kg car is 10.0 m from the same end. How much weight must each end of the bridge support?

12. An auto transmission of mass 165 kg is located 1.00 m from one end of a 2.50-m bench. What weight must each end of the bench support?

13. A bar 8.00 m long supports masses of 20.0 kg on the left end and 40.0 kg on the right end. At what distance from the 40.0-kg mass must the bar be supported for the bar to balance?

14. Two painters, each of mass 75.0 kg, stand on a 12.0-m scaffold, 6.00 m apart and 3.00 m from each end. They share a paint container of mass 21.0 kg in the middle of the scaffold. What weight must be supported by each of the ropes secured to the ends of the scaffold?

15. Two painters, one of mass 75.0 kg and the other 90.0 kg, stand on a 12.00-m scaffold, 6.00 m apart and 3.00 m from each end. They share a paint container of mass 21.0 kg in the middle of the scaffold. What weight must be supported by each of the ropes secured at the end of the scaffold?

16. Two painters stand on a 10.00-m scaffold. One, of mass 65.0 kg, stands 2.00 m from one end. The other, of mass 95.0 kg, stands 4.00 m from the other end. They share a paint container of mass 18.0 kg located between the two and

2.50 m from the larger person. What weight must be supported by each of the ropes secured at the ends of the scaffold?

17. An auto differential with a mass of 76.0 kg is 1.00 m from the end of a 2.22-m workbench. What mass must each end of the workbench support?

18. A chop saw with a mass of 12.60 kg is 0.75 m from one end of a worktable that is 2.00 m long. What mass must each end of the table support?

7.5 Center of Gravity

In Section 7.4 we disregarded the weight of the plank in the painter example. In practice, the weight of the plank or bridge is extremely important. The weight of a bridge being designed must be known in order to use materials of sufficient strength to support the bridge and the traffic and not collapse. An important concept in this kind of problem is center of gravity. *The **center of gravity** of any body is that point at which all of its weight can be considered to be concentrated.* A body such as a brick or a uniform rod has its center of gravity at its middle or center. The center of gravity of something like an automobile, however, is not at its center or middle because its weight is not evenly distributed throughout. Its center of gravity is located nearer the heavy engine.

The center of gravity of an irregularly shaped uniform thin plate is the point at which it can be supported as in Fig. 7.43(a). The center of gravity of a uniform thin plate can also be found by suspending it from a point and using a vertical chalk line with a suspended weight as shown in Fig. 7.43(b). The center of gravity is the point of intersection of any two or more such chalk lines as in Fig. 7.43(c).

You have probably had the experience of carrying a long board by yourself. If the board was not too heavy, you could carry it yourself by balancing it at its middle (Fig. 7.44). You didn't have to hold up both ends. You applied the principle of center of gravity and balanced the board at that point.

We shall represent the weight of a body by a vector through its center of gravity. We use a vector to show the weight (force due to gravity) of the body (Fig. 7.45). It is placed through the center of gravity to show that all the weight may be considered concentrated at that point. If the center of gravity is not at the middle of the body, its location will be given. In solving problems, the weight of the plank or bridge is represented like the other forces by a vector, which in the case of weight is through the center of gravity of the object.

Figure 7.43 The center of gravity of a uniform thin plate can also be found by suspending it from a point and using a vertical chalkline with a suspended weight, as shown in (b). The center of gravity is the point of intersection of any two or more such chalklines, as in (c).

Center of gravity

(a)   (b)   (c)

Figure 7.44 Support of the board at its center of gravity

Figure 7.45 Weight can be represented by a vector through the center of gravity.

(a)   (b)

EXAMPLE 1

A carpenter stands 2.00 ft from one end of a 6.00-ft scaffold that is uniform and weighs 20.0 lb. If the carpenter weighs 165 lb, how much weight must each end support?

1. **Sketch:**

Since the plank is uniform, its center of gravity is at the middle.

2. $\Sigma F = F_1 + F_2 = 165\ \text{lb} + 20.0\ \text{lb}$

3. Choose the left end as the point of rotation and label it A.

4. $\Sigma \tau_{\text{clockwise}} = (165\ \text{lb})(2.00\ \text{ft}) + (20.0\ \text{lb})(3.00\ \text{ft})$

5. $\Sigma \tau_{\text{counterclockwise}} = (F_2)(6.00\ \text{ft})$

6. $(165\ \text{lb})(2.00\ \text{ft}) + (20.0\ \text{lb})(3.00\ \text{ft}) = (F_2)(6.00\ \text{ft})$

7. $F_2 = \dfrac{33\overline{0}\ \text{lb ft} + 60.0\ \text{lb ft}}{6.00\ \text{ft}} = \dfrac{39\overline{0}\ \text{lb ft}}{6.00\ \text{ft}} = 65.0\ \text{lb}$

8. $F_1 + 65.0\ \text{lb} = 165\ \text{lb} + 20.0\ \text{lb}$

    $F_1 = 165\ \text{lb} + 20.0\ \text{lb} - 65.0\ \text{lb}$

    $= 12\overline{0}\ \text{lb}$

We can also use this method to find the magnitude and position of a parallel force vector that produces equilibrium.

Find the magnitude, direction, and placement (from point A) of a parallel vector F_6 that will produce equilibrium in the parallel force diagram in Fig. 7.46.

EXAMPLE 2

Figure 7.46

1. $\Sigma F = F_1 + F_5 + F_6 = F_2 + F_3 + F_4$
        $2850\text{ N} + 39\overline{0}0\text{ N} + F_6 = 2110\text{ N} + 4180\text{ N} + 3090\text{ N}$
        $F_6 = 2630\text{ N (up)}$

2.   Choose A' instead of A as the point of rotation to make the torque arm zero for F_1. Also, let x be the distance of F_6 from point A'.

3. $\Sigma\tau_{\text{clockwise}} = (2110\text{ N})(2.25\text{ m}) + (4180\text{ N})(3.75\text{ m}) + (3090\text{ N})(5.10\text{ m})$

4. $\Sigma\tau_{\text{counterclockwise}} = (39\overline{0}0\text{ N})(7.10\text{ m}) + (2630\text{ N})(x)$

5. $\Sigma\tau_{\text{clockwise}} = \Sigma\tau_{\text{counterclockwise}}$

6. $(2110\text{ N})(2.25\text{ m}) + (4180\text{ N})(3.75\text{ m}) + (3090\text{ N})(5.10\text{ m}) = (39\overline{0}0\text{ N})(7.10\text{ m})$
                                                 $+ (2630\text{ N})(x)$
                                   $3.23\text{ m} = x \text{ (from } A')$
                                   or $4.23\text{ m} = x \text{ (from } A)$

Thus, the equilibrium vector is 2630 N (up) placed at 4.23 m from point A.
. . . .

TRY THIS ACTIVITY

Center of Gravity

Stand with your back and heels touching a wall. Without moving your feet, bend over and touch your legs below your knees. Using the concept of center of gravity, explain what happened.

PROBLEMS 7.5

Solve each problem using the methods outlined in this chapter.

1.   Solve for F_1: $30.0F_1 = (14.0)(18.0) + (25.0)(17.0)$
2.   Solve for F_w: $(12.0)(15.0) + 45.0F_w = (21.0)(65.0) + (22.0)(32.0)$
3.   Two workers carry a uniform 15.0-ft plank that weighs 22.0 lb (Fig. 7.47). A load of blocks weighing 165 lb is located 7.00 ft from the first worker. What force must each worker exert to hold up the plank and load?

SKETCH

$$12\ cm^2 \quad w$$

4.0 cm

DATA

$A = 12\ cm^2,\ l = 4.0\ cm,\ w = ?$

BASIC EQUATION

$A = lw$

WORKING EQUATION

$w = \dfrac{A}{l}$

SUBSTITUTION

$w = \dfrac{12\ cm^2}{4.0\ cm} = 3.0\ cm$

Figure 7.47

4. Juan and Pablo carry a load weighing 720 N on a pole between them. (a) If the pole is 2.0 m long and the load is 0.50 m from Pablo, what force does each person support? Neglect the weight of the pole. (b) If the weight of the 120-N pole is considered, what force does each person support?

5. A wooden beam is 3.30 m long and has its center of gravity 1.30 m from one end. If the beam weighs 15$\overline{0}$0 N, what force is needed to support each end?

6. An auto engine weighs 65$\overline{0}$ lb and is located 4.00 ft from one end of a 10.0-ft workbench. If the bench is uniform and weighs 75.0 lb, what weight must each end of the bench support?

7. A bridge across a country stream weighs 89,200 N. A large truck stalls 4.00 m from one end of the 9.00-m bridge. What weight must each of the piers support if the truck weighs 98,$\overline{0}$00 N?

8. A window washer's scaffold 12.0 ft long and weighing 75.0 lb is suspended from each end. One washer weighs 155 lb and is 3.00 ft from one end. The other washer is 4.00 ft from the other end. If the force supported by the end near the first washer is 20$\overline{0}$ lb, how much does the second washer weigh?

9. A porch swing weighs 29.0 lb. It is 4.40 ft long and has a dog weighing 14.0 lb sleeping on it 1.90 ft from one end and a 125-lb person sitting 1.00 ft from the other end. What weight must the support ropes on each end hold up?

10. A wooden plank is 5.00 m long and supports a 75.0-kg block 2.00 m from one end. If the plank is uniform with mass 30.0 kg, how much force is needed to support each end?

11. A bridge has a mass of 1.60×10^4 kg, is 21.0 m long, and has a 35$\overline{0}$0-kg truck 7.00 m from one end. What force must each end of the bridge support?

12. A uniform steel beam is 5.00 m long and weighs 360 N. What force is needed to lift one end?

13. A wooden pole is 4.00 m long, weighs 315 N, and has its center of gravity 1.50 m from one end. What force is needed to lift each end?

14. A bridge has a mass of 2.60×10^4 kg, is 32.0 m long, and has a 35$\overline{0}$0-kg truck 15.0 m from one end. What force must each end of the bridge support?

15. An auto engine of mass 295 kg is located 1.00 m from one end of a 4.00-m workbench. If the uniform bench has a mass of 45.0 kg, what weight must each end of the bench support?

16. A 125-kg horizontal beam is supported at each end. A 325-kg mass rests one fourth of the way from one end. What weight must be supported at each end?

17. The sign shown in Fig. 7.48 is 4.00 m long, weighs 155$\overline{0}$ N, and is made of uniform material. A weight of 245 N hangs 1.00 m from the end. Find the tension in each support cable.

18. The uniform bar in Fig. 7.49 is 5.00 m long and weighs 975 N. A weight of 255 N is attached to one end while a weight of 375 N is attached 1.50 m from the other end. (a) Find the tension in the cable. (b) Where should the cable be tied to lift the bar and its weights so that the bar hangs in a horizontal equilibrium position?

Figure 7.48

Figure 7.49

Find the magnitude, direction, and placement (from point A) of a parallel vector F_6 that will produce equilibrium in each force diagram.

19. $F_1 = 125\bar{0}$ N

20.

Glossary

Center of Gravity The point of any body at which all of its weight can be considered to be concentrated. (p. 191)

Compression A force produced by forces pushing inward on the ends of an object (p. 173)

Concurrent Forces Two or more forces applied to, or acting at, the same point. (p. 164)

Equilibrant Force The force that, when applied at the same point as the resultant force, produces equilibrium. (p. 172)

Equilibrium An object is said to be in equilibrium when the net force acting on it is zero. A body that is in equilibrium is either at rest or moving at a constant velocity. (p. 170)

First Condition of Equilibrium The sum of all parallel forces on a body in equilibrium must be zero. (p. 185)

Resultant Force The sum of the forces applied at the same point; the single force that has the same effect as two or more forces acting together. (p. 164)

Second Condition of Equilibrium The sum of the clockwise torques on a body in equilibrium must be equal to the sum of the counterclockwise torques about any point. (p. 187)

Statics The study of objects that are in equilibrium. (p. 170)

Tension A stretching force produced by forces pulling outward on the ends of an object. (p. 173)

Torque The tendency to produce change in rotational motion; equal to the applied force times the length of the torque arm. (p. 181)

Formulas

7.1   To find the resultant vector $\mathbf{F_R}$ of two or more vectors:
  (a) Find the x- and y-components of each vector and add the components.
  (b) Find angle A as follows:

$$\tan A = \frac{|\text{sum of } y\text{-components}|}{|\text{sum of } x\text{-components}|} = \frac{|\mathbf{F}_{Ry}|}{|\mathbf{F}_{Rx}|}$$

  Determine the quadrant of the angle from the signs of the sum of the x- and y-components.
  (c) Find the magnitude of $\mathbf{F_R}$ using the Pythagorean theorem:

$$|\mathbf{F_R}| = \sqrt{|\mathbf{F}_{Rx}|^2 + |\mathbf{F}_{Ry}|^2}$$

7.2   Condition for equilibrium in one dimension:

$$\mathbf{F}_+ = \mathbf{F}_-$$

where $\mathbf{F}_+$ is the sum of the forces acting in one direction (call it the positive direction) and $\mathbf{F}_-$ is the sum of the forces acting in the opposite (negative) direction. Conditions for equilibrium in two dimensions:

  (a) The sum of x-components = 0; that is, $\mathbf{A}_x + \mathbf{B}_x + \mathbf{C}_x = 0$; and
  (b) The sum of y-components = 0; that is, $\mathbf{A}_y + \mathbf{B}_y + \mathbf{C}_y = 0$.

To solve equilibrium problems:

1. Draw a force diagram from the point at which the unknown forces act.
2. Find the x- and y-components of each force.
3. Substitute the components in the equations:

$$\text{sum of } x\text{-components} = 0$$
$$\text{sum of } y\text{-components} = 0$$

4. Solve for the unknowns. This may involve two simultaneous equations.

7.3 $\tau = Fr$

7.4   *First condition of equilibrium:* The sum of all parallel forces on an object must be zero.

$$\Sigma F = 0$$

Second condition of equilibrium: The sum of the clockwise torques on an object must equal the sum of the counterclockwise torques.

$$\Sigma \tau_{\text{clockwise}} = \Sigma \tau_{\text{counterclockwise}}$$

Review Questions

1. Concurrent forces act at
   (a) two or more different points.   (b) the same point.   (c) the origin.
2. The resultant force is
   (a) the last force applied.
   (b) the single force that has the same effect as two or more forces acting together.
   (c) equal to either diagonal when using the parallelogram method to add vectors.
3. A moving object
   (a) can be in equilibrium.   (b) is never in equilibrium.
   (c) has no force being applied.
4. The study of an object in equilibrium is called
   (a) dynamics.   (b) astronomy.   (c) statics.   (d) biology.
5. Torque is
   (a) applied force in rotational motion.
   (b) the length of the torque arm.
   (c) applied force times the length of the torque arm.
   (d) none of the above.
6. The first condition of equilibrium states that
   (a) all parallel forces must be zero.
   (b) all perpendicular forces must be zero.
   (c) all frictional forces must be zero.
7. In the second condition of equilibrium,
   (a) clockwise and counterclockwise torques are unequal.
   (b) clockwise and counterclockwise torques are equal.
   (c) there are no torques.
8. The center of gravity of an object
   (a) is always at its geometric center.
   (b) does not have to be at the geometric center.
   (c) exists only in symmetrical objects.
9. Is motion produced every time a force is applied to an object?
10. What is the relationship between opposing forces on a body that is in equilibrium?
11. Define *equilibrium*.
12. In what direction does the force due to gravity always act?
13. What may be said about concurrent forces whose sum of x-components equals zero and whose sum of y-components equals zero?
14. What is a force diagram?
15. Is the length of the pedal necessarily the true length of the torque arm in pedaling a bicycle?
16. In your own words, explain the second condition of equilibrium.
17. What is the primary consideration in the selection of a point of rotation in an equilibrium problem?
18. List three examples from daily life in which you use the concept of center of gravity.
19. Is the center of gravity of an object always at its geometric center?
20. On a 3.00-m scaffold of uniform mass, with supports at each end, there is a pile of bricks 0.500 m from one end. Which support will exert a greater force: the one closer to the bricks or the one farther away?

Review Problems

1. Find the sum of the following forces acting at the same point in a straight line: 345 N (right); 108 N (right); 481 N (left); 238 N (left); 303 N (left).

2. Forces of 275 lb and 225 lb act at the same point. (a) What is the magnitude of the maximum net force the two forces can exert together? (b) What is the magnitude of the minimum net force the two forces can exert together?

Find the sum of each set of vectors. Give angles in standard position.

3.

F_R
$F_1 = 1500$ N
$F_2 = 3400$ N

4.

$F_1 = 5080$ lb
F_R
$35.0°$
$F_2 = 4440$ lb

5.

F_3 F_2
$45.0°$
F_1

$|\mathbf{F}_1| = |\mathbf{F}_2| = |\mathbf{F}_3| = 54{,}600$ N

6. Forces of $\mathbf{F}_1 = 1250$ N, $\mathbf{F}_2 = 625$ N, and $\mathbf{F}_3 = 1850$ N are applied at the same point. The angle between $\mathbf{F}_1$ and $\mathbf{F}_2$ is 120.0° and the angle between $\mathbf{F}_2$ and $\mathbf{F}_3$ is 30.0°. $\mathbf{F}_2$ is between $\mathbf{F}_1$ and $\mathbf{F}_3$. Find the resultant force.

7. Eight people are involved in a tug-of-war. The blue team members pull with forces of 220 N, 340 N, 180 N, and 560 N. Three members of the red team pull with forces of 250 N, 160 N, and 420 N. With what force must the fourth person pull to maintain equilibrium?

8. A bridge has a weight limit of 14.0 tons. What is the maximum weight an 8.0-ton truck can carry across and still maintain equilibrium?

9. The x-components of three vectors are $\mathbf{F}_x$, 375 units, and 150 units. If their sum is equal to zero, what is $\mathbf{F}_x$?

10. If $\mathbf{W}_y = 60\overline{0}$ N and $\mathbf{W}_x = 90\overline{0}$ N, what are the magnitude and direction of the resultant $\mathbf{W}$?

Find forces $\mathbf{F}_1$ and $\mathbf{F}_2$ that produce equilibrium in each force diagram.

11.

F_2
F_1
1650 N
$23.0°$

12.

F_1
$65.0°$
F_2
4750 lb

13. Find the tension in the cable and the compression in the support of the sign shown in Fig. 7.50.
14. Find the tension in each cable in Fig. 7.51.
15. Find the tension in each cable in Fig. 7.52.

Figure 7.50

Figure 7.51

Figure 7.52

16. Find the tension and the compression in Fig. 7.53.
17. A man is changing a flat tire using a tire iron that is 50.0 cm long. If he exerts a force of 53.0 N, how much torque (in N m) does he produce?
18. A torque of 81.0 lb ft is produced by a torque arm of 3.00 ft. What force is being applied?
19. A hanging sign has mass $20\overline{0}$ kg. If the tension in one support cable is 1080 N, what is the tension in the other support cable?
20. A scaffold supports a bricklayer and bricks weighing 450 lb. If the force in one end support is 290 lb, what is the supporting force in the other?
21. Two ladders at the ends of a scaffold support a mass of 90.0 kg each. An 80.0 kg worker is on the scaffold with a pile of bricks. Find the mass of the bricks.
22. How far from the light end of a 68.0-cm bat would its center of gravity be if it is one fourth of the length of the bat from the heavy end?
23. A bridge has mass $80\overline{0}0$ kg. If a $32\overline{0}0$-kg truck stops in the middle of the bridge, what mass must each pier support?
24. If the truck in Problem 23 stops 7.00 m from one end of the 26.0-m bridge, what weight must each end support?
25. A uniform 2.20-kg steel bar with length 2.70 m is suspended on each end by a chain. A 40.0-kg person hangs 70.0 cm from one end while a 55.0-kg person hangs 50.0 cm from the other end. Another person pushes up halfway between the two persons with a force of 127 N. How much weight does each chain support?
26. Find the vertical force needed to support the 4.00-m-long uniform beam weighing 3475 N as shown in Fig. 7.54.
27. A horizontal cable supports the boom of a crane. If the tension in the cable is $45\overline{0}0$ N and the crane is supporting a $75\overline{0}0$-N girder, what is the compression in the boom?

Figure 7.53

Figure 7.54

APPLIED CONCEPTS

1. Archeologists in Egypt are attempting to open a 2.57-m-wide door in the tomb of Tutankhamen. Miles, one of the rookie archeologists, pushes perpendicularly on the door with a force of 894 N. (a) How much torque does Miles exert on the door? (b) As the door opens, Miles continues to push with the same force but is at an angle of 30.0° to the face of the door. How much torque does Miles apply to the door? (c) What could Miles do to continually exert the same amount of torque on the door?

2. Sean and Greg are on a job site standing on two beams 11.0 ft apart. They need to lift their crate of tools midway between them with ropes up 33.5 ft to where they are working. (a) What is the angle between the ropes when the crate is on the ground? (b) How much force do Sean and Greg need to exert on the ropes when lifting the 115-lb crate off the ground? (c) How much force do both Sean and Greg need to exert when the crate is 5.75 ft below them? (d) Explain why the force to lift the crate changes as it moves closer to them.

3. Maria has severe arthritis and can apply a maximum force of only 25.5 N when opening her front door. (a) How much more torque would Maria be able to apply if she purchased a lever-style door opener with a handle 12.7 cm long compared to her conventional doorknob with a radius of 3.74 cm? (b) How much less force would she need to apply to the new door handle in order to maintain the original amount of torque she had with her old doorknob?

4. Krista's flagpole bracket is mounted at an angle of 45.0° to the wall and is designed to support a maximum torque of 40.0 N m. (a) If the new flag and pole that Krista bought have a mass of 8.75 kg and their center of mass is located 0.750 m from the bracket, will the bracket support the flag and pole? (b) What could Krista do to the angle of the bracket to reduce the torque?

Figure 7.55

5. Luisa, whose mass is 45.0 kg, is standing at the end of a 5.50-m diving board, which has a mass of 35.7 kg (Fig. 7.55). (a) What force must the bracket and fulcrum exert on the diving board? (b) Where would Luisa stand to have both the fulcrum and the bracket apply an upward force?

CHAPTER 8

WORK AND ENERGY

Objectives

The major goals of this chapter are to enable you to:

1. Distinguish between the common and technical definitions of work.
2. Analyze how power is used and described in technical applications.
3. Relate kinetic and potential energy to the law of conservation of mechanical energy.

Work, power, and energy are common terms used to describe changes in physical activity. In science each of these terms has a limited definition. Work, for example, is accomplished only where there is movement of the object on which the applied force acts. Effort alone is insufficient.

We will now study how the scientist and engineer use work, power, and energy; how they are related; and how they differ from their everyday meanings.

8.1 Work

What is work? The common idea of the definition of work is quite different from the technical definition. We often associate work with physical or mental effort that leads to fatigue. The technical definition of work is more limited. If we try to lift a heavy crate that does not budge [Fig. 8.1(a)], we would probably say that we have done work because we strained our muscles and feel tired, but in a technical sense, no work was done because the crate did not move. Work would be done *on* the crate if we were to push it across the floor [Fig. 8.1(b)]. In this case, work was done *by* us, and work was done *on* the crate.

Figure 8.1 Work is done only in the case shown in part (b).

(a) Crate is not moved.　　(b) Crate is moved.

The technical meaning of work requires that work must be done *by* one object *on* another object. When a stake is driven into the ground (Fig. 8.2), work is done *by* the moving sledgehammer and work is done *on* the stake. When a bulldozer pushes dirt and rocks (Fig. 8.3), work is done *by* the bulldozer and work is done *on* the dirt and rocks.

Figure 8.2 Work is done by one object on another.

(a)　　　　　　　　　(b)

Figure 8.3 Work is done *by* the bulldozer *on* the dirt and rocks.

Courtesy of fotolia © ftfoxfoto

The previous examples show a limited meaning of work: Work is done when a force acts through a distance. The physical definition of **work** is even narrower: *Work is the product of the force in the direction of the motion and the magnitude of the displacement.*

$$W = Fs$$

where　W = work
　　　　F = force applied *in the direction of the motion*
　　　　s = magnitude of the displacement

Now, let us apply our technical definition of work to our unsuccessful effort to lift the crate. We applied a force by lifting on the crate but were unable to move it. Therefore, the displacement was zero, and the product of the force and the displacement must also be zero. So, no work was done.

It is important to remember that displacement is a vector describing both quantity and direction. Vector concepts therefore also apply to work. If the applied force is not in the direction of the motion of the object, we must determine the vector component of the force that is in the direction of the motion of the object. This critical concept is illustrated in Example 2.

From the equation for work, we can determine the units for work. In the metric system, force is expressed in newtons and the magnitude of the displacement in metres:

$$\text{work} = \text{force} \times \text{displacement} = \text{newton} \times \text{metre} = \text{N m}$$

James P. Joule (1818–1889),

physicist, was born in England. In a series of experiments from 1843 to 1878, he showed that heat is a form of energy and established the mechanical equivalent of heat, which formed the basis of the theory of the conservation of energy. He also worked with Lord Kelvin on temperature changes in gases, which led to the founding of the refrigeration industry. The unit of work, the joule, is named after him.

This unit (N m) has a special name in honor of **James P. Joule**. It is the joule (J) (pronounced jo͞ol):

$$1 \text{ N m} = 1 \text{ joule} = 1 \text{ J}$$

In the U.S. system, force is expressed in pounds (lb) and displacement in feet (ft):

$$\text{work} = \text{force} \times \text{displacement} = \text{pounds} \times \text{feet} = \text{ft lb}$$

The U.S. unit of work is called the foot-pound.

Work is *not* a vector quantity because it has no particular direction. It is a scalar and has only magnitude.

Find the amount of work done by a worker lifting 225 N of bricks to a height of 1.75 m as shown in Fig. 8.4.

EXAMPLE 1

Data:

$$F = 225 \text{ N}$$
$$s = 1.75 \text{ m}$$
$$W = ?$$

Basic Equation:

$$W = Fs$$

Working Equation: Same

Substitution:

$$W = (225 \text{ N})(1.75 \text{ m})$$
$$= 394 \text{ N m} \quad \text{or} \quad 394 \text{ J}$$

Figure 8.4

1.75 m

· · · · · · · · · · · · · · · ·

A worker pushes a 350-lb cart a distance of 3$\overline{0}$ ft by exerting a constant force of 4$\overline{0}$ lb as shown in Fig. 8.5. How much work does the person do?

EXAMPLE 2

Figure 8.5

350 lb

4$\overline{0}$ lb

3$\overline{0}$ ft

Data:

$$F = 4\overline{0} \text{ lb}$$
$$s = 3\overline{0} \text{ ft}$$
$$W = ?$$

Basic Equation:

$$W = Fs$$

Working Equation: Same

Substitution:

$$W = (4\overline{0} \text{ lb})(3\overline{0} \text{ ft})$$
$$= 1200 \text{ ft lb}$$

· · · · · · · · · · · · · · · · ·

Note: In Example 2 the cart weighs 350 lb but $F = 4\overline{0}$ lb. (Recall that the weight of an object is the measure of its gravitational attraction to the earth and is represented by a vertical vector pointing down to the center of the earth.) There is no vertical motion in the direction of the gravitational force. Therefore, the weight of the box is not the force used to determine the amount of work being done.

Work is being done by the worker pushing the pallet. Exerting a force of $4\overline{0}$ lb results in a displacement in the direction of the applied force. The work done is the product of this force ($4\overline{0}$ lb) and the displacement ($3\overline{0}$ ft) in the direction the force is applied.

Recall that the definition of work states that work is the product of the *force in the direction of the motion* and the magnitude of the displacement. To determine the work when the force is not applied in the direction of the motion, consider a block being pulled by a rope with a force **F** that makes an angle θ with level ground as shown in Fig. 8.6. First, draw the horizontal component $\mathbf{F}_x$ and complete the right triangle. Note that $\mathbf{F}_x$ is the force in the direction of the motion. From the right triangle we have

Figure 8.6

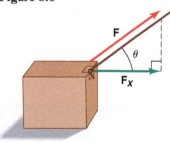

$$\cos\theta = \frac{\text{side adjacent to } \theta}{\text{hypotenuse}} = \frac{|\mathbf{F}_x|}{|\mathbf{F}|}$$

Or,

$$|\mathbf{F}_x| = |\mathbf{F}| \cos\theta$$

That is, when the applied force is not in the direction of the motion, the work done is

$$W = Fs \cos\theta$$

where W = the work done
F = the applied force
s = the displacement
θ = the angle between the applied force and the direction of the motion

EXAMPLE 3

A person pulls a sled along level ground a distance of 15.0 m by exerting a constant force of 215 N at an angle of 30.0° with the ground (Fig. 8.7). How much work does she do?

Figure 8.7

Data:

$$F = 215 \text{ N}$$
$$s = 15.0 \text{ m}$$
$$\theta = 30.0°$$
$$W = ?$$

Basic Equation:

$$W = Fs \cos \theta$$

Working Equation: Same

Substitution:

$$W = (215 \text{ N})(15.0 \text{ m}) \cos 30.0°$$
$$= 2790 \text{ N m}$$
$$= 2790 \text{ J} \quad (1 \text{ N m} = 1 \text{ J})$$

· · · · · · · · · · · · · · · · · ·

EXAMPLE 4

Juan and Sonja use a push mower to mow a lawn. Juan, who is taller, pushes at a constant force of 33.1 N on the handle at an angle of 55.0° with the ground. Sonja, who is shorter, pushes at a constant force of 23.2 N on the handle at an angle of 35.0° with the ground. Assume they each push the mower $30\overline{0}0$ m. Who does more work and by how much?

Sketch:

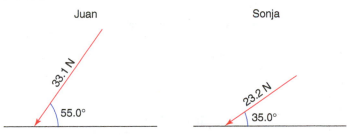

Data:

$F = 33.1$ N	$F = 23.2$ N
$s = 30\overline{0}0$ m	$s = 30\overline{0}0$ m
$\theta = 55.0°$	$\theta = 35.0°$
$W = ?$	$W = ?$

Basic Equation:

$W = Fs \cos \theta$	$W = Fs \cos \theta$

Working Equation: Same   Same

Substitution:

$$W = (33.1 \text{ N})(30\overline{0}0 \text{ m}) \cos 55.0° \qquad W = (23.2 \text{ N})(30\overline{0}0 \text{ m}) \cos 35.0°$$
$$= 57,\overline{0}00 \text{ N m} \qquad\qquad = 57,\overline{0}00 \text{ N m}$$
$$= 57,\overline{0}00 \text{ J} \quad (1 \text{ N m} = 1 \text{ J}) \qquad = 57,\overline{0}00 \text{ J}$$

They do the same amount of work. However, Juan must exert more energy because he pushes into the ground more than Sonja, who pushes more in the direction of the motion.

· · · · · · · · · · · · · · · ·

EXAMPLE 5

Find the amount of work done in vertically lifting a steel beam with mass 750 kg at uniform speed a distance of 45 m.
    Here the force is the weight of the beam.

Data:

$$F = mg$$
$$m = 750 \text{ kg}$$
$$g = 9.80 \text{ m/s}^2$$
$$s = 45 \text{ m}$$
$$W = ?$$

Basic Equation:

$$W = Fs = mgs$$

Working Equation: Same

Substitution:

$$W = (750 \text{ kg})(9.80 \text{ m/s}^2)(45 \text{ m})$$
$$= 3.3 \times 10^5 \text{ (kg m/s}^2)(\text{m})$$
$$= 3.3 \times 10^5 \text{ N m} \qquad (1 \text{ N} = 1 \text{ kg m/s}^2)$$
$$= 3.3 \times 10^5 \text{ J} \qquad (1 \text{ J} = 1 \text{ N m})$$

Do you see that 330 kJ would also be an acceptable answer?

· · · · · · · · · · · · · · · ·

SKETCH

12 cm² | w

4.0 cm

DATA

$A = 12 \text{ cm}^2, l = 4.0 \text{ cm}, w = ?$

BASIC EQUATION

$A = lw$

WORKING EQUATION

$w = \dfrac{A}{l}$

SUBSTITUTION

$w = \dfrac{12 \text{ cm}^2}{4.0 \text{ cm}} = 3.0 \text{ cm}$

PROBLEMS 8.1

1. Given: $F = 10.0 \text{ N}$
        $s = 3.43 \text{ m}$
        $W = ?$

   Given: $F = 1850 \text{ N}$
        $s = 625 \text{ m}$
        $\theta = 37.5°$
        $W = ?$

   Given: $F = 25,700 \text{ N}$
        $s = 238 \text{ m}$
        $W = 5.57 \times 10^6 \text{ J}$
        $\theta = ?$

2. Given: $F = 125 \text{ N}$
        $s = 4875 \text{ m}$
        $W = ?$

4. Given: $W = 697 \text{ ft lb}$
        $s = 976 \text{ ft}$
        $F = ?$

6. Given: $F = ma$
        $m = 16.0 \text{ kg}$
        $a = 9.80 \text{ m/s}^2$
        $s = 13.0 \text{ m}$
        $W = ?$

7. How much work is required for a mechanical hoist to lift a 90$\overline{0}$0-N automobile to a height of 1.80 m for repairs?

8. A hay wagon is used to move bales from the field to the barn. The tractor pulling the wagon exerts a constant force of 350 lb. The distance from field to barn is $\frac{1}{2}$ mi. How much work (ft lb) is done in moving one load of hay to the barn?

9. A worker lifts 75 concrete blocks a distance of 1.50 m to the bed of a truck. Each block has a mass of 4.00 kg. How much work is done to lift all the blocks to the truck bed?

10. The work required to lift eleven 94.0-lb bags of cement from the ground to the back of a truck is 4340 ft lb. What is the distance from the ground to the bed of the truck?

11. How much work is done in lifting 450 lb of cement 75 ft above the ground?

12. How much work is done lifting a 20$\overline{0}$-kg wrecking ball 6.50 m above the ground?

13. A gardener pushes a mower a distance of 90$\overline{0}$ m in mowing a yard. The handle of the mower makes an angle of 40.0° with the ground. The gardener exerts a force of 35.0 N along the handle of the mower (Fig. 8.8). How much work does the gardener do in mowing the lawn?

Figure 8.8

(a)   (b)

14. A traveler is pulling a suitcase at an angle of 40.0° with the horizontal (Fig. 8.9). If she exerts a force of 5.00 lb along the handle, how much work does the traveler do in pulling the suitcase $\frac{1}{4}$ mi?

Figure 8.9

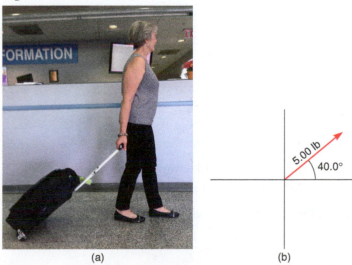

(a)   (b)

15. A crate is pulled 675 ft across a warehouse floor by a worker using a rope that makes an angle of 50.0° with the floor. If 375 lb is exerted on the rope, how much work is done in pulling the crate across the floor?

16. A man pulls a sled a distance of 231 m. The rope attached to the sled makes an angle of 30.0° with the ground. The man exerts a force of 775 N on the rope. How much work does the man do in pulling the sled?

17. A tractor tows a barge through a canal with a towrope that makes an angle of 21° with the bank of the canal. If the tension in the rope is 12,000 N, how much work is done in moving the barge 550 m?

18. Two tractors tow a barge through a canal; each tractor uses a towrope that makes an angle of 21° with the bank of the canal. If the tension in each rope is 12,000 N, how much work is done in moving the barge 550 m?

19. Two students push a dune buggy 35.0 m across a lot. The force required is 825 N. How much work is done?

20. After a rain, the force necessary to push the dune buggy in Problem 19 through the mud is doubled. How does the amount of work done by the students change?

21. A delivery person carries a 215-N box up stairs 4.20 m vertically and 6.80 m horizontally. (a) How much work does the delivery person do? (b) How much work does the delivery person do in carrying the box down the stairs?

22. A crate is pulled by a force of 628 N across the floor by a worker using a rope making an angle of 46.0° with the floor. If the crate is pulled 15.0 m, how much work does the force on the rope do?

23. A laborer pushes a wheelbarrow weighing $20\overline{0}$ N at 25.0° above the horizontal. If he pushes it a distance of 25.0 m, how much work is done?

24. An end loader lifts a $10\overline{0}0$-N bucket of gravel 1.75 m above the ground. How much work is done?

8.2 Power

While the technical concept of work is important, in industry and everyday life we are always concerned with how fast something happens. If we introduce time as another factor along with the amount of work done, we have the more familiar concept of power. **Power** is *the rate of doing work*; that is,

$$P = \frac{W}{t}$$

where P = power
$\quad\quad\ \ W$ = work
$\quad\quad\ \ t$ = time

The units of power are familiar to most of us. In the metric system, the unit of power is the *watt*:

$$P = \frac{W}{t} = \frac{Fs}{t} = \frac{\text{N m}}{\text{s}} = \frac{\text{J}}{\text{s}} = \text{watt}$$

Power is often expressed in kilowatts and megawatts:

1000 watts (W) = 1 kilowatt (kW)

1,000,000 watts = 1 megawatt (MW)

In the U.S. system, the unit of power is either ft lb/s or horsepower:

$$P = \frac{W}{t} = \frac{Fs}{t} = \frac{\text{ft lb}}{\text{s}}$$

Horsepower (hp) is a unit defined by **James Watt**:

1 horsepower (hp) = 550 ft lb/s = 33,000 ft lb/min

James Watt (1736–1819),

engineer and inventor, was born in Scotland. He made fundamental improvements to steam engines and is credited for several related inventions. The term horsepower was first used by him; the SI unit of power, the watt, is named after him.

One horsepower is equivalent to moving a force of 550 lb a distance of 1 ft in 1 s or moving 1 lb a distance of 550 ft in 1 s.

Note: Since the above is a definition, treat any conversion factor as an exact number, which does not affect the number of significant digits in any calculation.

TRY THIS ACTIVITY

Human Horsepower

How much equivalent horsepower do you possess? Using a tape measure, a stopwatch, and a flight of stairs, determine the horsepower in your legs when climbing a flight of stairs. First, measure the vertical height of the stairs. Use this distance and your weight to determine the work required to move your body's weight up the stairs. Clock the time it takes to walk and to run up the stairs. Then, find and compare the power. Convert your power to horsepower. How does your running horsepower compare to the horsepower of a lawnmower or a car?

EXAMPLE 1

A freight elevator with operator weighs $50\overline{0}0$ N. If it is raised to a height of 15.0 m in 10.0 s, how much power is developed?

Data:

$$F = 50\overline{0}0 \text{ N}$$
$$s = 15.0 \text{ m}$$
$$t = 10.0 \text{ s}$$
$$P = ?$$

Basic Equations:

$$P = \frac{W}{t} \quad \text{and} \quad W = Fs$$

Working Equation:

$$P = \frac{Fs}{t}$$

Substitution:

$$P = \frac{(50\overline{0}0 \text{ N})(15.0 \text{ m})}{10.0 \text{ s}}$$
$$= 75\overline{0}0 \text{ N m/s}$$

EXAMPLE 2

The power expended in lifting an 825-lb girder to the top of a building $10\overline{0}$ ft high is 10.0 hp. How much time is required to raise the girder?

Data:

$$F = 825 \text{ lb}$$
$$s = 10\overline{0} \text{ ft}$$
$$P = 10.0 \text{ hp}$$
$$t = ?$$

Basic Equations:

$$P = \frac{W}{t} \quad \text{and} \quad W = Fs$$

Working Equation:

$$t = \frac{W}{P} = \frac{Fs}{P}$$

Substitution:

$$t = \frac{(825 \text{ lb})(10\bar{0} \text{ ft})}{10.0 \text{ hp}}$$

$$= \frac{(825 \text{ lb})(10\bar{0} \text{ ft})}{10.0 \text{ hp}} \times \frac{1 \text{ hp}}{550 \frac{\text{ft lb}}{\text{s}}}$$

$$= 15.0 \text{ s}$$

$$\boxed{\frac{\text{lb ft}}{\text{hp}} \times \frac{\text{hp}}{\frac{\text{ft lb}}{\text{s}}} = \frac{\text{lb ft}}{\text{hp}} \times \left(\text{hp} \div \frac{\text{ft lb}}{\text{s}} \right) = \frac{\text{lb ft}}{\text{hp}} \times \left(\text{hp} \times \frac{\text{s}}{\text{ft lb}} \right) = \text{s}}$$

Note: We use a conversion factor to obtain time units.

················

EXAMPLE 3

The mass of a large steel wrecking ball is $200\bar{0}$ kg. What power is used to raise it to a height of 40.0 m if the work is done in 20.0 s?

Data:

$$m = 200\bar{0} \text{ kg}$$
$$s = 40.0 \text{ m}$$
$$t = 20.0 \text{ s}$$
$$P = ?$$

Basic Equations:

$$P = \frac{W}{t} \quad \text{and} \quad W = Fs$$

Working Equation:

$$P = \frac{Fs}{t}$$

Substitution: Note that we cannot directly substitute into the working equation because our data are given in terms of *mass* and we must find *force* to substitute in $P = Fs/t$. The force is the weight of the ball:

$$F = mg = (200\bar{0} \text{ kg})(9.80 \text{ m/s}^2) = 19,600 \text{ kg m/s}^2 = 19,600 \text{ N}$$

Then

$$P = \frac{Fs}{t} = \frac{(19,600 \text{ N})(40.0 \text{ m})}{20.0 \text{s}}$$

$$= 39,200 \text{ N m/s}$$

$$= 39,200 \text{ W} \quad \text{or} \quad 39.2 \text{ kW}$$

················

EXAMPLE 4

A machine is designed to perform a given amount of work in a given amount of time. A second machine does the same amount of work in half the time. Find the power of the second machine compared with the first.

Data (for the second machine given in terms of the first):

$$W = W$$

$$t = \tfrac{1}{2}t = \frac{t}{2}$$

$$P = ?$$

Basic Equation:

$$P = \frac{W}{t}$$

Working Equation: Same

Substitution:

$$P = \frac{W}{\dfrac{t}{2}} = W \div \frac{t}{2} = W \times \frac{2}{t} = 2\left(\frac{W}{t}\right) = 2P$$

Thus, the power is doubled when the time is halved.

· · · · · · · · · · · · · · · · · ·

EXAMPLE 5

A motor is capable of developing 10.0 kW of power. How large a mass can it lift 75.0 m in 20.0 s?

Data:

$$P = 10.0 \text{ kW} = 10,\overline{0}00 \text{ W}$$

$$s = 75.0 \text{ m}$$

$$t = 20.0 \text{ s}$$

$$F = ?$$

Basic Equations:

$$P = \frac{W}{t} \quad \text{and} \quad W = Fs \quad \text{or} \quad P = \frac{Fs}{t}$$

Working Equation:

$$F = \frac{Pt}{s}$$

Substitution:

$$F = \frac{(10,\overline{0}00 \text{ W})(20.0 \text{ s})}{75.0 \text{ m}}$$

$$= 2670 \frac{\text{W s}}{\text{m}} \times \frac{1 \text{ N m/s}}{1 \text{ W}} \qquad (1 \text{ W} = 1 \text{ J/s} = 1 \text{ N m/s})$$

$$= 2670 \text{ N}$$

Next, change the weight to mass as follows:

Data:

$$F = 2670 \text{ N}$$
$$g = 9.80 \text{ m/s}^2$$
$$m = ?$$

Basic Equation:

$$F = mg$$

Working Equation:

$$m = \frac{F}{g}$$

Substitution:

$$m = \frac{2670 \text{ N}}{9.80 \text{ m/s}^2} \times \frac{1 \text{ kg m/s}^2}{1 \text{ N}} \qquad (1 \text{ N} = 1 \text{ kg m/s}^2)$$
$$= 272 \text{ kg}$$

EXAMPLE 6

A pump is needed to lift $15\overline{0}0$ L of water per minute a distance of 45.0 m. What power, in kW, must the pump be able to deliver? (1 L of water has a mass of 1 kg.)

Data:

$$m = 15\overline{0}0 \text{ L} \times \frac{1 \text{ kg}}{1 \text{ L}} = 15\overline{0}0 \text{ kg}$$
$$s = 45.0 \text{ m}$$
$$t = 1 \text{ min} = 60.0 \text{ s}$$
$$g = 9.80 \text{ m/s}^2$$
$$P = ?$$

Basic Equations:

$$P = \frac{W}{t}, \quad W = Fs, \quad \text{and} \quad F = mg, \quad \text{or} \quad P = \frac{mgs}{t}$$

Working Equation:

$$P = \frac{mgs}{t}$$

Substitution:

$$P = \frac{(15\overline{0}0 \text{ kg})(9.80 \text{ m/s}^2)(45.0 \text{ m})}{60.0 \text{ s}}$$
$$= 1.10 \times 10^4 \text{ kg m}^2/\text{s} \quad \left(1 \text{ W} = \frac{1 \text{ J}}{\text{s}} = \frac{1 \text{ N m}}{\text{s}} = \frac{1 \text{ (kg m/s}^2)(\text{m})}{\text{s}} = 1 \text{ kg m}^2/\text{s}\right)$$
$$= 1.10 \times 10^4 \text{ W} \times \frac{1 \text{ kW}}{10^3 \text{ W}}$$
$$= 11.0 \text{ kW}$$

PROBLEMS 8.2

1. Given: $W = 132$ J
   $t = 7.00$ s
   $P = ?$

2. Given: $P = 231$ ft lb/s
   $t = 14.3$ s
   $W = ?$

3. Given: $P = 75.0$ W
   $W = 40.0$ J
   $t = ?$

4. Given: $W = 55.0$ J
   $t = 11.0$ s
   $P = ?$

5. The work required to lift a crate is $31\overline{0}$ J. If the crate is lifted in 25.0 s, what power is developed?

6. When a $36\overline{0}0$-lb automobile runs out of gas, it is pushed by its unhappy driver and a friend a quarter of a mile (0.250 mi). To keep the car rolling, they must exert a constant force of 175 lb. (a) How much work do they do? (b) If it takes them 15.0 min, how much power do they develop? (c) Expressed in horsepower, how much power do they develop?

7. An electric golf cart develops 1.25 kW of power while moving at a constant speed. (a) Express its power in horsepower. (b) If the cart travels $20\overline{0}$ m in 35.0 s, what force is exerted by the cart?

8. How many seconds would it take a 7.00-hp motor to raise a 475-lb boiler to a platform 38.0 ft high?

9. How long would it take a $95\overline{0}$-W motor to raise a 360-kg mass to a height of 16.0 m?

10. A $15\overline{0}0$-lb casting is raised 22.0 ft in 2.50 min. Find the required horsepower.

11. What is the rating in kW of a 2.00-hp motor?

12. A wattmeter shows that a motor is drawing $22\overline{0}0$ W. What horsepower is being delivered?

13. A 525-kg steel beam is raised 30.0 m in 25.0 s. How many kilowatts of power are needed?

14. How long would it take a 4.50-kW motor to raise a 175-kg boiler to a platform 15.0 m above the floor?

15. A 475-kg pre-stressed concrete beam is to be raised 10.0 m in 24.0 s. How many kilowatts of power are needed for the job?

16. A 50.0-kg welder is to be raised 15.0 m in 12.0 s. How many kilowatts of power are needed for the job?

17. An escalator is needed to carry 75 passengers per minute a vertical distance of 8.0 m. Assume that the mass of each passenger is $7\overline{0}$ kg. (a) What is the power (in kW) of the motor needed? (b) Express this power in horsepower. (c) What is the power (in kW) of the motor needed if 35% of the power is lost to friction and other losses?

18. A pump is needed to lift $75\overline{0}$ L of water per minute a distance of 25.0 m. What power (in kW) must the pump be able to deliver? (1 L of water has a mass of 1 kg.)

19. A machine is designed to perform a given amount of work in a given amount of time. A second machine does twice the same amount of work in half the time. Find the power of the second machine compared with the first.

20. A certain machine is designed to perform a given amount of work in a given amount of time. A second machine does 2.5 times the same amount of work in one third the time. Find the power of the second machine compared with the first.

21. A motor on an escalator is capable of developing 12 kW of power. (a) How many passengers of mass 75 kg each can it lift a vertical distance of 9.0 m per min, assuming no power loss? (b) What power, in kW, motor is needed to move the same number of passengers at the same rate if 45% of the actual power developed by the motor is lost to friction and heat loss?

22. A pump is capable of developing 4.00 kW of power. How many litres of water per minute can be lifted a distance of 35.0 m? (1 L of water has a mass of 1 kg.)

23. A pallet weighing 575 N is lifted a distance of 20.0 m vertically in 10.0 s. What power is developed in kilowatts?

SKETCH

$$\boxed{12 \text{ cm}^2} \quad w$$

4.0 cm

DATA

$A = 12$ cm^2, $l = 4.0$ cm, $w = ?$

BASIC EQUATION

$A = lw$

WORKING EQUATION

$w = \dfrac{A}{l}$

SUBSTITUTION

$w = \dfrac{12 \text{ cm}^2}{4.0 \text{ cm}} = 3.0$ cm

24. A pallet is loaded with bags of cement; the total weight of 875 N is lifted 21.0 m vertically in 11.0 s. What power in kilowatts is required to lift the cement?
25. A bundle of steel reinforcing rods weighing 175 N is lifted 32.0 m in 16.0 s. What power in kilowatts is required to lift the steel?
26. An ironworker carries a 7.50-kg toolbag up a vertical ladder on a high-rise building under construction. (a) After 30.0 s, he is 8.20 m above his starting point. How much work does the worker do on the toolbag? (b) If the worker weighs 645 N, how much work does he do in lifting himself and the toolbag? (c) What is the average power developed by the worker?

Figure 8.10 Here the counter-weight working with the electric motor balances the load in the elevator.

8.3 Energy

Energy is defined as the ability to do work. There are many forms of energy, such as mechanical, electrical, thermal, fluid, chemical, atomic, and sound (acoustic) energy.

The mechanical energy of a body or a system is due to its position, its motion, or its internal structure. There are two kinds of mechanical energy: potential energy and kinetic energy. **Potential energy** is the stored energy of a body due to its internal characteristics or its position. **Kinetic energy** is the energy due to the mass and the velocity of a moving object.

Internal potential energy is determined by the nature or condition of the substance; for example, gasoline, a compressed spring, or a stretched rubber band has internal potential energy due to its internal characteristics. **Gravitational potential energy** is determined by the position of an object relative to a particular reference level; for example, a rock lying on the edge of a cliff, the raised counterweight on an elevator (Fig. 8.10), or a raised pile driver has potential energy due to its position. Each weight has the ability to do work because of the pull of gravity on it. The unit of energy is the joule (J) in the metric system and the foot-pound (ft lb) in the U.S. system.

In Section 8.1 we discussed the technical definition of work. Work and energy are related, as we now have defined energy as the ability to do work. While internal potential energy is more difficult to measure, gravitational potential energy can be used to illustrate the relationship between the two.

Recall that work is equal to the force applied in the direction of motion times the displacement ($W = Fs$). Gravitational potential energy is measured by multiplying the weight of an object (mass times acceleration of gravity) times the height of the object relative to a particular reference level. So the work done in raising the object a distance h is mgh, or $W = Fs = mgh$.

The formula for gravitational potential energy is

$$E_p = mgh$$

Figure 8.11 Work done in raising the crate gives it potential energy.

where E_p = potential energy
m = mass
g = 9.80 m/s² or 32.2 ft/s²
h = height above reference level

In position 1 in Fig. 8.11, the crate is at rest on the floor. It has no ability to do work because it is in its lowest position. To raise the crate to position 2, work must be done to lift it. In the raised position, however, it now has stored ability to do work (by falling to the floor). Its E_p (potential energy) can be calculated by multiplying the mass of the crate times acceleration of gravity (g) times height above reference level (h). Note that we can calculate the potential energy of the crate with respect to any level we choose. Here we have chosen the floor as the zero or lowest reference level.

EXAMPLE 1

A wrecking ball of mass $20\bar{0}$ kg is poised 4.00 m above a concrete platform whose top is 2.00 m above the ground. (a) With respect to the platform, what is the potential energy of the ball? (b) With respect to the ground, what is the potential energy of the ball?

Sketch:

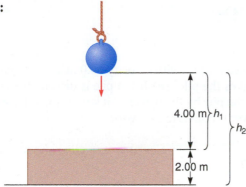

4.00 m h_1
h_2
2.00 m

Data:

$$m = 20\bar{0} \text{ kg}$$
$$h_1 = 4.00 \text{ m}$$
$$h_2 = 6.00 \text{ m}$$
$$E_p = ?$$

Basic Equation:

$$E_p = mgh$$

Working Equation: Same

(a) Substitution:

$$E_p = (20\bar{0} \text{ kg})(9.80 \text{ m/s}^2)(4.00 \text{ m})$$
$$= 7840 \frac{\text{kg m}^2}{\text{s}^2} \times \frac{1 \text{ J}}{\text{kg m}^2/\text{s}^2} \quad [1 \text{ J} = 1 \text{ N m} = 1 \text{ (kg m/s}^2)(\text{m}) = 1 \text{ kg m}^2/\text{s}^2]$$
$$= 7840 \text{ J} \quad \text{(which also indicates the amount of work done by gravity on a falling object)}$$

(b) Substitution:

$$E_p = (20\bar{0} \text{ kg})(9.80 \text{ m/s}^2)(6.00 \text{ m})$$
$$= 11,800 \frac{\text{kg m}^2}{\text{s}^2} \times \frac{1 \text{ J}}{\text{kg m}^2/\text{s}^2}$$
$$= 11,800 \text{ J}$$

Kinetic energy is due to the mass and the velocity of a moving object and is given by the formula

$$E_k = \tfrac{1}{2} mv^2$$

where E_k = kinetic energy
   m = mass of moving object
   v = velocity of moving object

A pile driver (Fig. 8.12) shows the relation of energy of motion to useful work. The energy of the driver is its kinetic energy as it hits. When the driver strikes the pile,

Figure 8.12 Energy of motion becomes useful work in the pile driver.

Courtesy of fotolia © sergey makarenko

work is done on the pile, and it is forced into the ground. The depth it goes into the ground is determined by the force applied to it. The force applied is determined by the energy of the driver. If all the kinetic energy of the driver is converted to useful work, then

$$\tfrac{1}{2}mv^2 = Fs$$

EXAMPLE 2

A pile driver with mass 10,$\overline{0}$00 kg strikes a pile with velocity 10.0 m/s. (a) What is the kinetic energy of the driver as it strikes the pile? (b) If the pile is driven 20.0 cm into the ground, what force is applied to the pile by the driver as it strikes the pile? Assume that all the kinetic energy of the driver is converted to work.

Sketch:

$m = 10,\overline{0}00$ kg

$v = 10.0$ m/s

$F = ?$

$s = 0.200$ m

Data:

$$m = 1.00 \times 10^4 \text{ kg}$$
$$v = 10.0 \text{ m/s}$$
$$s = 20.0 \text{ cm} = 0.200 \text{ m}$$
$$F = ?$$

(a) **Basic Equation:**

$$E_k = \tfrac{1}{2}mv^2$$

Working Equation: Same

Substitution:

$$E_k = \tfrac{1}{2}(1.00 \times 10^4 \text{ kg})(10.0 \text{ m/s})^2$$

$$= 5.00 \times 10^5 \frac{\text{kg m}^2}{\text{s}^2} \times \frac{1 \text{ J}}{\text{kg m}^2/\text{s}^2} \quad [1 \text{ J} = 1 \text{ N m} = 1 \text{ (kg m/s}^2)(\text{m}) = 1 \text{ kg m}^2/\text{s}^2]$$

$$= 5.00 \times 10^5 \text{ J} \quad \text{or} \quad 50\overline{0} \text{ kJ}$$

(b) **Basic Equation:**

$$E_k = W = Fs$$

Working Equation:

$$F = \frac{E_k}{s} \quad [\text{Use } E_k \text{ from part (a).}]$$

Substitution:

$$F = \frac{5.00 \times 10^5 \text{ J}}{0.200 \text{ m}} \times \frac{1 \text{ N m}}{1 \text{ J}} \quad (1 \text{ J} = 1 \text{ N m})$$

$$= 2.50 \times 10^6 \text{ N}$$

A 60.0-g bullet is fired from a gun with 3150 J of kinetic energy. Find its velocity.

EXAMPLE 3

Data:

$E_k = 3150$ J

$m = 60.0$ g $= 0.0600$ kg

$v = ?$

Basic Equation:

$E_k = \frac{1}{2}mv^2$

Working Equation:

$v = \sqrt{\dfrac{2E_k}{m}}$

Substitution:

$v = \sqrt{\dfrac{2(3150 \text{ J})}{0.0600 \text{ kg}} \times \dfrac{1 \text{ kg m}^2/\text{s}^2}{1 \text{ J}}}$ $[1 \text{ J} = 1 \text{ N m} = 1 \text{ (kg m/s}^2)(\text{m}) = 1 \text{ kg m}^2/\text{s}^2]$

$= 324$ m/s

A compressed or stretched spring contains internal potential energy. If a coiled spring is stretched, it will, up to a certain limit, return to its original shape when released. It is described as being elastic. There is, however, an elastic limit, that is, a point at which the spring will not return to its original form. If the spring is stretched too far, it remains distorted. Every solid has an elastic limit.

Robert Hooke (1635–1703) found that the amount of stretch in elastic solids is directly proportional to the distorting force if the elastic limit is not exceeded. This is the principle upon which a spring balance operates. Hooke measured this relationship in connection with the invention of a spring balance for a clock. Hooke's law is discussed in detail in Section 12.2.

Stretching the spring increases the internal potential energy of the spring in proportion to the force applied to it. Within the elastic limit, doubling the force stretching the spring doubles its internal potential energy. The same principle applies if the distorting force is compressing a spring. The internal potential energy is increased in direct proportion to the compressing force, within its elastic limit.

We have discussed only two types of energy—kinetic and potential. Keep in mind that energy exists in many other forms—chemical, atomic, electrical, sound, and heat. These forms and the conversion of energy from one form to another will be studied later.

PROBLEMS 8.3

1. Given: $m = 11.4$ kg
   $g = 9.80$ m/s^2
   $h = 22.0$ m
   $E_p = ?$

2. Given: $m = 3.50$ kg
   $g = 9.80$ m/s^2
   $h = 15.0$ m
   $E_p = ?$

3. Given: $m = 4.70$ kg
   $v = 9.60$ m/s
   $E_k = ?$

4. Given: $E_p = 93.6$ J
   $g = 9.80$ m/s^2
   $m = 2.30$ kg
   $h = ?$

SKETCH

$$12 \text{ cm}^2 \quad w$$

4.0 cm

DATA

$A = 12 \text{ cm}^2, l = 4.0 \text{ cm}, w = ?$

BASIC EQUATION

$A = lw$

WORKING EQUATION

$w = \dfrac{A}{l}$

SUBSTITUTION

$w = \dfrac{12 \text{ cm}^2}{4.0 \text{ cm}} = 3.0 \text{ cm}$

5. A truck with mass $95\overline{0}$ slugs is driven 55.0 mi/h. (a) What is its velocity in ft/s? (b) What is its kinetic energy?

6. A bullet with mass 12.0 g travels 415 m/s. Find its kinetic energy. (*Hint:* Convert 12.0 g to kg.)

7. A bicycle and rider together have a mass of 7.40 slugs. If the kinetic energy is 742 ft lb, find the velocity.

8. A crate of mass 475 kg is raised to a height 17.0 m above the floor. What potential energy has it acquired with respect to the floor?

9. A tank of water containing $250\overline{0}$ L of water is stored on the roof of a building. (a) Find its potential energy with respect to the floor, which is 12.0 m below the roof. (b) Find its potential energy with respect to the basement, which is 4.0 m below the first floor.

10. The potential energy of a girder, after being lifted to the top of a building, is 5.17×10^5 ft lb. If its mass is 173 slugs, how high is the girder?

11. A 30.0-g bullet is fired from a gun and possesses 1750 J of kinetic energy. Find its velocity.

12. The Hoover Dam is 726 ft high. Find the potential energy of 1.00 million ft³ of water at the top of the dam. (1 ft³ of water weighs 62.4 lb.)

13. A 250-kg part falls from a plane and hits the ground at 150 km/h. Find its kinetic energy.

14. A meteorite is a solid composed of stone and/or metal material from outer space that passes through the atmosphere and hits the earth's surface. Find the kinetic energy of a meteorite with mass 250 kg that hits the earth at 25 km/s.

15. Water is pumped at $25\overline{0}$ m³/min from a lake into a tank 65.0 m above the lake. (a) What power, in kW, must be delivered by the pump? (b) What horsepower rating does this pump motor have? (c) What is the increase in potential energy of the water each minute?

16. Oil is pumped at 25.0 m³/min into a tank 10.0 m above the ground (1 L of oil has a mass of 0.68 kg). (a) What power, in kW, must be delivered by the pump? (b) What is the increase in potential energy of the oil after 10.0 min? (c) Find the increase in potential energy of the oil after 10.0 min if the tank is 5.00 m above the ground.

17. If the velocity of an object is doubled, by what factor is its kinetic energy increased?

18. If the kinetic energy of an object is doubled, by what factor is its velocity increased?

19. A 4.20-g slug is shot from a rifle at 965 m/s. (a) What is the kinetic energy of the slug? (b) How much work is done on the slug if it starts from rest? (c) Find the average force on the slug if the work is done over a distance of 0.750 m. (d) If the slug comes to rest after penetrating 1.50 cm into metal, what is the magnitude of the average force it exerts?

20. A window washer with mass 90.0 kg first climbs 45.0 m upward to the top of a building, then from the top goes down 85.0 m to the ground. (a) What is the potential energy of the window washer at the top of the building, using his initial position as the reference level? (b) Find the potential energy of the window washer at ground level with respect to his initial position.

21. A painter weighing $63\overline{0}$ N climbs to a height of 5.00 m on a ladder. (a) How much work does she do in climbing the ladder? (b) What is the increase in gravitational potential energy of the painter? (c) Where does the energy come from to cause this increase in potential energy?

8.4 Conservation of Mechanical Energy

Kinetic and potential energy are related by the **law of conservation of mechanical energy**.

> **LAW OF CONSERVATION OF MECHANICAL ENERGY**
> The sum of the kinetic energy and the potential energy in a system is constant if no resistant forces do work.

A pile driver shows this energy conservation. When the driver is at its highest position, the potential energy is maximum and the kinetic energy is zero [Fig. 8.13(a)]. Its potential energy is

$$E_p = mgh$$

and its kinetic energy is

$$E_k = \tfrac{1}{2} mv^2 = \tfrac{1}{2} m(0)^2 = 0$$

Figure 8.13

(a)   (b)

When the driver hits the top of the pile [Fig. 8.13(b)], it has its maximum kinetic energy; its potential energy is

$$E_p = mgh = mg(0) = 0$$

Since the total energy in the system must remain constant, the maximum potential energy must equal the maximum kinetic energy:

$$\max E_p = \max E_k$$
$$mgh = \tfrac{1}{2} mv^2$$

Solving for the velocity of the driver as it hits the pile when the initial velocity of the driver is zero gives

$$v = \sqrt{2gh}$$

where $v =$ velocity
  $g = 9.80$ m/s^2 or 32.2 ft/s^2
  $h =$ height above the reference level

EXAMPLE 1

A pile driver falls freely from a height of 3.50 m above a pile. What is its velocity as it hits the pile?

Data:

$$h = 3.50 \text{ m}$$
$$g = 9.80 \text{ m/s}^2$$
$$v = ?$$

Basic Equation:

$$v = \sqrt{2gh}$$

Working Equation: Same

Substitution:

$$v = \sqrt{2(9.80 \text{ m/s}^2)(3.50 \text{ m})}$$
$$= 8.28 \text{ m/s} \qquad \boxed{\sqrt{\text{m}^2/\text{s}^2} = \text{m/s}}$$

· · · · · · · · · · · · · · · · ·

The conservation of mechanical energy can also be illustrated by considering a swinging pendulum where there is no resistance involved. Pull the pendulum over to the right side so that it makes an angle of 65° with the vertical [Fig. 8.14(a)]. At this point, the pendulum contains its maximum potential energy and its minimum kinetic energy (zero). Note that a larger maximum potential energy is possible when an initial deflection of greater than 65° is made.

Figure 8.14 Kinetic and potential energy changes in a pendulum

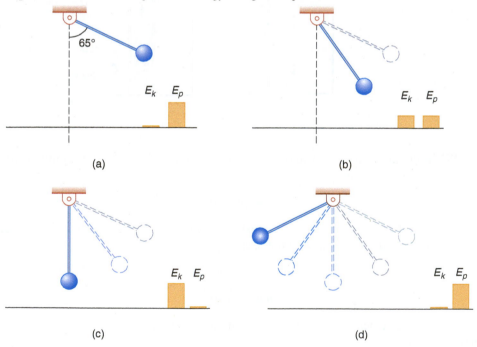

(a)

(b)

(c)

(d)

An instant later, the pendulum has lost some of its potential or stored energy, but it has gained in kinetic energy due to its motion [Fig. 8.14(b)]. At the bottom of its arc of swing [Fig. 8.14(c)], its potential energy is zero and its kinetic energy is maximum (its velocity is maximum). The kinetic energy of the pendulum then causes it to swing upward to the left. As it completes its swing [Fig. 8.14(d)], its kinetic energy is decreasing and its potential energy is increasing. That is, its kinetic energy is changing to potential energy.

According to the law of conservation of mechanical energy, the sum of the kinetic energy and the potential energy of the pendulum at any instant is a constant. Assuming no resistant forces, such as friction or air resistance, the pendulum would swing uniformly "forever."

EXAMPLE 2

Drop a 5.000-kg mass from a hot air balloon 400.0 m above the ground. Find its kinetic energy, its potential energy, and the sum of the kinetic energy and the potential energy in 1-s intervals until the mass hits the ground. (Assume no air resistance.)

Data:

$$m = 5.000 \text{ kg}$$
$$g = 9.80 \text{ m/s}^2$$

In the accompanying table, fill in each column as follows:

In column (1), list the times, t, in 1.000-s increments until the mass hits the ground.

In column (2), use $s = \frac{1}{2}gt^2$ to find the distance, s, the mass has fallen from the balloon at each time, t, rounded to the nearest 0.1 m.

In column (3), use $v = \sqrt{2gh} = \sqrt{2gs}$ to find its velocity at each time, t, rounded to the nearest 0.01 m/s.

In column (4), use $E_k = \frac{1}{2}mv^2$ to find the kinetic energy at each time, t, rounded to the nearest 10 J.

In column (5), use $h = 400.0 \text{ m} - s$ to find the height above the ground at each time, t, rounded to the nearest 0.1 m.

In column (6), use $E_p = mgh$ to find the potential energy at each time, t, rounded to the nearest 10 J.

In column (7), find the sum of the E_k and E_p columns at each time, t.

(1) t (s)	(2) s (m)	(3) v (m/s)	(4) E_k (J)	(5) h (m)	(6) E_p (J)	(7) Total (J)
0.000	0.0	0.00	0	400.0	19,6̄00	19,6̄00
1.000	4.9	9.80	240	395.1	19,360	19,6̄00
2.000	19.6	19.60	960	380.4	18,640	19,6̄00
3.000	44.1	29.40	2,160	355.9	17,440	19,6̄00
4.000	78.4	39.20	3,840	321.6	15,760	19,6̄00
5.000	122.5	49.00	6,0̄00	277.5	13,6̄00	19,6̄00
6.000	176.4	58.80	8,640	223.6	10,960	19,6̄00
7.000	240.1	68.60	11,760	159.9	7,840	19,6̄00
8.000	313.6	78.40	15,370	86.4	4,230	19,6̄00
9.000	396.9	88.20	19,450	3.1	150	19,6̄00
9.035	400.0	88.54	19,6̄00	0.0	0	19,6̄00

As you can see from the table, the sum of the kinetic energy and the potential energy at each time, t, is constant according to the *law of conservation of mechanical energy*.

Figure 8.15   Work is done to take the roller coaster car from the beginning of the ride to the top of the structure to give it potential energy. After it leaves the peak of the ride, the sum of its potential energy and its kinetic energy is equal at each point of the ride.

A roller coaster is an excellent example of the law of conservation of mechanical energy. Figure 8.15 shows a roller coaster with various points marked for discussion. First, work is done to take the roller coaster car from the beginning of the ride to the top of the structure (Point P) to give it potential energy. (Assume $v = 0$ at P for this discussion.) After it reaches the peak of the ride, the sum of its potential energy and its kinetic energy is equal at each point of the ride. That is,

Point P		Point Q		Point R	
Kinetic	Potential	Kinetic	Potential	Kinetic	Potential
0	$+ \quad mgh_p$	$= \quad \frac{1}{2}mv_Q^2$	$+ \quad mgh_Q$	$= \quad \frac{1}{2}mv_R^2$	$+ \quad 0$

In studying collisions (Chapter 6), we learned that whenever objects collide in the absence of any external forces, the total momentum of the objects before the collision equals the total momentum after the collision. This was an example of the law of conservation of momentum.

> In an *elastic* collision, the total kinetic energy in the system *before* the collision equals the total kinetic energy in the system *after* the collision.

That is,

$$\text{total kinetic energy}_{\text{before collision}} = \text{total kinetic energy}_{\text{after collision}}$$

Let's check this principle in Example 1 in Section 6.2.
Kinetic energy before collision:

$$\tfrac{1}{2}m_1v_1^2 + \tfrac{1}{2}m_2v_2^2 = \tfrac{1}{2}(0.600 \text{ kg})(9.00 \text{ m/s})^2 + \tfrac{1}{2}(0.300 \text{ kg})(-8.00 \text{ m/s})^2$$
$$= 33.9 \text{ kg m}^2/\text{s}^2 = 33.9 \text{ N m} = 33.9 \text{ J}$$

Kinetic energy after collision:

$$\tfrac{1}{2}m_1(v_1')^2 + \tfrac{1}{2}m_2(v_2')^2 = \tfrac{1}{2}(0.600 \text{ kg})(-2.33 \text{ m/s})^2 + \tfrac{1}{2}(0.300 \text{ kg})(14.66 \text{ m/s})^2$$
$$= 33.9 \text{ kg m}^2/\text{s}^2 = 33.9 \text{ N m} = 33.9 \text{ J}$$

TRY THIS ACTIVITY

Backyard Physics

Croquet players use the conservation of energy when their opponent's ball is touching their own ball and they want to knock their opponent's ball away (Fig. 8.16). To do so, a player will place his foot on his (red) ball and strike his ball with the mallet. Since the stepped-on ball remains stationary, the kinetic energy from the mallet gets transferred to the opponent's (green) ball, causing it to travel.

Figure 8.16   Step firmly on the red ball so that it will not move when it is struck with the croquet mallet. Observe what happens to the green ball.

PHYSICS CONNECTIONS

Landing on an Aircraft Carrier

Landing any airplane is a complicated procedure. Furthermore, placing a fast-moving jet on the deck of a ship in the middle of the ocean is one of the most complicated and dangerous tasks for a Navy pilot. Although the length of an aircraft carrier is over 300 m (approximately three football fields in length), its landing strip for the jets is only half that length. Even if the pilot can place the jet on the landing strip, he or she is not able to stop the jet in such a short distance without applying an external force on it.

Upon approaching the carrier, the pilot extends a mechanism called a tail-hook from the rear of the jet (Fig. 8.17(a)). This tail-hook is used to catch one of the four hydraulically controlled arresting cables on the deck of the aircraft carrier. Once the tail-hook catches one of the cables, the hydraulic cabling system extends and applies a force to slow the aircraft (Fig. 8.17(b)). The force from the cable, performed over the distance that the cable extends, works to bring the aircraft to a quick, safe stop. The hydraulic cabling system absorbs the kinetic energy of the moving plane. If the pilot misses all four arresting cables, he or she must increase the throttle to remain safely airborne after leaving the deck of the aircraft carrier and try again.

Figure 8.17(a) A fighter jet's tail-hook is positioned to latch onto one of the aircraft carrier's four arresting cables.

Figure 8.17(b) The tail-hook latches onto an arresting cable, causing the jet to transfer its kinetic energy to the cable system.

PROBLEMS 8.4

SKETCH

```
┌─────────┐
│ 12 cm²  │ w
└─────────┘
   4.0 cm
```

DATA
$A = 12\ cm^2, l = 4.0\ cm, w = ?$

BASIC EQUATION
$A = lw$

WORKING EQUATION
$w = \dfrac{A}{l}$

SUBSTITUTION
$w = \dfrac{12\ cm^2}{4.0\ cm} = 3.0\ cm$

1. A pile driver falls a distance of 2.50 m before hitting a pile. Find its velocity as it hits the pile.

2. A sky diver jumps out of a plane at a height of 50$\overline{0}$0 ft. If her parachute does not open until she reaches 10$\overline{0}$0 ft, what is her velocity at that point if air resistance is disregarded?

3. A piece of shattered glass falls from the 82nd floor of a building, 27$\overline{0}$ m above the ground. What is the velocity of the glass when it hits the ground, if air resistance is disregarded?

4. A 10.0-kg mass is dropped from a hot air balloon at a height of 325 m above the ground. Find its speed at points 30$\overline{0}$ m, 20$\overline{0}$ m, and 10$\overline{0}$ m above the ground and as it hits the ground.

5. A 0.175-lb ball is thrown upward with an initial velocity of 75.0 ft/s. What is the maximum height reached by the ball?

6. A pile driver falls a distance of 1.75 m before hitting a pile. Find its velocity as it hits the pile.

7. A sandbag is dropped from a hot air balloon at a height of 125 m above the ground. Find its velocity as it hits the ground. Disregard air resistance.

8. An ironworker drops a hammer 5.25 m to the ground. What is its speed as it hits the ground?

9. A box is dropped 3.60 m to the ground. What is its speed as it hits the ground?

10. A piece of broken glass with mass 15.0 kg falls from the side of a building 8.00 m above the street. (a) What is the kinetic energy of the glass as it hits the street? (b) What is the speed of the glass as it hits the street?

11. A ball is thrown downward from the top of a building at a speed of 75 ft/s. Find its velocity as it hits the ground 475 ft below. Disregard air resistance.

12. Find the maximum height reached by a ball thrown upward at a velocity of 95 ft/s.

13. A 4.000-kg mass is dropped from a hot air balloon 300.0 m above the ground. Find its kinetic energy, its potential energy, and the sum of the kinetic energy and the potential energy in 1-s intervals until the mass hits the ground. (Assume no air resistance.)

14. A 2.00-kg projectile is fired vertically upward with an initial velocity of 98.0 m/s. Find its kinetic energy, its potential energy, and the sum of its kinetic and potential energies at each of the following times:

   (a)   the instant of its being fired   (e) $t = 10.00$ s
   (b) $t = 1.00$ s   (f) $t = 12.00$ s
   (c) $t = 2.00$ s   (g) $t = 15.00$ s
   (d) $t = 5.00$ s   (h) $t = 20.00$ s

8.5 Alternative Energy Sources

The pile driver in Section 8.3 provides the massive steel driver with enough gravitational potential energy to repeatedly drive the pile or post into the ground. As discussed, this is an example of gravitational potential energy being transformed into kinetic energy and subsequently into useful work. However, doing work on the driver by lifting it into the air also takes energy. Typically, a gasoline- or diesel-powered engine produces the kinetic energy that is needed to lift the driver into the air.

Unfortunately, most useful machines that we utilize to do work for us use fossil fuels as the source of energy. Fossil fuels have become more expensive to obtain, refine, and distribute to consumers. In addition, many scientists argue that the long-term impact of using such products has been detrimental to our planet's environment.

Scientists have been developing other methods of harnessing energy from non–fossil fuel sources to provide cheaper, renewable sources of energy and reduce our dependence on fossil fuels. **Alternative energy** is defined as a source of energy that does not originate from fossil fuels and does not produce significant undesirable by-products that could harm the environment. Alternative energy sources include, but are not limited to, hydroelectric, wind, geothermal, and solar energy.

Hydroelectric energy takes advantage of flowing water to generate electricity. Dams built to create reservoirs have a control gate that opens toward the bottom of the dam. The escaping water spins a series of turbine blades that are connected to electrical generators. The generators create electrical energy that is distributed throughout the electrical power grid.

Wind energy is used to generate electricity in much the same manner that flowing water is utilized in hydroelectric power plants (Fig. 8.18). In the case of wind, giant windmill-like fan blades rotate and are connected to electrical generators that convert the kinetic energy of the turning fan blades into useful electrical energy. More and more wind farms are being established in windy land areas and offshore sites.

Figure 8.18   A field of modern, electricity-generating wind turbines

Courtesy of fotolia © JJAVA

Figure 8.19   Solar panels in a residential application

Geothermal energy takes advantage of the intense heat from the earth's molten core to heat underground water reservoirs. Drilling into these underground reservoirs allows steam to escape and be directed into turbines that generate electricity that can be distributed to consumers.

Solar energy uses the sun's light and a series of silicone-based solar arrays to create electrical energy (Fig 8.19). The generated electrical energy is stored in a series of batteries that can be distributed as electrical power when needed by the electrical device that is connected to the battery. Solar energy is often used to provide electricity for small and remote electrical devices such as street lights, navigational shipping lights, and orbiting satellites. Some power companies also allow their consumers to place solar panels on their roofs. Consumers can then sell the excess electricity they generate back to the power company.

The potential for alternative energy replacing or reducing our dependence on fossil fuels is promising. The United States alone needs five times more energy than it can produce through fossil fuel production. At the same time, the earth receives 10 times more energy from the sun annually than the earth's entire supply of fossil fuel could ever produce. To limit dependence on fossil fuels, scientists, policy makers, and entrepreneurs continually seek out scientific methods and develop financial incentives to increase our use of alternative sources of energy.

PHYSICS CONNECTION

Wind Energy

Wind power has become one of the largest sources of new electricity generation. New technology enables the production of zero-emissions wind power at an increasingly affordable cost. Wind projects do not use any fuel, so they are unaffected by increases in fuel prices. It is estimated that wind could supply as much as 20% of our national energy consumption. A typical 1-MW turbine generates enough electricity for 300 homes.

The wind causes large fan blades (each as long as a semi-truck) to turn electrical generators to generate electricity, which is then carried to the nation's power grid through underground transmission lines. To be commercially viable, however, the wind industry relies on government tax credits and grants to fund research and development of wind farms. Wind-farm developers compete in the marketplace to acquire leases from landowners in choice wind-farm locations. Criticism of the industry centers around government subsidies and the imposing visual presence of a wind-farm. While the appearance of a wind-farm is clearly subjective to the beholder, there is no doubt that wind power can deliver zero-emissions electricity in commercial quantities.

PHYSICS ON THE JOB

Wind Turbine Technician

Wind turbine technicians, otherwise known as windtechs, are responsible for all aspects of a wind turbine alternative energy generator. Although the technology focuses on alternative energy production, the maintenance of the turbine requires a strong background in mechanical, hydraulic, and electrical systems. In particular, windtechs spend the majority of maintenance work keeping the nacelle in proper working condition. The nacelle is the horizontally aligned section located directly behind the rotors. It contains the gearbox, rotor shaft, and brake assembly.

A wind turbine technician typically requires an associate's degree and engages in additional on-the-job training with the specific manufacturer of the wind turbine system. Becoming a windtech can be a good opportunity for electricians because a major component of the technology involved in the turbine is the electrical system. Finally, a windtech must not be afraid to work in confined spaces high above the ground (Fig. 8.20)!

Figure 8.20   Wind turbine technologists

Courtesy of fotolia © mediagram

Glossary

Energy The ability to do work. There are many forms of energy, such as mechanical, electrical, thermal, fluid, chemical, atomic, and sound. (p. 214)

Gravitational Potential Energy The energy determined by the position of an object relative to a particular reference level. (p. 214)

Internal Potential Energy The energy determined by the nature or condition of a substance. (p. 214)

Kinetic Energy The energy due to the mass and the velocity of a moving object. (p. 214)

Law of Conservation of Mechanical Energy The sum of the kinetic energy and the potential energy in a system is constant if no resistant forces do work. (p. 219)

Potential Energy The stored energy of a body due to its internal characteristics or its position. (p. 214)

Power The rate of doing work (work divided by time). (p. 208)

Work The product of the force in the direction of motion and the magnitude of the displacement. (p. 202)

Formulas

8.1 $W = Fs$

$W = Fs \cos \theta$

8.2 $P = \dfrac{W}{t}$

8.3 $E_p = mgh$

$E_k = \frac{1}{2} mv^2$

8.4 $v = \sqrt{2gh}$

Review Questions

1. Work is done when
    (a) a force is applied.   (b) a person tries unsuccessfully to move a crate.
    (c) force is applied and an object is moved.
2. Power
    (a) is work divided by time.   (b) is measured in newtons.
    (c) is time divided by work.   (d) none of the above.
3. A large boulder at rest possesses
    (a) potential energy.   (b) kinetic energy.   (c) no energy.
4. A large boulder rolling down a hill possesses
    (a) potential energy.   (b) kinetic energy.
    (c) no energy.   (d) both kinetic and potential energy.
5. With no air resistance and no friction, a pendulum would
    (a) not swing.   (b) swing for a short time.   (c) swing forever.
6. Can work be done by a moving object on itself?
7. Has a man swinging a sledgehammer done work if he misses the stake at which he is swinging?
8. Develop the units associated with work from the components of the definition: work = force × displacement.
9. Is work a vector quantity?
10. Is work being done on a boulder by gravity?
11. Is work being done by the weight of a grandfather clock?
12. How could the power developed by a man pushing a stalled car be measured?
13. How does water above a waterfall possess potential energy?
14. What are two devices possessing gravitational potential energy?
15. Is kinetic energy dependent on time?
16. At what point is the kinetic energy of a swinging pendulum at a maximum?
17. At what point is the potential energy of a swinging pendulum at a maximum?
18. Is either kinetic or potential energy a vector quantity?
19. Can an object possess both kinetic and potential energy at the same time?
20. Why is a person more likely to be severely injured by a bolt falling from the fourth floor of a job site than by one falling from the second floor?

Review Problems

1. How many joules are in one kilowatt-hour?
2. An endloader holds $15\overline{0}0$ kg of sand 2.00 m off the ground for 3.00 min. How much work does it do?
3. How high can a 10.0-kg mass be lifted by $100\overline{0}$ J of work?
4. A 40.0-kg pack is carried up a $25\overline{0}0$-m-high mountain in 10.0 h. How much work is done?
5. Find the average power output in Problem 4 in (a) watts; (b) horsepower.
6. A 10.0-kg mass has a potential energy of 10.0 J when it is at what height?
7. A 10.0-lb weight has a potential energy of 20.0 ft lb at what height?
8. At what speed does a 1.00-kg mass have a kinetic energy of 1.00 J?
9. At what speed does a 10.0-N weight have a kinetic energy of 1.00 J?
10. What is the kinetic energy of a $30\overline{0}0$-lb automobile moving at 55.0 mi/h?
11. What is the potential energy of an 80.0-kg diver standing 3.00 m above the water?
12. What is the kinetic energy of a 0.020-kg bullet having a velocity of 550 m/s?
13. What is the potential energy of an 85.0-kg high jumper clearing a 2.00-m bar?
14. A worker pulls a crate 10.0 m by exerting a force of $30\overline{0}$ N. (a) How much work does the worker do? (b) How much work does the worker do pulling the crate a distance of 10.0 m by exerting the same force at an angle of 20.0° with the horizontal?
15. A hammer falls from a scaffold on a building 50.0 m above the ground. Find its speed as it hits the ground.

SKETCH

12 cm² w

4.0 cm

DATA

$A = 12$ cm², $l = 4.0$ cm, $w = ?$

BASIC EQUATION

$A = lw$

WORKING EQUATION

$w = \dfrac{A}{l}$

SUBSTITUTION

$w = \dfrac{12\ \text{cm}^2}{4.0\ \text{cm}} = 3.0$ cm

APPLIED CONCEPTS

1. Rosita needs to purchase a sump pump for her basement. (a) If the pump must carry 10.0 kg of water to a height of 2.75 m each minute, what minimum wattage pump is needed? (b) What three main factors determine power for a sump pump?

2. A roller coaster designer must carefully balance the desire for excitement and the need for safety. The most recent design is shown in Fig. 8.21. (a) If a 355-kg roller coaster car has zero velocity on the top of the first hill, determine its potential energy. (b) What is the velocity of the roller coaster car at the specified locations in the design? (c) Explain the relationship between velocity and the position on the track throughout the ride. (Consider the track to be frictionless.)

Figure 8.21

3. A 22,500-kg Navy fighter jet flying 235 km/h must catch an arresting cable to land safely on the runway strip of an aircraft carrier. (a) How much energy must the cable absorb to stop the fighter jet? (b) If the cable allows the jet to move 115 m before coming to rest, what is the average force that the cable exerts on the jet? (c) If the jet were given more than 115 m to stop, how would the force applied by the cable change?

4. The hydroelectric plant at the Itaipu Dam, located on the Parana River between Paraguay and Brazil, uses the transfer of potential to kinetic energy of water to generate electricity. (a) If 1.00×10^6 gallons of water (3.79×10^6 kg) flows down 142 m into the turbines each second, how much power does the hydroelectric power plant generate? (For comparison purposes, the Hoover Dam generates 1.57×10^6 W of power.) (b) How much power could the plant produce if the Itaipu Dam were twice its actual height? (c) Explain why the height of a dam is important for hydroelectric power plants.

5. A 1250-kg wrecking ball is lifted to a height of 12.7 m above its resting point. When the wrecking ball is released, it swings toward an abandoned building and makes an indentation of 43.7 cm in the wall. (a) What is the potential energy of the wrecking ball at a height of 12.7 m? (b) What is its kinetic energy as it strikes the wall? (c) If the wrecking ball transfers all of its kinetic energy to the wall, how much force does the wrecking ball apply to the wall? (d) Why should a wrecking ball strike a wall at the lowest point in its swing?

CHAPTER 9

ROTATIONAL MOTION

Objectives

The major goals of this chapter are to enable you to:

1. Distinguish among rectilinear, curvilinear, and rotational motion.
2. Find angular displacement, velocity, and acceleration.
3. Use conservation of angular momentum to describe rotational motion.
4. Find centripetal force.
5. Find power in rotational systems.
6. Analyze how gears, gear trains, and pulleys are used to transfer rotational motion.

The concepts of displacement, velocity, acceleration, vectors, and forces in a straight line also apply to motion in a curved path and rotational motion.

9.1 Measurement of Rotational Motion

Until now we have considered only motion in a straight line, called **rectilinear motion**. Technicians are often faced with many problems with motion along a curved path or with objects rotating about an axis. Although these kinds of motion are similar, we must distinguish between them.

Motion along a curved path is called **curvilinear motion**. A satellite in orbit around the earth is an example of curvilinear motion [Fig. 9.1(a)].

Figure 9.1

(a) Curvilinear motion of an orbiting satellite

(b) Rotational motion of earth spinning on its axis

(c) Rotational motion of a wheel spinning on its axle

Rotational motion occurs when the body itself is spinning. Examples of rotational motion are the earth spinning on its axis, a turning wheel, a turning driveshaft, and the turning shaft of an electric motor [Fig. 9.1(b) and (c)].

We can see a wheel turn, but to gather useful information about its motion, we need a system of measurement. There are three basic systems of defining angle measurement. The **revolution** is one complete rotation of a body. This unit of measurement of rotational motion is the number of rotations—how many times the object goes around. The unit of rotation (most often used in industry) is the revolution (rev). A second system of angular measurement divides the circle of rotation into 360 degrees (360° = 1 rev). One **degree** is 1/360 of a complete revolution.

Figure 9.2

When $s = r$, $\theta = 1$ rad

The **radian** (rad), which is approximately 57.3° or exactly $\left(\dfrac{360}{2\pi}\right)^{\circ}$, is a third angular unit of measurement. A radian is defined as that angle with its vertex at the center of a circle whose sides cut off an arc on the circle equal to its radius (Fig. 9.2), where $s = r$ and $\theta = 1$ rad.

By definition, the radian measure of an angle is the ratio of the length of arc, s, that the angle intercepts on a circle to the length of its radius, r; that is,

$$\theta = \frac{s}{r}$$

where θ = angle determined by s and r
  s = length of the arc of the circle
  r = radius of the circle

Technically, angle θ measured in radians is defined as the ratio of two lengths: the lengths of the arc and the radius of a circle. Since the length units in the ratio cancel, the radian is a dimensionless unit. As a matter of convenience, "rad" is often used to show radian measurement. A useful relationship is 2π rad equals one revolution. Therefore,

$$1 \text{ rev} = 360° = 2\pi \text{ rad}$$

You may need to use this conversion between systems of measurement.

EXAMPLE 1

Convert the angle 10π rad (a) to rev and (b) to degrees.

Using 1 rev = 360° = 2π rad, form conversion factors so that the old units are in the denominator and the new units are in the numerator.

(a) $\theta = (10\pi \text{ rad})\left(\dfrac{1 \text{ rev}}{2\pi \text{ rad}}\right)$

$= 5$ rev

(b) $\theta = (10\pi \text{ rad})\left(\dfrac{360°}{2\pi \text{ rad}}\right)$

$= 1800°$

Angular displacement is the angle through which any point on a rotating body moves. Note that on any rotating body, all points on that body move through the same angle in any given amount of time—even though each may travel different linear distances. Point A on the flywheel shown in Fig. 9.3 travels much farther than point B (along a curved line), but during one revolution both travel through the same angle (have equal angular displacements).

In the automobile industry, technicians are concerned with the *rate* of rotational motion. Recall that in the linear system, velocity is the rate of motion (displacement/time). Similarly, **angular velocity** in the rotational system is the rate of angular displacement. Angular velocity (designated ω, the Greek lowercase letter omega) is usually measured in rev/min (rpm) for relatively slow rotations (e.g., automobile engines) and rev/s or rad/s for high-speed instruments. We use the term *angular velocity* when referring to a vector that includes the direction of rotation. We use the term *angular speed* in referring to a magnitude when the direction of rotation is either not known or not important.

Figure 9.3 The angular displacements of points A and B on the flywheel are always the same.

$$\omega = \text{angular velocity} = \frac{\text{number of revolutions}}{\text{time}} = \frac{\text{angular displacement}}{\text{time}}$$

Written as a formula,

$$\omega = \frac{\theta}{t}$$

where

ω = angular velocity or speed (rad/s)	or ω = angular velocity or speed (rev/min)
θ = angle (in radians)	θ = angle (in revolutions)
t = time (in seconds)	t = time (in minutes)

EXAMPLE 2

A motorcycle wheel turns $36\overline{0}0$ times while being ridden for 6.40 min. What is the angular speed in rev/min?

Data:

$$t = 6.40 \text{ min}$$

$$\text{number of revolutions} = 36\overline{0}0 \text{ rev}$$

$$\omega = ?$$

Basic Equation:

$$\omega = \frac{\theta}{t}$$

Working Equation: Same

Substitution:

$$\omega = \frac{3600 \text{ rev}}{6.40 \text{ min}}$$

$$= 563 \text{ rev/min or } 563 \text{ rpm}$$

. . . .

Formulas for linear speed of a rotating point on a circle and angular speed are related as follows. We know

$$(1) \quad \theta = \frac{s}{r} \qquad (2) \quad v = \frac{s}{t} \qquad (3) \quad \omega = \frac{\theta}{t}$$

Therefore, combining and substituting s/r for θ in (3), we obtain

$$\omega = \frac{s/r}{t}$$

$$\omega(r) = \frac{(s/r)(r)}{t} \qquad \text{Multiply both sides by } r.$$

$$\omega r = \frac{s}{t}$$

$$\omega r = v \qquad \text{Recall that } v = s/t.$$

Thus,

$$\boxed{v = \omega r}$$

where v = linear velocity of a point on the circle
ω = angular speed
r = radius

EXAMPLE 3

A wheel of 1.00 m radius turns at $10\overline{0}0$ rpm. (a) Express the angular speed in rad/s. (b) Find the angular displacement in 2.00 s. (c) Find the linear speed of a point on the rim of the wheel.

Sketch:

1.00 m

(a) Data:

$\omega = 10\overline{0}0$ rpm   (change to rad/s)

$$\omega = 10\overline{0}0 \; \frac{\text{rev}}{\text{min}} \times \frac{2\pi \text{ rad}}{1 \text{ rev}} \times \frac{1 \text{ min}}{60 \text{ s}} = 105 \text{ rad/s}$$

(b) Data:

$$\omega = 105 \text{ rad/s}$$
$$t = 2.00 \text{ s}$$
$$\theta = ?$$

Basic Equation:

$$\omega = \frac{\theta}{t}$$

Working Equation:

$$\theta = \omega t$$

Substitution:

$$\theta = (105 \text{ rad/s})(2.00 \text{ s})$$
$$= 21\overline{0} \text{ rad}$$

(c) Data:

$$\omega = 105 \text{ rad/s}$$
$$r = 1.00 \text{ m}$$
$$v = ?$$

Basic Equation:

$$v = \omega r$$

Working Equation: Same

Substitution:

$$v = (105 \text{ rad/s})(1.00 \text{ m})$$
$$= 105 \text{ m/s}$$

> (rad/s)(m) = m/s because the rad is a dimensionless unit.

• • • • • • • • • • • • • • • • • •

A device called a *stroboscope* or strobe light may be used to measure or check the speed of rotation of a shaft or other machinery part. Repeating motion is "slowed down" so it can be observed more easily. The light flashes rapidly, and the rate of flash can be adjusted to coincide with the rotation of a point or points on the rotating object. Knowing the rate of flashing will also then reveal the rate of rotation. A slight variation in the rate of rotation and flash will cause the observed point to appear to move either forward or backward as the stagecoach wheels in old western movies sometimes appear to do. Figure 9.4 shows a stroboscopic motion time-lapse photo of a tennis ball as it approaches the ground after being struck by a racket.

Figure 9.4

Photo courtesy of fotolia © Gregory Johnston

In linear motion, we found a change in velocity results in an acceleration. Similarly, in rotational motion, changing the rate of rotation involves a change in angular velocity and results in an angular acceleration. For uniformly accelerated rotational motion, **angular acceleration** is the rate of change of angular velocity. That is,

$$\alpha = \frac{\omega}{t}$$

where α = angular acceleration
ω = change in angular velocity
t = time

The equations for uniformly accelerated linear motion in Section 4.3 may easily be transformed into the corresponding equations for uniformly accelerated rotational motion by substituting θ for s, ω for v, and α for a.

Linear Motion	**Rotational Motion**
$s = v_{avg}t$	$\theta = \omega_{avg}t$
$s = v_i t + \frac{1}{2}at^2$	$\theta = \omega_i t + \frac{1}{2}\alpha t^2$
$v_{avg} = \dfrac{v_f + v_i}{2}$	$\omega_{avg} = \dfrac{\omega_f + \omega_i}{2}$
$v_f = v_i + at$	$\omega_f = \omega_i + \alpha t$
$a = \dfrac{v_f - v_i}{t}$	$\alpha = \dfrac{\omega_f - \omega_i}{t}$
$s = \frac{1}{2}(v_f + v_i)t$	$\theta = \frac{1}{2}(\omega_f + \omega_i)t$
$2as = v_f^2 - v_i^2$	$2\alpha\theta = \omega_f^2 - \omega_i^2$

where s = linear displacement
v_f = final linear velocity
v_i = initial linear velocity
v_{avg} = average linear velocity
a = linear acceleration
t = time

where θ = angular displacement
ω_f = final angular velocity
ω_i = initial angular velocity
ω_{avg} = average angular velocity
α = angular acceleration
t = time

While these rotational motion equations are somewhat intuitive, their derivations are beyond the scope of this text.

EXAMPLE 4

A rotating pulley 24.0 cm in diameter is rotating at an initial angular speed of 30.5 rad/s. The speed is steadily increased to 41.5 rad/s within 6.30 s. (a) Find the pulley's angular acceleration. (b) Find the final linear speed of a point on its rim.

(a) Data:

$$\omega_i = 30.5 \text{ rad/s}$$
$$\omega_f = 41.5 \text{ rad/s}$$
$$t = 6.30 \text{ s}$$
$$\alpha = ?$$

Basic Equation:

$$\omega_f = \omega_i + \alpha t$$

Working Equation:

$$\alpha = \frac{\omega_f - \omega_i}{t}$$

Substitution:

$$\alpha = \frac{41.5 \text{ rad/s} - 30.5 \text{ rad/s}}{6.30 \text{ s}}$$

$$= 1.75 \text{ rad/s}^2$$

(b) Data:

$$\omega = 41.5 \text{ rad/s}$$
$$r = 12.0 \text{ cm}$$
$$v = ?$$

Basic Equation:

$$v = \omega r$$

Working Equation: Same

Substitution:

$$v = (41.5 \text{ rad/s})(12.0 \text{ cm})$$

$$= 498 \text{ cm/s}$$

· · · · · · · · · · · · · · · ·

TRY THIS ACTIVITY

Global Rotational Physics

On a globe depicting the earth, place a push pin or a small mark on northern Canada, another on Florida, and another on Ecuador. Rotate the globe about its central axis. Of the places that you marked, where would a person experience the greatest linear velocity as the earth is rotating? Where would a person experience the greatest angular velocity?

Finally, what would happen to a person's angular velocity if the rotational speed of the earth changed from one rotation per day to two rotations per day?

PROBLEMS 9.1

1. Convert $6\frac{1}{2}$ revolutions
   (a) to radians.
   (b) to degrees.
2. Convert 2880°
   (a) to revolutions.
   (b) to radians.
3. Convert 25π rad
   (a) to revolutions.
   (b) to degrees.
4. Convert 12.0 revolutions
   (a) to radians.
   (b) to degrees.

Find the angular speed in Problems 5–10.

5. Number of revolutions = 525
   $t = 3.42$ min
   $\omega =$ _____ rpm

6. Number of revolutions = 7360
   $t = 37.0$ s
   $\omega =$ _____ rev/s

7. Number of revolutions = 4.00
   $t = 3.00$ s
   $\omega =$ _____ rad/s

8. Number of revolutions = 325
   $t = 5.00$ min
   $\omega =$ _____ rpm

9. Number of revolutions = 6370
   $t = 18.0$ s
   $\omega =$ _____ rev/s

10. Number of revolutions = 6.25
    $t = 5.05$ s
    $\omega =$ _____ rad/s

11. Convert 675 rad/s to rpm.

12. Convert 285 rpm to rad/s.

13. Convert 136 rpm to rad/s.

14. Convert 88.4 rad/s to rpm.

15. A motor turns at a rate of 11.0 rev/s. Find its angular speed in rpm.

16. A rotor turns at a rate of 180 rpm. Find its angular speed in rev/s.

17. A rotating wheel completes one revolution in 0.150 s. Find its angular speed
    (a) in rev/s.   (b) in rpm.   (c) in rad/s.

18. A rotor completes 50.0 revolutions in 3.25 s. Find its angular speed
    (a) in rev/s.   (b) in rpm.   (c) in rad/s.

19. A flywheel rotates at 1050 rpm. (a) How long (in s) does it take to complete one revolution? (b) How many revolutions does it complete in 5.00 s?

20. A wheel rotates at 36.0 rad/s. (a) How long (in s) does it take to complete one revolution? (b) How many revolutions does it complete in 8.00 s?

21. A shaft of radius 8.50 cm rotates 7.00 rad/s. Find its angular displacement (in rad) in 1.20 s.

22. A wheel of radius 0.240 m turns at 4.00 rev/s. Find its angular displacement (in rev) in 13.0 s.

23. A pendulum of length 1.50 m swings through an arc of 5.0°. Find the length of the arc through which the pendulum swings.

24. An airplane circles an airport twice while 5.00 mi from the control tower. Find the length of the arc through which the plane travels.

25. A wheel of radius 27.0 cm has an angular speed of 47.0 rpm. Find the linear speed (in m/s) of a point on its rim.

26. A belt is placed around a pulley that is 30.0 cm in diameter and rotating at 275 rpm. Find the linear speed (in m/s) of the belt. (Assume no belt slippage on the pulley.)

27. A flywheel of radius 25.0 cm is rotating at 655 rpm. (a) Express its angular speed in rad/s. (b) Find its angular displacement (in rad) in 3.00 min. (c) Find the linear distance traveled (in cm) by a point on the rim in one complete revolution. (d) Find the linear distance traveled (in m) by a point on the rim in 3.00 min. (e) Find the linear speed (in m/s) of a point on the rim.

28. An airplane propeller with blades 2.00 m long is rotating at 1150 rpm. (a) Express its angular speed in rad/s. (b) Find its angular displacement in 4.00 s. (c) Find the linear speed (in m/s) of a point on the end of the blade. (d) Find the linear speed (in m/s) of a point 1.00 m from the end of the blade.

29. An automobile is traveling at 60.0 km/h. Its tires have a radius of 33.0 cm. (a) Find the angular speed of the tires (in rad/s). (b) Find the angular displacement of the tires in 30.0 s. (c) Find the linear distance traveled by a point on the tread in 30.0 s. (d) Find the linear distance traveled by the automobile in 30.0 s.

30. Find the angular speed (in rad/s) of the following hands on a clock.
    (a) Second hand   (b) Minute hand   (c) Hour hand

31. A bicycle wheel of diameter 30.0 in. rotates twice each second. Find the linear velocity of a point on the wheel.

32. A point on the rim of a flywheel with radius 1.50 ft has a linear velocity of 30.0 ft/s. Find the time for it to complete 4π rad.

33. The earth rotates on its axis at an angular speed of 1 rev/24 h. Find the linear speed (in km/h) (a) of Singapore, which is nearly on the equator. (b) of Houston, which is approximately 30.0° north latitude. (c) of Minneapolis, which is approximately 45.0° north latitude. (d) of Anchorage, which is approximately 60.0° north latitude.

34. A truck tire rotates at an initial angular speed of 21.5 rad/s. The driver steadily accelerates, and after 3.50 s the tire's angular speed is 28.0 rad/s. What is the tire's angular acceleration during its linear acceleration?

35. Find the angular acceleration of a radiator fan blade as its angular speed increases from 8.50 rad/s to 15.4 rad/s in 5.20 s.

36. A wheel of radius 20.0 cm starts from rest and makes 6.00 revolutions in 2.50 s. (a) Find its angular velocity in rad/s. (b) Find its angular acceleration. (c) Find the final linear speed of a point on the rim of the wheel. (d) Find the linear acceleration of a point on the rim of the wheel.

37. A circular disk 30.0 cm in diameter is rotating at 275 rpm and then uniformly stops within 8.00 s. (a) Find its angular acceleration. (b) Find the initial linear speed of a point on its rim. (c) How many revolutions does the disk make before it stops?

38. A rotating flywheel of diameter 40.0 cm uniformly accelerates from rest to 250 rad/s in 15.0 s. (a) Find its angular acceleration. (b) Find the linear velocity of a point on the rim of the wheel after 15.0 s. (c) How many revolutions does the wheel make during the 15.0 s?

NEW TECHNOLOGIES

Football Helmet Technology

Figure 9.5 Football helmets collide in tackles

Courtesy of fotolia © Olexandr

The Biomedical Engineering Department at Virginia Tech University and the School of Biomedical Engineering and Science at Wake Forest University have focused attention on the high-tech world of football helmets. Most football helmets protect very well against skull fractures. Yet focus has shifted to the forces exerted on the head and the internal bruising deep inside the brain's white matter that strains nerve cells and the nerve fibers. This type of injury is called a concussion and repeated concussions can lead to permanent neurological problems later in life.

Researchers at Virginia Tech are using Head Impact Telemetry Systems, a series of accelerometers positioned inside helmets. They are discovering that linear acceleration, resulting in a one-dimensional blow to the head/brain system, is not nearly as damaging as a blow resulting in rotational acceleration to the head/brain system. Most hits are off-center and cause rotational acceleration of the head/brain. Thus, the nerves and the nerve fibers are twisting and turning inside the head, causing additional strain on the system and resulting in a more serious concussion.

Helmets need to protect against not only linear forces, but also rotational forces as well. Many manufacturers have redesigned helmets to provide for additional padding in high-impact areas of the head, but new technology is combating the rotational force issue. A Swedish company, Multidirectional Impact Protection System, claims that a plastic layer, when fit snugly on a player's head between the helmet padding and the skull, allows the head to "float" a bit when struck by a rotational force. The helmet rotates, but the head/brain system does not rotate as much, reducing the twisting and turning inside the head. The manufacturer claims to have reduced brain rotation by as much as 40%. Currently, the product is found in bicycle and ski helmets for the general consumer but is not yet used in football helmets (Fig. 9.5).

9.2 Angular Momentum

Recall that in Section 5.1 we saw that *inertia* is the property of a body that causes it to remain at rest if it is at rest or to continue moving with a constant velocity unless an unbalanced force acts upon it. Similarly, *rotational inertia,* called the **moment of inertia**, is the property of a rotating body that causes it to continue to turn until a torque causes it to change its rotational motion. A freely spinning wheel on an upside-down bicycle continues to spin after you stop hand cranking the pedals because of its rotational inertia.

In Section 7.3, we saw that a *torque* is produced when a force is applied to produce a rotation ($\tau = F_r$). To increase the speed of the spinning bicycle wheel above, additional torque must be applied by increasing the applied force, *F*. The angular acceleration of a rotating body is found to be directly proportional to the torque applied to it. This applied torque can be expressed as follows:

$$\tau = I\alpha$$

where τ = applied torque
   I = moment of inertia (rotational inertia)
   α = angular acceleration

This is the rotational equivalent of Newton's second law of motion and applies to a rigid body rotating about a fixed axis.

The moment of inertia, *I*, is a measure of the rotational inertia of a body. The rotational inertia is determined by the mass of the rotating object and how far away that mass is from its axis of rotation. Figure 9.6 shows two cylinders of equal mass, one solid and one hollow. The hollow cylinder has more rotational inertia because its mass is concentrated farther from its axis of rotation. A flywheel is a mechanical device, which is usually a heavy metal rotating wheel attached to a drive shaft with most of its weight concentrated at its circumference. A small motor can slowly increase the speed of a flywheel to store up kinetic energy; then, the small motor uses this inertia to perform a task for which it is ordinarily too small. A flywheel is also used to minimize rotational variations due to fluctuations in load and applied torque. Figure 9.7 shows a flywheel that is used to produce a steady rotation where the applied force of the piston is intermittent in this two-cylinder engine.

Figure 9.6   Even though the two cylinders are of equal mass, the hollow cylinder has more rotational inertia because its mass is concentrated farther from its axis of rotation.

Figure 9.7   The large flywheel on this 1913 Case 30–60 tractor produces steady rotation of the crankshaft between fuel ignition cycles in its two-cylinder engine.

In Section 6.1 we saw that (linear) *momentum* is a measure of the amount of inertia and motion an object has or of the difficulty in bringing a moving body to rest. The formula for linear momentum is $p = mv$. There we studied applications involving linear motion and found the impulse is the change in linear momentum, $Ft = mv_f - mv_i$. Similarly, **angular momentum** for a rotating body about a fixed axis is defined as

$$L = I\omega$$

where L = angular momentum
I = moment of inertia (rotational inertia)
ω = angular velocity

Note the comparison with linear dynamics:

angular momentum = (moment of inertia) $\times$ (angular velocity)

linear momentum = (mass, a measure of inertia) $\times$ (linear velocity)

Furthermore, the *angular impulse* is the change in angular momentum.

$$\tau t = I\omega_f - I\omega_i$$

where τ = torque
t = time
I = moment of inertia
ω_f = final angular velocity
ω_i = initial angular velocity

Compare the following pairs of equations for linear motion and rotational motion:

Linear Motion	Rotational Motion
$F = ma$	$\tau = I\alpha$
$p = mv$	$L = I\omega$
$Ft = mv_f - mv_i$	$\tau t = I\omega_f - I\omega_i$

where F = applied force
m = mass (inertia)
a = linear acceleration
p = linear momentum
v = linear velocity
t = time
v_f = final linear velocity
v_i = initial linear velocity

where τ = applied torque
I = moment of inertia
α = angular acceleration
L = angular momentum
ω = angular velocity
t = time
ω_f = final angular velocity
ω_i = initial angular velocity

Conservation of Angular Momentum

In Section 6.1 we learned from the law of conservation of momentum that the total linear momentum ($p = mv$) of a system remains unchanged unless an external force acts on it. Similarly, the **law of conservation of angular momentum** states that the total angular momentum ($L = I\omega$) of a system remains unchanged unless an external torque acts on it.

LAW OF CONSERVATION OF ANGULAR MOMENTUM
The angular momentum of a system remains unchanged unless an external torque acts on it.

A spinning ice skater is an interesting example. When the skater's arms are extended, the rotational inertia, I, is relatively large and the angular velocity, ω, is relatively small. Often at the end of a spin, the skater pulls his or her arms tight to the body, resulting in a much faster spin (larger angular velocity) because of a much smaller rotational inertia, I. When a rotating body contracts, its angular velocity, ω, increases; and when a rotating body expands, its angular velocity decreases. This phenomenon is the result of the conservation of angular momentum.

Similarly, gymnasts and divers generate their spins (torque) from a solid base or a diving board after which the angular momentum remains unchanged. The usual somersaults and twists result from making variations in their rotational inertia. Astronauts must also learn to control their spins as they maneuver their bodies to work in space.

TRY THIS ACTIVITY

Spinning Non–Ice Skaters

Ice skaters are able to control their rotational speed by changing their rotational inertia. To do this without a pair of skates, sit in an office chair that rotates freely and hold a couple of weights or heavy books close to your chest. Have someone spin the chair and gradually increase your rotational speed. After reaching your optimum rotational speed, stretch out your arms, holding the weights in your hands. What happens to your rotational speed as you move your arms from close to your chest to an outstretched position? Try this a few times. Explain how you can change your rotational speed like ice skaters do.

9.3 Centripetal Force

Figure 9.8 Rock on a string being swung in a circle. The centripetal force is directed toward the center.

Newton's laws of motion apply to motion along a curved path as well as in a straight line. Recall that a moving body tends to continue in a straight line because of inertia. If we are to cause the body to move in a circle, we must constantly apply a force perpendicular to the line of motion of the body. A simple example is a rock on the end of a string being swung in a circle (Fig. 9.8). By Newton's first law, the rock tends to go in a straight line but the string exerts a constant force on the rock perpendicular to this line of travel. The resulting path of the rock is a circle [Fig. 9.9(a)]. The force of the string on the rock is the *centripetal* (toward the center) force. The **centripetal force** acting on a body in circular motion causes it to move in a circular path. This force is exerted toward the center of the circle. If the string should break, however, there would no longer be a centripetal force acting on the rock, which would fly off tangent to the circle [Fig. 9.9(b)].

Figure 9.9 Centripetal force on a rock being swung in a circle

(a)   (b)

A familiar example of centripetal force is the spin cycle of an automatic washing machine. The wet clothes are spun in the tub after the washing cycle is completed. The restrained clothes continue in a circular motion while the water is spun away through holes in the tub. Centripetal force toward the agitator holds the clothes on the spinning tub. The unrestrained water flows outward and is drained away.

The equation for finding the centripetal force on any body moving along a curved path is

$$F = \frac{mv^2}{r}$$

where F = centripetal force
m = mass of the body
v = velocity of the body
r = radius of curvature of the path of the body

TRY THIS ACTIVITY

Whirling a Bucket of Water

Hold a bucket with some water in your hand and quickly swing it vertically in a circular motion. Why does the water not fall out of the bucket when the bucket is upside down? Gradually decrease the speed of the whirling bucket while measuring the approximate rotational speed of the bucket. Measure the radius of the circular arc from your shoulder to the bottom of the bucket, the mass of the bucket and water, and the rotational speed of the bucket to find the minimum centripetal force needed to keep the water in the bucket. Next, double the amount of water in the bucket and repeat the experiment. How does the amount of water influence the rotational speed needed to keep the water in the bucket?

An automobile of mass 1640 kg rounds a curve of radius 25.0 m with a velocity of 15.0 m/s (54.0 km/h). What centripetal force is exerted on the automobile while rounding the curve?

EXAMPLE 1

Sketch:

15.0 m/s

$r = 25.0$ m

Data:

$m = 1640$ kg
$v = 15.0$ m/s
$r = 25.0$ m
$F = ?$

Basic Equation:

$$F = \frac{mv^2}{r}$$

Working Equation: Same

Substitution:

$$F = \frac{(1640 \text{ kg})(15.0 \text{ m/s})^2}{25.0 \text{ m}}$$

$$= 14,800 \text{ kg m/s}^2 \qquad (\text{Recall: } 1 \text{ N} = 1 \text{ kg m/s}^2)$$

$$= 14,800 \text{ N}$$

. . . .

SKETCH

12 cm² | w
4.0 cm

DATA
$A = 12 \text{ cm}^2, l = 4.0 \text{ cm}, w = ?$

BASIC EQUATION
$A = lw$

WORKING EQUATION
$w = \dfrac{A}{l}$

SUBSTITUTION
$w = \dfrac{12 \text{ cm}^2}{4.0 \text{ cm}} = 3.0 \text{ cm}$

PROBLEMS 9.3

1. Given: $m = 64.0$ kg
   $v = 34.0$ m/s
   $r = 17.0$ m
   $F = $ _____ N

2. Given: $m = 11.3$ slugs
   $v = 3.00$ ft/s
   $r = 3.24$ ft
   $F = $ _____ lb

3. Given: $F = 25\overline{0}0$ lb
   $v = 47.6$ ft/s
   $r = 72.0$ ft
   $m = $ _____ slugs

4. Given: $F = 587$ N
   $v = 0.780$ m/s
   $m = 67.0$ kg
   $r = $ _____ m

5. Given: $F = 602$ N
   $m = 63.0$ kg
   $r = 3.20$ m
   $v = $ _____ m/s

6. Given: $m = 37.5$ kg
   $v = 17.0$ m/s
   $r = 3.75$ m
   $F = $ _____ N

7. Given: $F = 75.0$ N
   $v = 1.20$ m/s
   $m = 10\overline{0}$ kg
   $r = $ _____ m

8. Given: $F = 80.0$ N
   $m = 43.0$ kg
   $r = 17.5$ m
   $v = $ _____ m/s

9. An automobile of mass 117 slugs follows a curve of radius 79.0 ft with a speed of 49.3 ft/s. What centripetal force is exerted on the automobile while it is rounding the curve?

10. Find the centripetal force exerted on a 7.12-kg mass moving at a speed of 2.98 m/s in a circle of radius 2.72 m.

11. The centripetal force on a car of mass $80\overline{0}$ kg rounding a curve is 6250 N. If its speed is 15.0 m/s, what is the radius of the curve?

12. The centripetal force on a runner is 17.0 lb. If the runner weighs 175 lb and his speed is 14.0 mi/h, find the radius of the curve.

13. An automobile with mass 1650 kg is driven around a circular curve of radius $15\overline{0}$ m at 80.0 km/h. Find the centripetal force of the road on the automobile.

14. A cycle of mass 510 kg rounds a curve of radius $4\overline{0}$ m at 95 km/h. What is the centripetal force on the cycle?

15. What is the centripetal force exerted on a rock with mass 3.2 kg moving at 3.5 m/s in a circle of radius 2.1 m?

16. What is the centripetal force on a $150\overline{0}$-kg vehicle driven around a curve of radius 35.0 m at 60.0 km/h?

17. What is the centripetal force on a $75\overline{0}$-kg vehicle rounding a curve of radius 40.0 m at 30.0 km/h?

18. A truck with mass 215 slugs rounds a curve of radius 53.0 ft with a speed of 62.5 ft/s. (a) What centripetal force is exerted on the truck while rounding the curve? (b) How does the centripetal force change when the velocity is doubled? (c) What is the new force?

19. A 225-kg dirt bike is rounding a curve with linear velocity of 35 m/s and an angular speed of 0.25 rad/s. Find the centripetal force exerted on the bike.
20. A 55,000-kg truck rounds a curve at 62.0 km/h. If the radius of the curve is 38.0 m, what is the centripetal force on the truck?
21. The radius of a curve is 27.5 m. What is the centripetal force on a 10,000-kg truck going around it at 35.0 km/h?

9.4 Power in Rotational Systems

One of the most important aspects of rotational motion is the power developed. Recall that torque was discussed in Section 7.3. Power, however, must be considered whenever an engine or motor is used to turn a shaft. Some common examples are drive trains and winches (Fig. 9.10).

Figure 9.10   This driveshaft connects the engine transmission with the axle to supply power to the wheels and other components of this tractor. (Courtesy of Deere & Company, Moline, IL, USA)

Earlier we learned that

$$\text{power} = \frac{\text{force} \times \text{displacement}}{\text{time}} = \frac{\text{work}}{\text{time}}$$

in the linear system. In the rotational system

$$P = \frac{(\text{torque})(\text{angular displacement})}{\text{time}}$$
$$= (\text{torque})(\text{angular velocity})$$
$$= \tau\omega$$

Recall that angular displacement is the angle through which a shaft is turned. In the metric system, angular displacement must be expressed in radians (1 rev = 2π radians). Substituting symbols and units, we have, in watts (W),

$$P = \tau\omega$$

$$= (\text{N m})\left(\frac{1}{\text{s}}\right) = \frac{\text{N m}}{\text{s}} = \frac{\text{J}}{\text{s}} = \text{W}$$

To find the power in kilowatts (kW), multiply the number of watts by the conversion factor

$$\frac{1\ \text{kW}}{1000\ \text{W}}$$

Note: In problem solving, the radian unit is a dimensionless unit; ω is expressed with the unit /s.

EXAMPLE 1

How many watts of power are developed by a mechanic tightening bolts using 50.0 N m of torque at a rate of 2.50 rad/s? How many kW?

Data:

$$\tau = 50.0 \text{ N m}$$
$$\omega = 2.50/s$$
$$P = ?$$

Basic Equation:

$$P = \tau\omega$$

Working Equation: Same

Substitution:

$$P = (50.0 \text{ N m})(2.50/s)$$
$$= 125 \text{ N m/s}$$
$$= 125 \text{ W} \quad (1 \text{ W} = 1 \text{ N m/s})$$

To find the power in kW:

$$125 \text{ W} \times \frac{1 \text{ kW}}{1000 \text{ W}} = 0.125 \text{ kW}$$

·················

In the U.S. system, we measure angular displacement by multiplying the number of revolutions by 2π:

$$\text{angular displacement} = (\text{number of revolutions})(2\pi)$$

For the rotational system

$$\text{power} = \frac{(\text{torque})(2\pi \text{ revolutions})}{\text{time}}$$

When time is in minutes,

$$\text{power} = \text{torque} \times 2\pi \times \frac{\text{rev}}{\text{min}} \times \frac{1 \text{ min}}{60 \text{ s}}$$

$$\text{power in } \frac{\text{ft lb}}{\text{s}} = \text{torque in lb ft} \times \frac{\text{number of revolutions}}{\text{min}} \times 0.105 \underbrace{\frac{\text{min}}{\text{rev s}}}_{2\pi \times \frac{1 \text{ min}}{60 \text{ s}}}$$

Another common unit of power is the horsepower (hp). The conversion factor between $\frac{\text{ft lb}}{\text{s}}$ and hp is

$$\text{power in hp} = \text{power in } \frac{\text{ft lb}}{\text{s}} \times \frac{\text{hp}}{550 \text{ ft lb/s}}$$

What power (in ft lb/s) is developed by an electric motor with torque 5.70 lb ft and speed 425 rpm?

EXAMPLE 2

Data:

$$\tau = 5.70 \text{ lb ft}$$
$$\omega = 425 \text{ rpm}$$
$$P = ?$$

Basic Equation:

$$P = \text{torque} \times \frac{\text{rev}}{\text{min}} \times 0.105 \frac{\text{min}}{\text{rev s}}$$

Working Equation: Same

Substitution:

$$P = (5.70 \text{ lb ft})\left(425\frac{\text{rev}}{\text{min}}\right)\left(0.105 \frac{\text{min}}{\text{rev s}}\right)$$

$$= 254 \text{ ft lb/s}$$

What horsepower is developed by a racing engine with torque 545 lb ft at $65\overline{0}0$ rpm? First, find power in ft lb/s and then convert to hp.

EXAMPLE 3

Data:

$$\tau = 545 \text{ lb ft}$$
$$\omega = 65\overline{0}0 \text{ rpm}$$
$$P = ?$$

Basic Equation:

$$P = \text{torque} \times \frac{\text{rev}}{\text{min}} \times 0.105 \frac{\text{min}}{\text{rev s}}$$

Working Equation: Same

Substitution:

$$P = (545 \text{ lb ft})\left(65\overline{0}0 \frac{\text{rev}}{\text{min}}\right)\left(0.105 \frac{\text{min}}{\text{rev s}}\right)$$

$$= 372{,}000 \frac{\text{ft lb}}{\text{s}} \times \frac{1 \text{ hp}}{550 \frac{\text{ft lb}}{\text{s}}}$$

$$= 676 \text{ hp}$$

SKETCH

12 cm^2 | w

4.0 cm

DATA
$A = 12 \text{ cm}^2, l = 4.0 \text{ cm}, w = ?$

BASIC EQUATION
$A = lw$

WORKING EQUATION
$w = \dfrac{A}{l}$

SUBSTITUTION
$w = \dfrac{12 \text{ cm}^2}{4.0 \text{ cm}} = 3.0 \text{ cm}$

PROBLEMS 9.4

1. Given: $\tau = 125$ lb ft
   $\omega = 555$ rpm
   $P = \underline{\quad}$ ft lb/s

2. Given: $\tau = 39.4$ N m
   $\omega = 6.70/s$
   $P = \underline{\quad}$ W

3. Given: $\tau = 372$ lb ft
   $\omega = 264$ rpm
   $P = \underline{\quad}$ hp

4. Given: $\tau = 65\overline{0}$ N m
   $\omega = 45.0/s$
   $P = \underline{\quad}$ kW

5. Given: $P = 8950$ W
   $\omega = 4.80/s$
   $\tau = \underline{\quad}$

6. Given: $P = 650$ W
   $\tau = 540$ N m
   $\omega = \underline{\quad}$

7. What horsepower is developed by an engine with torque $40\overline{0}$ lb ft at $450\overline{0}$ rpm?

8. What torque must be applied to develop 175 ft lb/s of power in a motor if $\omega = 394$ rpm?

9. Find the angular velocity of a motor developing 649 W of power with torque 131 N m.

10. A high-speed industrial drill develops 0.500 hp at $160\overline{0}$ rpm. What torque is applied to the drill bit?

11. An engine has torque of 550 N m at 8.3 rad/s. What power in watts does it develop?

12. Find the angular velocity of a motor developing 33.0 N m/s of power with a torque of 6.0 N m.

13. What power (in hp) is developed by an engine with torque 524 lb ft
    (a) at $300\overline{0}$ rpm?   (b) at $600\overline{0}$ rpm?

14. Find the angular velocity of a motor developing 650 W of power with a torque of 130 N m.

15. A drill develops 0.500 kW of power at $180\overline{0}$ rpm. What torque is applied to the drill bit?

16. What power is developed by an engine with torque $75\overline{0}$ N m applied at $450\overline{0}$ rpm?

17. A tangential force of 150 N is applied to a flywheel of diameter 45 cm to maintain a constant angular velocity of 175 rpm. How much work is done per minute?

18. Find the power developed by an engine with a torque of $120\overline{0}$ N m applied at $200\overline{0}$ rpm.

19. Find the power developed by an engine with a torque of $160\overline{0}$ N m applied at $150\overline{0}$ rpm.

20. Find the power developed by an engine with torque 1250 N m applied at $500\overline{0}$ rpm.

21. Find the angular velocity of a motor developing $100\overline{0}$ W of power with a torque of $15\overline{0}$ N m.

22. A motor develops 0.75 kW of power at $200\overline{0}$ revolutions per $1\overline{0}$ min. What torque is applied to the motor shaft?

23. What power is developed when a tangential force of 175 N is applied to a flywheel of diameter 86 cm, causing it to have an angular velocity of 36 revolutions per 6.0 s?

24. What power is developed when a tangential force of $25\overline{0}$ N is applied to a wheel 57.0 cm in diameter with an angular velocity of 25.0 revolutions in 13.0 s?

25. An engine develops 1.50 kW of power at 10,$\overline{0}$00 revolutions per 5.00 min. What torque is applied to the engine's crankshaft?

26. A mechanic tightens engine bolts using 45.5 N m of torque at a rate of 2.75 rad/s. How many watts of power are used to tighten the bolts?

27. An ag mechanic tightens implement bolts using 52.5 N m of torque at a rate of 2.25 rad/s. What power does the mechanic develop in tightening the bolts?

9.5 Transferring Rotational Motion

Suppose that two disks are touching each other as in Fig. 9.11. Disk *A* is driven by a motor and turns disk *B* (wheel) by making use of the friction between them. The relationship between the diameters of the two disks and their number of revolutions is

$$D \cdot N = d \cdot n$$

where D = diameter of the driver disk
         d = diameter of the driven disk
         N = number of revolutions of the driver disk
         n = number of revolutions of the driven disk

Figure 9.11 In the self-propelled lawn mower, disk *A*, driven by the motor, turns disk *B*, the wheel, which results in the mower moving along the ground.

The relationship between the circumference and the diameter of a wheel is given by $C = \pi d$. If two rotating wheels are touching, the linear distance a point on each wheel travels must be the same. The smaller wheel must turn faster for the two distances traveled to be equal. Therefore, the diameter of the larger wheel times its number of revolutions equals the diameter of the smaller wheel times its number of revolutions.

However, using two disks to transfer rotational motion is not very efficient due to slippage that may occur between them. The most common ways to prevent disk slippage are placing teeth on the edge of each disk and connecting the disks with a belt. Therefore, instead of using disks, we use gears or belt-driven pulleys to transfer this motion. The teeth on the gears eliminate the slippage; the belt connecting the pulleys helps reduce the slippage and provides for distance between rotating centers (Fig. 9.12).

We can change the equation $D \cdot N = d \cdot n$ to the form $D/d = n/N$ by dividing both sides by dN. The left side indicates the ratio of the diameters of the disks. If the ratio is 2, this means that the larger disk must have a diameter two times the diameter of the smaller disk. The same ratio would apply to gears and pulleys. The ratio of the diameters of the gears must be 2 to 1, and the ratio of the diameters of the pulleys must be 2 to 1. In fact, the ratio of the number of teeth on the gears must be 2 to 1.

The right side of the equation indicates the ratio of the number of revolutions of the two disks. If the ratio is 2, this means that the smaller disk makes two revolutions

Figure 9.12 Gears and pulleys are used to reduce slippage in transferring rotational motion.

(a)   (b)   (c)

while the larger disk makes one revolution. The same would be true for gears and for pulleys connected by a belt.

Gears and pulleys are used to increase or reduce the angular velocity of a rotating shaft or wheel. When two gears or pulleys are connected, the speed at which each turns compared to the other is inversely proportional to the diameter of that gear or pulley. The larger the diameter of a pulley or gear, the slower it turns. The smaller the diameter of a pulley or gear, the faster it will turn when connected to a larger one.

9.6 Gears

Gears are used to transfer rotational motion from one gear to another. The gear that causes the motion is called the *driver gear*. The gear to which the motion is transferred is called the *driven gear*.

There are many different sizes, shapes, and types of gears. Some examples are shown in Fig. 9.13. For any type of gear, we use one basic formula:

Figure 9.13 Examples of different types of gears, some of which may have a diameter of up to 120 in. Some gears are very small, depending on the application. (Photos courtesy of Circle Gear and Machine Company, Inc., Cicero, IL)

(a) Spur gears
Courtesy of Circle Gear and Machine Co., Inc

(b) Helical gears
Courtesy of Circle Gear and Machine Co., Inc

Figure 9.13   (*Continued*)

(c) Herringbone gears
Courtesy of Circle Gear and Machine Co., Inc

(d) Worm and worm gears
Courtesy of Circle Gear and Machine Co., Inc

(e) Straight bevels
Courtesy of Circle Gear and Machine Co., Inc

(f) Spiral bevels
Courtesy of Circle Gear and Machine Co., Inc

$$T \cdot N = t \cdot n$$

where T = number of teeth on the driver gear
    N = number of revolutions of the driver gear
    t = number of teeth on the driven gear
    n = number of revolutions of the driven gear

The above gear equation is somewhat intuitive. When two gears are meshed, the teeth fit together. For any given rotation of one gear, the same number of teeth on each gear must make contact with each other. Therefore, since the number of contacts on each gear must be equal, the smaller gear turns more times for each revolution of the larger gear. The driver gear turns the driven gear.

A driver gear has 30 teeth. How many revolutions does the driven gear with 20 teeth make while the driver makes one revolution?

EXAMPLE 1

Data:

$T = 30$ teeth $t = 20$ teeth
$N = 1$ revolution $n = ?$

Basic Equation:

$$T \cdot N = t \cdot n$$

Working Equation:

$$n = \frac{T \cdot N}{t}$$

Substitution:

$$n = \frac{(30 \text{ teeth})(1 \text{ rev})}{20 \text{ teeth}}$$

$$= 1.5 \text{ rev}$$

· · · · · · · · · · · · · · · ·

EXAMPLE 2

A driven gear of 70 teeth makes 63.0 revolutions per minute (rpm). The driver gear makes 90.0 rpm. What is the number of teeth required for the driver gear?

Data:

$$N = 90.0 \text{ rpm}$$
$$t = 70 \text{ teeth}$$
$$n = 63.0 \text{ rpm}$$
$$T = ?$$

Basic Equation:

$$T \cdot N = t \cdot n$$

Working Equation:

$$T = \frac{t \cdot n}{N}$$

Figure 9.14 Meshed gears

A   B

Substitution:

$$T = \frac{(70 \text{ teeth})(63.0 \text{ rpm})}{90.0 \text{ rpm}}$$

$$= 49 \text{ teeth}$$

· · · · · · · · · · · · · · · ·

Figure 9.15 Gear train of three gears

A   B

Gear Trains

When two gears mesh (Fig. 9.14), they turn in opposite directions. If gear *A* turns clockwise, gear *B* turns counterclockwise. If gear *A* turns counterclockwise, gear *B* turns clockwise. If a third gear is inserted between the two (Fig. 9.15), then gears *A* and *B* are rotating in the same direction. This third gear is called an *idler*. A **gear train** is a series of gears that transfers rotational motion from one gear to another.

When the number of shafts in a gear train is odd (such as 1, 3, 5, ...), the first gear and the last gear rotate in the same direction. When the number of shafts is even, the gears rotate in opposite directions.

When a complex gear train is considered, the relationship between revolutions and number of teeth is still present. This relationship is: The number of revolutions of the first driver times the product of the number of teeth of all the driver gears equals the number of revolutions of the final driven gear times the product of the number of teeth on all the driven gears. That is,

$$NT_1T_2T_3T_4 \cdots = nt_1t_2t_3t_4 \cdots$$

where N = number of revolutions of first driver gear
T_1 = teeth on first driver gear
T_2 = teeth on second driver gear
T_3 = teeth on third driver gear
T_4 = teeth on fourth driver gear
n = number of revolutions of last driven gear
t_1 = teeth on first driven gear
t_2 = teeth on second driven gear
t_3 = teeth on third driven gear
t_4 = teeth on fourth driven gear

Determine the relative motion of gears A and B in Fig. 9.16.

EXAMPLE 3

Figure 9.16

(a)

(b)

(c)

(d)

Find the number of revolutions per minute of gear D in Fig. 9.17 if gear A rotates at 20.0 rpm. Gears A and C are drivers and gears B and D are driven.

EXAMPLE 4

Data:

$N = 20.0$ rpm $t_1 = 45$ teeth
$T_1 = 30$ teeth $t_2 = 60$ teeth
$T_2 = 15$ teeth $n = ?$

Figure 9.17

60 teeth   D
45 teeth   B
30 teeth   A
C
15 teeth

Basic Equation:

$$NT_1T_2 = nt_1t_2$$

Working Equation:

$$n = \frac{NT_1T_2}{t_1t_2}$$

Substitution:

$$n = \frac{(20.0 \text{ rpm})(30 \text{ teeth})(15 \text{ teeth})}{(45 \text{ teeth})(60 \text{ teeth})}$$

$$= 3.33 \text{ rpm}$$

· · · · · · · · · · · · · · · ·

EXAMPLE 5

Find the rpm of gear D in the train shown in Fig. 9.18. Gears A and C are drivers and gears B and D are driven.

Figure 9.18

D
A   B
C
48 teeth
15 teeth
30 teeth
60 teeth at 1600 rpm

Data:

$$N = 16\overline{0}0 \text{ rpm} \qquad t_1 = 30 \text{ teeth}$$
$$T_1 = 60 \text{ teeth} \qquad t_2 = 48 \text{ teeth}$$
$$T_2 = 15 \text{ teeth} \qquad n = ?$$

Basic Equation:

$$NT_1T_2 = nt_1t_2$$

Working Equation:

$$n = \frac{NT_1T_2}{t_1t_2}$$

Substitution:

$$n = \frac{(16\overline{0}0 \text{ rpm})(60 \text{ teeth})(15 \text{ teeth})}{(30 \text{ teeth})(48 \text{ teeth})}$$

$$= 10\overline{0}0 \text{ rpm}$$

· · · · · · · · · · · · · · · ·

EXAMPLE 6

In the gear train shown in Fig. 9.19, find the speed in rpm of gear A.

Figure 9.19

20 teeth   45 teeth   10 teeth
B   30 teeth
A   C   12 teeth at 225 rpm
D   E

Data:

$$t_1 = 45 \text{ teeth} \qquad T_1 = 20 \text{ teeth}$$
$$t_2 = 10 \text{ teeth} \qquad T_2 = 45 \text{ teeth}$$
$$t_3 = 12 \text{ teeth} \qquad T_3 = 30 \text{ teeth}$$
$$n = 225 \text{ rpm} \qquad N = ?$$

Gear B is both a driver and a driven gear.

Basic Equation:

$$NT_1T_2T_3 = nt_1t_2t_3$$

Working Equation:

$$N = \frac{nt_1t_2t_3}{T_1T_2T_3}$$

Substitution:

$$N = \frac{(225 \text{ rpm})(45 \text{ teeth})(10 \text{ teeth})(12 \text{ teeth})}{(20 \text{ teeth})(45 \text{ teeth})(30 \text{ teeth})}$$
$$= 45.0 \text{ rpm}$$

· · · · · · · · · · · · · · · · ·

In a gear train, when a gear is both a driver gear and a driven gear, it may be omitted from the computation.

The problem in Example 6 could have been worked as follows because gear B is both a driver and a driven gear.

EXAMPLE 7

Basic Equation:

$$NT_1T_3 = nt_2t_3$$

Figure 9.20

Working Equation:

$$N = \frac{nt_2t_3}{T_1T_3}$$

Substitution:

$$N = \frac{(225 \text{ rpm})(10 \text{ teeth})(12 \text{ teeth})}{(20 \text{ teeth})(30 \text{ teeth})}$$
$$= 45.0 \text{ rpm}$$

· · · · · · · · · · · · · · · ·

The mechanical pocket watch in Fig. 9.20 shows a number of interacting gears transferring rotational energy.

PHYSICS CONNECTIONS

Bicycle Gears

The gearing system on a bicycle allows a cyclist to choose how much force he or she would like to exert when riding a bicycle. What gear ratio should be used when riding uphill, downhill, or on level ground? Prior to the use of gears on bicycles, a rider needed to sit directly above the front wheel in order to pedal. Gears, chains, and other advances have made the modern bicycle much more comfortable and efficient.

To find the gear ratio of a bicycle's gearing system, divide the number of teeth on the rear, driven gear by the number of teeth on the front, driver gear. When the number of teeth on the front gear is larger, the gear ratio is less than 1 and the rear wheel turns faster than the pedals. This results in high speeds for going down slight inclines or for level ground and allows the cyclist to pedal fewer revolutions while traveling a greater distance. When the number of teeth in the rear gear is larger, the gear ratio is greater than 1. Here, the pedals turn faster than the rear wheel. This results in low speeds for going up large hills and allows the cyclist to pedal more revolutions while traveling a shorter distance but exerting a more manageable leg force. The following table illustrates the differences:

Number of Teeth on Front Gear (Driver)	Number of Teeth on Rear Gear (Driven)	Gear Ratio (Driven/Driver)	Number of Teeth on Front Gear (Driver)	Number of Teeth on Rear Gear (Driven)	Gear Ratio (Driven/Driver)
44	11	1/4	15	30	2/1
When the front gear turns once, the back gear turns four times (good for traveling at high speeds).			When the front gear turns once, the back gear turns only $\frac{1}{2}$ a rotation (good for reducing the force needed to pedal while going uphill).		

Most beginner bicycles simply connect the chain on the front gear to the rear gear with no option for changing the gear ratios. Children must stand up to pedal with more force. Geared bicycles, including road and mountain bikes, are engineered with a variety of gear ratios to allow a cyclist to travel more easily over many terrains (Fig. 9.21).

Figure 9.21   (a) Beginner bicycle

(b) Mountain bike

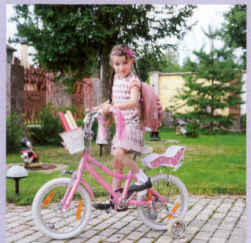

Courtesy of fotolia © Alinute

Courtesy of fotolia © Luis Viegas

PROBLEMS 9.6

Fill in the blanks.

	Number of Teeth		rpm	
	Driver	Driven	Driver	Driven
1.	16	48	156	_____
2.	36	24	_____	225
3.	18	_____	72.0	54.0
4.	_____	64	148	55.5
5.	48	36	_____	276
6.	16	12	144	_____

7. A driver gear has 36 teeth and makes 85.0 rpm. Find the rpm of the driven gear with 72 teeth.

8. A motor turning at 1250 rpm is fitted with a gear having 54 teeth. Find the speed of the driven gear if it has 45 teeth.

9. A gear running at $25\overline{0}$ rpm meshes with another revolving at $10\overline{0}$ rpm. If the smaller gear has 30 teeth, how many teeth does the larger gear have?

10. A driver gear with 40 teeth makes 154 rpm. How many teeth must the driven gear have if it makes $22\overline{0}$ rpm?

11. Two gears have a speed ratio of 4.2 to 1. If the smaller gear has 15 teeth, how many teeth does the larger gear have?

12. What size gear should be meshed with a 15-tooth pinion to achieve a speed reduction of 10 to 3?

13. A driver gear has 72 teeth and makes 162 rpm. Find the rpm of the driven gear with 81 teeth.

14. A driver gear with 60 teeth makes 1600 rpm. How many teeth must the driven gear have if it makes 480 rpm?

15. What size gear should be meshed with a 20-tooth pinion to achieve a speed reduction of 3 to 1?

16. A motor turning at $150\overline{0}$ rpm is fitted with a gear having 60 teeth. Find the speed of the driven gear if it has 40 teeth.

17. The larger of two gears in a clock has 36 teeth and turns at a rate of 0.50 rpm. How many teeth does the smaller gear have if it rotates at 1/30 rev/s?

18. How many revolutions does an 88-tooth gear make in 10.0 min when it is meshed with a 22-tooth pinion rotating at 44 rpm?

If gear *A* turns in a clockwise motion, determine the motion of gear *B* in each gear train.

SKETCH

12 cm²   *w*

4.0 cm

DATA

$A = 12 \text{ cm}^2, l = 4.0 \text{ cm}, w = ?$

BASIC EQUATION

$A = lw$

WORKING EQUATION

$w = \dfrac{A}{l}$

SUBSTITUTION

$w = \dfrac{12 \text{ cm}^2}{4.0 \text{ cm}} = 3.0 \text{ cm}$

19.

20.

21.

22.

23.

24.

25.

26.

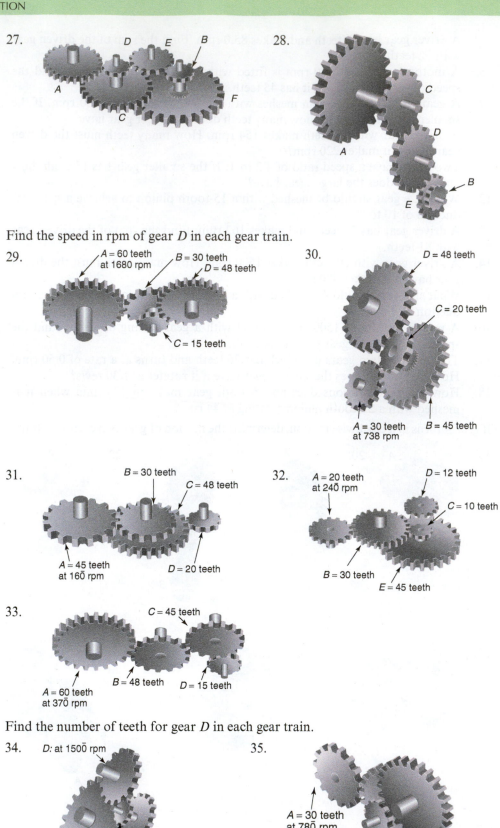

27. A D E B C F

28. C D A B E

Find the speed in rpm of gear D in each gear train.

29. $A = 60$ teeth at 1680 rpm, $B = 30$ teeth, $D = 48$ teeth, $C = 15$ teeth

30. $D = 48$ teeth, $C = 20$ teeth, $A = 30$ teeth at 738 rpm, $B = 45$ teeth

31. $B = 30$ teeth, $C = 48$ teeth, $A = 45$ teeth at $16\bar{0}$ rpm, $D = 20$ teeth

32. $A = 20$ teeth at $24\bar{0}$ rpm, $D = 12$ teeth, $C = 10$ teeth, $B = 30$ teeth, $E = 45$ teeth

33. $C = 45$ teeth, $A = 60$ teeth at $37\bar{0}$ rpm, $B = 48$ teeth, $D = 15$ teeth

Find the number of teeth for gear D in each gear train.

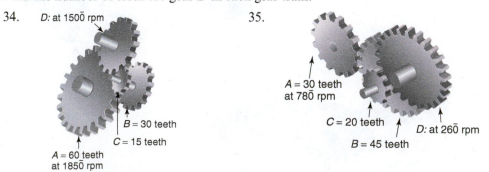

34. D: at $150\bar{0}$ rpm, $B = 30$ teeth, $C = 15$ teeth, $A = 60$ teeth at $185\bar{0}$ rpm

35. $A = 30$ teeth at $78\bar{0}$ rpm, $C = 20$ teeth, $B = 45$ teeth, D: at $26\bar{0}$ rpm

36.

A = 45 teeth
at 160 rpm

B = 30 teeth

D: at 576 rpm

C = 48 teeth

37.

D: at 1125 rpm

E = 45 teeth

C = 10 teeth

B = 30 teeth

A = 20 teeth
at 250 rpm

38.

D: at 555 rpm

C = 45 teeth

A = 60 teeth
at 370 rpm

B = 48 teeth

39. Find the direction of rotation of gear B if gear A is turned counterclockwise in Problems 22 through 28.

40. Find the effect of doubling the number of teeth on gear A in Problem 38.

9.7 Pulleys Connected with a Belt

Pulleys connected with a belt are used to transfer rotational motion from one shaft to another (Fig. 9.22). Assuming no slippage, the linear speed of any point on the belt equals the linear speed of any point on the rim of each pulley as the belt travels around each pulley. The larger the pulley, the larger is its circumference: $C = \pi d$. The larger the circumference of the pulley, the longer a point on the belt stays in contact with the pulley. The smaller the circumference of the pulley, the shorter a point on the belt stays in contact with the pulley, which causes the smaller pulley to rotate faster than the larger pulley. Two pulleys connected with a belt have a relationship similar to gears. Assuming no slippage, when two pulleys are connected,

Figure 9.22 A single belt drives several components of this engine.

Courtesy of fotolia © Boggy

$$D \cdot N = d \cdot n$$

where D = diameter of the driver pulley
        N = number of revolutions per minute of the driver pulley
        d = diameter of the driven pulley
        n = number of revolutions per minute of the driven pulley

The preceding equation may be generalized in the same manner as for gear trains as follows:

$$ND_1 D_2 D_3 \cdots = n d_1 d_2 d_3 \cdots$$

EXAMPLE 1

Find the speed in rpm of pulley A shown in Fig. 9.23.

Data:

$$D = 6.00 \text{ in.}$$
$$d = 30.0 \text{ in.}$$
$$n = 35\overline{0} \text{ rpm}$$
$$N = ?$$

Figure 9.23
Driver diameter = 6.00 in.

Diameter = 30.0 in.
at 35$\overline{0}$ rpm

Basic Equation:

$$D \cdot N = d \cdot n$$

Working Equation:

$$N = \frac{dn}{D}$$

Substitution:

$$N = \frac{(30.0 \text{ in.})(35\overline{0} \text{ rpm})}{6.00 \text{ in.}}$$
$$= 1750 \text{ rpm}$$

. . . .

When two pulleys are connected with an open-type belt, the pulleys turn in the same direction. When two pulleys are connected with a cross-type belt, the pulleys turn in opposite directions. See Fig. 9.24.

Figure 9.24   Crossing a belt reverses direction.

Open type

Cross type

(a) Pulleys rotate in
same direction.

(b) Pulleys rotate in
opposite directions.

PROBLEMS 9.7

Find each missing quantity using $D \cdot N = d \cdot n$.

	D	N	d	n
1.	18.0	$150\overline{0}$	12.0	___
2.	36.0	___	9.00	972
3.	12.0	$18\overline{0}0$	6.00	___
4.	___	2250	9.00	1125
5.	49.0	1860	___	$62\overline{0}$

6. A driver pulley of diameter 6.50 in. revolves at 1650 rpm. Find the speed of the driven pulley if its diameter is 26.0 in.
7. A driver pulley of diameter 25.0 cm revolves at $12\overline{0}$ rpm. At what speed will the driven pulley turn if its diameter is 48.0 cm?
8. One pulley of diameter 36.0 cm revolves at $60\overline{0}$ rpm. Find the diameter of the second pulley if it rotates at $36\overline{0}$ rpm.
9. One pulley rotates at $45\overline{0}$ rpm. The diameter of the second pulley is 15.0 in. and rotates at 675 rpm. Find the diameter of the first pulley.
10. A pulley with a radius of 10.0 cm rotates at $12\overline{0}$ rpm. The radius of the second pulley is 15.0 cm; find its rpm.

Determine the direction of pulley B in each pulley system.

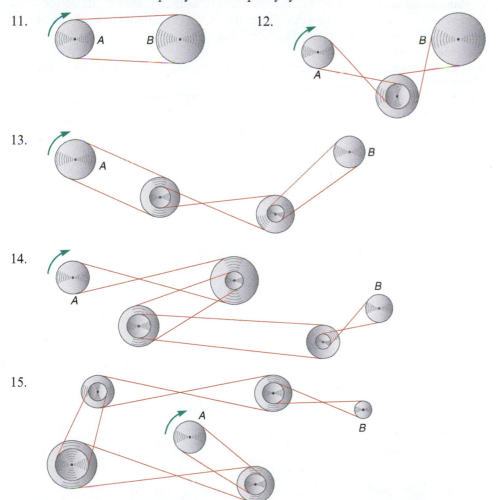

11.

12.

13.

14.

15.

16. What size pulley should be placed on a countershaft turning $15\overline{0}$ rpm to drive a grinder with a 12.0-cm pulley that is to turn at $120\overline{0}$ rpm?

Glossary

Angular Acceleration The rate of change of angular velocity (change in angular velocity/time). (p. 236)

Angular Displacement The angle through which any point on a rotating body moves. (p. 233)

Angular Momentum For a rotating body about a fixed axis, the angular momentum is the product of the moment of inertia and the angular velocity of the body. (p. 241)

Angular Velocity The rate of angular displacement (angular displacement/time). (p. 233)

Centripetal Force The force acting on a body in circular motion that causes it to move in a circular path. This force is exerted toward the center of the circle. (p. 242)

Curvilinear Motion Motion along a curved path. (p. 232)

Degree An angular unit of measure. Defined as 1/360 of one complete revolution. (p. 232)

Gear Train A series of gears that transfers rotational motion from one gear to another. (p. 252)

Law of Conservation of Angular Momentum The angular momentum of a system remains unchanged unless an outside torque acts on it. (p. 241)

Moment of Inertia Rotational inertia; the property of a rotating body that causes it to continue to turn until a torque causes it to change its rotational motion. (p. 240)

Radian An angular unit of measurement. Defined as that angle with its vertex at the center of a circle whose sides cut off an arc on the circle equal to its radius. Equal to $(360°/2\pi)$ or approximately $57.3°$. (p. 232)

Rectilinear Motion Motion in a straight line. (p. 231)

Revolution A unit of measurement in rotational motion; one complete rotation of a body. (p. 232)

Rotational Motion Spinning motion of a body. (p. 232)

Formulas

9.1 $\quad \theta = \dfrac{s}{r}$

$1 \text{ rev} = 360° = 2\pi \text{ rad}$

$\omega = \dfrac{\theta}{t}$

$v = \omega r$

$\alpha = \dfrac{\omega}{t}$

$\theta = \omega_{avg} t$

$\theta = \omega_i t + \frac{1}{2}\alpha t^2$

$\omega_{avg} = \dfrac{\omega_f + \omega_i}{2}$

$\omega_f = \omega_i + \alpha t$

$\alpha = \dfrac{\omega_f - \omega_i}{t}$

$\theta = \frac{1}{2}(\omega_f + \omega_i)t$

$2\alpha\theta = \omega_f^2 - \omega_i^2$

9.2 $\tau = I\alpha$

$L = I\omega$

$\tau t = I\omega_f - I\omega_i$

9.3 $F = \dfrac{mv^2}{r}$

9.4 $P = \tau\omega$

$\text{power in } \dfrac{\text{ft lb}}{\text{s}} = \text{torque in lb ft} \times \dfrac{\text{number of revolutions}}{\text{min}} \times 0.105 \dfrac{\text{min}}{\text{rev s}}$

$\text{power in hp} = \text{power in } \dfrac{\text{ft lb}}{\text{s}} \times \dfrac{\text{hp}}{550 \text{ ft lb/s}}$

9.5 $D \cdot N = d \cdot n$

9.6 $T \cdot N = t \cdot n$

$NT_1 T_2 T_3 T_4 \cdots = nt_1 t_2 t_3 t_4 \cdots$

9.7 $D \cdot N = d \cdot n$

$ND_1 D_2 D_3 \cdots = nd_1 d_2 d_3 \cdots$

Review Questions

1. Angular velocity is measured in
    (a) revolutions/minute.   (b) radians/second.
    (c) revolutions/second.   (d) all of the above.
2. Power in the rotational system
    (a) is found in the same way as in the linear system.
    (b) is found differently than it is in the linear system.
    (c) cannot be determined.   (d) is always a constant.
3. A gear train has 13 directly connected gears. The first and last gears will
    (a) rotate in opposite directions.   (b) rotate in the same direction.
    (c) not rotate.
4. Distinguish between curvilinear motion and rotational motion.
5. Name the two types of measurement of rotation.
6. In your own words, define *radian*.
7. What is angular displacement? In what units is it measured?
8. How is linear velocity of a point on a circle related to angular velocity?
9. How do equations for uniformly accelerated rotational motion compare with those for uniformly accelerated linear motion?
10. A girl jumping from a high platform into a pool tucks her body into a tight ball to complete two somersaults before extending her body as she enters the water. Her body rotates much more quickly during the somersaults than during her extension into the water. This is an example of what law?
11. Is the tangent to a circle always perpendicular to the radius?
12. Will inertia tend to keep a moving body following a curved path?
13. Explain the relationship between the number of teeth on two interlocking gears and their relative number of revolutions.
14. How does the presence of an idler gear affect the relationship between a driver gear and a driven gear in a gear train?
15. When the number of directly connected gears in a gear train is four, do the first and last gears in the train rotate in the same or in opposite directions?

16. Why can a gear that is both a driver gear and a driven gear be omitted from a computation?
17. How do pulley combination equations compare to gear train equations?
18. If a large pulley and a small pulley are connected with a belt, which will turn faster?
19. How do we know the belt connecting two pulleys travels at the same rate while in contact with the different-size pulleys?

Review Problems

SKETCH

12 cm² | w

4.0 cm

DATA

$A = 12 \text{ cm}^2, l = 4.0 \text{ cm}, w = ?$

BASIC EQUATION

$A = lw$

WORKING EQUATION

$w = \dfrac{A}{l}$

SUBSTITUTION

$w = \dfrac{12 \text{ cm}^2}{4.0 \text{ cm}} = 3.0 \text{ cm}$

1. Convert 13 revolutions to (a) radians and (b) degrees.
2. A bicycle wheel turns 25π rad during 45 s. Find the angular velocity of the wheel.
3. A lawn tractor tire turns at 65.0 rpm and has a radius of 13.0 cm. Find the linear speed of the tractor in m/s.
4. A model plane pulls into a tight curve of a radius of 25.0 m. The $30\overline{0}$-g plane is traveling at 90.0 km/h. What is the plane's centripetal force?
5. A 0.950-kg mass is spun in a circle on a string of radius 60.0 cm. If its centripetal force is 12.0 N, at what velocity does it travel?
6. A girl riding her bike creates a torque of 1.20 lb ft with an angular speed of 45.0 rpm. How much power does she produce?
7. A motor generates $30\overline{0}$ W of power. The torque necessary is 50.0 N m. Find the angular velocity.
8. Two rollers are side by side, with the large one turning the small one. The diameter of the small one is 2.00 cm and it turns at 15.0 rpm. The large roller has a diameter of 5.00 cm. How many revolutions does it make in one minute?
9. A clock is driven by a series of gears. The first gear has 30 teeth and rotates 60.0 times a minute. The second gear rotates at 90.0 rpm. How many teeth does it have?
10. Two gears have 13 and 26 teeth, respectively. The first gear turns at 115 rpm. How many times per minute does the second gear rotate?
11. A gear train has 17 directly connected gears. Do the first and last gears rotate in the same direction?
12. A pulley of diameter 14.0 cm is driven by an electric motor to revolve 75.0 rpm. The pulley drives a second one of diameter 10.0 cm. How many revolutions does the second pulley make in 1.00 min?
13. A pulley of diameter 5.00 cm is driven at $10\overline{0}$ rpm. Find the diameter of a second pulley if it is driven by the first at $25\overline{0}$ rpm.
14. If gear C turns counterclockwise, in what direction does gear F turn?

15. Find the speed in rpm of gear *D*.   16.   Find the number of teeth in gear *D*.

B = 30 teeth
C = 48 teeth
A = 45 teeth at 200 rpm
D = 20 teeth

A = 30 teeth at 800 rpm
D
C = 20 teeth
300 rpm
B = 50 teeth

APPLIED CONCEPTS

1. As part of their training, NASA astronauts are placed in large rotating machines to see how well their bodies can withstand various "*g*" forces. (One *g* equals the gravitational force of gravity at the earth's surface.) (a) What rotational speed is needed in a device with radius 6.25 m to allow a 75.8-kg astronaut to experience a force that is twice his normal weight or "2 *g*"? (b) What rotational speed is needed to achieve a "4 *g*" effect? (c) Would the rotational speeds found in (a) and (b) change if the astronaut had a different mass?

2. Waterwheels are used to convert kinetic energy from falling water into useful mechanical energy. (a) If an average of 10.0 kg of water flows onto a waterwheel every second, how much torque does the water exert on a waterwheel of radius 2.45 m? (b) The torque exerted by the water causes the wheel to move with an angular acceleration of 0.593 rad/s^2. Using this information, find the rotational inertia of the water wheel. (c) If you were the designer of this waterwheel, what are two things you could do to increase the rotational inertia of the waterwheel?

3. In Chapter 8 (Fig. 8.15), you were asked to calculate the velocity of a frictionless roller coaster car at various locations around the track. To make sure the loops are safe, you must make sure the track can support between two and five times the normal weight (that is, between 2 *g* and 5 *g*) of a car. (a) How many *g*'s does the 355-kg roller coaster car experience in each loop? (b) Are the loops safe? If not, explain what can be done to make the loops safe.

4. A hairpin turn on a concrete racetrack has a radius of 117 ft. (a) Since the frictional force between the road and the tires acts as the centripetal force, find the vehicle's maximum possible speed around the turn. (The coefficient of starting friction between rubber and concrete is 2.00.) (b) What would be the maximum velocity around the next turn if it has double the radius of the first turn? (c) Explain how the radius of a turn influences the centripetal force and the maximum velocity for a vehicle.

5. (a) How much power does a motorcycle need to produce 72.0 lb ft of torque while the engine crankshaft rotates at 3000 rpm? (b) When shifted into a higher gear, the same amount of power rotates the engine crankshaft at 5000 rpm. How much torque does the engine produce in this higher gear? (c) Using your knowledge of power, torque, and angular speed, explain why a motorcycle should be in a low gear when climbing a steep hill.

CHAPTER 10
SIMPLE MACHINES

Objectives

The major goals of this chapter are to enable you to:

1. Determine how energy is transferred using simple machines.
2. Analyze the efficiency and mechanical advantages of simple machines.
3. Distinguish among the three types of levers and the mechanical advantage of each.
4. Analyze the mechanical advantage of the wheel-and-axle, the pulley, the inclined plane, and the screw.

Machines may be used to transfer energy, multiply force, or multiply speed. We use them to obtain a mechanical advantage—to do something more efficiently that would be difficult or impossible without mechanical help. In this chapter we examine simple machines and their usefulness in technology.

10.1 Machines and Energy Transfer

A **machine** is an object, device, or system that is used to transfer energy from one place to another and allows work to be done that could not otherwise be done or could not be done as easily. For example, a nail driven into a board might be impossible to remove using only your bare hands. However, using a claw hammer makes the removal easy. The leverage gained by using a simple machine—the lever—allows you to gain a substantial advantage (called mechanical advantage) in performing the task of removing the nail. By using a pulley system [Fig. 10.1(a)], one person can easily lift an engine from an automobile. Pliers [Fig. 10.1(b)] allow a person to cut a wire or turn a nut with the strength of his or her hand.

Figure 10.1   Simple machines used to transfer energy

(a)   (b)

Simple machines are useful in three ways: (1) Machines can be used to multiply force. A car jack [Fig. 10.2(a)] (a lever) can be used to lift a much heavier weight than could be lifted otherwise. (2) Machines can be used to multiply speed. The gears of a bicycle (wheel-and-axle) [Fig. 10.2(b)] propel the rider much faster than he or she could go without aid from the machine. A single simple machine may be used to multiply force or speed, but not both. (3) Machines can be used to only change direction. When we use a single fixed pulley on a flag pole to raise a flag [Fig. 10.2(c)], the only advantage we get is the change in direction. (We pull the rope down, and the flag goes up.)

A **simple machine** is any one of six mechanical devices in which an applied force results in useful work (Fig. 10.3). All other machines—no matter how complex—are combinations of two or more of these simple machines.

In every machine we are concerned with two forces—effort and resistance. The **effort** is the force applied *to* the machine. The **resistance** is the force overcome *by* the machine. The popular phrase "You never get something for nothing" is particularly applicable to simple machines. A machine may greatly increase force or speed, but there is always a price to be paid. Work must be conserved ($W = Fs$). In the case of

Figure 10.2   Simple machines may be used to multiply force or speed or to change direction.

(a)   (b)   (c)

A car jack is used to   Gears are used to multiply speed.   A pulley is used to
multiply force.   change direction.

Figure 10.3 Six simple machines

1. Lever
2. Wheel and axle
3. Pulley
4. Inclined plane
5. Screw
6. Wedge

a lever, a much larger weight (force) can be lifted a short distance from the fulcrum, but the length of the lever arm must be increased. You have probably experienced this daily, from using a can or bottle opener, to using a long screwdriver to pry open a lid. In Fig. 10.4, a person applies $3\overline{0}$ lb on the jack handle to produce a lifting force of $60\overline{0}$ lb on the car. The effort force is $3\overline{0}$ lb. The resistance force is $60\overline{0}$ lb.

Figure 10.4 This lever multiplies force.

$3\overline{0}$ lb

$60\overline{0}$ lb

Jack

(a)

(b)

You can see in this example of a car jack that a smaller force applied to a longer distance can raise a much larger weight (force) but only by a distance that must be correspondingly much shorter. Work must be conserved, so the effort force times the effort distance must equal the resistance force times the resistance distance. This relationship is expressed as the **law of simple machines**.

LAW OF SIMPLE MACHINES

resistance force × resistance distance = effort force × effort distance

So a mechanical advantage can be gained using a simple machine, but the total amount of work done remains the same (is conserved).

Mechanical Advantage and Efficiency

We have all used a simple machine to make a task easier or to enable us to lift something or do something that we could not do without it. We use mathematics to describe and calculate just how much advantage using a machine can give us. This **mechanical advantage** (MA) of a simple machine is described as the ratio of the resistance force to the effort force. By formula,

$$\text{MA} = \frac{\text{resistance force}}{\text{effort force}}$$

The MA of the jack in Fig. 10.4 is found as follows:

$$\text{MA} = \frac{\text{resistance force}}{\text{effort force}} = \frac{60\overline{0}\ \text{lb}}{3\overline{0}\ \text{lb}} = \frac{2\overline{0}}{1}$$

This MA means that, for each pound applied by the person, he or she lifts $2\overline{0}$ pounds. Note that MA has no units. Why?

Each time a machine is used, part of the energy or effort applied to the machine is lost due to friction (Fig. 10.5). (The effect of friction is discussed in Section 10.9.) The **efficiency** of a machine is the ratio of the work output to the work input. By formula,

$$\text{efficiency} = \frac{\text{work output}}{\text{work input}} \times 100\% = \frac{F_{\text{output}} \times s_{\text{output}}}{F_{\text{input}} \times s_{\text{input}}} \times 100\%$$

Figure 10.5   Some work is always lost to friction.

10.2   The Lever

A **lever** consists of a rigid bar free to turn on a pivot called a **fulcrum** (Fig. 10.6). The mechanical advantage is the ratio of the effort arm (s_E) to the resistance arm (s_R):

$$\text{MA}_{\text{lever}} = \frac{\text{effort arm}}{\text{resistance arm}} = \frac{s_E}{s_R}$$

Figure 10.6   Mechanical advantage of
the lever: $MA = \dfrac{s_E}{s_R}$

The **effort arm** is the distance from the effort force to the fulcrum. The **resistance arm**
is the distance from the fulcrum to the resistance force. The three types or classes of levers
are shown in Fig. 10.7. The law of simple machines as applied to levers (basic equation) is

$$F_R \cdot s_R = F_E \cdot s_E$$

where F_R = resistance force
$\quad\quad s_R$ = length of resistance arm
$\quad\quad F_E$ = effort force
$\quad\quad s_E$ = length of effort arm

Figure 10.7   Three classes of levers

First class: The fulcrum is between the resistance force
(F_R) and the effort force (F_E).

Second class: The resistance force (F_R) is between the
fulcrum and the effort force (F_E).

Third class: The effort force (F_E) is between the fulcrum
and the resistance force (F_R).

TRY THIS ACTIVITY

Pulling Nails

Hammer two identical nails into a piece of wood so that the heads are slightly above the wood. Wedge the jaws of the hammer beneath the head of the first nail and hold the handle close to the head of the hammer. Before removing the nail, note the resistance distance, the position of the fulcrum, and the effort distance. Note the amount of effort force necessary to remove the first nail. Next, hold the end of the handle as you remove the second nail. Explain why simple machines, such as hammers pulling nails, are able to reduce the effort force.

A bar is used to raise a $12\overline{0}0$-N stone. The pivot is placed 30.0 cm from the stone. The worker pushes 2.50 m from the pivot. What is the mechanical advantage? What force is exerted?

EXAMPLE 1

Sketch:

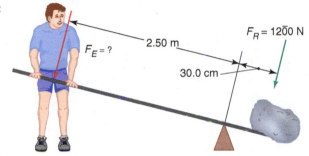

$F_R = 12\overline{0}0$ N

$F_E = ?$

2.50 m

30.0 cm

First, find the MA:

$$MA_{lever} = \frac{s_E}{s_R} = \frac{2.50 \text{ m}}{0.300 \text{ m}} = \frac{8.33}{1}$$

To find the force:

Data:

$$s_E = 2.50 \text{ m}$$
$$s_R = 30.0 \text{ cm} = 0.300 \text{ m}$$
$$F_R = 12\overline{0}0 \text{ N}$$
$$F_E = ?$$

Basic Equation:

$$F_R \cdot s_R = F_E \cdot s_E$$

Working Equation:

$$F_E = \frac{F_R \cdot s_R}{s_E}$$

Substitution:

$$F_E = \frac{(12\overline{0}0 \text{ N})(0.300 \text{ m})}{2.50 \text{ m}}$$
$$= 144 \text{ N}$$

A wheelbarrow 1.20 m long has a $90\overline{0}$-N load 40.0 cm from the axle. What is the MA? What force is needed to lift the wheelbarrow?

EXAMPLE 2

Sketch:

$F_R = 90\bar{0}$ N $F_E = ?$

40.0 cm

1.20 m

First, find the MA.

$$\text{MA} = \frac{s_E}{s_R} = \frac{1.20 \text{ m}}{0.400 \text{ m}} = \frac{3.00}{1}$$

To find the force:

Data:

$$s_E = 1.20 \text{ m}$$
$$s_R = 40.0 \text{ cm} = 0.400 \text{ m}$$
$$F_R = 90\bar{0} \text{ N}$$
$$F_E = ?$$

Basic Equation:

$$F_R \cdot s_R = F_E \cdot s_E$$

Working Equation:

$$F_E = \frac{F_R \cdot s_R}{s_E}$$

Substitution:

$$F_E = \frac{(90\bar{0} \text{ N})(0.400 \text{ m})}{1.20 \text{ m}}$$
$$= 30\bar{0} \text{ N}$$

EXAMPLE 3

The MA of a pair of pliers is 6.0/1. A force of 8.0 lb is exerted on the handle. What force is exerted on a wire in the pliers?

MA = 6.0/1 means that for each pound of force applied on the handle, 6.0 lb is exerted on the wire. Therefore, if a force of 8.0 lb is applied on the handle, a force of (6.0)(8.0 lb) or 48 lb is exerted on the wire.

PHYSICS CONNECTION

The Human Body—A Complex Machine

The human arm is a classic example of a simple machine. Figure 10.8(a) illustrates the forces and lever arms that are in place while lifting a weight. The elbow's hinge joint acts as a relatively low-friction fulcrum. The bicep is the muscle that exerts the effort force, and the barbell is the resistance force. According to the equation for mechanical advantage, when the resistance force is farther from the fulcrum than the effort force, the mechanical advantage is less than 1. Instead of reducing the force needed to lift the object, the bicep actually exerts more force than the weight of the barbell.

Figure 10.8 (a) The bicep moves very little while exerting a large force to lift the heavy barbell. (b) The parts of the arm act as several simple machines to throw a ball with a high velocity.

(a)

(b)

Machines with mechanical advantages less than 1, such as the human arm, are useful because the effort force does not have to move far in order for the weight to move large distances. The bicep muscle contracts a relatively small distance, whereas the barbell moves a larger distance. When the arm is used to throw a ball, the muscles exert a large force over a small distance, whereas the end of the forearm moves a relatively large distance with a high velocity. As seen in Fig. 10.8(b), the arm acts like several levers when throwing a ball. Although the force needed to throw the ball is greater than the weight of the ball, the great distance covered by the ball in a short period of time translates into a high velocity of the ball.

PROBLEMS 10.2

Given $F_R \cdot s_R = F_E \cdot s_E$, find each missing quantity.

	F_R	F_E	s_R	s_E
1.	20.0 N	5.00 N	3.70 cm	___ cm
2.	___ N	176 N	49.2 cm	76.3 cm
3.	37.0 N	12.0 N	___ cm	112 cm
4.	23.4 lb	9.80 lb	___ in.	53.9 in.
5.	119 N	___ N	29.7 cm	67.4 cm

Given $MA_{lever} = \dfrac{F_R}{F_E}$, find each missing quantity.

	MA	F_R	F_E
6.	___	20.0 N	5.00 N
7.	___	23.4 lb	9.80 lb
8.	7.00	119 N	___ N
9.	4.00	___ lb	12.2 lb
10.	___	37.0 N	12.0 N

Given $MA_{lever} = \dfrac{s_E}{s_R}$, find each missing quantity.

	MA	s_R	s_E
11.	___	49.2 cm	76.3 cm
12.	7.00	29.7 in.	___ in.
13.	___	29.7 cm	67.4 cm
14.	4.00	___ cm	67.4 cm

15. A pole is used to lift a car that fell off a jack (Fig. 10.9). The pivot is 2.00 ft from the car. Two people together exert 275 lb of force 8.00 ft from the pivot. (a) What force is applied to the car? (Disregard the weight of the pole.) (b) Find the MA.

SKETCH

12 cm²	w

4.0 cm

DATA

$A = 12$ cm², $l = 4.0$ cm, $w = ?$

BASIC EQUATION

$A = lw$

WORKING EQUATION

$w = \dfrac{A}{l}$

SUBSTITUTION

$w = \dfrac{12 \text{ cm}^2}{4.0 \text{ cm}} = 3.0$ cm

Figure 10.9

Figure 10.10

16. A bar is used to lift a $10\overline{0}$-kg block of concrete. The pivot is 1.00 m from the block. (a) If the worker pushes down on the other end of the bar a distance of 2.50 m from the pivot, what force (in N) must the worker apply? (b) Find the MA.

17. A wheelbarrow 6.00 ft long is used to haul a $18\overline{0}$-lb load. (a) How far from the wheel is the load placed so that a person can lift the load with a force of 45.0 lb? (b) Find the MA.

18. (a) Find the force, F_E, pulling up on the beam holding the sign shown in Fig. 10.10. (b) Find the MA.

10.3 The Wheel-and-Axle

The **wheel-and-axle** consists of a large wheel attached to an axle so that both turn together (Fig. 10.11). Other examples include a doorknob and a screwdriver with a thick handle.
The law of simple machines as applied to the wheel-and-axle (basic equation) is

$$F_R \cdot r_R = F_E \cdot r_E$$

where F_R = resistance force
r_R = radius of resistance force
F_E = effort force
r_E = radius of effort force

Note: Although s is generally used to designate distance, r is used when the distance is a radius.

Figure 10.11   Examples of the wheel-and-axle

Winch
(a)

Bicycle sprocket
and pedal assembly
(b)

Steering wheel
(c)

r_R = Resistance radius
r_E = Effort radius

(d)

A winch has a handle that turns in a radius of 30.0 cm. The radius of the drum or axle is 10.0 cm. Find the force required to lift a bucket weighing $50\overline{0}$ N (Fig. 10.12).

EXAMPLE 1

Data:

$$F_R = 50\overline{0} \text{ N}$$
$$r_E = 30.0 \text{ cm}$$
$$r_R = 10.0 \text{ cm}$$
$$F_E = ?$$

Figure 10.12

$r_E = 30.0$ cm
$r_R = 10.0$ cm

$50\overline{0}$ N

Basic Equation:

$$F_R \cdot r_R = F_E \cdot r_E$$

Working Equation:

$$F_E = \frac{F_R \cdot r_R}{r_E}$$

Substitution:

$$F_E = \frac{(50\overline{0} \text{ N})(10.0 \text{ cm})}{30.0 \text{ cm}}$$

$$= 167 \text{ N}$$

The mechanical advantage (MA) of the wheel-and-axle is the ratio of the radius of the effort force to the radius of the resistance force.

$$MA_{\text{wheel-and-axle}} = \frac{\text{radius of effort force}}{\text{radius of resistance force}} = \frac{r_E}{r_R}$$

Calculate the MA of the winch in Example 1.

EXAMPLE 2

$$MA_{\text{wheel-and-axle}} = \frac{r_E}{r_R} = \frac{30.0 \text{ cm}}{10.0 \text{ cm}} = \frac{3.00}{1}$$

PROBLEMS 10.3

Given $F_R \cdot r_R = F_E \cdot r_E$, find each missing quantity.

	F_R	F_E	r_R	r_E
1.	20.0 N	5.30 N	3.70 cm	____ cm
2.	$37\overline{0}$ N	$12\overline{0}$ N	____ m	1.12 m
3.	____ N	175 N	49.2 cm	76.3 cm
4.	23.4 lb	9.80 lb	____ in.	53.9 in.
5.	1190 N	____ N	29.7 cm	67.4 cm

Given $MA_{\text{wheel-and-axle}} = \dfrac{r_E}{r_R}$, find each missing quantity.

	MA	r_E	r_R
6.	7.00	119 mm	____ mm
7.	4.00	____ in.	12.2 in.
8.	____	49.2 cm	31.7 cm
9.	3.00	61.3 cm	____ cm
10.	____	67.4 mm	29.7 mm

SKETCH

$\boxed{12 \text{ cm}^2}$ w

4.0 cm

DATA

$A = 12 \text{ cm}^2, l = 4.0 \text{ cm}, w = ?$

BASIC EQUATION

$A = lw$

WORKING EQUATION

$w = \dfrac{A}{l}$

SUBSTITUTION

$w = \dfrac{12 \text{ cm}^2}{4.0 \text{ cm}} = 3.0 \text{ cm}$

11. A wheel with radius 75.0 cm is attached to an axle of radius 13.6 cm. What force must be applied to the rim of the wheel to raise a 1000-N weight?
12. An axle of radius 12.0 cm is used with a wheel of radius 62.0 cm. What force must be applied to the rim of the wheel to lift a weight of 975 N?
13. The radius of the axle of a winch is 3.00 in. The length of the handle (radius of wheel) is 1.50 ft. (a) What weight will be lifted by an effort of 73.0 lb? (b) Find the MA.
14. A wheel of radius of 70.0 cm is attached to an axle of radius 20.0 cm. (a) What force must be applied to the rim of the wheel to raise a weight of $15\overline{0}0$ N? (b) Find the MA. (c) What weight can be lifted if a force of 575 N is applied?
15. The diameter of the wheel of a wheel-and-axle is 10.0 cm. (a) If a force of 475 N is raised by applying a force of 142 N, find the diameter of the axle. (b) Find the MA.
16. Two persons use a large winch to raise a mass of 470 kg. The radius of the wheel is 48 cm and the radius of the axle is 4.0 cm. (a) What force is required to lift the load? (b) Find the MA of the large winch. (c) If the efficiency of the large winch is 60% and each person exerts the same force, how much force must each apply?

10.4 The Pulley

A **pulley** is a grooved wheel that turns readily on an axle and is supported in a frame. It can be fastened to a fixed object or to the resistance that is to be moved. A **fixed pulley** is a pulley fastened to a fixed object [Fig. 10.13(a)]. A **movable pulley** is fastened to the object to be moved [Fig. 10.13(b)]. A pulley system consists of combinations of fixed and movable pulleys [Fig. 10.13(c)–(e)].

Figure 10.13 Pulleys and pulley systems

One fixed
(a)

One movable
(b)

One fixed
One movable
(c)

Three fixed
Two movable
(d)

Three fixed
Three movable
(e)

Figure 10.14 Law of simple machines applied to pulleys:
$F_R \cdot s_R = F_E \cdot s_E$

The law of simple machines as applied to pulleys (Fig. 10.14) is

$$F_R \cdot s_R = F_E \cdot s_E$$

Here, s refers to the distance moved. From the preceding equation,

$$\frac{F_R}{F_E} = \frac{s_E}{s_R} = MA_{pulley}$$

However, when one continuous cord is used, this ratio reduces to the number of strands holding the resistance in the pulley system. Therefore,

$$MA_{pulley} = \text{number of strands holding the resistance}$$

This result may be explained as follows: When a weight is supported by two strands, each individual strand supports one half of the total weight. Thus, the MA = 2. If a weight is supported by three strands, each individual strand supports one third of the total weight. Thus, the MA = 3. If a weight is supported by four strands, each individual strand supports one fourth of the total weight. In general, when a weight is supported by n strands, each individual strand supports $\frac{1}{n}$ of the total weight. Thus, the MA = n.

Stated another way, the resistance force, F_R, is spread equally among the supporting strands. Thus, $F_R = nT$, where n is the number of strands holding the resistance and T is the tension in each supporting strand.

The effort force, F_E, equals the tension, T, in each supporting strand. The equation may then be written

$$MA_{pulley} = \frac{F_R}{F_E} = \frac{nT}{T} = n$$

Note: The mechanical advantage of the pulley does not depend on the diameter of the pulley.

A number of examples are shown in Fig. 10.15.

Figure 10.15   Mechanical advantage of pulleys and pulley systems

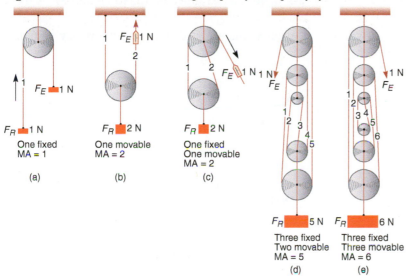

Draw two different sets of pulleys, each with an MA of 4.

EXAMPLE 1

Sketch:

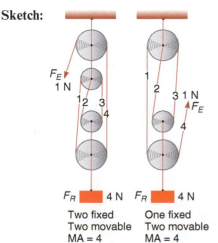

EXAMPLE 2

What effort will lift a resistance of 480 N in the pulley systems in Example 1?

Data:

$$MA_{pulley} = 4$$
$$F_R = 480 \text{ N}$$
$$F_E = ?$$

Basic Equation:

$$MA_{pulley} = \frac{F_R}{F_E}$$

Working Equation:

$$F_E = \frac{F_R}{MA_{pulley}}$$

Substitution:

$$F_E = \frac{480 \text{ N}}{4}$$
$$= 120 \text{ N}$$

EXAMPLE 3

If the resistance moves 7.00 ft, what is the effort distance of the pulley system in Example 1?

Data:

$$MA_{pulley} = 4$$
$$s_R = 7.00 \text{ ft}$$
$$s_E = ?$$

Basic Equation:

$$MA_{pulley} = \frac{s_E}{s_R}$$

Working Equation:

$$s_E = s_R(MA_{pulley})$$

Substitution:

$$s_E = (7.00 \text{ ft})(4)$$
$$= 28.0 \text{ ft}$$

EXAMPLE 4

The pulley system in Fig. 10.16 is used to raise a 650-lb object 25 ft. What is the mechanical advantage? What force is exerted?

$$MA_{pulley} = \text{number of strands holding the resistance}$$
$$= 5$$

To find the force exerted:

Data:

$$MA_{pulley} = 5$$
$$F_R = 650 \text{ lb}$$
$$F_E = ?$$

Basic Equation:

$$MA_{pulley} = \frac{F_R}{F_E}$$

Working Equation:

$$F_E = \frac{F_R}{MA_{pulley}}$$

Substitution:

$$F_E = \frac{650 \text{ lb}}{5}$$
$$= 130 \text{ lb}$$

Figure 10.16

$F_E = ?$

F_R 650 lb

Three fixed
Two movable
MA = ?

PROBLEMS 10.4

Find the mechanical advantage of each pulley system.

1.

F_E

F_R

2.

F_E

F_R

3.

F_E

F_R

4.

F_E

F_R

5.

F_E

F_R

6.

F_E

F_R

7.

F_E

F_R

8.

F_E

F_R

SKETCH

| 12 cm² | w |

4.0 cm

DATA
$A = 12 \text{ cm}^2, l = 4.0 \text{ cm}, w = ?$

BASIC EQUATION
$A = lw$

WORKING EQUATION
$w = \frac{A}{l}$

SUBSTITUTION
$w = \frac{12 \text{ cm}^2}{4.0 \text{ cm}} = 3.0 \text{ cm}$

Draw each pulley system for Problems 9–14.

9. One fixed and two movable. Find the system's MA.
10. Two fixed and two movable with an MA of 5.
11. Three fixed and three movable with an MA of 6.

12. Four fixed and three movable. Find the system's MA.
13. Four fixed and four movable with an MA of 8.
14. Three fixed and four movable with an MA of 8.
15. What is the MA of a single movable pulley?
16. (a) What effort will lift a $25\overline{0}$-lb weight by using a single movable pulley? (b) If the weight is moved 15.0 ft, how many feet of rope are pulled by the person exerting the effort?
17. A system consisting of two fixed pulleys and two movable pulleys has a mechanical advantage of 4. (a) If a force of 97.0 N is exerted, what weight is raised? (b) If the weight is raised 20.5 m, what length of rope is pulled?
18. A $40\overline{0}$-lb weight is lifted 30.0 ft. (a) Using a system of one fixed and two movable pulleys, find the effort force and effort distance. (b) If an effort force of 65.0 N is applied through an effort distance of 13.0 m, find the weight of the resistance and the distance it is moved.
19. Can an effort force of 75.0 N lift a 275-N weight using the pulley system in Problem 3?
20. (a) What effort will lift a 1950-N weight using the pulley system in Problem 5? (b) If the weight is moved 3.00 m, how much rope must be pulled through the pulley system by the person exerting the force?
21. Complete the following pulley system mechanical advantage chart, which lists two possible arrangements of fixed and movable pulleys for each given mechanical advantage.

Pulleys	Mechanical Advantage (MA)							
	1	2	3	4	5	6	7	8
Fixed	1	1	2					
Movable	0	1	1					
Fixed		0	1					
Movable		1	1					

22. Can you arrange a pulley system containing 10 pulleys and obtain a mechanical advantage of 12? Why or why not?

10.5   The Inclined Plane

An **inclined plane** is a plane surface set at an angle from the horizontal used to raise objects that are too heavy to lift vertically. Gangplanks, chutes, and ramps are all examples of the inclined plane (Fig. 10.17). The work done in raising a resistance using the inclined plane equals the resistance times the height. This must also equal the work input, which can be found by multiplying the effort times the length of the plane.

$$F_R \cdot s_R = F_E \cdot s_E \quad \text{(law of machines)}$$

$$F_R \cdot \text{height of plane} = F_E \cdot \text{length of plane}$$

Figure 10.17   Inclined plane

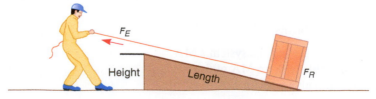

From the preceding equation,

$$\frac{F_R}{F_E} = \frac{\text{length of plane}}{\text{height of plane}} = MA_{\text{inclined plane}}$$

A worker is pushing a box weighing $15\overline{0}0$ N up a ramp 6.00 m long onto a platform 1.50 m above the ground. What is the mechanical advantage? What effort is applied?

EXAMPLE 1

Sketch:

6.00 m

1.50 m

$$MA_{\text{inclined plane}} = \frac{\text{length of plane}}{\text{height of plane}} = \frac{6.00\ \cancel{m}}{1.50\ \cancel{m}} = 4.00$$

To find the effort force:

Data:

$$F_R = 15\overline{0}0\ \text{N}$$
$$MA_{\text{inclined plane}} = 4.00$$
$$F_E = ?$$

Basic Equation:

$$MA_{\text{inclined plane}} = \frac{F_R}{F_E}$$

Working Equation:

$$F_E = \frac{F_R}{MA_{\text{inclined plane}}}$$

Substitution:

$$F_E = \frac{15\overline{0}0\ \text{N}}{4.00}$$
$$= 375\ \text{N}$$

....

Find the length of the shortest ramp that can be used to push a $60\overline{0}$-lb resistance onto a platform 3.50 ft high by exerting a force of 72.0 lb.

EXAMPLE 2

Data:

$$F_R = 60\overline{0}\ \text{lb}$$
$$F_E = 72.0\ \text{lb}$$
$$\text{height} = 3.50\ \text{ft}$$
$$\text{length} = ?$$

Basic Equation:

$$F_R \cdot \text{height} = F_E \cdot \text{length}$$

Working Equation:

$$\text{length} = \frac{F_R \cdot \text{height}}{F_E}$$

Substitution:

$$\text{length} = \frac{(60\overline{0} \text{ lb})(3.50 \text{ ft})}{72.0 \text{ lb}}$$

$$= 29.2 \text{ ft}$$

. . . .

EXAMPLE 3

An inclined plane is 13.0 m long and 5.00 m high. What is its mechanical advantage and what weight can be raised by exerting a force of 375 N?

$$\text{MA}_{\text{inclined plane}} = \frac{\text{length of plane}}{\text{height of plane}} = \frac{13.0 \text{ m}}{5.00 \text{ m}} = 2.60$$

To find the weight of the resistance:

Data:

$$\text{MA}_{\text{inclined plane}} = 2.60$$
$$F_E = 375 \text{ N}$$
$$F_R = ?$$

Basic Equation:

$$\text{MA}_{\text{inclined plane}} = \frac{F_R}{F_E}$$

Working Equation:

$$F_R = (F_E)(\text{MA}_{\text{inclined plane}})$$

Substitution:

$$F_R = (375 \text{ N})(2.60)$$
$$= 975 \text{ N}$$

. . . .

PHYSICS CONNECTION

The Physics of Handicap Ramps

Handicap ramps are inclined planes. They make locations that are at different elevations accessible to people with physical disabilities. There are many factors related to physics that play a part in the design and construction of a wheelchair ramp (Fig. 10.18).

The most important factor in designing a wheelchair ramp is to make sure the ideal mechanical advantage of the incline is at least 12. The general guideline is that for each 1 in. of height, there should be a 12-in. "run" or horizontal component to the incline. However, this does not mean that if someone needs to go up 5 feet to a door, there should be a 60-ft-long incline. Imagine someone unable to continue upward along the incline and rolling back down a 60-ft inclined plane. That would be a very scary and dangerous situation for a person in a wheelchair. As a result, if the length is more than 30 ft, then a 5-ft-long landing must be placed somewhere in the incline.

A feature that is often overlooked is the landing at the top of the incline. A landing must be placed at the top of any wheelchair ramp to allow the wheelchair rider room to open a door without being on the incline.
Finally, frictional forces must be taken into consideration. Most manufacturers of wheelchair ramps use concrete with a brushed finish or aluminum with adhesive pads or ridges in the metal so the coefficient of friction between the ramp and the wheels prevents the wheels from losing traction as the wheelchair makes its way up the incline.

Figure 10.18   A school with appropriately sloped inclines and landings

PROBLEMS 10.5

Given $F_R \cdot \text{height} = F_E \cdot \text{length}$, find each missing quantity.

	F_R	F_E	Height of Plane	Length of Plane
1.	20.0 N	5.30 N	3.40 cm	_____ cm
2.	98$\overline{0}$0 N	2340 N	_____ m	3.79 m
3.	119 lb	_____ lb	13.2 in.	74.0 in.
4.	_____ N	1760 N	82.1 cm	3.79 m
5.	37$\overline{0}$0 N	12$\overline{0}$0 N	_____ cm	112 cm

Given $\text{MA}_{\text{inclined plane}} = \dfrac{\text{length of plane}}{\text{height of plane}}$, find each missing quantity.

	MA	Length of Plane	Height of Plane
6.	9.00	3.40 ft	_____ ft
7.	_____	3.79 m	0.821 m
8.	1.30	_____ ft	9.72 ft
9.	_____	74.0 cm	13.2 cm
10.	17.4	_____ in.	13.4 in.

11. An inclined plane is 10.0 m long and 2.50 m high. (a) Find its mechanical advantage. (b) A resistance of 727 N is pushed up the plane. What effort is needed? (c) An effort of 20$\overline{0}$ N is applied to push an 815-N resistance up the inclined plane. Is the effort enough?

12. A safe is loaded onto a truck whose bed is 5.50 ft above the ground. The safe weighs 538 lb. (a) If the effort applied is 14$\overline{0}$ lb, what length of ramp is needed? (b) What is the MA of the inclined plane? (c) Another safe weighing 257 lb is loaded onto the same truck. If the ramp is 21.1 ft long, what effort is needed?

13. A 3.00-m-long plank is used to raise a cooling unit 1.00 m. What is the MA of the ramp made by the plank?

SKETCH

12 cm²   w

4.0 cm

DATA

$A = 12 \text{ cm}^2, l = 4.0 \text{ cm}, w = ?$

BASIC EQUATION

$A = lw$

WORKING EQUATION

$w = \dfrac{A}{l}$

SUBSTITUTION

$w = \dfrac{12 \text{ cm}^2}{4.0 \text{ cm}} = 3.0 \text{ cm}$

Figure 10.19 The screw is an inclined plane wound around a cylinder. The hypotenuse of the triangular section of paper corresponds to the inclined plane (threads) of a screw as it is wound around the pencil.

14. A 2.75-m-long board is used to slide a compressor a vertical distance of 0.750 m. What is the MA of the ramp made by the board?
15. A resistance of 325 N is raised by using a ramp 5.76 m long and by applying a force of 75.0 N. (a) How high can it be raised? (b) Find the MA of the ramp.
16. A plank 12 ft long is used as an inclined plane to a platform 3.0 ft high. (a) What force must be used to push a load weighing 480 lb up the plank? (b) Find the MA of the inclined plane.
17. A pallet stacked with bags of cement weighing a total of 5500 N must be pushed up a 2.30-m incline to a 75.0-cm-high platform. What force must be applied to get the job done?
18. A nursery loading dock is 1.20 m above the ground. What force must be used to push a 255-N pallet of fertilizer up a 3.80-m incline to the platform?

10.6 The Screw

A **screw** is an inclined plane wrapped around a cylinder. To illustrate, cut a sheet of paper in the shape of a right triangle and wind it around a pencil as shown in Fig. 10.19. The jackscrew, wood screw, and auger are examples of this simple machine (Fig. 10.20). The distance a beam rises or the distance the wood screw advances into a piece of wood in one revolution is called the **pitch** of the screw. Therefore, the pitch of a screw is also the distance between two successive threads.

From the law of machines,

$$F_R \cdot s_R = F_E \cdot s_E$$

However, for advancing a screw with a screwdriver

$s_R = $ pitch of screw

$s_E = $ circumference of the handle of the screwdriver

or

$$s_E = 2\pi r$$

where r is the radius of the handle of the screwdriver. Therefore,

$$F_R \cdot \text{pitch} = F_E \cdot 2\pi r$$

Figure 10.20 Examples of screw-type simple machines.

Jackscrew

(a)

Wood screw

(b)

Drill bit for wood

(c)

Earth auger drills holes in ground

(d)

so

$$\frac{F_R}{F_E} = \frac{2\pi r}{\text{pitch}} = \text{MA}_{\text{screw}}$$

In the case of a jackscrew, *r is the length of the handle turning the screw and not the radius of the screw.*

Find the mechanical advantage of a jackscrew having a pitch of 25.0 mm and a handle radius of 35.0 cm.

EXAMPLE 1

Data:

$$\text{pitch} = 25.0 \text{ mm} = 2.50 \text{ cm}$$
$$r = 35.0 \text{ cm}$$
$$\text{MA}_{\text{screw}} = ?$$

Basic Equation:

$$\text{MA}_{\text{screw}} = \frac{2\pi r}{\text{pitch}}$$

Working Equation: Same

Substitution:

$$\text{MA}_{\text{screw}} = \frac{2\pi(35.0 \text{ cm})}{2.50 \text{ cm}}$$
$$= 88.0$$

What resistance can be lifted using the jackscrew in Example 1 if an effort of 203 N is exerted?

EXAMPLE 2

Data:

$$\text{MA}_{\text{screw}} = 88.0$$
$$F_E = 203 \text{ N}$$
$$F_R = ?$$

Basic Equation:

$$\text{MA}_{\text{screw}} = \frac{F_R}{F_E}$$

Working Equation:

$$F_R = (F_E)(\text{MA}_{\text{screw}})$$

Substitution:

$$F_R = (203 \text{ N})(88.0)$$
$$= 17,900 \text{ N}$$

EXAMPLE 3

A 19,400-N weight is raised using a jackscrew having a pitch of 5.00 mm and a handle length of 255 mm. What force must be applied?

Data:

$$\text{pitch} = 5.00 \text{ mm}$$
$$r = 255 \text{ mm}$$
$$F_R = 19,400 \text{ N}$$
$$F_E = ?$$

Basic Equation:

$$F_R \cdot \text{pitch} = F_E \cdot 2\pi r$$

Working Equation:

$$F_E = \frac{F_R(\text{pitch})}{2\pi r}$$

Substitution:

$$F_E = \frac{(19,400 \text{ N})(5.00 \text{ mm})}{2\pi (255 \text{ mm})}$$
$$= 60.5 \text{ N}$$

. . . .

SKETCH

12 cm² | w

4.0 cm

DATA
$A = 12$ cm², $l = 4.0$ cm, $w = ?$

BASIC EQUATION
$A = lw$

WORKING EQUATION
$w = \dfrac{A}{l}$

SUBSTITUTION
$w = \dfrac{12 \text{ cm}^2}{4.0 \text{ cm}} = 3.0$ cm

PROBLEMS 10.6

Given $F_R \cdot \text{pitch} = F_E \cdot 2\pi r$, find each missing quantity.

	F_R	F_E	Pitch	r
1.	20.7 N	5.30 N	3.70 mm	___ mm
2.	___ lb	17.6 lb	0.130 in.	24.5 in.
3.	234 N	9.80 N	___ mm	53.9 mm
4.	1190 N	___ N	2.97 mm	67.4 mm
5.	370 lb	12.0 lb	___ in.	11.2 in.

Given $\text{MA}_{\text{screw}} = \dfrac{2\pi r}{\text{pitch}}$, find each missing quantity.

	MA	r	Pitch
6.	7.00	34.0 mm	___ mm
7.	___	3.79 in.	0.812 in.
8.	9.00	___ in.	0.970 in.
9.	___	7.40 mm	1.32 mm
10.	13.0	___ mm	2.10 mm

11. A 3650-lb car is raised using a jackscrew having eight threads to the inch and a handle 15.0 in. long. (a) What effort must be applied? (b) What is the MA?

12. The mechanical advantage of a jackscrew is 97.0. (a) If the handle is 34.5 cm long, what is the pitch? (b) How much weight can be raised by applying an effort of 405 N to the jackscrew?

13. A wood screw with pitch 0.125 in. is advanced into wood using a screwdriver whose handle is 1.50 in. in diameter. (a) What is the mechanical advantage of the screw? (b) What is the resistance of the wood if 15.0 lb of effort is applied on the wood screw? (c) What is the resistance of the wood if 15.0 lb of effort is applied to the wood screw using a screwdriver whose handle is 0.500 in. in diameter?

14. The handle of a jackscrew is 60.0 cm long. (a) If the mechanical advantage is 78.0, what is the pitch? (b) How much weight can be raised by applying a force of $43\overline{0}$ N to the jackscrew handle?

10.7　The Wedge

A **wedge** is an inclined plane in which the plane is moved instead of the resistance. Examples are shown in Fig. 10.21.

Finding the mechanical advantage of a wedge is not practical because of the large amount of friction. A narrow wedge is easier to drive than a thick wedge. Therefore, the mechanical advantage depends on the ratio of its length to its thickness.

Figure 10.21　Inclined planes where the plane moves instead of the resistance

Nail

Wood chisel

Hatchet

10.8　Compound Machines

A **compound machine** is a combination of simple machines. Examples are shown in Fig. 10.22. In most compound machines, *the total mechanical advantage is the product of the mechanical advantage of each simple machine.*

$$MA_{\text{compound machine}} = (MA_1)(MA_2)(MA_3) \ldots$$

Figure 10.22　Compound machines multiply mechanical advantage.

Courtesy of fotolia © wormseed

A crate weighing $95\overline{0}0$ N is pulled up the inclined plane using the pulley system shown in Fig. 10.23. (a) Find the mechanical advantage of the total system. (b) What effort force (F_E) is needed?

> **EXAMPLE 1**

Figure 10.23

9500 N

10.0 m

1.50 m

F_E

(a) First, find the MA of the inclined plane.

$$MA_{\text{inclined plane}} = \frac{\text{length of plane}}{\text{height of plane}} = \frac{10.5 \text{ m}}{1.50 \text{ m}} = 7.00$$

The MA of the pulley system = 5 (the number of supporting strands).
The MA of the total system (compound machine) is

$$(MA_{\text{inclined plane}})(MA_{\text{pulley system}}) = (7.00)(5) = 35.0$$

(b) Data:

$$MA_{\text{compound machine}} = 35.0$$
$$F_R = 95\overline{0}0 \text{ N}$$
$$F_E = ?$$

Basic Equation:

$$MA_{\text{compound machine}} = \frac{F_R}{F_E}$$

Working Equation:

$$F_E = \frac{F_R}{MA_{\text{compound machine}}}$$

Substitution:

$$F_E = \frac{95\overline{0}0 \text{ N}}{35.0}$$
$$= 271 \text{ N}$$

PROBLEMS 10.8

1. The box shown in Fig. 10.24 being pulled up an inclined plane using the indicated pulley system (called a block and tackle) weighs 9790 N. If the inclined plane is 6.00 m long and the height of the platform is 2.00 m, find the mechanical advantage of this compound machine.

Figure 10.24

2. What effort force must be exerted to move the box to the platform in Problem 1?
3. Find the mechanical advantage of the compound machine shown in Fig. 10.25. The radius of the crank is 1.00 ft and the radius of the axle is 0.500 ft.

Figure 10.25

SKETCH

12 cm² | w

4.0 cm

DATA
$A = 12 \text{ cm}^2, l = 4.0 \text{ cm}, w = ?$

BASIC EQUATION
$A = lw$

WORKING EQUATION
$w = \dfrac{A}{l}$

SUBSTITUTION
$w = \dfrac{12 \text{ cm}^2}{4.0 \text{ cm}} = 3.0 \text{ cm}$

4. If an effort of 30$\overline{0}$ lb is exerted, what weight can be moved up the inclined plane using the compound machine in Problem 3?

5. What effort is required to move a load of 1.50 tons up the inclined plane using the compound machine in Problem 3? (1 ton = 200$\overline{0}$ lb)

6. Find the mechanical advantage of the compound machine in Problem 1 if the inclined plane is 8.00 m long and 2.00 m high.

7. What effort force is needed to move a box of weight 250$\overline{0}$ N to the platform in Problem 6?

8. Find the mechanical advantage of the compound machine in Problem 3 if the radius of the crank is 40.0 cm, the radius of the axle is 12.0 cm, the length of the inclined plane is 12.0 m, and the height of the inclined plane is 50.0 cm.

9. If an effort of 45$\overline{0}$ N is exerted in Problem 8, what weight can be moved up the inclined plane?

10. What effort force (in N) is needed to move 2.50 metric tons up the inclined plane in Problem 8?

PHYSICS ON THE JOB

Heavy Equipment Maintenance Engineer

Heavy equipment maintenance engineers learn about the theory, functionality, repair, and operation of major construction equipment. In addition to mechanical skills and a complete working knowledge of diesel engine operations and maintenance, heavy equipment maintenance engineers must be knowledgeable in the area of complex machines, pneumatics, and hydraulics. Construction companies, heavy equipment dealers, and manufacturers typically employ heavy equipment maintenance engineers to keep their fleet of equipment operating efficiently.

A high school diploma, driver's license, and physical ability are necessary to become a heavy equipment maintenance engineer. Operating and maintaining more advanced equipment such as large cranes and pile drivers may require additional coursework and certification. Although a formal degree is not typically a requirement, many community and technical colleges offer certification and/or associate programs in heavy equipment maintenance (Fig. 10.26).

Figure 10.26 Heavy equipment system technologist

Courtesy of fotolia © Nomad-Soul

10.9 The Effect of Friction on Simple Machines

Our initial discussion of mechanical advantage disregarded the fact that each time a machine is used, part of the energy or effort applied to the machine is lost due to friction. When we disregard the effect of friction, we have 100% efficiency and the mechanical advantage is called **ideal mechanical advantage** (IMA). **Actual mechanical advantage** (AMA) does take the effect of friction into account and recognizes that true efficiency is never 100% but varies with the type of machine.

The *law of simple machines* has been stated for a particular machine and in the general case in the previous sections. Each has been stated in terms of what is called *ideal mechanical advantage* (IMA), in which we have 100% efficiency. Actually, in

Figure 10.27 Friction has been disregarded in our study of pulleys.

1200 N

every machine energy is lost through heat to overcome friction. This lost energy decreases the efficiency of the machine; that is, more work must be put into a machine than is gotten out of the machine. This lost energy is heat energy, which results in machine wear or even burning out of certain parts of the machine.

Throughout this chapter we have been discussing simple machines, mechanical advantage, resistance force, effort force, resistance distance, and effort distance in the ideal case while disregarding friction. For example, in the pulley system in Fig. 10.27, we find the IMA is 4 to 1. Ideally, it takes 300 N of effort force to lift the resistant force of 1200 N; that is, it ideally takes 1 N of force to raise 4 N of weight. However, it actually takes 400 N of effort to lift the 1200-N weight; the *actual mechanical advantage* (AMA) is then 3 to 1; that is, it actually takes 1 N of force to raise 3 N of weight.

In general, the actual mechanical advantage is found by the following formula:

$$\text{AMA} = \frac{F_R}{F_E} = \frac{\text{resistance force}}{\text{effort force}}$$

In Section 5.3, we studied the effects of sliding friction. The actual effects of sliding friction are substantial in inclined plane problems. In Example 1 of Section 10.5, the inclined plane (repeated in Fig. 10.28) has an IMA of 4 to 1. Actually, it takes 545 N of effort to move the 15$\overline{0}$0-N box up the ramp. Therefore, the AMA is

$$\text{AMA} = \frac{F_R}{F_E} = \frac{15\overline{0}0 \text{ N}}{545 \text{ N}} = 2.75$$

That is, it actually takes 1.00 N of force to push 2.75 N up the ramp.

Figure 10.28

6.00 m   1.50 m

Other simple machines are likewise subject to the effects of friction, which reduce their efficiency. For the wheel-and-axle (Section 10.3) the ideal mechanical advantage is the radius of the effort force r_E divided by the radius of the resistance force r_R. The actual mechanical advantage may only be about 60% of the ideal mechanical advantage.

EXAMPLE 1

The ideal mechanical advantage of a winch is 8.0. If the actual mechanical advantage of the winch is 5.0, what is the efficiency of the machine?

Data:

$$\text{IMA} = 8.0$$
$$\text{AMA} = 5.0$$
$$\text{efficiency} = ?$$

Basic Equation:

$$\text{efficiency} = \frac{\text{AMA}}{\text{IMA}}$$

Working Equation: Same

Substitution:

$$\text{efficiency} = \frac{5.0}{8.0}$$

$$= 0.63$$

$$= 63\%$$

· · · · · · · · · · · · · · · · ·

The ideal mechanical advantage of the screw is equal to $\dfrac{2\pi r}{\text{pitch}}$. The actual mechanical advantage is considerably less because of the substantial effect of friction. Remember that the screw is really a form of inclined plane wrapped around a cylinder. Its actual mechanical advantage may be less than 30%.

If the efficiency of a jackscrew is 25% and the actual mechanical advantage is 95, what is the ideal mechanical advantage of the jackscrew?

> **EXAMPLE 2**

Data:

$$\text{efficiency} = 25\%$$

$$\text{AMA} = 95$$

$$\text{IMA} = ?$$

Basic Equation:

$$\text{efficiency} = \frac{\text{AMA}}{\text{IMA}}$$

Working Equation:

$$\text{IMA} = \frac{\text{AMA}}{\text{efficiency}}$$

Substitution:

$$\text{IMA} = \frac{95}{0.25}$$

$$= 380$$

· · · · · · · · · · · · · · · · ·

The lever can be the most efficient machine. Very little friction may be involved with the lever, depending upon the application. The ideal mechanical advantage of the lever is equal to the length of the effort arm s_E divided by the length of the resistance arm s_R. A lever may be more than 90% efficient.

If the efficiency of a lever is 92% and the ideal mechanical advantage is $2\overline{0}$, what is the actual mechanical advantage of the lever?

> **EXAMPLE 3**

Data:

$$\text{efficiency} = 92\%$$

$$\text{IMA} = 2\overline{0}$$

$$\text{AMA} = ?$$

Basic Equation:

$$\text{efficiency} = \frac{\text{AMA}}{\text{IMA}}$$

Working Equation:

$$\text{AMA} = (\text{efficiency})(\text{IMA})$$

Substitution:

$$\text{AMA} = (0.92)(2\overline{0})$$
$$= 18$$

· · · · · · · · · · · · · · · · ·

The **efficiency** of a machine may also be described as the ratio of actual mechanical advantage to ideal mechanical advantage as follows:

$$\text{efficiency} = \frac{\text{actual mechanical advantage (AMA)}}{\text{ideal mechanical advantage (IMA)}}$$

TRY THIS ACTIVITY

Reducing Friction

Observe the amount of effort force required to drag a heavy book up an inclined plane. Before repeating this process, place several small wooden dowels or marbles under the book. Drag the book up the incline again and observe the amount of effort force required. How do the wooden dowels or marbles reduce the effort force required to move the book up the incline? The ancient Egyptians may have used large dowels to reduce the effort force to drag massive stone blocks to the tops of pyramids.

PROBLEMS 10.9

Round all results to two significant digits.

1. If the IMA of a ramp is 4.0 and its AMA is 3.2, what is its efficiency?
2. If the AMA of a wheel-and-axle is 9.0 and its IMA is 13, what is its efficiency?
3. If the efficiency of a screw is 32% and its IMA is $4\overline{0}$, what is its AMA?
4. If the efficiency of lever is 94% and its AMA is 12, what is its IMA?
5. If the IMA of an inclined plane is 6.0 and its AMA is 4.8, what is its efficiency?
6. If the AMA of a ramp is 4.6 and its IMA is 6.0, what is its efficiency?
7. If the IMA of a pulley is 12 and its AMA is 9.0, what is its efficiency?
8. If the IMA of a screw is $6\overline{0}$ and its AMA is 26, what is its efficiency?
9. If the efficiency of a pulley is 82% and its IMA is 22, what is its AMA?
10. A wheel-and-axle has an efficiency of 65%. If its AMA is 16, what is its IMA?

Glossary

Actual Mechanical Advantage The ratio of the resistance force to the actual effort force. (p. 289)

Compound Machine A combination of simple machines. Its total mechanical advantage is the product of the mechanical advantage of each simple machine. (p. 287)

Efficiency The ratio of the work output to the work input of a machine. (p. 269)

Effort The force applied to a machine. (p. 267)

Effort Arm The distance from the effort force to the fulcrum of a lever. (p. 270)

Fixed Pulley A pulley that is fastened to a fixed object. (p. 276)

Fulcrum A pivot about which a lever is free to turn. (p. 269)

Ideal Mechanical Advantage The ratio of the resistance force to the effort force (assuming 100% efficiency and disregarding friction). (p. 289)

Inclined Plane A plane surface set at an angle from the horizontal used to raise objects that are too heavy to lift vertically. (p. 280)

Law of Simple Machines Resistance force × resistance distance = effort force × effort distance. (p. 268)

Lever A rigid bar free to turn on a pivot called a fulcrum. (p. 269)

Machine An object or system that is used to transfer energy from one place to another and allows work to be done that could not otherwise be done or could not be done as easily. (p. 266)

Mechanical Advantage The ratio of the resistance force to the effort force. (p. 269)

Movable Pulley A pulley that is fastened to the object to be moved. (p. 276)

Pitch The distance a screw advances in one revolution of the screw. Also the distance between two successive threads. (p. 284)

Pulley A grooved wheel that turns readily on an axle and is supported in a frame. (p. 276)

Resistance The force overcome by a machine. (p. 267)

Resistance Arm The distance from the resistance force to the fulcrum of a lever. (p. 270)

Screw An inclined plane wrapped around a cylinder. (p. 284)

Simple Machine Any one of six mechanical devices in which an applied force results in useful work. The six simple machines are the lever, the wheel-and-axle, the pulley, the inclined plane, the screw, and the wedge. (p. 267)

Wedge An inclined plane in which the plane is moved instead of the resistance. (p. 287)

Wheel-and-Axle A large wheel attached to an axle so that both turn together. (p. 274)

Formulas

10.1 resistance force × resistance distance = effort force × effort distance

$$MA = \frac{\text{resistance force}}{\text{effort force}}$$

$$\text{efficiency} = \frac{\text{work output}}{\text{work input}} \times 100\% = \frac{F_{\text{output}} \times s_{\text{output}}}{F_{\text{input}} \times s_{\text{input}}} \times 100\%$$

10.2 $MA_{\text{lever}} = \dfrac{\text{effort arm}}{\text{resistance arm}} = \dfrac{s_E}{s_R}$

$F_R \cdot s_R = F_E \cdot s_E$

10.3 $MA_{\text{wheel-and-axle}} = \dfrac{\text{radius of effort force}}{\text{radius of resistance force}} = \dfrac{r_E}{r_R}$

$F_R \cdot r_R = F_E \cdot r_E$

10.4 MA_{pulley} = number of strands holding the resistance

$$MA_{pulley} = \frac{s_E}{s_R}$$

10.5 $MA_{inclined\ plane} = \dfrac{length\ of\ plane}{height\ of\ plane}$

$F_R \cdot height = F_E \cdot length$

10.6 $MA_{screw} = \dfrac{2\pi r}{pitch}$

$F_R \cdot pitch = F_E \cdot 2\pi r$

10.8 $MA_{compound\ machine} = (MA_1)(MA_2)(MA_3) \cdots$

10.9 $AMA = \dfrac{F_R}{F_E} = \dfrac{resistance\ force}{effort\ force}$

$efficiency = \dfrac{actual\ mechanical\ advantage\ (AMA)}{ideal\ mechanical\ advantage\ (IMA)}$

Review Questions

1. Which of the following is not a simple machine?
   (a) Pulley   (b) Lever
   (c) Wedge   (d) Automobile
2. The force applied to the machine is the
   (a) effort.   (b) frictional.
   (c) horizontal.   (d) resistance.
3. Efficiency is
   (a) the same as mechanical advantage.   (b) a percentage.
   (c) impossible to determine.
4. A second-class lever has
   (a) two fulcrums.   (b) two effort arms.
   (c) two resistance arms.   (d) a resistance arm shorter than
                                           the effort arm.
5. A pulley has eight strands holding the resistance. The mechanical advantage is
   (a) 4.   (b) 8.   (c) 16.   (d) 64.
6. The mechanical advantage of a compound machine
   (a) is the sum of the MA of each simple machine.
   (b) is the product of the MA of each simple machine.
   (c) cannot be found.   (d) is none of the above.
7. Cite three examples of machines used to multiply speed.
8. What name is given to the force overcome by the machine?
9. State the law of simple machines in your own words.
10. What is the term used for the ratio of the resistance force to the effort force?
11. What is the term used for the ratio of the amount of work obtained from a machine to the amount of work put into the machine?
12. Does a friction-free machine exist?
13. What is the pivot point of a lever called?
14. In your own words, state how to find the MA of a lever.
15. Which type of lever do you think would be most efficient?
16. State the law of simple machines as it is applied to levers.
17. Where is the fulcrum located in a third-class lever?
18. In your own words, explain the law of simple machines as applied to the wheel-and-axle.
19. Does the MA of a wheel-and-axle depend on the force applied?
20. Describe the difference between a fixed pulley and a movable pulley.

21. Does the MA of a pulley depend on the radius of the pulley?
22. How can you find the MA of an inclined plane?
23. In your own words, describe the pitch of a screw.
24. How does the MA of a jackscrew differ from the MA of a screwdriver?

Review Problems

1. A girl uses a lever to lift a box. The box has a resistance force of $25\overline{0}$ N while she exerts an effort force of 125 N. What is the mechanical advantage of the lever?

2. A bicycle requires 1575 N m of input but only puts out 1150 N m of work. What is the bicycle's efficiency?

3. A lever uses an effort arm of 2.75 m and has a resistance arm of 72.0 cm. What is the lever's mechanical advantage?

4. Two people are on a teeter-totter. One person exerts a force of 540 N and is 2.00 m from the fulcrum. If they are to remain balanced, how much force does the other person exert if she is (a) also 2.00 m from the fulcrum? (b) 3.00 m from the fulcrum?

5. A wheel-and-axle has an effort force of 125 N and an effort radius of 17.0 cm. (a) If the resistance force is 325 N, what is the resistance radius? (b) Find the mechanical advantage.

6. What is the mechanical advantage of a pulley system having 12 strands holding the resistance?

7. A pulley system has a mechanical advantage of 5. What is the resistance force if an effort of 135 N is exerted?

8. An inclined plane has a height of 1.50 m and a length of 4.50 m. (a) What effort must be exerted to pull up an 875-N box? (b) What is the mechanical advantage?

9. What height must a 10.0-ft-long inclined plane be to lift a $100\overline{0}$-lb crate with $23\overline{0}$ lb of effort?

10. A screw has a pitch of 0.0200 cm. An effort force of 29.0 N is used to turn a screwdriver whose handle diameter is 36.0 mm. What is the maximum resistance force?

11. A 945-N resistance force is overcome with a 13.5-N effort using a screwdriver whose handle is 24.0 mm in diameter. What is the pitch of the screw?

12. Find the mechanical advantage of a jackscrew with a 1.50-cm pitch and a handle 36.0 cm long.

13. A courier uses a bicycle with rear wheel radius 35.6 cm and gear radius 4.00 cm. If a force of 155 N is applied to the chain, the wheel rim moves 14.0 cm. (a) If the efficiency is 95.0%, what is the ideal mechanical advantage of the wheel and gear? (b) What is the actual mechanical advantage of the wheel and gear? (c) What is the force on the pavement applied by the wheel?

14. (a) If the gear radius is doubled on the courier's bicycle in Problem 13, how does the mechanical advantage change? (b) How far did the courier move the chain to produce the 14.0-cm linear movement of the rim?

15. A farmer uses a pulley system to raise a 225-N bale 16.5 m. A 129-N force is applied by pulling a rope 33.0 m. What is the mechanical advantage of the pulley system?

16. A laborer uses a lever to raise a 1250-N rock a distance of 13.0 cm by applying a force of 225 N. If the efficiency of the lever is 88.7%, how far does the laborer have to move his end of the lever?

SKETCH

12 cm²	w

4.0 cm

DATA
$A = 12$ cm², $l = 4.0$ cm, $w = ?$

BASIC EQUATION
$A = lw$

WORKING EQUATION
$w = \dfrac{A}{l}$

SUBSTITUTION
$w = \dfrac{12 \text{ cm}^2}{4.0 \text{ cm}} = 3.0 \text{ cm}$

Find the mechanical advantage of each pulley system.

17.

18.

19. (a) Find the mechanical advantage of the compound machine in Fig. 10.29. The radius of the crank is 32.0 cm and the radius of the axle is 8.00 cm. (b) If an effort force of 75 N is applied to the handle of the crank, what force can be moved up the inclined plane?

Figure 10.29

20. If an effort force of 45 N is applied to a simple machine and moves a resistance of 270 N, what is the actual mechanical advantage?

APPLIED CONCEPTS

1. In the third century BC, Archimedes said, "Give me a long enough lever and a firm fulcrum and I will lift the earth." (a) Using the moon as a fixed fulcrum, how long a lever would Archimedes have needed to lift the earth? (b) What would be the mechanical advantage of such a simple machine? (Disregard the mass of the lever. Assume the weight of the earth is 5.85×10^{25} N, Archimedes' weight is 858 N, and the distance from the earth to the moon is 3.84×10^8 m.)

2. (a) What is the mechanical advantage of the fishing pole in Fig. 10.30? (b) If the fisherman's hand moves forward 30.5 cm, how far will the tip of the fishing pole move? (c) How fast will the tip of the fishing pole move if the forward motion is completed in 0.554 s? (d) In general, what is the benefit of using a simple machine that has a mechanical advantage less than 1.0?

Figure 10.30

$F_{upper\ hand}$

2.13 m

25.5 cm

Fulcrum
(lower hand)

3. A snowblower auger has a radius of 7.75 in. and a pitch of 6.80 in. (a) What is the mechanical advantage of this auger system? (b) How much effort force is needed for the auger to throw 23.4 lb of snow? (c) What happens to the mechanical advantage if the pitch is reduced?

4. Aaron, a bicycle mechanic, is studying the mechanical advantage of the rear wheel. The rear wheel's radius is 14.0 in. and the chain is currently on a gear with radius 1.65 in. (Fig. 10.31). (a) If the chain applies a force of 54.5 lb on the wheel, what is the mechanical advantage of the wheel? (b) What is the resistance force? (c) How many inches of chain must be used to turn the wheel one complete rotation? (d) Finally, under what circumstances is it beneficial for a bicycle to have a mechanical advantage less than 1.0?

Figure 10.31

r_{wheel} = 14.0 in.

F = 54.5 lb

r_{gear} = 1.65 in.

Figure 10.32

0.265 m

0.568 m

1.87 m

r = 0.075 m

5. Willie is using a wheelbarrow (Fig. 10.32) to move 345 lb of patio blocks. (a) What is the mechanical advantage of the wheelbarrow handles? (b) What is the mechanical advantage of the wheelbarrow's wheel-and-axle? (c) How can the design of the wheelbarrow be altered to increase the mechanical advantage?

CHAPTER 11

UNIVERSAL GRAVITATION AND SATELLITE MOTION

Objectives

The major goals of this chapter are to enable you to:

1. Describe how gravitation acts between all objects with mass.
2. Calculate what your weight would be on various planets.
3. Explain the concept of a gravitational field.
4. Express how objects are able to orbit the earth.
5. Determine the connection between orbiting the earth and being in free fall.

What allows thousands of satellites to orbit the earth? Why does vertical motion on the moon appear to take place in slow motion? In this chapter we will discover how we can send massive objects into orbit and why a person weighs less on the moon than on the earth. In this chapter we appreciate the importance of understanding the difference between mass and weight.

11.1 Universal Gravitation

What makes an astronaut weigh less on the moon than on the earth? **Newton's law of universal gravitation** states that all objects that have mass are attracted to one another by a gravitational force. Newton determined that the greater the mass of two objects, the stronger is the attractive gravitational force between them. He also discovered that as objects move away from each other, the attraction between them diminishes dramatically. Newton's law of universal gravitation is

$$F_G = G\frac{m_1 m_2}{r^2}$$

where F_G = gravitational force between the two objects
  $G = 6.67 \times 10^{-11}$ N m^2/kg$^2 = 3.44 \times 10^{-8}$ lb ft^2/slug^2 (universal gravitational constant)
  m_1 = mass of the first object
  m_2 = mass of the second object
  r = distance between the centers of mass of the two objects.

The universal gravitational constant, G, is an experimentally derived quantity. This constant is difficult to measure because gravity is such a relatively weak force. It is, however, essential to Newton's law of universal gravitation and Einstein's general theory of relativity. It was actually first measured by Henry Cavendish as discussed below.

The law of universal gravitation makes it possible to calculate an object's weight on any planet; the gravitational force exerted between the person sitting next to you in class and you; and the gravitational forces between planets, moons, and stars. Table 11.1 gives the average radius, the mass, and the mean distance from the sun of various objects in our solar system.

Table 11.1   Table of Planetary Data

Object	Average Radius (m)	Mass (kg)	Mean Sun to Planet Distance (m)
Sun	6.96×10^8	1.99×10^{30}	—
Mercury	2.44×10^6	3.30×10^{23}	5.79×10^{10}
Venus	6.05×10^6	4.87×10^{24}	1.08×10^{11}
Earth	6.38×10^6	5.97×10^{24}	1.50×10^{11}
Mars	3.40×10^6	6.42×10^{23}	2.28×10^{11}
Jupiter	7.15×10^7	1.90×10^{27}	7.78×10^{11}
Saturn	6.03×10^7	5.68×10^{26}	1.43×10^{12}
Uranus	2.56×10^7	8.68×10^{25}	2.87×10^{12}
Neptune	2.48×10^7	1.02×10^{26}	4.50×10^{12}
Pluto*	1.19×10^6	1.30×10^{22}	5.87×10^{12}

*Pluto is now classified as a dwarf planet.
Data courtesy of NASA

Other information:

  earth to moon distance = 3.84×10^8 m
  moon's radius   = 1.74×10^6 m
  moon's mass   = 7.35×10^{22} kg

EXAMPLE 1

Compare (a) the gravitational force between Vince and the earth to (b) the gravitational force between Vince and Matt, who is sitting 2.34 m away from Vince. Both Matt and Vince have a mass of 85.5 kg.

Data:

$$m_{earth} = 5.97 \times 10^{24} \text{ kg}$$

$$m_{Matt} = 85.5 \text{ kg}$$

$$m_{Vince} = 85.5 \text{ kg}$$

$$r_{Vince-earth} = 6.38 \times 10^6 \text{ m}$$

$$r_{Vince-Matt} = 2.34 \text{ m}$$

Basic Equation:

$$F_G = G \frac{m_1 m_2}{r^2}$$

Working Equation: Same

Substitution:

(a)
$$F_G = \left(6.67 \times 10^{-11} \frac{\text{N m}^2}{\text{kg}^2} \right) \frac{(85.5 \text{ kg})(5.97 \times 10^{24} \text{ kg})}{(6.38 \times 10^6 \text{ m})^2}$$

$$= 836 \text{ N}$$

(b)
$$F_G = \left(6.67 \times 10^{-11} \frac{\text{N m}^2}{\text{kg}^2} \right) \frac{(85.5 \text{ kg})(85.5 \text{ kg})}{(2.34 \text{ m})^2}$$

$$= 8.90 \times 10^{-8} \text{ N}$$

The gravitational force between Vince and Matt is virtually undetectable compared to the gravitational force between Vince and the earth.

· · · · · · · · · · · · · · · · · ·

Henry Cavendish (1731–1810),

English physicist and chemist, was born in France. He was the first person to experimentally determine the universal gravitational constant used in Isaac Newton's law of universal gravitation (Fig. 11.1). He also found that water was composed of hydrogen and oxygen and estimated the density of the earth.

In 1798, Henry Cavendish attached a pair of 2-in. lead balls to the ends of a light horizontal rod, which was suspended by a string onto which a mirror was mounted as shown in Fig. 11.1. When this pair of lead balls was brought near two more massive

Figure 11.1 The Cavendish experiment

(a)

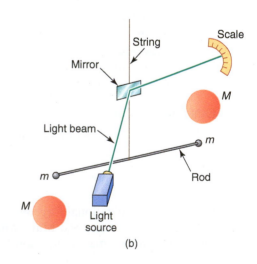

(b)

12-in. balls, the gravitational force caused the smaller balls to move toward the more massive balls, which caused the string and mirror to twist. The tiny movement was measured by reflecting a narrow light beam onto a scale. By carefully measuring the very small twisting motion of the suspended balls, Cavendish was able to determine G within 1% of today's accepted value.

One of Newton's preliminary studies of universal gravitation involved the moon and a falling apple. (a) Find the force between an apple ($m = 0.153$ kg) and the earth. (b) Compare this to the force between the moon and the earth, two objects that are quite massive yet far away from each other. See Table 11.1.

EXAMPLE 2

Data:

$$m_{apple} = 0.153 \text{ kg}$$
$$m_{moon} = 7.35 \times 10^{22} \text{ kg}$$
$$m_{earth} = 5.97 \times 10^{24} \text{ kg}$$
$$r_{apple-earth} = 6.38 \times 10^6 \text{ m}$$
$$r_{moon-earth} = 3.84 \times 10^8 \text{ m}$$

Basic Equation:

$$F_G = G\frac{m_1 m_2}{r^2}$$

Working Equation: Same

Substitutions:

(a)
$$F_G = \left(6.67 \times 10^{-11} \frac{\text{N m}^2}{\text{kg}^2}\right) \frac{(0.153 \text{ kg})(5.97 \times 10^{24} \text{ kg})}{(6.38 \times 10^6 \text{ m})^2}$$
$$= 1.50 \text{ N}$$

(b)
$$F_G = \left(6.67 \times 10^{-11} \frac{\text{N m}^2}{\text{kg}^2}\right) \frac{(7.35 \times 10^{22} \text{ kg})(5.97 \times 10^{24} \text{ kg})}{(3.84 \times 10^8 \text{ m})^2}$$
$$= 1.98 \times 10^{20} \text{ N}$$

The force between the earth and the moon is much greater than the force between the earth and the apple. Although the moon is much farther away from the earth than the apple is, the moon's much larger mass more than makes up for the difference in the distances.

What would be the gravitational force exerted on a 65.0-kg person on Jupiter?

EXAMPLE 3

Data:

$$m_{person} = 65.0 \text{ kg}$$
$$m_{Jupiter} = 1.90 \times 10^{27} \text{ kg}$$
$$r_{Jupiter} = 7.15 \times 10^7 \text{ m}$$

Basic Equation:

$$F_G = G\frac{m_1 m_2}{r^2}$$

Working Equation: Same

Substitution:

$$F_G = \left(6.67 \times 10^{-11}\ \frac{\text{N m}^2}{\text{kg}^2}\right)\frac{(65.0\ \text{kg})\ (1.90 \times 10^{27}\ \text{kg})}{(7.15 \times 10^7\ \text{m})^2}$$

$$= 1610\ \text{N}$$

As these examples show, the gravitational force becomes stronger as the mass of the objects increases and as the distance between them decreases.

PROBLEMS 11.1

SKETCH

12 cm² | w

4.0 cm

DATA

$A = 12\ \text{cm}^2, l = 4.0\ \text{cm}, w = ?$

BASIC EQUATION

$A = lw$

WORKING EQUATION

$w = \dfrac{A}{l}$

SUBSTITUTION

$w = \dfrac{12\ \text{cm}^2}{4.0\ \text{cm}} = 3.0\ \text{cm}$

1. Compare the gravitational force that (a) the earth exerts on an 84.3-kg person and (b) the force that the sun exerts on the same person.
2. Find the gravitational force between the sun and the earth.
3. Find the gravitational force between the sun and Mercury.
4. Find the gravitational force between the sun and Jupiter.
5. Find the gravitational force between the sun and Pluto.
6. Explain why the gravitational force between the sun and Jupiter is greater than the gravitational force between the sun and the earth even though the sun and the earth are much closer to one another than are the sun and Jupiter.
7. A satellite is orbiting 3.22×10^5 m above the surface of the earth. If the mass of the satellite is 3.80×10^4 kg, what is the weight or gravitational force exerted on the satellite by the earth?
8. If the satellite in Problem 7 is orbiting at twice its original distance from the earth, what would be the weight or gravitational force exerted on the satellite by the earth?
9. What is the gravitational force exerted between an electron ($m = 9.11 \times 10^{-31}$ kg) and a proton ($m = 1.67 \times 10^{-27}$ kg) in a hydrogen atom where the distance between the electron and proton is 5.3×10^{-9} m?
10. The Apollo 16 lunar module had a mass of 4240 kg. Using Newton's law of universal gravitation, find its weight (a) on the earth and (b) on the moon.

11.2 Gravitational Fields

Isaac Newton's contemporaries did not welcome Newton's laws. The thought that the earth could exert a force on the moon almost 240,000 mi away without touching it seemed outrageous to many at the time. Until then, it was thought that most objects needed to be physically touching in order to have an effect on one another.

The area around a massive body where an object experiences a gravitational force is called a **gravitational field**. By introducing the concept of a gravitational field, physicists have been able to determine that an object does not need to be in physical contact with another object to exert a force on it (Fig. 11.2). The change in velocity caused by the gravitational field is known as acceleration due to gravity.

The gravitational field around the surface of the earth is quite strong. However, if one travels out into space a bit, the distance alone affects the intensity of the field. That is, the acceleration due to gravity decreases significantly, as shown in Fig. 11.3.

Figure 11.2   Gravitational field lines represent the strength of gravity around the earth. The closer the lines, the stronger is the gravitational attraction, as shown by the force vectors.

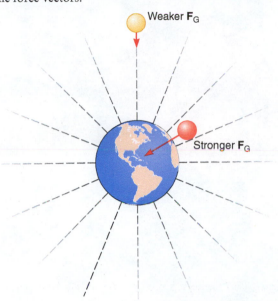

Weaker **F**$_G$

Stronger **F**$_G$

Figure 11.3   The gravitational force depends on the object's distance from the center of the earth. For example, if an object weighs 100 lb on the earth's surface, its weight becomes dramatically less the farther away the object is located.

$F = 10\overline{0}$ lb $\frac{1}{4}F = 25.0$ lb $\frac{1}{9}F = 11.1$ lb $\frac{1}{16}F = 6.25$ lb $\frac{1}{25}F = 4.00$ lb

r $2r$ $3r$ $4r$ $5r$

The concept of a gravitational field inspired Albert Einstein to establish a theory that would describe a warped space-time field in our universe. This fascinating, yet conceptually challenging, theory will be covered in Chapter 24.

11.3 Satellite Motion

An **orbit** is the path taken by an object during its revolution around another object, such as the path of the moon or a satellite about the earth or of a planet about the sun. We have seen pictures of astronauts weightless and appearing to "float" in space as they orbit the earth (Fig. 11.4). The reason is because of gravity, not the lack of gravity, at that point in space.

In Chapter 4, we learned that when an object is dropped or falls from an elevated point on earth, nothing is pushing back up on the object, so it appears weightless (Fig. 11.5). Gravity may be pulling the object down, but there is no normal force to support the object. In essence, when astronauts orbit the earth, they are in a constant state of free fall.

A person in free fall feels weightless, just like the astronauts. However, as the person continues to fall, he or she will crash to the earth. Newton realized that the moon and other orbiting objects must sustain a large enough horizontal velocity to remain in orbit and avoid crashing to the earth.

Figure 11.4   Although an astronaut in orbit feels weightless, there is still a gravitational force acting on the astronaut. Here astronaut Joseph R. Tanner (right) stands fixed on the end of Discovery's remote manipulator system arm and aims a camera at the solar array panels of the Hubble Space Telescope as astronaut Gregory J. Harbaugh assists.

Courtesy of NASA/Johnson Space Center

TRY THIS ACTIVITY

Weightless Water

Place two small holes on opposite sides near the bottom of a Styrofoam cup. Cover the holes with your thumb and finger and fill the cup with water. While standing on a stepladder or a table, grasp the cup near the top with your other hand. Uncover the holes just long enough to see the water streaming out. Then drop the cup. What happens to the streams of water as the cup and water fall? How is this activity similar to what an astronaut orbiting the earth experiences? How is it different?

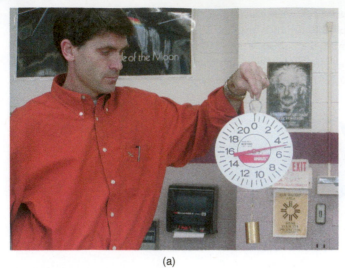

(a)

Figure 11.5   (a) When supported, the 0.50-kg mass weighs 4.90 N. (b) When no longer supported, the object appears weightless and the scale reads 0 N.

(b)

Figure 11.6   Newton's conceptual thought experiment

An object launched beyond the earth's atmosphere and given a horizontal velocity moves sideways while falling at the same time. Traveling horizontally at the appropriate velocity while falling due to gravity allows the space shuttle, astronauts, and other objects to continually miss the surface of the earth and achieve orbit (Fig. 11.6). An object traveling horizontally at less than the appropriate velocity will fall back to the earth.

Isaac Newton conceived this concept several centuries before the development of rockets and artificial satellites and predicted that orbiting the earth could theoretically be done. Newton developed a thought experiment where, on top of an unrealistically tall mountain and above the frictional forces of air resistance, a cannonball would be shot horizontally. And, the greater its initial speed, the greater its range. In fact, its range would also increase due to the curvature of the earth. Then, when the cannonball's path curved too much as shown, the cannonball would simply orbit the entire earth and strike the backside of the cannon! Newton made the connection between satellite motion and the orbit of the moon around the earth. The velocity of satellites orbiting the earth or any other planet can be calculated by combining centripetal motion and the law of universal gravitation:

$$F_c = \frac{mv^2}{r}, \qquad F_G = G\frac{m_1m_2}{r^2}$$

Solve the centripetal force equation for velocity,

$$v = \sqrt{\frac{F_c r}{m}}$$

Since the centripetal force holding a satellite in orbit is actually the gravitational force between the earth and the satellite, F_G can be substituted for F_c:

$$v = \sqrt{\frac{Gm_1 m_2 r}{r^2 m_1}} \qquad \text{(Cancel like terms.)}$$

The mass of the satellite (m_1) is the term that is canceled in the last equation. As a result, the mass of the object being orbited (m_2) is the only mass that determines the orbital velocity of the satellite and is included in the following simplified formula:

$$v = \sqrt{\frac{Gm}{r}}$$

where v = velocity of the satellite
$G = 6.67 \times 10^{-11}$ N m^2/kg^2 = 3.44×10^{-8} lb ft^2/slug^2
m = mass of the object being orbited
r = distance from the center of the earth to the satellite

The time or period, T, to orbit the earth or other celestial body is

$$T = 2\pi \sqrt{\frac{r^3}{Gm}} \qquad \text{(The derivation of this equation is beyond the scope of this text.)}$$

Note that just as acceleration due to gravity is independent of the mass of falling objects, the velocity and the period of the object are independent of the mass of the satellite as well.

If a satellite orbits the earth at $20\overline{0}0$ km above sea level, (a) how fast will the orbiting satellite travel and (b) how long will it take to orbit the earth once?

EXAMPLE 1

Data:

$r_{\text{earth}-\text{satellite}} = 6.38 \times 10^6$ m $+ 2.00 \times 10^6$ m $= 8.38 \times 10^6$ m

$m_{\text{earth}} = 5.97 \times 10^{24}$ kg

Basic Equations:

$$v = \sqrt{\frac{Gm}{r}}$$

$$T = 2\pi \sqrt{\frac{r^3}{Gm}}$$

Working Equations: Same

Substitutions:

(a) $v = \sqrt{\dfrac{(6.67 \times 10^{-11} \text{ N m}^2/\text{kg}^2)(5.97 \times 10^{24} \text{ kg})}{8.38 \times 10^6 \text{ m}}}$

$= 6890$ m/s $\boxed{\sqrt{\dfrac{(\text{N m}^2/\text{kg}^2)(\text{kg})}{\text{m}}} = \sqrt{\dfrac{(\text{kg m/s}^2)(\text{m}^2/\text{kg}^2)(\text{kg})}{\text{m}}} = \sqrt{\text{m}^2/\text{s}^2} = \text{m/s}}$

(b) $T = 2\pi \sqrt{\dfrac{(8.38 \times 10^6 \text{ m})^3}{(6.67 \times 10^{-11} \text{ N m}^2/\text{kg}^2)(5.97 \times 10^{24} \text{ kg})}}$

$= 7640$ s $= 2.12$ h $\boxed{\sqrt{\dfrac{\text{m}^3}{(\text{N m}^2/\text{kg}^2)(\text{kg})}} = \sqrt{\dfrac{\text{m}^3}{(\text{kg m/s}^2)(\text{m}^2/\text{kg}^2)(\text{kg})}} = \sqrt{\text{s}^2} = \text{s}}$

EXAMPLE 2

An asteroid orbits the sun 8.35×10^{11} m from the sun. (a) How fast must the asteroid travel to maintain its orbit around the sun? (b) How long will it take the asteroid to orbit the sun?

Data:

$$r_{\text{sun}-\text{asteroid}} = 8.35 \times 10^{11} \text{ m}$$
$$m_{\text{sun}} = 1.99 \times 10^{30} \text{ kg}$$

Basic Equations:

$$v = \sqrt{\frac{Gm}{r}}$$

$$T = 2\pi\sqrt{\frac{r^3}{Gm}}$$

Working Equations: Same

Substitutions:

(a)
$$v = \sqrt{\frac{(6.67 \times 10^{-11} \text{ N m}^2/\text{kg}^2)(1.99 \times 10^{30} \text{ kg})}{8.35 \times 10^{11} \text{ m}}}$$
$$= 1.26 \times 10^4 \text{ m/s}$$

(b)
$$T = 2\pi\sqrt{\frac{(8.35 \times 10^{11} \text{ m})^3}{(6.67 \times 10^{-11} \text{ N m}^2/\text{kg}^2)(1.99 \times 10^{30} \text{ kg})}}$$
$$= 4.16 \times 10^8 \text{ s} = 13.2 \text{ yr}$$

PHYSICS CONNECTIONS

Satellite Orbits

Thousands of functioning artificial satellites currently orbit the earth. These satellites have many functions such as tracking weather patterns; collecting scientific data about the earth and space; and transmitting telephone, television, and Internet communications. Our society has become dependent upon satellites over the last 50 years; much of our use of technology today would not be possible without satellites orbiting the earth.

Several types of orbits allow artificial satellites to maintain a stable orbit (Fig. 11.7). Communication and weather satellites need to remain over a fixed point on the earth. A television satellite dish on a house always points at a particular satellite in a stationary orbit and common weather images are also taken by satellites in a stationary orbit. A stationary orbit is known as a geosynchronous orbit, and satellites in such an orbit are positioned 22,400 mi (36,000 km) above the equator.

Imaging and intelligence satellites need to take clear and detailed pictures of the entire earth. Such satellites are placed in low-altitude, north–south polar orbits. The earth rotates underneath the satellites, allowing pictures and images to be collected for most of the earth.

Finally, the most versatile orbit for satellites is the asynchronous orbit. The space shuttle, scientific research satellites, and global positioning satellites need to be in an orbit that allows them to pass over various locations on earth at different times. These satellites typically orbit at varying altitudes ranging from 200 to 12,000 mi above the surface of the earth.

Figure 11.7 Three types of orbits for artificial satellites

PROBLEMS 11.3

1. The moon orbits 3.84×10^8 m from the earth. How fast does the moon travel?
2. The moon orbits 3.84×10^8 m from the earth. How long does it take to orbit the earth?
3. Find the orbital velocity for Mercury as it orbits the sun.
4. Find the orbital velocity for the earth as it orbits the sun.
5. Find the orbital velocity for Saturn as it orbits the sun.
6. Find the orbital velocity for Uranus as it orbits the sun.
7. Find the time Mercury takes to orbit the sun.
8. Find the time the earth takes to orbit the sun.
9. Find the time Saturn takes to orbit the sun.
10. Find the time Uranus takes to orbit the sun.

SKETCH

$$12 \text{ cm}^2 \quad w$$

4.0 cm

DATA

$A = 12 \text{ cm}^2, l = 4.0 \text{ cm}, w = ?$

BASIC EQUATION

$A = lw$

WORKING EQUATION

$w = \dfrac{A}{l}$

SUBSTITUTION

$w = \dfrac{12 \text{ cm}^2}{4.0 \text{ cm}} = 3.0 \text{ cm}$

NEW TECHNOLOGIES

Space Launch Systems—NASA's Powerful Rocket Technology

NASA has set its sights on the human exploration of Mars. In order to transport equipment to Mars to support human life, an entirely new propulsion system needs to be devised to overcome the gravitational force of earth and propel a spacecraft and its payload to Mars as well as to the International Space Station and moon.

NASA claims its Space Launch System (SLS) will be sustainable, affordable, and a safe launching mechanism that will carry American astronauts beyond earth's orbit patterns. Standing taller than the massive Saturn V rocket, the SLS, a 130-metric-ton rocket, will be capable of providing 20% more thrust than its predecessor that transported astronauts to the moon. The propulsion system consists mainly of a liquid hydrogen and oxygen propulsion system, similar to the system used on the space shuttle. The SLS will be able to carry a payload of 77 tons, nearly three times the payload capacity of the retired space shuttle.

The rocket consists of three main stages. Advanced booster rockets generate initial power to launch massive payloads into space. The main section or upper stage will generate superior space propulsion for escaping the earth's gravitation pull. The upper stage will also house additional power to maneuver and travel into deep space.

Two missions are slated for the SLS. Exploration Mission 1 will be a remotely operated Orion spacecraft that will travel to an orbit around the moon and return the Orion back to earth. Exploration Mission 2 will launch a crew of four astronauts further into space than the United States has ever traveled before. The hope is that the SLS will eventually lead NASA and its partners with a human payload to Mars (Fig. 11.8).

Figure 11.8 Artist's rendering of the SLS rocket on the launch pad

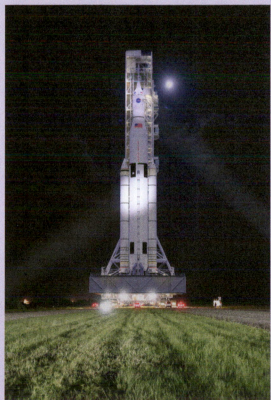

Courtesy of NASA Headquarters

Glossary

Gravitational Field The area around a massive body in which an object experiences a gravitational force. The more massive and closer an object is to that body, the stronger is the gravitational field. (p. 302)

Newton's Law of Universal Gravitation All objects that have mass are attracted to one another by a gravitational force. (p. 298)

Orbit The path taken by an object during its revolution around another object, such as the path of the moon or a satellite about the earth or of a planet about the sun. (p. 303)

Formulas

11.1 $F_G = G\dfrac{m_1 m_2}{r^2}$

11.3 $v = \sqrt{\dfrac{Gm}{r}}$

$T = 2\pi\sqrt{\dfrac{r^3}{Gm}}$

Review Questions

1. What type of force is related to the mass of objects?
   (a) Electric force   (b) Strong force
   (c) Magnetic force   (d) Gravitational force
2. As the distance increases between two objects, the gravitational force between the objects
   (a) increases.   (b) decreases.   (c) remains constant.
3. As the mass of two objects increases, the gravitational force between the objects
   (a) increases.   (b) decreases.   (c) remains constant.
4. The mass of a satellite is increased. In order to maintain the same orbital period, its distance from the earth must
   (a) increase.   (b) decrease.   (c) remain constant.
5. As the distance increases between a satellite and the earth, what happens to the time it takes to complete an orbit?
   (a) Increases   (b) Decreases   (c) Remains constant
6. Explain why the gravitational force that exists between the person sitting next to you in class and you is much less than the gravitational force that exists between the earth and you.
7. What would happen to your weight on earth if the radius of the earth doubled, but its mass stayed the same?
8. What would happen to your weight on earth if the mass of the earth doubled, but its radius stayed the same?
9. Explain how a satellite in orbit is in a constant state of free fall yet does not crash to the earth.
10. According to Isaac Newton, how is the motion of a falling apple different from the motion of the moon orbiting the earth?
11. Most planets actually have slightly elliptical orbits around the sun. What is the force exerted on a planet at its perigee (the point closest to the sun) compared to the force at its apogee (the point farthest from the sun)?
12. Does the mass of a satellite influence the time it takes to orbit the earth?

Review Problems

1. Two 0.300-kg apples are 25.0 cm apart from one center to the other. Find the gravitational attraction between the two apples.
2. Two 65.0-kg people are standing 1.00 m apart. What is the attractive gravitational force between them?
3. Find the weight of a 65.0-kg person on the earth (in newtons and pounds).
4. Find the weight of a 65.0-kg person on Jupiter (in newtons and pounds).
5. Find the weight of a 65.0-kg person on Pluto (in newtons and pounds).
6. If the moon orbited at one half the present distance to the earth, what would be the orbiting time for the moon?
7. If the moon orbited at twice the present distance to the earth, what would be the orbiting time for the moon?
8. If the moon orbited at four times the present distance to the earth, what would be the orbiting time for the moon?
9. Using Newton's law of universal gravitation, find the amount of gravitational force on an 85.0-kg astronaut on the launch pad.
10. If an 85.0-kg astronaut in a space shuttle orbits the earth 362 km above sea level, what is the amount of gravitational force between the astronaut and the earth?

SKETCH

12 cm² | w

4.0 cm

DATA
$A = 12 \text{ cm}^2, l = 4.0 \text{ cm}, w = ?$

BASIC EQUATION
$A = lw$

WORKING EQUATION
$w = \dfrac{A}{l}$

SUBSTITUTION
$w = \dfrac{12 \text{ cm}^2}{4.0 \text{ cm}} = 3.0 \text{ cm}$

A P P L I E D   C O N C E P T S

1. The gravitational differences between the earth and Mars is a factor that engineers and scientists must consider before sending an astronaut to the "Red Planet." (a) What is the acceleration due to gravity on Mars? (b) If an 85.0-kg astronaut landed on Mars, how much would the astronaut weigh there compared to his weight on the earth? (c) Based on the acceleration due to gravity on Mars, what might happen to the strength of the astronaut's muscles while the astronaut is away from earth?

2. (a) How far from the center of the earth must a person be located in space so that his or her weight would be the same as when he or she is standing on the moon? (b) How many earth radii would this location be from the center of the earth?

3. A geosynchronous communication satellite orbits at a fixed point above the earth's surface. The earth takes 24.0 h to rotate on its axis. (a) Find the communication satellite's orbiting altitude from the center of the earth and from the surface of the earth. (b) Find its linear speed.

4. Flight engineers for the Apollo Lunar Orbiter placed the orbiter 15$\overline{0}$ km above the surface of the moon. What was the Orbiter's (a) linear velocity and (b) period as it orbited the moon? (c) How would increasing the altitude affect the Orbiter's velocity and period?

5. (a) What is the gravitational force on a 65.7-kg space shuttle astronaut orbiting the earth 427 km above the surface of the earth? (b) Explain why astronauts orbiting the earth experience weightlessness even though the earth continually applies a force on their bodies.

CHAPTER 12

MATTER

Objectives

The major goals of this chapter are to enable you to:

1. Describe the properties of matter.
2. Apply Hooke's law.
3. Describe the properties of solids, liquids, and gases.
4. Solve density and specific gravity problems.
5. Calculate the amount of stress on objects.

Technology is used to take raw materials and shape, refine, mold, and transform them into products useful to our society. To do this, it is necessary to have some understanding of the nature and properties of matter and basic characteristics of its various forms of solids, liquids, and gases. Once these materials take shape and become useful products, they are subjected to various stresses and strains that can ultimately result in structural failure. We will show how to analyze these stresses and strains so that materials can be evaluated before being put to use.

12.1 Properties of Matter

What are the building blocks of matter? First, **matter** is anything that occupies space and has mass. Suppose that we take a cube of sugar and divide it into two pieces. Then we divide a resulting piece into another two pieces. Can we continue this process indefinitely and get smaller and smaller particles of sugar each time? No, at some point we will arrive at the building blocks of sugar.

An **element** is a substance that cannot be separated into simpler substances. A **compound** is a substance containing two or more elements.

A **molecule** is the smallest particle of an element that can exist in a free state and still retain the characteristics of that element or compound. Most simple molecules are about 3×10^{-10} m in diameter. An **atom** is the smallest particle of an element that can exist in a stable or independent state. The molecules of elements consist of one atom or two or more similar atoms; the molecules of compounds consist of two or more different atoms.

What do we get if we divide the sugar molecule? The resulting particles are carbon, hydrogen, and oxygen atoms. Models of water and sugar molecules are shown in Fig. 12.1. Not all atoms are the same size. The hydrogen atom is the smallest, with diameter 6×10^{-11} m and mass 1.67×10^{-27} kg. Uranium is one of the heaviest atoms, 3.96×10^{-25} kg.

Figure 12.1

Hydrogen atom

Oxygen atom

Hydrogen atom

(a) The water molecule is composed of two hydrogen atoms and one oxygen atom (H_2O).

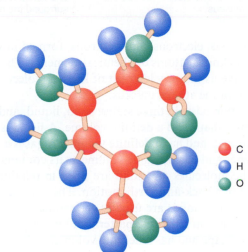

C

H

O

(b) The sugar (glucose) molecule is composed of six carbon atoms, twelve hydrogen atoms, and six oxygen atoms ($C_6H_{12}O_6$).

What happens if an atom is subdivided? Several constituent particles of the atom have been discovered. Of these, the three most important are (a) the **proton**, a particle with a positive charge, (b) the **electron**, a particle with a negative charge, and (c) the **neutron**, a particle that does not have an electric charge. Although there are smaller particles constituting the atom, we limit our discussion here to these three. Table 12.1 provides some basic information about these three particles.

Table 12.1   Properties of Atomic Particles

Particle	Mass	Diameter	Charge
Proton	1.673×10^{-27} kg	8.2×10^{-16} m	$+1$
Electron	9.109×10^{-31} kg	0^*	-1
Neutron	1.675×10^{-27} kg	8.2×10^{-16} m	0

*The zero size of an electron is due to its lack of any internal structure.

Models of the hydrogen atom and the carbon atom are shown in Fig. 12.2. The **nucleus** is the center part of an atom made up of protons and neutrons, while the electrons surround the nucleus. Electrons do not really orbit the nuclei like planets around the sun; they are more like a cloud around a given nucleus. The atoms of the nuclei are held together by strong nuclear forces. The molecules are held together by electrical forces.

Figure 12.2

(a) The hydrogen atom is composed of a nucleus which contains one proton. Its one electron surrounds the nucleus.

(b) The carbon atom is composed of a nucleus which contains six protons and six neutrons. Its six electrons surround the nucleus.

Ernest Rutherford (1871–1937),

physicist, was born in New Zealand. He developed the concept of the atomic nucleus, proposed in 1903 that radioactivity results from the disintegration of atoms, and received the Nobel Prize for Chemistry in 1908.

Protons, electrons, and neutrons, formed in various combinations, give us more than 100 known atoms or chemical elements. The atoms, formed in various combinations, give us the very long list of known molecules. **Ernest Rutherford** developed the concept of the atomic nucleus.

Matter exists in three states: solid, liquid, and gas. A **solid** is a substance that has a definite shape and a definite volume. A **liquid** is a substance that takes the shape of its container and has a definite volume. A **gas** is a substance that takes the shape of its container and has the same volume as its container.

The molecules of a solid are fixed in relation to each other [Fig. 12.3(a)]. They vibrate in a back-and-forth motion. They are so close that a solid can be compressed only slightly. Solids are usually crystalline substances, meaning that their molecules are arranged in a definite pattern and in fixed positions. This is why a solid tends to hold its shape and has a definite volume.

Figure 12.3

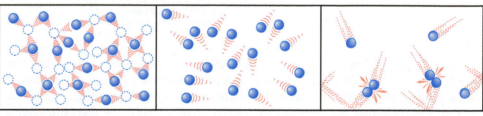

(a) Solid molecules vibrate in fixed positions.

(b) Liquid molecules flow over each other.

(c) Gas molecules move rapidly in all directions and collide.

The molecules of a liquid are not fixed in relation to each other [Fig. 12.3(b)]. They normally move in a flowing type of motion but are so close together that they are practically incompressible and have a definite volume. Because the molecules move in a smooth flowing motion and not in any fixed manner, a liquid takes the shape of its container.

The molecules of a gas are not fixed in relation to each other and move rapidly in all directions, colliding with each other [Fig. 12.3(c)]. They are much farther apart than molecules in a liquid, and they are extremely far apart when compared to the distance between molecules in solids. The movement of the molecules is limited only by the container. Therefore, a gas takes the shape of its container. Because the molecules are far apart, a gas can easily be compressed, and it has the same volume as its container.

12.2 Properties of Solids

Solids have a definite shape and a definite volume and have molecules that are usually arranged in a definite pattern. The following properties are common to most solids.

Cohesion and Adhesion

The molecules of a solid are held together by large internal molecular forces. **Cohesion** is the force of attraction between like molecules. The cohesive forces hold the closely packed molecules of a solid together and keep its shape and volume from being easily changed. Cohesion can also be shown by grinding and polishing the surfaces of two like solids and then sliding their surfaces together. For example, take two pieces of polished plate glass and slide them together. Try to pull them apart. The force of attraction of the like molecules of the two pieces of glass makes it difficult to pull them apart.

Adhesion is the force of attraction between different or unlike molecules. Common examples include glue and wood, adhesive tape and skin, and tar and shoe soles.

Tensile Strength

The **tensile strength** of a solid is a measure of its resistance to being pulled apart. That is, the tensile strength of a solid is a measure of its cohesive forces between adjacent molecules. The tensile strength of a rod or wire is found by putting it in a machine that pulls the rod or wire until it breaks (Fig. 12.4). The tensile strength is the ratio

$$\frac{\text{force required to break the rod or wire}}{\text{cross-sectional area of the rod or wire}}$$

Figure 12.4   This tensile testing system measures the tensile strength of the rod while pulling it apart.

Photo courtesy of Instron®

(a)

Pull

Pull

(b)

Hardness

The **hardness** of a solid is a measure of the internal resistance of its molecules to being forced farther apart or closer together. More commonly, we classify the hardness of a solid in terms of its difficulty in being scratched using a "scratch test." The given material is scratched in a certain way. Its scratch is then compared with a series of standard scratches of materials that form an arbitrary hardness scale from very soft solids to the hardest known substance, diamond.

The **Brinell method**, named after **Johan Brinell**, is a common industrial method used to measure the hardness of a metal. A machine is used to press a 10-mm hardened chrome-steel ball with the same force as an equivalent mass of 3000 kg into the metal being tested (Fig. 12.5). The diameter of the resulting impression is used as a measure of the metal's hardness. The Brinell value or number is the ratio

$$\frac{\text{mass placed on the object (in kg)}}{\text{surface area of the impression (in mm}^2\text{)}}$$

Johan August Brinell (1849–1925),

engineer and metallurgist, was born in Sweden. He invented the Brinell machine for measuring hardness of alloys and metals.

Figure 12.5

Courtesy of Clive Streeter © Dorling Kindersley

(a) The Brinell hardness testing machine determines the hardness of a metal.

(b) The size of the dent determines the hardness of the metal.

This value is compared with a scale of the accepted hardnesses of given metals. The larger the Brinell number, the harder is the metal.

Steel can be hardened by heating it to a very high temperature, then suddenly cooling it by putting it in water. However, it then becomes brittle. This cooled steel can then be tempered (toughened) by reheating it and allowing it to cool slowly. As the steel cools, it loses hardness and gains toughness. If the steel cools down slowly and completely, we say that it is *annealed*. Annealed steel is soft and tough but not brittle.

Ductility

Ductility is the property of a metal that enables it to be drawn through a die to produce a wire. As the rod is pulled through the die, its diameter is decreased and its length is increased as it becomes a wire (Fig. 12.6).

Figure 12.6 Ductility: the ability of a metal to be drawn into a wire

Die

Wire before being pulled through die

Wire after being pulled through die

Malleability

Malleability is the property of a metal that enables it to be hammered and rolled into a sheet. As the metal is hammered or rolled as in Fig. 12.7, its shape or thickness is changed. During this process, the atoms slide over each other and change positions. The cohesive forces are relatively strong; thus, the atoms do not become widely separated during their rearrangement and the resulting shape remains relatively stable.

Figure 12.7   Malleability: the ability of a metal to be rolled into a sheet

Elasticity

An object becomes deformed when outside forces change its shape or size. **Elasticity** is a measure of a deformed object's ability to return to its original size and shape once the outside forces are removed. When the solid is being deformed, sometimes the molecules attract each other and sometimes they repel each other. For instance, try to pull a rubber ball apart (Fig. 12.8). You will notice that the ball stretches out of shape.

Figure 12.8   Elasticity in a rubber ball

(a)   (b)   (c)

However, when you release the pulling force, the ball returns to its original shape because the molecules, being farther apart than normal, attract each other. If you squeeze the ball, it will again become out of shape. Now release the pressure, and the ball will again return to its original shape because the molecules, being too close together, repel each other. Therefore, we can see that when molecules are slightly pulled out of position, they attract each other. When they are pressed too close together, they repel each other.

Most solids have the property of elasticity; however, some are only slightly elastic. For example, wood and Styrofoam are two solids whose elasticity is small.

Not every elastic object returns to its original shape after being deformed. If too large a deforming force is applied, an object may become deformed permanently. Take a spring [Fig. 12.9(a)] and pull it apart by a moderate amount [Fig. 12.9(b)]. When you let it go, it should return to its original shape. Next, pull the spring apart as far as you can [Fig. 12.9(c)]. When you let it go this time, it will probably not return to its original shape. The **elastic limit** of a solid is the point beyond which a deformed object cannot return to its original shape. The spring's molecules were pulled far enough apart that they slid past one another beyond the point at which the original molecular forces could return the spring to its original shape. If the applied force is great enough, the spring breaks apart [Fig. 12.9(d)].

Figure 12.9

(a) Spring before stretching

(b) Spring stretched near its elastic limit

(c) Spring stretched beyond its elastic limit

(d) Spring stretched much beyond its elastic limit ... break occurs!

Stress is the ratio of the outside applied force, which tends to cause a distortion, to the area over which the force acts. In other words,

$$\text{stress} = \frac{\text{applied force}}{\text{area over which the force acts}}$$

or

$$S = \frac{F}{A}$$

where S = stress, usually in N/m^2 (Pa) or lb/in^2 (psi)
F = force applied, N or lb, perpendicular to the surface to which it is applied
A = area, m^2 or in^2

Since the SI metric unit for force is the newton (N) and the unit for area is the square metre (m^2), the corresponding pressure unit is N/m^2. This unit is given the special name *pascal* (Pa), named after **Blaise Pascal**, who made important discoveries in science and mathematics.

$$1 \text{ N/m}^2 = 1 \text{ Pa}$$

Imagine a brick weighing 12.0 N first lying on its side on a table and then standing on one end (Fig. 12.10). The weight of the brick is the same no matter what its position, so the total force (the weight of the brick) on the table is the same in both cases.

Figure 12.10   The weight of the brick is constant, but the stress on the table in part (b) is greater.

However, the position of the brick does make a difference in the stress exerted on the table. In which case is the stress greater? When standing on end, the brick exerts a greater stress on the table because the area of contact on the end is *smaller* than on the side. Using $S = F/A$, find the stress in each case.

Case 1	Case 2
$F = 12.0$ N	$F = 12.0$ N
$A = 8.00 \text{ cm} \times 16.0 \text{ cm} = 128 \text{ cm}^2$	$A = 6.00 \text{ cm} \times 8.00 \text{ cm} = 48.0 \text{ cm}^2$
$S = \dfrac{F}{A} = \dfrac{12.0 \text{ N}}{128 \text{ cm}^2} \times \left(\dfrac{100 \text{ cm}}{1 \text{ m}}\right)^2$	$S = \dfrac{F}{A} = \dfrac{12.0 \text{ N}}{48.0 \text{ cm}^2} \times \left(\dfrac{100 \text{ cm}}{1 \text{ m}}\right)^2$
$= 938 \text{ N/m}^2 = 938 \text{ Pa}$	$= 2500 \text{ N/m}^2 = 2500 \text{ Pa}$

This shows that when the same force is applied to a smaller area, the stress is greater. From the discussion so far, would you rather a woman step on your foot with a flat-heel shoe or with a pointed-heel shoe? (See Fig. 12.11.) This is a serious issue for the aircraft industry. They must design and construct airplane floors light in weight

Figure 12.11   The stress exerted on the floor by the heel in part (b) is greater because the weight of the woman rests on a smaller area.

(a)   (b)

but strong enough to withstand the stress of pointed-heel shoes. For example, if a 160-lb woman rests her weight on a 4.0-in^2 heel, the stress is

$$S = \frac{F}{A} = \frac{160 \text{ lb}}{4.0 \text{ in}^2} = 4\bar{0} \text{ lb/in}^2$$

But if she rests her weight on a pointed heel of $\frac{1}{4}$ in^2, which is a common area of a pointed heel, the stress is

$$S = \frac{F}{A} = \frac{160 \text{ lb}}{\frac{1}{4} \text{ in}^2} = 640 \text{ lb/in}^2$$

A similar comparison may be shown using metric units. If a 65-kg woman rests her weight on a 25-cm^2 heel, the stress is

$$S = \frac{F}{A} = \frac{(65 \text{ kg})(9.80 \text{ m/s}^2)}{25 \text{ cm}^2 \times \left(\dfrac{1 \text{ m}}{100 \text{ cm}}\right)^2} = 2.5 \times 10^5 \text{ N/m}^2 \qquad (1 \text{ N} = 1 \text{ kg m/s}^2)$$

$$= 2.5 \times 10^5 \text{ Pa} = 250 \text{ kPa}$$

If she rests her weight on a pointed heel of 1.0 cm^2, the stress is

$$S = \frac{F}{A} = \frac{(65 \text{ kg})(9.80 \text{ m/s}^2)}{1.0 \text{ cm}^2 \times \left(\dfrac{1 \text{ m}}{100 \text{ cm}}\right)^2} = 6.4 \times 10^6 \text{ N/m}^2$$

$$= 6.4 \times 10^6 \text{ Pa} = 6400 \text{ kPa}$$

Since the pascal is a relatively small unit, the kilopascal (kPa) is a commonly used unit of pressure.

Five basic types of stresses are as follows:

Tension is a stress caused by two forces acting directly opposite each other. This stress tends to cause objects to become longer and thinner. An example of such a stress is that on the rope in a tug-of-war (Fig. 12.12). The rope has one team's force pulling one way and another team's force pulling in the opposite direction. If the rope is not strong enough to withstand the tension, it could ultimately stretch beyond its elastic limit and break.

Figure 12.12 The rope in a tug-of-war competition is in constant tension.

Courtesy of fotolia © micromonkey

Compression is a stress caused by two forces acting directly toward each other. This stress tends to cause objects to become shorter and thicker. An example of compression is the stress present in a supporting column (Fig. 12.13). A load is pushing down on the column, while the ground is applying a force pushing up on the column. As a result, the pillar compresses.

Figure 12.13 A column under the New Clark Bridge crossing the Mississippi River is in compression.

Shearing is a stress caused by two forces applied in parallel, opposite directions. In Fig. 12.14, the table pushing the book to the left counteracts the force of the hand

Figure 12.14   A book being pushed in this way is undergoing shear.

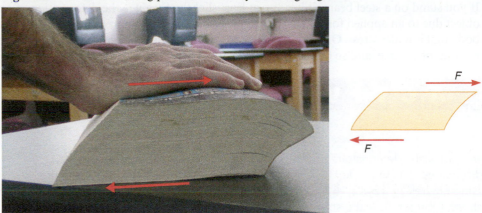

pushing the book to the right. The normally rectangular shape of the book is altered. Scissors use shearing to cut paper.

Torsion is a stress related to a twisting motion. Torsion occurs when two torques act in opposite directions. This type of stress severely compromises the strength of most materials. An example of torsion is the stress on a bolt or a screw as it is being tightened (Fig. 12.15).

Figure 12.15   The twisting of the bolt in one direction is counteracted by the force of the wood resisting the turning motion.

Bending consists of both tension and compression stresses. It occurs when a force is placed on an object and this causes it to sag. An example of bending is caused by a person sitting on a board (Fig. 12.16). The top section of the board is being pushed together, in compression, while the bottom section of the board is being pulled apart, in tension.

Figure 12.16   A beam that is bending

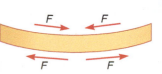

Whenever a stress is applied to an object, the object is changed minutely, at least. If you stand on a steel beam, it bends—at least slightly. **Strain** is the deformation of an object due to an applied force. That is, strain is the relative amount of deformation of a body that is under stress. Or strain is *change in length per unit of length*, change in volume per unit of volume, and so on. Strain is a direct and necessary consequence of stress.

TRY THIS ACTIVITY

Stresses

Foam is a flexible material that can easily demonstrate the various types of stresses on solid materials. Using a permanent marker, draw lines at 1.0-in. intervals along the top and bottom sides of a rectangular piece of foam (Fig. 12.17). Using Figures 12.12–12.16 as a guide, apply the appropriate forces to the foam to simulate the five types of stress. Observe how the stresses affect the spacing of the drawn lines. Describe how the strength of building materials such as concrete, wood, and steel is affected when the materials are subjected to the five types of stress.

Figure 12.17

EXAMPLE 1

A steel column in a building has a cross-sectional area of $25\overline{0}0$ cm^2 and supports a weight of 1.50×10^5 N. Find the stress on the column.

Data:

$$A = 25\overline{0}0 \text{ cm}^2 \times \left(\frac{1 \text{ m}}{100 \text{ cm}}\right)^2 = 0.250 \text{ m}^2$$
$$F = 1.50 \times 10^5 \text{ N}$$
$$S = ?$$

Basic Equation:

$$S = \frac{F}{A}$$

Working Equation: Same

Substitution:

$$S = \frac{1.50 \times 10^5 \text{ N}}{0.250 \text{ m}^2}$$
$$= 6.00 \times 10^5 \text{ N/m}^2$$
$$= 6.00 \times 10^5 \text{ Pa} \quad \text{or} \quad 60\overline{0} \text{ kPa}$$

Hooke's Law

One of the most basic principles related to the elasticity of solids is **Hooke's law**, named after **Robert Hooke**.

Robert Hooke (1635–1703),

chemist and physicist, was born in England. He formulated the law governing elasticity (Hooke's law), invented the balance spring for watches, worked with and made important observations with the telescope and the microscope, and formulated the theory of planetary movement.

HOOKE'S LAW

The ratio of the force applied to an object to its change in length (resulting in its being stretched or compressed by the applied force) is constant as long as the elastic limit has not been exceeded.

Or stated another way: Stress is directly proportional to strain as long as the elastic limit has not been exceeded. (See Fig. 12.18.) In equation form,

$$\frac{F}{\Delta l} = k$$

where F = applied force
   Δl = change in length
   k = elastic constant

Note: Δ (the Greek letter delta) is often used in mathematics and science to mean "change in."

Figure 12.18   Graph of Hooke's law showing behavior within and beyond the elastic limit

A force of 5.00 N is applied to a spring whose elastic constant is 0.250 N/cm. Find its change in length.

EXAMPLE 2

Sketch:

5.00 N

Data:

$F = 5.00$ N
$k = 0.250$ N/cm
$\Delta l = ?$

Basic Equation:

$$\frac{F}{\Delta l} = k$$

Working Equation:

$$\Delta l = \frac{F}{k}$$

Substitution:

$$\Delta l = \frac{5.00 \text{ N}}{0.250 \text{ N/cm}}$$

$$= 20.0 \text{ cm}$$

$$\frac{\text{N}}{\text{N/cm}} = \text{N} \div \frac{\text{N}}{\text{cm}} = \cancel{\text{N}} \cdot \frac{\text{cm}}{\cancel{\text{N}}} = \text{cm}$$

EXAMPLE 3

A force of 3.00 lb stretches a spring 12.0 in. What force is required to stretch the spring 15.0 in.?

Sketch:

12.0 in.

15.0 in.

3.00 lb

? lb

Data:

$$F_1 = 3.00 \text{ lb}$$
$$l_1 = 12.0 \text{ in.}$$
$$l_2 = 15.0 \text{ in.}$$
$$F_2 = ?$$

Basic Equation:

$$\frac{F}{\Delta l} = k$$

Working Equations:

$$\frac{F}{\Delta l} = k \quad \text{and} \quad F = k(\Delta l)$$

Substitution: There are two substitutions, one to find k and one to find the second force F_2:

$$\frac{3.00 \text{ lb}}{12.0 \text{ in.}} = k$$
$$0.250 \text{ lb/in.} = k$$

$$F_2 = (0.250 \text{ lb/in.})(15.0 \text{ in.})$$
$$= 3.75 \text{ lb}$$

· · · · · · · · · · · · · · · · · ·

EXAMPLE 4

A support column is compressed 3.46×10^{-4} m under a weight of 6.42×10^5 N. How much is the column compressed under a weight of 5.80×10^6 N?

First find k:

Data:

$$F_2 = 6.42 \times 10^5 \text{ N}$$
$$\Delta l_2 = 3.46 \times 10^{-4} \text{ m}$$
$$k = ?$$

Basic Equation:

$$\frac{F_2}{\Delta l_2} = k$$

Working Equation: Same

Substitution:

$$k = \frac{6.42 \times 10^5 \text{ N}}{3.46 \times 10^{-4} \text{ m}}$$
$$= 1.86 \times 10^9 \text{ N/m}$$

Then:

Data:

$$k = 1.86 \times 10^9 \text{ N/m}$$
$$F_1 = 5.80 \times 10^6 \text{ N}$$
$$\Delta l_1 = ?$$

Basic Equation:

$$\frac{F_1}{\Delta l_1} = k$$

Working Equation:

$$\Delta l_1 = \frac{F_1}{k}$$

Substitution:

$$\Delta l_1 = \frac{5.80 \times 10^6 \text{ N}}{1.86 \times 10^9 \text{ N/m}}$$
$$= 3.12 \times 10^{-3} \text{ m or } 3.12 \text{ mm}$$

PHYSICS CONNECTIONS

Development of Materials

Engineers choose materials based on two main factors—cost and strength. The strength of a material is chosen according to its proposed use. For example, will the material be used underneath a roadway to support cars and trucks, or will it be used to suspend a roadway? Advances in the science and engineering of materials have helped in the development of more economical bridges that span greater and greater distances.

Ancient Roman, Mesopotamian, and Chinese bridge builders used stone to create their arch bridges. Stone was abundant and very strong under compression forces. However, the need for lighter and stronger materials sparked technological improvements in building materials during the Industrial Revolution of the 18th and 19th centuries. During that time, an English engineer, Abraham Darby, built the first cast-iron arch bridge (Fig. 12.19). The low cost and high compression strength of iron made it a much better alternative to the expensive and heavy stone arch. In later years, however, problems with cast-iron bridges became evident, as the brittle nature of iron under tension and shear stresses caused fractures in cast-iron structures.

In the late 19th century, an American engineer, James Eads, constructed the first steel arch bridge over the Mississippi River. The steel used in this bridge was not much stronger under compression than iron but was more than twice as strong as iron under tension and shearing stresses. Steel is used for beams and cables in almost all bridges today.

Although concrete is extremely weak under shearing and tension stresses, it is used to combat compression stresses. To improve concrete's ability to withstand tension and shearing stresses, stretched steel wire mesh or cables are embedded in the concrete. After the concrete dries, the tension in the wire mesh or cables is released, placing the concrete in a permanent state of compression. Therefore, when a tension force is placed on the prestressed concrete, the concrete itself remains under compression from the steel mesh or cables.

Figure 12.19

Photo courtesy of Andy Crawford/Dorling Kindersley, Ltd

(a) Darby's cast-iron bridge, constructed at Coalbrookdale, England, in 1779

Tension

When the concrete is dry, tension is released, resulting in compression.

Tension

(b) Pre-stressed concrete requires tension on the cables while the concrete dries. After the concrete dries, the tension is released, placing the concrete in compression.

PROBLEMS 12.2

SKETCH

$$12 \text{ cm}^2 \quad w$$
$$4.0 \text{ cm}$$

DATA
$A = 12 \text{ cm}^2, l = 4.0 \text{ cm}, w = ?$

BASIC EQUATION
$A = lw$

WORKING EQUATION
$w = \dfrac{A}{l}$

SUBSTITUTION
$w = \dfrac{12 \text{ cm}^2}{4.0 \text{ cm}} = 3.0 \text{ cm}$

1. A packing crate 2.50 m × 0.80 m × 0.45 m weighs 1.41×10^5 N. Find the stress (in kPa) exerted by the crate on the floor in each of its three possible positions.
2. A packing crate 2.50 m × 20.0 cm × 30.0 cm has a mass of 975 kg. Find the stress (in kPa) exerted by the crate on the floor in each of its three possible positions.
3. A spring is stretched 24.0 in. by a force of 54.0 lb. (a) How far will it stretch if a force of 104 lb is applied? (b) What weight will stretch the spring 9.00 in.?
4. A 17.0-N force stretches a wire 0.650 cm. (a) What force will stretch a similar piece of wire 1.87 cm? (b) A force of 21.3 N is applied to a similar piece of wire. How far will it stretch?
5. A force of 36.0 N stretches a spring 18.0 cm. Find the spring constant (in N/m).
6. A force of 5.00 N is applied to a spring whose spring constant is 0.250 N/cm. Find its change in length (in cm).
7. Each vertical steel column of an office building supports 1.30×10^5 N and is compressed 5.90×10^{-3} cm. (a) Find the compression in each column if a weight of 5.50×10^5 N is supported. (b) If the compression of each steel column is 0.0710 cm, what weight is supported by each column?
8. Each vertical steel column of an office building supports $30,\overline{0}00$ lb and is compressed 0.00234 in. Find the compression of each column if a weight of 125,000 lb is supported.
9. If the compression of each steel column in Problem 8 is 0.0279 in., what weight is supported by each column?

10. A coiled spring is stretched 40.0 cm by a 5.00-N weight. (a) How far is it stretched by a 15.0-N weight? (b) What weight will stretch the spring 60.0 cm?

11. A 12,000-N load is hanging from a steel cable that is 10.0 m long and 16.0 mm in diameter. Find the stress.

12. A rectangular cast-iron column 25.0 m × 25.0 cm × 5.00 m supports a weight of 6.80×10^6 N. Find the stress in pascals on the column.

13. In a Hooke's law experiment, the following weights were attached to a spring, resulting in the following elongations:

Weight (N)	Elongation (cm)
50	2.0
75	3.3
105	4.2
125	5.0
150	6.0
175	7.4
225	9.5
275	11.1

(a) Plot the graph of weight versus elongation and draw the best straight line through the data. (b) From the graph, what weight corresponds to an elongation of 7.5 cm? (c) From the graph, what elongation corresponds to a weight of 220 N? (d) From the graph, determine the spring constant.

14. What was the original length of a spring with spring constant 96.0 N/m that is stretched to 28.0 cm by a 15.0-N weight?

15. A bar weighing 26.0 N is supported horizontally on each end by two hanging springs, each 15.0 cm long, with spring constants 0.970 N/cm and 1.45 N/cm, respectively. The bar is 6.00 m long and has a center of mass 2.00 m from the spring with constant 0.970 N/cm. How far does each spring stretch?

16. A firefighter weighs 725 N. She wears shoes that each cover an area of 206 cm². (a) What is the average stress she applies to the ground on which she is standing? (b) How does the stress change if she stands on only one foot?

17. Two identical wires are 125 cm and 375 cm long, respectively. The first wire is broken by a force of 489 N. What force is needed to break the other?

18. The cross-sectional area of a wire is 2.50×10^{-3} cm² and its tensile strength is 1.00×10^5 N/cm². What force will break the wire?

19. A spring having a force constant of 1.25 N/cm is stretched through a distance of 11.5 cm. How much work is required to stretch the spring?

12.3 Properties of Liquids

As noted previously, a liquid is a substance that has a definite volume and takes the shape of its container. The molecules move in a flowing motion yet are so close together that it is very difficult to compress a liquid. Most liquids share the following common properties.

Cohesion and Adhesion

Cohesion, the force of attraction between like molecules, causes a liquid like molasses to be sticky. Adhesion, the force of attraction between unlike molecules, causes the molasses to also stick to your finger. In the case of water, its adhesive forces are greater than its cohesive forces. Put a glass in water and pull it out. Some water remains on the glass.

In the case of mercury, the opposite is true. Mercury's cohesive forces are greater than its adhesive forces. If glass is submerged in mercury and then pulled out, virtually no mercury remains on the glass.

A liquid whose adhesive forces are greater than its cohesive forces tends to wet any surface that comes in contact with it. A liquid whose cohesive forces are greater than its adhesive forces tends to leave objects dry or not wet when any surface comes in contact with it.

Figure 12.20 (a) The surface tension of water will support a needle. (b) Adding soap reduces the surface tension and the needle sinks.

(a) Water (b) Soap added

Surface Tension

Surface tension is the ability of the surface of a liquid to act like a thin, flexible film. The ability of the surface of water to support a needle is an example. The water's surface acts like a thin, flexible surface film. The surface tension of water can be reduced by adding soap to the water (Fig. 12.20). Soaps are added to laundry water to decrease the surface tension of water so that the water more easily penetrates the fibers of the clothes being washed.

Surface tension causes a raindrop to hold together and a small drop of mercury to keep an almost spherical shape. A liquid drop suspended in space is spherical. A falling raindrop's shape is due to the friction with the air (Fig. 12.21).

Figure 12.21 (a) Surface tension causes a drop of mercury to be more spherical. (b) Surface tension causes a liquid drop to hold together. (c) The shape of a falling raindrop is due to the friction with air.

(a) Mercury drop on a surface (b) Liquid drop in space (c) Falling raindrop

Viscosity

Viscosity is the internal friction of a fluid caused by molecular attraction, which makes it resist a tendency to flow. The greater the molecular attraction, the greater is the liquid's internal friction and viscosity. For example, it takes more force to move a block of wood through oil than through water. This is because oil is more viscous than water.

If a liquid's temperature is increased, its viscosity decreases. For example, the viscosity of oil in a car engine before it is started on a winter morning at −10.0°C is greater than after the engine has been running for an hour (Fig. 12.22).

Higher viscosity does not mean higher density. For example, oil is more viscous, but water is denser. Therefore, oil floats on water.

Figure 12.22 Cold oil is more viscous than hot oil.

Cold oil Hot oil

Capillary Action

Capillary action is the behavior of liquids that causes the liquid level in very small-diameter tubes to be different than that in larger-diameter tubes. A liquid keeps the same level in connected tubes filled with it if the tubes have a large enough diameter (Fig. 12.23). In tubes of different very small diameters, water does not stand at the same level. The smaller the diameter, the higher the water rises (Fig. 12.24). Mercury

Figure 12.23   A liquid keeps the same level in tubes of large enough diameter that are connected. Neither the shape nor the size of the containers makes a difference.

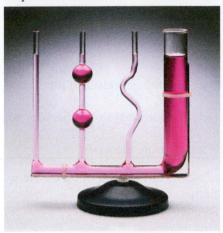

Courtesy of Richard Megna © Fundamental Photographs, NYC

Figure 12.24   Liquids rise in a thin tube because of surface tension and adhesion. The height of the rise depends on the surface tension, the diameter of the tube, and the density of the liquid. Here you can see the varying heights of colored water in thin tubes of various diameters.

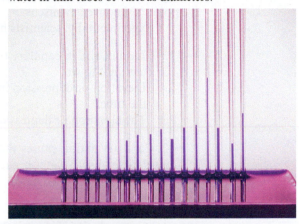

Courtesy of Ken Kay © Fundamental Photographs, NYC

does not stand at the same level either, but instead of rising up the narrower tube, the mercury level falls or is depressed in the smaller-diameter tube. The smaller the diameter, the lower mercury's level is depressed (Fig. 12.25).

Figure 12.25   Comparison of the capillary action of water and mercury: In very thin tubes, water stands at higher levels, and the water surface is concave. In very thin tubes, mercury stands at lower levels, and the mercury surface is convex. [The tube diameters and differences in liquid levels are exaggerated in (a) and (b).]

Water surface is concave

Water

(a)

Mercury surface is convex

(b)

Capillary action is due to both the adhesion of the liquid molecules with the tube and the surface tension of the liquid. For water, the adhesive forces are greater than the cohesive molecular forces. Thus, water creeps up the sides of the tube and produces a concave water surface. The surface tension of the water tends to flatten the concave surface. Together, these two forces raise the water up the tube until it is counterbalanced by the weight of the water column itself.

In the case of mercury, the cohesive molecular forces are greater than the adhesive forces and produce a convex mercury surface. The surface tension of the mercury tends to further hold down its level. This crescent-shaped surface of a liquid column in a tube, whether concave or convex, is called a **meniscus**. To measure the height of a liquid in a tube, measure to the lowest point of a concave meniscus or to the highest point of a convex meniscus.

Experimentally, scientists have found that:

1. Liquids rise in capillary tubes they tend to wet and are depressed in tubes they tend not to wet.
2. Elevation or depression in the tube is inversely proportional to the diameter of the tube.
3. The elevation or depression decreases as the temperature increases.

Capillary action causes the rise of oil (or kerosene) in the wick of an oil lamp. Towels also absorb water because of capillary action.

PHYSICS CONNECTIONS

Oil Viscosity

One of the most important factors in choosing the correct motor oil for an automobile is its viscosity grade (Fig. 12.26). The viscosity grade is indicated by a numbering system established by the Society of Automotive Engineers (S.A.E.), such as 10W-30. The first number followed by the "W" indicates that the oil remains viscous and does not lose its lubricating abilities at low temperatures. A 5W weight oil is guaranteed to remain viscous at −25°C, a 10W weight oil is guaranteed to remain viscous at −20°C, a 15W weight oil is guaranteed to remain viscous at −15°C, and so on.

The second number in the viscosity grade indicates that the oil has had polymers added to it to prevent the oil from losing its viscosity as its temperature increases. When the temperature increases, the polymers stretch into long molecular chains that prevent the oil from shearing apart in the engine. This number indicates how the oil will behave when it reaches typical engine operating temperatures. In other words, a 10W-30 oil remains viscous at −20°C, but when heated in the engine, it behaves like the 30W oil would when it is cold. A 5W-20 oil remains viscous at −25°C, but when heated in the engine, it behaves like the 20W oil would when it is cold.

Figure 12.26   Filling the engine with the motor oil of the correct viscosity is essential for maintaining a well-lubricated engine.

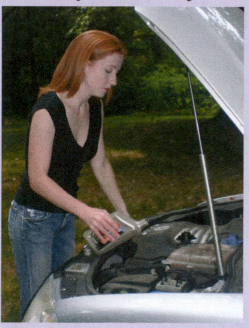

Although the addition of polymers to the engine oil is beneficial in achieving viscosity at large temperature ranges, those polymers also form deposits in the engine. A large range between the two numbers, such as 10W-40, indicates that more polymers are present in the oil as compared to a smaller range, such as 10W-20. If there are more polymers, then there is also an increased risk of thermal breakdown and the formation of unwanted deposits in the engine. As a result, a narrower range between the two numbers is typically desired, but consumers should refer to the recommendations from the car's manufacturer when they select the proper engine oil.

Synthetic motor oil is increasingly being used as a lubricant for superior engine wear protection. Synthetic oils perform better than conventional oil at very high and very low operating temperatures. Conventional oil is refined crude oil. Synthetic oil is refined but further broken down into its basic molecules. It is then rebuilt and added to, giving it even better lubricating characteristics than conventional motor oil.

12.4 Properties of Gases

Expansion is a property of a gas in which the rapid random movement of its molecules causes the gas to completely occupy the volume of its container.

Diffusion is the process by which molecules of a gas mix with the molecules of a solid, a liquid, or another gas. If you remove the cap from a can of gasoline, you soon smell the fumes. The air molecules and the gasoline molecules mix throughout the room because of diffusion.

A balloon inflates due to the pressure of the air molecules on its inside surface. This pressure is caused by the bombardment on the walls by the moving molecules. The pressure may be increased by increasing the number of molecules by blowing more air into the balloon. Pressure may also be increased by heating the air molecules already in the balloon. Heat increases the velocity of the molecules.

The behavior of liquids and gases is often very similar. **A fluid** is a substance that takes the shape of its container. The term is used when discussing principles and behaviors common to both liquids and gases.

12.5 Density

Density is a property of all three states of matter. **Mass density**, D_m, is defined as mass per unit volume. **Weight density**, D_w, is defined as weight per unit volume, or

$$D_m = \frac{m}{V} \qquad\qquad D_w = \frac{F_w}{V}$$

where D_m = mass density D_w = weight density
$\quad\quad\;\; m$ = mass F_w = weight
$\quad\quad\;\;\, V$ = volume V = volume

Although mass density and weight density can be expressed in both the metric system and the U.S. system, mass density is usually given in the metric units kg/m^3 and weight density is usually given in the U.S. units lb/ft^3 (Table 12.2).

The mass density of water is $10\overline{0}0$ kg/m^3; that is, 1 cubic metre of water has a mass of $10\overline{0}0$ kg. The weight density of water is 62.4 lb/ft^3; that is, 1 cubic foot of water weighs 62.4 lb.

In nearly all forms of matter, the density usually decreases as the temperature increases and increases as the temperature decreases. Water does not follow the usual pattern of increasing density at lower temperatures; ice is actually less dense than liquid water. This phenomenon is discussed more fully in Section 14.7.

Note: Conversion factors must often be used to obtain the desired units.

Table 12.2 Densities for Various Substances

Substance	Mass Density (kg/m³)	Weight Density (lb/ft³)
Solids		
Aluminum	2,700	169
Brass	8,700	540
Concrete	2,300	140
Copper	8,890	555
Cork	240	15
Ice	917	57
Iron	7,800	490
Lead	11,300	708
Wood, white pine	420	26
Liquids		
Alcohol	790	49.4
Gasoline	680	42.0
Mercury	13,600	846
Oil	870	54.2
Seawater	1,025	64.0
Water	1,000	62.4
Gases*	At 0°C and 1 atm pressure	At 32°F and 1 atm pressure
Air	1.29	0.081
Ammonia	0.760	0.047
Carbon dioxide	1.96	0.123
Carbon monoxide	1.25	0.078
Helium	0.178	0.011
Hydrogen	0.0899	0.0056
Nitrogen	1.25	0.078
Oxygen	1.43	0.089
Propane	2.02	0.126

*The density of a gas is found by pumping the gas into a container, measuring its volume and mass or weight, and then using the appropriate density formula.

EXAMPLE 1

Find the weight density of a block of wood 3.00 in. × 4.00 in. × 5.00 in. with weight 0.700 lb.

Sketch:

5.00 in.

3.00 in.

4.00 in.

Data:

$$l = 4.00 \text{ in.}$$
$$w = 3.00 \text{ in.}$$
$$h = 5.00 \text{ in.}$$
$$F_w = 0.700 \text{ lb}$$
$$D_w = ?$$

Basic Equations:

$$V = lwh \quad \text{and} \quad D_w = \frac{F_w}{V}$$

Working Equations: Same

Substitutions:

$$V = (4.00 \text{ in.})(3.00 \text{ in.})(5.00 \text{ in.})$$
$$= 60.0 \text{ in}^3$$

$$D_w = \frac{0.700 \text{ lb}}{60.0 \text{ in}^3}$$

$$= 0.0117 \frac{\text{lb}}{\text{in}^3} \times \left(\frac{12 \text{ in.}}{1 \text{ ft}} \right)^3$$

$$= 20.2 \text{ lb/ft}^3$$

· · · · · · · · · · · · · · · · ·

Find the mass density of a ball bearing with mass 22.0 g and radius 0.875 cm. ◄ **EXAMPLE 2**

Data:

$$r = 0.875 \text{ cm}$$
$$m = 22.0 \text{ g}$$
$$D_m = ?$$

Basic Equations:

$$V = \tfrac{4}{3} \pi r^3 \quad \text{and} \quad D_m = \frac{m}{V}$$

Working Equations: Same

Substitutions:

$$V = \tfrac{4}{3} \pi (0.875 \text{ cm})^3$$
$$= 2.81 \text{ cm}^3$$

$$D_m = \frac{22.0 \text{ g}}{2.81 \text{ cm}^3}$$
$$= 7.83 \text{ g/cm}^3$$

$$= 7.83 \frac{\text{g}}{\text{cm}^3} \times \left(\frac{100 \text{ cm}}{1 \text{ m}} \right)^3 \times \frac{1 \text{ kg}}{10^3 \text{ g}} = 7830 \text{ kg/m}^3$$

· · · · · · · · · · · · · · · · ·

Find the weight density of a gallon of water weighing 8.34 lb. ◄ **EXAMPLE 3**

Data:

$$F_w = 8.34 \text{ lb}$$
$$V = 1 \text{ gal} = 231 \text{ in}^3$$
$$D_w = ?$$

Basic Equation:

$$D_w = \frac{F_w}{V}$$

Working Equation: Same

Substitution:

$$D_w = \frac{8.34 \text{ lb}}{231 \text{ in}^3}$$

$$= 0.0361 \frac{\text{lb}}{\text{in}^3} \times \left(\frac{12 \text{ in.}}{1 \text{ ft}}\right)^3$$

$$= 62.4 \text{ lb/ft}^3$$

EXAMPLE 4

Find the weight density of a can of oil (1 quart) weighing 1.90 lb.

Data:

$$V = 1 \text{ qt} = \tfrac{1}{4} \text{ gal} = \tfrac{1}{4}(231 \text{ in}^3) = 57.8 \text{ in}^3$$
$$F_w = 1.90 \text{ lb}$$
$$D_w = ?$$

Basic Equation:

$$D_w = \frac{F_w}{V}$$

Working Equation: Same

Substitution:

$$D_w = \frac{1.90 \text{ lb}}{57.8 \text{ in}^3}$$

$$= 0.0329 \frac{\text{lb}}{\text{in}^3} \times \left(\frac{12 \text{ in.}}{1 \text{ ft}}\right)^3$$

$$= 56.9 \text{ lb/ft}^3$$

EXAMPLE 5

A quantity of gasoline weighs 5.50 lb with weight density 42.0 lb/ft^3. Find its volume.

Data:

$$D_w = 42.0 \text{ lb/ft}^3$$
$$F_w = 5.50 \text{ lb}$$
$$V = ?$$

Basic Equation:

$$D_w = \frac{F_w}{V}$$

Working Equation:

$$V = \frac{F_w}{D_w}$$

Substitution:

$$V = \frac{5.50 \text{ lb}}{42.0 \text{ lb/ft}^3}$$

$$= 0.131 \text{ ft}^3$$

The density of an irregular solid (rock) cannot be found directly because of the difficulty of finding its volume. However, we could find the amount of water the solid displaces, which is the same as the volume of the irregular solid. In Fig. 12.27, the volume of water in the small beaker equals the volume of the rock.

Figure 12.27   The volume of the rock can be found by measuring the volume of the liquid it displaces.

(a)   (b)

The volume of the rock = the volume of the water displaced into the overflow beaker.

A rock of mass 10.8 kg displaces $32\overline{0}0$ cm^3 of water. What is the mass density of the rock?

EXAMPLE 6

Data:

$$m = 10.8 \text{ kg}$$
$$V = 32\overline{0}0 \text{ cm}^3$$
$$D_m = ?$$

Basic Equation:

$$D_m = \frac{m}{V}$$

Working Equation: Same

Substitution:

$$D_m = \frac{10.8 \text{ kg}}{32\overline{0}0 \text{ cm}^3} \times \left(\frac{100 \text{ cm}}{1 \text{ m}}\right)^3$$
$$= 3380 \text{ kg/m}^3$$

A rock displaces 3.00 gal of water and has a weight density of 156 lb/ft^3. What is its weight?

EXAMPLE 7

Data:

$$D_w = 156 \text{ lb/ft}^3$$
$$V = 3.00 \text{ gal}$$
$$F_w = ?$$

Basic Equation:

$$D_w = \frac{F_w}{V}$$

Working Equation:

$$F_w = D_w V$$

Substitution:

$$F_w = 156 \, \frac{\text{lb}}{\text{ft}^3} \times 3.00 \, \text{gal} \times \frac{231 \, \text{in}^3}{1 \, \text{gal}} \times \left(\frac{1 \, \text{ft}}{12 \, \text{in.}} \right)^3$$

$$= 62.6 \, \text{lb}$$

To compare the densities of two materials, we compare each with the density of water. The **specific gravity** of any material is the ratio of the density of the material to the density of water. That is,

$$\text{specific gravity (sp gr)} = \frac{D_{\text{material}}}{D_{\text{water}}}$$

Note that specific gravity is a unitless quantity.

EXAMPLE 8

The density of iron is 7830 kg/m^3. Find its specific gravity.

Data:

$$D_{\text{material}} = 7830 \, \text{kg/m}^3$$
$$D_{\text{water}} = 10\overline{0}0 \, \text{kg/m}^3$$
$$\text{sp gr} = ?$$

Basic Equation:

$$\text{sp gr} = \frac{D_{\text{material}}}{D_{\text{water}}}$$

Working Equation: Same

Substitution:

$$\text{sp gr} = \frac{7830 \, \text{kg/m}^3}{10\overline{0}0 \, \text{kg/m}^3}$$

$$= 7.83$$

This means that iron is 7.83 times as dense as water, and thus it sinks in water.

EXAMPLE 9

The density of oil is 54.2 lb/ft^3. Find its specific gravity.

$$D_{\text{material}} = 54.2 \, \text{lb/ft}^3$$
$$D_{\text{water}} = 62.4 \, \text{lb/ft}^3$$
$$\text{sp gr} = ?$$

Basic Equation:

$$\text{sp gr} = \frac{D_{\text{material}}}{D_{\text{water}}}$$

Working Equation: Same

Substitution:

$$\text{sp gr} = \frac{54.2 \text{ lb/ft}^3}{62.4 \text{ lb/ft}^3}$$
$$= 0.869$$

This means oil is 0.869 times as dense as water and thus it floats on water.

Figure 12.28   A hydrometer measures the density of a liquid.

In general, the specific gravity of

water = 1

a material denser than water > 1

a material less dense than water < 1

When we check the antifreeze in a radiator in winter, we are really finding the specific gravity of the liquid. Specific gravity is a ratio comparison of the density of a substance to that of water. Because the density of antifreeze is different from the density of water, we find the concentration of antifreeze (and thus the amount of protection from freezing) by measuring the specific gravity of the solution in the radiator.

A **hydrometer** is a sealed glass tube weighted at one end so that it floats vertically in a liquid (Fig. 12.28). It sinks in the liquid until it displaces an amount of liquid equal to its own weight. The densities of the displaced liquids are inversely proportional to the depths to which the tube sinks. That is, the greater the density of the liquid, the less the tube sinks; the less the density of the liquid, the more the tube sinks. A hydrometer usually has a scale inside the tube and is calibrated so that it floats in water at the 1.000 mark. Anything with a specific gravity greater than 1 sinks in water. A substance with a specific gravity less than 1 floats in water; its specific gravity indicates the fractional volume that is under water.

Hydrometers are commonly used to measure the specific gravities of battery acid and antifreeze in radiators (Fig. 12.29). In a lead storage battery, the electrolyte is a solution of sulfuric acid and water, and the specific gravity of the solution varies with the amount of charge of the battery. Table 12.3 gives common specific gravities of conditions of a lead storage battery. Table 12.4 gives various specific gravities and the corresponding temperatures below which the antifreeze and water solution will freeze.

Figure 12.29   A common hydrometer

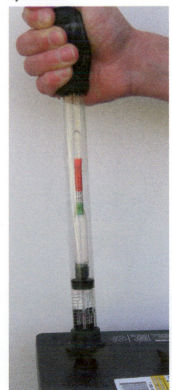

Table 12.3   Specific Gravities for a Lead Storage Battery

Condition	Specific Gravity
New (fully charged)	1.30
Old (discharged)	1.15

Table 12.4   Specific Gravities for Antifreeze and Water Solution

Temperature (°C)	Specific Gravity
−1.24	1.00
−2.99	1.01
−6.89	1.02
−19.82	1.05
−44.83	1.07
−51.23	1.08

One other factor must be considered in the use of the hydrometer—that of temperature. Significant differences in readings will occur over a range of temperatures. Specific gravities of some common liquids at room temperature are given in Table 12.5.

Table 12.5   Specific Gravities of Common Liquids at Room Temperature (20°C or 68°F)

Liquid	Specific Gravity
Benzene	0.9
Ethyl alcohol	0.79
Gasoline	0.68
Kerosene	0.82
Mercury	13.6
Seawater	1.025
Sulfuric acid	1.84
Turpentine	0.87
Water	1.000

TRY THIS ACTIVITY

Pour some water into a clear glass. Then pour in some cooking oil. Which liquid is less dense? (See Fig. 12.30.)

Figure 12.30

Figure 12.31   Fluids of different densities float on top of each other.

Courtesy of Stephen Oliver © Dorling Kindersley

Oil and other fluids of different densities float on top of each other (Fig. 12.31, contrasting colors have been added for illustration purposes).

PROBLEMS 12.5

Express mass density in kg/m^3 and weight density in lb/ft^3.

1. Find the mass density of a chunk of rock of mass 215 g that displaces a volume of 75.0 cm^3 of water.
2. A block of wood is 55.9 in. × 71.1 in. × 25.4 in. and weighs 1810 lb. Find its weight density.
3. If a block of wood of the size in Problem 2 has a weight density of 30.0 lb/ft^3, what does it weigh?
4. Find the volume (in cm^3) of 1350 g of mercury.
5. Find the volume (in cm^3) of 1350 g of cork.
6. Find the volume (in m^3) of 1350 g of nitrogen at 0°C and 1 atm pressure.
7. A block of gold 9.00 in. × 8.00 in. × 6.00 in. weighs 302 lb. Find its weight density.

8. A cylindrical piece of copper is 9.00 in. tall and 1.40 in. in radius. How much does it weigh?
9. A piece of aluminum of mass 6.24 kg displaces water that fills a container 12.0 cm × 12.0 cm × 16.0 cm. Find its mass density.
10. If 1.00 pint of turpentine weighs 0.907 lb, what is its weight density?
11. Find the mass density of gasoline if 106 g occupies 155 cm³.
12. How much does 1.00 gal of gasoline weigh?
13. Determine the volume (in m³) of 3045 kg of oil.
14. How many ft³ will 573 lb of water occupy?
15. If 20.4 in³ of linseed oil weighs 0.694 lb, what is its weight density?
16. If 108 in³ of ammonia gas weighs 0.00301 lb, what is its weight density?
17. Find the volume of 3.00 kg of propane at 0°C and 1 atm pressure.
18. Granite has a mass density of 2650 kg/m³. Find its weight density in lb/ft³.
19. Find the mass density of a metal block 18.0 cm × 24.0 cm × 8.00 cm with mass 9.76 kg.
20. Find the mass (in kg) of 1.00 m³ of (a) water, (b) gasoline, (c) copper, (d) mercury, and (e) air at 0°C and 1 atm pressure.
21. What size tank (in litres) is needed for $10\overline{0}0$ kg of (a) water? (b) gasoline? (c) mercury?
22. Copper has a mass density of 8890 kg/m³. Find its mass density in g/cm³.
23. What is the mass of gasoline in a 1250-litre gas tank?
24. If racing alcohol has a mass density of 790 kg/m³, what mass will a 1250-litre tank hold?
    Use Table 12.2 to find the specific gravity of each material.
25. Ice
26. Concrete
27. Iron
28. Air
29. Gasoline
30. Cork
31. The specific gravity of material X is 0.82. Does it sink in or float on water?
32. The specific gravity of material Y is 1.7. Does it sink in or float on water?
33. The specific gravity of material Z is 0.52. Does it sink in or float on gasoline?
34. The specific gravity of material W is 11.5. Does it sink in or float on mercury?
35. A proton has mass 1.67×10^{-27} kg and diameter 8.2×10^{-16} m. Find its specific gravity.
36. Find the mass density of a 315-g object that displaces 0.275 m³ of water.
37. What is the mass density of a $50\overline{0}$-g block that displaces 215 cm³ of water?

SKETCH

12 cm² w

4.0 cm

DATA
$A = 12$ cm², $l = 4.0$ cm, $w = ?$

BASIC EQUATION
$A = lw$

WORKING EQUATION
$w = \dfrac{A}{l}$

SUBSTITUTION
$w = \dfrac{12 \text{ cm}^2}{4.0 \text{ cm}} = 3.0 \text{ cm}$

Glossary

Adhesion The force of attraction between different or unlike molecules. (p. 313)
Atom The smallest particle of an element that can exist in a stable or independent state. (p. 311)
Bending Consists of both tension and compression stresses. It occurs when a force is placed on a beam causing it to sag. (p. 319)
Brinell Method Common industrial method used to measure the hardness of a metal. (p. 313)
Capillary Action The behavior of liquids that causes the liquid level in very small-diameter tubes to be different than in larger-diameter tubes. This behavior is due both to adhesion of the liquid molecules to the tube and to the surface tension of the liquid. (p. 326)

Cohesion The force of attraction between like molecules; holds the closely packed molecules of a solid together. (p. 313)

Compound A substance containing two or more elements. (p. 310)

Compression A stress caused by two forces acting directly toward each other. This stress tends to cause objects to become shorter and thicker. (p. 318)

Diffusion The process by which molecules of a gas mix with the molecules of a solid, a liquid, or another gas. (p. 329)

Ductility A property of a metal that enables it to be drawn through a die to produce a wire. (p. 314)

Elastic Limit The point beyond which a deformed object cannot return to its original shape. (p. 315)

Elasticity A measure of a deformed object's ability to return to its original size and shape once the deforming force is removed. (p. 315)

Electron One of the particles that makes up atoms; has a negative charge. (p. 311)

Element A substance that cannot be separated into simpler substances. (p. 310)

Expansion Property of a gas in which the rapid random movement of its molecules causes the gas to completely occupy the volume of its container. (p. 329)

Fluid A substance that takes the shape of its container; either a liquid or a gas. (p. 329)

Gas A substance that takes the shape of its container and has the same volume as its container. (p. 312)

Hardness A measure of the internal resistance of the molecules of a solid being forced farther apart or closer together. (p. 313)

Hooke's Law A principle of elasticity in solids: The ratio of the force applied to an object to its change in length (resulting in its being stretched or compressed by the applied force) is constant as long as the elastic limit has not been exceeded. (p. 320)

Hydrometer A sealed glass tube weighted at one end so that it floats vertically in a liquid; an instrument used to determine specific gravity. (p. 335)

Liquid A substance that takes the shape of its container and has a definite volume. (p. 312)

Malleability A property of a metal that enables it to be hammered and rolled into a sheet. (p. 315)

Mass Density The mass per unit volume of a substance. (p. 329)

Matter Anything that occupies space and has mass. (p. 310)

Meniscus The crescent-shaped surface of a liquid column in a tube. (p. 328)

Molecule The smallest particle of a substance that exists in a stable and independent state. (p. 310)

Neutron One of the particles that makes up atoms; does not carry an electric charge. (p. 311)

Nucleus The center part of an atom made up of protons and neutrons. (p. 311)

Proton One of the particles that makes up atoms; has a positive charge. (p. 311)

Shearing A stress caused by two forces applied in parallel, opposite directions. (p. 318)

Solid A substance that has a definite shape and a definite volume. (p. 312)

Specific Gravity The ratio of the density of any material to the density of water. (p. 334)

Strain The deformation of an object due to an applied force. (p. 320)

Stress The ratio of an outside applied distorting force to the area over which the force acts. (p. 316)

Surface Tension The ability of the surface of a liquid to act like a thin, flexible film. (p. 326)

Tensile Strength A measure of a solid's resistance to being pulled apart. (p. 313)

Tension A stress caused by two forces acting directly opposite each other. This stress tends to cause objects to become longer and thinner. (p. 318)

Torsion A stress related to a twisting motion. This type of stress severely compromises the strength of most materials. (p. 319)

Viscosity The internal friction of a fluid caused by molecular attraction, which makes it resist a tendency to flow. (p. 326)

Weight Density The weight per unit volume of a substance. (p. 329)

Formulas

12.2 $S = \dfrac{F}{A}$

$\dfrac{F}{\Delta l} = k$

12.5 $D_m = \dfrac{m}{V}$

$D_w = \dfrac{F_w}{V}$

specific gravity (sp gr) $= \dfrac{D_{material}}{D_{water}}$

Review Questions

1. The most important particles that make up atoms include which of the following?
    (a) Neutron   (b) Molecule   (c) Electron
    (d) Hydrogen   (e) Proton

2. Matter exists in which of the following?
    (a) Gas   (b) Neutrons   (c) Electrons
    (d) Solid   (e) Liquid

3. The common industrial method used to measure the hardness of a metal is
    (a) the Bernoulli method.   (b) Hooke's method.
    (c) the capillary method.   (d) the Brinell method.
    (e) none of the above.

4. Density is a property of
    (a) gases.   (b) liquids.   (c) solids.
    (d) all of the above. (e) none of the above.

5. The process by which molecules of a gas mix with the molecules of a solid, a liquid, or a gas is called
    (a) expansion.   (b) contraction.   (c) capillary action.
    (d) diffusion.   (e) none of the above.

6. Capillary action refers to
    (a) the mixing of molecules of different types.
    (b) the behavior of liquids in small tubes.
    (c) the attractive force between molecules.
    (d) stretching beyond the elastic limit.

7. The relationship of the change in length of a stretched or compressed object to the force causing the change is given by
    (a) Pascal's law.   (b) Brinell's law.   (c) the elastic limit.
    (d) Hooke's law.   (e) none of the above.

8. The ability of the surface of water to support a needle is an example of
    (a) mass density.   (b) Hooke's law.   (c) diffusion.
    (d) stress.   (e) surface tension.

9. In your own words, describe the difference between mass density and weight density.

10. Would the mass density of an object be the same if the object were on the moon rather than on the earth? Would the weight density be the same?

11. In your own words, describe capillary action.

12. What is the difference between adhesion and cohesion?

13. Give one example of the effect of surface tension that is not described in this book.

14. The mass of a proton is approximately _____ times the mass of an electron.

15. The applied force divided by the area over which the force acts is called ____.
16. In your own words, state Hooke's law.
17. The commonly used unit of stress in the metric system is the ____.
18. Describe how to find the specific gravity of an object.
19. What is the ratio of mass to volume called?
20. What is friction in liquids called?
21. A spring that has been permanently deformed is said to have been deformed past its ____ ____.
22. List the three states of matter.
23. Distinguish between a molecule and an atom.
24. Distinguish between a neutron and a proton.
25. List the five basic stresses.
26. Explain how a hydrometer measures the charge in a lead storage battery. Does the temperature affect the measurement?

Review Problems

PROBLEM SOLVING

SKETCH

12 cm² | w
4.0 cm

DATA

$A = 12 \text{ cm}^2, l = 4.0 \text{ cm}, w = ?$

BASIC EQUATION

$A = lw$

WORKING EQUATION

$w = \dfrac{A}{l}$

SUBSTITUTION

$w = \dfrac{12 \text{ cm}^2}{4.0 \text{ cm}} = 3.0 \text{ cm}$

1. A force of 32.5 N stretches a wire 0.470 cm. What force will stretch a similar piece of wire 2.39 cm?
2. A force of 7.33 N is applied to a spring whose spring constant is 0.298 N/cm. Find its change in length.
3. Each vertical steel column of an office building supports 42,100 lb and is compressed 0.0258 in. What is the spring constant of the steel? Find the compression if the weight were 51,700 lb.
4. A rectangular cast-iron column 16.0 cm × 16.0 cm × 4.50 m supports a weight of 7.95×10^6 N. Find the stress on the top of the column.
5. Find the weight density of a block of metal 7.00 in. × 6.50 in. × 8.00 in. that weighs 425 lb.
6. A cylindrical piece of aluminum is 4.25 cm tall and 1.95 cm in radius. How much does it weigh?
7. A piece of metal has a mass of 8.36 kg. If it displaces water that fills a container 9.34 cm × 10.0 cm × 10.0 cm, what is the mass density of the metal?
8. A block of wood is 27.7 in. × 36.3 in. × 12.4 in. and weighs 602 lb. Find its weight density.
9. Find the volume (in cm³) of 759 g of mercury.
10. Find the volume (in m³) of 1970 g of hydrogen at 0°C and 1 atm.
11. Find the mass of 1510 m³ of oxygen at 0°C and 1 atm.
12. Find the weight of 951 ft³ of water.
13. Find the weight density of a block of material 4.27 in. × 3.87 in. × 5.44 in. that weighs 0.982 lb.
14. Find the weight density of 2.00 quarts of liquid weighing 3.67 lb.
15. A quantity of liquid weighs 4.65 lb with a weight density of 39.8 lb/ft³. What is its volume?
16. The density of a metal is 694 kg/m³. Find its specific gravity.
17. A solid displaces 4.30 gal of water and has a weight density of 135 lb/ft³. What is its weight?
18. Find the mass of a rectangular gold bar 4.00 cm × 6.00 cm × 20.00 cm. The mass density of gold is 19,300 kg/m³.
19. Find the mass density of a chunk of rock using only a scale and knowing the following information: mass of rock is 225 g; mass of water the rock displaces is 75.9 g.
20. The specific gravity of an unknown substance is 0.80. Will it float on or sink in gasoline?

APPLIED CONCEPTS

1. Instead of carrying a full-size spare tire, many automobiles are equipped with a significantly smaller spare tire with a warning that the car should not exceed a speed of 50 mi/h. (a) How much pressure does the smaller tire withstand if it must support one fourth of a 1650-kg minivan and its area of contact with the road measures 8.26 cm × 15.9 cm? (b) Find the pressure on a full-size spare tire if its area of contact with the road measures 15.2 cm × 21.0 cm. (c) Why is it important to limit the speed of the smaller spare tire?

2. Observe the warped lines on asphalt pavement in front of a stoplight (Fig. 12.32). (a) What type of stress causes such curved lines? (b) Why are the warped lines, as seen in the photograph, more noticeable at stop signs and stoplights than on other sections of road?

Figure 12.32

3. Raul weighs 235 lb and is able to float in seawater but not in fresh water. Find his volume assuming that his density is the average density of seawater and fresh water.

4. A tanker truck with a cylindrical container 11.3 m long and 2.85 m in diameter transports several types of liquids. How much mass does it carry if it is transporting a full load of (a) water, (b) oil, and (c) gasoline?

5. Every morning Shakira weighs herself on a bathroom spring scale. Today, the scale reads 134 lb. (a) If the spring in the bathroom scale compresses 0.752 in., find its spring constant. (b) If her husband steps on the scale and the spring compresses 1.13 in., how much does he weigh? (c) Why is there a limit to how much weight a spring scale can measure?

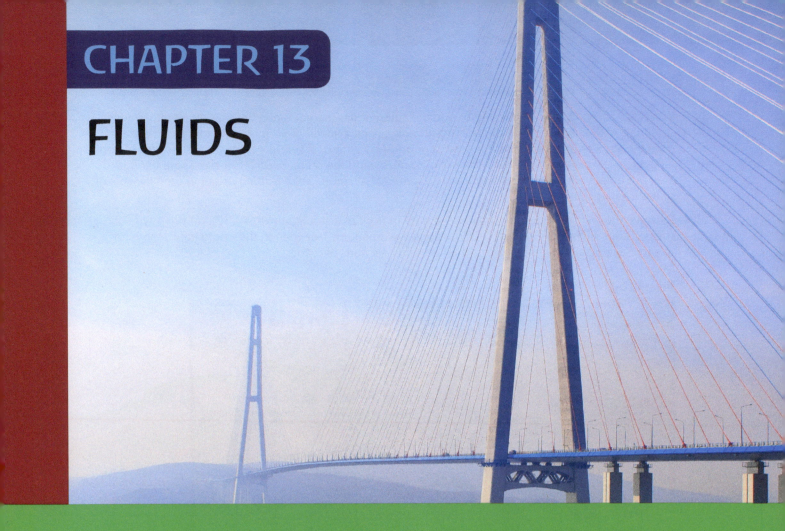

CHAPTER 13

FLUIDS

Objectives

The major goals of this chapter are to enable you to:

1. Describe the behavior of fluids.
2. Determine pressure using the hydraulic principle.
3. Distinguish between gauge pressure and absolute pressure.
4. Calculate buoyancy using Archimedes' principle.
5. Analyze fluid flow and Bernoulli's principle.

Substances that flow are called fluids. Because liquids and gases behave in much the same manner, they will be presented together. From ships to airplanes to automobile brake systems to balloons, the utilization of fluids and their properties is important to technology and everyday life. We will now study hydraulics, hydrostatic pressure, buoyancy, and flow.

13.1 Hydrostatic Pressure

Since liquids and gases behave in much the same manner, they are often studied together as fluids. The gas and water piped to your home are fluids having several common characteristics.

Pressure is the force applied per unit area. **Hydrostatic pressure** is the pressure a liquid at rest exerts on a submerged object (Fig. 13.1). As you probably know, the pressure on an object increases as the water depth increases. Liquids differ from solids in that they exert force in all directions, whereas solids exert only a downward force due to gravity.

Figure 13.1   (a) The pressure on a submerged object is exerted by the liquid in all directions. The pressure at ocean-floor level is so great that very strong containers must be built to withstand the enormous pressures. (b) When a force is exerted on a liquid in a container, the force is exerted uniformly in all directions. Note that the liquid squirts out of the holes uniformly in all directions.

(a)

(b)

The pressure in a liquid depends *only* on the depth and weight density of the liquid and *not* on the surface area. Because the pressure exerted by water increases with depth, dams are built much thicker at the base than at the top (Fig. 13.2).

Figure 13.2   Dams must be built much thicker at the base because the pressure exerted by the water increases with depth.

(a)

Courtesy of fotolia © Zelijko Radojko

Schematic of basic dam

(b)

Courtesy of fotolia © Graphies.theque

TRY THIS ACTIVITY

Water Pressure

The next time you go swimming in a pool, swim toward the bottom. As you swim deeper, notice the gradual change in pressure on your eardrums. Explain how the change in water pressure on your eardrums can be used to describe hydrostatic pressure. How would the pressure change if the pool were full of salt water?

This general pressure principle may also be illustrated with a pile of bricks. The weight of the pile and the pressure on the ground clearly increase as the pile is made taller. The force (weight) applied to the same area at the bottom of the pile increases the pressure at the bottom of the pile with the addition of each brick. Likewise, the pressure on a submerged object increases the farther it is pushed down due to the weight of the additional water above it.

To find the pressure at a given depth in a liquid, use the formula

$$P = hD_w$$

where P = pressure
   h = height (or depth)
   D_w = weight density of the liquid

One way to increase water pressure is to raise a water storage tank above the ground. This same formula applies except h is the sum of the depth of the water in the tank and the distance the bottom of the water tank is above the ground or some other reference such as a building. This is why community water storage tanks are placed on towers or on the tops of large hills.

TRY THIS ACTIVITY

Water Pressure and Depth

Figure 13.3

Punch three small holes near the top, the middle, and the bottom on the side of a large can with its lid removed. Next, fill the can with water and compare the streams of water spurting out each hole (Fig. 13.3). Pressure in a liquid also depends on the weight density of the liquid. Pressure on an object 50 m below the surface of fresh water is less than the pressure at the same depth in salt water because fresh water is less dense than salt water.

EXAMPLE 1

Find the pressure at the bottom of a water-filled drum 4.00 ft high.

Sketch:

4.00 ft

Data:

$$h = 4.00 \text{ ft}$$
$$D_w = 62.4 \text{ lb/ft}^3$$
$$P = ?$$

Basic Equation:

$$P = hD_w$$

Working Equation: Same

Substitution:

$$P = (4.00 \text{ ft})\left(62.4 \frac{\text{lb}}{\text{ft}^3}\right)$$

$$= 25\bar{0}\frac{\text{lb}}{\text{ft}^2} \times \left(\frac{1 \text{ ft}}{12 \text{ in.}}\right)^2$$

$$= 1.74 \text{ lb/in}^2$$

Note: The pressure depends only on the height, not the width or area of the container.

EXAMPLE 2

Find the depth in a lake at which the pressure is 105 lb/in².

Data:

$$P = 105 \text{ lb/in}^2$$
$$D_w = 62.4 \text{ lb/ft}^3$$
$$h = ?$$

Basic Equation:

$$P = hD_w$$

Working Equation:

$$h = \frac{P}{D_w}$$

Substitution:

$$h = \frac{105 \text{ lb/in}^2}{62.4 \text{ lb/ft}^3}$$

$$= 1.68\frac{\text{ft}^3}{\text{in}^2}$$

$$\boxed{\frac{\text{lb/in}^2}{\text{lb/ft}^3} = \frac{\text{lb}}{\text{in}^2} \div \frac{\text{lb}}{\text{ft}^3} = \frac{\text{lb}}{\text{in}^2} \times \frac{\text{ft}^3}{\text{lb}} = \frac{\text{ft}^3}{\text{in}^2}}$$

$$= 1.68\frac{\frac{\text{ft}^3}{\text{ft}^3}}{\text{in}^2} \times \left(\frac{12 \text{ in.}}{1 \text{ ft}}\right)^2$$

$$= 242 \text{ ft}$$

EXAMPLE 3

Find the height of a water column where the pressure at the bottom of the column is 40\bar{0} kPa and the weight density of water is 980\bar{0} N/m³.

Data:

$$P = 40\bar{0} \text{ kPa}$$
$$D_w = 980\bar{0} \text{ N/m}^3$$
$$h = ?$$

Basic Equation:

$$P = hD_w$$

Working Equation:

$$h = \frac{P}{D_w}$$

Substitution:

$$h = \frac{40\overline{0}\text{ kPa}}{980\overline{0}\text{ N/m}^3}$$

$$= \frac{40\overline{0}\text{ kPa}}{980\overline{0}\text{ N/m}^3} \times \frac{10^3\text{ N/m}^2}{1\text{ kPa}} \quad (Recall: 1\text{ kPa} = 10^3\text{ N/m}^2)$$

$$= 40.8\text{ m}$$

$$\boxed{\frac{\text{kPa}}{\text{N/m}^3} \times \frac{\text{N/m}^2}{\text{kPa}} = \text{N/m}^2 \div \text{N/m}^3 = \frac{\text{N}}{\text{m}^2} \times \frac{\text{m}^3}{\text{N}} = \text{m}}$$

· · · · · · · · · · · · · · · · ·

Total Force Exerted by Liquids

The *total force* exerted by a liquid on a horizontal surface (such as the bottom of a barrel) depends on the area of the surface, the depth of the liquid, and the weight density of the liquid. By formula,

$$F_t = AhD_w$$

where F_t = total force
   A = area of bottom or horizontal surface
   h = height or depth of the liquid
   D_w = weight density

EXAMPLE 4

Find the total force on the bottom of a rectangular tank 10.0 ft by 5.00 ft by 4.00 ft deep filled with water.

Sketch:

4.00 ft

10.0 ft

5.00 ft

Data:

$$A = lw = (10.0\text{ ft})(5.00\text{ ft}) = 50.0\text{ ft}^2$$
$$h = 4.00\text{ ft}$$
$$D_w = 62.4\text{ lb/ft}^3$$
$$F_t = ?$$

Basic Equation:

$$F_t = AhD_w$$

Working Equation: Same

Substitution:

$$F_t = (50.0 \text{ ft}^2)(4.00 \text{ ft})\left(62.4\frac{\text{lb}}{\text{ft}^3}\right)$$

$$= 12{,}500 \text{ lb}$$

· · · · · · · · · · · · · · · · · · ·

The total force on a vertical surface F_s (such as the *side* of a tank) is found by using half the vertical height (average height):

$$F_s = \tfrac{1}{2} AhD_w$$

where A is the area of the side or vertical surface.

Find the total force on the small side of the rectangular tank in Example 4.

EXAMPLE 5

Data:

$$A = lw = (5.00 \text{ ft})(4.00 \text{ ft}) = 20.0 \text{ ft}^2$$
$$h = 4.00 \text{ ft}$$
$$D_w = 62.4 \text{ lb/ft}^3$$
$$F_s = ?$$

Basic Equation:

$$F_s = \tfrac{1}{2} AhD_w$$

Working Equation: Same

Substitution:

$$F_s = \tfrac{1}{2}(20.0 \text{ ft}^2)(4.00 \text{ ft})(62.4 \text{ lb/ft}^3)$$
$$= 25\overline{0}0 \text{ lb}$$

· · · · · · · · · · · · · · · · · ·

Find the total force on the side of a water-filled cylindrical tank 3.00 m high with radius 5.00 m. The weight density of water is $98\overline{0}0$ N/m³.

EXAMPLE 6

Sketch:

5.00 m

3.00 m

Data:

$$A = 2\pi rh = 2\pi(5.00 \text{ m})(3.00 \text{ m})$$ **Note:** The area of the vertical surface is the lateral surface of the cylinder.

$$h = 3.00 \text{ m}$$
$$D_w = 98\overline{0}0 \text{ N/m}^3$$
$$F_s = ?$$

Basic Equation:

$$F_s = \tfrac{1}{2} AhD_w$$

Working Equation: Same

Substitution:

$$F_s = \tfrac{1}{2} [2\pi(5.00 \text{ m})(3.00 \text{ m})](3.00 \text{ m})(98\overline{0}0 \text{ N/m}^3)$$

$$= 1.39 \times 10^6 \text{ N}$$

PHYSICS CONNECTIONS

Hydroelectric Power

More than 2000 dams in the United States have been constructed to satisfy two specific functions. First, dams are often used to increase the capacity for fresh water reservoirs. The fresh water is typically used to supply consumers with fresh drinking water or serve as a recreational park facility. Second, many dams are equipped with a system of turbines and generators that take advantage of the hydrostatic pressure of the water in the reservoir to generate electricity for consumers along the local electrical power grid.

As discussed in Section 13.1 and in the two associated Try This activities, water pressure increases with increasing depth of the water. As indicated in Fig. 13.4, the intake opening in the dam is placed deep enough in the dam to take advantage of the hydrostatic water pressure at that depth, but not too deep so that sediment from the bottom of the reservoir enters into the penstock channel, which channels the water to the turbines. As the "Water Pressure and Depth" Try This Activity indicates, the speed of the water exiting through the penstock in the dam will be greater than if the penstock was placed closer to the top of the dam.

Figure 13.4 Cross section of a typical hydroelectric power plant

Once the water has passed through the penstock, the water strikes a series of curved turbine blades that turn a rotor inside an electrical power generator. A series of moving magnets on the rotor causes electrons to move through coils of wire, which in turn generates electricity. (The process of creating electricity in this manner is discussed in Chapter 17.) The electricity is then distributed through electrical power lines to the electrical grid for consumer use.

One of the most famous hydroelectric power plants is located within the Hoover Dam (Fig. 13.5). The American Society of Civil Engineers considers the dam, completed in 1936, one of the Seven Modern Engineering Marvels of the World. Although the top of the concrete dam is only 50 ft thick, the bottom of the dam is 660 ft thick to accommodate the tremendous hydrostatic pressure and house the four penstocks that allow water to travel through the lower section of the dam. Seventeen turbines and generators are used to generate four billion kilowatt-hours of electrical power annually. The hydroelectric power is then distributed to consumers throughout the Nevada, Arizona, and California power grid.

Figure 13.5 The Hoover Dam and hydroelectric power plant

Courtesy of fotolia © irisphoto1

PROBLEMS 13.1

1. Find the pressure (in lb/in^2) at the bottom of a tower with water 50.0 ft deep.
2. Find the height of a column of water where the pressure at the bottom of the column is 20.0 lb/in^2.
3. Find the density of a liquid that exerts a pressure of 0.400 lb/in^2 at a depth of 42.0 in.
4. (a) Find the total force on the bottom of a water-filled circular cattle tank 0.750 m high with radius 1.30 m where the weight density of water is 9800 N/m^3. (b) Find the total force on the side of the tank.
5. What must the water pressure be to supply water to the third floor of a building (35.0 ft up) with a pressure of 40.0 lb/in^2 at that level?
6. A small rectangular tank 5.00 in. by 9.00 in. is filled with mercury. (a) If the total force on the bottom of the tank is 165 lb, how deep is the mercury? (Weight density of mercury = 0.490 lb/in^3.) (b) Find the total force on the larger side of the tank.
7. Find the water pressure (in kPa) at the 25.0-m level of a water tower containing water 50.0 m deep.
8. Find the height of a column of water where the pressure at the bottom is 115 kPa.
9. What is the height of a column of water if the pressure at the bottom of the column is 95.0 kPa?
10. What is the mass density of a liquid that exerts a pressure of 180 kPa at a depth of 30.0 m?
11. What is the mass density of a liquid that exerts a pressure of 178 kPa at a depth of 24.0 m?
12. (a) Find the total force on the bottom of a cylindrical gasoline storage tank 15.0 m high with radius 23.0 m. (b) Find the total force on the side of the tank.

SKETCH

12 cm^2 w

4.0 cm

DATA

$A = 12$ cm^2, $l = 4.0$ cm, $w = ?$

BASIC EQUATION

$A = lw$

WORKING EQUATION

$w = \dfrac{A}{l}$

SUBSTITUTION

$w = \dfrac{12 \text{ cm}^2}{4.0 \text{ cm}} = 3.0 \text{ cm}$

13. What must the water pressure be to supply the second floor (18.0 ft up) with a pressure of 50.0 lb/in^2 at that level?
14. Find the water pressure at ground level to supply water to the third floor of a building 8.00 m high with a pressure of 325 kPa at the third-floor level.
15. What pressure must a pump supply to pump water up to the 30th floor of a skyscraper with a pressure of 25 lb/in^2? Assume that the pump is located on the first floor and that there are 16.0 ft between floors.
16. A submarine is submerged to a depth of 3550 m in the Pacific Ocean. What air pressure (in kPa) is needed to blow water out of the ballast tanks?

A filled water tower sits on the top of the highest hill in a town (use Fig. 13.6 for Problems 17–21). The cylindrical tower has a radius of 12.0 m and a height of 50.0 m.

Figure 13.6

Water tower

32.5 m

Fire station

School

8.20 m

17. Find the total force on the bottom of the water tower.
18. Find the total force on the sides of the water tower.
19. Find the pressure (in kPa) on the bottom of the water tower.
20. What is the water pressure (in kPa) at the fire station?
21. What is the water pressure (in kPa) at the school?
22. A cylindrical grain bin 24.0 ft in diameter is filled with corn whose weight density is 45.1 lb/ft^3. How tall can the bin be for the floor to support 94.0 lb/in^2 of pressure?

13.2 Hydraulic Principle

Put a stopper in one end of a metal pipe. Fill it with a fluid such as water. Then put a second stopper into the open end. Put the pipe in a horizontal position as in Fig. 13.7 and push on stopper *A*. What happens? Stopper *B* is pushed out. This illustrates a basic principle of hydraulics: The liquid in the pipe transmits the pressure from one stopper or piston to another without measurable loss. This is known as the **hydraulic principle** or **Pascal's principle**.

Figure 13.7 Pressure on a confined liquid is transmitted in all directions without measurable loss.

A   *B*

HYDRAULIC PRINCIPLE (PASCAL'S PRINCIPLE)

The pressure applied to a confined liquid is transmitted without measurable loss throughout the entire liquid to all inner surfaces of the container.

The rear-wheel hydraulic brake system of a front-wheel-drive automobile (Fig. 13.8) is an application of Pascal's principle. When the driver pushes the brake pedal, the pressure on the piston in the master cylinder is transmitted through the brake fluid to the two pistons in the brake cylinder. This transmitted pressure then forces the brake-cylinder pistons to push the brake shoes against the brake drum and stop the automobile. Releasing the brake pedal releases the pressure on the pistons in the brake cylinder. The spring pulls the brake shoes away from the brake drum, which allows the wheels to turn freely again.

Figure 13.8   Hydraulic brake system of an automobile

The hydraulic jack, lift, and press are applications of hydraulics being used as a simple machine to multiply force. If we apply a force to the small piston of the hydraulic lift in Fig. 13.9, the pressure is transmitted without measurable loss in all directions. The reason for this is the virtual noncompressibility of liquids. The *pressure* on the large piston is the same as the pressure on the small piston; however, the *total force* on the large piston is greater because of its larger surface area.

Figure 13.9   Hydraulic lift

Although it may seem we are acquiring a "free" increase in force by being able to lift an automobile using a very small applied force, that is really not the case. The principle is similar to the mechanical advantage of simple machines; we can multiply force but at the expense of distance. The lift, although able to move a large weight, can only move it a relatively short distance.

EXAMPLE 1

From the diagram of the hydraulic jack in Fig. 13.10, find (a) the pressure on the small piston. (b) the pressure on the large piston. (c) the total force on the large piston. (d) the mechanical advantage of the jack.

Figure 13.10 Hydraulic jack

Data:

$$F_1 = 15.0 \text{ lb}$$
$$A_{\text{small piston}} = A_1 = 3.00 \text{ in}^2$$
$$A_{\text{large piston}} = A_2 = 75.0 \text{ in}^2$$
$$P_1 = ?$$
$$P_2 = ?$$
$$F_2 = ?$$
$$\text{MA} = ?$$

(a) Basic Equation:

$$P_1 = \frac{F_1}{A_1}$$

Working Equation: Same

Substitution:

$$P_1 = \frac{15.0 \text{ lb}}{3.00 \text{ in}^2}$$
$$= 5.00 \text{ lb/in}^2$$

(b) Applying Pascal's Principle:

$$P_2 = P_1 = 5.00 \text{ lb/in}^2$$

(c) Basic Equation:

$$P_2 = \frac{F_2}{A_2}$$

Working Equation:

$$F_2 = P_2 A_2$$

Substitution:

$$F_2 = \left(5.00 \frac{\text{lb}}{\text{in}^2}\right)(75.0 \text{ in}^2)$$

$$= 375 \text{ lb}$$

(d) Basic Equation:

$$\text{MA} = \frac{F_R}{F_E}$$

Working Equation: Same

Substitution:

$$\text{MA} = \frac{375 \text{ lb}}{15.0 \text{ lb}}$$

$$= 25.0$$

The small piston of a hydraulic press has an area of 10.0 cm². If the applied force is 50.0 N, what must the area of the large piston be to exert a pressing force of 4800 N?

EXAMPLE 2

Data:

$$A_1 = 10.0 \text{ cm}^2$$
$$F_1 = 50.0 \text{ N}$$
$$F_2 = 4800 \text{ N}$$
$$A_2 = ?$$

Basic Equations:

$$P_1 = \frac{F_1}{A_1}, \quad P_2 = \frac{F_2}{A_2}, \quad \text{and since } P_1 = P_2, \quad \frac{F_1}{A_1} = \frac{F_2}{A_2}$$

Working Equation:

$$A_2 = \frac{A_1 F_2}{F_1}$$

Substitution:

$$A_2 = \frac{(10.0 \text{ cm}^2)(4800 \text{ N})}{50.0 \text{ N}}$$

$$= 960 \text{ cm}^2$$

PROBLEMS 13.2

1. The area of the small piston in a hydraulic jack is 0.750 in². The area of the large piston is 3.00 in². If a force of 15.0 lb is applied to the small piston, what weight can be lifted by the large one?
2. The mechanical advantage of a hydraulic press is 25. What applied force is necessary to produce a pressing force of 2400 N?
3. Find the mechanical advantage of a hydraulic press that produces a pressing force of 8250 N when the applied force is 375 N.
4. The mechanical advantage of a hydraulic press is 18. What applied force is necessary to produce a pressing force of 990 lb?

SKETCH

12 cm² $\quad$ w

4.0 cm

DATA

$A = 12$ cm², $l = 4.0$ cm, $w = ?$

BASIC EQUATION

$A = lw$

WORKING EQUATION

$w = \dfrac{A}{l}$

SUBSTITUTION

$w = \dfrac{12 \text{ cm}^2}{4.0 \text{ cm}} = 3.0$ cm

5. Find the mechanical advantage of a hydraulic press that produces a pressing force of 1320 N when the applied force is 55.0 N.

6. The small piston of a hydraulic press has an area of 8.00 cm². If the applied force is 25.0 N, find the area of the large piston to exert a pressing force of 3600 N.

7. The MA of a hydraulic jack is 250. What force must be applied to lift an automobile weighing 12,000 N?

8. The small piston of a hydraulic press has an area of 4.00 in². If the applied force is 10.0 lb, what must the area of the large piston be to exert a pressing force of 865 lb?

9. The MA of a hydraulic jack is 420. Find the weight of the heaviest automobile that can be lifted by an applied force of 55 N.

10. The mechanical advantage of a hydraulic jack is 450. Find the weight of the heaviest automobile that can be lifted by an applied force of 60.0 N.

11. The pistons of a hydraulic press have radii of 2.00 cm and 12.0 cm, respectively. (a) What force must be applied to the smaller piston to exert a force of 5250 N on the larger? (b) What is the pressure (in N/cm²) on each piston? (c) What is the mechanical advantage of the press?

12. The small circular piston of a hydraulic press has an area of 8.00 cm². If the applied force is 25.0 N, what must the area of the large piston be to exert a pressing force of 3650 N?

13. The large piston on a hydraulic lift has radius 40.0 cm. The small piston has radius 5.00 cm to which a force of 75.0 N is applied. (a) Find the force exerted by the large piston. (b) Find the pressure on the large piston. (c) Find the pressure on the small piston. (d) Find the mechanical advantage of the lift. (e) What happens when the area of the small piston is half as large? (f) What happens when the radius of the small piston is half as large?

14. In a hydraulic system a 20.0-N force is applied to the small piston with cross-sectional area 25.0 cm². What weight can be lifted by the large piston with cross-sectional area 50.0 cm²?

15. If the diameter of the larger piston in Problem 14 is doubled, how is the weight able to be lifted changed?

16. If a dentist's chair weighs 1600 N and is raised by a large piston with cross-sectional area 75.0 cm², what force must be exerted on a small piston of cross-sectional area 3.75 cm² to lift the chair?

17. A hydraulic jack whose piston has a cross-sectional area of 115 cm² supports a pickup truck weighing 1.20×10^4 N. Compressed air is used to apply a force on the second piston with cross-sectional area 25.0 cm². How large must this force be to support the truck?

18. Compressed air in a car lift applies a force to a piston with radius 5.00 cm. This pressure is transmitted through a hydraulic system to a second piston with radius 15.0 cm. (a) How much force must the compressed air exert to lift a vehicle weighing 1.33×10^4 N? (b) What pressure produces the lift?

19. The small piston of an automobile lift has an area of 12.0 cm². If the applied force on the lift is 75.0 N, find the area of the large piston needed to exert a lifting force of 9800 N.

20. If the lifting force of a hydraulic truck jack is 19,600 N and the ratio of the area of the large piston to the small piston is 150 to 1, what is the applied force on the jack?

13.3 Air Pressure

Since air has weight, as does any fluid, it exerts pressure. The atmosphere exerts pressure on objects on the surface of the earth. This atmospheric pressure can be illustrated by using a bell jar with a hole in the top over which a thin rubber membrane is stretched, as in Fig. 13.11(a). As air is pumped out of the bell jar, the inside air pressure

Figure 13.11 Effects of air pressure

Air pressure

Thin rubber membrane

To vacuum pump

Air pressure

Air pressure is reduced

(a)

(b)

Figure 13.12 The barometer measures air pressure.

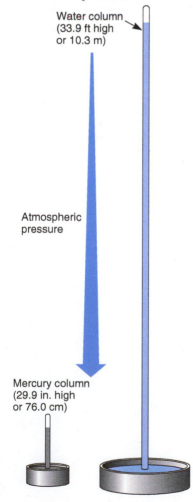

Water column (33.9 ft high or 10.3 m)

Atmospheric pressure

Mercury column (29.9 in. high or 76.0 cm)

is reduced by removing a number of air molecules. Thus, there are fewer molecular bombardments on the inside surface of the rubber membrane than there are on its outside surface. The outside air pressure, now greater than the inside air pressure, pushes the rubber membrane down into the bell jar. When a straw is used to drink, the air pressure inside the straw is reduced [Fig. 13.11(b)]. As a result, the outside air pressure is higher than the pressure in the straw, which forces the fluid up the straw.

In Section 13.1 we saw that the pressure on a submerged body increases as the body goes deeper into the liquid. Some creatures live near the bottom of the ocean, where the pressure of the water is so great that it would collapse any human body and most submarines, but through the process of evolution, such creatures have adapted to this tremendous pressure. Similarly, we on the earth live at the bottom of a fluid, air, that is several miles deep. The pressure from this fluid is normally 14.7 lb/in^2 or 101.32 kPa at sea level. We do not feel this pressure because it normally is almost the same from all directions and also because living bodies maintain an internal pressure that balances the external pressure. **Atmospheric pressure** is the pressure caused by the weight of the air in the atmosphere. Air pressure and the amount of air decrease with altitude. Mountain climbers often must use oxygen tanks to help them breathe at high altitudes. Aircraft must have pressurized cabins for passengers to be able to breathe and function. When the air pressure becomes unequal, its force becomes quite evident in the form of wind. This wind may be a cool summer breeze or the tremendous concentrated force of a tornado.

What is the *pressure of our atmosphere* equivalent to? Experiments have shown that the atmosphere supports a column of water 33.9 ft or 10.3 m high in a tube from which the air has been removed. The atmosphere supports 29.9 in. or 76.0 cm of mercury in a similar tube (Fig. 13.12). This is not surprising; mercury is 13.6 times as dense as water and

$$\frac{1}{13.6} \times 33.9 \text{ ft} \times \frac{12 \text{ in.}}{1 \text{ ft}} = 29.9 \text{ in.}$$

The height of the mercury column in a barometer is independent of the width (or diameter or cross-sectional area) of the barometer tube. This "inches of mercury" measurement has been standard for many years on TV weather programs but is increasingly being replaced by the metric standard measurement in kilopascals (kPa).

The pressure of the atmosphere can be expressed in terms of the pressure of an equivalent column of mercury. Air pressure at sea level is normally 29.9 in. or 76.0 cm or $76\overline{0}$ mm of mercury. How do we arrive at the 14.7 lb/in^2 measurement? In the case of mercury, the height of the column is 29.9 in. or 2.49 ft. Its density is 13.6×62.4 lb/ft^3 or 849 lb/ft^3. Therefore,

$$P = hD_w$$

$$P = 2.49 \text{ ft} \times 849 \frac{\text{lb}}{\text{ft}^3} \times \left(\frac{1 \text{ ft}}{12 \text{ in.}}\right)^2$$

$$= 14.7 \text{ lb/in}^2 \text{ at standard temperature}$$

A pressure of 2 atm is equivalent to 29.4 lb/in^2 or 202.64 kPa. If the pressure is $\frac{1}{2}$ atm at a given point in the atmosphere, it is 7.35 lb/in^2 or 50.66 kPa.

An interesting demonstration illustrates atmospheric pressure by using it to collapse a metal can. A can that can be tightly capped is placed on a burner with its cap off. A small amount of water is placed in the can and heated until it boils. The cap is then secured, and the can is removed from the heat source. The steam has pushed air from the can, and any remaining air has expanded from being heated. When the sealed can cools and the steam inside condenses, the pressure in the can is reduced. The greater pressure of the atmosphere outside the can crushes the can (Fig. 13.13). The same concept has been used for generations to seal jars of food in home canning, but the jars are strong enough to not collapse.

Figure 13.13   (a) The can is heated with its cap removed. (b) The cap of the can is twisted tight, and the can is removed from the heat. As the sealed can cools, the greater pressure of the atmosphere slowly crushes the can, as shown.

(a)   (b)

Different types of gauges read either atmospheric or gauge pressure. When we purchase bottled gas, the amount of gas and its density vary with the pressure. If the pressure is low, the amount of gas in the bottle is low. If the pressure is high, the bottle is "nearly full." The *gauge* that is usually used for checking the pressure in bottles and tires shows a reading of zero at normal atmospheric pressure. The pressure of the atmosphere is not included in this reading. Thus, **gauge pressure** is the amount of air pressure excluding the normal atmospheric pressure, as when one uses a tire gauge. The actual pressure, called **absolute pressure**, is the gauge pressure reading plus the normal atmospheric pressure, 101.32 kPa or 14.7 lb/in², as when one uses an atmospheric barometer. That is,

absolute pressure = gauge pressure + atmospheric pressure

or

$$P_{abs} = P_{ga} + P_{atm}$$

where $P_{atm} = 101.32$ kPa or 14.7 lb/in² at standard temperature.

Note: Atmospheric pressure has not been included in our previous pressure calculations.

What is the absolute pressure in a tire inflated to 32.0 lb/in² (a) in lb/in²? (b) in kPa?

EXAMPLE 1

(a) Data:

$$P_{ga} = 32.0 \text{ lb/in}^2$$
$$P_{atm} = 14.7 \text{ lb/in}^2$$
$$P_{abs} = ?$$

Basic Equation:

$$P_{abs} = P_{ga} + P_{atm}$$

Working Equation: Same

Substitution:

$$P_{abs} = 32.0 \text{ lb/in}^2 + 14.7 \text{ lb/in}^2$$
$$= 46.7 \text{ lb/in}^2$$

(b) We use the conversion factor:

$$101.32 \text{ kPa} = 14.7 \text{ lb/in}^2$$

Therefore,

$$P_{abs} = 46.7 \text{ lb/in}^2 \times \frac{101.32 \text{ kPa}}{14.7 \text{ lb/in}^2}$$
$$= 322 \text{ kPa}$$

PROBLEMS 13.3

1. Change 815 kPa to lb/in².
2. Change 64.3 lb/in² to kPa.
3. Change 42.5 lb/in² to kPa.
4. Change 215 kPa to lb/in².
5. Find the pressure of (a) 3 atm (in kPa), (b) 2 atm (in kPa), (c) 6 atm (in lb/in²), (d) 5 atm (in kPa), (e) $\frac{1}{3}$ atm (in kPa), and (f) $\frac{1}{4}$ atm (in kPa).
6. A barometer in the Rocky Mountains reads 516 mm of mercury. Find this pressure (a) in kPa and (b) in lb/in².
7. Find the absolute pressure in a bicycle tire with a gauge pressure of 485 kPa.
8. Find the absolute pressure of a motorcycle tire with a gauge pressure of 255 kPa.
9. Find the gauge pressure of a tire with an absolute pressure of 45.0 lb/in².
10. Find the gauge pressure of a tire with an absolute pressure of 425 kPa.
11. Find the absolute pressure of a tire gauge that reads 205 kPa.
12. Find the absolute pressure of a tank whose gauge pressure reads 362 lb/in².
13. Find the gauge pressure of a tank whose absolute pressure is 1275 kPa.
14. Find the gauge pressure of a tank whose absolute pressure is 218 lb/in².

SKETCH

12 cm²   w

4.0 cm

DATA
$A = 12$ cm², $l = 4.0$ cm, $w = ?$

BASIC EQUATION
$A = lw$

WORKING EQUATION
$w = \dfrac{A}{l}$

SUBSTITUTION
$w = \dfrac{12 \text{ cm}^2}{4.0 \text{ cm}} = 3.0$ cm

15.   Find the absolute pressure of a cycle tire with gauge pressure 3.00×10^5 Pa.
16.   Find the absolute pressure in a hydraulic jack with a small piston of area 23.0 cm^2 when a force of 125 N is applied. The area of the large piston is 46.0 cm^2.

13.4 Buoyancy

Archimedes was one of the first to study fluids and formulated what is now called **Archimedes' principle**.

> **ARCHIMEDES' PRINCIPLE**
> Any object placed in a fluid apparently loses weight equal to the weight of the displaced fluid.

Archimedes (287–212 B.C.),

Greek mathematician, was born in Syracuse. He is remembered for the construction of siege-engines against the Romans, Archimedes' screw (still used for raising water), discovering the principle for the buoyant force of a floating body, and founding the science of hydrostatics. In mathematics, he discovered the formulas for the areas and volumes of spheres, cylinders, parabolas, and other plane and solid figures.

In Fig. 13.14, the toy boat floats on water. Note that the weight of the boat equals the weight of the water it displaces. Three people in a boat displace more water than only one person in the boat. The boat rides lower due to the increased weight.

Figure 13.14   Weight of object = weight of displaced water

Similarly, a loaded ship displaces more water than an empty ship and rides lower due to its increased weight (Fig. 13.15).

Figure 13.15   (a) This loaded barge rides low in the water because it displaces an amount of water equal to its weight. (b) This empty barge rides high in the water because its weight is much less than when full and thus it displaces much less water.

(a)   (b)

What happens to the weight of a brick or some other object that sinks in water? First, weigh the brick in air. Then, lower the brick under the water and weigh it again. It weighs less. The difference between the two weights is the buoyant (upward) force of the water. That is, the **buoyant force** is the upward force exerted on a submerged or partially submerged object. This is illustrated in the following example.

EXAMPLE 1

A solid concrete block 15.0 cm × 20.0 cm × 10.0 cm weighs 67.6 N in air (Fig. 13.16). When lowered into water, it weighs 38.2 N. Find the buoyant force. Using the scale readings, we see that the buoyant force is 67.6 N − 38.2 N = 29.4 N.

Figure 13.16   Weight in water = weight in air – buoyant force (weight of displaced water)

(a)

38.2 N weight in water

29.4 N buoyant force
(weight of displaced water)

67.6 N weight in air

(b)

By calculation, the volume of the displaced water is

$$V = lwh$$
$$V = (15.0 \text{ cm})(20.0 \text{ cm})(10.0 \text{ cm})$$
$$= 30\overline{0}0 \text{ cm}^3 = 3.00 \times 10^{-3} \text{ m}^3$$

The mass of the displaced water is

$$m = D_m V$$
$$m = (100\overline{0} \text{ kg/m}^3)(3.00 \times 10^{-3} \text{ m}^3)$$
$$= 3.00 \text{ kg}$$

The weight of the displaced water is then

$$F_w = mg$$
$$F_w = (3.00 \text{ kg})(9.80 \text{ m/s}^2)$$
$$= 29.4 \text{ N}   (1 \text{ N} = 1 \text{ kg m/s}^2)$$

which equals the buoyant force.

A submarine uses these same principles to dive and rise in the ocean. A submarine has ballast tanks that are filled with seawater, making it heavier than the volume of water it displaces, and so it descends. For the submarine to rise, the water in the ballast tanks is blown out by compressed air. The submarine then becomes lighter than the volume of water it displaces and so it rises.

Early human cultures made their boats of wood, which has densities less than that of water. How can a barge made of steel, which is approximately eight times as dense as water, float? The barge must be built so that it displaces at least its own weight plus its cargo load equivalent of water. The wider the barge and the deeper it is immersed, the more water is displaced and the greater is the buoyant force exerted on the barge by the water.

Imagine a loaded fishing boat coming in with the day's catch and just barely clearing a bridge on its way to the dock. After it is emptied of its cargo and heads back to sea, will it have more or less clearance under the bridge? Use Fig. 13.15 to explain your answer.

EXAMPLE 2

A rectangular boat is 4.00 m wide, 8.00 m long, and 3.00 m deep. (a) How many m^3 of water will it displace if the top stays 1.00 m above the water? (b) What load (in newtons) will the boat contain under these conditions if the empty boat weighs 8.60×10^4 N in dry dock?

(a)   The volume of water displaced by the boat is

$$V = lwh$$
$$V = (8.00 \text{ m})(4.00 \text{ m})(2.00 \text{ m})$$
$$= 64.0 \text{ m}^3$$

(b)   The load of the boat is the buoyant force of the displaced water ($D_w V$) minus the weight of the boat in dry dock.

$$(98\overline{0}0 \text{ N/m}^3)(64.0 \text{ m}^3) - (8.60 \times 10^4 \text{ N}) = 5.41 \times 10^5 \text{ N}$$

Note: The weight density of water is $98\overline{0}0$ N/m^3.

. . . .

Archimedes' principle applies to gases as well as liquids. Lighter-than-air craft (such as the Goodyear blimps) operate on this principle. Since they are filled with helium, which is lighter than air, the buoyant force on them causes them to be supported by the air. Being "submerged" in the air, a blimp is buoyed up by the weight of the air it displaces, which equals the buoyant force of the air on the balloon.

PROBLEMS 13.4

1.   A metal alloy weighs 81.0 lb in air and 68.0 lb when under water. Find the buoyant force of the water.
2.   A piece of metal weighs 67.0 N in air and 62.0 N in water. Find the buoyant force of the water.
3.   A rock weighs 25.7 N in air and 21.8 N in water. What is the buoyant force of the water?
4.   A metal bar weighs 455 N in air and 437 N in water. What is the buoyant force of the water?
5.   A rock displaces 1.21 ft^3 of water. What is the buoyant force of the water?
6.   A metal displaces 16.8 m^3 of water. Find the buoyant force of the water.
7.   A metal casting displaces 327 cm^3 of water. Find the buoyant force of the water.
8.   A piece of metal displaces 657 cm^3 of water. Find the buoyant force of the water.
9.   A metal casting displaces 2.12 ft^3 of alcohol. Find the buoyant force of the alcohol.
10.   A metal cylinder displaces 515 cm^3 of gasoline. Find the buoyant force of the gasoline.

SKETCH

12 cm^2 w

4.0 cm

DATA
$A = 12 \text{ cm}^2, l = 4.0 \text{ cm}, w = ?$

BASIC EQUATION
$A = lw$

WORKING EQUATION
$w = \dfrac{A}{l}$

SUBSTITUTION
$w = \dfrac{12 \text{ cm}^2}{4.0 \text{ cm}} = 3.0 \text{ cm}$

11. A 75.0-kg rock lies at the bottom of a pond. Its volume is 3.10×10^4 cm^3. How much force is needed to lift the rock?

12. A 125-lb rock lies at the bottom of a pond. Its volume is 0.800 ft^3. How much force is needed to lift the rock?

13. A flat-bottom river barge is 30.0 ft wide, 85.0 ft long, and 15.0 ft deep. (a) How many ft^3 of water will it displace while the top stays 3.00 ft above the water? (b) What load in tons will the barge contain under these conditions if the empty barge weighs $16\overline{0}$ tons in dry dock?

14. A flat-bottom river barge is 12.0 m wide, 30.0 m long, and 6.00 m deep. (a) How many m^3 of water will it displace while the top stays 1.00 m above the water? (b) What load (in newtons) will the barge contain under these conditions if the empty barge weighs 3.55×10^6 N in dry dock?

15. What is the volume (in m^3) of the water displaced by a submerged air tank that is acted on by a buoyant force of 7.50×10^4 N?

16. A lifeguard swims with her head just above the water. What is the volume of the submerged part of her body if she weighs $60\overline{0}$ N

17. An underwater camera weighing 1250 N in air is submerged and supported by a tether line. If the volume of the camera is 8.30×10^{-2} m^3, what is the tension in the line?

13.5 Fluid Flow

Think about the motion of water flowing down a fast-moving mountain stream that contains boulders and rapids and about the motion of the air during a thunderstorm or during a tornado. These types of motion are complex, indeed. Our discussion will focus on the simpler examples of fluid flow.

Streamline flow, also known as *laminar flow*, is the smooth flow of a fluid through a tube (Fig. 13.17). By smooth flow we mean that all particles of the fluid follow the same uniform path. **Turbulent flow**, also known as *nonlaminar flow*, is the erratic, unpredictable flow of a fluid resulting from excessive speed of the flow or sudden changes in direction or size of the tube or pipe.

(a)

(b)

(c)

Figure 13.17   Streamline flow of a fluid through a smooth tube or pipe is shown in (a) and (b). Water flowing in a creek or a mountain stream over and around rocks, resulting in sudden changes in direction and speed, is a common example of turbulent flow as in (c).

The **flow rate** of a fluid is the volume of fluid flowing past a given point in a pipe per unit time. Assume that we have a streamline flow through a straight section of pipe at speed v. During a time interval of t, each particle of fluid travels a distance vt. If A is the cross-sectional area of the pipe, the volume of fluid passing a given point during the time interval t is vtA. Thus, the flow rate, Q, is given by

$$Q = \frac{vtA}{t}$$

or

$$Q = vA$$

where Q = flow rate
v = speed of the fluid through the tube or pipe
A = cross-sectional area of the tube or pipe

EXAMPLE 1

Water flows through a fire hose of diameter 6.40 cm at a speed of 5.90 m/s. Find the flow rate of the fire hose in L/min.

Data:

$$v = 5.90 \text{ m/s}$$
$$r = 3.20 \text{ cm} = 0.0320 \text{ m}$$
$$A = \pi r^2 = \pi (0.0320 \text{ m})^2$$
$$Q = ?$$

Basic Equation:

$$Q = vA$$

Working Equation: Same

Substitution:

$$Q = \left(5.90 \frac{\text{m}}{\text{s}}\right) \pi (0.0320 \text{ m})^2 \times \frac{10^3 \text{ L}}{1 \text{ m}^3} \times \frac{60 \text{ s}}{1 \text{ min}}$$
$$= 1140 \text{ L/min}$$

EXAMPLE 2

Water flows through a hose of diameter 2.50 in. at a rate of 24.0 gal/min. Find its speed in ft/s as it leaves the end of the hose.

Data:

$$A = \pi r^2 = \pi (1.25 \text{ in.})^2 = 4.91 \text{ in}^2$$
$$Q = 24.0 \text{ gal/min}$$
$$v = ?$$

Basic Equation:

$$Q = vA$$

Working Equation:

$$v = \frac{Q}{A}$$

Substitution:

$$v = \frac{24.0 \text{ gal/min}}{4.91 \text{ in}^2} \times \frac{231 \text{ in}^3}{1 \text{ gal}} \times \frac{1 \text{ ft}}{12 \text{ in.}} \times \frac{1 \text{ min}}{60 \text{ s}}$$
$$= 1.57 \text{ ft/s}$$

The overall volume rate of flow of water in a stream or creek is relatively constant; that is, the volume of water passing through the various sections is constant. Thus, the speed

Figure 13.18 For an incompressible fluid, the flow rate is constant throughout.

$$Q = A_1v_1 = A_2v_2 = A_3v_3$$

of the water increases within sections where the stream is narrow and decreases within sections where the stream is broad.

For an incompressible fluid, the flow rate is constant throughout the pipe. If the cross-sectional area of the pipe changes and streamline flow is maintained, the flow rate is the same all along the pipe. That is, as the cross-sectional area increases, the velocity decreases, and vice versa (Fig. 13.18).

TRY THIS ACTIVITY

The Venturi Hose

Turn on the water to a garden hose that does not have a nozzle. Note the speed of the water as it flows out of the hose. Without turning the faucet, pinch the hose just behind the metal coupling, allowing only a small space for the water to flow. Describe what happens to the speed of the water as it now comes out of the hose. Explain why the speed changes although the faucet remained the same.

What happens to the pressure as the cross-sectional area of the pipe changes? This concept can be illustrated by use of a Venturi meter (Fig. 13.19), named after **Giovanni Battista Venturi**. Here the vertical tubes act like pressure gauges; the higher the column, the higher the pressure. As you can see, the higher the speed, the lower the pressure, and vice versa. This change in pressure of a fluid in streamline flow was first explained by **Daniel Bernoulli**.

Figure 13.19 A Venturi meter shows that the higher the speed of a fluid through a tube, the lower the pressure; the lower the speed of a fluid, the higher the pressure.

Do you recall walking into the wind in winter on a city street lined with several tall buildings when the wind seemed stronger than usual? The wind actually was stronger because it was acting as a fluid as illustrated using the Venturi meter principle in Fig. 13.19. That is, the speed of the wind increased as it was forced to flow between the tall buildings along the street.

BERNOULLI'S PRINCIPLE

For the horizontal flow of a fluid through a tube, the sum of the pressure and energy of motion (kinetic energy) per unit volume of the fluid is constant.

One application of **Bernoulli's principle** involves a small engine carburetor (Fig. 13.20). The volume of airflow is determined by the position of the butterfly valve. As the air flows through the throat, the air gains speed and loses pressure. The pressure in the fuel bowl equals the pressure above the throat. Due to the difference in

Giovanni Battista Venturi (1746–1822),

physicist, was born in Italy. He worked on the flow of fluids and is remembered for his discovery of the Venturi effect, the decrease of the pressure in a fluid in a pipe as the diameter is gradually reduced. The effect has many applications, such as in the carburetor and fluid-flow measuring instruments.

Daniel Bernoulli (1700–1782),

mathematician, was born in The Netherlands. He worked on trigonometry, mechanics, vibrating systems, and hydrodynamics (leading to the kinetic theory of gases) and developed Bernoulli's principle.

Figure 13.20   (a) Schematic of a small engine carburetor (b) Top view of a butterfly valve on a carburetor

(a)   (b)

Figure 13.21   Air flowing past an airplane wing with its more-rounded curved upper surface creates lift.

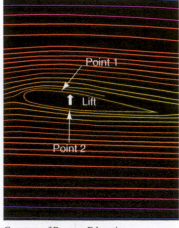

Courtesy of Pearson Education

pressure between the fuel bowl and the throat, gasoline is drawn into and is mixed with the airstream. The reduced pressure in the throat also helps the gasoline to vaporize.

Another application of Bernoulli's principle involves airplane travel. Figure 13.21 shows the flow of air rushing past the wing of an airplane. The velocity, v_1, of the air above the wing is greater than the velocity of the air below, v_2, because it has farther to travel in a given time. Thus, the pressure at point 2 is greater, which causes lift on the wing.

An airplane requires a longer distance for takeoff in summer than in winter. Why? Hot air is less dense and has fewer air molecules to lift the plane as it moves down the runway.

TRY THIS ACTIVITY

Low Pressure

Tear a sheet of paper in half and position each half as shown in Fig. 13.22. What do you think will happen to the positions of the papers when a continuous stream of air is blown between them? Will the sheets separate or move closer to one another? Test your prediction by blowing a continuous stream of air between the two sheets of paper. Was your prediction correct? Use Bernoulli's principle to explain the behavior of the paper.

Figure 13.22

Figure 13.23   Bernoulli's principle explains why a baseball curves.

Two other examples that illustrate Bernoulli's principle are a curving baseball and a paint spray gun. The rotating baseball in Fig. 13.23 drags a layer of air immediately next to its surface. This reduces the speed of air at the bottom of the ball and increases the speed at the top of the ball, causing the pressure to drop more at the top than at the bottom so that the ball curves as shown.

When the air in a spray gun is accelerated through a narrowing in the line, the pressure is reduced, and paint is drawn into the airstream and is then forced from the gun.

PHYSICS CONNECTIONS

Blood Pressure

A fluid exerts pressure on the walls of its closed container. As a fluid, blood exerts pressure on the walls of the heart, arteries, vessels, and capillaries that make up the circulatory system. When the diameter of the arteries and vessels is narrowed, an increased pressure is exerted on the walls of the blood vessels.

A sphygmomanometer (Fig. 13.24) is a device used to measure a person's blood pressure. The sphygmomanometer cuff is placed around the upper arm, inflated, and then deflated while the meter measures the pressure of blood passing through that section of the arm. As seen in this chapter, liquid pressure is dependent on the depth of the fluid. Since a sphygmomanometer cannot be placed around the heart and the depth of the fluid must be at the same level as the heart, the upper arm becomes a convenient location to measure blood pressure. When a person is lying down, the depth of the fluid throughout the body is roughly the same, so that the pressure can be measured on any limb of the body.

When standing on one's head, one notices that the blood vessels in the head experience a great deal of pressure. As a result, the blood vessels on the sides of the temple protrude. The vessels in the feet and legs typically withstand much higher pressures than the vessels in the head.

Figure 13.24 A person having his blood pressure measured. Notice that the sphygmomanometer is at the same level as the heart.

Courtesy of Michal Heron, Pearson Education/PH College

PROBLEMS 13.5

1. Water flows through a hose of diameter 3.90 cm at a velocity of 5.00 m/s. Find the flow rate of the hose in L/min.
2. Water flows through a 15.0-cm fire hose at a rate of 5.00 m/s. (a) Find the rate of flow through the hose in L/min. (b) How many litres pass through the hose in 30.0 min?
3. Water flows from a pipe at 650 L/min. (a) What is the diameter (in cm) of the pipe if the velocity of the water is 1.5 m/s? (b) Find the velocity (in m/s) of the water if the diameter of the pipe is 20.0 cm.
4. Water flows through a pipe of diameter 8.00 cm at 45.0 m/min. Find the flow rate (a) in m³/min and (b) in L/s.
5. A pump is rated to deliver 50.0 gal/min. Find the velocity of water in (a) a 6.00-in.-diameter pipe and (b) a 3.00-in.-diameter pipe.
6. What size pipe needs to be attached to a pump rated at 36.0 gal/min if the desired velocity is 10.0 ft/min? Give the inner diameter in inches.
7. What is the diameter of a pipe in which water travels 32.0 m in 16.0 s and has a flow rate of 2620 L/min?
8. A garden hose is used to fill a bucket in 30.0 s. If you cover part of the hose nozzle so the speed of the water leaving the hose doubles, how long does it take to fill the bucket?

SKETCH

12 cm² | w

4.0 cm

DATA
$A = 12$ cm², $l = 4.0$ cm, $w = ?$

BASIC EQUATION
$A = lw$

WORKING EQUATION
$w = \dfrac{A}{l}$

SUBSTITUTION
$w = \dfrac{12 \text{ cm}^2}{4.0 \text{ cm}} = 3.0 \text{ cm}$

9. A liquid flows through a pipe with a diameter of 0.50 m at a speed of 4.20 m/s. What is the rate of flow in L/min?

10. A pipe system with a radius of 0.060 m has a liquid flowing through it at a speed of 3.96 m/s. What is the rate of flow in L/min?

PHYSICS ON THE JOB

Aerospace Technician or Avionics Technician

Aerospace technicians are involved in the development, installation, and maintenance of all aspects of aircraft and aerospace systems. Aerospace technicians must be acutely aware of the properties specifically related to fluid dynamics and the properties of gases and its interaction with mechanical systems. Technicians are often on the front lines of calibrating and testing equipment and checking equipment failure.

A minimum of an associate's degree in engineering technology is needed to qualify as a technician and a bachelor's degree in an engineering field is typically required to become an aerospace engineer. Courses in mathematics, physics, drafting and computer skills are necessary, as are courses that promote hands-on technical repair and problem solving (Fig. 13.25).

Source: Bureau of Labor Statistics, www.bls.gov/ooh/

Figure 13.25 Avionics technicians

Courtesy of fotolia © Christian42

Glossary

Absolute Pressure The actual air pressure given by the gauge reading plus the normal atmospheric pressure. (p. 356)

Archimedes' Principle Any object placed in a fluid apparently loses weight equal to the weight of the displaced fluid. (p. 358)

Atmospheric Pressure The pressure caused by the weight of the air in the atmosphere. (p. 355)

Bernoulli's Principle For the horizontal flow of a fluid through a tube, the sum of the pressure and energy of motion (kinetic energy) per unit volume of the fluid is constant. (p. 363)

Buoyant Force The upward force exerted on a submerged or partially submerged object. (p. 359)

Flow Rate The volume of fluid flowing past a given point in a pipe per unit time. (p. 361)

Gauge Pressure The amount of air pressure excluding the normal atmospheric pressure. (p. 356)

Hydraulic Principle (Pascal's Principle) The pressure applied to a confined liquid is transmitted without measurable loss throughout the entire liquid to all inner surfaces of the container. (p. 350)

Hydrostatic Pressure The pressure a liquid at rest exerts on a submerged object. (p. 342)

Pressure The force applied per unit area. (p. 342)

Streamline Flow The smooth flow of a fluid through a tube. (p. 361)
Turbulent Flow The erratic, unpredictable flow of a fluid resulting from excessive speed of the flow or sudden changes in direction or size of the tube or pipe. (p. 361)

Formulas

13.1 $P = hD_w$
$F_t = AhD_w$
$F_s = \frac{1}{2} AhD_w$
13.3 $P_{abs} = P_{ga} + P_{atm}$
13.5 $Q = vA$

Review Questions

1. The force applied to a unit area is called
   (a) strain.
   (b) total force.
   (c) pressure.
   (d) none of the above.
2. The statement that the pressure applied to a confined liquid is transmitted without measurable loss throughout the entire liquid to all inner surfaces of the container is called
   (a) Hooke's law.
   (b) Pascal's principle.
   (c) Archimedes' principle.
   (d) none of the above.
3. For an incompressible fluid, the flow rate is
   (a) equal for all surfaces.
   (b) constant throughout the pipe.
   (b) greater for the larger parts of the pipe.
   (d) none of the above.
4. Bernoulli's principle states that for horizontal flow of a fluid through a tube, the sum of the pressure and energy of motion per unit volume is
   (a) increasing with time.
   (b) decreasing with time.
   (c) constant.
   (d) none of the above.
5. Bernoulli's principle explains
   (a) curving baseballs.
   (b) the hydraulic principle.
   (c) absolute pressure.
   (d) buoyant forces.
   (e) none of the above.
6. What is the metric unit for pressure?
7. In your own words, define *pressure*.
8. In your own words, state how to find the force exerted on the vertical side of a rectangular water tank.
9. In your own words, state the hydraulic principle.
10. Describe why a ship floats.
11. Describe how a rotating baseball follows a curved path.
12. How does an airplane wing provide lift?
13. What is the difference between streamline and turbulent flow?
14. Give an example of how Archimedes' principle applies to gases.
15. Describe the difference between absolute and gauge pressure. Which do you use when you measure the pressure in your automobile tires?
16. Is the pressure on a small piston different from the pressure on a large piston in the same hydraulic system? Are the forces on the two pistons the same?
17. On what does the total force exerted by a liquid on a horizontal surface depend?
18. Why must the thickness of a dam be greater at the bottom than at the top?

19. Is the hydraulic piston in the master brake cylinder in an automobile larger or smaller than the piston in the brake cylinder at the wheels? Why?
20. Would a drinking straw work in space, where there is no gravity? Explain.

Review Problems

SKETCH

| 12 cm² | w |

4.0 cm

DATA
$A = 12 \text{ cm}^2, l = 4.0 \text{ cm}, w = ?$

BASIC EQUATION
$A = lw$

WORKING EQUATION
$w = \dfrac{A}{l}$

SUBSTITUTION
$w = \dfrac{12 \text{ cm}^2}{4.0 \text{ cm}} = 3.0 \text{ cm}$

1. Find the pressure (in kPa) at the bottom of a water-filled drum 3.24 m high.
2. Find the depth in a lake at which the pressure is 197 lb/in².
3. Find the height of a water column when the pressure at the bottom of the column is 297 kPa.
4. What is the total force exerted on the bottom of a rectangular tank 8.67 ft × 4.83 ft × 3.56 ft deep?
5. Find the water pressure (in kPa) at a point 35.0 m from the bottom of a 55.0-m-tall full water tower.
6. Find the total force on the bottom of a cylindrical water tower 55.0 m high and 7.53 m in radius.
7. Find the total force on the side of a cylindrical water tower 55.0 m high and 7.53 m in radius.
8. Find the total force on the side of a rectangular water trough 1.25 m high × 1.55 m × 2.95 m.
9. What must the water pressure (in kPa) be on the ground to supply a water pressure of 252 N/cm² on the third floor, which is 9.00 m above the ground?
10. What water pressure must a pump that is located on the first floor supply to have water on the 25th floor of a building with a pressure of 26 lb/in²? Assume that the distance between floors is 16.0 ft.
11. A submarine is submerged to a depth of 3150 ft in the Atlantic Ocean. What air pressure (in kPa) is needed to blow water out of the ballast tanks?
12. The area of the large piston in a hydraulic jack is 4.75 in². The area of the small piston is 0.564 in². (a) What force must be applied to the small piston if a weight of 650 lb is to be lifted? (b) What is the mechanical advantage of the hydraulic jack?
13. The MA of a hydraulic jack is 324. What force must be applied to lift an automobile weighing 11,500 N?
14. The pistons of a hydraulic press have radii of 0.543 cm and 3.53 cm, respectively. (a) What force must be applied to the smaller piston to exert a force of 4350 N on the larger? (b) What is the pressure (in N/cm²) on each piston? (c) What is the mechanical advantage of the press?
15. Find the absolute pressure in a bicycle tire with a gauge pressure of 202 kPa.
16. Find the gauge pressure of a tire with an absolute pressure of 655 kPa.
17. Find the gauge pressure of a tank whose absolute pressure is 314 lb/in².
18. A rock weighs 55.4 N in air and 52.1 N in water. Find the buoyant force on the rock.
19. A metal displaces 643 cm³ of water. Find the buoyant force of the water.
20. A rock displaces 314 cm³ of alcohol. Find the buoyant force on the rock.
21. A flat-bottom barge is 22.3 ft wide, 87.5 ft long, and 16.5 ft deep. (a) How many ft³ of water will it displace while the top stays 3.20 ft above the water? (b) What load in tons will the barge contain if the barge weighs 157 tons in dry dock?
22. Water flows through a hose of diameter 3.00 cm at a velocity of 4.43 m/s. Find the flow rate of the hose in L/min.
23. Water flows through a 13.0-cm-diameter fire hose at a rate of 4.53 m/s. (a) What is the rate of flow through the hose in L/min? (b) How many litres pass through the hose in 25.0 min?
24. (a) What is the weight density of a liquid that exerts a total force of 433 N on the sides of a 3.00-m-tall cylindrical tank? The radius of the tank is 0.913 m, and it is filled to 1.75 m. (b) What liquid might this be?

APPLIED CONCEPTS

1. An aquarium's main tank holds $20\overline{0},000$ gal or 758 m³ of salt water. (a) What is the lateral surface area of the glass wall if the cylindrical saltwater tank is 14.5 m tall? (b) What is the force applied to the vertical glass surface? (c) Steel bands are often placed around aquariums to reinforce the glass walls. Explain how the steel bands should be spaced toward the bottom of the tank.

2. The piston in a master cylinder has a radius of 0.570 in. and the radius of each of the two brake cylinder pistons is 1.75 in. (a) How much pressure is created in the master cylinder if a driver quickly applies 45.5 lb of force to the automobile's brake pedal? (b) Given the area of the two brake cylinder pistons, what is the force applied to each of the brake drums? (See Fig. 13.8.)

3. A crane that can lift a maximum of $90\overline{0}0$ N is preparing to lift and move an underwater concrete mooring. The 1.25 m × 1.25 m × 0.450 m concrete block is located in seawater. Verify that the crane is strong enough to lift the mooring by finding (a) its volume, (b) its dry-dock weight, (c) the water's buoyant force, and (d) the force needed to lift the mooring while it is under water. (Refer to Table 12.2 for the mass density of concrete and of water.)

4. Wind tunnels are used to measure the aerodynamic properties of prototype models. (a) If a fan generates a wind speed of 25.0 mi/h inside an 8.75-ft² section of a wind tunnel, what is the wind speed as the air enters the narrower, 4.35-ft² section of the wind tunnel? (b) Explain why it is often windier on city streets surrounded by tall buildings than in more open areas.

5. A flexible hose with inside radius 0.250 in. leads to a shower fixture with 15 holes, each with radius 3.13×10^{-2} in. (a) If a faucet is opened to allow the water to move through the pipes at 3.94 ft/s, what is the speed of the water as it comes out of the holes? (b) What are two ways to increase the speed of the water from a shower fixture without opening the faucet farther? Explain.

CHAPTER 14

TEMPERATURE AND HEAT TRANSFER

Objectives

The major goals of this chapter are to enable you to:

1. Distinguish between temperature and heat.
2. Express temperature using different scales.
3. Analyze heat transfer applications.
4. Determine final temperature using the method of mixtures.
5. Relate heat transfer to the expansion of solids and liquids.
6. Find the heat required for change of phase of solids, liquids, and gases.

Almost all forms of technology have concerns about temperature and heat transfer. The concern may be direct, as in refrigeration, or indirect, as in the thermal expansion of highways. Being able to measure heat transfer can mean the difference between the success and failure of many things, from steam heat to space travel.

14.1 Temperature

An understanding of temperature and heat and their differences is very important in many applications in technology. The automotive, truck, and heavy equipment industries are very concerned with the heat energy released by the fuel mixture in an engine cylinder's combustion chamber (Fig. 14.1). While the contained explosion drives the piston, the excess heat produced must be transferred to the atmosphere.

This is accomplished in air-cooled small engines by cooling fins and in larger engines by liquid cooling systems. Allowing the engine temperature to become too hot causes the metal parts to excessively expand and freeze up or warp, permanently damaging the engine.

Basically, **temperature** is a measure of the hotness or coldness of an object. Temperature could be measured in a simple way by using your hand to sense the hotness or coldness of an object. However, the range of temperatures that your hand can withstand is too small, and your hand is not precise enough to measure temperature adequately. Therefore, other methods are used for measuring temperature.

Certain properties of matter vary with their temperature. For example, when objects are heated, they give off light of different colors. When an object is heated, in the absence of chemical reactions, it first gives off red light. As it is heated more, it appears white.

Chemical reactions sometimes cause different colors. When carbon steel is heated and exposed to air, several colors are observed before the rod appears red. This is due to a chemical reaction involving the carbon. If we could measure the color of the light, we could then determine the temperature. Although this works only for high temperatures, it is used in the production of metal alloys. The temperature of hot molten metals is determined this way.

Another property of matter that we use to find temperature is the change in volume of a liquid or a solid as its temperature changes. The liquid in glass thermometers is an example. This type of thermometer [Fig. 14.2(a)] consists of a hollow glass bulb and a hollow glass tube joined together. A small amount of liquid such as alcohol is placed in the bulb. The air is removed from the tube. When the liquid is heated, it expands and rises up the glass tube. The height to which the liquid rises indicates the temperature.

The glass thermometer is standardized by marking two points on the glass that indicate the liquid level at two known temperatures. The temperatures used are the *freezing point* of water and the *boiling point* of water at sea level. The distance between these marks is then divided up into equal segments called *degrees*.

Digital display thermometers [Fig. 14.2(b)] measure the electrical resistance of the metal tip at the end of the thermometer. As the temperature of the metal increases, the resistance increases. A microchip inside the housing of the thermometer converts the electrical resistance of the metal to a temperature reading on the digital display.

We will study the four temperature scales shown in Fig. 14.3. The common metric temperature scale is the **Celsius scale** with freezing point 0°C and boiling point 100°C. To write a temperature, we write the number followed by the degree symbol

Figure 14.1   Force on a piston produced by hot expanding gas

Figure 14.2   (a) Common glass thermometer

(a)

Courtesy of fotolia © rachwai

Figure 14.2   (b) Digital display thermometer

(b)

Courtesy of fotolia © zozulinskyi

Figure 14.3   Four basic temperature scales

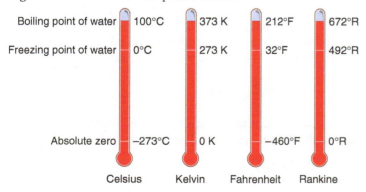

	Celsius	Kelvin	Fahrenheit	Rankine
Boiling point of water	100°C	373 K	212°F	672°R
Freezing point of water	0°C	273 K	32°F	492°R
Absolute zero	−273°C	0 K	−460°F	0°R

Anders Celsius (1701–1744),

astronomer, was born in Sweden. He devised the centigrade scale of temperature in 1742. The Celsius scale (formerly the centigrade scale) is named after him.

Gabriel Daniel Fahrenheit (1686–1736),

physicist, was born in Poland. He invented the alcohol thermometer in 1709 and the mercury thermometer in 1714.

(°) followed by the capital letter of the scale used. Temperatures below zero on a scale are written as negative numbers. Thus, 20° below zero on the Celsius scale is written as $-20°$C.

The common U.S. temperature scale is the **Fahrenheit scale** with freezing point 32°F and boiling point 212°F. The relationship between Fahrenheit temperatures (T_F) and Celsius temperatures (T_C) is given by

$$T_C = \frac{5}{9}(T_F - 32°)$$

$$T_F = \frac{9}{5}T_C + 32°$$

where T_C = Celsius temperature
T_F = Fahrenheit temperature

EXAMPLE 1

The human body average temperature is 98.6°F. What is it in degrees Celsius?

Data:

$$T_F = 98.6°F$$
$$T_C = ?$$

Basic Equation:

$$T_C = \frac{5}{9}(T_F - 32°)$$

Working Equation: Same

Substitution:

$$T_C = \frac{5}{9}(98.6° - 32°)$$

$$= \frac{5}{9}(66.6°)$$

$$= 37.0°C$$

Lord Kelvin (Sir William Thomson) (1824–1907),

mathematician and physicist, was born in Belfast, Ireland. He helped develop the law of conservation of energy and the absolute temperature scale (now named the Kelvin scale), did fundamental research in thermodynamics, presented the dynamic theory of heat, developed theorems for the mathematical analysis of electricity and magnetism, and designed several kinds of electrometers.

Sometimes it is necessary to use the *absolute temperature scales*, which are the Kelvin scale and the Rankine scale. These are called absolute scales because 0 on either scale refers to the lowest limit of temperature, called *absolute zero*.

The **Kelvin scale** is the metric absolute temperature scale on which absolute zero is 0 K and is closely related to the Celsius scale. The relationship is*

$$T_K = T_C + 273$$

The **Rankine scale** is the U.S. absolute temperature scale on which absolute zero is 0°R and is closely related to the Fahrenheit scale. The relationship is

$$T_R = T_F + 46\overline{0}°$$

William Rankine (1820–1872),

engineer and scientist, was born in Scotland. He is noted for his work on the steam engine, machinery, shipbuilding, applied mechanics, the new science of thermodynamics, and the theories of elasticity and of waves.

*The degree symbol (°) is not used when writing a temperature on the Kelvin scale.

Change 18°C to Kelvin.

EXAMPLE 2

Data:

$$T_C = 18°C$$
$$T_K = ?$$

Basic Equation:

$$T_K = T_C + 273$$

Working Equation: Same

Substitution:

$$T_K = 18 + 273$$
$$= 291 \text{ K}$$

· · · · · · · · · · · · · · · ·

Change 535°R to degrees Fahrenheit.

EXAMPLE 3

Data:

$$T_R = 535°R$$
$$T_F = ?$$

Basic Equation:

$$T_R = T_F + 46\bar{0}°$$

Working Equation:

$$T_F = T_R - 46\bar{0}°$$

Substitution:

$$T_F = 535° - 46\bar{0}°$$
$$= 75°F$$

· · · · · · · · · · · · · · · ·

PROBLEMS 14.1

Find each temperature as indicated.

1. $T_F = 77°F$, $T_C = $ _____
2. $T_F = 113°F$, $T_C = $ _____
3. $T_F = 257°F$, $T_C = $ _____
4. $T_C = 15°C$, $T_F = $ _____
5. $T_C = 145°C$, $T_F = $ _____
6. $T_C = 35°C$, $T_F = $ _____
7. $T_F = 1\bar{0}°F$, $T_C = $ _____
8. $T_F = 2\bar{0}°F$, $T_C = $ _____
9. $T_C = 95°C$, $T_F = $ _____
10. $T_F = -5\bar{0}°F$, $T_C = $ _____
11. $T_C = 25°C$, $T_K = $ _____
12. $T_F = -45°F$, $T_R = $ _____
13. $T_K = 406 \text{ K}$, $T_C = $ _____
14. $T_C = 75°C$, $T_K = $ _____
15. $T_C = -5\bar{0}°C$, $T_K = $ _____
16. $T_K = 175 \text{ K}$, $T_C = $ _____
17. $T_K = 600\bar{0} \text{ K}$, $T_C = $ _____
18. The melting point of pure iron is 1505°C. What Fahrenheit temperature is this?
19. The melting point of mercury is −38.0°F. What Celsius temperature is this?
20. A welding white heat is approximately 140\bar{0}°C. Find this temperature expressed in degrees Fahrenheit.

SKETCH

12 cm²   w

4.0 cm

DATA
$A = 12$ cm², $l = 4.0$ cm, $w = ?$

BASIC EQUATION
$A = lw$

WORKING EQUATION
$w = \dfrac{A}{l}$

SUBSTITUTION
$w = \dfrac{12 \text{ cm}^2}{4.0 \text{ cm}} = 3.0 \text{ cm}$

PROBLEM SOLVING

21.   The temperature in a crowded room is 85°F. What is the Celsius reading?
22.   The temperature of an iced tea drink is 5°C. What is the Fahrenheit reading?
23.   The boiling point of liquid nitrogen is −196°C. What is the Fahrenheit reading?
24.   The melting point of ethyl alcohol is −179°F. What is the Celsius reading?

During the forging and heat-treating of steel, the color of heated steel is used to determine its temperature. Complete the following table, which shows the color of heat-treated steel and the corresponding approximate temperatures in degrees Celsius and Fahrenheit. (Round to three significant digits.)

	Color	°C	°F
25.	White	——	220̄0
26.	Yellow	110̄0	——
27.	Orange	——	1725
28.	Cherry red	718	——
29.	Dark red	635	——
30.	Faint red	——	90̄0
31.	Pale blue	31̄0	——

14.2 Heat

Figure 14.4   Friction causes a rise in temperature of the drill and plate.

When a hole is drilled in a metal block (Fig. 14.4), it becomes very hot. As the drill does mechanical work on the metal, the temperature of the metal increases. How can we explain this? Note the difference between the metal at low temperatures and at high temperatures. At high temperatures, the atoms in the metal vibrate more rapidly than at low temperatures. Their velocity is higher at high temperatures, and thus their kinetic energy ($E_k = \frac{1}{2} mv^2$) is greater. To raise the temperature of a material, we must speed up the atoms; that is, we must add energy to them. **Heat** is a form of internal kinetic and potential energy contained in an object associated with the motion of its atoms or molecules and may be transferred from an object at a higher temperature to one at a lower temperature.

Drilling a hole in a metal block causes a temperature increase. As the drill turns, it collides with atoms of the metal, causing them to speed up. This mechanical work done on the metal has caused an increase in the energy of the atoms. For this reason, any friction between two surfaces results in a temperature rise of the materials.

Since heat is a form of energy, we could measure it in joules or ft lb, which are energy units. However, before it was known that heat is a form of energy, special units for heat were developed, which are still in use. These units are the calorie and the kilocalorie in the metric system and the Btu (British thermal unit) in the U.S. system. The **kilocalorie** (kcal) is the amount of heat necessary to raise the temperature of 1 kg of water 1°C. *Note:* The precise definition is based on the amount of heat needed to raise the temperature of 1 kg of water from 14.5°C to 15.5°C; however, the variation for each 1°C change in temperature is so minimal that it can be ignored for all practical purposes. The **Btu** is the amount of heat (energy) necessary to raise the temperature of 1 lb of water 1°F. The **calorie** (cal) is the amount of heat (energy) necessary to raise the temperature of 1 g of water 1°C. *Note:* One food calorie is the same as 1 kcal.

To lower the temperature of a substance, we need to remove some of the heat, the energy of motion of the molecules. When we have removed all the heat possible (when the molecules are moving as slowly as possible), we have reached **absolute zero**, the lowest possible temperature. Lower temperatures cannot be reached because all the heat has been removed. However, there is no upper limit on temperature because we can always add more heat (energy) to a substance to increase its temperature.

As mentioned before, heat and work are somehow related. James Prescott Joule determined by experiments the relationship between heat and work, called the **mechanical equivalent of heat.** He found that

1. 1 cal of heat is produced by 4.19 J of work.
2. 1 kcal of heat is produced by 4190 J of work.
3. 1 Btu of heat is produced by 778 ft lb of work.

The following are some examples in which heat is converted into useful work:

1. **In our bodies.** When food is oxidized, heat energy is produced, which can be converted into muscular energy, which in turn can be turned into work. Experiments have shown that only about 25% of the heat energy from our food is converted into muscular energy. That is, our bodies are about 25% efficient.
2. **By burning gases.** When a gas is burned, the gas expands and builds up a tremendous pressure that may convert heat to work by exerting a force to move a piston in an engine or turn the blades of a turbine. Since the burning of the fuel occurs within the cylinder or turbine, such engines are called *internal combustion engines*.
3. **By steam.** Heat from burning oil, coal, or wood may be used to generate steam. When water changes to steam under normal atmospheric pressure, it expands about 1700 times. When confined to a boiler, the pressure exerts a force against the piston in a steam engine or against the blades of a steam turbine. Since the fuel burns outside the engine, most steam engines or steam turbines are *external combustion engines*.

Technically, what is the difference between temperature and heat? *Temperature* is a measure of the hotness or coldness of an object. *Heat* is the total thermal energy (kinetic and potential) that can be transferred from an object at a higher temperature to one at a lower temperature. There are two basic ways of changing the temperature of an object:

1. By doing work *on* the object, such as the work done by the drill on the metal block in Fig. 14.4.
2. By supplying energy *to* the object, such as mechanical, chemical, or electrical energy.

Find the amount of work (in J) that is equivalent to 4850 cal of heat.   ◄——— **EXAMPLE 1**

$$4850 \text{ cal} \times \frac{4.19 \text{ J}}{1 \text{ cal}} = 20{,}300 \text{ J} \quad \text{or} \quad 20.3 \text{ kJ}$$

· · · · · · · · · · · · · · ·

How much work must a person do to offset eating a 775-calorie breakfast?   ◄——— **EXAMPLE 2**
   First, note that one food calorie equals one kilocalorie.

$$775 \text{ kcal} \times \frac{4190 \text{ J}}{1 \text{ kcal}} = 3.25 \times 10^6 \text{ J} \quad \text{or} \quad 3.25 \text{ MJ}$$

· · · · · · · · · · · · · · ·

A given coal gives off 7150 kcal/kg of heat when burned. How many joules of work   ◄——— **EXAMPLE 3**
result from burning one metric ton, assuming that 35.0% of the heat is lost?
   First, note that one metric ton equals 1000 kg.

$$7150 \frac{\text{kcal}}{\text{kg}} \times \frac{4190 \text{ J}}{\text{kcal}} \times 1000 \text{ kg} \times 0.350 = 1.05 \times 10^{10} \text{ J}$$

· · · · · · · · · · · · · · ·

SKETCH

12 cm²	w

4.0 cm

DATA
$A = 12$ cm², $l = 4.0$ cm, $w = ?$

BASIC EQUATION
$A = lw$

WORKING EQUATION
$w = \dfrac{A}{l}$

SUBSTITUTION
$w = \dfrac{12 \text{ cm}^2}{4.0 \text{ cm}} = 3.0$ cm

PROBLEMS 14.2

1. Find the amount of heat in cal generated by 95 J of work.
2. Find the amount of heat in kcal generated by 7510 J of work.
3. Find the amount of work that is equivalent to 1550 Btu.
4. Find the amount of work that is equivalent to 3850 kcal.
5. Find the mechanical work equivalent (in J) of 765 kcal of heat.
6. Find the mechanical work equivalent (in J) of 8550 cal of heat.
7. Find the heat equivalent (in Btu) of 3.46×10^6 ft lb of work.
8. Find the heat equivalent (in kcal) of 7.63×10^5 J of work.
9. How much work must a person do to offset eating a piece of cake containing 625 cal?
10. How much work must a person do to offset eating a $2\overline{0}0$-g bag of potato chips if 28 g of chips contain 150 cal?
11. A fuel yields 1.15×10^4 cal/g when burned. How many joules of work are obtained by burning $10\overline{0}0$ g of the fuel?
12. A racing fuel produces 1.60×10^4 cal/g when burned. If $50\overline{0}$ g of the fuel is burned, how many joules of work are produced?
13. A given gasoline yields 1.15×10^4 cal/g when burned. How many joules of work are obtained by burning 875 g of gasoline?
14. A coal sample yields 1.25×10^4 Btu/lb. How many foot-pounds of work result from burning 1.00 ton of this coal?
15. Natural gas burned in a gas turbine has a heating value of 1.10×10^5 cal/g. If the turbine is 24.0% efficient and 2.50 g of gas is burned each second, find (a) how many joules of work are obtained and (b) the power output in kilowatts.
16. Find the amount of heat energy that must be produced by the body to be converted into muscular energy and then into $10\overline{0}0$ ft lb of work. Assume that the body is 25% efficient.
17. What is the mechanical work equivalent in $50,\overline{0}00$ kcal of heat produced by a stationary diesel engine?
18. An industrial engine produces $38,\overline{0}00$ kcal of heat. What is the mechanical work equivalent of the heat produced?

14.3 Heat Transfer

Figure 14.5 Transfer of heat from a warmer to a colder area

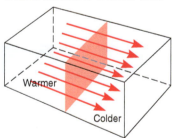

The movement of heat from a hot engine to the air is necessary to keep the engine from overheating. The heat produced by a furnace must be transferred to the various rooms in a house. The movement of heat is a major technical application.

The transfer of heat from one object to another is always from the warmer object to the colder one or from the warmer part of an object to a colder part (Fig. 14.5). There are three methods of heat transfer: *conduction, convection*, and *radiation*. **Conduction** is the heat transfer from a warmer part of a substance to a cooler part as a result of molecular collisions, which cause the slower-moving molecules to move faster. Conduction is the usual method of heat transfer in solids. When one end of a metal rod is heated, the molecules in that end move faster than before. These molecules collide with other molecules and cause them to move faster also. In this way, the heat is transferred from one end of the metal to the other (Fig. 14.6). Another example of conduction is the transfer of the excess heat produced in the combustion chamber of an engine through the engine block into the coolant (Fig. 14.7).

The conduction of heat through some materials is better than through others. A poor conductor of heat is called an *insulator*. A list of good conductors and poor conductors is given in Table 14.1.

Figure 14.6   Heat flows by conduction in the metal rod.

Heat flow

Metal rod

Figure 14.7   Heat conduction in an automobile engine cooling system

Table 14.1   Good and Poor Heat Conductors

Good Heat Conductors	Poor Heat Conductors
Copper	Asbestos
Aluminum	Glass
Steel	Wood
	Air
	Snow
	Fiberglass
	Cellulose

Sitting on aluminum bleachers on a cold day feels colder than sitting on wood bleachers at the same temperature because aluminum is a much better heat conductor than wood. Wood is a good insulator and is often used as handles for pots and pans. In winter, animals find shelter in snow banks by making snow holes because snow is a poor heat conductor. Snow causes not only the ground heat, but also the animals' body heat to be retained. Before health concerns were realized, asbestos was once widely used in insulating buildings because it is such a poor heat conductor.

Convection is the heat transfer by the movement of warm molecules from one region of a gas or a liquid to another. The wind carries heat along with it. The coolant in an engine carries hot antifreeze from the engine block to the radiator by a convection process. Heat transfer by the wind is a natural convection process. Heat transfer by the engine coolant is a forced convection process because it depends on a pump.

A dramatic illustration of the difference between convection and conduction can be shown by restraining some ice at the bottom of a test tube with some steel wool and filling the tube with water. When the water in the top of the tube is heated with a flame, it will boil without melting the ice. The poor conductivity of water keeps the less dense boiling water at the top and any heat transfer to the ice must be by conduction.

Convection currents are caused by the expansion of liquids or gases as they are heated. This expansion makes the hot gas or liquid less dense than the surrounding fluid. The lighter fluid is then forced upward by the heavier, surrounding fluid, which then flows in to replace it (Fig. 14.8). This type of behavior occurs in a fireplace as hot air goes up the chimney and is replaced by cool air from the adjacent room. The cool air draft, as this is called, is eventually supplied from outside air. This is why a fireplace is not very effective in heating a house. An airtight woodburning

Figure 14.8 A fireplace draws room air up into the chimney.

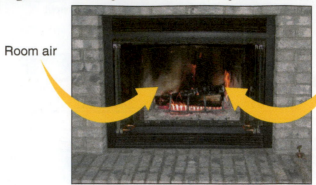

Room air

Room air

stove, however, draws little air from the inside of a house and is therefore much more efficient at heating the house.

Conduction and convection require the presence of matter. All life on earth depends on the transfer of energy from the sun; this energy is transferred through nearly empty space. **Radiation** is heat transfer through energy being transmitted in the forms of rays, waves, or particles. Put your hand several inches from a hot iron (Fig. 14.9). The heat you feel is not transferred by conduction because air is a poor conductor. It is not transferred by convection because the hot air rises. This heat transfer is through radiation. This radiant heat is similar to light and passes through air, glass, and the vacuum of space. The energy that comes to us from the sun is in the form of radiant energy. At night, heat in the ground is radiated into the air. Dark objects absorb radiant heat and light objects reflect radiant heat. This is why we feel cooler on a hot day in light-colored clothing than in dark clothing (Fig. 14.10).

Figure 14.9 Heat radiation

Figure 14.10 Dark objects absorb more radiant heat than light objects.

Heat flow calculations are important because of concern about energy conservation. All three of the heat-transfer mechanisms discussed here must be considered in any estimation of heat loss from a building. In addition, infiltration losses that arise from leakage through cracks and openings near doors, windows, and other such areas must also be considered. Infiltration losses are a form of convective transfer. Heat loss by conduction must also be considered. We will discuss methods for calculating heat loss by conduction.

The equations describing the flow of heat through an object are very similar to those for the flow of electricity, which are developed in later chapters. The driving potential for heat flow is the temperature difference between the hot and cold sides of the object. Heat flow is similar to the flow of electrical charge.

The ability of a material to transfer heat by conduction is called its **thermal conductivity**. Metals are good conductors of heat. Glass and air are poor conductors. The rate at which heat is transferred through an object depends on the following factors:

1. The thermal conductivity
2. The cross-sectional area through which the heat flows
3. The thickness of the material
4. The temperature difference between the two sides of the material

The total amount of heat transferred is given by the equation

$$Q = \frac{KAt(T_2 - T_1)}{L}$$

where Q = heat transferred in J or Btu
   K = thermal conductivity (from Table 14.2)
   A = cross-sectional area
   t = total time
   T_2 = temperature of the hot side
   T_1 = temperature of the cool side
   L = thickness of the material

Table 14.2 gives the thermal conductivities of some common materials.

Table 14.2   Thermal Conductivities

Substance	J/(s m °C)	Btu/(ft °F h)
Air	0.025	0.015
Aluminum	230	140
Brass	120	68
Brick/concrete	0.84	0.48
Cellulose fiber (loose fill)	0.039	0.023
Copper	380	220
Corkboard	0.042	0.024
Glass	0.75	0.50
Gypsum board (sheetrock)	0.16	0.092
Mineral wool	0.045	0.026
Plaster	0.14	0.083
Polystyrene foam	0.035	0.020
Polyurethane (expanded)	0.024	0.014
Steel	45	26
Water	0.56	0.32

Find the heat flow in an 8.0-h period through a 36 in. × 36 in. pane of glass (0.125 in. thick) if the temperature of the inner surface of the glass is 65°F and the temperature of the outer surface is 15°F.

EXAMPLE 1

Data:

$K = 0.50$ Btu/(ft °F h)

$A = 36$ in. × 36 in. $= 3.0$ ft × 3.0 ft $= 9.0$ ft^2

$t = 8.0$ h

$T_2 = 65°F$

$T_1 = 15°F$

$L = 0.125$ in. $\times \left(\dfrac{1 \text{ ft}}{12 \text{ in.}} \right) = 0.0104$ ft

$Q = ?$

Basic Equation:

$$Q = \frac{KAt(T_2 - T_1)}{L}$$

Working Equation: Same

Substitution:

$$Q = \frac{[0.50 \text{ Btu/(ft °F h)}](9.0 \text{ ft}^2)(8.0 \text{ h})(65°F - 15°F)}{0.0104 \text{ ft}}$$

$$= 1.7 \times 10^5 \text{ Btu} \qquad \boxed{\dfrac{\left(\dfrac{\text{Btu}}{\cancel{\text{ft}}\, \cancel{°F}\, \cancel{\text{h}}}\right) \text{ft}^2 \, \cancel{\text{h}} \, \cancel{°F}}{\cancel{\text{ft}}} = \text{Btu}}$$

· · · · · · · · · · · · · · · ·

The insulation value of construction material is often expressed in terms of the *R value*, which indicates the ability of the material to resist the flow of heat and uses U.S. units. The *R* value is inversely proportional to the thermal conductivity and directly proportional to the thickness. Low thermal conductivity is characteristic of good insulators. This is described by the equation

$$R = \frac{L}{K}$$

where R = R value (in ft^2 °F/Btu/h)
  K = thermal conductivity
  L = thickness of the material (in ft)

EXAMPLE 2 ➤ Calculate the *R* value of 6.0 in. of mineral wool insulation.

Data:

$L = 6.0$ in. $= 0.50$ ft
$K = 0.026$ Btu/(ft °F h)
$R = ?$

Basic Equation:

$$R = \frac{L}{K}$$

Working Equation: Same

Substitution:

$$R = \frac{0.50 \text{ ft}}{0.026 \text{ Btu/(ft°F h)}}$$

$$= 19 \text{ ft}^2 \text{ °F/Btu/h} \qquad \boxed{\dfrac{\text{ft}}{\dfrac{\text{Btu}}{\text{ft °F h}}} = \text{ft} \div \dfrac{\text{Btu}}{\text{ft °F h}} = \text{ft} \cdot \dfrac{\text{ft °F h}}{\text{Btu}} = \dfrac{\text{ft}^2 \text{ °F h}}{\text{Btu}} = \dfrac{\text{ft}^2 \text{ °F}}{\text{Btu/h}}}$$

This result could also have been written *R*-19. There is no equivalent in the metric system.

· · · · · · · · · · · · · · · ·

PHYSICS CONNECTIONS

Insulation

The purpose of insulating a home is to limit the amount of heat transfer through the walls and ceiling. Insulation limits heat transfer via conduction, convection, and radiation. Different insulating materials and methods are used. Less dense materials are often more effective in preventing heat transfer. Attic insulation must be fluffy and not compressed to be less dense. Molecules farther apart make fewer collisions, which results in less transfer of heat energy to a cooler environment. Other examples of less dense materials limiting heat transfer via conduction are argon gas placed between double-pane windows and evacuated linings in vacuum bottles. Argon has very low density; evacuated linings contain almost no air and therefore few molecules to transfer heat.

Fiberglass is one of the most popular types of insulation. The glass fiber is a poor conductor of heat. In addition, the air pockets in the fluffy fiberglass insulation prevent convection currents from transferring energy between the molecules and transferring heat (Fig. 14.11).

Figure 14.11 (a) Insulation is made of fiberglass, a poor conductor of heat. The fluffiness of the insulation creates air pockets to also eliminate convection currents. (b) The evacuated chamber inside a vacuum bottle prevents heat from transferring through the sides of the bottle.

Courtesy of fotolia © Roman Milert

(a)

(b)

PROBLEMS 14.3

1. Find the R value of a pane of 0.125-in.-thick glass.
2. Find the R value of a brick wall 4.0 in. thick.
3. Find the R value of 0.50-in.-thick sheetrock.
4. Find the thermal conductivity of a piece of building material 0.25 in. thick that has an R value of 1.6 ft^2 °F/Btu/h.
5. Find the R value of 0.50-in.-thick corkboard.
6. The dimensions of a rectangular building are $2\overline{0}$ ft × $10\overline{0}$ ft. The average outer wall temperature is $2\overline{0}$°F and the average inner wall temperature is 55°F. Find the amount of heat conducted through the walls of the building in 24 h if the R value of the walls is 11 ft^2 °F/Btu/h.
7. Find the heat flow during 30.0 days through a glass window of thickness 0.20 in. with area 15 ft^2 if the average outer surface temperature is 25°F and the average inner glass surface temperature is $5\overline{0}$°F.
8. Find the heat flow in 30.0 days through a 0.25-cm-thick steel plate with cross section 45 cm × 75 cm. Assume a temperature differential of 95°C.

SKETCH

12 cm^2 $\quad w$

4.0 cm

DATA

$A = 12$ cm^2, $l = 4.0$ cm, $w = ?$

BASIC EQUATION

$A = lw$

WORKING EQUATION

$w = \dfrac{A}{l}$

SUBSTITUTION

$w = \dfrac{12 \text{ cm}^2}{4.0 \text{ cm}} = 3.0 \text{ cm}$

9. Find the heat flow in 75 s through a steel rod of length 85 cm and diameter 0.50 cm if the temperature of the hot end of the rod is $11\bar{0}°C$ and the temperature of the cool end is −25°C.

10. Find the heat flow in 15 min through a 0.10-cm-thick copper plate with cross-sectional area 150 cm² if the temperature of the hot side is $99\bar{0}°C$ and the temperature of the cool side is 5°C.

11. Find the heat flow in 24 h through a refrigerator door 30.0 in. × 58.0 in. insulated with cellulose fiber 2.0 in. thick. The temperature inside the refrigerator is 38°F. Room temperature is 72°F.

12. Find the heat flow in 30.0 days through a freezer door 30.0 in. × 58.0 in. insulated with cellulose fiber 2.0 in. thick. The temperature inside the freezer is $−1\bar{0}°F$. Room temperature is 72°F.

13. Find the heat flow in 24 h through a refrigerator door 76.0 cm × 155.0 cm insulated with cellulose fiber 5.0 cm thick. The temperature inside the refrigerator is 3°C. Room temperature is 21°C.

14. Find the heat flow in 30.0 days through a freezer door 76.0 cm × 155.0 cm insulated with cellulose fiber 5.0 cm thick. The temperature inside the freezer is −18°C. Room temperature is 21°C.

15. Find the heat flow through the sides of an 18-cm-tall glass of ice water in 45 s. The glass is 6.00 mm thick; the temperature inside is 28.0°C. The temperature outside is 43.3°C and the radius is 7.0 cm.

14.4 Specific Heat

If we placed a piece of steel and a pan of water in the direct summer sunlight, we would find that the water becomes only slightly warmer whereas the steel gets quite hot. Why should one get so much hotter than the other? If equal masses of steel and water were placed over the same flame for 1 min, the temperature of the steel would increase almost 10 times more than that of the water. The water has a greater capacity to absorb heat.

Because water has a much higher capacity for storing energy than most common materials, it is very useful in cooling systems in engines and power plants. This property of water affects the climate of many places. Cities on large lakes, like Chicago, are warmer in the winter and cooler in the summer near the lake because of the high heat capacity of water. High temperatures in summer and low temperatures in winter in the middle of large continents are largely due to the absence of large bodies of water. Europe is warmer in the winter than mid-Canada because warm air from over the Atlantic Ocean is blown by prevailing westerly winds over the land.

The specific heat of a substance is a measure of its capacity to absorb or give off heat per degree change in temperature. This property of water to absorb or give off large amounts of heat makes it an effective substance for transferring heat in industrial processes.

The **specific heat** of a substance is the amount of heat necessary to change the temperature of 1 kg of it 1°C (1 lb of it 1°F in the U.S. system). By formula,

$$c = \frac{Q}{m\Delta T} \quad \text{(metric)} \qquad c = \frac{Q}{w\Delta T} \quad \text{(U.S.)}$$

To find the amount of heat added or taken away from a substance to produce a certain temperature change, we use

$$Q = cm\Delta T \quad \text{(metric)} \qquad Q = cw\Delta T \quad \text{(U.S.)}$$

where c = specific heat
   Q = heat
   m = mass
   w = weight
   ΔT = change in temperature

A list of specific heats is given in Table 15 of Appendix C.

TRY THIS ACTIVITY

Cool Floors

A dramatic example of heat conduction is often experienced on cold winter mornings. While standing with your bare feet on a cold tile floor, note how quickly heat is transferred from your feet to the tile. Then, stand in a doorway with one bare foot on tile and one bare foot on wood and note the difference in the rate at which heat is transferred. What are the general characteristics that determine the heat capacity for your floors? Why are mats commonly placed on bathroom floors?

How many kilocalories of heat must be added to 10.0 kg of steel to raise its temperature $15\overline{0}°C$?

EXAMPLE 1

Data:

$$m = 10.0 \text{ kg}$$
$$\Delta T = 15\overline{0}°C$$
$$c = 0.115 \text{ kcal/kg °C} \qquad \text{(from Table 15 of Appendix C)}$$
$$Q = ?$$

Basic Equation:

$$Q = cm\Delta T$$

Working Equation: Same

Substitution:

$$Q = \left(0.115 \frac{\text{kcal}}{\text{kg °C}}\right)(10.0 \text{ kg})(15\overline{0}°C)$$
$$= 173 \text{ kcal}$$

How many joules of heat must be absorbed to cool 5.00 kg of water from 75.0°C to 10.0°C?

EXAMPLE 2

Data:

$$m = 5.00 \text{ kg}$$
$$\Delta T = 75.0°C - 10.0°C = 65.0°C$$
$$c = 4190 \text{ J/kg °C} \qquad \text{(from Table 15 of Appendix C)}$$
$$Q = ?$$

Basic Equation:

$$Q = cm\Delta T$$

Working Equation: Same

Substitution:

$$Q = \left(4190 \frac{\text{J}}{\text{kg °C}}\right)(5.00 \text{ kg})(65.0°C)$$
$$= 1.36 \times 10^6 \text{ J} \qquad \text{or} \qquad 1.36 \text{ MJ}$$

SKETCH

```
┌──────────┐
│  12 cm²  │ w
└──────────┘
   4.0 cm
```

DATA

$A = 12 \text{ cm}^2, l = 4.0 \text{ cm}, w = ?$

BASIC EQUATION

$A = lw$

WORKING EQUATION

$w = \dfrac{A}{l}$

SUBSTITUTION

$w = \dfrac{12 \text{ cm}^2}{4.0 \text{ cm}} = 3.0 \text{ cm}$

PROBLEMS 14.4

Find Q for each material.

1. Steel, $w = 3.00$ lb, $\Delta T = 50\overline{0}°$F, $Q =$ _____ Btu
2. Copper, $m = 155$ kg, $\Delta T = 170°$C, $Q =$ _____ kcal
3. Water, $w = 19.0$ lb, $\Delta T = 200°$F, $Q =$ _____ Btu
4. Water, $m = 25\overline{0}$ g, $\Delta T = 17.0°$C, $Q =$ _____ cal
5. Ice, $m = 5.00$ kg, $\Delta T = 2\overline{0}°$C, $Q =$ _____ J
6. Steam, $w = 5.00$ lb, $\Delta T = 40°$F, $Q =$ _____ Btu
7. Aluminum, $m = 79.0$ g, $\Delta T = 16°$C, $Q =$ _____ cal
8. Brass, $m = 750$ kg, $\Delta T = 125°$C, $Q =$ _____ J
9. Steel, $m = 1250$ g, $\Delta T = 50.0°$C, $Q =$ _____ J
10. Aluminum, $m = 85\overline{0}$ g, $\Delta T = 115°$C, $Q =$ _____ kcal
11. Water, $m = 80\overline{0}$ g, $\Delta T = 80.0°$C, $Q =$ _____ kcal
12. Lead, $m = 475$ kg, $\Delta T = 245°$C, $Q =$ _____ J
13. How many Btu of heat must be added to 1200 lb of copper to raise its temperature from $10\overline{0}°$F to $45\overline{0}°$F?
14. How many Btu of heat are given off by $50\overline{0}$ lb of aluminum when it cools from $65\overline{0}°$F to 75°F?
15. How many kcal of heat must be added to 1250 kg of copper to raise its temperature from 25°C to 275°C?
16. How many joules of heat are absorbed by an electric freezer in lowering the temperature of 1850 g of water from 80.0°C to 10.0°C?
17. How many joules of heat are required to raise the temperature of $75\overline{0}$ kg of water from 15.0°C to 75.0°C?
18. How many kilocalories of heat must be added to $75\overline{0}$ kg of steel to raise its temperature from 75°C to $30\overline{0}°$C?
19. How many joules of heat are given off when 125 kg of steel cools from 1425°C to 82°C?
20. A 525-kg steam boiler is made of steel and contains 315 kg of water at 40.0°C. Assuming that 75% of the heat is delivered to the boiler and water, how many kilocalories are required to raise the temperature of both the boiler and water to 100.0°C?
21. Find the initial temperature of a 49.0-N cube of zinc, 16.0 cm on a side, that gives off 3.36×10^5 J of heat while cooling to 80.0°C.
22. A coolant lowers the temperature 13°C in a steel engine weighing 16,250 N and running at a temperature of $11\overline{0}°$C. What is the heat reduction (in joules) in the steel engine?
23. A block of iron with mass 0.400 kg is heated to 325°C from 295°C. How much heat is absorbed by the iron?
24. A block of copper is heated from 20.0°C to 80.0°C. How much heat is absorbed by the copper if its mass is 60.0 g?
25. The cooling system of a truck engine contains 20.0 L of water. (1 L of water has a mass of 1 kg.) (a) If the engine is run until 845 kJ of heat is added, what is the change in temperature of the water? (b) In the winter, the system was filled with 20.0 L of ethyl alcohol with density 0.800 g/cm³. If the ethyl alcohol absorbed the same 845 kJ of heat, what would be the increase in temperature of the alcohol? (c) Would ethyl alcohol or water be a better coolant? Why?

14.5 Method of Mixtures

When two substances at different temperatures are mixed together, heat flows from the warmer body to the cooler body until they reach the same temperature (Fig. 14.12). This is known as thermal equilibrium or the **method of mixtures**. Part of the heat lost by the warmer body is transferred to the cooler body and part is lost to the surrounding objects or the air. In most cases almost all the heat is transferred to the cooler body. We assume here that all the heat lost by the warmer body equals the heat gained by the cooler body. The amount of heat lost or gained by a body is

Figure 14.12 Heat flows from the warmer substance to the cooler.

$$Q = cm\Delta T \quad \text{or} \quad Q = cw\Delta T$$

By formula,

$$Q_{\text{lost}} = Q_{\text{gained}}$$
$$c_l m_l(T_l - T_f) = c_g m_g(T_f - T_g)$$

where the subscript *l* refers to the warmer body, which *loses* heat; the subscript *g* refers to the cooler body, which *gains* heat; and T_f is the final temperature of the mixture.

A 10.0-lb piece of hot copper is dropped into 30.0 lb of water at $5\overline{0}°F$. If the final temperature of the mixture is 65°F, what was the initial temperature of the copper?

EXAMPLE 1

Data:

$w_l = 10.0$ lb	$w_g = 30.0$ lb
$c_l = 0.093$ Btu/lb °F	$c_g = 1.00$ Btu/lb °F
$T_l = ?$	$T_g = 5\overline{0}°F$
$T_f = 65°F$	

Basic Equation:

$$c_l w_l(T_l - T_f) = c_g w_g(T_f - T_g)$$

Working Equation:

$$T_l = \frac{c_g w_g}{c_l w_l}(T_f - T_g) + T_f$$

Substitution:

$$T_l = \frac{(1.00 \text{ Btu/lb °F})(30.0 \text{ lb})}{(0.093 \text{ Btu/lb °F})(10.0 \text{ lb})}(65°F - 5\overline{0}°F) + 65°F$$

$$= 550°F$$

Some find it easier to find T_l using a second method. Substitute the data directly into the basic equation. Then solve for T_l as follows:

$$\left(0.093\frac{\text{Btu}}{\text{lb °F}}\right)(10.0 \text{ lb})(T_l - 65°F) = \left(1.00\frac{\text{Btu}}{\text{lb °F}}\right)(30.0 \text{ lb})(65°F - 5\overline{0}°F)$$

$$0.93T_l \text{ Btu/°F} - 6\overline{0} \text{ Btu} = 450 \text{ Btu}$$

$$0.93T_l \text{ Btu/°F} = 510 \text{ Btu}$$

$$T_l = \frac{510 \text{ Btu}}{0.93 \text{ Btu/°F}}$$

$$T_l = 550°F$$

If $20\overline{0}$ g of steel at $22\overline{0}°C$ is added to $50\overline{0}$ g of water at 10.0°C, find the final temperature of this mixture.

EXAMPLE 2

Data:

$c_l = 0.115$ cal/g °C	$c_g = 1.00$ cal/g °C
$m_l = 20\overline{0}$ g	$m_g = 50\overline{0}$ g
$T_l = 22\overline{0}°C$	$T_g = 10.0°C$
	$T_f = ?$

Basic Equation:

$$c_l m_l (T_l - T_f) = c_g m_g (T_f - T_g)$$

Working Equation:

$$T_f = \frac{c_l m_l T_l + c_g m_g T_g}{c_l m_l + c_g m_g}$$

Substitution:

$$T_f = \frac{(0.115 \text{ cal/g °C})(20\overline{0} \text{ g})(22\overline{0}°C) + (1.00 \text{ cal/g °C})(50\overline{0} \text{ g})(10.0°C)}{(0.115 \text{ cal/g °C})(20\overline{0} \text{ g}) + (1.00 \text{ cal/g °C})(50\overline{0} \text{ g})}$$

$$= 19.2°C$$

To find T_f by the second method, substitute the data directly into the basic equation. Then, solve for T_f as follows:

$$\left(0.115 \frac{\text{cal}}{\text{g °C}}\right)(20\overline{0} \text{ g})(22\overline{0}°C - T_f) = \left(1.00 \frac{\text{cal}}{\text{g °C}}\right)(50\overline{0} \text{ g})(T_f - 10.0°C)$$

$$5060 \text{ cal} - 23.0 \frac{\text{cal}}{°C} T_f = 50\overline{0} \frac{\text{cal}}{°C} T_f - 50\overline{0}0 \text{ cal}$$

$$10,060 \text{ cal} = 523 \frac{\text{cal}}{°C} T_f$$

$$\frac{10,060 \text{ cal}}{523 \text{ cal/°C}} = T_f$$

$$19.2°C = T_f$$

. . . .

SKETCH

12 cm² | w

4.0 cm

DATA
$A = 12$ cm², $l = 4.0$ cm, $w = ?$

BASIC EQUATION
$A = lw$

WORKING EQUATION
$w = \dfrac{A}{l}$

SUBSTITUTION
$w = \dfrac{12 \text{ cm}^2}{4.0 \text{ cm}} = 3.0 \text{ cm}$

PROBLEMS 14.5

Refer to Table 15 of Appendix C.

1. A 2.50-lb piece of steel is dropped into 11.0 lb of water at 75.0°F. The final temperature is 84.0°F. What was the initial temperature of the steel?
2. Mary mixes 5.00 lb of water at 20\overline{0}°F with 7.00 lb of water at 65.0°F. Find the final temperature of the mixture.
3. A 25\overline{0}-g piece of tin at 99°C is dropped in 10\overline{0} g of water at 1\overline{0}°C. If the final temperature of the mixture is 2\overline{0}°C, what is the specific heat of the tin?
4. How many grams of water at 2\overline{0}°C are necessary to change 80\overline{0} g of water at 9\overline{0}°C to 5\overline{0}°C?
5. A 159-lb piece of aluminum at 50\overline{0}°F is dropped into 40\overline{0} lb of water at 6\overline{0}°F. What is the final temperature?
6. A 42.0-lb piece of steel at 67\overline{0}°F is dropped into 10\overline{0} lb of water at 75.0°F. What is the final temperature of the mixture?
7. If 1250 g of copper at 20.0°C is mixed with 50\overline{0} g of water at 95.0°C, find the final temperature of the mixture.
8. If 50\overline{0} g of brass at 20\overline{0}°C and 30\overline{0} g of steel at 15\overline{0}°C are added to 90\overline{0} g of water in an aluminum pan of mass 15\overline{0} g both at 20.0°C, find the final temperature, assuming no loss of heat to the surroundings.
9. The following data were collected in the laboratory to determine the specific heat of an unknown metal:

Mass of copper calorimeter	153 g
Specific heat of calorimeter	0.092 cal/g °C
Mass of water	275 g

Specific heat of water	1.00 cal/g °C
Mass of metal	236 g
Initial temperature of water and calorimeter	16.2°C
Initial temperature of metal	99.6°C
Final temperature of calorimeter, water, and metal	22.7°C

Find the specific heat of the unknown metal. *Note:* A calorimeter is usually a metal cup inside another metal cup that is insulated by the air between them (Fig. 14.13).

Figure 14.13   Apparatus for measuring the specific heat of a metal by the method of mixtures

10. The following data were collected in the laboratory to determine the specific heat of an unknown metal:

Mass of aluminum calorimeter	132 g
Specific heat of calorimeter	$92\overline{0}$ J/kg °C
Mass of water	285 g
Specific heat of water	4190 J/kg °C
Mass of metal	215 g
Initial temperature of water and calorimeter	12.6°C
Initial temperature of metal	99.1°C
Final temperature of calorimeter, water, and metal	18.6°C

Find the specific heat of the unknown metal.

11. Determine the original temperature of a $56\overline{0}$-g piece of lead placed in a 165-g brass calorimeter that contains 325 g of water. The initial temperature of the water and calorimeter was 18.0°C. The final temperature of the lead, calorimeter, and water is 31.0°C.

12. How much heat must be absorbed by its surroundings to cool a 565-g cube of iron from 100.0°C to 20.0°C?

13. How much water at 0°C would be needed to cool the iron in Problem 12 to 20.0°C?

14. The specific heat of water is 4190 J/kg °C. The specific heat of steel is 481 J/kg °C. Why may you burn your tongue on hot coffee but not on the spoon when both are at the same temperature?

14.6 Expansion of Solids

Most solids expand when heated and contract when cooled. They expand or contract in all three dimensions—length, width, and thickness. When a solid is heated, the expansion is due to the increased length of the vibrations of the atoms and molecules. This results in the solid expanding in all directions. This increase in volume results in a decrease in weight density, which was discussed in Chapter 12. Engineers, technicians, and designers must know the effects of thermal expansion. You have no doubt heard of highway pavements buckling on a hot summer day (Fig. 14.14). Bridges are built with special joints that allow for expansion and contraction of the bridge deck (Fig. 14.15). Similarly, TV towers, pipelines, and buildings must be designed and built to allow for this expansion and contraction.

Figure 14.14   Thermal expansion causes pavement to sometimes break up (buckle) in the summer. Air temperatures in the mid-90s (°F) or low 30s (°C) can easily translate into pavement temperatures in the 125°F or 52°C range. As the pavement absorbs heat, it expands and can reach the point where the concrete buckles across the traffic lanes.

Courtesy of Louis Haasis, District 6 Operations Engineer, Illinois Department of Transportation, Springfield, IL

There are some advantages to solids expanding. A bimetallic strip is made by fusing two different metals together side by side as illustrated in Fig. 14.16. When heated, the brass expands more than the steel, which makes the strip curve. If the bimetallic

Figure 14.15   Thermal expansion joint on a bridge

Courtesy of fotolia © peacepics

Figure 14.16   When the thin bimetal strip is heated, it bends because of unequal expansion of the two metals. Here brass expands more than steel when heated, making the strip bend toward the steel side.

Courtesy of Frank LaBua © Pearson Education/PH College

strip is cooled below room temperature, the brass will contract more than the steel, forcing the strip to curve in the opposite direction. The thermostat operates on this principle. As shown in Fig. 14.17, the basic parts of a thermostat are a bimetallic strip on the right and a regular metal strip on the left. The bimetallic strip of brass and steel bends with the temperature. The regular metal strip is moved by hand to set the temperature desired. This particular bimetallic strip is made and placed so that it bends to the left when cooled. As a result, when it comes in contact with the strip on the left, it completes a circuit, which turns on the furnace. When the room warms to the desired temperature, the bimetallic strip moves back to the right, which opens the contacts and shuts off the heat. Bimetallic strips are in spiral form in some thermostats (Fig. 14.18).

Figure 14.17 Simple thermostat

(a) Circuit open   (b) Circuit closed

Figure 14.18 These inner workings of a thermostat show the bimetal coil that expands and contracts as the temperature changes to activate the mercury bulb switch.

Linear Expansion

The amount that a solid expands depends on the following:

1. *Material.* Different materials expand at different rates. Steel expands at a rate less than that of brass.
2. *Length of the solid.* The longer the solid, the larger is the expansion. A 20.0-cm steel rod will expand twice as much as a 10.0-cm steel rod.
3. *Amount of change in temperature.* The greater the change in temperature, the greater is the expansion.

This can be written as a formula:

$$\Delta l = \alpha \, l \Delta T$$

where Δl = change in length
    α = a constant called the **coefficient of linear expansion***
    l = original length
    ΔT = change in temperature

Table 14.3 lists the coefficients of linear expansion for some common solids.

Table 14.3   Coefficients of Linear Expansion

Material	α (metric)	α (U.S.)
Aluminum	$2.3 \times 10^{-5}/\text{C}°$	$1.3 \times 10^{-5}/\text{F}°$
Brass	$1.9 \times 10^{-5}/\text{C}°$	$1.0 \times 10^{-5}/\text{F}°$
Concrete	$1.1 \times 10^{-5}/\text{C}°$	$6.0 \times 10^{-6}/\text{F}°$
Copper	$1.7 \times 10^{-5}/\text{C}°$	$9.5 \times 10^{-6}/\text{F}°$
Glass	$9.0 \times 10^{-6}/\text{C}°$	$5.1 \times 10^{-6}/\text{F}°$
Pyrex	$3.0 \times 10^{-6}/\text{C}°$	$1.7 \times 10^{-6}/\text{F}°$
Steel	$1.3 \times 10^{-5}/\text{C}°$	$6.5 \times 10^{-6}/\text{F}°$
Zinc	$2.6 \times 10^{-5}/\text{C}°$	$1.5 \times 10^{-5}/\text{F}°$

*Defined as change in unit length of a solid when its temperature is changed 1 degree.

Comparing the coefficients of linear expansion of common glass and Pyrex, we can see that Pyrex expands and contracts approximately one third as much as glass. This is why it is used in cooking and chemical laboratories.

EXAMPLE 1

A steel railroad rail is 40.0 ft long at 0°F. How much will it expand when heated to $10\overline{0}$°F?

Data:

$$l = 40.0 \text{ ft}$$
$$\Delta T = 10\overline{0}°F - 0°F = 10\overline{0}°F$$
$$\alpha = 6.5 \times 10^{-6}/F°$$
$$\Delta l = ?$$

Basic Equation:

$$\Delta l = \alpha l \Delta T$$

Working Equation: Same

Substitution:

$$\Delta l = (6.5 \times 10^{-6}/F°)(40.0 \text{ ft})(10\overline{0}°F)$$
$$= 0.026 \text{ ft} \quad \text{or} \quad 0.31 \text{ in.}$$

· · · · · · · · · · · · · · · · ·

Pipes that undergo large temperature changes are installed to allow for expansion and contraction (Fig. 14.19).

Figure 14.19 (a) The zigzag arrangement of these geothermal pipes in Iceland protects the integrity of the pipes despite major temperature variations.

Figure 14.19 (b) Another arrangement of geothermal pipes allows for their expansion and contraction as the temperature changes.

(a)

(b)

Courtesy of fotolia © Graham Prentice

EXAMPLE 2

What allowance for expansion must be made for a steel pipe $12\overline{0}$ m long that handles coolants and must undergo temperature changes of $20\overline{0}$°C?

Data:

$$\alpha = 1.3 \times 10^{-5}/C°$$
$$l = 12\overline{0} \text{ m}$$
$$\Delta T = 20\overline{0}°C$$
$$\Delta l = ?$$

Basic Equation:

$$\Delta l = \alpha l \Delta T$$

Working Equation: Same

Substitution:

$$\Delta l = (1.3 \times 10^{-5}/\text{C}°)(12\overline{0} \text{ m})(20\overline{0} °\text{C})$$
$$= 0.31 \text{ m} \quad \text{or} \quad 31 \text{ cm}$$

Area and Volume Expansion of Solids

Solids expand in width and thickness as well as in length when heated. The area of a hole cut out of a metal sheet will expand in the same way as the surrounding material. To allow for this expansion the following formulas are used:

> Area expansion: $\Delta A = 2\alpha A \, \Delta T$
> Volume expansion: $\Delta V = 3\alpha V \, \Delta T$

where A = original area
       V = original volume

The top of a circular copper disk has an area of 64.2 in² at $2\overline{0}°$F. What is the change in area when the temperature is increased to $15\overline{0}°$F?

EXAMPLE 3

Data:

$$\alpha = 9.5 \times 10^{-6}/\text{F}°$$
$$A = 64.2 \text{ in}^2$$
$$\Delta T = 15\overline{0}°\text{F} - 2\overline{0}°\text{F} = 13\overline{0}°\text{F}$$
$$\Delta A = ?$$

Basic Equation:

$$\Delta A = 2\alpha A \, \Delta T$$

Working Equation: Same

Substitution:

$$\Delta A = 2(9.5 \times 10^{-6}/\text{F}°)(64.2 \text{ in}^2)(13\overline{0}° \text{ F})$$
$$= 0.16 \text{ in}^2$$

A section of concrete in a bridge support measures 6.00 m × 12.0 m × 30.0 m at 38°C. What allowance for change in volume is necessary for a temperature drop to −15°C?

EXAMPLE 4

Data:

$$V = (6.00 \text{ m})(12.0 \text{ m})(30.0 \text{ m}) = 2160 \text{ m}^3$$
$$\alpha = 1.1 \times 10^{-5}/\text{C}°$$
$$\Delta T = 38°\text{C} - (-15°\text{C}) = 53°\text{C}$$

Basic Equation:

$$\Delta V = 3\alpha V \Delta T$$

Working Equation: Same

Substitution:

$$\Delta V = 3(1.1 \times 10^{-5}/\mathrm{C}°)(2160 \text{ m}^3)(53°\text{C})$$
$$= 3.8 \text{ m}^3$$

· · · · · · · · · · · · · · · · · ·

SKETCH

12 cm² | w

4.0 cm

DATA

$A = 12$ cm², $l = 4.0$ cm, $w = ?$

BASIC EQUATION

$A = lw$

WORKING EQUATION

$w = \dfrac{A}{l}$

SUBSTITUTION

$w = \dfrac{12 \text{ cm}^2}{4.0 \text{ cm}} = 3.0$ cm

PROBLEMS 14.6

1. Find the increase in length of copper tubing 200.0 ft long at 40.0°F when it is heated to 200.0°F.
2. Find the increase in length of a zinc rod 50.0 m long at 15.0°C when it is heated to 130.0°C.
3. Find the increase in length of 300.00 m of copper wire when its temperature changes from 14°C to 34°C.
4. A steel pipe 8.25 m long is installed at 45°C. Find the decrease in length when coolants at −6$\overline{0}$°C pass through the pipe.
5. A steel tape measures 200.00 m at 15°C. What is its length at 55°C?
6. A brass rod 1.020 m long expands 3.0 mm when it is heated. Find the temperature change.
7. The road bed on a bridge 500.0 ft long is made of concrete. What allowance is needed for temperatures of −4$\overline{0}$°F in winter and 14$\overline{0}$°F in summer?
8. An aluminum plug has a diameter of 10.003 cm at 40.0°C. At what temperature will it fit precisely into a hole of constant diameter 10.000 cm?
9. The diameter of a steel drill at 45°F is 0.750 in. Find its diameter at 375°F.
10. A brass ball with diameter 12.000 cm is 0.011 cm too large to pass through a hole in a copper plate when the ball and plate are at a temperature of 20.0°C. What is the temperature of the ball when it will just pass through the plate, assuming that the temperature of the plate does not change? What is the temperature of the plate when the ball will just pass through, assuming that the temperature of the ball does not change?
11. A brass cylinder has a cross-sectional area of 482 cm² at −5°C. Find its change in area when heated to 95°C.
12. The volume of the cylinder in Problem 11 is 4820 cm³ at 240.0°C. Find its change in volume when cooled to −75.0°C.
13. An aluminum pipe has a cross-sectional area of 88.40 cm² at 15°C. What is its cross-sectional area when the pipe is heated to 155°C?
14. A steel pipe has a cross-sectional area of 127.20 in² at 25°F. What is its cross-sectional area when the pipe is heated to 175°F?
15. A glass plug has a volume of 60.00 cm³ at 12°C. What is its volume at 76°C?
16. The diameter of a hole drilled through brass at 21°C measures 6.500 cm. Find the diameter and area of the hole when the brass is heated to 175°C.
17. Steel rails 15.000 m long are laid at 10.0°C. How much space should be left between them if they are to just touch at 35.0°C?
18. Steel beams 60.000 ft long are placed in a highway overpass to allow for expansion and contraction. The temperature range allowance is −3$\overline{0}$°F to 13$\overline{0}$°F. (a) Find the space allowance (in inches) between the beams at −3$\overline{0}$°F if the beams touch at 13$\overline{0}$°F. (b) Find the space allowance between the beams if placed at 75°F and touch at 13$\overline{0}$°F.

19. The spaces between 13.00-m steel rails are 0.711 cm at −15°C. If the rails touch at 35.5°C, what is the coefficient of linear expansion?
20. A section of concrete dam is a rectangular solid 20.0 ft × 50.0 ft × 80.0 ft at 115°F. What allowance for change in volume is necessary for a temperature of −15°F?
21. A glass ball has a radius of 12.000 cm at 6.0°C. Find its change in volume when the temperature is increased to 81.0°C.
22. Find the final height of a concrete column that is 1.250 m × 1.250 m × 4.250 m at 0.0°C when the column is heated to 45.0°C.
23. What is the final volume of a glass right circular cylinder with original height 1.200 m and radius 30.00 cm that is heated from 13.0°C to 56.0°C?
24. A metal bar at 21.0°C is 2.6000 m long. If the bar is heated to 93.0°C, its change in length is 3.40 mm. What is the coefficient of linear expansion of the bar?

14.7 Expansion of Liquids

Liquids also generally expand when heated and contract when cooled. The thermometer is made using this principle. When a glass thermometer is placed under your tongue, the heat from your mouth causes the liquid in the bottom of the thermometer to expand. The liquid is then forced to rise up the thin calibrated tube (Fig. 14.20). Similarly, when the gasoline tank on a car is filled to capacity on a hot summer day and then parked in a hot parking lot, it overflows. The cold gas from the underground storage tanks expands as it warms up in the car's tank. An automobile radiator filled to the brim with cold water would likewise overflow as the engine heats the water. The formula for *volume expansion of liquids* is

Figure 14.20 Liquid expansion in a thermometer

Courtesy of fotolia © rachwai

$$\Delta V = \beta V \, \Delta T$$

where β = coefficient of volume expansion for liquids
       V = original volume

Table 14.4 lists the coefficients of volume expansion for some common liquids.

Table 14.4 Coefficients of Volume Expansion

Liquid	β (metric)	β (U.S.)
Acetone	$1.49 \times 10^{-3}/C°$	$8.28 \times 10^{-4}/F°$
Alcohol, ethyl	$1.12 \times 10^{-3}/C°$	$6.62 \times 10^{-4}/F°$
Carbon tetrachloride	$1.24 \times 10^{-3}/C°$	$6.89 \times 10^{-4}/F°$
Mercury	$1.8 \times 10^{-4}/C°$	$1.0 \times 10^{-4}/F°$
Petroleum	$9.6 \times 10^{-4}/C°$	$5.33 \times 10^{-4}/F°$
Turpentine	$9.7 \times 10^{-4}/C°$	$5.39 \times 10^{-4}/F°$
Water	$2.1 \times 10^{-4}/C°$	$1.17 \times 10^{-4}/F°$

EXAMPLE 1

If petroleum at 0°C occupies $25\overline{0}$ L, what is its volume at $5\overline{0}$°C?

Data:

$$\beta = 9.6 \times 10^{-4}/C°$$
$$V = 25\overline{0} \text{ L}$$
$$\Delta T = 5\overline{0}°C$$
$$\Delta V = ?$$

Basic Equation:

$$\Delta V = \beta V \Delta T$$

Working Equation: Same

Substitution:

$$\Delta V = (9.6 \times 10^{-4}/\mathrm{C}°)(25\overline{0}\text{ L})(5\overline{0}°\mathrm{C})$$
$$= 12\text{ L}$$
$$\text{volume at } 5\overline{0}°\text{C} = V + \Delta V$$
$$= 25\overline{0}\text{ L} + 12\text{ L} = 262\text{ L}$$

. . . .

EXAMPLE 2

Find the increase in volume of 18.2 in³ of water when the water is heated from $4\overline{0}°$F to $18\overline{0}°$F.

Data:

$$\beta = 1.17 \times 10^{-4}/\text{F}°$$
$$V = 18.2\text{ in}^3$$
$$\Delta T = 18\overline{0}°\text{F} - 4\overline{0}°\text{F} = 14\overline{0}°\text{F}$$
$$\Delta V = ?$$

Basic Equation:

$$\Delta V = \beta V \Delta T$$

Working Equation: Same

Substitution:

$$\Delta V = (1.17 \times 10^{-4}/\text{F}°)(18.2\text{ in}^3)(14\overline{0}°\text{ F})$$
$$= 0.298\text{ in}^3$$

. . . .

Figure 14.21 Expansion of water in change from liquid to solid

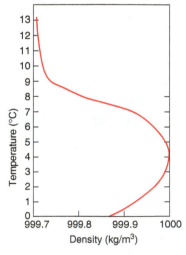

Courtesy of Clive Streeter
© Dorling Kindersley

Figure 14.22 Change in density of water with change in temperature

Density (kg/m³)

Expansion of Water

Water is unusual in its expansion characteristics. Recall the mound in the middle of each ice cube in an ice cube tray (Fig. 14.21). This is evidence of the expansion of water during its change of state from liquid to solid form.

Nearly all liquids are the most dense at their lowest temperature before a change of phase to become solids. As the temperature drops, the molecular motion slows and the substance becomes denser. Water does not follow this general rule. Because of its unusual structural characteristics, water is most dense at 4°C or 39.2°F instead of 0°C or 32°F. A graph of its change in density with increase in temperature is shown in Fig. 14.22. As ice melts and the water temperature is slightly increased, there are still groups of molecules that have the open crystallographic structure of ice, which is less dense than water. As the water is heated to 4°C, these groupings disappear and the water becomes denser. Above 4°C, water then expands normally as the temperature is raised.

This unique behavior of water is critically important in lakes that freeze in winter. If, as in most liquids, water were most dense at its freezing point, the coldest water would settle to the bottom and the lake would freeze from the bottom up. Any living creatures would be killed. Fortunately, the most dense water at 4°C is at the bottom and the less dense water at 0°C is above it, so ice forms at the surface, the water below remains liquid, and the lake freezes from the top down.

When ice melts at 0°C or 32°F, the water formed *contracts* as the temperature is raised to 4°C or 39.2°F. Then it begins to *expand,* as do most other liquids.

PHYSICS CONNECTIONS

Hot Fuel

The U.S. government has been investigating the possibility of regulations on the sale of "hot fuel." Petroleum, like all liquids, expands when its temperature increases; however, the amount of energy in a given amount of fuel does not increase as its volume expands. Fuel pricing is based on a price per gallon at a temperature of 60°F. Fuel at the distribution center is typically pumped from 60°F underground storage tanks into tanker trucks that transport the fuel to local gas stations, where the fuel tends to heat up and expand.

Figure 14.23   On a summer day, the consumer receives less energy per gallon of gas than on a cool day.

During the summer, the temperature of fuel often reaches 90°F as it is stored at local gas stations and as it enters an automobile's gas tank (Fig. 14.23). As a result, a consumer receives 1 gallon of expanded gasoline, which does not have as much energy as contained in the gasoline stored in the 60°F underground storage tanks.

Let's put the formula for the volume expansion of liquids to work. One gallon is equal to 231 in^3. When gas is pumped into a tank at 90°F, the gas becomes less dense and occupies a volume of 235 in^3. However, the pump still only pumps 231 in^3 into the tank, shortchanging the consumer of the energy contained in the missing 4 in^3 of petroleum.

The impact on consumers of the sale of hot fuel is greater in warm climates than in cooler, northern climates. In the United States it is estimated that consumers are short-changed by as much as $1.5 billion over the course of a summer as a result of the sale of hot fuel.

PROBLEMS 14.7

1. A quantity of carbon tetrachloride occupies 625 L at 12°C. Find its volume at 48°C.
2. Some mercury occupies 157 in^3 at $-3\overline{0}$°F. What is its change in volume when heated to $9\overline{0}$°F?
3. Some petroleum occupies 11.7 m^3 at -17°C. Find its volume at 28°C.
4. Find the increase in volume of 35 L of acetone heated from 28°C to 38°C.
5. Some water at $18\overline{0}$°F occupies $378\overline{0}$ ft^3. What is its volume at 122°F?
6. A $120\overline{0}$-L tank of petroleum is completely filled at 9°C. How much spills over if the temperature rises to 45°C?
7. Find the increase in volume of 215 cm^3 of mercury when its temperature increases from $1\overline{0}$°C to 25°C.
8. Find the decrease in volume of $200\overline{0}$ ft^3 of alcohol in a railroad tank car if the temperature drops from 75°F to 54°F.

SKETCH

12 cm^2 *w*

4.0 cm

DATA

$A = 12$ cm^2, $l = 4.0$ cm, $w = ?$

BASIC EQUATION

$A = lw$

WORKING EQUATION

$w = \dfrac{A}{l}$

SUBSTITUTION

$w = \dfrac{12 \text{ cm}^2}{4.0 \text{ cm}} = 3.0 \text{ cm}$

9. A gasoline service station owner receives a truckload of 34,00$\overline{0}$ L of gasoline at 32°C. It cools to 15°C in the underground tank. At 75 cents/L, how much money is lost as a result of the contraction of the gasoline?

10. A Pyrex container is completely filled with 275 cm^3 of mercury at 10.0°C. How much mercury spills over when heated to 75.0°C?

11. What was the temperature of 18$\overline{0}$ mL of acetone before it was heated to 98°C and increased to a volume of 20$\overline{0}$ mL?

12. What is the increase in volume of 12$\overline{0}$0 L of petroleum as it warms from 3.0°C to 25.0°C?

13. Five hundred litres of petroleum at 4.0°C is heated to 30.0°C. What is its increase in volume?

14.8 Change of Phase

Many industries are concerned with a change of phase in the materials they use. In foundries the principal activity is to change solid metals to liquid, pour the liquid metal into molds, and allow it to become solid again (Fig. 14.24). **Change of phase** (sometimes called *change of state*) is a change in a substance from one form of matter (solid, liquid, or gas) to another.

Figure 14.24   Molten iron at about 2900°F is poured from a bucket into an open mold.

Courtesy of fotolia © brozova

Fusion

The change of phase from solid to liquid is called **melting** or **fusion**. The change from liquid to solid is called **freezing** or **solidification**. Most solids have a crystalline structure and a definite melting point at any given pressure. Melting and solidification of these substances occur at the same temperature. For example, water at 0°C (32°F) changes to ice and ice changes to water at the same temperature. There is no temperature change during change of phase. Ice at 0°C changes to water at 0°C. Only a few substances, such as butter and glass, have no particular melting temperature but change phase gradually.

Although there is no temperature change during a change of phase, *there is a transfer of heat*. A melting solid *absorbs* heat and a solidifying liquid *gives off* heat. When 1 g of ice at 0°C melts, it absorbs 8$\overline{0}$ cal of heat. Similarly, when 1 g of water freezes at 0°C, ice at 0°C is produced, and 8$\overline{0}$ cal of heat is released.

When 1 kg of ice at 0°C melts, it absorbs $8\bar{0}$ kcal of heat. Similarly, when 1 kg of water freezes at 0°C, ice at 0°C is produced and $8\bar{0}$ kcal of heat is released.

Or when 1 kg of ice at 0°C melts, it absorbs 335 kilojoules (kJ) of heat. Then when 1 kg of water freezes at 0°C, ice at 0°C is produced and 335 kJ of heat is released.

When 1 lb of ice at 32°F melts, it absorbs 144 Btu of heat. Similarly, when 1 lb of water freezes at 32°F, ice at 32°F is produced and 144 Btu of heat is released.

The amount of heat required to melt 1 g or 1 kg or 1 lb of a substance is called its **heat of fusion,** designated L_f.

$$L_f = \frac{Q}{m} \quad \text{(metric)} \qquad L_f = \frac{Q}{w} \quad \text{(U.S.)}$$

where L_f = heat of fusion (see Table 15 in Appendix C)
Q = quantity of heat
m = mass of substance (metric system)
w = weight of substance (U.S. system)

If 1340 kJ of heat is required to melt 4.00 kg of ice at 0°C into water at 0°C, what is the heat of fusion of water?

EXAMPLE 1

Data:

$$Q = 1340 \text{ kJ}$$
$$m = 4.00 \text{ kg}$$
$$L_f = ?$$

Basic Equation:

$$L_f = \frac{Q}{m}$$

Working Equation: Same

Substitution:

$$L_f = \frac{1340 \text{ kJ}}{4.00 \text{ kg}}$$
$$= 335 \text{ kJ/kg}$$

. . . .

heat of fusion (water) = $8\overline{0}$ cal/g, or $8\overline{0}$ kcal/kg, or 335 kJ/kg, or 144 Btu/lb

A very interesting (and delicious) change-of-phase activity is to make home-made ice cream. A sealed container with a mixture of milk, egg, vanilla, and sugar is submerged in a mixture of rock salt and crushed ice. The salt causes the ice to rapidly melt, which requires heat, while the ice changes phase from solid to liquid. Most of this heat is transferred from the ice cream mixture, which hardens into ice cream.

Vaporization

The change of phase from a liquid to a gas or vapor is called **vaporization**. A kettle of boiling water (Fig. 14.25) vividly shows this change of phase as the steam evaporates and leaves the liquid. Note that vaporization requires that heat be supplied; in this case heat is required to boil the water. The reverse process (change from a gas to a liquid) is called **condensation**. As steam condenses in radiators (Fig. 14.26), large amounts of heat are released.

Figure 14.25 Heat supplied to boiling water changes liquid water into steam—the gas form of water.

Figure 14.26 A large amount of heat is released by condensation of steam in a radiator.

Courtesy of fotolia © nikkytok

At the point of condensation, the vapor becomes *saturated*; that is, the vapor cannot hold any more moisture. For example, water vapor is always present in some amount in the earth's atmosphere. The weather term **relative humidity** is the ratio

of the actual amount of vapor in the atmosphere to the amount of vapor required to reach 100% of saturation at the existing temperature. As the air temperature decreases without change in pressure or vapor content, the relative humidity increases until it reaches 100% at saturation. The temperature at which saturation is reached is called the **dew point**. Once saturation is reached and the temperature continues to decrease, condensation occurs in the form of dew, fog, mist, clouds, and rain or other forms of precipitation.

While a liquid is boiling, the temperature of the liquid does not change. However, there is a transfer of heat. A liquid being vaporized (boiled) *absorbs* heat. As a vapor condenses, heat is given off.

The amount of heat required to vaporize 1 g or 1 kg or 1 lb of a liquid is called its **heat of vaporization**, designated L_v. So when 1 g of water at $10\overline{0}°C$ changes to steam at $10\overline{0}°C$, it absorbs $54\overline{0}$ cal; when 1 g of steam at $10\overline{0}°C$ condenses to water at $10\overline{0}°C$, $54\overline{0}$ cal of heat is given off. The tremendous amount of heat released accounts for the potential for far more serious burns from steam than from hot water.

When 1 kg of water at $10\overline{0}°C$ changes to steam at $10\overline{0}°C$, it absorbs $54\overline{0}$ kcal of heat. Similarly, when 1 kg of steam at $10\overline{0}°C$ condenses to water at $10\overline{0}°C$, $54\overline{0}$ kcal of heat is given off.

Or when 1 kg of water at $10\overline{0}°C$ changes to steam at $10\overline{0}°C$, it absorbs 2.26 MJ $(2.26 \times 10^6$ J) of heat. Then, when 1 kg of steam at $10\overline{0}°C$ condenses to water at $10\overline{0}°C$, 2.26 MJ of heat is given off.

When 1 lb of water at 212°F changes to steam at 212°F, $97\overline{0}$ Btu of heat is absorbed; when 1 lb of steam at 212°F condenses to water at 212°F, $97\overline{0}$ Btu of heat is given off.

$$L_v = \frac{Q}{m} \quad \text{(metric)} \qquad L_v = \frac{Q}{w} \quad \text{(U.S.)}$$

where L_v = heat of vaporization (see Table 15 in Appendix C)
  Q = quantity of heat
  m = mass of substance (metric system)
  w = weight of substance (U.S. system)

EXAMPLE 2

If 135,000 cal of heat is required to vaporize $25\overline{0}$ g of water at $10\overline{0}°C$, what is the heat of vaporization of water?

Data:

$$Q = 135,000 \text{ cal}$$
$$m = 25\overline{0} \text{ g}$$
$$L_v = ?$$

Basic Equation:

$$L_v = \frac{Q}{m}$$

Working Equation: Same

Substitution:

$$L_v = \frac{135,000 \text{ cal}}{25\overline{0} \text{ g}}$$
$$= 54\overline{0} \text{ cal/g}$$

heat of vaporization (water) = $54\overline{0}$ cal/g, or $54\overline{0}$ kcal/kg, or 2.26 MJ/kg, or $97\overline{0}$ Btu/lb

EXAMPLE 3

If 15.8 MJ of heat is required to vaporize 18.5 kg of ethyl alcohol at 78.5°C (its boiling point), what is the heat of vaporization of ethyl alcohol?

Data:

$$Q = 15.8 \text{ MJ}$$
$$m = 18.5 \text{ kg}$$
$$L_v = ?$$

Basic Equation:

$$L_v = \frac{Q}{m}$$

Working Equation: Same

Substitution:

$$L_v = \frac{15.8 \text{ MJ}}{18.5 \text{ kg}}$$
$$= 0.854 \text{ MJ/kg or } 854 \text{ kJ/kg or } 8.54 \times 10^5 \text{ J/kg}$$

Figures 14.27 through 14.29 show the heat gained by one unit of ice at a temperature below its melting point as it warms to its melting point, changes to water, warms to its boiling point, changes to steam, and then is heated above its boiling point in joules, Btu, and calories. Note that during each change of phase there is no temperature change. Recall the basic shape of these graphs because we will use it to find the amount of heat gained or lost when a quantity of material goes through one or both changes of phase. Refer to Fig. 14.30 to do such problems. See Table 15 of Appendix C for the heat constants of some common substances.

Figure 14.27

Heat gained by one kilogram of ice at –20°C as it is converted to steam at 120°C

Figure 14.28

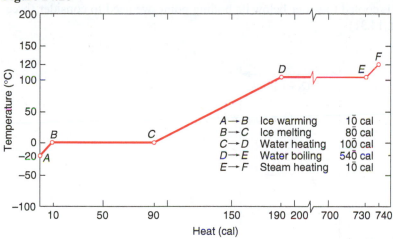

Heat gained by one gram of ice at –20°C as it is converted to steam at 120°C

Figure 14.29

Heat gained by one pound of ice at –20°F as it is converted to steam at 250°F

Figure 14.30   Graph of heat transfer during change of phase

EXAMPLE 4

How many Btu of heat are released when 4.00 lb of steam at 222°F is cooled to water at 82°F?

To find the amount of heat released when steam at a temperature above its vaporization point is cooled to water below its boiling point, we need to consider three amounts (see Fig. 14.31):

Figure 14.31

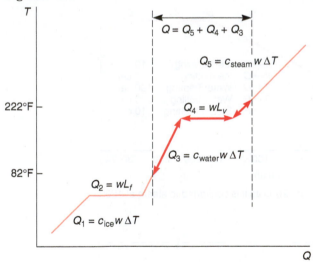

$Q_5 = c_{\text{steam}} w \Delta T$ (amount of heat released as the steam changes temperature from 222°F to 212°F)

$Q_4 = w L_v$ (amount of heat released as the steam changes to water)

$Q_3 = c_{\text{water}} w \Delta T$ (amount of heat released as the water changes temperature from 212°F to 82°F)

So the total amount of heat released is

$$Q = Q_5 + Q_4 + Q_3$$

Data:

$$w = 4.00 \text{ lb}$$
$$T_i \text{ of steam} = 222°F$$
$$T_f \text{ of water} = 82°F$$
$$Q = ?$$

Basic Equation:

$$Q = Q_5 + Q_4 + Q_3$$

Working Equation:

$$Q = c_{\text{steam}} w \Delta T + w L_v + c_{\text{water}} w \Delta T$$

Substitution:

$$Q = \left(0.48 \frac{\text{Btu}}{\text{lb} \,°\text{F}}\right)(4.00 \text{ lb})(10°\text{F}) + (4.00 \text{ lb})\left(970 \frac{\text{Btu}}{\text{lb}}\right)$$

$$+ \left(1.00 \frac{\text{Btu}}{\text{lb} \,°\text{F}}\right)(4.00 \text{ lb})(130°\text{F})$$

$$= 4420 \text{ Btu}$$

How many joules of heat are needed to change 3.50 kg of ice at −15.0°C to steam at 120.0°C?

EXAMPLE 5

Sketch:

Data:

$$m = 3.50 \text{ kg}$$
$$T_i \text{ of ice} = -15.0°\text{C}$$
$$T_f \text{ of steam} = 120.0°\text{C}$$
$$Q = ?$$

Basic Equation:

$$Q = Q_1 + Q_2 + Q_3 + Q_4 + Q_5$$

Working Equation:

$$Q = c_{\text{ice}} m \Delta T + m L_f + c_{\text{water}} m \Delta T + m L_v + c_{\text{steam}} m \Delta T$$

Substitution:

$$Q = \left(2100\,\frac{J}{kg\,°C}\right)(3.50\text{ kg})(15.0°C) + (3.50\text{ kg})\left(335\,\frac{kJ}{kg}\right) \times \frac{10^3\,J}{1\text{ kJ}} \quad \text{(Change to joules.)}$$

$$+ \left(4190\,\frac{J}{kg\,°C}\right)(3.50\text{ kg})(100.0°C) + (3.50\text{ kg})\left(2.26\,\frac{MJ}{kg}\right) \times \frac{10^6\,J}{1\text{ MJ}}$$

$$+ \left(2\overline{0}00\,\frac{J}{kg\,°C}\right)(3.50\text{ kg})(20.0°C)$$

$$= 1.080 \times 10^7\,J \quad \text{or} \quad 10.80\text{ MJ}$$

PHYSICS CONNECTIONS

Geothermal Electrical Generation

Areas prone to earthquakes or volcanic activity often have the ability to harness geothermal energy and convert it into electrical energy. Seismic activity in these areas causes cracks in the earth's crust that allows molten lava to escape from the core of the earth and make its way up toward the surface. Molten lava escaping from the core sometimes encounters underground water reservoirs and heats the water to high temperatures—enough to cause vaporization.

Figure 14.32   The Geysers geothermal energy plant

Geophysicists drill into such underground reservoirs to take advantage of the tremendous heat that is generated from vaporization. The escaping steam travels up through the well and strikes turbine blades that spin and generate electricity.

One of the world's largest geothermal power plants is The Geysers in northern California (Fig. 14.32). It is located on the edge of the North American tectonic plate and comprises a series of 22 geothermal plants and 350 geothermal wells. At its peak power production, The Geysers generated electrical energy for 1.8 million people in the Northern California electrical grid. However, geophysicists have noticed that electrical generation has decreased since the late 1980s. Scientists believe that the underground reservoirs are beginning to dry up, thus producing less high-pressure steam to spin the turbines in the power plant. Plans are underway to inject water into the underground reservoirs to bring steam levels back to their pre-1980 levels.

One of the most progressive countries with regard to geothermal energy production is Iceland (Fig. 14.33). The government of Iceland has been on the forefront of geothermal electrical production and generates over half of its electrical energy needs from this source. Iceland has taken advantage of its location on top of the area where the North American tectonic plate meets the Eurasian tectonic plate. Seismic activity between these plates has resulted in the formation of natural geothermal structures such as geysers, hot springs, and volcanoes throughout the island country. Policy makers and scientists have taken advantage of this abundant, alternative, renewable energy source.

Figure 14.33   Swimmers relax in the "Blue Lagoon" runoff pond at the Svartsengi geothermal power plant in Iceland.

Evaporation as a Cooling Process

Evaporation is the process by which high-energy molecules of a liquid continually leave its surface. This change of a liquid to a gas helps keep your body cool. When you become too warm, your sweat glands produce water, which evaporates from your skin. As the water evaporates, your body loses heat at the rate of $\frac{Q}{m} = L_v$. In a cool summer breeze, the perspiration evaporates more rapidly, which cools you faster. On a hot, humid day, you tend to remain hot because the perspiration does not evaporate as quickly.

Years ago when a person, especially a child, had a very high body temperature, the common medical practice was to rub the body with rubbing alcohol because it quickly evaporates from the skin. As it evaporates, it removes heat from the body. This practice is not recommended now because we know the alcohol is also absorbed in the body.

During evaporation the molecules of a liquid continually leave the surface of the liquid. Some molecules have enough energy to leave, freeing themselves from the liquid's surface. The rest, not having enough energy to leave, fall back and remain as part of the liquid. The rate of evaporation of a liquid depends on the following:

1. *Amount of surface area.*   The larger the surface area, the greater is the number of molecules that have a chance to escape from the surface.
2. *Temperature.*   The higher the temperature, the higher is the molecular energy of the molecules, which allows more molecules to escape.
3. *Surface currents.*   Air currents blowing over the liquid's surface remove many of the molecules that have evaporated before they fall back into the liquid, which is why a cool summer breeze feels so good.
4. *Volatility.*   The **volatility** of a liquid is a measure of its ability to vaporize. Examples of highly volatile liquids are rubbing alcohol and gasoline. The more volatile the liquid, the greater is its rate of evaporation.
5. *Pressure on or above the liquid.*   The lower the pressure, the greater is the rate of evaporation. Under a partial vacuum, there are fewer molecules available with which the liquid molecules may collide, allowing for a higher rate of escape and a higher rate of evaporation.
6. *Humidity.*   If the liquid is exposed to the atmosphere, lower humidity values will provide for greater evaporation.

Heat Pump

A **heat pump** is a device that warms or cools by transferring heat from a lower-temperature source to a higher-temperature source, or vice versa. It is often used for heating in the winter and cooling in the summer.

A heat pump contains a vapor, usually called a *refrigerant*, that is easily condensed to a liquid when under pressure. The liquid refrigerant gives up the heat gained during compression to the higher-temperature source. The liquid is then released to a low-pressure part of the heat pump, where it quickly evaporates and takes its heat of vaporization from the lower-temperature heat source. The vapor is then compressed again, and the cycle repeats. Work is done on the vapor for the heat pump to transfer the heat from a lower-temperature source to a higher-temperature one.

In winter, a house may be heated by a heat pump that extracts heat from the outside air and transfers it into the house. In summer, the heat pump is reversed and extracts heat from the inside air and transfers it to the outdoors (Fig. 14.34).

Refrigerators and freezers are forms of heat pumps. Heat is transferred from the appliance to the room.

Figure 14.34   The heat pump is used to heat a home in winter and cool it in summer.

Effects of Pressure and Impurities on Change of Phase

Automobile cooling systems present important problems concerning change of phase. Most substances contract on solidifying. However, water and a few other substances expand. The tremendous force exerted by this expansion is shown by the number of cracked automobile blocks and burst radiators suffered by careless motorists every winter.

Impurities in water tend to *lower* the freezing point. Alcohol has a lower freezing point than water and is used in some types of antifreeze. By mixing water with antifreeze in the cooling system, an unknowing person can raise the freezing point of the coolant, allowing it to freeze and crack the engine block. Automobile engines may also be ruined in winter by overheating if the water in the radiator is frozen, preventing the engine from being cooled by circulation in the system.

An increase in the pressure on a liquid *raises* the boiling point. Automobile manufacturers utilize this fact by pressurizing their cooling systems and thereby raising the boiling point of the coolant used.

A decrease in the pressure on a liquid *lowers* the boiling point. Frozen concentrated orange juice is produced by subjecting the pure juice to very low pressures at which the water in the juice is evaporated. Then the consumer must restore the lost water before serving the juice.

SKETCH

12 cm² | w

4.0 cm

DATA

$A = 12 \text{ cm}^2, l = 4.0 \text{ cm}, w = ?$

BASIC EQUATION

$A = lw$

WORKING EQUATION

$w = \dfrac{A}{l}$

SUBSTITUTION

$w = \dfrac{12 \text{ cm}^2}{4.0 \text{ cm}} = 3.0 \text{ cm}$

PROBLEMS 14.8

1. How many calories of heat are required to melt 14.0 g of ice at 0°C?
2. How many pounds of ice at 32°F can be melted by the addition of 635 Btu of heat?
3. How many Btu of heat are required to vaporize 11.0 lb of water at 212°F?
4. How many grams of steam in a boiler at $10\overline{0}$°C can be condensed to water at $10\overline{0}$°C by the removal of 1520 cal of heat?
5. How many calories of heat are required to melt 320 g of ice at 0°C?
6. How many calories of heat are given off when 3250 g of steam is condensed to water at $10\overline{0}$°C?
7. How many joules of heat are required to melt 20.0 kg of ice at 0°C?
8. How many kilocalories of heat are required to melt 20.0 kg of ice at 0°C?
9. How many joules of heat need to be removed to condense 1.50 kg of steam at $10\overline{0}$°C?

10. How many litres of water at 100°C are vaporized by the addition of 5.00 MJ of heat?

11. How many Btu of heat are required to melt 33.0 lb of ice at 32°F and to raise the temperature of the melted ice to 72°F?

12. How many Btu of heat are released when 20.0 lb of water at 80°F is cooled to 32°F and then frozen in an ice plant?

13. How many Btu of heat are required to change 9.00 lb of ice at 10°F to steam at 232°F?

14. How many calories of heat are released when 200 g of steam at 120°C is changed to ice at -12°C?

15. How many kilocalories of heat are required to melt 50.0 kg of ice at 0°C and to raise the temperature of the melted ice to 20°C?

16. How many joules of heat are required to melt 15.0 kg of ice at 0°C and to raise the temperature of the melted ice to 75°C?

17. How many joules of heat need to be removed from 1.25 kg of steam at 115°C to condense it to water and cool the water to 50°C?

18. How many kcal of heat are needed to vaporize 5.00 kg of water at 100°C and raise the temperature of the steam to 145°C?

19. How many calories of heat are needed to change 625 g of ice at -24.0°C to steam at 132.0°C?

20. How many Btu of heat must be withdrawn from 5.65 lb of steam at 236.0°F to change it to ice at 12.0°F?

21. How many kilocalories of heat are needed to change 143 N of ethyl alcohol at 65.0°C to vapor?

22. How many joules of heat does 620 g of mercury require to go from a solid at -38.9°C to vapor?

Glossary

Absolute Zero The lowest possible temperature. (p. 374)

Btu (British thermal unit) The amount of heat (energy) necessary to raise the temperature of 1 lb of water 1°F. (p. 374)

Calorie The amount of heat necessary to raise the temperature of 1 g of water 1°C. (p. 374)

Celsius Scale The metric temperature scale on which ice melts at 0° and water boils at 100°. (p. 371)

Change of Phase (sometimes called *change of state*) A change in a substance from one form of matter (solid, liquid, or gas) to another. (p. 396)

Coefficient of Linear Expansion A constant that indicates the amount by which a solid expands or contracts when its temperature is changed 1 degree. (p. 389)

Condensation The change of phase from gas or vapor to a liquid. (p. 398)

Conduction A form of heat transfer from a warmer part of a substance to a cooler part as a result of molecular collisions, which cause the slower-moving molecules to move faster. (p. 376)

Convection A form of heat transfer by the movement of warm molecules from one region of a gas or a liquid to another. (p. 377)

Dew Point The temperature at which air becomes saturated with water vapor and condensation occurs. (p. 399)

Evaporation The process by which high-energy molecules of a liquid continually leave its surface. (p. 405)

Fahrenheit Scale The U.S. temperature scale on which ice melts at 32° and water boils at 212°. (p. 372)

Freezing The change of phase from liquid to solid. Also called *solidification*. (p. 396)

Fusion The change of phase from solid to liquid. Also called *melting*. (p. 396)

Heat A form of internal kinetic and potential energy contained in an object associated with the motion of its atoms or molecules and which may be transferred from an object at a higher temperature to one at a lower temperature. (p. 374)

Heat of Fusion The heat required to melt 1 g or 1 kg or 1 lb of a substance. (p. 397)

Heat of Vaporization The amount of heat required to vaporize 1 g or 1 kg or 1 lb of a liquid. (p. 399)

Heat Pump A device that warms or cools by transferring heat. It contains a vapor (refrigerant) that is easily condensed to a liquid when under pressure. Produces heat during compression and cooling during vaporization. (p. 405)

Kelvin Scale The metric absolute temperature scale on which absolute zero is 0 K and the units are the same as on the Celsius scale. (p. 372)

Kilocalorie The amount of heat necessary to raise the temperature of 1 kg of water 1°C. (p. 374)

Mechanical Equivalent of Heat The relationship between heat and mechanical work. (p. 375)

Melting The change of phase from solid to liquid. Also called *fusion*. (p. 396)

Method of Mixtures When two substances at different temperatures are mixed together, heat flows from the warmer body to the cooler body until they reach the same temperature. Part of the heat lost by the warmer body is transferred to the cooler body and to surrounding objects. If the two substances are well insulated from surrounding objects, the heat lost by the warmer body is equal to the heat gained by the cooler body. (p. 384)

Radiation A form of heat transfer through energy being radiated or transmitted in the forms of rays, waves, or particles. (p. 378)

Rankine Scale The U.S. absolute temperature scale on which absolute zero is 0°R and the degree units are the same as on the Fahrenheit scale. (p. 372)

Relative Humidity Ratio of the actual amount of vapor in the atmosphere to the amount of vapor required to reach 100% of saturation at the existing temperature. (p. 398)

Solidification The change of phase from liquid to solid. Also called *freezing*. (p. 396)

Specific Heat The amount of heat necessary to change the temperature of 1 kg of a substance by 1°C in the metric system or 1 lb of a substance by 1°F in the U.S. system. (p. 382)

Temperature A measure of the hotness or coldness of an object. (p. 371)

Thermal Conductivity The ability of a material to transfer heat by conduction. (p. 378)

Vaporization The change of phase from liquid to a gas or vapor. (p. 398)

Volatility A measure of a liquid's ability to vaporize. The more volatile the liquid, the greater is its rate of evaporation. (p. 405)

Formulas

14.1 $T_C = \frac{5}{9}(T_F - 32°)$

$T_F = \frac{9}{5}T_C + 32°$

$T_K = T_C + 273$
$T_R = T_F + 46\overline{0}°$

14.3 $Q = \dfrac{KAt(T_2 - T_1)}{L}$

$R = \dfrac{L}{K}$

14.4 $Q = cm\Delta T$
$Q = cw\Delta T$

14.5 $Q_{lost} = Q_{gained}$
$c_l m_l (T_l - T_f) = c_g m_g (T_f - T_g)$

14.6 $\Delta l = \alpha l \Delta T$
$\Delta A = 2\alpha A \Delta T$
$\Delta V = 3\alpha V \Delta T$

14.7 $\Delta V = \beta V \Delta T$

14.8 $L_f = \dfrac{Q}{m}$ $L_f = \dfrac{Q}{w}$

$L_v = \dfrac{Q}{m}$ $L_v = \dfrac{Q}{w}$

Review Questions

1. Which of the following are methods of heat transfer?
   (a) Convection   (b) Conduction   (c) Temperature
   (d) Radiation   (e) Potential energy
2. Which of the following are good conductors of heat?
   (a) Air   (b) Copper   (c) Steel
   (d) Aluminum   (e) Brick   (f) Mineral wool
3. The amount that a solid expands when heated depends on
   (a) the type of material.   (b) the length of the solid.
   (b) the density of the solid.   (d) the amount of temperature change.
   (e) all of the above.
4. The rate of evaporation from the surface of a liquid depends on
   (a) the temperature.   (b) the volatility of the liquid.
   (c) the mass density of the liquid. (d) the air pressure above the liquid.
5. The amount of heat required to melt 1 kg of a solid is called its
   (a) heat of vaporization.   (b) mass density.
   (c) weight density.   (d) heat of fusion.
   (e) volume expansion.
6. The operation of a simple thermostat depends on
   (a) the mechanical equivalent of heat. (b) specific heat.
   (c) thermal expansion.   (d) the R value.
7. In your own words, describe the method of mixtures.
8. What is the mechanical equivalent of heat in the U.S. system?
9. Which other temperature scale is closely related to the Fahrenheit scale?
10. Which other temperature scale is closely related to the Celsius scale?
11. Distinguish between the Celsius and Fahrenheit temperature scales.
12. Distinguish between heat and temperature.
13. Give three examples of the conversion of heat into useful work.
14. Give three examples of the conversion of work into heat.
15. Should you wear light- or dark-colored clothing on a hot, sunny summer day? Explain.
16. Does the area of a hole cut out of a metal block increase or decrease as the metal is heated? Explain.
17. (a) At what temperature does water have its highest density?
    (b) How does water differ from other liquids in this regard?
    (c) What is the impact of this unique characteristic of water?
18. Which would cool a hot object better: 10 kg of water at 0°C or 10 kg of ice at 0°C? Explain.

19. Steam can cause much more severe burns than hot water. Explain.
20. Why are ice cubes often observed to have a slight mound on the top of the cube?
21. In your own words, describe each method of heat transfer.
22. Describe why automotive cooling systems are designed to operate at elevated pressures.
23. Explain how a heat pump works to heat in the winter and cool in the summer.

Review Problems

SKETCH

12 cm² | w

4.0 cm

DATA
$A = 12 \text{ cm}^2, l = 4.0 \text{ cm}, w = ?$

BASIC EQUATION
$A = lw$

WORKING EQUATION
$w = \dfrac{A}{l}$

SUBSTITUTION
$w = \dfrac{12 \text{ cm}^2}{4.0 \text{ cm}} = 3.0 \text{ cm}$

1. Change 344 K to degrees Celsius.
2. Change 24°C to Kelvin.
3. Change 5110°C to degrees Fahrenheit.
4. Change 635°F to degrees Celsius.
5. Find the amount of heat in cal generated by 43.0 J of work.
6. Find the amount of heat in kcal generated by 6530 J of work.
7. Find the amount of work equivalent to 435 Btu.
8. Find the heat flow during 4.10 h through a glass window of thickness 0.15 in. with an area of 33 ft² if the average outer surface temperature is 22°F and the average inner glass surface is 48°F.
9. Find the heat flow in 25.0 days through a freezer door 80.0 cm × 144 cm insulated with cellulose fiber 4.0 cm thick. The temperature inside the freezer is −14°C. Room temperature is 22°C.
10. How many Btu of heat must be added to 835 lb of steel to raise its temperature from 20.0°F to 455°F?
11. How many kcal of heat must be added to 148 kg of aluminum to raise its temperature from 21.5°C to 485°C?
12. A 161-kg steam boiler is made of steel and contains 1127 N of water at 8.9°C. How much heat is required to raise the temperature of both the boiler and the water to $10\overline{0}$°C?
13. A 3.80-lb piece of copper is dropped into 8.35 lb of water at 48.0°F. The final temperature is 98.2°F. What was the initial temperature of the copper?
14. A 355-g piece of metal at 48.0°C is dropped into 111 g of water at 15.0°C. If the final temperature of the mixture is 32.5°C, what is the specific heat of the metal?
15. A brass rod 45.2 cm long expands 0.734 mm when heated. Find the temperature change.
16. The length of a steel rod at 5°C is 12.500 m. What is its length when heated to 154°C?
17. The diameter of a hole drilled through aluminum at 22°C is 7.50 mm. Find the diameter and the area of the hole at 89°C.
18. A steel ball has a radius of 1.54 cm at 35°C. Find its change in volume when the temperature is increased to 84.5°C.
19. Find the increase in volume of 44.8 L of acetone when it is heated from 37.0°C to 75.5°C.
20. What is the decrease in volume of 3450 ft³ of alcohol in a railroad tank car if the temperature drops from 87.0°F to 33.0°F?
21. How many kcal of heat are required to vaporize 21.5 kg of water at $10\overline{0}$°C?
22. How many Btu of heat are required to melt 8.35 lb of ice at 32°F?
23. How many kcal of heat must be withdrawn from 4.56 kg of steam at 125°C to change it to ice at −44.5°C?
24. How many joules of heat are required to change 336 g of ethyl alcohol from a solid at −117°C to a vapor at 78.5°C?

APPLIED CONCEPTS

1. A polystyrene foam cover prevents an ice–water mixture from absorbing heat from 25.3°C air outside a cooler. If the cover is 54.5 cm × 37.8 cm in cross section and 5.25 cm thick, how much heat will transfer through the foam in 60.0 min?

2. Every winter a local recreation department fills a parking lot with water so that the water will freeze and people can ice skate on it. Although the weather does not get cold enough for the water to freeze, the water temperature drops from 4.30°C to 0.00°C. If the water is 12.7 cm deep and the parking lot is 24.5 m × 33.8 m, how much heat energy does the water release into the air?

3. (a) What is the heat of fusion for the water in Problem 2 to freeze into a solid block of ice? (b) What is the total energy needed for the water at 4.30°C to change to ice at 0.00°C? (c) How much additional heat is removed from the ice to lower its temperature to −3.43°C?

4. Pedro, a contractor, is trying to choose between purchasing a steel or an aluminum 50-ft tape measure. (a) If he wants the tape measure that expands and contracts less, which tape measure should he purchase? Consider that the tape measures were calibrated at 68.0°F in the factory. (b) How much can he expect his new tape measure to contract if the temperature outside is 18.5°F?

5. In anticipation of winter snowstorms, Jamal fills his 2.50-gal gas can at the local gas station. (a) If the temperature is 65.3°F on the day he fills the gas can, what volume of gas will Jamal have when the temperature drops to 10.5°F? (b) If the gas cost $1.97/gal, how much money does Jamal lose?

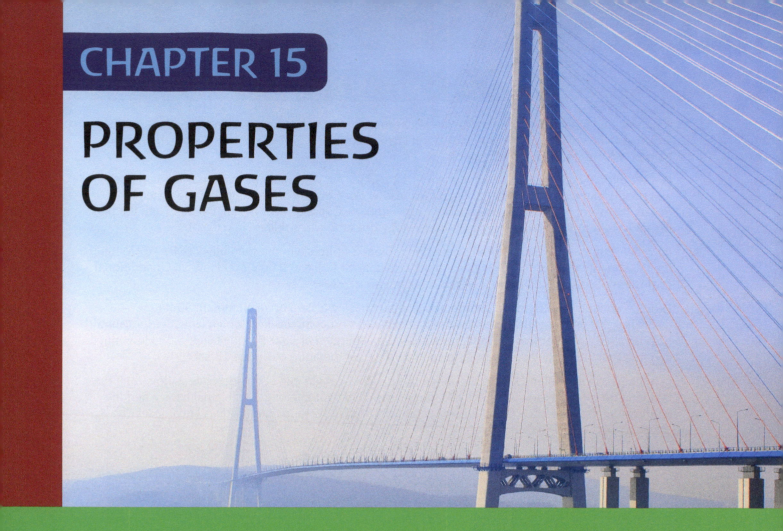

CHAPTER 15

PROPERTIES OF GASES

Objectives

The major goals of this chapter are to enable you to:

1. Use Charles's law to determine thermal expansion.

2. Apply Boyle's law to calculate volume changes of gases.

3. Relate gas density to pressure and temperature.

Properties of gases are related to the temperature and the pressure under which the gas is contained. Gas laws concern the behavior of "ideal" gases at standard conditions of temperature and pressure. The focus of this chapter is on the behavior of gases as temperature and pressure are varied.

15.1 Charles's Law

Before making a long summer trip, you notice that the tires are low. You stop at a gas station around the corner and add air to the 28-lb/in^2 gauge pressure reading. Later in the afternoon you stop for gas. Since your tires were low that morning, you decide to check them again. Now you notice that they look a bit larger, and the gauge pressure is 40 lb/in^2. What happened? When a gas is heated, the increased kinetic energy causes the volume to increase, the pressure to increase, or both to increase.

To study these concepts we will use the idea of an "ideal gas." When the density of a gas is sufficiently low, the pressure, volume, and temperature of the gas tend to be related in a rather simple way. These relationships are fairly accurate over a broad range of temperatures and pressures in real gases. They also apply over a broad range of different gases. We will therefore consider the wide variety of gases and their behaviors by using the ideal gas model. **Charles's and Boyle's laws** mathematically describe these relationships.

CHARLES'S LAW

If the pressure on a gas is constant, the volume is directly proportional* to its absolute (Kelvin or Rankine) temperature.

By formula,

$$\frac{V}{T} = \frac{V'}{T'} \quad \text{or} \quad VT' = V'T$$

where
- V = original volume
- T = original temperature
- V' = final volume
- T' = final temperature

Jacques Charles (1746–1823),

physicist, was born in France. He became famous for making the first manned ascent by hydrogen balloon in 1783. His interest in gases led him to formulate Charles's law.

A gas occupies $45\overline{0}$ cm^3 at $3\overline{0}°$C. At what temperature will the gas occupy $48\overline{0}$ cm^3?

EXAMPLE 1

Data:

$V = 45\overline{0}$ cm^3

$T = 3\overline{0}° + 273 = 303$ K   (**Note:** We must use Kelvin temperature.)

$V' = 48\overline{0}$ cm^3

$T' = ?$

Basic Equation:

$$\frac{V}{T} = \frac{V'}{T'}$$

Working Equation:

$$T' = \frac{TV'}{V}$$

Substitution:

$$T' = \frac{(303 \text{ K})(48\overline{0} \text{ cm}^3)}{45\overline{0} \text{ cm}^3}$$

$$= 323 \text{ K} \quad \text{or} \quad 5\overline{0}°\text{C}$$

·············

---

*"Directly proportional" means that as temperature increases, volume increases, and as temperature decreases, volume decreases.

EXAMPLE 2　→　At $4\bar{0}°F$, some helium occupies 15.0 ft³. What will be its volume at $9\bar{0}°F$?

Data:

$$V = 15.0 \text{ ft}^3$$

$$T = 4\bar{0}° + 46\bar{0}° = 50\bar{0}°R \qquad (\textit{Note: We must use Rankine temperature.})$$

$$T' = 9\bar{0}° + 46\bar{0}° = 55\bar{0}°R$$

$$V' = ?$$

Basic Equation:

$$\frac{V}{T} = \frac{V'}{T'}$$

Working Equation:

$$V' = \frac{VT'}{T}$$

Substitution:

$$V' = \frac{(15.0 \text{ ft}^3)(55\bar{0}°R)}{50\bar{0}°R}$$

$$= 16.5 \text{ ft}^3$$

....

TRY THIS ACTIVITY

Mylar® Balloons

Place a room-temperature, helium-filled Mylar® balloon in a significantly colder environment, such as outside in the cold winter or in a freezer. As the temperature of the helium decreases, observe what happens to the volume of the balloon. After it has been outside several minutes, bring the balloon back into the room-temperature environment. Observe the change that occurs. Then, place the balloon in a warmer area such as near a hot radiator. Why does the volume of the balloon change in different temperature environments?

PROBLEMS 15.1

1. Change 15°C to K.
2. Change −14°C to K.
3. Change 317 K to °C.
4. Change 235 K to °C.
5. Change 72°F to °R.
6. Change −55°F to °R.
7. Change $55\bar{0}°R$ to °F.
8. Change 375°R to °F.

Use $\dfrac{V}{T} = \dfrac{V'}{T'}$ to find each quantity:

9. $T = 315$ K, $V' = 225$ cm³, $T' = 275$ K, find V.
10. $T = 615°R$, $V = 60.3$ in³, $T' = 455°R$, find V'.
11. $V = 20\bar{0}$ ft³, $T' = 95°F$, $V' = 25\bar{0}$ ft³, find T.
12. $V = 19.7$ L, $T = 51°C$, $V' = 25.2$ L, find T'.
13. Some gas occupies a volume of 325 m³ at 41°C. What is its volume at 94°C?
14. Some oxygen occupies 275 in³ at 35°F. Find its volume at 95°F.
15. Some methane occupies 1575 L at 45°C. Find its volume at 15°C.
16. Some helium occupies $120\bar{0}$ ft³ at 70°F. At what temperature will its volume be $60\bar{0}$ ft³?

17. Some nitrogen occupies 14,300 cm³ at 25.6°C. What is the temperature when its volume is 10,250 cm³?

18. Some propane occupies 1270 cm³ at 18.0°C. What is the temperature when its volume is 1530 cm³?

19. Some carbon dioxide occupies 34.5 L at 49.0°C. Find its volume at 12.0°C.

20. Some oxygen occupies 28.7 ft³ at 11.0°F. Find its temperature when its volume is 18.5 ft³.

21. A balloon contains 26.0 L of hydrogen at 40.0°F. What is the Kelvin temperature change needed to make the balloon expand to 36.0 L?

22. Using Charles's law, determine the effect (a) on the temperature of a gas when the volume is doubled, (b) on the temperature of a gas when the volume is tripled, and (c) on the volume when temperature is doubled. (d) Explain the relationship between the volume and the temperature.

23. If 38.0 L of hydrogen is heated to $11\overline{0}$°C and expands to 90.0 L, what was the original temperature?

24. A tank contains 3.00 L of acetylene at 4.00°C. Find the volume at 12.0°C.

25. A hot air balloon contains 147 m³ of air at 19.0°C. What is the volume of air at 32.0°C?

26. A tank with 139 L of propane is cooled from 91.0°C to 37.0°C. Find the original volume.

27. A $200\overline{0}$ L fuel tank filled with propane at 21°C is cooled to 9°C. What is the new volume of propane?

28. A propane nurse tank is left on a job site overnight. If the tank was filled in the heat of the day at 19°C and cools overnight to 3°C, what is the new propane volume if the tank was filled with $140\overline{0}$ litres?

29. A propane tank now containing $25\overline{0}$ L of propane was cooled from 35.0°C to 5.0°C. What was the original volume of the propane?

30. A tank with $50\overline{0}$ L of propane is heated from 17.0°C to 31.0°C. What is the new volume of the propane in the tank?

SKETCH

12 cm²   w
4.0 cm

DATA
$A = 12$ cm², $l = 4.0$ cm, $w = ?$

BASIC EQUATION
$A = lw$

WORKING EQUATION
$w = \dfrac{A}{l}$

SUBSTITUTION
$w = \dfrac{12 \text{ cm}^2}{4.0 \text{ cm}} = 3.0 \text{ cm}$

15.2 Boyle's Law

Next, we study Boyle's law.

BOYLE'S LAW

If the temperature of a gas is constant, the volume is inversely proportional* to the absolute pressure.

By formula,

$$\frac{V}{V'} = \frac{P'}{P} \quad \text{or} \quad VP = V'P'$$

where V = original volume
V' = final volume
P = original pressure
P' = final pressure

Note: The pressure must be expressed in terms of *absolute pressure.*

Robert Boyle (1627–1691),

chemist, was born in Ireland. Working with Robert Hooke as his assistant, he performed experiments on air, vacuums, combustion, and respiration. He developed Boyle's law in 1662. He also did research on properties of acids and alkalis, specific gravity, crystallography, and refraction.

---

*"Inversely proportional" means that as volume increases, pressure decreases, and as volume decreases, pressure increases.

EXAMPLE 1

Some oxygen occupies $50\overline{0}$ in^3 at an absolute pressure of 40.0 lb/in^2 (psi). What is its volume at an absolute pressure of $10\overline{0}$ psi?

Data:

$$V = 50\overline{0} \text{ in}^3$$
$$P = 40.0 \text{ psi}$$
$$P' = 10\overline{0} \text{ psi}$$
$$V' = ?$$

Basic Equation:

$$\frac{V}{V'} = \frac{P'}{P}$$

Working Equation:

$$V' = \frac{VP}{P'}$$

Substitution:

$$V' = \frac{(50\overline{0} \text{ in}^3)(40.0 \text{ psi})}{10\overline{0} \text{ psi}}$$
$$= 20\overline{0} \text{ in}^3$$

· · · · · · · · · · · · · · ·

EXAMPLE 2

Some nitrogen occupies 20.0 m^3 at a gauge pressure of 274 kPa. Find the absolute pressure when its volume is 30.0 m^3.

Data:

$$V = 20.0 \text{ m}^3$$
$$P_{abs} = 274 \text{ kPa} + 101 \text{ kPa} = 375 \text{ kPa} \qquad \text{(See Section 13.3.)}$$
$$V' = 30.0 \text{ m}^3$$
$$P' = ?$$

Basic Equation:

$$\frac{V}{V'} = \frac{P'}{P}$$

Working Equation:

$$P' = \frac{VP}{V'}$$

Substitution:

$$P = \frac{(20.0 \text{ m}^3)(375 \text{ kPa})}{30.0 \text{ m}^3}$$
$$= 25\overline{0} \text{ kPa}$$

· · · · · · · · · · · · · · ·

Density and Pressure

If the pressure of a given amount (constant volume) of gas is increased, its density increases as the gas molecules are forced closer together. (Recall that density is discussed in Section 12.5.) Also, if the pressure is decreased, the density decreases. That is, the *density of a gas is directly proportional to its pressure* as long as there is no change in state. In equation form,

$$\frac{D}{D'} = \frac{P}{P'} \quad \text{or} \quad DP' = D'P$$

where D = original density
D' = final density
P = original pressure (absolute)
P' = final pressure (absolute)

Some amount of carbon dioxide has a density of 1.60 kg/m³ at an absolute pressure of 95.0 kPa. What is the density when the pressure is decreased to 80.0 kPa?

EXAMPLE 3

Data:

$$D = 1.60 \text{ kg/m}^3$$
$$P = 95.0 \text{ kPa}$$
$$P' = 80.0 \text{ kPa}$$
$$D' = ?$$

Basic Equation:

$$\frac{D}{D'} = \frac{P}{P'}$$

Working Equation:

$$D' = \frac{DP'}{P}$$

Substitution:

$$D' = \frac{(1.60 \text{ kg/m}^3)(80.0 \text{ kPa})}{95.0 \text{ kPa}}$$

$$= 1.35 \text{ kg/m}^3$$

· · · · · · · · · · · · · · · ·

A gas has a density of 2.00 kg/m³ at a gauge pressure of 16$\overline{0}$ kPa. What is the density at a gauge pressure of 30$\overline{0}$ kPa?

EXAMPLE 4

Data:

$$D = 2.00 \text{ kg/m}^3$$
$$P = 16\overline{0} \text{ kPa} + 101 \text{ kPa} = 261 \text{ kPa}$$
$$P' = 30\overline{0} \text{ kPa} + 101 \text{ kPa} = 401 \text{ kPa}$$
$$D' = ?$$

Basic Equation:

$$\frac{D}{D'} = \frac{P}{P'}$$

Working Equation:

$$D' = \frac{DP'}{P}$$

Substitution:

$$D' = \frac{(2.00 \text{ kg/m}^3)(401 \text{ kPa})}{261 \text{ kPa}}$$

$$= 3.07 \text{ kg/m}^3$$

PROBLEMS 15.2

Use $\frac{V}{V'} = \frac{P'}{P}$ or $\frac{D}{D'} = \frac{P}{P'}$ to find each quantity. (All pressures are absolute unless otherwise stated.)

1. $V' = 315 \text{ cm}^3$, $P = 101 \text{ kPa}$, $P' = 85.0 \text{ kPa}$; find V.
2. $V = 45\bar{0} \text{ L}$, $V' = 70\bar{0} \text{ L}$, $P = 75\bar{0} \text{ kPa}$; find P'.
3. $V = 76.0 \text{ m}^3$, $V' = 139 \text{ m}^3$, $P' = 41.0 \text{ kPa}$; find P.
4. $V = 439 \text{ in}^3$, $P' = 38.7 \text{ psi}$, $P = 47.1 \text{ psi}$; find V'.
5. $D = 1.80 \text{ kg/m}^3$, $P = 108 \text{ kPa}$, $P' = 125 \text{ kPa}$; find D'.
6. $D = 1.65 \text{ kg/m}^3$, $P = 87.0 \text{ kPa}$, $D' = 1.85 \text{ kg/m}^3$; find P'.
7. $P = 51.0 \text{ psi}$, $P' = 65.3 \text{ psi}$, $D' = 0.231 \text{ lb/ft}^3$; find D.
8. Some air at 22.5 psi occupies $140\bar{0} \text{ in}^3$. What is its volume at 18.0 psi?
9. Some nitrogen at a pressure of 110.0 kPa occupies 185 m^3. Find its pressure if its volume is changed to 225 m^3.
10. Some methane at 185.0 kPa occupies 65.0 L. What is its volume at a pressure of 95.0 kPa?
11. Some carbon dioxide has a density of 3.75 kg/m^3 at 815 kPa. What is its density if the pressure is decreased to 725 kPa?
12. Some oxygen has a density of 1.75 kg/m^3 at normal atmospheric pressure. What is its pressure (in kPa) when the density is changed to 1.45 kg/m^3?
13. Some methane at $50\bar{0} \text{ kPa}$ gauge pressure occupies $75\bar{0} \text{ m}^3$. What is its gauge pressure if its volume is $50\bar{0} \text{ m}^3$?
14. Some helium at 15.0 psi gauge pressure occupies 20.0 ft^3. Find its volume at 20.0 psi gauge pressure.
15. Some nitrogen at 80.0 psi gauge pressure occupies 13.0 ft^3. Find its volume at 50.0 psi gauge pressure.
16. Some carbon dioxide has a density of 6.35 kg/m^3 at 685 kPa gauge pressure. What is the density when the gauge pressure is 455 kPa?
17. Some propane has a density of 48.5 oz/ft^3 at 265 psi gauge pressure. What is the gauge pressure when the density is 30.6 oz/ft^3?
18. Some propane occupies 2.30 m^3 at a gauge pressure of 36.0 kPa. What is the new volume at a gauge pressure of 40.0 kPa?
19. A quantity of oxygen at a gauge pressure of 20.0 kPa occupies 2.60 m^3. What is the new volume at a gauge pressure of 26.0 kPa?
20. Some air occupies 4.5 m^3 at a gauge pressure of 46 kPa. What is the volume at a gauge pressure of 13 kPa?
21. Some oxygen at 87.6 psi (absolute) occupies 75.0 in^3. Find its volume if its absolute pressure is (a) doubled, (b) tripled, (c) halved.
22. A gas at $30\bar{0} \text{ kPa}$ (absolute) occupies 40.0 m^3. Find its absolute pressure if its volume is (a) doubled, (b) tripled, (c) halved.
23. A volume of 58.0 L of hydrogen is heated from 33°C to 68°C. If its original density is 4.85 kg/m^3 and its original absolute pressure is $12\bar{0} \text{ kPa}$, what is the resulting density?
24. Some argon gas is in a 42.0-L container at a pressure of $32\bar{0} \text{ kPa}$. What is the pressure if it is transferred to a container with a volume of 51.0 L?
25. A 2.00-L plastic bottle contains air at a pressure of 33.0 kPa. A person squeezes the bottle, resulting in a pressure of 57.0 kPa. What is the new volume of the container?

SKETCH

12 cm²	w

4.0 cm

DATA
$A = 12 \text{ cm}^2$, $l = 4.0 \text{ cm}$, $w = ?$

BASIC EQUATION
$A = lw$

WORKING EQUATION
$w = \frac{A}{l}$

SUBSTITUTION
$w = \frac{12 \text{ cm}^2}{4.0 \text{ cm}} = 3.0 \text{ cm}$

26. The 3.25-cm^3 volume of air remaining in a water balloon is reduced to 2.75 cm^3 upon impact when the air pressure is 48.3 kPa. (a) Is the original pressure more or less than that upon impact? (b) Find the original pressure.

27. A mass of 1.31 kg of neon is in a 3.00-m^3 container at a pressure of 121 kPa. When the pressure is reduced to 97.4 kPa, what is the final density?

28. The air density in a tractor tire is 1.40 kg/m^3 at a pressure of 314 kPa. (a) As the air pressure increases to $70\overline{0}$ kPa, does the final density of the air increase or decrease? (b) What is the resulting density of the air?

29. An unknown gas is in a tank at 13.3 kPa. (a) If the density of the gas changes from 1.45 kg/m^3 to 1.35 kg/m^3, will the resulting pressure be higher or lower than the original pressure? (b) Find the resulting pressure.

15.3 Charles's and Boyle's Laws Combined

Most of the time it is very difficult to keep the pressure constant or the temperature constant. In this case we combine Charles's law and Boyle's law as follows:

$$\frac{VP}{T} = \frac{V'P'}{T'} \quad \text{or} \quad VPT' = V'P'T$$

Note: Both pressure and temperature must be absolute.

We have 5.00 m^3 of acetylene at 4.00°C at $16\overline{0}$ kPa (absolute). What is the pressure if its volume is changed to 4.00 m^3 at 30.0°C?

EXAMPLE 1

Data:

$$V = 5.00 \text{ m}^3$$
$$P = 16\overline{0} \text{ kPa}$$
$$T = 4.00° + 273 = 277 \text{ K}$$
$$V' = 4.00 \text{ m}^3$$
$$T' = 30.0° + 273 = 303 \text{ K}$$
$$P' = ?$$

Basic Equation:

$$\frac{VP}{T} = \frac{V'P'}{T'}$$

Working Equation:

$$P' = \frac{VPT'}{TV'}$$

Substitution:

$$P' = \frac{(5.00 \text{ m}^3)(16\overline{0} \text{ kPa})(303 \text{ K})}{(277 \text{ K})(4.00 \text{ m}^3)}$$

$$= 219 \text{ kPa}$$

The gas laws are reasonably accurate except at very low temperatures and under extreme pressures. A commonly used reference in gas laws is called **standard temperature and pressure** (STP). Standard temperature is the freezing point of water, 0°C or 32°F. Standard pressure is equivalent to atmospheric pressure, 101.32 kPa or 14.7 lb/in^2.

Vapor Pressure and Humidity

When a liquid evaporates, molecules of the liquid pass from its surface into the air above the liquid. This increase in the number of molecules in the air causes an increase in the pressure above the liquid. This increase in pressure is called *vapor pressure*.

Water boiling is an example of the change of phase of a liquid to a gas. Bubbles of gas form below the surface of the water, rise to the surface, and escape into the air. Recall that boiling water remains at a constant 100°C temperature at standard pressure. Water can, however, have its temperature raised above 100°C and be kept from boiling in a pressure cooker. This pot has a heavy, tight-fitting lid with a steam escape valve. As the water in the pot begins to boil, the pressure above the water builds up and interferes with further boiling. This way, the boiling temperature of the water can be raised above 100°C, allowing for fast cooking of a roast or other foods. Likewise, at a higher altitude, water boils at a lower temperature. In the mile-high city of Denver, water boils at 95°C, making for longer cooking times. Many premixed baking foods provide alternate baking times for high altitudes.

Evaporation of water molecules into the air from a lake followed by cooling as the air passes over cooler land may create a fog. Fog is basically a cloud that forms near the ground. The returning of some water vapor molecules in the air to a liquid is an example of a process called *condensation*. Condensation occurs at a point in temperature called the *dew point*. Dew forms when moist air is cooled by the earth's surface. It can be closely reproduced in the laboratory by taking a glass container partly filled with water and adding pieces of ice. When the water and glass reach the dew point in temperature, water will condense from the air on the outside of the glass. The condensation on a can of cold soda is a familiar example, in which water vapor in the air is chilled upon contact with the cold can. The water molecules slow when cooled and change phase to liquid.

Condensation of steam can be very dangerous. Since steam gives up considerable energy when it changes from vapor to liquid (540 cal/g °C), getting burned by steam can be much worse than getting burned by water at the same temperature. This same concept explains why a steam heating system produces much more heat than a hot water system.

The maximum amount of water vapor that air will hold at a given temperature is *absolute humidity*. Any increase becomes rainfall. *Relative humidity*, in contrast, is the ratio of the amount of water vapor a sample of air holds to the maximum amount it can hold if it is saturated.

Another example of the effect of condensation on people can be seen by comparing the "dry heat" in Arizona with the humid air at the same temperature on the coast of the Gulf of Mexico. In Phoenix, where the humidity is very low, evaporation is substantially greater than condensation. In New Orleans, however, where the humidity is much higher, condensation is much greater and offsets evaporation, so people are more uncomfortable as condensation limits the cooling effect of the evaporation of a person's perspiration. For most people, a relative humidity of 50% to 60% at about 20°C is comfortable. When the relative humidity is higher, the air feels heavy and uncomfortable.

SKETCH

12 cm² | w

4.0 cm

DATA

$A = 12$ cm², $l = 4.0$ cm, $w = ?$

BASIC EQUATION

$A = lw$

WORKING EQUATION

$w = \dfrac{A}{l}$

SUBSTITUTION

$w = \dfrac{12 \text{ cm}^2}{4.0 \text{ cm}} = 3.0$ cm

PROBLEMS 15.3

Use $\dfrac{VP}{T} = \dfrac{V'P'}{T'}$ to find each quantity. (All pressures are absolute unless otherwise stated.)

1. $P = 825$ psi, $T = 575°$R, $V' = 1550$ in³, $P' = 615$ psi, $T' = 525°$R; find V.
2. $V = 50\overline{0}$ in³, $T = 50\overline{0}°$R, $V' = 80\overline{0}$ in³, $P' = 800$ psi, $T' = 45\overline{0}°$R; find P.
3. $V = 90\overline{0}$ m³, $P = 105$ kPa, $T = 30\overline{0}$ K, $P' = 165$ kPa, $T' = 265$ K; find V'.
4. $V = 18.0$ m³, $P = 112$ kPa, $V' = 15.0$ m³, $P' = 135$ kPa, $T' = 235$ K; find T.
5. $V = 532$ m³, $P = 135$ kPa, $T = 87°$C, $V' = 379$ m³, $P' = 123$ kPa; find T'.
6. We have $60\overline{0}$ in³ of oxygen at $150\overline{0}$ psi at 65°F. What is the volume at $120\overline{0}$ psi at $9\overline{0}°$F?
7. We have $80\overline{0}$ m³ of natural gas at 235 kPa at $3\overline{0}°$C. What is the temperature if the volume is changed to $120\overline{0}$ m³ at 215 kPa?
8. We have $140\overline{0}$ L of nitrogen at 135 kPa at 54°C. What is the temperature if the volume is changed to $80\overline{0}$ L at 275 kPa?

PHYSICS CONNECTIONS

Decompression Sickness: "The Bends"

When a scuba diver dives in deep water, he or she experiences a greater water pressure than when closer to the surface. According to Boyle's law, when the pressure is increased on an object, its volume decreases. In the case of a scuba diver, the volume of air inside the diver's lungs decreases as well. In order for the diver to breathe correctly, the air in the diver's tank must be highly pressurized and comprise 10% to 20% oxygen and 80% to 90% nitrogen.

Figure 15.1 A hyperbaric chamber

Courtesy of Shutterstock © Dallas Events Inc.

Under normal atmospheric pressure, nitrogen does not dissolve in blood. Under pressure, nitrogen's volume decreases, thereby increasing its solubility in blood. As the diver swims deeper, the pressure in the lungs increases, and the blood accepts the smaller nitrogen bubbles. If the scuba diver returns too quickly to the surface, the diver experiences a quickly reduced pressure on his or her body, which results in enlarged nitrogen bubbles in the bloodstream. The presence of nitrogen bubbles in the blood can cause generalized barotrauma, otherwise known as decompression sickness or "the bends."

Common symptoms of "the bends" include joint pain, dizziness, and numbness. The only way to prevent "the bends" is to slowly rise to the surface. This gradually reduces the pressure on the diver and the nitrogen and gives the nitrogen bubbles time to dissolve and prevents them from forming large bubbles in the blood.

In the mid-1800s, workers on the Brooklyn Bridge were the first individuals to be documented with "the bends." They worked under the Hudson River in high-pressure compartments where they would enter and exit several times per day without slowly decompressing. Symptoms were documented and the cause of this extremely painful condition was eventually discovered. Today, someone suffering from "the bends" is taken to decompression chambers, called hyperbaric chambers, where the patient breathes highly pressurized oxygen that is gradually decompressed to normal atmospheric pressure (Fig. 15.1).

9. An acetylene welding tank has a pressure of $20\overline{0}0$ psi at $4\overline{0}°F$. If the temperature rises to $9\overline{0}°F$, what is the new pressure?

10. What is the new pressure in Problem 9 if the temperature falls to $-3\overline{0}°F$?

11. An ideal gas occupies a volume of 5.00 L at STP. What is its gauge pressure (in kPa) if the volume is halved and its temperature increases to $40\overline{0}°C$?

12. An ideal gas occupies a volume of 5.00 L at STP. (a) What is its temperature if its volume is halved and its absolute pressure is doubled? (b) What is its temperature if its volume is doubled and its absolute pressure is tripled?

13. Some propane occupies 2.00 m^3 at 18.0°C at an absolute pressure of 3.50×10^5 N/m^2. (a) Find the absolute pressure (in kPa) at the same temperature when the volume is halved. (b) Find the new temperature when the absolute pressure is doubled and the volume is doubled. (c) Find the new volume when the absolute pressure is halved and the temperature is decreased to $-12.0°C$. (d) Find the new volume if the absolute pressure is 1.30×10^6 N/m^2 and the temperature is 31.0°C.

14. A balloon with volume $32\overline{0}0$ mL of xenon gas is at a gauge pressure of 122 kPa and a temperature of 27°C. What is the volume when the balloon is heated to 65°C and the gauge pressure is decreased to 112 kPa?

15. A 7.85-L helium-filled balloon experiences a change in both pressure from normal atmospheric pressure to 60.5 kPa and temperature from 24.0°C to 6.00°C. What is the resulting volume?

Glossary

Boyle's Law If the temperature of a gas is constant, the volume is inversely proportional to the absolute pressure, $V/V' = P'/P$. (p. 415)

Charles's Law If the pressure on a gas is constant, the volume is directly proportional to its Kelvin or Rankine temperature, $V/T = V'/T'$. (p. 413)

Standard Temperature and Pressure (STP) A commonly used reference in gas laws. Standard temperature is the freezing point of water. Standard pressure is equivalent to atmospheric pressure. (p. 419)

Formulas

15.1 Charles's law: $\dfrac{V}{T} = \dfrac{V'}{T'}$

15.2 Boyle's law: $\dfrac{V}{V'} = \dfrac{P'}{P}$

$$\frac{D}{D'} = \frac{P}{P'}$$

15.3 $\dfrac{VP}{T} = \dfrac{V' P'}{T'}$

Review Questions

1. The gas law that relates volume and temperature is called
   (a) Boyle's law.   (b) Hooke's law.
   (c) Charles's law.   (d) none of the above.

2. The gas law that relates volume and pressure is called
   (a) Boyle's law.   (b) Hooke's law.
   (c) Charles's law.   (d) none of the above.

3. If the temperature of a gas is constant and the volume is decreased, the pressure will
   (a) stay the same.   (b) decrease.
   (c) increase.   (d) increase or decrease, depending on the gas.

4. If the temperature of a gas is constant and the pressure is decreased, the volume will
   (a) stay the same.   (b) decrease.
   (c) increase.   (d) increase or decrease, depending on the gas.

5. If the pressure on a gas is constant and the temperature is decreased, the volume will
   (a) stay the same.   (b) decrease.
   (c) increase.   (d) increase or decrease, depending on the gas.

6. If the pressure on a gas is constant and the volume is decreased, the temperature will
   (a) stay the same.   (b) decrease.
   (c) increase.
   (d) increase or decrease, depending on the amount of gas.

7. Describe the conditions of standard temperature and pressure.
8. Describe what happens to the volume of a gas if its temperature and pressure increase.
9. Describe what happens to the temperature of a gas if its volume and pressure increase.
10. What causes the tendency of the volume and pressure of a gas to increase when it is heated?
11. What causes the tendency of the temperature of a gas to increase when it is compressed?
12. What causes the tendency of the pressure of a gas to decrease when the volume is increased?

Review Problems

All pressures are absolute unless otherwise stated.

1. A gas occupies 13.5 ft³ at 35.8°F. What will the volume of this gas be at 88.6°F if the pressure is constant?
2. A gas occupies 3.45 m³ at 18.5°C. What will the volume of this gas be at 98.5°C if the pressure is constant?
3. Some hydrogen occupies 115 ft³ at 54.5°F. What is the temperature when the volume is 132 ft³ if the pressure is constant?
4. Some carbon dioxide occupies 45.3 L at 38.5°C. What is the temperature when the volume is 44.2 L if the pressure is constant?
5. Some propane occupies 145 cm³ at 12.4°C. What is the temperature when the volume is 156 cm³ if the pressure is constant?
6. Some air at 276 kPa occupies 32.4 m³. What is its absolute pressure if its volume is doubled at constant temperature?
7. Some helium at 17.5 psi gauge pressure occupies 35.0 ft³. What is the volume at 32.4 psi if the temperature is constant?
8. Some carbon dioxide has a density of 6.35 kg/m³ at 685 kPa gauge pressure. What is the density when the gauge pressure is 355 kPa if the temperature is constant?
9. We have 435 in³ of nitrogen at 1340 psi gauge pressure at 75°F. What is the volume at 1150 psi gauge pressure at 45°F?
10. We have 755 m³ of carbon dioxide at 344 kPa at 25°C. Find the temperature if the volume is changed to 1330 m³ at 197 kPa.
11. A welding tank has a gauge pressure of 1950 psi at 38°F. (a) What is the new pressure if the temperature rises to 98°F? (b) What is the temperature if the gauge pressure falls to 1870 psi?
12. An ideal gas occupies a volume of 4.50 L at STP. What is its gauge pressure (in kPa) if the volume is halved and the absolute temperature is doubled?
13. An ideal gas occupies a volume of 5.35 L at STP. What is its gauge pressure (in kPa) if the volume is halved and the temperature is increased by 45.5°C?
14. A volume of 1120 L of helium at 40$\overline{0}$0 Pa is heated from 45°C to 77°C, increasing the pressure to 60$\overline{0}$0 Pa. What is the resulting volume of the gas?
15. In a 47-cm-tall cylinder of radius 7.0 cm, hydrogen of density 2.50 kg/m³ is at a gauge pressure of 327 kPa. What is the density when the absolute pressure is changed to 525 kPa?

SKETCH

12 cm² w

4.0 cm

DATA

$A = 12$ cm², $l = 4.0$ cm, $w = ?$

BASIC EQUATION

$A = lw$

WORKING EQUATION

$w = \dfrac{A}{l}$

SUBSTITUTION

$w = \dfrac{12 \text{ cm}^2}{4.0 \text{ cm}} = 3.0 \text{ cm}$

APPLIED CONCEPTS

1. Fran purchases a 1.85-ft³, helium-filled Mylar® balloon in a store with an inside temperature of 65.5°F. When she walks outside, the air temperature is only 11.0°F. (a) What happens to the balloon as Fran walks toward her car? (b) What is the volume of the balloon when the temperature of the helium reaches the air temperature? (c) What would have been the volume of the balloon if she had purchased it in the summer with an outside temperature of 101°F? (d) Depending on the season, what can the store clerk do to compensate for the outside temperature and its impact on the volume of the balloon?

2. An automobile tire is filled to an air pressure of 32 lb/in² at an air temperature of $4\overline{0}$°F. (a) After the car is driven for a period of time, the temperature of the air inside the tire increases to 145°F. If the volume of the tire remains constant, what is the pressure inside the tire? (b) If a person wants to maintain an air pressure of 32 lb/in² when the tire is warm, what initial air pressure should be put in the tire?

3. A 15.0-cm-long cylinder has a movable piston with radius 2.00 cm. (a) What is the volume of the cylinder? (b) If the initial air pressure is 101 kPa at 0.0°C, what is the air pressure inside the cylinder when the piston is pushed down 4.00 cm? (c) If the air temperature inside the compressed cylinder is heated to 20.0°C, what is the new air pressure?

4. A 0.0300-m³ steel tank containing helium is stored at 20.0°C under a pressure of 22.5×10^3 kPa. (a) If the helium is released into a 20.0°C, 101-kPa atmosphere, how much volume will the helium occupy? (b) How much volume will the released helium occupy if released into a −28.5°C, 285-Pa stratosphere?

5. A lightweight weather-collecting sensor is attached to a 2.50-L balloon. The balloon is released on a day when the temperature is 23.4°C and the atmospheric pressure is 1.35×10^5 Pa. The balloon will collect data until it reaches the stratosphere. (a) The weather sensor measures a temperature of −28.2°C and an atmospheric pressure of 285 Pa. What is the volume of the balloon? (b) Given the change in the volume of the balloon as it reaches the stratosphere, should the balloon be fully inflated when released from the ground?

CHAPTER 16

WAVE MOTION AND SOUND

Objectives

The major goals of this chapter are to enable you to:

1. Describe characteristics of mechanical waves.
2. Describe electromagnetic waves and the electromagnetic spectrum.
3. Analyze sound waves and explain the Doppler effect.

The importance of radio and television signals and other forms of electromagnetic waves cannot be overstated. Communication using these phenomena is the backbone of modern civilization. We will first study mechanical waves to see how waves in general behave and then study electromagnetic radiation and sound.

16.1 Characteristics of Waves

Energy may be transferred by the motion of particles. Electricity, for example, is conducted along a wire by the motion of electrons. Heat is conducted by the motion of atoms and molecules. Tides and winds are examples of transfer of energy by the motion of fluids.

Energy may also be transferred by *wave motion*. The sun's energy is transported to the earth by light waves. Radio waves are an illustration of energy transfer for communications. Light waves produced by lasers are being used for voice and data transmission in optical fibers. Sound waves are yet another method by which energy may be transferred. The source of all waves is something that vibrates. The frequency of the vibrating source and the frequency of the wave it produces are the same.

A **wave** is a disturbance that moves through a medium or through space. This disturbance may be a displacement of atoms away from their equilibrium or rest positions in an elastic medium, a pulse in a spring, a change in pressure of a gas, or a variation in light intensity. There is a transfer of energy in the direction of propagation of the disturbance for each type of wave. A leaf floating in a pond as a wave passes will bob up and down with the water but will not move along with the wave.

The elastic medium through which a wave travels or propagates is in many respects similar to a chain of particles connected by a series of springs like those shown in Fig. 16.1. If particle *A* is pulled to the left away from its equilibrium position, the neighboring spring exerts a force that tends to return *A* to its equilibrium position. The same spring exerts an equal but opposite force on particle *B*, which also tends to displace *B* to the left. As particle *B* moves to the left, the next particle experiences a force to the left, and so on, until each particle experiences a displacement. If particle *A* is returned to its equilibrium position, the other particles will return to their equilibrium positions at a later time.

Figure 16.1 Springs in series are a form of elastic medium.

If particle *A* is forced to oscillate about its equilibrium position, all the other particles will also oscillate about their equilibrium positions. The kinetic energy given to the first particle is transmitted to each successive particle in the system. Although energy is transferred through the connecting springs, there is no transfer of particles from position *A* to *E*. This energy transfer without matter transfer is typical of all types of wave motion.

Another type of wave motion is shown in Fig. 16.2. In this case the elastic medium is a long spring. If the left end of the spring is rapidly lifted up and then returned to its starting position, a crest is formed that travels to the right [Fig. 16.2(a)]. If the left end is displaced downward and rapidly returned to its original position, a trough is formed that travels to the right [Fig. 16.2(b)].

A **pulse** is a nonrepeated disturbance that carries energy through a medium or through space. If the pulse is repeated periodically, then a series of crests and troughs will travel through the medium, creating a traveling wave. A **transverse wave** is a disturbance in a medium in which the motion of the particles is perpendicular to the direction of the wave motion (Fig. 16.3). Water waves are another example of transverse waves. The **amplitude** of a wave is the maximum displacement of any part of the wave from its equilibrium, or rest, position.

If a spring is compressed at the left end as shown in Fig. 16.4 and then released, the compression will travel to the right. Similarly, if the spring is stretched, a rarefaction (the stretched portion of the spring) is formed that will propagate or travel to the right. In this case the particle motion is along the direction of the wave travel. A **longitudinal wave** is a disturbance in a medium in which the motion of the particles is along the direction of the wave travel. Sound is another example.

The **wavelength** λ is the minimum distance between particles that have the same displacement and are moving in the same direction (Fig. 16.5).

The **period** is the time required for a single wave to pass a given point. The **frequency** is the number of complete waves passing a given point per unit time. The common unit for frequency is the hertz (Hz) (named after **Heinrich Hertz**), where one oscillation per second is equal to 1 hertz (1 Hz = 1/s). Higher frequencies are measured in kilohertz (kHz), megahertz (MHz), and gigahertz (GHz). Radar and microwaves are measured in GHz, frequency-modulated (FM) radio waves are measured in MHz, and

Heinrich Hertz (1857–1894),

physicist, was born in Germany. His main work was on electromagnetic waves; he was the first to broadcast and receive radio waves. The unit of frequency, the hertz, is named after him.

Figure 16.2   Pulses in a long spring

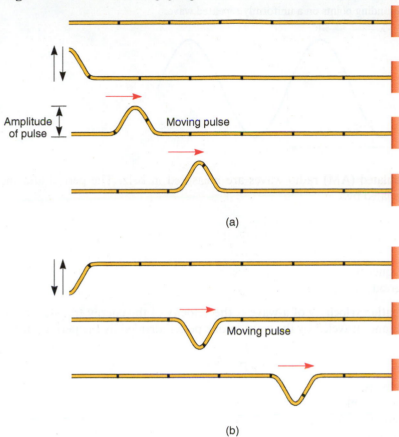

Amplitude of pulse

Moving pulse

(a)

Moving pulse

(b)

Figure 16.3   Transverse waves

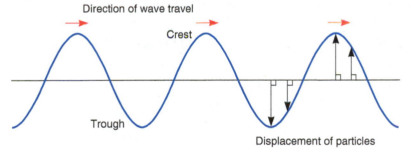

Direction of wave travel

Crest

Trough

Displacement of particles

Figure 16.4   Longitudinal wave in a spring

Direction of wave travel

Displacement of particles

Figure 16.5   The wavelength of a wave is the distance between successive corresponding points on a uniformly repeated wave.

amplitude-modulated (AM) radio waves are measured in kHz. The period and the frequency are related by

$$f = \frac{1}{T}$$

where f = frequency
  T = period

The **propagation velocity** v of a wave is the velocity of the energy transfer and is given by the distance traveled by the wave in one period divided by the period, or

$$v = \frac{\lambda}{T} = \lambda f$$

where v = velocity
  λ = wavelength
  T = period
  f = frequency

These relationships apply to sound, water, light, and all other waves.

EXAMPLE 1

Find the velocity of a wave with wavelength 2.5 m and frequency 44 Hz.

Data:

$$\lambda = 2.5 \text{ m}$$
$$f = 44 \text{ Hz} = 44/\text{s}$$
$$v = ?$$

Basic Equation:

$$v = \lambda f$$

Working Equation: Same

Substitution:

$$v = (2.5 \text{ m})(44/\text{s})$$
$$= 110 \text{ m/s}$$

Superposition of Waves

When two waves of a similar type pass through the same medium, a new wave is created by the **superposition of waves**. This new wave is the algebraic sum of the separate displacements of the individual waves (Fig. 16.6). **Constructive interference** occurs where the waves add together to form a larger displacement as at point A. **Destructive interference** occurs where the waves add together to form a smaller displacement as at point B.

Figure 16.6   Superposition of the two blue waves in (a) and (b) to form the new green wave in (c) by adding their displacements

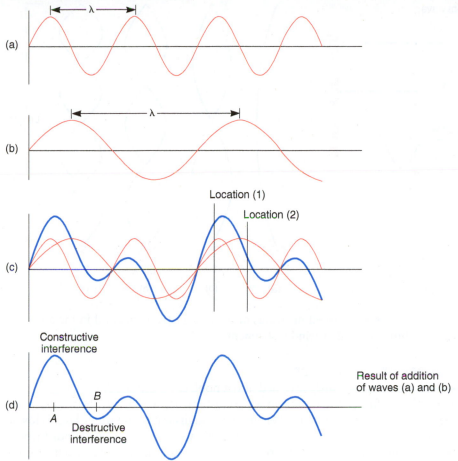

The addition is algebraic, so two positive amplitudes (that is, two amplitudes in the same direction) add together to make a larger amplitude in the same direction (constructive interference). Where the amplitudes are in opposite directions, the smaller counteracts the larger with a net amplitude in the direction of the larger (destructive interference). See location (1) of Fig. 16.6(c) for an illustration of constructive interference and location (2) for an illustration of destructive interference.

Standing Waves

When a transverse pulse reaches the end of a spring that is fastened to a rigid support (Fig. 16.7), the pulse is reflected along the spring with the displacement of the reflected wave opposite in direction to that of the incident wave. A traveling wave is also reflected at the rigid end of a spring, producing two waves moving in opposite directions. Reflections from a free end are not inverted.

In one special case two waves combine so that there is no propagation of energy along the wave. The wave displacements are constant and remain fixed in location. This is called a **standing wave** (Fig. 16.8) because the two waves of equal amplitude and wavelength do not appear to be traveling. The points of destructive interference and constructive interference remain in fixed positions. Figure 16.8 shows an example of a standing wave on a string that could produce sound on a musical instrument. Note that there is no motion of the string at the end points. Although there is no propagation of energy along the string, there may be energy transfer from the string to the air surrounding it, producing sound waves at the same frequency of oscillation as that of the vibrating string.

Figure 16.7 The reflected wave in the spring has the same amplitude as the incident wave.

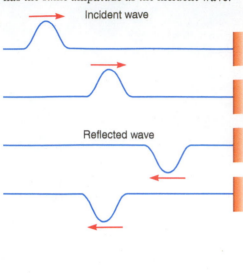

Figure 16.8 Standing waves in a string generated by increasing the frequency of vibration

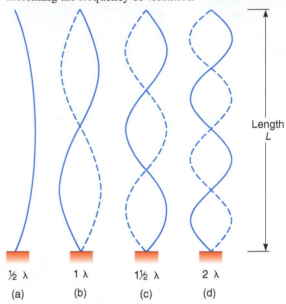

Standing waves are formed in strings of musical instruments and in the air in an organ pipe, a flute, and other wind instruments.

Interference and Diffraction

If two rocks are simultaneously dropped into a pool, each will produce a set of waves or ripples: [Fig. 16.9(a)]. Wherever two wave crests cross each other, the water height is higher than for either crest alone. Where two troughs cross, the water level is lower than for one alone. If a trough crosses a crest, the water level is nearly undisturbed. This is an example of wave **interference**. Constructive interference occurs when two crests or troughs meet, giving a larger disturbance than for either wave alone. Destructive interference occurs when a wave and a trough meet and cancel each other out. Those areas where waves cancel each other out are called *nodes* [Fig. 16.9(b)].

Figure 16.9 Interference of waves from two sources. The waves combine to form larger waves where two wave crests cross each other, and they cancel each other out where a wave crest and a trough meet.

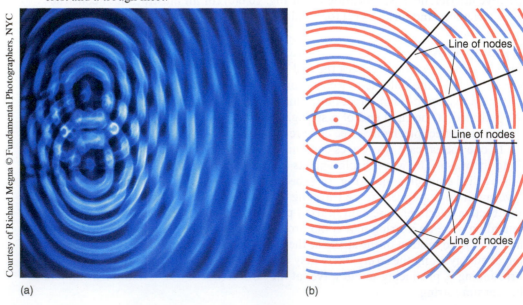

Interference of waves can occur for any type of wave, including sound, light, and radio waves. Some directional radio broadcast antennas rely on constructive interference between waves from different parts of the antenna to direct the signal in the desired direction. Destructive interference prevents the signal from propagating in undesired directions. "Dead spots" in an auditorium are caused by destructive interference of waves coming directly from the sound source with sound waves reflected from the walls, ceiling, or floor of the room. Proper choice of room shape and proper placement of materials that absorb sound well can lead to a room with good acoustical characteristics.

TRY THIS ACTIVITY

Interfering Waves

Place either a long Slinky® toy or a length of rope on a smooth, low-friction floor. With one person at each end of the rope, create a "wave pulse" toward each other. Observe what happens when the two waves interfere with each other. Explain what happens when two positive pulses interfere, two negative pulses interfere, and a positive and a negative pulse interfere with one another (Fig. 16.10).

Figure 16.10

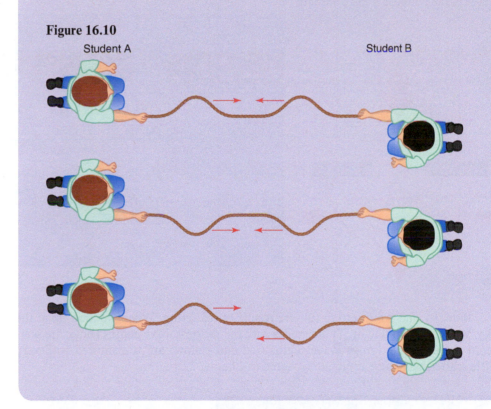

Student A   Student B

Diffraction is a property of a wave that describes its ability to bend around obstacles in its path. Water waves bend around the supports of a pier or a large rock (Fig. 16.11). Sound waves pass from one room through a door and spread into a second room (Fig. 16.12).

Wave diffraction is commonly observed only when the obstacle or opening is nearly the same size as the wavelength. Water waves and sound waves often have wavelengths in an easily observed range. Light, however, has a wavelength approximately 5×10^{-7} m. For this reason diffraction of light is not as easily recognized as diffraction of some other waves. The light waves from the sun when they encounter obstacles (air molecules in the atmosphere) are diffracted around them. The diffracted wave fronts are more or less spherical and are spread out, or *scattered*.

Figure 16.11 Water waves bend around obstacles and pass into the region behind (a) the pier supports or (b) a large rock. This property is called diffraction.

(a)   (b)

Figure 16.12 Diffraction of (a) sound waves entering a room through a door and (b) water waves passing through a small opening

(a)   (b)

Blue sky and a red setting sun are results of scattering. When we look up at the sky, we see scattered light. Those colors with the shortest wavelengths (blue) are scattered the most; the longer wavelengths (red and yellow) are transmitted with very little scattering. (These topics are covered in Chapter 22.)

PHYSICS CONNECTIONS

Wave Pools

Land-locked beach enthusiasts enjoy visiting water parks that generate large waves and allow them to body surf as if they are at the beach. Huge pumps, reservoirs, and pools are necessary to create such waves.

At Typhoon Lagoon in Disney World, a large pool shaped like a funnel holds over 2 million gallons of water (Fig. 16.13). Giant pumps and a large reservoir hold and then release 40,000 to 90,000 gal of water into the pool. The massive amount of water creates a surge in the pool that results in a high-amplitude wave. Water is lifted into the reservoir by a number of large pumps. Once the reservoir is full, giant valves are moved away from an opening in the bottom of the reservoir. Since water seeks to achieve its own level, the water rushes out of the bottom of the reservoir and into a narrow but deep section of the pool.

Once the giant surge of water enters the pool, the slope of the pool and the gradual increase in width of the pool promote the surf-like wave (Fig. 16.14). Once the wave passes and the water level flattens out, water is drawn out of the pool and back to the pumps so the process can be repeated.

Figure 16.13   Overhead view of a typical human-made wave pool

Figure 16.14   Swimmers enjoying a human-made wave at Typhoon Lagoon in Disney World

16.2 Electromagnetic Waves

An **electromagnetic wave** is a transverse wave resulting from a periodic disturbance in an electromagnetic field, which has an electric component and a magnetic component, each being perpendicular to the other and both perpendicular to the direction of travel. All electromagnetic waves travel with velocity $v = c =$ **speed of light** $= 3.00 \times 10^8$ m/s. So, for electromagnetic waves

$$c = \lambda f$$

where c = speed of light (3.00×10^8 m/s)
λ = wavelength
f = frequency

Note that λ and f are inversely proportional. That is, when the frequency increases, the wavelength decreases; and when the frequency decreases, the wavelength increases.

EXAMPLE 1

The FM band of a radio is centered around a frequency of $10\overline{0}$ megahertz (MHz). Find the length of an FM antenna if each arm must be a quarter-wavelength.

First, find the wavelength, λ.

Data:

$$f = 10\overline{0} \text{ MHz} = 10\overline{0} \times 10^6 \text{ Hz} = 1.00 \times 10^8/\text{s} \qquad (1 \text{ Hz} = 1/\text{s})$$
$$c = 3.00 \times 10^8 \text{ m/s}$$
$$\lambda = ?$$

Basic Equation:

$$c = \lambda f$$

Working Equation:

$$\lambda = \frac{c}{f}$$

Substitution:

$$\lambda = \frac{3.00 \times 10^8 \text{ m/s}}{1.00 \times 10^8/\text{s}}$$
$$= 3.00 \text{ m}$$

Therefore,

$$\frac{\lambda}{4} = \frac{3.00 \text{ m}}{4} = 0.750 \text{ m}$$

· · · · · · · · · · · · · · · · · ·

The **electromagnetic spectrum** is the entire range of electromagnetic waves classified according to frequency (Fig. 16.15). All electromagnetic waves travel at the same speed. The radio broadcast band is in the region of 1 MHz. The very high frequency (VHF) television band starts in the region of 50 MHz; the ultra high frequency (UHF) band is even higher. The highest-frequency waves generated by electronic oscillators are microwaves. Microwaves may be used to relay TV signals from a remote location to the studio transmitter.

Molecular and atomic oscillations create waves of even higher frequencies, including infrared, visible light, ultraviolet, and X-ray waves. Visible light is electromagnetic radiation in the range 4.3×10^{14} to 7×10^{14} Hz.

Figure 16.15   The electromagnetic spectrum

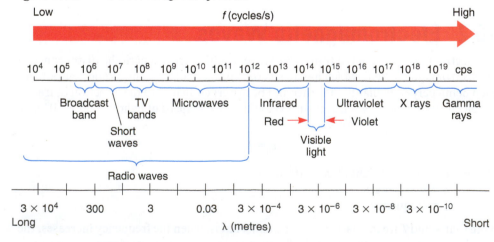

We are surrounded by and bombarded by a sea of electromagnetic radiation. The light we see is only a very tiny part of this radiation.

PHYSICS CONNECTIONS

Evading Radar

Radar (RAdio Detecting And Ranging) is used as a navigation, early-warning, and detection device. Radar transmitters send electromagnetic radio waves into the atmosphere to measure the time and the frequency of radio waves after they have reflected off objects. (See Section 16.4 for more information on the "Doppler effect.") The reflected radio waves provide the observer with the object's position, size, and velocity.

A futuristic French fighter plane known as Rafale uses a device to help the plane evade radar. The plane uses a technology called active cancellation. As radio waves from radar strike the Rafale, an inverted version of the reflected radio wave is emitted. When two identical but opposite waves meet, they cancel each other, which results in destructive interference. Through destructive interference, the producer of a radar signal will not see any reflected radio waves from the Rafale.

Stealth aircraft avoid radar detection because of their body shape and design. The odd shapes and angles of the stealth body are designed to deflect the radio waves around the plane instead of reflecting them to the producer of the radar signal. Some of the stealth material is also designed to absorb the radio waves (Fig. 16.16).

Figure 16.16 Stealth aircraft are designed to be invisible to radar.

Courtesy of Gary Ombler © Dorling Kindersley

PROBLEMS 16.2

1. Find the period of a wave whose frequency is $50\overline{0}$ Hz.
2. Find the frequency of a wave whose period is 0.550 s.
3. Find the velocity of a wave with wavelength 2.00 m and frequency $40\overline{0}$ Hz.
4. (a) What is the frequency of a light wave with wavelength 5.00×10^{-7} m and velocity 3.00×10^8 m/s? (b) Find the period of the wave.
5. What is the speed of a wave with frequency 3.50 Hz and wavelength 0.550 m?
6. Find the wavelength of water waves with frequency 0.650 Hz and velocity 1.50 m/s.
7. What is the wavelength of longitudinal waves in a coil spring with frequency 7.50 Hz and velocity 6.10 m/s?
8. A wave generator produces 20 pulses in 3.50 s. (a) What is its period? (b) What is its frequency?
9. Find the frequency of a wave produced by a generator that emits 30 pulses in 2.50 s.
10. What is the wavelength of an electromagnetic wave with frequency 50.0 MHz?
11. What is the frequency of an electromagnetic wave with a wavelength of 0.25 m?
12. What is the wavelength in metres of an electromagnetic wave with a frequency of 10^{14}/s?
13. Find the frequency of an electromagnetic wave with wavelength 1.50 m.
14. Find the wavelength of a wave traveling at 2.68×10^6 m/s with a period of 0.0125 s.
15. Find the wavelength of a wave traveling twice the speed of sound (speed of sound = 331 m/s) that is produced by an oscillator emitting 63 pulses every 8.3×10^{-6} min.
16. (a) Find the velocity of X rays emitted with wavelength 3.00×10^{-9} m and frequency 3.00×10^{18} Hz. (b) Find the period of the waves.
17. Find the velocity of microwaves having wavelength 0.750 m and frequency 2.75×10^{10} Hz.

SKETCH

12 cm² $\quad w$

4.0 cm

DATA
$A = 12$ cm², $l = 4.0$ cm, $w = ?$

BASIC EQUATION
$A = lw$

WORKING EQUATION
$$w = \frac{A}{l}$$

SUBSTITUTION
$$w = \frac{12 \text{ cm}^2}{4.0 \text{ cm}} = 3.0 \text{ cm}$$

16.3 Sound Waves

Daily we are exposed to many sounds and communicate with other people using the medium of sound. Whether the sound is pleasant music, a voice, or a loud siren, there are three requirements for the detection of sound in a physiological sense. There must be a source of sound, a medium (such as air) for transmitting it, and an ear to receive it. In a physical sense, sound is a vibratory disturbance in an elastic medium that may produce the sensation of sound. The frequency range over which the human ear responds is approximately 20 to 20,000 Hz. Ultrasonic waves have a frequency higher than 20,000 Hz.

Sound refers to those waves transmitted through a medium with frequencies capable of being detected by the human ear and is produced by a vibrating source. A ringing bell, the vibrating head of a drum, and a tuning fork (Fig. 16.17) are common examples of vibrating sources of sound. Other vibrating sources are not as easily recognized. Vibrating vocal cords produce speech. The notes of a clarinet originate with a vibrating reed. An auto horn uses an electrically driven vibrating diaphragm.

The most common medium for the propagation of sound is air. Sound will also propagate in solids or liquids. A vacuum will not carry sound. A mechanic may listen to the sounds of a running engine by placing one end of a metal rod against the engine and the other end against his or her ear. Sounds transmitted through water are utilized by passive sonar receivers aboard ships or submarines to identify other ships nearby.

Sound waves transmitted through the earth may be detected by an instrument called a *seismograph,* which can detect small motions of the earth's crust. An earthquake produces both longitudinal and transverse waves, which propagate with different velocities through the earth's crust. The distance to the source can be determined by measuring the time interval between the arrival of the two types of waves. Comparison of such data from seismographs at several points on the surface of the earth helps locate the epicenter of the earthquake. Waves can be intentionally set off by buried explosives. Reflections of these sound waves from different rock formations are recorded by seismographs. These recordings allow geologists to determine the underlying structure of the earth and predict the location of possible oil- or gas-producing regions.

Musical instruments utilize vibrating strings and vibrating columns of air to produce regular sounds. Wind instruments use an enclosed or partially enclosed column of air where waves are produced in the confined air by a pressure change (usually, blowing). Standing waves can then be produced in a column of proper length.

You may have watched distant lightning and noticed the time lapse before the sound of thunder reaches you. This is an example of the relatively slow speed of sound compared to the speed of light (3.00×10^8 m/s). The **speed of sound** in dry air at 1 atm pressure and 0°C is 331 m/s. Changes in humidity and temperature cause a variation in the speed of sound. The speed of sound increases with temperature at the rate of 0.61 m/s/°C. The speed of sound in dry air at 1 atm pressure is then given by

$$v = 331 \text{ m/s} + \left(0.61 \frac{\text{m/s}}{\text{°C}}\right)T$$

$$v = 1087 \text{ ft/s} + \left(1.1 \frac{\text{ft/s}}{\text{°F}}\right)(T - 32\text{°F})$$

where v = speed of sound in air
          T = air temperature

Figure 16.17 Sound waves produced by a vibrating tuning fork

EXAMPLE 1

Find the speed of sound in dry air at 1 atm pressure if the temperature is 23°C.

Data:

$$T = 23\text{°C}$$
$$v = ?$$

Basic Equation:

$$v = 331 \text{ m/s} + \left(0.61\frac{\text{m/s}}{\text{°C}}\right)T$$

Working Equation: Same

Substitution:

$$v = 331 \text{ m/s} + \left(0.61\frac{\text{m/s}}{\text{°C}}\right)(23\text{°C})$$

$$= 345 \text{ m/s}$$

· · · · · · · · · · · · · · ·

What is the time required for the sound from an explosion to reach an observer 1900 m away for the conditions of Example 1?

EXAMPLE 2

Data:

$$v = 345 \text{ m/s}$$
$$s = 1900 \text{ m}$$
$$t = ?$$

Basic Equation:

$$s = vt \qquad \text{(from Section 4.1)}$$

Working Equation:

$$t = \frac{s}{v}$$

Substitution:

$$t = \frac{1900 \text{ m}}{345 \text{ m/s}}$$

$$= 5.51 \text{ s}$$

$$\boxed{\frac{\text{m}}{\text{m/s}} = \text{m} \div \frac{\text{m}}{\text{s}} = \text{m} \cdot \frac{\text{s}}{\text{m}} = \text{s}}$$

· · · · · · · · · · · · · · ·

Sound propagates faster in a dense medium such as water than it does in a less dense medium such as air. A list of the speed of sound in various media is given in Table 16.1.

Table 16.1   Speed of Sound in Various Media

Medium	Speed m/s	ft/s
Aluminum	6,420	21,100
Brass	4,700	15,400
Steel	5,960	19,500
Granite	6,000	19,700
Alcohol	1,210	3,970
Water (25°C)	1,500	4,920
Air, dry (0°C)	331	1,090
Vacuum	0	0

All sounds have characteristics that we associate only with that sound. A siren is loud. A whisper is soft. Music can be loud or soft. The physical properties that differ for these sounds are intensity and frequency. The physiological characteristics of these sounds are loudness and pitch. Sound quality is related to the number of frequencies present. Pitch and frequency are closely related terms.

Intensity is the energy transferred by sound per unit time through unit area, thus $\frac{power}{area}$. **Loudness** refers to the strength of the sensation of sound heard by an observer and describes how strong or faint the sensation of sound seems. The ear does not respond equally to all frequencies. Sound must reach a certain intensity before it can be heard. The human ear normally detects sounds ranging in intensity from 10^{-12} W/m^2 (the threshold of hearing) to 10^0 W/m^2 or 1 (the threshold of pain). Levels of intensity are also measured on a logarithmic scale in decibels (dB); the unit "bel" is named after **Alexander Graham Bell**. Table 16.2 shows a range of familiar sounds.

Table 16.2   Range of Sound Levels and Intensities

Situation	Sound Level (dB)	Intensity (W/m^2)	Sound
Threshold of hearing	0	10^{-12}	Scarcely audible
Minimal sounds	10	10^{-11}	
Ticking watch	20	10^{-10}	
Whisper	30	10^{-9}	Faint
Leaves rustling, refrigerator	40	10^{-8}	
Average home, neighborhood street	50	10^{-7}	Moderate
Normal conversation, microwave	60	10^{-6}	
Car, alarm clock, city traffic	70	10^{-5}	Loud
Garbage disposal, vacuum cleaner, outboard motor, noisy restaurant	80	10^{-4}	Very loud
Factory, electric shaver, screaming child	85		
Lawn mower, passing motorcycle, convertible ride on a freeway	90	10^{-3}	
Hair dryer, diesel truck, chain saw, helicopter, subway train	100	10^{-2}	
Car horn, snowblower	110	10^{-1}	Deafening
Rock concert, prop plane	120	10^0	Painful
100 ft from jet engine, air raid siren	130	10^1	
Shotgun blast	140	10^2	

Source: Better Hearing Institute, *Self-Help for Hard of Hearing People*. Each increase of 10 dB is a tenfold increase in sound intensity; that is, 90 dB is 10 times noisier than 80 dB. The U.S. government advises wearing earplugs or other hearing protection for anyone exposed to 85 dB for a period of more than a few hours. Some hearing experts have found that hearing damage is done at sound levels as low as 70 dB. Federal regulations require a hearing conservation program in the workplace where employees are exposed to 85 dB or more during an 8-h work period.

The **pitch** of a sound is the effect of the frequency of sound waves on the ear. Higher-pitched sounds have a higher frequency than lower ones. The quality of sound can easily be determined by a casual listener. Irregular vibrations tend to produce noise, whereas regular vibrations in multiples of the fundamental vibration rate of an object produce sounds more pleasing to the human ear. These multiples of the fundamental tone are called harmonics. The quality of a sound depends on the number of harmonics produced and their relative intensities.

Sound waves from two sources reaching an observer at the same time will constructively or destructively interfere with each other. Identical waves from two separate sources such as stereo loudspeakers will enhance or reduce each other at different points in space. Sound waves of slightly different frequencies will also enhance or reduce each other at different points in time. The alternating periods of increasing and decreasing volume are called *beats*. The number of times per second the sound reaches a maximum is known as the *beat frequency* and is equal to the difference in frequencies of the two sound waves. This phenomenon is used for tuning musical instruments against a set of standard-frequency tuning forks. When the beat frequency is small, the instrument is well matched with the tuning fork.

The ear is the human organ used to detect sound. It consists of three sections: the outer ear, the middle ear, and the inner ear (Fig. 16.18). Sound travels through the ear canal, which ends at the eardrum. The sound waves cause the eardrum to vibrate. The sound energy is transferred through the three smallest bones in the body to a fluid in the inner ear, which excites tiny hairs located in the cochlea, called cilia. Cilia resonate in the fluid, sending nerve impulses to the brain in the form of electrical waves. The brain then interprets the sound and we are able to hear.

Figure 16.18 Sound waves move through all three regions of the ear and are sent to the brain in the form of electrical waves.

16.4 The Doppler Effect

The **Doppler effect** refers to the variation of the frequency heard when a source of sound and the ear are moving relative to each other. As an automobile or a train passes by at a high speed sounding its horn or whistle, the frequency or pitch of the sound drops noticeably as it passes the observer. This variation in pitch is called the Doppler effect, named after **Christian Johann Doppler**.

Water waves spread over the flat surface of the water. Sound and light waves, though, travel in three-dimensional space in all directions like an expanding balloon. Just as circular wave crests are closer together in front of a swimming duck, spherical sound or light wave crests ahead of a moving source are closer together than those behind the source and reach a receiver more frequently (Fig. 16.19).

Motion of a source of sound toward an observer increases the rate at which he or she receives the vibrations. The velocity of each vibration is the speed of sound whether the source is moving or not. Each vibration from an approaching source has a shorter distance to travel. The wavelength is shortened when the source is moving toward the observer and is lengthened when the source is moving away from the observer. The vibrations are therefore received at a higher frequency than they are sent. Similarly, sound waves from a receding source are received at a lower frequency than that at which they are sent (Fig. 16.20).

The apparent Doppler-shifted frequency for sound is given by the equation

$$f' = f\left(\frac{v}{v \pm v_s}\right)$$

where f' = Doppler-shifted frequency
  f = actual source frequency
  v = speed of sound
  v_s = speed of the source

The + sign in the denominator is used when the source is moving away from the observer. The − sign is used when the source is moving toward the observer.

Christian Johann Doppler (1803–1853),

physicist and mathematician, was born in Austria. He is best known for his explanation of the perceived frequency variation of sound and light waves relative to the motion of the source and detector, commonly known as the Doppler effect.

Figure 16.19 A source moving to the left creates higher-frequency sound waves ahead and lower-frequency sound waves behind.

Figure 16.20 As this car moves to the left, Observer A in front of the car hears the car horn at a higher pitch than the driver, whereas Observer B hears a lower pitch than the driver.

EXAMPLE 1

An automobile sounds its horn while passing an observer at 25 m/s. The actual horn frequency is $40\overline{0}$ Hz. Assume that the speed of sound is 345 m/s. (a) What is the frequency heard by the observer while the car is approaching? (b) What is the frequency heard when the car is leaving?

(a) Data:

$$f = 40\overline{0} \text{ Hz}$$
$$v = 345 \text{ m/s}$$
$$v_s = 25 \text{ m/s toward observer}$$
$$f' = ?$$

Basic Equation:

$$f' = f\left(\frac{v}{v - v_s}\right)$$

Working Equation: Same

Substitution:

$$f' = (40\overline{0} \text{ Hz})\left(\frac{345 \text{ m/s}}{345 \text{ m/s} - 25 \text{ m/s}}\right)$$
$$= 431 \text{ Hz}$$

(b) We simply change the sign from − to + in the basic equation of part (a). All other data remain the same. We then find

$$f' = (40\overline{0} \text{ Hz})\left(\frac{345 \text{ m/s}}{345 \text{ m/s} + 25 \text{ m/s}}\right)$$
$$= 373 \text{ Hz}$$

The Doppler effect can be easily demonstrated by two students and a toy horn. One is the observer and remains stationary. The other, with the horn, blows it while turning around. The change in pitch heard by the observer is the result of the Doppler effect.

The Doppler effect is usually experienced with sound waves. However, it is common to all waves. For visible light, the Doppler effect is seen as a change in color since the color of light is determined by its frequency. Astronomers use this principle in the study of the universe to determine whether heavenly bodies are approaching or moving away from us.

PHYSICS CONNECTIONS

Ultrasound

Ultrasound is used to examine tissue and liquid-based internal organs and systems without subjecting the patient to invasive procedures. Ultrasound is an extremely high-frequency sound wave ranging from 20 kHz to 5 MHz. A device called a transducer sends sound waves of a particular frequency and measures the frequency as the waves reflect off various media or organs. Since the speed of sound is dependent on the medium through which it travels, the ultrasound processor that takes the information from the device can determine the position and density of the tissue.

Ultrasound is often used for diagnostic purposes in place of X rays because it does not use radiation and is safer for the person and/or the fetus being examined. Along with monitoring the status of a fetus and generating deep heat for therapeutic purposes, ultrasound is also used to observe various human systems such as the nervous, urinary, reproductive, and circulatory systems. Ultrasound is not used to view bone structures because the high-frequency sound waves reflect well off only liquid-based objects. Recent technological improvements in ultrasound have led researchers to advances in three-dimensional imaging and Doppler ultrasound, where the movement of blood can be monitored using sound waves (Fig. 16.21).

Figure 16.21

Courtesy of Antonia Deutsch © Dorling Kindersley

(a) Ultrasound technology is used to visualize tissue inside the human body without subjecting the person to invasive procedures or harmful radiation.

(b) A three-dimensional ultrasound of a human fetus

PROBLEMS 16.4

1. Find the speed of sound in m/s at $1\overline{0}°C$ at 1 atm pressure in dry air.
2. Find the speed of sound in m/s at 35°C at 1 atm pressure in dry air.
3. Find the speed of sound in m/s at −23°C at 1 atm pressure in dry air.
4. How long will it take a sound to travel 21.0 m for the conditions of Problem 1?
5. How long will it take a sound to travel through $750\overline{0}$ m of water at 25°C?

SKETCH

$$\boxed{12\ cm^2} \quad w$$

4.0 cm

DATA

$A = 12\ cm^2,\ l = 4.0\ cm,\ w = ?$

BASIC EQUATION

$A = lw$

WORKING EQUATION

$w = \dfrac{A}{l}$

SUBSTITUTION

$w = \dfrac{12\ cm^2}{4.0\ cm} = 3.0\ cm$

6. A sound wave is transmitted through water from one submarine, is reflected off another submarine 15 km away, and returns to the sonar receiver on the first submarine. What is the round-trip transit time for the sound wave? Assume that the water temperature is 25°C.

7. A sonar receiver detects a reflected sound wave from another ship 3.52 s after the wave was transmitted. How far away is the other ship? Assume that the water temperature is 25°C.

8. A woman is swimming when she hears the underwater sound wave from an exploding ship across the harbor. She immediately lifts her head out of the water. The sound wave from the explosion propagating through the air reaches her 4.00 s later. How far away is the ship? Assume that the water temperature is 25°C and the air temperature is 23°C.

9. A train traveling at a speed of $4\overline{0}$ m/s approaches an observer at a station and sounds a $55\overline{0}$-Hz whistle. What frequency will be heard by the observer? Assume that the sound velocity in air is 345 m/s.

10. What frequency is heard by an observer who hears the $45\overline{0}$-Hz siren on a police car traveling at 35 m/s away from her? Assume that the velocity of sound in air is 345 m/s.

11. A car is traveling toward you at 40.0 mi/h. The car horn produces a sound at a frequency of $48\overline{0}$ Hz. What frequency do you hear? Assume that the sound velocity in air is 1090 ft/s.

12. A car is traveling away from you at 40.0 mi/h. The car horn produces a sound at a frequency of $48\overline{0}$ Hz. What frequency do you hear? Assume that the sound velocity in air is 1090 ft/s.

13. A jet airplane taxiing on the runway at 13.0 km/h is moving away from you. The engine produces a frequency of $66\overline{0}$ Hz in −6.0°C air. What frequency do you hear?

14. While snorkeling you hear a dolphin's sound as it approaches at 5.00 m/s. If the perceived frequency is $85\overline{0}$ Hz, what is the actual frequency being emitted?

15. Two construction workers stand 112 m apart. One strikes a steel beam with a hammer. How long does it take for the other to hear the sound? Assume the velocity of sound in air is 345 m/s.

16. What is the length of a brass pipe through which a sound wave is transmitted in 0.136 s?

17. A crop duster airplane flies overhead at 44.7 m/s. The frequency of the sound is 605 Hz. (a) What do you perceive as the frequency as it approaches? Assume the velocity of sound in air is 345 m/s. (b) What frequency do you hear as the plane flies away and has accelerated to 55.0 m/s?

18. Two iron workers are on a project 72.0 m apart. To get the other's attention, one worker strikes his lunch pail with his wrench. How long does it take the sound to travel through the air if the speed of sound under these conditions is 345 m/s?

19. A construction worker 30.0 m above the ground drops his hammer. How long does his shouted warning take to reach the ground if the speed of sound under these conditions is 345 m/s?

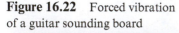

Figure 16.22 Forced vibration of a guitar sounding board

Courtesy of fotolia © Andrii Muzyka

16.5 Resonance

The **natural frequency** of an object, such as a tuning fork, is the frequency at which it vibrates when struck by another object, such as a rubber hammer. This frequency depends on the length and thickness of the tuning fork and the material from which it is made. Strings on a guitar also vibrate at a natural frequency. The sounding board of a guitar is forced to vibrate at the same frequency as the strings because of energy transfer from the strings to the sounding board (Fig. 16.22). This is an example of *forced vibration*. The natural frequency of the board is typically different from that of the strings or tuning fork. Because the area of the sounding board is large, energy transfer into sound waves is very efficient. Therefore, the vibrating string or tuning fork loses its energy or dies out more rapidly if in contact with a sounding board.

Consider two objects such as tuning forks with the same natural frequency that are set close together (Fig. 16.23). One is set into vibration and then stopped after a few seconds. It is found that the other tuning fork is weakly vibrating. The sound waves of the first fork cause the second to vibrate. This is called *sympathetic vibration* or *resonance*. **Resonance** occurs when the natural vibration rates of two objects are the same. Energy transfer into vibrations of the second fork is found to be much more efficient when both forks have the same frequency than when they have different frequencies. Large vibrations can be set up if the driving force is at the natural frequency of a system. Auto body rattles sometimes occur at certain speeds and disappear for small speed changes. Radio receivers operate on the principle of resonance. The natural frequency of vibration of electrical currents in a circuit may be tuned to that of an incoming radio signal, which is then amplified and converted into sound.

Figure 16.23   Identical tuning forks with resonant air column. Sound waves coming from the left fork cause the right fork to vibrate.

Resonant air column

Sound waves from left fork

TRY THIS ACTIVITY

Singing Wineglasses

Figure 16.24

Moisten the tip of your finger and gently rub it around the rim of a quality wineglass. At just the right speed, the friction between your finger and the wineglass causes the glass to vibrate at its natural frequency (Fig. 16.24). The glass forms a standing wave, begins to resonate, and produces the pure ringing sound of the wineglass. What can be done to the glass to change the frequency of the resonating sound?

The playing of a musical instrument demonstrates resonance. In wind instruments either the lips or a reed vibrates. Without an air column, however, no music is produced. The tube that makes up the instrument is necessary. Air in the tube vibrates

Figure 16.25 Resonance in an air column

Hammer

Tuning fork

Air column

Water

at the same frequency as the lips or reed (resonance) to produce the musical sound. The pitch of the sound is varied by changing the length of the resonating column of vibrating air. The length of the air column determines the resonant frequencies of the vibrating air. This is easily demonstrated by holding a tuning fork above a hollow tube and varying the length of the air column (Fig. 16.25). The sound is louder when the air column is in resonance with the tuning fork.

The tuning fork produces alternating high- and low-pressure variations in the air as it vibrates. Because of the movement of waves up and down the tube with accompanying constructive and destructive interference between the waves, resonance is found at one-fourth wavelengths in a closed tube. Resonance is found at multiples of $\lambda/4$. Open pipes also resonate. However, resonance in open pipes is at half-wavelength multiples ($\lambda/2$, λ, $3\lambda/2$, etc.).

16.6 Simple Harmonic Motion

Periodic motion occurs when an object moves repeatedly over the same path in equal time intervals. Attach a mass m to a spring suspended from a support [Fig. 16.26(a)]. Pull the mass down [Fig. 16.26(b)] and release it [Fig. 16.26(c)]. The mass moves up and down in periodic motion.

Simple harmonic motion is a type of linear motion in which the acceleration of an object is directly proportional to its displacement from its equilibrium position; the motion is always directed toward the equilibrium position. That is, the farther the spring is pulled down, the faster the spring moves when it is released, and the motion is always directed toward the equilibrium (rest) position. The mass on the spring in Fig. 16.26 is an example of an object in simple harmonic motion.

Next, we will compare simple harmonic motion and circular motion and discuss some of the corresponding terms. Assume that a Ferris wheel is rotating uniformly, with you in the only seat and with the sun directly overhead (Fig. 16.27).

Figure 16.26 Mass suspended by a spring from a support in simple harmonic motion

(a)   (b)   (c)

Now compare your position with the position of your shadow. When you are in position *a*, your shadow is in position *a'*; when you are in position *b*, your shadow is in position *b'*; and so on. As you complete one revolution on the Ferris wheel, your shadow makes one complete vibration (cycle) on the ground in simple harmonic motion. When your shadow is at *b'*, the *displacement* of your shadow is the distance *b'O'*, which is the distance from your shadow to the midpoint of its vibration, *O'*. In general, the **displacement** of an object in simple harmonic motion is its distance from its equilibrium, or rest, position. The **amplitude** of the vibration is the maximum displacement $O'P$ or $O'Q$, which is also the radius of the Ferris wheel. The **period** is the time required for one complete vibration—the time required for you to make one complete revolution on the Ferris wheel and the time required for your shadow to make one complete vibration on the ground. The **frequency** is the number of complete vibrations per unit of time or the number of complete revolutions that you make on the Ferris wheel per unit of time. The motion of the rider is represented as a special curve called a sine wave when the displacement of the rider is measured graphically over time as shown in Fig. 16.28. The frequency *f* equals the reciprocal of

Figure 16.27   The simple harmonic motion of the shadow of a person is shown on line *PQ*, where the person is rotating at uniform speed in a circle on a Ferris wheel. When the person is at position *a* on the Ferris wheel, his or her shadow is shown at position *a′* on line *PQ*. Shadow positions *b′*, *c′*, *d′*, and *e′* correspond to positions *b*, *c*, *d*, and *e*, respectively, of the person on the Ferris wheel.

Figure 16.28   Top view of the motion of a Ferris wheel rider is represented by a displacement vs. time graph. Positions *a′*, *b′*, *c′*, *d′*, and *e′* from Fig. 16.27 are placed on the *y*-axis.

the period *T*. That is, $f = \dfrac{1}{T}$. The equilibrium position of the shadow is the midpoint of its path, *O′*.

A **pendulum** consists of an object suspended so that it swings freely back and forth about a pivot (Fig. 16.29). Pendulums have been commonly used in clocks for many years. The motion of a pendulum, when the displacement is small, very closely approximates simple harmonic motion. There are three basic properties of a pendulum discovered by Galileo:

Figure 16.29   Free-swinging pendulum of length *l*

Pendulum

1. Its period is independent of its mass. (Air resistance is more affected by the size and shape of the bob than by its mass.)
2. Its period is independent of the amplitude when the arc is small, that is, when its arc is less than 10°.
3. Subject to these conditions, its period is given by

$$T = 2\pi\sqrt{\dfrac{l}{g}}$$

where   *T* = period (usually, in seconds)
   l = length of pendulum (m or ft)
   g = 9.80 m/s² or 32.2 ft/s²

The period of any pendulum depends only on its length and the acceleration of gravity. The longer the pendulum, the longer is the time for each complete swing or period. The less the gravitational attraction on a particular planet or moon, the larger is its period.

TRY THIS ACTIVITY

Swing Set Physics

A swing often serves as a good pendulum. Using a stopwatch, measure the period of a swing as it moves forward and backward. Vary the amplitude of the swing, the mass of the person sitting on it, and the length of the swing's chain. Which variable alters the period?

EXAMPLE 1

Find the length (in cm) of a pendulum with a period of 1.50 s.

Data:

$$T = 1.50 \text{ s}$$
$$g = 9.80 \text{ m/s}^2$$
$$l = ?$$

Basic Equation:

$$T = 2\pi\sqrt{\frac{l}{g}}$$

$$T^2 = 4\pi^2\frac{l}{g} \quad \text{Square both sides.}$$

$$gT^2 = 4\pi^2 l \quad \text{Multiply both sides by } g.$$

$$\frac{gT^2}{4\pi^2} = l \quad \text{Divide both sides by } 4\pi^2.$$

Working Equation:

$$l = \frac{gT^2}{4\pi^2}$$

Substitution:

$$l = \frac{(9.80 \text{ m/s}^2)(1.50 \text{ s})^2}{4\pi^2}$$
$$= 0.559 \text{ m} = 55.9 \text{ cm}$$

SKETCH

12 cm² | w

4.0 cm

DATA
$A = 12 \text{ cm}^2, l = 4.0 \text{ cm}, w = ?$

BASIC EQUATION
$A = lw$

WORKING EQUATION
$w = \dfrac{A}{l}$

SUBSTITUTION
$w = \dfrac{12 \text{ cm}^2}{4.0 \text{ cm}} = 3.0 \text{ cm}$

PROBLEMS 16.6

1. Find the length (in cm) of a pendulum with a period of 1.50 s.
2. Find the length (in ft) of a pendulum with a period of 3.00 s.
3. Find the period of a pendulum 1.25 m long.
4. Find the period of a pendulum 2.00 ft long.
5. Find the length (in in.) of a pendulum with a period of 2.25 s.
6. Find the length (in m) of a pendulum with a period of 0.700 s.
7. Find the period of a pendulum 18.0 in. long.
8. Find the period of a pendulum 35.0 cm long.
9. If you double the length of a pendulum, what happens to its period?
10. If you double the period of a pendulum, what happens to its length?
11. A grandfather clock has a 0.750-m pendulum. What is its period?
12. A grandfather clock has a pendulum with period 2.40 s. (a) Find the length of this pendulum. (b) Is the length of this pendulum longer or shorter than the length of the clock in Problem 11?
13. What is the period of a pendulum that is 0.25 m long?
14. What is the length of a pendulum with a period of 0.85 s?

Glossary

Amplitude The maximum displacement of any part of a wave or a vibration from its equilibrium, or rest, position. (pp. 426, 444)

Constructive Interference The superposition of waves to form a larger disturbance (wave) in a medium; occurs when two crests or troughs of superimposed waves meet. (p. 428)

Destructive Interference The superposition of waves to form a smaller disturbance (wave) in a medium. (p. 428)

Diffraction The property of a wave that describes its ability to bend around obstacles in its path. (p. 431)

Displacement The distance of an object in simple harmonic motion from its equilibrium, or rest, position. (p. 444)

Doppler Effect The variation of the frequency heard when a source of sound and the ear are moving relative to each other. (p. 439)

Electromagnetic Spectrum The entire range of electromagnetic waves classified according to frequency. (p. 434)

Electromagnetic Wave A transverse wave resulting from a periodic disturbance in an electromagnetic field, which has an electric component and a magnetic component, each being perpendicular to the other and both perpendicular to the direction of travel. (p. 433)

Frequency The number of complete waves passing a given point per unit time; the number of complete vibrations per unit time in simple harmonic motion. (pp. 426, 444)

Intensity The energy transferred by sound per unit time through unit area. (p. 438)

Interference The effect of two intersecting waves resulting in a loss of displacement in certain areas and an increase in displacement in others. (p. 430)

Longitudinal Wave A disturbance in a medium in which the motion of the particles is along the direction of the wave travel. (p. 426)

Loudness The strength of the sensation of sound to an observer. (p. 438)

Natural Frequency The frequency at which an object vibrates when struck by another object, such as a rubber hammer. (p. 442)

Pendulum An object suspended so that it swings freely back and forth about a pivot. (p. 445)

Period The time required for a single wave to pass a given point or the time required for one complete vibration of an object in simple harmonic motion. (pp. 426, 444)

Pitch The effect of the frequency of sound waves on the ear. (p. 438)

Propagation Velocity The velocity of energy transfer of a wave, given by the distance traveled by the wave in one period divided by the period. (p. 428)

Pulse Nonrepeated disturbance that carries energy through a medium or through space. (p. 426)

Resonance A sympathetic vibration of an object caused by the transfer of energy from another object vibrating at the natural frequency of vibration of the first object. (p. 443)

Simple Harmonic Motion A type of linear motion of an object in which the acceleration is directly proportional to its displacement from its equilibrium position and the motion is always directed to the equilibrium position. (p. 444)

Sound Those waves transmitted through a medium with frequencies capable of being detected by the human ear. (p. 436)

Speed of Light The speed at which light and other forms of electromagnetic radiation travel: 3.00×10^8 m/s in a vacuum. (p. 433)

Speed of Sound The speed at which sound waves travel in a medium: 331 m/s in dry air at 1 atm pressure and 0°C. (p. 436)

Standing Waves A special case of superposition of two waves when no energy propagation occurs along the wave. The wave displacements are constant and remain fixed in location. (p. 429)

Superposition of Waves The algebraic sum of the separate displacements of two or more individual waves passing through a medium. (p. 428)

Transverse Wave A disturbance in a medium in which the motion of the particles is perpendicular to the direction of the wave motion. (p. 426)
Wave A disturbance that moves through a medium or through space. (p. 426)
Wavelength The minimum distance between particles in a wave that have the same displacement and are moving in the same direction. (p. 426)

Formulas

16.1 $f = \dfrac{1}{T}$

$v = \lambda f$

16.2 $c = \lambda f$

16.3 $v = 331 \text{ m/s} + \left(0.61 \dfrac{\text{m/s}}{°\text{C}} \right) T$

$v = 1087 \text{ ft/s} + \left(1.1 \dfrac{\text{ft/s}}{°\text{F}} \right)(T - 32°\text{F})$

16.4 $f' = f\left(\dfrac{v}{v \pm v_s} \right)$

16.5 $T = 2\pi \sqrt{\dfrac{l}{g}}$

Review Questions

1. Which of the following are methods of energy transfer?
    (a) Conduction
    (b) Radiation
    (c) Wave motion
    (d) None of the above
2. The minimum distance between particles in a wave that have the same displacement and are moving in the same direction is called
    (a) the period.
    (b) the frequency.
    (c) the wavelength.
    (d) none of the above.
3. Which of the following refers to the time required for a single wave to pass a given point?
    (a) The period
    (b) The frequency
    (c) The wavelength
    (d) None of the above
4. Which of the following refers to the number of complete waves passing a given point per unit time?
    (a) The period
    (b) The frequency
    (c) The wavelength
    (d) None of the above
5. An example of a transverse wave is
    (a) a sound wave.
    (b) a water wave.
    (c) interference.
    (d) none of the above.
6. Which of the following is an example of longitudinal waves?
    (a) Sound waves
    (b) Water waves
    (c) Interference
    (d) None of the above
7. Which of the following are electromagnetic waves?
    (a) Sound
    (b) Water waves
    (c) Radar waves
    (d) X rays
    (e) All of the above
8. Explain the difference between interference and diffraction.

9.   Explain the difference between constructive and destructive interference.
10. If waves did not exhibit the property of diffraction, under what conditions would your stereo system sound different?
11. Give an example of diffraction of water waves.
12. What happens to the frequency of a vibrating string on a guitar if the length of the string is decreased?
13. Explain the difference between a wave and a pulse.
14. Give an example of a pulse.
15. What happens to the speed of sound when the temperature increases? Explain why this might happen.
16. Explain how a seismograph works.
17. How does the speed of sound differ in water and air? Explain the reason for this difference.
18. In your own words, explain the Doppler effect.
19. Distinguish between sympathetic and forced vibration.
20. In your own words, explain resonance.
21. State a reason that might explain why many stars appear to have their light shifted to the red (longer wavelength) part of the electromagnetic spectrum when viewed from the earth.
22. Distinguish between amplitude and displacement.
23. Distinguish between period and frequency.
24. Does the period of a pendulum depend on its mass, and if so, how?

Review Problems

1.   Find the period of a wave with frequency 355 kHz.
2.   Find the frequency of a wave with period 0.320 s.
3.   (a) What is the frequency of a light wave with wavelength 4.50×10^{-7} m and velocity 3.00×10^{8} m/s? (b) Find the period of the wave.
4.   Find the speed of a wave with frequency 8.97 Hz and wavelength 0.654 m.
5.   What is the wavelength of longitudinal waves in a coil spring with frequency 4.65 Hz and velocity 5.78 m/s?
6.   Find the frequency of a wave produced by a generator that emits 85 pulses in 1.3 s.
7.   What is the wavelength of an electromagnetic wave with frequency 65.5 MHz?
8.   Find the speed of sound in m/s at 85°C at 1 atm pressure in dry air.
9.   Find the speed of sound in m/s at −35°C at 1 atm pressure in dry air.
10. How long will it take a sound wave to travel through 1450 m of water at 25°C?
11. A sound wave is transmitted through water from one ship, is reflected off another ship 22 km away, and returns to the sonar receiver on the first ship. What is the round-trip transit time for the sound wave if the water temperature is 23°C?
12. A train traveling at a speed of 95 mi/h approaches an observer at a station and sounds a 525-Hz whistle. What frequency will be heard by the observer? Assume that the sound velocity in air is 1090 ft/s.
13. A car is traveling toward you at 95 km/h. The car horn produces a sound at frequency 4950 Hz. (a) What frequency do you hear? Assume that the sound velocity in air is 345 m/s. (b) What frequency do you hear if the car is traveling away from you?
14. What is the frequency of the sound waves being emitted from a train whistle while approaching at 45 m/s in air that is 11°C? The perceived frequency is 425 Hz.
15. The taillight on a car produces light with wavelength 5.00×10^{-7} m. What frequency do you observe when the car is departing at 24 m/s at 0°C?
16. A pendulum has a length of 0.450 m. What is its period?
17. A pendulum has a period of 0.700 s. Find the length of the pendulum in inches.

SKETCH

12 cm² — w

4.0 cm

DATA

$A = 12$ cm², $l = 4.0$ cm, $w = ?$

BASIC EQUATION

$A = lw$

WORKING EQUATION

$w = \dfrac{A}{l}$

SUBSTITUTION

$w = \dfrac{12 \text{ cm}^2}{4.0 \text{ cm}} = 3.0 \text{ cm}$

APPLIED CONCEPTS

1. The pendulum on a grandfather clock is calibrated so its period equals 1.00 s. (a) What is the length of the pendulum cable? (b) If the grandfather clock is moved to the moon where the acceleration due to gravity is 1.62 m/s², will the clock keep the correct time? (c) If not, what can be done to the pendulum to correct the problem?

Figure 16.30

Courtesy of Getty Images Inc., Hulton Archive Photos

2. The Tacoma Narrows Bridge, built across Puget Sound in Washington, formed a standing wave before it collapsed on November 7, 1940. As seen in the photo in Fig. 16.30, (a) how many full wavelengths were between the two towers, which were spaced 2800 ft apart? (b) If the frequency of the vibrations was 0.20 Hz, find the wave speed for this mechanical wave. (c) Find the wavelength of the bridge's standing wave if the frequency is doubled.

3. Maintaining strong AM radio reception when driving under overpasses or through tunnels is often difficult. FM radio frequencies typically are not as affected as AM frequencies. (a) Using the frequency range for AM radio, 550 kHz to 1650 kHz, and FM radio, 88 MHz to 108 MHz, find the range of wavelengths for both AM and FM bands. (b) In terms of wave diffraction, explain why AM radio waves have a more difficult time passing under overpasses and through tunnels.

4. Dave, a jet engine technician, is exposed to sound intensities of 10.3 W/m² ($12\overline{0}$ dB) at 30.4 m from a jet (Fig. 16.31). (a) What is the audible power produced by the jet engine? (b) Since the sound power produced by the engine remains constant and the sound propagates in all directions (spherically), what is the intensity of the sound 60.8 m from the engine? (c) Compare the intensity of the sound at 30.4 m to the intensity of the sound at 60.8 m from the engine.

Figure 16.31

5. The speed of an automobile can be determined using the Doppler effect by sounding the horn. (a) An automobile's horn produces a frequency of 765 Hz. How fast is the car traveling if a stationary microphone measures the horn's frequency as 836 Hz at a temperature of 23.3°C? (b) What is the frequency of the sound wave as the car moves away from the microphone?

CHAPTER 17

BASIC ELECTRICITY

Objectives

The major goals of this chapter are to enable you to:

1. Describe the nature of electric charges.
2. Distinguish conduction and induction.
3. Use Coulomb's law to find the force between charges.
4. Describe the characteristics of electricity.
5. Use Ohm's law to solve electric flow problems.
6. Use electrical symbols to describe circuits.
7. Find current, voltage, and resistance in simple circuits.
8. Describe the nature of cells and batteries.
9. Analyze circuits with cells in series and parallel.
10. Find electric power.

Electricity is the best means of transmitting energy for many purposes. We consider the basics of electricity, including electric charge, electric fields, static electricity, electric current and circuits, batteries, and electric power.

17.1 Electric Charges

Electrification was first studied 2500 years ago in ancient Greece. It was found that when an amber rod was rubbed with a wool cloth, the rod attracted small objects (Fig. 17.1). When two objects are rubbed together, they become charged.

When you slide rubber-soled shoes on a wool rug on a dry day, you become electrified. That is, you have acquired a static charge. This static charge is usually lost when you touch an object at a different potential as in Fig. 17.2. Part of the static charge may be lost when you touch an object, such as another person, to which charge is transferred.

Figure 17.1 Charged rod attracting bits of paper after being rubbed

To understand electricity, we need to know more about the structure of matter. We have seen that all matter is made up of atoms. These atoms are made of electrons, protons, and neutrons. Each **proton** has one unit of positive charge and each **electron** has one unit of negative charge. The **neutron** has no charge. The protons and neutrons are tightly packed into what is called the *nucleus*. Electrons may be thought of as small charged clouds that surround the nucleus of atoms (Fig. 17.3). An atom normally has the same number of electrons as protons and thus is uncharged. If an electron is removed, the atom is left with a *positive charge* (+), that is, an excess of protons. If an extra electron is added, the atom has a *negative charge* (−), that is, an excess of electrons. **Benjamin Franklin** first referred to electric charges as being positive and negative. The study of electric charges and the forces between them is referred to as *electrostatics*. As the name implies, *static electricity* is electricity that does not flow. *Current electricity* is the flow of electric charge through a conductor.

Figure 17.2 Stored charge being transferred

Figure 17.3

(a) Normal atom (uncharged)   (a) Atom with a positive charge   (a) Atom with a negative charge

Electron (−)   Proton (+)

When two materials are rubbed together, the atoms on the two surfaces move across each other and brush off electrons. The electrons are transferred from one surface to the other. One surface is then left with a *positive charge* and the other is *negative*. This process is called *electrification*.

T R Y   T H I S   A C T I V I T Y

Picking up Dust

Rub a furry or wool fabric against a plastic rod for approximately 15 to 20 s. Bring the rod into a dusty area or near a pile of sawdust. What happens to some of the dust or sawdust? Explain what is taking place in relation to the movement of charge.

The two types of electric charges can be observed indirectly by using an electroscope. A very simple electroscope is a ball of wood pith on a silk thread [Fig. 17.4(a)]. We can produce a charge on a hard rubber rod by rubbing it with a wool cloth. The rubber rod acquires a negative charge and the wool acquires a positive charge. The universal acceptance of the description of these charges establishes the convention of positive and negative charge in electric circuits. Now transfer some of this negative charge from the rubber rod to the pith ball [Fig. 17.4(b)]. The pith ball becomes negatively charged by **conduction**, a transfer of charge from one place to another. Another pith ball charged in the same way is repelled by the other pith ball [Fig. 17.4(c)]. This charge is *negative* (−).

Figure 17.4 Charging a simple electroscope

(a) Simple electroscope   (b) Charge transfer by conduction   (c) Like negative charges repel.

(d) Positive charge by conduction   (e) Unlike charges attract.   (f) Like positive charges repel.

Now rub a glass rod with silk. The glass rod acquires a positive charge and the silk acquires a negative charge. Transfer some of the positive charge from the glass rod to a pith ball [Fig. 17.4(d)]. This pith ball is attracted to the negatively charged pith ball [Fig. 17.4(e)]. The charge produced by glass and silk is called a *positive charge* (+). Two pith balls that are positively charged will repel each other [Fig. 17.4(f)].

17.2 Induction

Induction is a method of charging one object by bringing a charged object near to, but not touching, it. The leaf electroscope is more sensitive than the pith ball type and can also be used to show electrification or charging by induction. The leaf electroscope [Fig. 17.5(a)] usually consists of a metal rod with a metal ball on one end and two thin strips of gold foil leaf hanging from the other end. The delicate leaves are enclosed to protect them and the rod is insulated from the enclosure. When charge is placed on the leaves, they diverge because of the force of repulsion of their similar charge.

Electroscopes may be charged by *conduction* by touching a charged object to the metal ball. Electroscopes are charged by *induction* by bringing a charged object near to, but not touching, the metal ball. Like charges on the object are repelled in the electroscope to the leaves, which in turn repel each other and separate, and leave the ball with the unlike charges [Fig. 17.5(b) and (c)].

Figure 17.5 Charging an electroscope by induction

(a) Gold leaf electroscope

(b) Induced negative charges repelling

(c) Induced positive charges repelling

When the charged object is removed, the leaves close as the free electrons redistribute themselves over the ball, rod, and leaves. The electroscope has been only temporarily charged by induction. A residual charge by induction may be obtained by charging the electroscope temporarily as in Fig. 17.6(a) and then touching the electroscope with a neutral object (like your finger) while the charged object is still held close [Fig. 17.6(b)]. The neutral object provides a path for some of the induced charge on the electroscope to escape. When everything is removed, a residual charge by induction remains on the electroscope [Fig. 17.6(d)].

Figure 17.6 Residual charge by induction

| Leaves charged negatively by induction | Electrons repelled to earth | Electroscope with electron shortage | Positively charged electroscope |
| (a) | (b) | (c) | (d) |

In summary,

Like charges repel each other and unlike charges attract each other.

TRY THIS ACTIVITY

Repelled Balloons

Inflate two latex balloons, tie a piece of light string to each balloon, and suspend them from the ceiling or a horizontal rod so they are 1 ft or so apart. Rub each of the balloons with a furry fabric or your hair. Both balloons are now charged. Slowly reduce the distance between the two balloons. Look for evidence of an electric force. If you see such evidence, explain why the force is either attractive or repulsive. In addition, what are the two main factors that play a role in the strength of the electric force?

Lightning is simply a huge static electricity spark produced in the atmosphere by moving air masses that results in a tremendous discharge (Fig. 17.7). During a thunderstorm the negative charge buildup at the bottom of a layer of clouds induces a positive charge on the ground below. The ground is charged by induction, and lightning is the resulting electric discharge between the negatively charged clouds and the positively charged ground. Lightning also occurs as electric discharges between oppositely charged parts of clouds.

The lightning rod was invented by Benjamin Franklin, who observed that electricity from the air tended to accumulate on pointed objects. He found that a buildup of charge is decreased by installing a pointed rod or rods on the roof of a building and connecting the rod to the ground with a heavy conducting wire. A lightning rod not only prevents a large buildup of charge by induction, but also provides a conducting path to the ground for any sudden tremendous discharge of a lightning strike. This often results in protecting the building from damage.

The *van de Graaff generator*, named after **Robert van de Graaff**, is a laboratory machine that is used to produce static electricity and transfer it to a metal sphere by conduction. The van de Graaff consists of an electron source, a rubber belt driven by a motor, and a metal sphere supported by an insulating stand (Fig. 17.8). The electron source charges the rubber belt as it passes by, which carries the electrons up to the sphere and deposits them there. This builds up a high potential difference (several hundred thousand volts) between the sphere and the ground. Note that there is no charge on the inside surface of the sphere.

Figure 17.7   Lightning: static electric discharge

Courtesy of fotolia © yevgeniy

Figure 17.8   Van de Graaff generator

(a) Laboratory model

(b) Schematic

Collector comb
Metal sphere
Insulating support
Belt
Emitter comb
Source of electrons
Motor

(c) Student model

Courtesy of Pearson Learning Photo Studio

17.3   Coulomb's Law

The charge on a proton is denoted by the symbol e^+ and the charge on an electron by e^-. In the study of electricity, a much larger unit of charge is required and is called the *coulomb*, C, named after **Charles Coulomb**. The measurement of the unit charges gives $e^+ = +1.60 \times 10^{-19}$ C and $e^- = -1.60 \times 10^{-19}$ C. Thus, a collection of 6.25×10^{18} electrons, which is defined by

$$\frac{1}{1.60 \times 10^{-19}} = 6.25 \times 10^{18}$$

has a total charge of 1.00 C. Although this seems like a lot of electrons, it represents only the amount of charge that passes through a 100-W light bulb in about 1 s. We use q to denote amount of electric charge.

In 1789, Coulomb made a scientific study of the forces of attraction and repulsion between charged objects using a very sensitive torsion balance. From his experiments he determined the existence of an *inverse square law* for charged particles that can be used to calculate the forces of attraction or repulsion between charged objects.

Robert van de Graaff (1901–1967),

physicist, was born in Tuscaloosa, Alabama. He invented a constant-potential electrostatic generator (later known as the van de Graaff generator) and later developed this generator for use as a particle accelerator. He invented the insulating-core transformer in the late 1950s.

Charles Coulomb (1736–1806),

physicist, was born in France. He pioneered research in electricity and magnetism. The unit of measuring electricity, the coulomb, is named after him.

The inverse-square-law behavior was also found to apply to the much weaker gravitational force studied in Chapter 5. While the gravitational force is always attractive, the electric force can be either attractive or repulsive. These two forces are important because one holds the solar system together (gravity) and the other holds atoms and molecules together (electricity). Electric forces, however, are billions of times greater than the earth's gravitational force.

COULOMB'S LAW OF ELECTROSTATICS

The force between two point charges q_1 and q_2 is directly proportional to the product of their magnitudes and inversely proportional to the square of the distance separating them, r.

Figure 17.9   (a) Two like charges at distance r apart repel each other with force F. (b) Two unlike charges attract.

(a)

(b)

Figure 17.9(a) shows the repulsive force between two like (positive in this case) charges separated by a distance r. The attractive force between two unlike charges is shown in Fig. 17.9(b).

We use a *proportionality constant* k in writing Coulomb's law as an equation to take into account the air or other medium between the charges. Written in equation form, **Coulomb's law** becomes

$$F = \frac{kq_1q_2}{r^2}$$

where F = force of attraction or repulsion   (in newtons)
          $k = 9.00 \times 10^9$ N m^2/C^2 (k was found by experiment)
          q_1, q_2 = electric charges   (in coulombs)
          r = distance between the charges   (in metres)

The force between the charges is a vector quantity that acts on each charge.

TRY THIS ACTIVITY

Attracting Water

Turn on a faucet so only a very thin stream of water comes out of the tap. Charge a plastic rod, latex balloon, or plastic comb by rubbing it with a piece of wool or fur. Slowly bring the charged object toward the stream of water. Explain what happens to the stream of water as it flows from the tap.

EXAMPLE 1

Two charges, each with magnitude +6.50 μC, are separated by a distance of 0.200 cm. Find the force of repulsion between them.

Data:

$$q_1 = q_2 = +6.50 \ \mu C = +6.50 \times 10^{-6} \ C$$
$$r = 0.200 \ cm = 0.00200 \ m = 2.00 \times 10^{-3} \ m$$
$$k = 9.00 \times 10^9 \ N \ m^2/C^2$$
$$F = ?$$

Basic Equation:

$$F = \frac{kq_1q_2}{r^2}$$

Working Equation: Same

Substitution:

$$F = \frac{(9.00 \times 10^9 \text{ N m}^2/\text{C}^2)(6.50 \times 10^{-6} \text{ C})(6.50 \times 10^{-6} \text{ C})}{(2.00 \times 10^{-3} \text{ m})^2}$$

$$= 9.51 \times 10^4 \text{ N}$$

PROBLEMS 17.3

1. Two identical charges, each -8.00×10^{-5} C, are separated by a distance of 25.0 cm. What is the force of repulsion?
2. The force of repulsion between two identical positive charges is 0.800 N when the charges are 0.100 m apart. Find the value of each charge.
3. A charge of $+3.0 \times 10^{-6}$ C exerts a force of 940 N on a charge of $+6.0 \times 10^{-6}$ C. How far apart are the charges?
4. A charge of -3.0×10^{-8} C exerts a force of 0.045 N on a charge of $+5.0 \times 10^{-7}$ C. How far apart are the charges?
5. When a -9.0-μC charge is placed 0.12 cm from a charge q in a vacuum, the force between the two charges is 850 N. What is the value of q?
6. How far apart are two identical charges of $+6.00$ μC if the force between them is 25.0 N?
7. Three charges are located along the x-axis. Charge A ($+3.00$ μC) is located at the origin. Charge B ($+5.50$ μC) is located at $x = +0.400$ m. Charge C (-4.60 μC) is located at $x = +0.750$ m. (a) Find the total force (and direction) on charge B. (b) Find the total force (and direction) on charge A. (c) Find the total force (and direction) on charge C.
8. Three charges are located along the x-axis. Charge A ($+5.00$ μC) is located at the origin. Charge B ($+4.50$ μC) is located at $x = +0.650$ m. Charge C (-4.20 μC) is located at $x = +0.650$ m. Find the total force (and direction) on charge B.

17.4 Electric Fields

So far, we have discussed electrification due to the brushing of electrons from a surface. The concept of the electric field is also an important part of the study of static electricity. Two magnets may either attract or repel each other even though they may not be touching. This illustrates the idea of a "field": even though they are not touching, there is an invisible region around each magnet that affects the other magnet if that magnet is placed in the region of the first magnet.

In terms of static electricity, an **electric field** exists where an electric force (of attraction or repulsion) acts on a charge brought into the area. A charged balloon put on a wall illustrates this principle (Fig. 17.10). Note that the balloon attracts the wall even without physical contact. The balloon has acquired a negative charge through friction, but the wall surface acquires a positive charge produced by the electric field of the charged balloon. Such an invisible electric field is present around every charged object.

Static electricity can be a real hazard in industry as well as a curiosity and sometimes a nuisance in daily life. The electric spark from static electricity, particularly in synthetic fiber textile mills, is extremely dangerous. Also, some workers in cosmetic factories in which aerosol (spray) products are made with hydrocarbon (petroleum type) propellants are required to wear cotton clothes rather than those made with synthetic fibers.

Figure 17.10 Common electric field. Static electricity causes the balloon to adhere to the wall.

Figure 17.11 An electric field is created in the space around a charged object.

Electric fields can be measured by using test charges. Fields are represented using lines to show the direction and intensity of the field. Field lines do not really exist. They are just a means of providing a model of the field to visualize how the field is stronger where the lines are closer together and also to represent the direction of the field. Electric fields really do exist, though we study them by observing the effects they produce. Keep in mind that electric fields exist in three dimensions, although our drawings are only two-dimensional models (Fig. 17.11).

We can use the test charge to calculate the strength of a field but we do not yet know why charged bodies exert forces on each other. Still, we can detect and measure fields because fields produce forces on charges placed in the field.

A test charge creates an electric field about it in all directions. If a second charge is placed at some point in the field of the first charge, it interacts with the first charge. Gravitational fields exist around bodies like the earth and exert a force on nearby bodies, like people. We recall that $F_w = mg$, and we can find the gravitational field to be $g = F_w/m$. Similarly, an electric field can be described by

$$E = \frac{F}{q'}$$

where E = electric field
            F = force on a test charge placed in the field
            q' = test charge, measured in coulombs

The result is a vector quantity that is the magnitude of the electric field. By measuring the force on the test charge at different locations in the field, we can map the field and then represent it by using field lines for a model.

EXAMPLE 1

A positive test charge of 30.0 μC is placed in an electric field. The force on it is 0.600 N. What is the magnitude of the electric field at the location of the test charge?

Data:

$$q' = 30.0 \ \mu C = 3.00 \times 10^{-5} \ C$$
$$F = 0.600 \ N$$
$$E = ?$$

Basic Equation:

$$E = \frac{F}{q'}$$

Working Equation: Same

Substitution:

$$E = \frac{0.600 \ N}{3.00 \times 10^{-5} \ C}$$
$$= 2.00 \times 10^4 \ N/C$$

Electric fields are used for many applications in electronics and elsewhere. An ink-jet printer uses the deflection caused by an electric field on charged ink droplets to direct ink to the appropriate spot on the paper. In a similar way, the electron beam in an old-fashioned television picture tube is deflected to the correct spots on the screen to produce a picture (Fig. 17.12).

Figure 17.12   Electron beam deflection in an old-fashioned television

Energy can be transmitted by means of an electric field through empty space. When charges move in an electric field, they cause a disturbance that produces changes near the moving charges and results in the transfer of energy.

PHYSICS CONNECTIONS

Lightning Safety

The earth experiences about 25 lightning ground strikes every second. Each year in the United States approximately 400 people are struck by lightning, which results in 100 deaths. Many of these strikes take place on ball fields, golf courses, and other open areas.

Figure 17.13   The proper position to take if one is not able to get inside during a thunderstorm

The safest place to be during a lightning storm is inside a building or a car. The charge from the lightning strike gathers on the outside of the building or car because like electric charges repel and remain as far away from each other as possible. After accumulating on the exterior of the building or car, the charges move quickly to a ground. Although you may be safest inside a building or a car, you should stay away from electric appliances and plumbing fixtures that are connected to the ground and will act as conductors of electricity if the building is struck by lightning.

If you are not able to get inside a building or a car, you should take the following precautions during a lightning storm:

1. Move to the lowest section of ground and crouch down (Fig. 17.13). By reducing your height, you decrease the likelihood of becoming a lightning target. Do not let your hands or other parts of your unprotected body touch the ground. When lightning strikes the ground, it typically spreads out along the ground and can still reach you. If only your shoes are on the ground, the amount of current passing through your body will be reduced. If you are lying on the ground, the electric current could easily pass through your entire body, including your heart.
2. Remove all metal objects from your body. Metal on your body acts as a conductor of electricity and will attract lightning.
3. Avoid tall objects such as trees and hilltops and open areas like water and fields. Being the tallest object or standing under tall objects makes a person part of a giant lightning rod. Tall objects like trees and masts attract the charges from a thundercloud and lightning strikes.
4. If you are on a body of water, get back to shore as quickly as possible. If this is not possible, crouch down and move away from tall metal masts.

New advances in measuring electric fields are being used at ball fields. These devices measure the electric field intensity in the atmosphere and warn of possible lightning strikes. Since many lightning strikes occur at ball fields, such devices are becoming standard equipment at outdoor sporting events.

SKETCH

$A = 12\ cm^2$ w

4.0 cm

DATA

$A = 12\ cm^2$, $l = 4.0\ cm$, $w = ?$

BASIC EQUATION

$A = lw$

WORKING EQUATION

$w = \dfrac{A}{l}$

SUBSTITUTION

$w = \dfrac{12\ cm^2}{4.0\ cm} = 3.0\ cm$

PROBLEMS 17.4

1. An electric field has a positive test charge of 4.00×10^{-5} C placed on it. The force on it is 0.600 N. What is the magnitude of the electric field at the test charge location?
2. What is the field magnitude of an electric field in which a negative charge of 2.00×10^{-8} C experiences a force of 0.0600 N?
3. An electric field exerts a force of 2.50×10^{-4} N on a positive test charge of 5.00×10^{-4} C. Find the magnitude of the field at the charge location.
4. An electric field exerts a force of 3.00×10^{-4} N on a positive test charge of 7.50×10^{-4} C. Find the magnitude of the field at the charge location.
5. An electric field of magnitude 0.450 N/C exerts a force of 8.00×10^{-4} N on a test charge placed in the field. What is the magnitude of the test charge?
6. An electric field of magnitude 0.370 N/C exerts a force of 6.20×10^{-4} N on a test charge placed in the field. What is the magnitude of the test charge?
7. What force is exerted on a test charge of 3.86×10^{-5} C if it is placed in an electric field of magnitude 1.75×10^4 N/C?
8. What force is exerted on a test charge of 4.00×10^{-5} C if it is placed in an electric field of magnitude 3.00×10^6 N/C?

Figure 17.14 Current flowing in only one direction is direct current.

● Electron

Figure 17.15 Changing stored electric energy to mechanical energy for drilling

Courtesy of fotolia © Titi Matei

17.5 Simple Circuits

Electrons moving in a wire produce a current in the wire. When the electron current flows in only one direction (Fig. 17.14), it is called **direct current** (dc). Current that changes direction is called **alternating current** (ac). Alternating current will be considered in Chapter 19.

An electric current is a convenient and cost-effective means of transmitting energy. We all face many daily situations that require energy to do a particular task. To drill a hole in a metal stud (Fig. 17.15), energy must be supplied and transformed into mechanical energy to turn the drill bit. The problem is how to supply energy to the machine being used in a form that the machine can turn into useful work. Electricity is often the most satisfactory means of transmitting energy.

We begin our study of the use of electricity in transferring energy with the circuit of a simple flashlight (Fig. 17.16). An **electric circuit** is a conducting loop in which electrons carrying electric energy may be transferred from a suitable source to do useful work and returned to the source. Energy is stored in the battery. When the switch is closed, energy is transmitted to the light, and the light glows.

Figure 17.16 Simple electric circuit

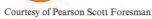

Dry cell

Switch

Light bulb

Courtesy of Pearson Scott Foresman

Current electricity is the flow of energized electrons through an electron carrier called a conductor (Fig. 17.17). The electrons move from the energy *source* (the battery, here) to the *load* (where the transmitted energy is turned into useful work). There they lose energy picked up in the source. We now consider each part of the circuit and determine its function.

The Source of Energy

The **source** is the object that supplies electric energy for the flow of electric charge (electrons) in a circuit. The dry cell (Fig. 17.18) is a device that converts chemical energy to electric energy. How the cell does this will be studied later in this chapter. Here,

Figure 17.17   Flow of energized electrons through a conductor

we simply state that, by chemical action, electrons are given energy in the cell. When energy is given to electrons in this manner, their electric potential energy is raised.

What does "electric potential energy" mean? The flow of charge in an electric circuit is often compared to the flow of water in a hydraulic system, as shown in Fig. 17.19(b). Water naturally flows from a position of high potential to a position of low potential and performs work in the process, such as turning a waterwheel or turbine. A pump is needed to return the water from its low-potential position to its high-potential position. There is a *difference in potential* due to its position. Work has been done in lifting it against gravity to the higher position. In a source of electric energy something similar happens. In the source (the battery), work is done on electrons that gives them potential energy. This potential difference between the high–potential energy energized electrons [at the negative (−) pole] and the low–potential energy electrons [at the positive (+) pole] causes the electrons to flow from one point (−) to the other (+) when connected [Fig. 17.19(a)].

Figure 17.18   Chemical energy is changed to electric energy in a dry cell.

Figure 17.19   The flow of charge in an electric circuit is often compared to the flow of water in a hydraulic system.

(a) Energized electrons flow from the source to the load, where energy is lost.

(b)

Think of this potential difference between the two points as an electric field set up between the poles that drives the electrons in the external circuit from the negative pole to the positive pole. The energized electrons collect at the source's negative pole, repel each other, and flow through the circuit to the positive pole. They lose their potential electric energy to the load.

The Conductor

A conductor carries or transfers the electric charge to the load [Fig. 17.19(a), the light bulb]. **A conductor** is a material (such as copper) through which an electric charge is readily transferred. Such materials have large numbers of free electrons (electrons that are free to move throughout the conductor). As high-energy electrons from the source pass through the conductor, they collide with other electrons in the conductor. These electrons then carry the energy farther along the wire until they collide again and transfer energy on through the wire.

Silver, copper, and aluminum are metals that allow electrons to pass freely through and thus are good conductors. Other metals offer more opposition to the flow of electrons and are poorer conductors (Fig. 17.20). Substances that do not allow electrons to pass readily are called **insulators**. Common insulators are rubber, wool, silk, glass, wood, distilled water, and dry air.

A small number of materials, called **semiconductors**, fall between conductors and insulators in their ability to conduct electric current. Their importance is due to the fact that these materials, under certain conditions, allow current to flow in one direction only. Silicon is a semiconductor used in transistors and integrated circuits (ICs). Semiconductors are neither good conductors nor good insulators in their pure form. However, they become excellent conductors or insulators when an impurity is added. Transistors are made by layering semiconductor materials together.

Selenium is a semiconductor used in the process of making photocopies. Charge built up on its surface will remain there in the dark. When a certain colored light shines on it, the charge will dissipate very quickly from the areas exposed to the light. A powder sticking only to the charged areas is transferred to a piece of paper, creating a photocopy.

A **superconductor** is a material that continuously conducts electric current without resistance when cooled to typically very low temperatures, usually near absolute zero. Heike Kamerlingh Onnes discovered superconductivity in 1911 shortly after he discovered how to liquefy helium gas. He determined that mercury metal lost its resistance to the flow of electricity at temperatures just below 4.2 K, the boiling point of helium. Scientists are currently finding materials in which superconductivity exists at higher temperatures. The ultimate goal is to find a material in which superconductivity exists at room temperature.

Figure 17.20

(a) Good conductor

(b) Poor conductor

The Load

In the load, electrons lose their energy. The **load** in a circuit converts the electric energy into other forms of energy or work. In a light bulb, electric energy is changed to light and heat (Fig. 17.21). An electric motor changes electric energy to mechanical energy. The load may be a complex motor or only a simple resistor with heat the only new form of energy. The electrons do not collect and remain in the load, but continue back to the low-energy side of the battery (+). There they are energized again for another trip through the circuit.

Figure 17.21   Electric energy is changed to light and heat in the load.

Current

The flow of electrons through a conductor is called **current**. We could count the electrons passing a point during a certain time to get the rate of flow. This is impractical because the flow of electrons is so large (about 10^{18}/s). To have a workable unit of electric charge, we define a charge of 6.25×10^{18} electrons as 1 *coulomb* (C). The *ampere* (A) is the rate of flow of 1 C of charge passing a point in 1 s. We define a unit for the rate of flow of charge as follows:

$$1 \text{ ampere (A)} = \frac{1 \text{ coulomb (C)}}{1 \text{ second (s)}}$$

As mentioned earlier, the charge carriers in metals are electrons. In some other conductors, such as electrolytes (conducting liquid solutions), the charge carriers may be positive or negative or both. An agreement must be made to determine which charge carriers should be assumed in our following discussions.

Note that positive charges flow in the opposite direction (toward the negative terminal) from that of negative charges (toward the positive terminal) when a battery is connected to a circuit (Fig. 17.22). A positive current moving in one direction is equivalent for almost all measurements to a current of negative charges flowing in the opposite direction.

In this book we assume that the charge carriers are positive, and we draw our current arrows in the direction that a positive charge would flow. This is the practice of the majority of engineers and technicians. If you encounter the negative-current convention, you should remember that a negative current flows in the opposite direction from that of a positive current. Regardless of the method used, the analysis of a situation by either method will give the correct result. Some of the rules discussed later, such as the right-hand rule for finding the direction of the magnetic field, will be different if the negative-current convention is used.

Figure 17.22 Current flow as positive current or negative current

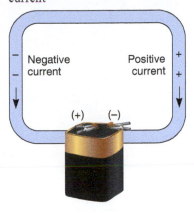

Voltage

We have seen that current flows in a circuit because of the difference in potential of the different points in the circuit. Work is done as a charge moves from one point to another in an electric field. Work is required to move a charge from one point to another when such points differ in electric potential. The *potential difference* between two points in an electric field is the work done per unit of charge as the charge is moved between two points. That is,

$$\text{potential difference} = \frac{\text{work}}{\text{charge}}$$

In *sources*, the raising of the potential energy of electrons that results in a potential difference across a source is called **emf** (*E*). In *circuits*, the lowering of the potential difference across a load is called **voltage drop**.

The *volt* (V), named after **Allessandro Volta**, is the unit of both emf and voltage drop. We define the volt as the potential difference between two points if 1 J of work is produced or used in moving 1 C of charge from one point to another:

$$1 \text{ volt (V)} = \frac{1 \text{ joule (J)}}{1 \text{ coulomb (C)}}$$

Resistance

Not all substances and not even all metals are good conductors of electricity. Those with few free electrons tend to have greater opposition to the flow of charge. This opposition to current flow is called **resistance**. The unit of resistance is the *ohm* (Ω). It is not a fundamental unit and is discussed in Section 17.6.

The resistance of a wire is determined by several factors. Among these are:

1. *Temperature.* An increase in temperature results in an increase in resistance in a wire, for most metals. Other materials, such as semiconductors, show a decrease in resistance with increasing temperature.
2. *Length.* Resistance varies directly with length. If we double the length of a given wire, the resistance is doubled [Fig. 17.23(a)].
3. *Cross-sectional area.* Resistance varies inversely with cross-sectional area. If we double the cross section of a wire, the resistance is *halved*. This is similar to water flowing through two pipes. It flows more easily through the larger pipe. (Note that doubling the radius of a wire [Fig. 17.23(b)] *more than* doubles the cross-sectional area: $A = \pi r^2$.)
4. *Material.* Resistance depends on the nature of the material. For example, copper is a better conductor than steel. The conducting characteristic of various materials is described by resistivity. **Resistivity** (ρ) is the resistance per unit length of a material with uniform cross section.

Figure 17.23

(a) Resistance varies directly with length.

(b) Doubling the radius more than doubles the cross-sectional area.

These factors are related by the equation

$$R = \frac{\rho l}{A}$$

where R = resistance
ρ = resistivity
l = length
A = cross-sectional area

EXAMPLE 1

Find the resistance of a copper wire 20.0 m long with cross-sectional area of 6.56×10^{-3} cm^2 at 20°C. The resistivity of copper at 20°C is 1.72×10^{-6} Ω cm.

Data:

$$l = 20.0 \text{ m} = 2.00 \times 10^3 \text{ cm}$$
$$A = 6.56 \times 10^{-3} \text{ cm}^2$$
$$\rho = 1.72 \times 10^{-6} \text{ Ω cm}$$
$$R = ?$$

Basic Equation:

$$R = \frac{\rho l}{A}$$

Working Equation: Same

Substitution:

$$R = \frac{(1.72 \times 10^{-6} \text{ Ω cm})(2.00 \times 10^3 \text{ cm})}{6.56 \times 10^{-3} \text{ cm}^2}$$

$$= 0.524 \text{ Ω}$$

SKETCH

12 cm²	w

4.0 cm

DATA
$A = 12$ cm^2, $l = 4.0$ cm, $w = ?$

BASIC EQUATION
$A = lw$

WORKING EQUATION
$w = \dfrac{A}{l}$

SUBSTITUTION
$w = \dfrac{12 \text{ cm}^2}{4.0 \text{ cm}} = 3.0$ cm

PROBLEMS 17.5

1. Find the resistance of 78.0 m of No. 20 aluminum wire at 20°C. ($\rho = 2.83 \times 10^{-6}$ Ω cm, $A = 2.07 \times 10^{-2}$ cm^2.)
2. Find the resistance of 315 ft of No. 24 copper wire with resistance 0.0262 Ω/ft.
3. Find the resistance per foot of No. 22 copper wire if $58\overline{0}$ ft has a resistance of 9.57 Ω.
4. At 77°F, $10\overline{0}$ ft of No. 18 copper wire has a resistance of 0.651 Ω. Find the resistance of $50\overline{0}$ ft of this wire.
5. Find the resistance of 475 m of No. 20 copper wire at 20°C. ($\rho = 1.72 \times 10^{-6}$ Ω cm, $A = 2.07 \times 10^{-2}$ cm^2.)

6. Find the resistance of $10\overline{0}$ m of No. 20 copper wire at 20°C. ($\rho = 1.72 \times 10^{-6} \ \Omega$ cm, $A = 2.07 \times 10^{-2}$ cm^2.)
7. Find the resistance of 50.0 m of No. 20 aluminum wire at 20°C. ($\rho = 2.83 \times 10^{-6} \ \Omega$ cm, $A = 2.07 \times 10^{-2}$ cm^2.)
8. Find the length of copper wire with resistance 0.0262 Ω/ft and total resistance 3.00 Ω.
9. Find the cross-sectional area of copper wire at 20°C that is 60.0 m long and has resistivity $\rho = 1.72 \times 10^{-6} \ \Omega$ cm and resistance 0.788 Ω.
10. Find the length of a copper wire with resistance 0.0262 Ω/ft and total resistance 5.62 Ω.

17.6 Ohm's Law

When a voltage is applied *across* a material that conducts electric current, the relationship between current *through* the material and voltage across it depends upon the type of material as shown in Fig. 17.24. The straight-line relationship shown in Fig. 17.24(a) is typical of many materials, including metal conductors. Other materials, such as semiconductors shown in Fig. 17.24(b), show a nonlinear relationship between I and V. For the materials with a straight-line relationship, the equation relating I and V was determined by **Georg Simon Ohm**. The relationship is called **Ohm's law** (see Fig. 17.25).

> **Georg Simon Ohm (1787–1854),**
>
> *physicist, was born in Bavaria. Ohm's law resulted from his work and research in electricity. The unit of the measure of electric resistance, the ohm, is named after him.*

Figure 17.24

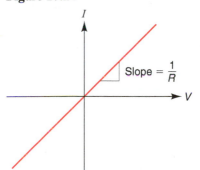

(a) Straight-line (linear) I–V characteristic (typical of resistors)

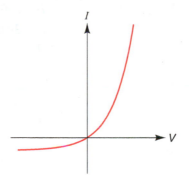

(b) Nonlinear I–V characteristic (typical of semiconductor diodes)

Figure 17.25 The relationship between the voltage *across* a resistance and the current *through* the resistance is described by Ohm's law.

Ohm's law

$$I = \frac{V}{R}$$

where I = current *through* the resistance
V = voltage drop *across* the resistance
R = resistance

Ohm's law can also be written

$$I = \frac{E}{R}$$

where E = emf of the source of electrical energy

Ohm's law applies to dc circuits containing linear resistors and those ac circuits containing only resistance. It may be applied to the whole circuit or to any part of it.

Ohm's law can aid us in understanding resistance. As we mentioned earlier, the ohm (Ω) is a derived unit. From Ohm's law,

$$I = \frac{V}{R}$$

Solving for R, we obtain

$$R = \frac{V}{I}$$

Substituting units, we obtain

$$\Omega = \frac{V}{A}$$

TRY THIS ACTIVITY

String of Lights

Attach a D-cell battery to a small 2.5-V or 3.5-V light bulb and observe the brightness of the light. Disconnect the circuit and attach another light bulb between the first bulb and the battery. Observe the relative brightness of the bulbs compared to when only one bulb was lit. Repeat the process with two or three additional bulbs. Using Ohm's law, explain what happened to the brightness of each bulb.

EXAMPLE 1

A heating element on an electric range operating on 240 V has a resistance of 30.0 Ω. What current does it draw?

Data:

$$E = 240 \text{ V}$$
$$R = 30.0 \text{ } \Omega$$
$$I = ?$$

Basic Equation:

$$I = \frac{E}{R}$$

Working Equation: Same

Substitution:

$$I = \frac{240 \text{ V}}{30.0 \text{ } \Omega}$$
$$= 8.0 \text{ V/}\Omega$$
$$= 8.0 \text{ A} \qquad \boxed{\frac{V}{\Omega} = A}$$

· · · · · · · · · · · · · · · · ·

EXAMPLE 2

A flashlight bulb is connected to two dry cells with an equivalent voltage of 3.0 V. If it draws 15 mA, what is its resistance?

Sketch:

$$R = ?$$
$$E = 3.0 \text{ V}$$
$$I = 15 \text{ mA}$$

Data:

$$E = 3.0 \text{ V}$$
$$I = 15 \text{ mA} = 15 \times 10^{-3} \text{ A} = 0.015 \text{ A}$$
$$R = ?$$

Basic Equation:

$$I = \frac{E}{R}$$

Working Equation:

$$R = \frac{E}{I}$$

Substitution:

$$R = \frac{3.0 \text{ V}}{0.015 \text{ A}}$$
$$= 2\overline{0}0 \text{ V/A}$$
$$= 2\overline{0}0 \text{ }\Omega$$

Electric shock is a very real hazard to electricians. High voltage does not carry the danger that a large current does. The resistance of the human body dry would normally be in the range of 100,000 to 500,000 Ω. If a person is standing in a bathtub, however, his or her resistance is greatly lowered by the water, and any electric appliance falling into the water could deliver sufficient current to cause death. A current as small as 0.070 A can cause serious damage to the nervous system and be fatal. Electricians can protect themselves by using only one hand in some circumstances where a wire must be grasped so that no circuit is completed. Another technique is to touch any questionable wire with the back of one's hand so that if there is any current present, an unexpected shock will not cause a muscular contraction that will keep the hand gripping the wire. In any attempted rescue of a shock victim, the first task is to clear the person from the electric supply with a piece of wood or other nonconductor to avoid the rescuer becoming a second victim.

Appliances today are usually supplied with a third prong on the electric plug to conduct any charge buildup on the appliance to the ground. Sometimes appliances are also made with insulating cases to achieve the same goal—to provide a path not through the human body for electricity to flow.

PROBLEMS 17.6

1. A heating element operates on 115 V. If it has a resistance of 24.0 Ω, what current does it draw?
2. A coffeepot operates on 12.0 V. If it draws 2.50 A, find its resistance.
3. An electric heater draws a maximum of 14.0 A. If its resistance is 15.7 Ω, on what voltage is it operating?
4. A heating coil operates on $22\overline{0}$ V. If it draws 15.0 A, find its resistance.
5. Find the resistance that draws 0.750 A on 115 V.
6. What current does a 75.0-Ω resistance draw on 115 V?
7. A heater operates on $22\overline{0}$ V. If it draws 12.5 A, what is its resistance?
8. What current does a 50.0-Ω resistance draw on 115 V?
9. What current does a 175-Ω resistance draw on $22\overline{0}$ V?
10. A heater draws 3.50 A on 115 V. What is its resistance?
11. (a) What current does a 150-Ω resistance draw on a $1\overline{0}$-V battery? (b) What voltage battery would produce three times the current in (a)? (c) What current would a 75-Ω resistor draw on the $1\overline{0}$-V battery?
12. A heater draws 4.25 A on 32.0 V. (a) What is the resistance of the heater? (b) What resistance heater would draw 8.50 A on 32.0 V?
13. Electric characteristics of all consumer electric devices must be shown on an attached plate. What is the resistance of an iron that discloses 6.40 A of current used on a $12\overline{0}$-V line?

SKETCH

12 cm² w

4.0 cm

DATA
$A = 12$ cm², $l = 4.0$ cm, $w = ?$

BASIC EQUATION
$A = lw$

WORKING EQUATION
$w = \dfrac{A}{l}$

SUBSTITUTION
$w = \dfrac{12 \text{ cm}^2}{4.0 \text{ cm}} = 3.0$ cm

14. What is the effective resistance of a television that draws 2.50 A on a 115-V line?
15. Find the current used by a stereo with resistance 65.0 Ω in a $12\bar{0}$-V system.
16. What is the current used by a microwave oven with resistance 20.0 Ω in a $12\bar{0}$-V system?

17.7 Series Circuits

In order to communicate about problems in electricity, technicians have developed a "picture language" of their own using symbols and diagrams. The circuit diagram is the most common and useful way to show a circuit. Note how each component (part) of the picture in Fig. 17.26(a) is represented by its symbol in the symbol diagram in its relative position in Fig. 17.26(b). The light bulb can be represented as a resistance. Then the circuit diagram appears as in Fig. 17.26(c). Some of the symbols used most often appear in Appendix C, Table 20.

Figure 17.26   A series circuit

There are two basic types of circuits: series and parallel. A fuse in a house is wired in series with the outlets. The outlets themselves are wired in parallel. A study of series and parallel circuits is basic to a study of electricity.

An electric circuit with only one path for the current to flow (Fig. 17.27) is called a **series circuit**. The current in a series circuit is the same throughout. That is, the current flows out of one resistance and into the next resistance. Therefore, the total current is the same as the current flowing through each resistance in the circuit.

Figure 17.27   Series circuit

$$\boxed{\begin{array}{c} \textit{SERIES} \\ I = I_1 = I_2 = I_3 = \cdots \end{array}}$$

where I = total current
  I_1 = current through R_1
  I_2 = current through R_2
  I_3 = current through R_3

In a series circuit, the emf of the source equals the sum of the separate voltage drops in the circuit (Fig. 17.27):

$$\boxed{\begin{array}{c} \textit{SERIES} \\ E = V_1 + V_2 + V_3 + \cdots \end{array}}$$

where E = emf of the source
  V_1 = voltage drop across R_1
  V_2 = voltage drop across R_2
  V_3 = voltage drop across R_3

The resistance of the conducting wires is very small and will be neglected here. The total resistance of a series circuit equals the sum of all the resistances in the circuit:

$$\boxed{\begin{array}{c} \textit{SERIES} \\ R = R_1 + R_2 + R_3 + \cdots \end{array}}$$

where R = total or equivalent resistance of the circuit
  R_1 = resistance of first load
  R_2 = resistance of second load
  R_3 = resistance of third load

The **equivalent resistance** is the single resistance that can replace a series and/or parallel combination of resistances in a circuit and provide the same current flow and voltage drop. The equivalent resistance of a series combination is larger than the resistance of any one of the resistances in series.

Find the total resistance of the circuit shown in Fig. 17.28.

EXAMPLE 1

Data:

$$R_1 = 7.00 \ \Omega$$
$$R_2 = 9.00 \ \Omega$$
$$R_3 = 21.0 \ \Omega$$
$$R = ?$$

Basic Equation:

$$R = R_1 + R_2 + R_3$$

Working Equation: Same

Substitution:

$$R = 7.00 \ \Omega + 9.00 \ \Omega + 21.0 \ \Omega$$
$$= 37.0 \ \Omega$$

Figure 17.28

EXAMPLE 2

Find the current in the circuit shown in Fig. 17.29.

Data:

Figure 17.29

$$R_1 = 5.00\ \Omega$$
$$R_2 = 13.0\ \Omega$$
$$R_3 = 12.0\ \Omega$$
$$R_4 = 96.0\ \Omega$$
$$E = 90.0\ \text{V}$$
$$I = ?$$

Basic Equations:

$$R = R_1 + R_2 + R_3 + R_4 \quad \text{and} \quad I = \frac{E}{R}$$

Working Equations: Same

Substitutions:

$$R = 5.00\ \Omega + 13.0\ \Omega + 12.0\ \Omega + 96.0\ \Omega$$
$$= 126.0\ \Omega$$

$$I = \frac{90.0\ \text{V}}{126.0\ \Omega}$$
$$= 0.714\ \text{A}$$

· · · · · · · · · · · · · · ·

EXAMPLE 3

Find the value of R_3 in the circuit shown in Fig. 17.30.

Data:

Figure 17.30

$$I = 3.00\ \text{A}$$
$$E = 115\ \text{V}$$
$$R_1 = 23.0\ \Omega$$
$$R_2 = 14.0\ \Omega$$
$$R_3 = ?$$

Basic Equations:

$$I = \frac{E}{R} \quad \text{and} \quad R = R_1 + R_2 + R_3$$

Working Equations:

$$R = \frac{E}{I} \quad \text{and} \quad R_3 = R - R_1 - R_2$$

Substitutions:

$$R = \frac{115\ \text{V}}{3.00\ \text{A}}$$
$$= 38.3\ \Omega$$

$$R_3 = 38.3\ \Omega - 23.0\ \Omega - 14.0\ \Omega$$
$$= 1.3\ \Omega$$

· · · · · · · · · · · · · · ·

Find the voltage drop across R_3 in Example 3.

EXAMPLE 4

Data:

$$I = I_3 = 3.00 \text{ A}$$
$$R_3 = 1.3 \text{ } \Omega$$
$$V_3 = ?$$

Basic Equation:

$$I_3 = \frac{V_3}{R_3}$$

Working Equation:

$$V_3 = I_3 R_3$$

Substitution:

$$V_3 = (3.00 \text{ A})(1.3 \text{ } \Omega)$$
$$= 3.9 \text{ V}$$

Table 17.1 summarizes the characteristics of series circuits.

Table 17.1   Characteristics of Series Circuits

	Series
Current	$I = I_1 = I_2 = I_3 = \cdots$
Equivalent resistance	$R = R_1 + R_2 + R_3 + \cdots$
Voltage	$E = V_1 + V_2 + V_3 + \cdots$

PROBLEMS 17.7

1. Three resistors of 2.00 Ω, 5.00 Ω, and 6.50 Ω are connected in series with a 24.0-V battery. Find the total resistance of the circuit.
2. Find the current in Problem 1.
3. Find the equivalent resistance in the circuit shown in Fig. 17.31.
4. Find the current through R_2 in Problem 3.
5. Find the current in the circuit shown in Fig. 17.32.
6. Find the voltage drop across R_1 in Problem 5.

SKETCH

12 cm² w

4.0 cm

DATA
$A = 12 \text{ cm}^2, l = 4.0 \text{ cm}, w = ?$

BASIC EQUATION
$A = lw$

WORKING EQUATION
$w = \dfrac{A}{l}$

SUBSTITUTION
$w = \dfrac{12 \text{ cm}^2}{4.0 \text{ cm}} = 3.0 \text{ cm}$

Figure 17.31

Figure 17.32

7. What emf is needed for the circuit shown in Fig. 17.33?
8. Find the voltage drop across R_3 in Problem 7.
9. Find the equivalent resistance in the circuit shown in Fig. 17.34.
10. Find R_3 in the circuit in Problem 9.

Figure 17.33

Figure 17.34

11. Find the values of R_1, R_2, and R_3 in Fig. 17.35.
12. Find the values of V_1, R_2, and V_3 in Fig. 17.36.

Figure 17.35

Figure 17.36

13. Find the values of R_1, V_2, and R_3 in Fig. 17.37.

Figure 17.37

17.8 Parallel Circuits

An electric circuit with more than one path for the current to flow (Fig. 17.38) is called a **parallel circuit**. All resistances connected in parallel have their ends connected to two common points (nodes) in the circuit (points A and B in Fig. 17.38).

Figure 17.38   Different ways to represent a parallel circuit

(a)   (b)

Figure 17.39 $I = I_1 + I_2 + I_3$

The current in a parallel circuit is divided among the branches of the circuit (Fig. 17.39). How it is divided depends on the resistance of each branch. The paths with the least resistance allow the largest currents to flow. Since the current divides, the current from the source equals the sum of the currents through each of the branches:

$$\textit{PARALLEL}$$
$$I = I_1 + I_2 + I_3 + \cdots$$

where I = total current in the circuit
  I_1 = current through R_1
  I_2 = current through R_2
  I_3 = current through R_3

Since the ends of all resistances in parallel are connected to the same common points (nodes) in the circuit, the voltage across each resistance is the same (Fig. 17.39):

$$\boxed{\begin{array}{c} PARALLEL \\ V_1 = V_2 = V_3 = \cdots \end{array}}$$

The emf of the source is the same as the voltage drop across each resistance in the circuit if there are no other (series) elements in the circuit (Fig. 17.40):

Figure 17.40 $E = V_1 = V_2 = V_3$

$$\boxed{\begin{array}{c} PARALLEL\ WITH\ VOLTAGE\ SOURCE \\ E = V_1 = V_2 = V_3 = \cdots \end{array}}$$

where E = emf of the source
  V_1 = voltage drop across R_1
  V_2 = voltage drop across R_2
  V_3 = voltage drop across R_3

Therefore, several different loads requiring the same voltage are connected in parallel.

The single resistance that would result in the same current flow and voltage drop as the combination of resistances is called the *equivalent resistance*. The equivalent resistance of a parallel circuit is less than the resistance of any single branch of the circuit. To find the equivalent resistance, use the formula

$$\boxed{\begin{array}{c} PARALLEL \\ \dfrac{1}{R} = \dfrac{1}{R_1} + \dfrac{1}{R_2} + \dfrac{1}{R_3} + \cdots \end{array}}$$

where R = equivalent resistance
  R_1 = resistance of R_1
  R_2 = resistance of R_2
  R_3 = resistance of R_3

If the parallel combination of resistances is replaced by a single resistance with the resistance R, the same current flows in the circuit. In the case where there are only two resistances in parallel, then

Figure 17.41   Resistor R in part (b) is equivalent to the pair of resistances R_1 and R_2 connected in parallel in part (a).

$$\frac{1}{R} = \frac{1}{R_1} + \frac{1}{R_2}$$

$$R = \frac{R_1 R_2}{R_1 + R_2}$$

(a)

(See Fig. 17.41.)

For comparison to parallel circuits, consider the water system shown in Fig. 17.42(a).

1.   The total amount of water flowing through $R_1 + R_2 + R_3$ equals the amount flowing through A or B.
2.   The water flowing past point A divides into the three branches R_1, R_2, and R_3.
3.   The larger pipes have *less* opposition to water flow than do the smaller pipes. Because R_1 has a larger cross-sectional area than R_2 or R_3, it has less opposition to the flow of water and therefore carries more water than R_2 or R_3.

(b)

Figure 17.42   A water system may be compared to a parallel electric circuit.

(a)

(b)

Similarly, in a parallel electric circuit as in Fig. 17.42(b):

1. The total amount of current flowing through $R_1 + R_2 + R_3$ equals the amount flowing through A or B.
2. The current flowing past point A divides into the three branches R_1, R_2, and R_3.
3. The smaller resistances have *less* opposition to current flow and therefore carry larger currents.

TRY THIS ACTIVITY

Parallel Bulbs

Attach a D-cell battery to a small 2.5-V or 3.5-V light bulb and observe the brightness of the light. Attach a second light bulb in parallel with the first. After adding a third bulb in parallel with the others, note the brightness of the bulbs. Why, when using the same battery, wires, and bulbs as in the "String of Lights" Try This Activity in Section 17.6 does the brightness of the bulbs differ from the bulbs in the series circuit?

EXAMPLE 1

Find the equivalent resistance of the circuit shown in Fig. 17.43.

Data:

Figure 17.43

$$R_1 = 7.00 \ \Omega$$
$$R_2 = 9.00 \ \Omega$$
$$R_3 = 12.0 \ \Omega$$
$$R = ?$$

Basic Equation:

$$\frac{1}{R} = \frac{1}{R_1} + \frac{1}{R_2} + \frac{1}{R_3}$$

Working Equation:

When using this formula, you should solve for the reciprocal of the unknown, then substitute.

Substitution:

$$\frac{1}{R} = \frac{1}{7.00 \ \Omega} + \frac{1}{9.00 \ \Omega} + \frac{1}{12.0 \ \Omega}$$
$$R = 2.96 \ \Omega$$

The key entry sequence on a scientific calculator for this calculation is

```
2.964705882
```

· · · · · · · · · · · · · · · · ·

Find the total current in the circuit shown in Fig. 17.44.

EXAMPLE 2

Data:

$$R_1 = 23.0 \ \Omega$$
$$R_2 = 14.0 \ \Omega$$
$$R_3 = 5.00 \ \Omega$$
$$E = 90.0 \ V$$
$$I = ?$$

Figure 17.44

First, find the equivalent resistance, R. Second, find the total current, I. To find R:

Basic Equation:

$$\frac{1}{R} = \frac{1}{R_1} + \frac{1}{R_2} + \frac{1}{R_3}$$

Working Equation: Same

Substitution:

$$\frac{1}{R} = \frac{1}{23.0 \ \Omega} + \frac{1}{14.0 \ \Omega} + \frac{1}{5.00 \ \Omega}$$

Using a calculator sequence as in Example 1, we find

$$R = 3.18 \ \Omega$$

To find I:

Basic Equation:

$$I = \frac{E}{R}$$

Working Equation: Same

Substitution:

$$I = \frac{90.0 \ V}{3.18 \ \Omega}$$
$$= 28.3 \ A$$

· · · · · · · · · · · · · · · · ·

Find the current through R_2 in Fig. 17.44 from Example 2.

EXAMPLE 3

Data:

$$R_2 = 14.0 \ \Omega$$
$$E = 90.0 \ V = V_2$$
$$I_2 = ?$$

Basic Equation:

$$I_2 = \frac{V_2}{R_2}$$

Working Equation: Same

Substitution:

$$I_2 = \frac{90.0 \text{ V}}{14.0 \text{ }\Omega}$$
$$= 6.43 \text{ A}$$

. . . .

EXAMPLE 4

Find the equivalent resistance and the value of R_3 in the circuit shown in Fig. 17.45.

Data:

Figure 17.45

$$E = 115 \text{ V}$$
$$I = 7.00 \text{ A}$$
$$R_1 = 38.0 \text{ }\Omega$$
$$R_2 = 49.0 \text{ }\Omega$$
$$R_3 = ?$$

First find R:

Basic Equation:

$$I = \frac{E}{R}$$

Working Equation:

$$R = \frac{E}{I}$$

Substitution:

$$R = \frac{115 \text{ V}}{7.00 \text{ A}}$$
$$= 16.4 \text{ }\Omega$$

To find R_3:

Basic Equation:

$$\frac{1}{R} = \frac{1}{R_1} + \frac{1}{R_2} + \frac{1}{R_3}$$

Working Equation:

$$\frac{1}{R_3} = \frac{1}{R} - \frac{1}{R_1} - \frac{1}{R_2}$$

Substitution:

$$\frac{1}{R_3} = \frac{1}{16.4 \text{ }\Omega} - \frac{1}{38.0 \text{ }\Omega} - \frac{1}{49.0 \text{ }\Omega}$$
$$R_3 = 70.2 \text{ }\Omega$$

The key entry sequence on a scientific calculator for this calculation is

$$\boxed{70.16727941}$$

. . . .

The characteristics of parallel circuits are summarized in Table 17.2.

Table 17.2   Characteristics of Parallel Circuits

	Parallel
Current	$I = I_1 + I_2 + I_3 + \cdots$
Resistance	$\dfrac{1}{R} = \dfrac{1}{R_1} + \dfrac{1}{R_2} + \dfrac{1}{R_3} + \cdots$
Voltage	$E = V_1 = V_2 = V_3 = \cdots$

Overloading is a concern with parallel circuits like those used in household wiring. As more load (bulbs, appliances, etc.) is added to a parallel circuit, the circuit draws more current. Although no change of current occurs in any individual branch, the current in the circuit as a whole increases. As the current increases, more and more heat is produced, which may cause a fire.

Circuit breakers or fuses (Fig. 17.46) are used to prevent overloading in circuits. Circuit breakers have magnets or bimetallic strips that open a switch when the current in the circuit becomes too large. A fuse has a metal strip that melts when the heat from a given amount of current passes through it. All fuses have a rating that describes the amount of current that can pass through the fuse before it melts or "blows" and breaks the circuit.

Figure 17.46   (a) Assorted fuses (b) Circuit breakers

Courtesy of Andy Crawford © Dorling Kindersley

Courtesy of Frank Labua © Pearson Education/PH College

(a)   (b)

PROBLEMS 17.8

1. (a) Find the equivalent resistance in the circuit shown in Fig. 17.47. (b) What is the total current in the circuit? (c) What is the current through R_1? (d) What is the current through R_2?

2. (a) Find I_2 (current through R_2) in the circuit shown in Fig. 17.48. (b) Find I_3. (c) Find I_1. (d) Find the total current in the circuit. (e) Find the equivalent resistance in the circuit.

Figure 17.47

$R_1 = 11.0 \ \Omega$

$R_2 = 7.00 \ \Omega$

$E = 50.0 \text{ V}$

Figure 17.48

$E = 10.0 \text{ V}$

R_1 R_2 R_3
2.00 Ω 5.00 Ω 8.00 Ω

3. (a) Find the resistance of R_3 in the circuit in Fig. 17.49. (b) What is the current through R_1? (c) What is the current through R_3?

4. (a) What is the equivalent resistance in the circuit shown in Fig. 17.50? (b) What emf is required for the circuit? (c) What is the voltage drop across each resistance? (d) What is the current through each resistance?

Figure 17.49

$I = 4.00 \text{ A}$
$E = 25.0 \text{ V}$

R_1 R_2 R_3
15.0 Ω 12.0 Ω

Figure 17.50

$I = 10.0 \text{ A}$

R_1 R_2 R_3 R_4
13.0 Ω 7.00 Ω 15.0 Ω 21.0 Ω

17.9 Compound Circuits

A *compound circuit* contains a combination of resistances in series and parallel arrangements. To simplify solving this kind of circuit, we apply the rules for series and parallel circuits to find an equivalent circuit that reduces to a circuit with one resistance.

EXAMPLE 1

Circuit *B* in Fig. 17.51 is equivalent to circuit *A*, where $R_4 = R_1 + R_2$. Then, circuit *C* is equivalent to circuit *B*, where $\dfrac{1}{R_5} = \dfrac{1}{R_3} + \dfrac{1}{R_4}$.

Figure 17.51   Circuit *A* can be replaced by circuit *C*, where R_5 is the equivalent resistance.

R_1 R_2

R_3

Circuit *A*

R_4

R_3

Circuit *B*

R_5

Circuit *C*

Circuit B in Fig. 17.52 is equivalent to circuit A, where $\dfrac{1}{R_4} = \dfrac{1}{R_2} + \dfrac{1}{R_3}$. Then, circuit C is equivalent to circuit B, where $R_5 = R_1 + R_4$.

EXAMPLE 2

Figure 17.52   Circuit A can be replaced by circuit C, where R_5 is the equivalent resistance.

Find the total current in the circuit shown in Fig. 17.53.

EXAMPLE 3

Sketch:

Figure 17.53

Circuit A is equivalent to the circuit in Fig. 17.53. Then, circuit B is equivalent to circuit A.

Data:

$$E = 115 \text{ V}$$
$$R_1 = 10.0 \ \Omega$$
$$R_2 = 32.0 \ \Omega$$
$$R_3 = 19.0 \ \Omega$$
$$R_4 = R_1 + R_2 = 10.0 \ \Omega + 32.0 \ \Omega = 42.0 \ \Omega$$
$$I = ?$$

First, find the equivalent resistance, R_5. Second, find the total current, I.
    To find R_5:

Basic Equation:

$$\frac{1}{R_5} = \frac{1}{R_3} + \frac{1}{R_4}$$

Working Equation: Same

Substitution:

$$\frac{1}{R_5} = \frac{1}{19.0 \ \Omega} + \frac{1}{42.0 \ \Omega}$$

$$R_5 = 13.1 \ \Omega$$

The key entry sequence on a scientific calculator for this calculation is

$$13.08196721$$

To find I:

Basic Equation:

$$I = \frac{E}{R_5}$$

Working Equation: Same

Substitution:

$$I = \frac{115 \text{ V}}{13.1 \text{ }\Omega}$$

$$= 8.78 \text{ A}$$

EXAMPLE 4

Find the equivalent resistance in the circuit shown in Fig. 17.54.

Sketch:

Circuit A Circuit B

Figure 17.54

Circuit A is equivalent to the circuit in Fig. 17.54, and circuit B is equivalent to circuit A.

Data:

$$R_1 = 7.00 \text{ }\Omega$$
$$R_2 = 35.0 \text{ }\Omega$$
$$R_3 = 25.0 \text{ }\Omega$$
$$R_4 = 20.0 \text{ }\Omega$$
$$R_5 = 62.0 \text{ }\Omega$$
$$E = 12\bar{0} \text{ V}$$
$$R_8 = ?$$

First, find R_6. Second, find R_7. Third, find the equivalent resistance, R_8. To find R_6:

Basic Equation:

$$\frac{1}{R_6} = \frac{1}{R_1} + \frac{1}{R_2} + \frac{1}{R_3}$$

Working Equation: Same

Substitution:

$$\frac{1}{R_6} = \frac{1}{7.00\ \Omega} + \frac{1}{35.0\ \Omega} + \frac{1}{25.0\ \Omega}$$

$$R_6 = 4.73\ \Omega$$

The key entry sequence on a scientific calculator for this calculation is

7 $\boxed{x^{-1}}$ $\boxed{+}$ 35 $\boxed{x^{-1}}$ $\boxed{+}$ 25 $\boxed{x^{-1}}$ $\boxed{=}$ $\boxed{x^{-1}}$ $\boxed{=}$

$$\boxed{\quad 4.72972973 \quad}$$

To find R_7:

Basic Equation:

$$\frac{1}{R_7} = \frac{1}{R_4} + \frac{1}{R_5}$$

Working Equation: Same

Substitution:

$$\frac{1}{R_7} = \frac{1}{20.0\ \Omega} + \frac{1}{62.0\ \Omega}$$

$$R_7 = 15.1\ \Omega$$

To find R_8:

Basic Equation:

$$R_8 = R_6 + R_7$$

Working Equation: Same

Substitution:

$$R_8 = 4.73\ \Omega + 15.1\ \Omega$$
$$= 19.83\ \Omega \text{ or } 19.8\ \Omega$$

· · · · · · · · · · · · · · · · ·

Find the total current in Example 4.

EXAMPLE 5

Data:

$$E = 12\overline{0}\ V$$
$$R_8 = 19.8\ \Omega$$
$$I = ?$$

Basic Equation:

$$I = \frac{E}{R_8}$$

Working Equation: Same

Substitution:

$$I = \frac{12\overline{0}\ V}{19.8\ \Omega}$$
$$= 6.06\ A$$

· · · · · · · · · · · · · · · · ·

Table 17.3 summarizes the characteristics of series and parallel circuits.

Table 17.3   Characteristics of Series and Parallel Circuits

	Series	Parallel
Current	$I = I_1 = I_2 = I_3 = \cdots$	$I = I_1 + I_2 + I_3 + \cdots$
Resistance	$R = R_1 + R_2 + R_3 + \cdots$	$\dfrac{1}{R} = \dfrac{1}{R_1} + \dfrac{1}{R_2} + \dfrac{1}{R_3} + \cdots$
Voltage	$E = V_1 + V_2 + V_3 + \cdots$	$E = V_1 = V_2 = V_3 = \cdots$

PROBLEMS 17.9

Use Fig. 17.55 in Problems 1 through 5.

1. (a) Which resistances are connected in parallel? (b) What is the equivalent resistance of the resistances connected in parallel?
2. Find the equivalent resistance of the entire circuit.
3. Find the current in R_1.
4. Find the voltage drop across R_1.
5. (a) Find the current through R_3. (b) Find the current through R_2.

Use Fig. 17.56 in Problems 6 through 12.

6. What is the equivalent resistance of the resistances connected in parallel?
7. Find the equivalent resistance of the circuit.
8. Find the current in R_1.
9. What is the voltage drop across the parallel part of the circuit?
10. Find the current through R_3.
11. Find the current through R_5.
12. What is the voltage drop across R_3?

Figure 17.55

Figure 17.56

Use Fig. 17.57 in Problems 13 through 20.

13. Find the equivalent resistance of the parallel arrangement in the upper branch.
14. Find the equivalent resistance of the parallel arrangement in the lower branch.
15. Find the equivalent resistance of the entire circuit.
16. What emf is required for the given current flow in the circuit?
17. Find the voltage drop across the parallel arrangement in the upper branch.
18. Find the voltage drop across R_4.
19. Find the voltage drop across R_6.
20. Find the current through R_6.

Figure 17.57

$R_2 = 24.0\ \Omega$ $R_3 = 6.00\ \Omega$
$R_1 = 5.00\ \Omega$
$I = 3.00\ A$
$R_4 = 15.0\ \Omega$
$R_5 = 3.00\ \Omega$
$R_6 = 10.0\ \Omega$

Figure 17.58

$R_1 = 5.00\ \Omega$
$R_2 = 20.0\ \Omega$
$R_4 = 6.00\ \Omega$
$R_3 = 5.00\ \Omega$
$E = 80.0\ V$
$R_5 = 10.00\ \Omega$

Use Fig. 17.58 in Problems 21 through 25.

21. Find the equivalent resistance in the circuit.
22. Find the current through R_5.
23. Find the voltage drop across R_5.
24. Find the voltage drop across R_4.
25. Find the current through R_2.

PHYSICS ON THE JOB

Electrical and Electronics Engineers

Electrical engineers are involved in the design, development, testing, and maintenance of all equipment as it relates to electricity. Electrical engineers need a strong foundation in basic electricity and electronic systems. In addition, an understanding of how electrical systems integrate into mechanical and chemical systems is important when working with a team of engineers. Finally, computer-aided drafting as it relates to electronics and general engineering concepts must be mastered in order to design and troubleshoot electrical components in mechanical systems.

Becoming an electrical or electronics engineer is considered one of the most rigorous engineering fields as it involves a mixture of theoretical as well as practical experiences in electronics. Drafting, trigonometry, algebra, calculus, and physics are necessary prerequisites for a bachelor's degree in electrical or electronics engineering. Many colleges and universities offer internship programs in conjunction with classroom experience as part of a degree program. Many of these programs lead to a master's degree in electrical engineering (Fig. 17.59).

Figure 17.59 Electronics engineer

17.10 Electric Instruments

In the laboratory we use several kinds of electric meters for measurements. Great care must be taken to avoid passing a large current through the meters. Meters are fragile instruments, and abuse will ruin them. A large current will burn out the meter. A **multimeter** is an instrument used to measure current flow, voltage drop, and resistance in ac circuits and dc circuits.

Digital instruments have more than one range on which readings are made (Fig. 17.60). Autorange meters adjust ranges automatically. The reading and use of the different modes of operation will be studied in the laboratory.

1. *Voltmeter.*   In this mode, the **voltmeter** instrument measures the difference in potential (voltage drop) between two points in a circuit. It should *always* be connected in *parallel* with the part of the circuit across which one wishes to measure the voltage drop (Fig. 17.61). The voltmeter is a high-resistance instrument and draws very little current.

2. *Ammeter.*   The **ammeter** mode measures the current flowing in a circuit. Therefore, it is connected in *series* in the circuit (Fig. 17.62). Since all the current flows through the meter, it has very low resistance in this mode so that its effect on the circuit will be as small as possible.

3. *Ohmmeter.*   The **ohmmeter** mode is used to measure the resistance of a circuit component. It should only be used when there is *no current* flowing in the circuit. The ohmmeter has a small battery as a built-in source of energy.

Figure 17.60   Digital multimeter

Figure 17.61   A voltmeter measures the difference in potential (voltage drop) across two points in an electric circuit.

Figure 17.62   An ammeter measures the current flowing in an electric circuit.

A small current provided by the ohmmeter is caused to flow through the component under test. The presence of another current or a complete circuit connected to the component under test will distort the resistance reading since the rest of the circuit may allow some of this test current to flow "around" the component under test. To avoid this problem, the component should be tested in isolation from any complete circuit (Fig. 17.63).

Another instrument is the *galvanometer*, a very sensitive instrument that is used to detect the presence and direction of *very small* currents.

Figure 17.63   An ohmmeter measures the resistance of a component in an electric circuit.

PROBLEMS 17.10

Using the formulas for series and parallel circuits, fill in the blanks in the tables shown opposite each circuit. In the blanks across from Battery under

    V: Write the emf of the battery.
    I: Write the total current in the circuit.
    R: Write the equivalent or total resistance of the entire circuit.

In the blanks across from R_1 under

    V: Write the voltage drop across R_1.
    I: Write the current flowing through R_1.
    R: Write the resistance of R_1.

In the blanks across from R_2, R_3, ... , fill in the appropriate numbers under V, I, and R. (Begin by looking for key information given in the table and work from there.)

SKETCH

12 cm² | w
4.0 cm

DATA
$A = 12$ cm^2, $l = 4.0$ cm, $w = ?$

BASIC EQUATION
$A = lw$

WORKING EQUATION
$w = \dfrac{A}{l}$

SUBSTITUTION
$w = \dfrac{12 \text{ cm}^2}{4.0 \text{ cm}} = 3.0$ cm

1.

	V	I	R
Battery	12.0 V	A	Ω
R_1	V	A	2.00 Ω
R_2	V	A	4.00 Ω

2.

	V	I	R
Battery	V	A	Ω
R_1	V	2.00 A	4.00 Ω
R_2	V	A	6.00 Ω
R_3	V	A	8.00 Ω

3.

	V	I	R
Battery	V	A	Ω
R_1	V	2.00 A	Ω
R_2	V	3.00 A	12.0 Ω
R_3	V	1.00 A	Ω

4.

	V	I	R
Battery	12.0 V	2.00 A	Ω
R_1	V	A	6.00 Ω
R_2	V	A	4.00 Ω
R_3	V	A	15.0 Ω

5.

	V	I	R
Battery	50.0 V	5.00 A	Ω
R_1	V	2.00 A	Ω
R_2	25.0 V	A	Ω
R_3	10.0 V	A	Ω
R_4	V	3.00 A	Ω

6.

	V	I	R
Battery	24.0 V	A	Ω
R_1	8.00 V	A	Ω
R_2	V	4.00 A	Ω
R_3	V	2.00 A	Ω

7.

	V	I	R
Battery	V	A	Ω
R_1	12.0 V	A	2.00 Ω
R_2	V	A	4.00 Ω
R_3	24.0 V	A	4.00 Ω
R_4	V	A	8.00 Ω

8.

	V	I	R
Battery	30.0 V	A	Ω
R_1	6.00 V	3.00 A	Ω
R_2	V	2.00 A	Ω
R_3	V	A	3.00 Ω
R_4	V	1.00 A	Ω
R_5	8.00 V	A	Ω
R_6	V	A	Ω

9.

	V	I	R
Battery	V	12.0 A	Ω
R_1	V	A	Ω
R_2	18.0 V	2.00 A	Ω
R_3	V	A	3.00 Ω
R_4	V	A	4.00 Ω
R_5	V	A	2.00 Ω
R_6	V	8.00 A	Ω
R_7	6.00 V	A	Ω

10.

	V	I	R
Battery	46.0 V	A	Ω
R_1	V	3.00 A	Ω
R_2	V	4.00 A	Ω
R_3	V	A	6.00 Ω
R_4	V	3.00 A	Ω
R_5	V	7.00 A	Ω

11.

	V	I	R
Battery	V	A	Ω
R_1	V	A	20.0 Ω
R_2	10.0 V	A	Ω
R_3	V	A	4.00 Ω
R_4	V	1.00 A	Ω
R_5	V	5.00 A	5.00 Ω
R_6	V	A	6.00 Ω

17.11 Voltage Sources

Lead Storage Cell

A battery is a group of cells connected together. Each cell consists of a positive plate and a negative plate in a conducting solution. These lead cells are **secondary cells**, which means that they are rechargeable. The passing of an electric current through the cell to restore the original chemicals is called **recharging**. Cells, such as the dry cell, that cannot be efficiently recharged are called **primary cells**. *Note:* The cells are all connected together in series (Fig. 17.64). Six storage cells of 2.0 V each connected in series give 12.0 V for an automobile storage battery.

Lead storage batteries are used to generate voltage in automobiles and in many other types of vehicles and machinery. Lead storage cells are made up of two kinds of lead plates (lead and lead oxide) submerged in a solution of distilled water and sulfuric acid (Fig. 17.65). This acid solution that produces large numbers of free electrons at the negative pole of a cell is called an **electrolyte**. The chemical action between the lead plates and the acid solution produces large numbers of free electrons at the negative (−) pole of the battery.

Figure 17.64 Cross-section cutaway of an automobile battery

Figure 17.65 Simple storage cell

Acid solution

These electrons have a large amount of electric potential energy, which is used in the load in the circuit (for instance, to operate headlights or to turn a starter motor). The work of **Luigi Galvani** led to the production of the electric battery.

As the electric energy is used in the load, the battery must be recharged. This is done by a generator or an alternator. Such devices provide an electric current to reverse the chemical reaction taking place in the battery. The recharging process extends the life of the battery, which would otherwise be very short.

Luigi Galvani (1737–1798),

physiologist, was born in Italy. He investigated the effects of electrostatic stimuli applied to muscle fiber. His work inspired his friend, Volta, and led to the production of the electric battery and initiated research into electrophysiology. The galvanometer is named after him.

Dry Cell

The dry cell is the most widely used primary cell. This kind of cell is used in flashlights and portable products such as cellular phones, notebook computers, drills, etc. A **dry cell** is a voltage-generating cell that consists of an electrolyte in the form of a chemical paste and two electrodes of unlike materials, one of which reacts chemically with the electrolyte to provide energized electrons. The carbon–zinc dry cell is made of a carbon rod, which is the positive (+) terminal or pole, and a zinc can, which acts as the negative (−) terminal (Fig. 17.66). In between is a paste of chemicals and water that reacts with the terminals to provide energized electrons. These cells are available in a wide range of sizes. Common battery voltages range from 1.5 to 9 V. To achieve 9 V requires a series stack of six cells.

The dry cell, as well as the lead cell, has resistance within the cell itself that opposes the movement of the electrons. This is called the **internal resistance**, r, of the cell. Every cell has internal resistance. Because current flows in the cell, the emf of the cell is reduced by the voltage drop across the internal resistance (Fig. 17.67). The voltage applied to the external circuit is then

$$V = E - Ir$$

where $V =$ voltage applied to the circuit
   $E =$ emf of the cell
   $I =$ current through the cell
   $r =$ internal resistance of the cell

Figure 17.66 Dry cell batteries

Figure 17.67 Effect of the internal resistance of a dry cell

When the current or the voltage available from a single cell is inadequate for a particular job, we usually connect two or more cells in a parallel or series arrangement.

Alkaline cells resemble carbon–zinc cells but are five to eight times longer lasting. An alkaline cell has a highly porous zinc anode, which oxidizes more readily than does the carbon–zinc cell's anode. Its electrolyte is a strong alkali solution called *potassium hydroxide*. This compound conducts electricity inside the cell very well and enables the alkaline cell to deliver relatively high currents with greater efficiency than that of carbon–zinc cells.

Nickel–cadmium, metal hydride, and lithium batteries are now also in common use. Their advantage over other types of dry cells is that they are rechargeable. Lithium batteries have greater capacity than nickel–cadmium or hydride batteries.

Another type of dc power source is the solar cell. It is commonly made from a semiconductor material (typically amorphous silicon or gallium arsenide). Many small calculators operate with power supplied by solar cells.

17.12   Cells in Series and Parallel

To connect cells in series, the positive terminal of one is connected to the negative terminal of the next cell. This procedure is continued until the desired number of cells is connected (Fig. 17.68). The rules for cells connected in series and parallel are similar to those for resistances.

Figure 17.68

Series-connected dry cells

Circuit diagram for three cells in series

CELLS IN SERIES

1. The current in the circuit equals the current in any single cell:
$$I = I_1 = I_2 = I_3 = \cdots$$

2. The internal resistance of the battery equals the sum of the individual internal resistances of the cells:
$$r = r_1 + r_2 + r_3 + \cdots$$

3. The emf of the battery equals the sum of the emfs of the individual cells:
$$E = E_1 + E_2 + E_3 + \cdots$$

EXAMPLE 1

Two 6.00-V cells with internal resistance of 0.100 Ω each are connected in series to form a battery with a current of 0.750 A in each cell (Fig. 17.69).

Figure 17.69

Series

(a)　What is the emf of the battery?
(b)　Find the internal resistance of the battery.
(c)　Find the current in the external circuit.

(a)　$E = E_1 + E_2 = 6.00\text{ V} + 6.00\text{ V} = 12.00\text{ V}$　　(Rule 3)
(b)　$r = r_1 + r_2 = 0.100\ \Omega + 0.100\ \Omega = 0.200\ \Omega$　　(Rule 2)
(c)　$I = 0.750\text{ A}$　　(Rule 1)

. . . .

To connect cells in parallel, the positive terminals of all the cells are connected together and the negative terminals are all connected together (Fig. 17.70).

Figure 17.70

Parallel-connected dry cells

Circuit diagram for
three cells in parallel

A common example of the use of cells in parallel is the practice of jump-starting a car that has a dead battery (Fig. 17.71). It is not common to find cells hooked up in parallel because a mismatch of output voltages could cause problems. The leads from the external circuit may be connected to any positive and negative terminals. (The external circuit is all of the circuit *outside* the battery or cell.)

Figure 17.71　How to properly make connections to jump-start a car
with a dead battery

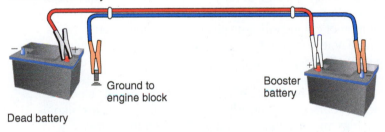

Dead battery

Ground to
engine block

Booster
battery

CELLS IN PARALLEL

1.　The total current equals the sum of the individual currents in each cell:

$$I = I_1 + I_2 + I_3 + \cdots$$

2.　The internal resistance equals the resistance of one cell divided by the number of cells:*

$$r = \frac{r \text{ of one cell}}{\text{number of cells}}$$

3.　The emf of the battery equals the emf of any single cell:

$$E = E_1 = E_2 = E_3 = \cdots$$

----

*This formula works only when all the cells have the same internal resistance. Otherwise, a formula similar to that for resistors in parallel must be used.

EXAMPLE 2

Four cells, each 1.50 V and having an internal resistance of 0.0500 Ω, are connected in parallel to form a battery with a current output of 0.250 A in each cell (Fig. 17.72).

Figure 17.72

Parallel cells   Circuit diagram

(a)   What is the emf of the battery?
(b)   Find the internal resistance of the battery.
(c)   Find the current in the external circuit.

(a) $E = 1.50$ V   (Rule 3)

(b) $r = \dfrac{r \text{ of one cell}}{\text{number of cells}} = \dfrac{0.0500 \ \Omega}{4} = 0.0125 \ \Omega$ (Rule 2)

(c) $I = I_1 + I_2 + I_3 + I_4 = 0.250$ A $+ 0.250$ A $+ 0.250$ A $+ 0.250$ A $= 1.000$ A
       (Rule 1)

PHYSICS CONNECTIONS

Ice Block Air Conditioning

In an effort to limit the strain on power grids and lower the cost of cooling, several companies have been using ice blocks to cool office buildings in New York City. This method of cooling also reduces the amount of pollutants in the skies over New York City and the surrounding area.

Instead of using a conventional central air conditioning system that runs constantly throughout the day to keep the building cool, the Met Life Tower in New York City uses three cooling rooms in the basement to freeze 800 gal of water in each of 64 tanks. A solution of chilled ethylene glycol is pumped through pipes in the tanks to freeze the water at night. During the day, the ice melts in the tanks and cools the ethylene glycol solution in the pipes. The solution is then blown through the building and cools the air as it encounters heat exchange coils throughout the building.

The environmental results and savings are impressive. The Met Life building has lowered its electric usage by nearly 2.15 million kilowatt-hours over the course of a year. Although the initial costs are significant, companies are finding that the long-term savings and environmental benefits are tremendous.

PROBLEMS 17.12

1.   A cell has an emf of 1.50 V and an internal resistance of 0.0450 Ω. If there is 0.250 A in the cell, what voltage is applied to the external circuit?
2.   The voltage applied to a circuit is 11.8 V when the current through the battery is 0.500 A. If the internal resistance of the battery is 0.150 Ω, what is the emf of the battery?
3.   The emf of a battery is 12.0 V. If the internal resistance is 0.300 Ω and the voltage applied to the circuit is 11.6 V, what is the current through the battery?
4.   Three 1.50-V cells, each with an internal resistance of 0.0500 Ω, are connected in series to form a battery with a current of 0.850 A in each cell. (a) Find the current in the external circuit. (b) What is the emf of the battery? (c) Find the internal resistance of the battery.

5. Five 9.00-V cells, each with an internal resistance of 0.100 Ω and a current output of 0.750 A, are connected in parallel to form a battery in a certain circuit. (a) Find the current in the external circuit. (b) What is the emf of the battery? (c) Find the internal resistance of the battery.
6. Find the current in the circuit shown in Fig. 17.73.
7. Find the current in the circuit shown in Fig. 17.74.

Figure 17.73

$E_1 = 1.50$ V
$R_1 = 7.40$ Ω
$E_2 = 1.50$ V

Figure 17.74

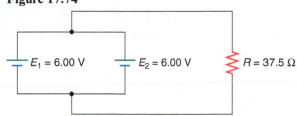

$E_1 = 6.00$ V $E_2 = 6.00$ V $R = 37.5$ Ω

8. If the current in the circuit in Fig. 17.75 is 1.20 A, what is the value of R_2?
9. Find the current in the circuit shown in Fig. 17.76.
10. Find the total resistance in the circuit shown in Fig. 17.76.

Figure 17.75

$E_1 = 9.00$ V $E_2 = 9.00$ V R_2

$R_1 = 6.00$ Ω

Figure 17.76

$E = 1.50$ V

$R_1 = 10.0$ Ω $R_2 = 15.0$ Ω

$E = 1.50$ V

17.13 Electric Power

Tremendous quantities of energy are used by industry and sold by power companies. The rate of consuming energy is called **power**. The unit of power is the watt. One *watt* (W) is the power generated by a current of 1 A flowing because of a potential difference of 1 V. A volt is a joule/coulomb (J/C); an ampere is a coulomb/second (C/s). Their product is

$$VA = \frac{J}{C} \cdot \frac{C}{s} = \frac{J}{s}$$

Thus, 1 W = 1 J/s.
Hence, power is

$$P = VI$$

where P = power (watts)
V = voltage drop
I = current

This equation applies to components of dc circuits and to whole dc circuits as well as to ac circuits with resistance only.
Recalling Ohm's law, $I = V/R$, we find two other equations for power:
Given

$$P = VI$$

substitute for V using $V = IR$ to obtain

$$P = (IR)I = I^2R$$

$$\boxed{P = I^2R}$$

Note from the following unit analysis that amps squared times ohms gives watts:

$$\boxed{A^2\,\Omega = A^2 \cdot \frac{V}{A} = AV = \frac{C}{s} \cdot \frac{J}{C} = \frac{J}{s} = W}$$

Also, given

$$P = I^2R$$

substitute

$$I = \frac{V}{R}$$

to get

$$P = \left(\frac{V}{R}\right)^2 R = \frac{V^2}{R^2} \cdot R$$

$$\boxed{P = \frac{V^2}{R}}$$

EXAMPLE 1

A soldering iron draws 7.50 A in a 115-V circuit. What is its wattage rating?

Data:

$$I = 7.50 \text{ A}$$
$$V = 115 \text{ V}$$
$$P = ?$$

Basic Equation:

$$P = VI$$

Working Equation: Same

Substitution:

$$P = (115 \text{ V})(7.50 \text{ A})$$
$$= 863 \text{ W}$$

Therefore, a soldering iron drawing 7.50 A in a 115-V circuit has a rating of 863 W.

· · · · · · · · · · · · · · · · ·

EXAMPLE 2

A hand drill draws 4.00 A and has a resistance of 14.6 Ω. What power does it use?

Data:

$$I = 4.00 \text{ A}$$
$$R = 14.6 \text{ } \Omega$$
$$P = ?$$

Basic Equation:

$$P = I^2 R$$

Working Equation: Same

Substitution:

$$P = (4.00 \text{ A})^2 (14.6 \ \Omega)$$
$$= 234 \text{ W}$$

Thus, a drill that draws 4.00 A with a resistance of 14.6 Ω has a rating of 234 W.

· · · · · · · · · · · · · · · ·

Since the watt is a relatively small unit, the kilowatt (1 kW = 1000 W) is commonly used in industry.

Although we speak of "paying our power bill," what power companies actually sell is **energy** in a form of work delivered to an electric component or appliance. Energy is sold in kilowatt-hours (kWh). The amount of energy consumed is equal to the power used times the time it is used. Therefore,

$$\boxed{\text{energy} = \text{power} \times \text{time}}$$

or

$$\text{energy (in kWh)} = (VI)t$$
$$\text{number of kWh} = VIt$$

when V is in volts, I is in amperes, and t is time in hours. Note that electric energy can be expressed in other units (joules), but kilowatt-hours is commonly used. This equation is useful in finding the cost of electric energy. Cost is measured in cents per kilowatt-hour. The cost of operating an electric device may be found as follows:

$$\text{cost} = \text{energy} \times \text{cost per unit energy}$$

$$\text{cost} = (\text{kWh})\left(\frac{\text{cents}}{\text{kWh}}\right)$$

$$\boxed{\text{cost (in cents)} = \text{power (in W)} \times \text{hours} \times \frac{1 \text{ kW}}{1000 \text{ W}} \times \frac{\text{cents}}{\text{kWh}}}$$

↑   conversion factor

An iron is rated at 550 W. How much would it cost to operate it for 2.50 h at \$0.08/kWh?

EXAMPLE 3

Data:

$$P = 550 \text{ W}$$
$$t = 2.50 \text{ h}$$
$$\text{rate} = \$0.08/\text{kWh}$$
$$\text{cost} = ?$$

Basic Equation:

$$\text{cost} = Pt\left(\frac{\text{kW}}{1000 \text{ W}}\right)\left(\frac{\text{cents}}{\text{kWh}}\right)$$

Working Equation: Same

Substitution:

$$\text{cost} = (550 \text{ W})(2.50 \text{ h})\left(\frac{\text{kW}}{1000 \text{ W}}\right)\left(\frac{\$0.08}{\text{kWh}}\right)$$

$$= \$0.11$$

Remember that the source of electrons in a circuit is the conducting circuit material itself. You may be able to buy an empty water hose, but you cannot buy an "empty" wire. If you plug in an appliance, *energy* flows from the outlet to the appliance, not electrons. Energy is carried by the electric field and causes motion in the electrons already in the appliance. The power company sells the energy; the appliance supplies the electrons.

SKETCH

12 cm² | w

4.0 cm

DATA

$A = 12 \text{ cm}^2$, $l = 4.0 \text{ cm}$, $w = ?$

BASIC EQUATION

$A = lw$

WORKING EQUATION

$w = \dfrac{A}{l}$

SUBSTITUTION

$w = \dfrac{12 \text{ cm}^2}{4.0 \text{ cm}} = 3.0 \text{ cm}$

PROBLEMS 17.13

1. A heater draws 8.70 A on a $11\overline{0}$-V line. What is its wattage rating?
2. What power is needed for a sander that draws 3.50 A and has a resistance of 6.70 Ω?
3. How many amperes will a 75.0-W lamp draw on a $11\overline{0}$-V line?
4. Find the resistance of the lamp in Problem 3.
5. How many amperes will a $75\overline{0}$-W lamp draw on a $11\overline{0}$-V circuit?
6. Find the cost to operate the lamp in Problem 5 for 40.0 h if the cost of energy is $0.07 per kWh.
7. Six 50.0-W bulbs are operated for 25.0 h on a 115-V circuit. If energy costs $0.075 per kWh, find the cost of operating them.
8. A small furnace uses 3.00 kW of power. If the cost of operation of the furnace is $3.84 for a 24.0-h period, what is the cost of energy per kWh?
9. Will a 20.0-A fuse blow if a $100\overline{0}$-W hair dryer, a $120\overline{0}$-W electric skillet, and an $110\overline{0}$-W toaster are all used at once on a $11\overline{0}$-V line?
10. How long could you operate a $100\overline{0}$-W soldering iron for $0.50 if the cost of energy is $0.075/kWh?
11. Find the cost of operating a 1.50-A motor on a $11\overline{0}$-V circuit for 2.00 h at $0.08/kWh.
12. Find the cost of operating a 2.50-A motor on a $11\overline{0}$-V circuit for 3.00 h at $0.07/kWh.
13. Find the cost of operating a 3.00-A motor on a $11\overline{0}$-V circuit for 2.00 h at $0.07/kWh.
14. How many amperes will a $6\overline{0}$-W lamp draw on a $11\overline{0}$-V line?
15. Using the following table, list two different combinations of bulbs and appliances that could be used on a $2\overline{0}$-A circuit breaker in a $11\overline{0}$-V system without causing the circuit breaker to trip.

Appliance	Power Rating
Light bulb (60 W)	60 W
Fluorescent bulb (40 W)	40 W
12″ TV	55 W
Projection TV	1500 W
Personal computer	550 W
Hand drill	400 W
Microwave oven	1000 W

16. Using the preceding table, list two different combinations that could be used on a $3\overline{0}$-A circuit breaker in a $11\overline{0}$-V system without causing the circuit breaker to trip.
17. Find the power output of a cell phone charger that delivers $40\overline{0}$ mA of current at 5.90 V dc.
18. A power supply for electronic devices delivers 1.10 A of current at 4.40 V. What is its power output?

19. At what rate does a light bulb convert electric energy to light if it draws 0.500 A of current on a $12\overline{0}$-V outlet?

20. What power is used by a light that draws 2.00 A from a 12.0-V battery?

21. How much electric energy (in joules) is delivered in 5.00 min to an electric motor that draws 0.500 A from a 6.00-V battery?

22. A car has a 12.0-V battery. If the current through the starter is $21\overline{0}$ A, what electric energy (in joules) is delivered to the starter in 10.0 s?

23. (a) How much power does a television use if it draws 2.00 A on a $12\overline{0}$-V line? (b) What energy in kWh does the television use in 30 days if it is used an average of 7.00 h/day? (c) Find the cost of operating the television for 30 days if the cost of energy is \$0.11/kWh.

24. An electric heater is used 5.00 h each day. (a) If it draws 15.0 A on a $12\overline{0}$-V line, how much power does it use? (b) In 30 days, how much energy in kWh does the heater use? (c) At \$0.11/kWh, what does it cost to operate the heater for 30 days?

25. A digital timer is used on a 115-V line. (a) If the resistance of the timer is $12,\overline{0}00\ \Omega$, how much current does it draw? (b) How much power does the timer use? (c) What does it cost to operate the timer at \$0.09/kWh for 30 days?

26. A current of $23\overline{0}$ A is delivered to a truck starter motor. What electric energy (in joules) is used in the 12.0-V system in 2.00 s?

27. A job site generator delivers 205 A in 15.0 s in a 12.0-V system. What energy (in joules) is expended in 2.50 s?

NEW TECHNOLOGIES

Tesla High-Performance Electric Automobile

Tesla is a unique automobile as it is considered to be the first-of-its-kind, fully electric and high-performance electric vehicle. Prior to the Tesla, electric cars were hybrid vehicles that relied on a combination of electric power supplemented by gasoline. Until now, pure electric cars were not considered high-performance model vehicles.

The brakes in a Tesla not only slow down the vehicle, but also utilize the energy transfer between brakes and brake pads to create additional energy for the vehicle. This regenerative braking system shifts the kinetic energy typically lost in braking and instead transfers it back to electrical energy for the automobile.

The Tesla Model S has designed a massive 1,000-lb bank of high-density, rechargeable lithium-ion batteries, similar to those used in laptops or other mobile devices. Tesla suggests charging the vehicle overnight in a home garage by plugging the vehicle into a standard 110 V or 220 V circuit. Charging speeds vary between 29 mi of driving range per charging hour at 110 V to 58 mi of driving range per charging hour using a 220 V circuit. Most impressively, Tesla supercharge stations, located at Tesla dealerships and other locations around the nation, can charge up to 340 mi of range per charging hour and can replenish half the battery in 30 minutes (Fig. 17.77)!

Figure 17.77 (a) Tesla Model S electrical vehicle

Figure 17.77 (b) Tesla charging station

Glossary

Alternating Current Current that changes direction. (p. 460)

Ammeter An instrument that measures the current flowing in a circuit. (p. 484)

Conduction A transfer of charge from one place to another. (p. 452)

Conductor A material through which an electron charge is readily transferred. (p. 462)

Coulomb's Law The force between two point charges is directly proportional to the product of their magnitudes and inversely proportional to the square of the distance between them. (p. 456)

Current The flow of charge that passes through a conductor. (p. 462)

Direct Current Current that flows in one direction. (p. 460)

Dry Cell A voltage-generating cell that consists of a chemical paste and two electrodes of unlike materials, one of which reacts chemically with the electrolyte. (p. 487)

Electric Circuit A conducting loop in which electrons carrying electric energy may be transferred from a suitable source to do useful work and returned to the source. (p. 460)

Electric Field An electric field exists where an electric force acts on a charge brought into the area. (p. 457)

Electrolyte An acid solution that produces large numbers of free electrons at the negative pole of a cell. (p. 487)

Electron A negatively charged particle found in every atom. (p. 452)

emf The potential difference across a source. (p. 463)

Energy Work delivered to an electric component or appliance (power × time). (p. 493)

Equivalent Resistance The single resistance that can replace a series and/or parallel combination of resistances in a circuit and provide the same current flow and voltage drop. (p. 469)

Induction A method of charging one object by bringing a charged object near to, but not touching, it. (p. 453)

Internal Resistance The resistance within a cell that opposes movement of the electrons. (p. 487)

Insulator A substance that does not allow electric current to flow through it readily. (p. 462)

Load The object in a circuit that converts electric energy into other forms of energy or work. (p. 462)

Multimeter An instrument used to measure current flow, voltage drop, and resistance. (p. 483)

Neutron A neutral particle found in the nucleus of most atoms. (p. 452)

Ohm's Law When a voltage is applied across a resistance in an electric circuit, the current equals the voltage drop across the resistance divided by the resistance, $I = V/R$. (p. 465)

Ohmmeter An instrument that measures the resistance of a circuit component. (p. 484)

Parallel Circuit An electric circuit with more than one path for the current to flow. The current is divided among the branches of the circuit. (p. 472)

Power Energy per unit time consumed in a circuit. (p. 491)

Primary Cell A cell that cannot be recharged. (p. 487)

Proton A positively charged particle found in the nucleus of every atom. (p. 452)

Recharging The passing of an electric current through a secondary cell to restore the original chemicals. (p. 487)

Resistance The opposition to current flow. (p. 463)

Resistivity The resistance per unit length of a material with uniform cross section. (p. 463)

Secondary Cell A rechargeable type of cell. (p. 487)

Semiconductors A small number of materials that fall between conductors and insulators in their ability to conduct electric current. (p. 462)

Series Circuit An electric circuit with only one path for the current to flow. The current in a series circuit is the same throughout. (p. 468)

Source The object that supplies electric energy for the flow of electric charge (electrons) in a circuit. (p. 460)

Superconductor A material that continuously conducts electric current without resistance when cooled to typically very low temperatures, often near absolute zero. (p. 462)

Voltage Drop The potential difference across a load in a circuit. (p. 463)

Voltmeter An instrument that measures the difference in potential (voltage drop) between two points in a circuit. (p. 484)

Formulas

17.3   Coulomb's law: $F = \dfrac{kq_1 q_2}{r^2}$

17.4 $E = \dfrac{F}{q'}$

17.5 $R = \dfrac{\rho l}{A}$

17.6   Ohm's law: $I = \dfrac{V}{R}$

17.7   *Characteristics of Series Circuits*

Current	$I = I_1 = I_2 = I_3 = \cdots$
Resistance	$R = R_1 + R_2 + R_3 + \cdots$
Voltage	$E = V_1 + V_2 + V_3 + \cdots$

17.8   *Characteristics of Parallel Circuits*

Current	$I = I_1 + I_2 + I_3 + \cdots$
Resistance	$\dfrac{1}{R} = \dfrac{1}{R_1} + \dfrac{1}{R_2} + \dfrac{1}{R_3} + \cdots$
Voltage	$E = V_1 = V_2 = V_3 = \cdots$

17.11 $V = E - Ir$

17.12   *Cells in series:*   *Cells in parallel:*

$I = I_1 = I_2 = I_3 = \cdots$ $I = I_1 + I_2 + I_3 + \cdots$

$r = r_1 + r_2 + r_3 + \cdots$ $r = \dfrac{r \text{ of one cell}}{\text{number of cells}}$

$E = E_1 + E_2 + E_3 + \cdots$ $E = E_1 = E_2 = E_3 = \cdots$

17.13 $P = VI$

$P = I^2 R$

$P = \dfrac{V^2}{R}$

energy = power × time

cost (in cents) = power (in W) × hours × $\dfrac{1 \text{ kW}}{1000 \text{ W}}$ × $\dfrac{\text{cents}}{\text{kWh}}$

Review Questions

1. The atomic particle that carries a positive charge is
   - (a) the neutron.
   - (b) the proton.
   - (c) the electron.
   - (d) none of the above.
2. The atomic particle that carries a negative charge is
   - (a) the neutron.
   - (b) the proton.
   - (c) the electron.
   - (d) none of the above.
3. The process by which an object becomes charged when it comes in contact with a charged object is called
   - (a) induction.
   - (b) electrification.
   - (c) conduction.
   - (d) none of the above.
4. The process by which an object becomes permanently charged when it comes near a charged object requires that the first object be
   - (a) an insulator.
   - (b) touched by another object.
   - (c) a conductor.
   - (d) none of the above.
5. The resistance of a wire is dependent on all of the following except
   - (a) temperature.
   - (b) cross-sectional area.
   - (c) length.
   - (d) material.
   - (e) voltage.
6. Which of the following are good electric conductors?
   - (a) Aluminum
   - (b) Wood
   - (c) Glass
   - (d) Distilled water
   - (e) Silver
7. The total resistance in a circuit containing resistors connected in series is given by
   - (a) the sum of the individual resistances.
   - (b) the sum of the inverse of the individual resistances.
   - (c) the sum of the currents.
   - (d) the sum of the voltages.
8. The current in a parallel circuit is given by
   - (a) the sum of the inverse currents.
   - (b) the sum of the voltages.
   - (c) the sum of the currents in the branches.
   - (d) none of the above.
9. The emf of a battery with cells connected in series equals the sum of the
   - (a) internal resistances of the cells.
   - (b) emf of the individual cells.
   - (c) current in the individual cells.
10. The current in a battery with cells connected in series equals the
    - (a) internal resistances of the cells.
    - (b) emf of the individual cells.
    - (c) current in the individual cells.
11. The current in a battery with cells connected in parallel equals the
    - (a) current in one cell.
    - (b) internal resistance.
    - (c) sum of the currents in each cell.
12. Examples of dry cells include
    - (a) lead–zinc cells.
    - (b) nickel–cadmium cells.
    - (c) carbon–zinc cells.
    - (d) fuel cells.
13. In your own words, describe how materials can become charged by electrification.
14. What particles make up an atom?
15. What particles are located in the nucleus (center) of an atom?
16. Where are electrons located in an atom?
17. What are the two types of charge? What atomic particle carries each type of charge?
18. Describe the process of charging an electroscope by conduction.
19. Describe the process of charging an electroscope by induction.
20. In your own words, describe Coulomb's law of electrostatics.
21. Describe an electric field.
22. Describe lightning.
23. The flow of electrons through a conductor is called _____.
24. (a) The unit of current is the _____. (b) The unit of emf is the _____. (c) The unit of resistance is the _____.

25. What effect does doubling the diameter of a wire have on the wire's resistance?
26. In your own words, explain Ohm's law.
27. Differentiate between a series and a parallel circuit.
28. Differentiate between the equivalent resistance in a series circuit and a parallel circuit.
29. In using an electric instrument, with what range should you start when making a measurement?
30. Explain how a parallel water system compares to a parallel electric circuit.
31. How does the current change in a circuit if the resistance increases by a factor of 2?
32. How does the current change in a circuit if the voltage is increased by a factor of 2?
33. How would the resistance of a wire change if the length were to be increased by a factor of 2?
34. Explain the concept of electric potential.
35. Explain the transfer of energy that occurs in a circuit that includes a dry cell and two lamps in series.
36. Distinguish between a primary and a secondary cell.
37. Explain recharging.
38. Describe the function of an electrolyte.
39. In your own words, describe the manner in which a secondary cell produces electric energy.
40. What is the effect of the internal resistance of a cell?
41. The unit of electric power is the _____.
42. In your own words, explain the relationship among power, voltage, and current.
43. Do we pay the utility company for our power use or our energy use? Explain.
44. Explain the relationship among power, voltage, and resistance.
45. If the current in a circuit is increased by a factor of 2 and the voltage stays constant, how does the power change?
46. If the resistance in a circuit decreases by a factor of 2 and the voltage stays constant, how does the power change?
47. If the voltage and current in a circuit each decrease by a factor of 2, how does the power change?
48. If the current increases in a circuit by a factor of 2 and the voltage stays constant, how would the cost of operating the circuit change?

Review Problems

1. Two charges, each $-4.50 \ \mu C$, are 0.150 cm apart. Find their repulsive force.
2. The repulsive force between two identical negative charges is 0.750 N when they are 0.100 m apart. Find the amount of each charge.
3. A charge of 2.50×10^{-8} C exerts a force of 0.0250 N on a second charge of 5.00×10^{-7} C. How far apart are the charges?
4. A positive test charge of $2.50 \ \mu C$ is placed in an electric field. The force on it is 0.500 N. What is the magnitude of the electric field at the location of the test charge?
5. Find the magnitude of the electric field in which a negative charge of 1.50×10^{-8} C experiences a force of 0.0500 N.
6. What force is exerted on a test charge of 4.25×10^{-5} C if it is placed in an electric field of magnitude 2.50×10^4 N/C?
7. Find the resistance of 85.5 m of No. 20 aluminum wire ($\rho = 2.83 \times 10^{-6} \ \Omega$ cm, $A = 2.07 \times 10^{-2} \ \text{cm}^2$).
8. At 75°F, 12$\overline{0}$ ft of wire has a resistance of 0.743 Ω. Find the resistance of 56$\overline{0}$ ft of this wire.
9. Find the resistance of 134 m of No. 20 copper wire at 20°C ($\rho = 1.72 \times 10^{-6} \ \Omega$ cm, $A = 2.07 \times 10^{-2} \ \text{cm}^2$).
10. Find the length of a copper wire with resistance 0.0273 Ω/ft and total resistance 3.97 Ω.

SKETCH

```
┌──────────┐
│  12 cm²  │  w
└──────────┘
   4.0 cm
```

DATA
$A = 12 \ \text{cm}^2, l = 4.0 \ \text{cm}, w = ?$

BASIC EQUATION
$A = lw$

WORKING EQUATION
$w = \dfrac{A}{l}$

SUBSTITUTION
$w = \dfrac{12 \ \text{cm}^2}{4.0 \ \text{cm}} = 3.0 \ \text{cm}$

Figure 17.78

$R_1 = 11.5\ \Omega$ $R_2 = 16.5\ \Omega$
$E = ?\ V$
$R_3 = 2.43\ \Omega$
$R_4 = 5.76\ \Omega$
$I = 1.34\ A$

Figure 17.79

$R_1 = 11.3\ \Omega$
$E = 24.5\ V$
$R_2 = 1.45\ \Omega$
R_3
$I = 1.51\ A$

11. Find the cross-sectional area of copper wire at 20°C that is 55.4 m long and has resistivity $\rho = 1.79 \times 10^{-6}\ \Omega$ cm and resistance 0.943 Ω.

12. A heating element operates on 115 V. If it has a resistance of 15.4 Ω, what current does it draw?

13. A heating coil operates on $22\overline{0}$ V. If it draws 8.75 A, find its resistance.

14. What current does a 234-Ω resistance draw on 115 V?

15. Four resistors of 3.40 Ω, 6.54 Ω, 8.32 Ω, and 1.34 Ω are connected in series with a 12.0-V battery. Find the total resistance of the circuit.

16. Find the current in Problem 15.

17. Find the emf in the circuit shown in Fig. 17.78.

18. Find the equivalent resistance in the circuit shown in Fig. 17.79.

19. Find R_3 in the circuit in Problem 18.

20. Find the equivalent resistance in the circuit shown in Fig. 17.80.

21. Find the current in Fig. 17.80.

22. Find the current through R_1 in Fig. 17.80.

23. Find the current through R_2 in Fig. 17.80.

24. Find the equivalent resistance in the circuit of Fig. 17.81.

25. Find the current through R_3 in Fig. 17.81.

26. Find the current through R_1 in Fig. 17.81; through R_2.

Figure 17.80

$R_1 = 5.65\ \Omega$
$E = 11\overline{0}\ V$
$R_2 = 4.32\ \Omega$

Figure 17.81

$R_1 = 4.32\ \Omega$
$E = 50.0\ V$
$R_2 = 6.35\ \Omega$
$R_3 = 2.43\ \Omega$

Figure 17.82

$R_1 = 4.35\ \Omega$
$R_4 = 0.765\ \Omega$
$R_2 = 3.75\ \Omega$
$R_3 = 1.56\ \Omega$
$E = 75.5\ V$
$R_5 = 3.42\ \Omega$

27. Find the equivalent resistance in Fig. 17.82.

28. Find the current in R_5 in Fig. 17.82.

29. Find the voltage drop across R_5 in Fig. 17.82.

30. Find the current in R_1 in Fig. 17.82.

31. Find the voltage drop across R_1 in Fig. 17.82.

32. Using the formulas for series and parallel circuits, fill in the blanks in the table below for the circuit of Fig. 17.83.

Figure 17.83

R_1
R_3
R_4
R_2
R_6
R_5

	V	I	R
Battery	35.0 V	A	Ω
R_1	5.00 V	2.75 A	Ω
R_2	V	1.95 A	Ω
R_3	V	A	2.80 Ω
R_4	V	0.97 A	Ω
R_5	7.50 V	A	Ω
R_6	V	A	Ω

33. A cell has an emf of 1.44 V and an internal resistance of 0.0550 Ω. If there is 0.135 A in the cell, what voltage is applied to the external circuit?

34. The voltage applied to a circuit is 12.0 V when the current through the battery is 0.858 A. If the internal resistance of the battery is 0.245 Ω, what is the emf?

35. Six 6.00-V cells, each with an internal resistance of 0.0987 Ω and current output of 0.658 A, are connected in parallel to form a battery in a certain circuit. (a) What is the current in the external circuit? (b) What is the emf of the battery? (c) What is the internal resistance of the battery?

36.   Find the current in the circuit shown in Fig. 17.84.
37.   Find the total resistance in the circuit shown in Fig. 17.85.

Figure 17.84   **Figure 17.85**

38.   What power is needed for a drill that draws 2.45 A and has a resistance of 6.55 Ω on a $11\overline{0}$-V circuit?
39.   How many amperes will a $15\overline{0}$-W light bulb draw on a $11\overline{0}$-V circuit?
40.   What is the cost to operate the lamp in Problem 39 for 135 h if the cost of energy is $0.05/kWh?
41.   If the cost of energy is $0.043/kWh, how long could you operate a motor that draws 0.40 A on a 110-V line for $0.45?
42.   How many amperes will a $10\overline{0}$-W lamp draw on a 110-V line?

APPLIED CONCEPTS

1.   A hydrogen atom contains one electron and one proton. (a) Find the electric force between the proton and the electron. (The distance between an electron and a proton in a hydrogen atom is 5.29×10^{-11} m. A proton's positive charge has the same magnitude as the negative charge of an electron.) (b) What is the gravitational force between a proton and an electron? (c) If the electric force is stronger than the gravitational force, why don't people accelerate toward charged combs, balloons, and other electrically charged objects?

2.   A rod with charge -4.31×10^{-8} C is held 10.3 cm above a piece of sawdust with mass 3.12×10^{-5} kg. (a) How much does the sawdust weigh? (b) How much electrostatic force must the rod apply to lift the sawdust? (c) If the charged rod lifts the sawdust, what is the electric charge of the sawdust? (d) What happens to the strength of the electric field as the sawdust comes closer to the rod?

3.   Hairdryers work by blowing heat that is generated by exposed, high-resistance wires. (a) If a woman using a hairdryer comes in direct contact with the $11\overline{0}$-V wires, what is the current that travels through her if she is dry and has a resistance of $50\overline{0},000$ Ω? (b) If she is wet, the resistance of her body can drop as low as $10\overline{0}$ Ω. What is the current that passes through her body at this resistance? (Death can result when a current of 0.07 A flows through a person's body for more than 1 s.)

4.   A $100\overline{0}$-W microwave, a 40.0-W fluorescent light bulb, and a $55\overline{0}$-W computer are plugged into a $12\overline{0}$-V parallel circuit. (a) What is the current passing through each appliance in the parallel circuit? (b) Find the resistance of each appliance.

5.   A $70\overline{0}$-W toaster is plugged into a $11\overline{0}$-V outlet. When turned on, the current flowing through the 6.00-m strand of nichrome wire heats the wire and toasts the bread. As the designer of the toaster, what radius do you choose for the nichrome wire? (Assume the temperature of the wire does not affect the resistance. $\rho_{nichrome} = 1.00 \times 10^{-6}$ Ω m.)

CHAPTER 18

MAGNETISM

Objectives

The major goals of this chapter are to enable you to:

1. Describe the nature of magnetism and the magnetic effect of electric current.
2. Describe how induced magnetism and electric current are related.
3. Distinguish between generators and motors, and describe the principles that apply to both.

Magnetism and electricity are closely related. Electromagnetism is at the heart of the generation of electric power. We consider how generators and motors use these principles in the production and consumption of electricity.

18.1   Introduction to Magnetism

Many devices that use or produce electric energy depend on the relation of magnetism and electric current. Motors and meters are designed to use the fact that electric currents in wires behave like magnets. Generators produce electric current due to the movement of wires near very large magnets.

In this chapter we investigate the basic properties of magnets and the relation between electric current and magnetism. In later sections on generators, motors, and transformers, we will use the basic principles of magnetism that are developed here.

Figure 18.1   Magnetic materials attract iron and steel.

Figure 18.2   A suspended magnet will rotate to line up north and south.

Metals with the ability to attract pieces of iron or steel are said to be **magnetic** (Fig. 18.1). Iron ore that is naturally magnetic is called *lodestone*.

Artificial magnets can be made from iron, steel, and several special alloys such as Permalloy™ and alnico. We will discuss the process of creating artificial magnets later. Materials that can be made into magnets are called *magnetic materials*. Most materials are nonmagnetic (examples are wood, aluminum, copper, and zinc).

Forces Between Magnets

Suppose that a bar magnet is suspended by a string so that it is free to rotate. It will rotate until one end points north and the other south (Fig. 18.2). The end that points north is called the north-seeking pole, or *north* (N) *pole*. The other end is the south-seeking pole, or *south* (S) *pole*.

If the north pole of another bar magnet is brought near the north pole of this magnet, the two like poles will repel [Fig. 18.3(a)]. The south pole of one magnet will attract the north pole of the other [Fig. 18.3(b)]. In summary:

Like magnetic poles repel each other, and unlike magnetic poles attract each other.

Figure 18.3   (a) Like poles repel each other. (b) Unlike poles attract each other.

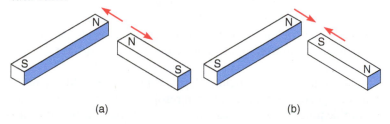

(a)   (b)

Figure 18.4   Simple compass

A **compass** (Fig. 18.4) is simply a small magnetic needle that is free to rotate on a bearing.

Magnetic Fields of Force

A **magnetic field** is a field of force near a magnetic pole or near an electric current that can be detected using another magnet. We can represent this field of force by drawing lines that indicate the direction of the force exerted on a north pole placed in the field. The field of a bar magnet can be mapped by moving a small compass around the magnet as shown in Fig. 18.5(a). These resulting lines are called **flux lines** (lines of force). The flux lines can also be found by sprinkling iron filings near a magnet [Fig. 18.5(b)].

Figure 18.5

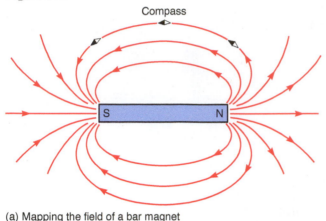

(a) Mapping the field of a bar magnet

Courtesy of Stephen Oliver © Dorling Kindersley

(b) Iron filings and plotting compasses reveal the flux lines of magnetic force around a bar magnet.

Figure 18.6 Flux lines of (a) unlike poles and (b) like poles near each other shown by iron filings

(a)   (b)

Figure 18.7 A magnetic field is three-dimensional.

The fields of combinations of magnets can also be found in this way (Fig. 18.6). Although the iron filing patterns are two-dimensional, the magnetic field around a magnet is actually a three-dimensional field, as shown in Fig. 18.7.

Magnetic flux is the total number of magnetic lines of force entering or leaving the pole of a magnet. The symbol for magnetic flux is the Greek letter ϕ. One magnetic line of force is given the unit maxwell (Mx). A more useful unit is 10^8 Mx or one weber (Wb).

The earth's three-dimensional magnetic field is depicted in Fig. 18.8. Many puzzling aspects of the earth's magnetic field have not been resolved. The north magnetic pole and the north geographic pole (sometimes called *true north*) are at different locations. The axis of rotation and the magnetic field axis are slightly different and change approximately 10 min of arc each year. Even more puzzling, scientific evidence indicates that the earth's magnetic field reverses completely every few hundred thousand years without significantly affecting the earth's rotational or orbital motions.

Magnetic poles are similar to the behavior of electric charges in many ways, but there is a very important difference. Electric charges can be isolated, but magnetic poles cannot. A cluster of electrons doesn't need to be accompanied by a cluster of protons. However, a north magnetic pole never exists without the presence of a south magnetic pole. If a magnet is broken in half, each piece still has a north and a south pole.

Figure 18.8 The earth's magnetic field is similar to that of other magnets. The geographic and magnetic poles do not coincide; similarly, the geographic and magnetic equators differ. The needle on a compass points to the north magnetic pole.

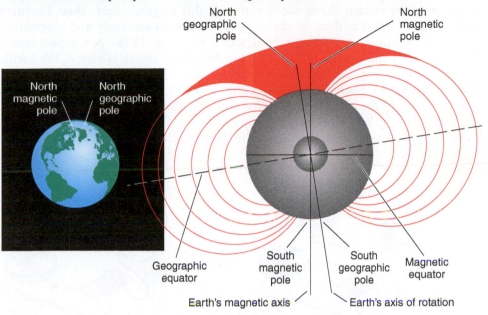

NEW TECHNOLOGIES

NASA's Magnetospheric Multiscale Mission (MMS)

The earth's magnetic field, the magnetosphere, is a flowing entity that fluctuates in strength and position depending on the current state of "space weather." Energy from the sun interacts with the magnetosphere causing magnetic lines to connect and disconnect from the earth, thus releasing massive amounts of energy. These disturbances can cause electrical flow issues on earth, expose orbiting astronauts to unwanted radiation from the sun, and damage satellites.

NASA's Magnetospheric Multiscale mission (MMS) consists of launching four identical spacecraft that will gather data to collect a more precise, three-dimensional view of the magnetosphere and better understand how it behaves. Three sets of instruments on each spacecraft will measure magnetic field strength, plasma, and energetic particles. By studying the reconnection of magnetic fields around the earth, scientists will be better able to predict and respond to electrical disruptions on earth caused by solar weather. In addition, scientists are seeking to more fully understand our magnetosphere so they may better understand the impact that magnetic field lines and solar weather have on other planes in our solar system (Fig. 18.9).

Figure 18.9 (a) The four spacecraft that make up NASA's Magnetospheric Multiscale mission (MMS)

Courtesy of NASA

Figure 18.9 (b) Artistic rendition of earth's magnetosphere

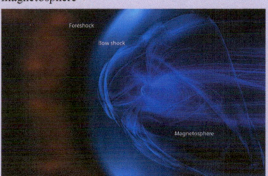

Courtesy of NASA

18.2 Magnetic Effects of Current

Hans Christian Oersted (1777–1851),

physicist, was born in Denmark. He was the first to discover the connection between electricity and magnetism in 1820.

When a current passes through a conductor, it forms a magnetic field. A compass placed near the current shows the direction of this magnetic field. **Hans Christian Oersted** was the first to discover the connection between electricity and magnetism. We can show this by connecting a battery to a wire (Fig. 18.10). A compass needle is placed under the wire as in Fig. 18.10(a). When the switch is closed, the needle is deflected as in Fig. 18.10(b). If the terminals of the battery are reversed, the needle is deflected in the opposite direction [Fig. 18.10(c)]. When the compass needle is placed on top of the conductor, the direction of deflection is reversed in each case [Fig. 18.10(d)–(f)]. When the current in a wire flows in a given direction, the flux lines point in one direction below the wire and in the opposite direction above the wire.

Figure 18.10   Magnetic effects of electric current

(a)   Compass below conductor, switch open

(b)   Compass below conductor, switch closed

(c)   Compass below conductor, switch closed. Battery terminals reversed.

(d)   Compass above conductor, switch open

(e)   Compass above conductor, switch closed

(f)   Compass above conductor, switch closed. Battery terminals reversed.

Andre Ampère (1775–1836),

mathematician and physicist, was born in France. His work became the basis of the science of electrodynamics following Oersted's discovery in 1820 of the magnetic effects of electric currents. The SI unit of electric current, the ampere or amp, is named after him.

The field actually curves around the straight current-carrying wire [Fig. 18.11(a)]. Iron filings on a sheet of paper perpendicular to a current-carrying wire show the shape of the field. The magnetic field is stronger for large currents than for small currents. The direction of the field near a current in a straight wire is shown in Fig. 18.11(b) and given by **Ampère's rule:**

AMPÈRE'S RULE

Hold the wire in your right hand with your thumb extended in the direction of the current. Your fingers circle the wire in the direction of the flux lines.

Figure 18.11   (a) Field around a current-carrying wire
(b) Direction of the field around a current-carrying wire

(a)   (b)

The magnetic field near a long current-carrying wire, measured in units of teslas, is circular about the wire and given by Ampère's law:

$$B = \frac{\mu_0 I}{2\pi R}$$

where B = magnetic field
I = current through the wire
R = perpendicular distance from the center of the wire
$\mu_0 = 4\pi \times 10^{-7}$ T m/A

The magnetic field, B, has the unit *tesla* (T), named after **Nikola Tesla**, and is defined in terms of electric current by the constant μ_0, the permeability constant. The value of μ_0 is not experimentally determined but is an assigned value that explicitly defines magnetic field in terms of electric current.

The magnetic field, B, may also be described in terms of magnetic flux density, or the number of magnetic lines of force passing through a given area:

$$\text{magnetic flux density} = B = \frac{\phi}{A} = \frac{\text{magnetic flux}}{\text{area}}$$

Note that magnetic flux, ϕ, includes the total area, whereas magnetic flux density describes a magnetic field in a specified area. The unit tesla (T) = 1 W/m². If the area is measured in square centimetres, the unit Gauss = 1 Mx/cm² may be used.

Nikola Tesla (1856–1943),

electrical engineer and inventor, was born in Croatia. He invented the electromagnetic motor that became the basis for most alternating-current machinery and sold his patents to Westinghouse in 1888. He then produced important inventions involving high-frequency electricity: the Tesla coil and a resonant air-core transformer. His alternating current system illuminated the Chicago World's Fair in 1893 and led to the construction of the Niagara Falls hydroelectric generating plant in 1896. His later inventions included wireless transmission of electricity and radio-controlled craft. He also worked on pulsed radar, harnessing solar power, and radio communications to other planets.

EXAMPLE 1

A power line carrying $40\overline{0}$ A is 9.00 m above a transit used by a surveying student [Fig. 18.12(a)].

(a)   Find the magnetic field formed by the power-line current above the transit.
(b)   If the earth's horizontal component of magnetic field is 5.20×10^{-5} T at that location, what error could be introduced in the angular measurement? (Assume that the power line runs north–south.)

(a)   **Data:**

$$I = 40\overline{0} \text{ A}$$
$$R = 9.00 \text{ m}$$
$$\mu_0 = 4\pi \times 10^{-7} \text{ T m/A}$$
$$B = ?$$

Figure 18.12

B_E = magnetic field of earth

B = magnetic field of power line

B_R = resultant magnetic field

(a)   (b)

Basic Equation:

$$B = \frac{\mu_0 I}{2\pi R}$$

Working Equation: Same

Substitution:

$$B = \frac{(4\pi \times 10^{-7} \text{ T m/A})(40\overline{0} \text{ A})}{2\pi(9.00 \text{ m})}$$

$$= 8.89 \times 10^{-6} \text{ T}$$

Therefore, the magnetic field from the power line is 8.89×10^{-6} T. With the current from south to north, Ampère's rule shows the direction of B to be east to west.

(b)   The angle that the resultant vector [the earth's field plus the wire's field ($B_E + B$)] makes with B_E would be the angular error θ.

Data:

$$B_E = 5.20 \times 10^{-5} \text{ T} \quad \text{earth's component}$$
$$B = 8.89 \times 10^{-6} \text{ T} \quad \text{[from part (a)]}$$
$$\theta = ?$$

Basic Equation:

$$\tan \theta = \frac{B}{B_E}$$

Working Equation: Same

Substitution:

$$\tan \theta = \frac{8.89 \times 10^{-6} \text{ T}}{5.20 \times 10^{-5} \text{ T}} = 0.171$$

$$\theta = 9.7°$$

The bearing on the surveying student's transit could be in error by 9.7° because of the power line.

· · · · · · · · · · · · · · · · · ·

Magnetic Field of a Loop

If a wire is bent into a loop, the magnetic field lines become bunched up or concentrated in the center of the loop. A second and subsequent loops increase the intensity of the magnetic field in the loop even more.

To determine the direction of the flux lines of a current in a loop, use Ampère's rule as shown in Fig. 18.13. If several loops are made into a tight spiral as shown in Fig. 18.14, the flux lines add to form the field shown. A coil of tightly wrapped wire is called a **solenoid**. The left side of this solenoid acts like a south magnetic pole. The right side acts like a north magnetic pole. This polarity could be found by using a compass. The rule for finding the polarity of a solenoid is the following:

> Hold the solenoid in your right hand so that your fingers circle it in the same direction as the current. The direction your thumb points indicates the northern orientation of the solenoid (Fig. 18.15).

Figure 18.13   Magnetic field around (a) a straight wire and (b) a loop in a current-carrying wire

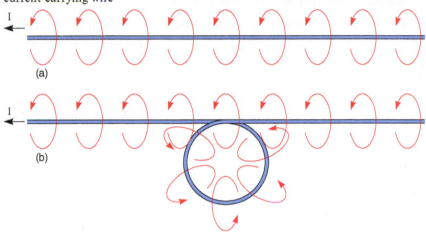

Figure 18.14   (a) Magnetic field of a coil (b) Iron filings show the magnetic lines of force from the magnetic field of a coil of red wire charged with an electric current. The coil of wire passes repeatedly through a white surface. (c) Compasses also represent a magnetic field around a coil of wire.

(a)

(b)

Courtesy of Manfred Kage/Getty Images

(c)

Courtesy of Andy Crawford/Dorling Kindersley, Ltd

Figure 18.15   Polarity of a solenoid

For a long coil that is tightly turned, the field strength at its center is

$$B = \mu_0 In$$

where B = magnetic field in the region at the center of the solenoid
μ_0 = permeability constant, $4\pi \times 10^{-7}$ T m/A
I = current through the solenoid
n = number of turns per unit length of solenoid

The longer the solenoid in relation to its radius, the more uniform the magnetic field is inside the solenoid; for an infinitely long solenoid, the value of B is uniform throughout.

EXAMPLE 2

Find the magnetic field at the center of a solenoid that is 0.425 m long and 0.0750 m in diameter, and has three layers of $85\overline{0}$ turns each, when 0.250 A flows throughout.

Data:

$$I = 0.250 \text{ A}$$

$$n = \frac{3 \times 85\overline{0} \text{ turns}}{0.425 \text{ m}} = 60\overline{0}0 \text{ turns/m}$$

$$\mu_0 = 4\pi \times 10^{-7} \text{ T m/A}$$

$$B = ?$$

Note: The length, 42.5 cm, can be considered "long" compared with the radius, 3.75 cm (about 11 times longer).

Basic Equation:

$$B = \mu_0 In$$

Working Equation: Same

Substitution:

$$B = (4\pi \times 10^{-7} \text{ T m/A})(0.250 \text{ A})(60\overline{0}0/\text{m})$$
$$= 1.88 \times 10^{-3} \text{ T}$$

PROBLEMS 18.2

1. Find the magnetic field at 0.250 m from a long wire carrying a current of 15.0 A.
2. Find the magnetic field at 0.500 m from a long wire carrying a current of 7.50 A.
3. What is the current in a wire if the magnetic field is 5.75×10^{-6} T at a distance of 2.00 m from the wire?

4. A power line runs north–south carrying 675 A and is 5.00 m above a transit used by a surveyor. (a) What is the magnetic field at the transit because of the power-line current? (b) If the earth's horizontal component of magnetic field is 5.20×10^{-5} T, what error is introduced in the surveyor's angular measurement?

5. Find the magnetic field at 0.350 m from a long wire carrying a current of 3.00 A.

6. Find the current in a wire if the magnetic field is 3.50×10^{-6} T at a distance of 2.50 m from the wire.

7. A solenoid has $100\overline{0}$ turns of wire, is 0.320 m long, and carries a current of 5.00 A. What is the magnetic field at the center of the solenoid? Assume that its length is long in comparison with its diameter.

8. A solenoid has $300\overline{0}$ turns of wire and is 0.350 m long. What current is required to produce a magnetic field of 0.100 T at the center of the solenoid? Assume that its length is long in comparison with its diameter.

9. A small solenoid is 0.150 m in length and 0.0150 m in diameter, and has $60\overline{0}$ turns of wire. What current is required to produce a magnetic field of 1.25×10^{-3} T at the center of the solenoid?

10. A solenoid has $250\overline{0}$ turns of wire and is 0.200 m long. What current is required to produce a magnetic field of 0.100 T at the center of the solenoid? Assume that its length is long in comparison with its diameter.

11. A long solenoid has $100\overline{0}$ turns and is 0.250 m long. If the wire carries a current of 2.50 A, what is the strength of the magnetic field at its center?

12. A small solenoid 0.100 m in length has $100\overline{0}$ turns of wire. What current is required to produce a magnetic field of 1.30×10^{-3} T at the center of the solenoid?

13. An auto mechanic wants to use a solenoid she found on a car starter. If the solenoid is 0.150 m in length and has $75\overline{0}$ turns of wire, what current is required to produce a magnetic field of 1.50×10^{-3} T at its center?

14. An earthmover requires a solenoid with $250\overline{0}$ turns and whose length is 0.150 m. What current is required to produce a magnetic field of 0.100 T at the center of the solenoid?

18.3 Induced Magnetism and Electromagnets

Induced magnetism is produced in a magnetic material such as iron when the magnetic material is placed in a magnetic field, such as that produced in the core of a current-carrying solenoid (Fig. 18.16). An **electromagnet** consists of a solenoid and a magnetic core. When a current is passed through the solenoid, the magnetic fields of the atoms in the magnetic material line up to produce a strong magnetic field. When the current through the coil is turned off, the strength of the induced magnet decreases, but some remains. When the core is removed, a magnetic field remains in the core. In materials such as soft iron, very little magnetic field remains in the core after the current flow stops. In other materials, such as steel, alnico, and Permalloy™, a much stronger field remains. The latter materials are used for permanent magnets. However, they are undesirable for use as a core in an induction motor. Soft-iron cores are often used for this application because less energy is required to reverse the polarity of the induced magnetic field.

A magnet can be thought to consist of many atoms, each behaving like a small magnet. In each atom, the electrons orbit about the nucleus and each electron spins about its own axis, producing small current loops that generate magnetic fields. In most materials, these current loops are arranged so that their magnetic fields point in different directions. The result is that the magnetism of one loop (atom) is canceled out by those of its neighbors.

In magnetic materials, the atomic magnetic fields line up with each other in regions called *magnetic domains* when no field is present [Fig. 18.17(a)]. Each domain

Figure 18.16 Simple electromagnet

Magnetic core

I

Figure 18.17 Magnetic fields

(a) Unmagnetized iron (no field)

S   N

(b) Slightly magnetized iron (weak field)

S   N

(c) Strongly magnetized iron (strong field)

has a magnetic field direction to which most atoms in the domain are aligned. When no magnetic field is present, the orientations of the domains are random. However, when an external field is present, the domains tend to line up with the field, causing the domain boundaries to shift [Fig. 18.17(b)]. In high electric fields, nearly all the material is aligned [Fig. 18.17(c)]. When the external magnetic field is removed, some materials such as the alloy alnico retain the aligned domains, creating a permanent magnet.

TRY THIS ACTIVITY

Lifting Paper Clips

Using a large nail, an insulated wire, and a 3.0- to 9.0-V battery, make an electromagnet that will pick up as many paper clips as possible. Without adding more batteries, what can be done to the electromagnet to allow it to pick up more paper clips?

18.4 Induced Current

An **induced current** is produced in a circuit by motion of the circuit through the flux lines of a magnetic field. The strength depends on the strength of the magnetic field and on the rate at which the flux lines are cut by moving the magnet or wire. Increasing the strength of the field or increasing the rate at which the flux lines are cut also causes the current to increase.

While the magnet shown in Fig. 18.18(a) is moving downward, the galvanometer indicates that a current flows through the wire. If the magnet shown in Fig. 18.18(b) is moving upward, the induced current is in the opposite direction.

A current also flows in the wire if the circuit is closed, the wire is moved, and the magnet is stationary [Fig. 18.18(c)]. The current is produced by the relative motion of the magnet and wire. In commercial generators, magnets are spun inside a set of coils of wire. The current is increased by replacing the single wire with a coil of many turns. For example, tripling the number of turns triples the induced emf [Fig. 18.18(d)].

Electromagnets can be made strong enough to lift very heavy loads, as shown in Fig. 18.19.

Figure 18.18   Induced current in a wire

(a) Magnet moving downward

(b) Magnet moving upward causes the current to flow in the opposite direction.

(c) Wire moving downward

(d) Coil moving downward increases the current flow in same direction.

Figure 18.19   (a) A crane with an electromagnet lifts scrap metal from a pile in a salvage yard.

Figure 18.19   (b) The scrap metal drops when the current through the coil of the electromagnet is turned off.

Courtesy of fotolia © petaran

(a)

Courtesy of fotolia © petaran

(b)

TRY THIS ACTIVITY

Factors Influencing Current

Attach a solenoid to an ammeter to read the current that is flowing through a coil of wire. Then move a bar magnet back and forth inside the coil of wire while observing the current on the ammeter. Instead of increasing the number of coils as discussed in the text, move the magnet in and out of the solenoid at a greater rate of speed. In addition, put one magnet on top of another while moving the magnet back and forth in the solenoid. What impact, if any, do these actions have on the current flowing through the solenoid wire?

PHYSICS CONNECTIONS

Speakers and Electromagnets

A conventional speaker consists of three major components: (a) an amplifier or some other signal source, (b) a cone that vibrates the air to produce a sound wave, and (c) a permanent magnet surrounding a coil of electric wire attached to the back of the cone.

For the speaker to create a sound wave, the amplifier must send an alternating current signal through the wire. As the alternating current travels through the wire, it induces an alternating magnetic field around the coil. The wire's induced magnetic field is repeatedly attracted to and repelled from the magnetic field of the permanent magnet. This repeated motion causes the cone to vibrate at the frequency determined by the amplifier or signal generator (Fig. 18.20).

Figure 18.20 The cross section of a conventional speaker

Permanent magnet

Coil of wire

N

S

To receiver

Speaker cone

18.5 Generators

The induction of an emf in a coil of wire can be used to produce electrical power. A **generator** consists of a coil of wire rotating in a magnetic field, which produces an induced current in the coil, converting mechanical energy into electric energy. This is the simplest kind of generator to build in the laboratory, so we will study its operation here and compare it to commercial generators, where the magnets (electromagnets) are rotated.

The current produced by rotating the wire through the magnetic field is called an alternating current (ac). As side A of the current loop in Fig. 18.21(a) passes

Figure 18.21   Direction of current flow in a rotating wire in a magnetic field

(a)   (b)

downward by the north pole, the induced current is in one direction. As side *A* (same side of the rotating loop) passes upward by the south pole as in Fig. 18.21(b), the induced current is in the opposite direction. The result is an alternating current induced in the rotating wire. As side *B* (the other side of the rotating loop) passes upward and then downward, the current in it also alternates.

The direction of current flow in a wire as it rotates between the north and south magnet poles is illustrated in Fig. 18.22. In beginning position (1), the current is zero. As the wire passes down through position (2), the current builds and reaches a maximum value in position (3). The current then becomes smaller through position (4) until it is zero again in position (5).

As the wire begins to pass back up through the magnetic field in position (6), the current begins to build and reaches a maximum value in the opposite direction in position (7). It then becomes smaller in position (8) and falls to zero again as it reaches position (1) again. The cycle is then repeated for the next rotation of the wire through the magnetic field.

A graph of the induced current showing the relative magnitude of the current for the changing positions of the wire is shown in Fig. 18.23.

Figure 18.22   Induced current in a wire rotating in a magnetic field

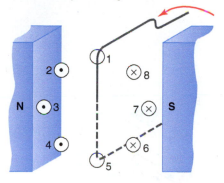

• Current out of page
× Current into page

Figure 18.23   Graph of induced current as the wire changes position in a magnetic field

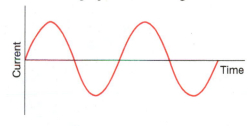

One cycle is produced by one rotation of the wire. The time required for one cycle depends on the rotational speed of the coil. If the coil rotates 60 times each second, an alternating current of frequency 60 hertz (cycles per second) is produced. The current produced in the coil is conducted by brushes on slip rings to the external circuit as shown in Fig. 18.24. The rotating coil is called the **rotor** or **armature**, and the *field magnets* are called the **stator**.

The generator does not actually create electric energy; it changes the mechanical energy of rotation into electric energy. The energy to turn the rotor may be supplied by water falling down a waterfall, a diesel engine, or a steam turbine.

Power companies use large commercial ac generators to produce the current they need to supply to their customers. These generators work similarly to the generator discussed here, but they have many coils and use electromagnets instead of permanent magnets. The large generators used by electric power companies can produce voltages as large as 13,000 V and currents up to 10 A. The alternator used in automobiles is an ac generator that produces voltages about 13 V and currents up to 40 A.

Figure 18.24   Current in a generator

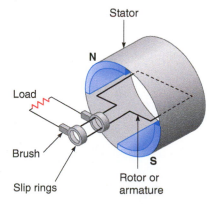

dc Generators

By the use of a special device called a commutator, the ac generator can be used to produce direct current. The **commutator** is a split ring that replaces the slip rings as shown in Fig. 18.25(a). When side *A* of the coil passes upward along the north pole,

Figure 18.25 Current in a dc generator

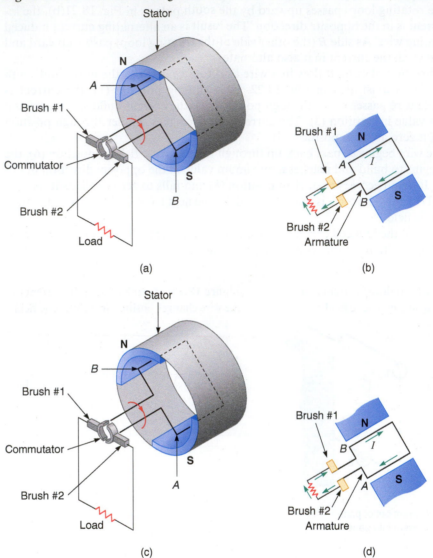

(a)

(b)

(c)

(d)

the induced current flows in the direction shown in Fig. 18.25(b) and is picked up by brush 1. The current in the external circuit is also shown.

When side *B* of the coil passes upward along the north pole, the induced current flows in the direction shown and is picked up by brush 1 [Fig. 18.25(c)]. The current in the external circuit is in the same direction as it was when *A* passed along the north pole [Fig. 18.25(d)]. Thus, this is a direct current.

The current produced by this dc generator does not have the same value at all times. A graph of the induced current is shown in Fig. 18.26(a). Commercial dc generators that are used for industrial purposes contain many coils. The output current has almost the same value at all times due to the large number of coils [Fig. 18.26(b)].

Figure 18.26 Induced current in a dc generator

(a) One coil

(b) Many coils

18.6 The Motor Principle

We have seen that like poles of magnets repel each other. A magnet that is pivoted will spin due to the repulsion of another magnet nearby, as shown in Fig. 18.27(a). We can construct an electromagnet by wrapping wire around an iron core and passing a current through the wire [Fig. 18.27(b)]. The north pole of the electromagnet will be repelled by a north pole of another magnet. The electromagnet will turn until its south pole is next to the north pole of the permanent magnet [Fig. 18.27(c)].

If we could suddenly change the polarity of the electromagnet (often called the *armature*), the magnet would repel the north pole and the electromagnet would continue to spin [Fig. 18.27(d)]. If a dc current supply is used [Fig. 18.27(e)], this change can be made by using a commutator (split ring) to change the direction of the current in the electromagnet. Changing the direction of the current flowing through the coil of the electromagnet changes the poles.

Figure 18.27   Motor principle

(a)   (b)   (c)

(d)   (e)

As the current changes direction, the electromagnet spins due to the repulsion of like poles. A shaft may be connected to the electromagnet so that the rotational motion can be used to do work. In a **motor** the armature rotates in the magnetic field of the stator when a current is passed through the armature and converts electric energy to mechanical energy and thus performs the reverse function of a generator (Fig. 18.28).

If ac current is supplied to the electromagnet, slip rings are used instead of a commutator. The use of alternating current makes the commutator unnecessary. The changes in direction are supplied by the ac current itself (Fig. 18.29).

Figure 18.28 A motor performs the reverse function of a generator.

Figure 18.29 Use of ac makes a commutator unnecessary.

Generator

Motor

Industrial Motors

Commercial motors operate in the same way as the motors just discussed. However, they usually use electromagnets in place of the permanent magnets and are much more complex. Slip rings are not necessary in ac motors. The current in the rotating electromagnet can be induced in the same way a current is induced in a generator.

Motors can be designed for many different purposes. Heavily loaded motors need certain types of starters. The torque and power outputs can be greatly varied by differences in design. Three types of ac motors are discussed here.

1. The **universal motor** (Fig. 18.30) can be run on either ac or dc power. Slip rings are used in this type of motor as in dc motors. This motor is often used in small hand tools and appliances.

2. The **induction motor** (Fig. 18.31), a widely used, heavy-duty ac motor, depends on a rotating magnetic field for its operation. Instead of the rotating electromagnet being connected to a power source by slip rings, the current in the electromagnet is induced by a moving magnetic field created by an ac current. Three pairs of poles and a three-phase current cause the stator to rotate and induce currents in the rotor.

Figure 18.30 Universal motor used in an electric drill

Courtesy of fotolia © manusy

Figure 18.31 Induction motor

Courtesy of Bodine Electric Company, Chicago, IL

3. The **synchronous motor** (Fig. 18.32) is very similar to the slip-ring ac motor discussed earlier. The rotating electromagnet is supplied with current through slip rings. The speed of rotation of a synchronous motor is constant, depends on the number of coils, and is directly proportional to the frequency of its ac power supply. The synchronous motor will work only when operated with an ac power source of the frequency for which it is designed. The word *synchronous* is derived from *syn*, meaning same, and *chrona*, or time. Synchronous motors are used to operate clocks and other devices needing accurate speed control.

Figure 18.32

(a) This disassembled permanent magnet motor is a synchronous motor

Courtesy of gettyimages © Bosca78

(b) Synchronous motor

PHYSICS CONNECTIONS

Superconducting Electromagnets and Magnetic Resonance Imaging

Superconducting electromagnets are used to accelerate subatomic particles in particle accelerators for nuclear and particle research (Fig. 18.33). The magnetically levitated vehicle, or maglev, now being developed uses electromagnets to levitate the vehicle above a track and virtually eliminates any friction between the vehicle and the track (Fig. 18.34; also see Fig. 0.11). Air resistance remains a key vehicle design factor.

Magnetic resonance imaging (MRI) is used in medical facilities to provide incredibly clear images of tissues and organs inside the human body without invasive procedures or X-ray exposure. Superconducting electromagnets align hydrogen atoms in the body. After radio waves hit the protons of the hydrogen atoms, the protons return to their previous pattern while emitting electromagnetic signals. Sensors pick up those signals and computers then record and analyze the resulting effects on the different tissues and provide a computer image "slice" of the tissue or organ (Fig. 18.35).

Figure 18.33 Components of a particle accelerator used for particle physics experiments

Courtesy of CERN/European Organization for Nuclear Research

Figure 18.34 A maglev system pushes a spacecraft along a rail for testing.

Courtesy of NASA

Figure 18.35 (a) Magnetic resonance imaging (MRI) machine

Figure 18.35 (b) Magnetic resonance images on monitors

(a)

Courtesy of fotolia © uwimages

(b)

Courtesy of fotolia © beerkoff

18.7 Magnetic Forces on Moving Charged Particles

A charged particle that is not moving will produce no interaction with a magnetic field. However, if the charged particle moves through a magnetic field, it is deflected by the magnetic field. The deflection force is greatest when the direction is perpendicular to the magnetic field lines.

The force that acts on a moving charged particle does not act in a direction between the sources of interaction but is perpendicular to both the magnetic field and the direction of the charged particle (Fig. 18.36). This behavior is used to spread electrons on a TV tube to produce a TV picture.

Figure 18.36 Deflection of an electron beam by a magnetic field

TRY THIS ACTIVITY

Moving Electrons

Take a bar magnet and move it close to the front of an old computer monitor, black and white television, or oscilloscope. (Do not risk using a new television set.) What happens to the image on the screen? Why does the magnet influence the electrons' placement on the screen and appear to warp the image?

Glossary

Ampère's Rule To find the direction of a magnetic field near a current and a straight wire, hold the wire in your right hand with your thumb extended in the direction of the current. Your fingers circle the wire in the direction of the flux lines. (p. 506)

Armature The rotating coil or electromagnet in a generator. (p. 515)

Commutator A device in an ac generator that produces a direct current. Composed of a split ring that replaces the slip rings in an ac generator and produces a direct current in the circuit connected to the split ring of the generator. (p. 515)

Compass A small magnetic needle that is free to rotate on a bearing. (p. 503)

Electromagnet A combination of a solenoid and a magnetic material, such as iron, in the core of the solenoid. When a current is passed through the solenoid, the magnetic fields of the atoms in the magnetic material line up to produce a strong magnetic field. (p. 511)

Flux Lines Lines indicating the direction of the magnetic field near a magnetic pole. (p. 504)

Generator An apparatus consisting of a coil of wire rotating in a magnetic field. A current is induced in the coil, converting mechanical energy into electric energy. (p. 514)

Induced Current A current produced in a circuit by motion of the circuit through the flux lines of a magnetic field. (p. 512)

Induced Magnetism Magnetism produced in a magnetic material such as iron when the material is placed in a magnetic field, such as that produced in the core of a current-carrying solenoid. (p. 511)

Induction Motor An ac motor with an electromagnetic current induced by the moving magnetic field of the ac current. (p. 518)

Magnetic Property of metals or other materials that can attract iron or steel. (p. 503)

Magnetic Field A field of force near a magnetic pole or a current that can be detected using a magnet. (p. 503)

Motor A device that is composed of an armature and a stator. When a current is passed through the armature, the armature rotates in the magnetic field of the stator and converts electric energy to mechanical energy. (p. 517)

Rotor The rotating coil in a generator. (p. 515)

Solenoid A coil of tightly wrapped wire. Commonly used to create a strong magnetic field by passing current through the wire. (p. 509)

Stator The field magnets in a generator. (p. 515)

Synchronous Motor An ac motor whose speed of rotation is constant and is directly proportional to the frequency of its ac power supply. (p. 518)

Universal Motor A motor that can be run on either ac or dc power. (p. 518)

Formulas

18.2 $B = \dfrac{\mu_0 I}{2\pi R}$

$B = \mu_0 I n$

Review Questions

1. The presence of a magnetic force field may be detected by using
   (a) a compass.
   (b) iron filings.
   (c) a magnet.
   (d) all of the above.
2. The deflection of a compass needle placed near a current-carrying wire shows
   (a) the magnetic field of the sun.
   (b) the magnetic field of the wire.
   (c) the electric field.
3. Ampère's rule relates
   (a) the strength of a magnetic field to the magnetic pole.
   (b) the direction of a magnetic field surrounding a current-carrying wire.
   (c) the direction of a magnetic field near a bar magnet.
   (d) none of the above.
   (e) all of the above.
4. The unit used to express the strength of a magnetic field is the _____.
5. Describe how a strong magnetic field can be produced in a solenoid.
6. Describe how to determine the direction of a magnetic field in a solenoid.
7. Describe how a magnetic field is induced by a current-carrying coil surrounding a core of magnetic material.
8. Describe how a generator produces current.
9. Describe the function of a commutator.
10. Describe how a motor works.
11. What is a synchronous motor, and how does it work?
12. Distinguish between a universal motor and an induction motor.
13. Distinguish between an armature and a stator.
14. Describe how an electromagnet works.
15. If the current in a solenoid is increased by a factor of 2, how does the magnetic field change?
16. If the radius of a solenoid decreases by a factor of 2, how does the magnetic field change?
17. If the number of turns per inch in a solenoid were increased by a factor of 4, how would the magnetic field change?
18. Describe how to find the flux lines near a bar magnet.
19. How is alternating current produced by a generator?

Review Problems

1. Find the magnetic field at 0.255 m from a long wire carrying a current of 1.38 A.
2. Find the magnetic field at 0.365 m from a long wire carrying a current of 8.95 A.
3. What is the current in a wire if the magnetic field is 4.75×10^{-6} T at a distance of 1.75 m from the wire?
4. A solenoid has $20\overline{0}0$ turns of wire, is 0.452 m long, and carries a current of 4.55 A. What is the magnetic field at the center of the solenoid?
5. A solenoid has 2750 turns of wire and is 0.182 m long. What current is required to produce a magnetic field of 0.235 T at the center of the solenoid?
6. A power line running north–south carrying $50\overline{0}$ A is 7.00 m above a transit used by a surveyor. What error is induced in the compass used by the surveyor? (Assume the earth's horizontal component of magnetic field is 5.20×10^{-5} T.)

PROBLEM SOLVING

SKETCH

12 cm² | w

4.0 cm

DATA
$A = 12$ cm², $l = 4.0$ cm, $w = ?$

BASIC EQUATION
$A = lw$

WORKING EQUATION
$w = \dfrac{A}{l}$

SUBSTITUTION
$w = \dfrac{12 \text{ cm}^2}{4.0 \text{ cm}} = 3.0 \text{ cm}$

APPLIED CONCEPTS

1. A ship's compass is mistakenly placed 8.35 cm away from a wire carrying a current of 8.25 A. (a) What is the strength of the wire's magnetic field on the compass? (b) The strength of the earth's magnetic field is 5.20×10^{-5} T. How far from the wire must the compass be mounted so that it only experiences a magnetic field of 5.20×10^{-7} T $\left(\frac{1}{100}\right.$ of the magnetic field of the earth$\left.\right)$ due to the wire?

2. Figure 18.10 shows a compass near a current-carrying wire. (a) What is the strength of the magnetic field when the compass needle is placed 1.03 cm under the wire? Assume the voltage of the battery is 6.00 V and the resistance of the wire is 15.0 Ω. (b) Compare the strength of the earth's magnetic field to the strength of the magnetic field of the wire.

3. A coaxial cable consists of an inner conducting wire encased in an insulating material surrounded by a conducting metal braid and another insulating sheath. The inner wire carries current in one direction, while the outer braid carries current in the opposite direction. Using Ampère's right-hand rule, explain why there is no significant magnetic field outside the coaxial cable. (Coaxial cables are often used to transmit cable TV and other audiovisual signals.)

4. Figure 18.37 shows a picture tube diagram that uses two solenoids to guide electrons as they move through the magnetic field. (a) What is the orientation of the magnetic field for the two solenoids? (b) As a result of the magnetic field, will the electrons travel toward a, b, or c? (Use Fig. 18.36 as a reference.)

Figure 18.37

5. A copper wire 20.0 m long with radius 4.57×10^{-3} m and $\rho = 1.72 \times 10^{-8}$ Ω cm is used to form a solenoid that will produce a magnetic field. (a) What is the wire's resistance? (b) If the wire is connected to a 4.50-V battery, what is the current that passes through the wire? (c) If the solenoid is designed to have an internal radius of 3.25 cm, how many single-layer coils will be produced? (d) What will be the length of the solenoid if the coils touch one another? (e) Find the strength of the solenoid's magnetic field.

CHAPTER 19

ALTERNATING CURRENT ELECTRICITY

Objectives

The major goals of this chapter are to enable you to:

1. Describe the nature and characteristics of alternating current.
2. Describe the use of transformers in changing voltage.
3. Apply inductance and inductive reactance in circuits.
4. Use capacitance and capacitive reactance in circuits.

Ordinary household current is alternating current. The current produced by rotating a loop through a magnetic field is alternating current and constantly changing. We now consider the nature and characteristics of alternating current, transformation of voltage, and various devices used in electric circuits.

19.1 What is Alternating Current?

In Chapter 18 we learned that the current produced by rotating a loop of wire through a magnetic field alternates in direction. This current, called an *alternating current*, is used more frequently in industry and in everyday living than direct current because the transmission of power over long distances is more efficient with alternating current than with direct current.

As its name implies, **alternating current** (ac) is current that flows in one direction in a conductor, then changes direction and flows in the other direction. Electrons in the circuit move first in one direction and then in the opposite direction, alternating back and forth about relatively fixed positions. The direction of flow changes many times in 1 s. Ordinary household current is 60-Hz current. This means that the voltage goes through 60 complete cycles, positive to negative and back again, each second.

Every time the current repeats itself—flows, changes direction, flows, and changes direction—it goes through one *cycle* (Fig. 19.1). The reason for this alternation is the way current comes from electric generators. The emf and current produced

Figure 19.1   Alternating current changes direction as it flows in a conductor.

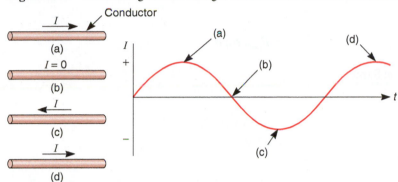

Figure 19.2   Alternating current does not alternate instantly between maximum values in each direction but builds to maximum values as in part (a). Current in a computer digital circuit varies as shown in part (b).

(a)

(b)

by a generator do not alternate instantly between maximum values in each direction, but they build up to maximum values and then decrease, change direction, and build to maximum values in the other direction (Fig. 19.2).

Direct current is usually a steady flow at a constant value. Graphically, it can be represented as shown in Fig. 19.3. Alternating current, however, is constantly changing. To graphically represent ac, we must show that it builds up and drops off. This can be demonstrated by the curve shown in Fig. 19.4, called a *sine curve*. We form the curve by rotating a vector **V** about a point and plotting the vertical components of **V**. Rotating **V** through 360° graphs one cycle.

Figure 19.3   Direct current does not alternate but is steady.

Direct current

Figure 19.4   Graph of one cycle of alternating current

The graph in Fig. 19.4 shows one complete cycle of the ac-current curve. A graph of ac voltage is also a sine curve. The current and voltage of ac are constantly changing. **Instantaneous current** is the current at any instant of time; **instantaneous voltage** is the voltage at any instant of time. We can find the value of current or voltage at any instant by using the fact that the graph of each makes a sine curve (Fig. 19.5).

Figure 19.5   Instantaneous values of current and voltage

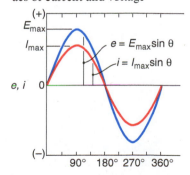

$$i = I_{max} \sin \theta$$
$$e = E_{max} \sin \theta$$

where i = instantaneous current (current at any instant)
I_{max} = maximum instantaneous current
θ = angle measured from the beginning of the cycle (see Fig. 19.4)
e = instantaneous voltage
E_{max} = maximum instantaneous voltage

EXAMPLE 1

The maximum voltage in an alternating current is 75 V. Find the instantaneous voltage at $\theta = 35°$.

Data:

$$E_{max} = 75 \text{ V}$$
$$\theta = 35°$$
$$e = ?$$

Basic Equation:

$$e = E_{max} \sin \theta$$

Working Equation: Same

Substitution:

$$e = (75 \text{ V})(\sin 35°)$$
$$= 43 \text{ V}$$

· · · · · · · · · · · · · · · ·

Both curves in Fig. 19.6(a) show e and i reaching a maximum at the same time and falling to zero at the same time. When this occurs, they are "in phase." When there is only resistance in the circuit, e and i are in phase. Later we will study some ac components that will cause e and i to be out of phase [see Fig. 19.6(b)].

Figure 19.6

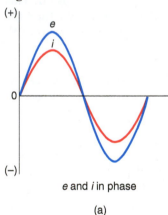

e and i in phase

(a)

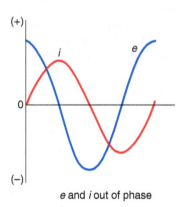

e and i out of phase

(b)

Effective Values of ac

A direct measurement of ac is difficult because it is constantly changing. The most useful value of ac is based on its heating effect and is called its effective value. The **effective value** of an alternating current (sometimes called rms, root-mean-square, value) is the number of amperes that produces the same amount of heat in a resistance as an equal number of amperes of a steady direct current.

The numerical factors in the following equations are derived from an average of the sine-wave time variation of the ac current. When this time average is taken, the factors $\sqrt{2} = 1.414$ and $1/\sqrt{2} = 0.707$ are found.

$$I = 0.707 \, I_{max}$$
$$I_{max} = 1.41 \, I$$

where I = effective value of current
        I_{max} = maximum instantaneous current

The current supplied to a woodworking shop is rated at 10.0 A. What is the maximum value of the current supplied?

EXAMPLE 2

Data:

$$I = 10.0 \text{ A}$$
$$I_{max} = ?$$

Basic Equation:

$$I_{max} = 1.41 \, I$$

Working Equation: Same

Substitution:

$$I_{max} = 1.41 \, (10.0 \text{ A})$$
$$= 14.1 \text{ A}$$

. . . .

The effective value for ac voltage may be expressed similarly:

$$E = 0.707 \, E_{max}$$
$$E_{max} = 1.41 \, E$$

where E = effective value of voltage
     E_{max} = maximum instantaneous voltage

When we say a house is wired for 120 V, we are using the effective value of the voltage. Actually, the voltage varies between +170 V and −170 V during each cycle.

Unless otherwise stated, ac voltage and current are *always* expressed in terms of effective, or rms, values.

PROBLEMS 19.1

1. What is the maximum voltage in an ac circuit in which the instantaneous voltage at $\theta = 35.0°$ is 27.0 V?
2. The instantaneous voltage in an ac circuit at $\theta = 65.0°$ is 82.0 V. What is the maximum voltage?
3. If the maximum ac voltage on a line is 165 V, what is the instantaneous voltage at $\theta = 45.0°$?
4. The maximum current in an ac circuit is 8.00 A. Find the instantaneous current at $\theta = 60.0°$.
5. The instantaneous current in an ac circuit is 6.50 A at $\theta = 45.0°$. Find the maximum current.
6. What is the maximum voltage in an ac circuit where the instantaneous voltage at $\theta = 51.0°$ is 14.5 V?
7. If the maximum ac voltage on a line is 145 V, what is the instantaneous voltage at $\theta = 35.0°$?
8. The maximum current in an ac circuit is 5.75 A. What is the instantaneous current at $\theta = 80.0°$?
9. Find the maximum current in an ac circuit where the instantaneous current at $\theta = 45.0°$ is 4.00 A.
10. The instantaneous voltage in an ac circuit at $\theta = 55.0°$ is $45\overline{0}$ V. Find the maximum voltage.
11. If $I_{max} = 4.59$ A and $I = 4.32$ A, what is θ?
12. If $I = 1.23$ A and $I_{max} = 3.41$ A, what is θ?

SKETCH

12 cm² w

4.0 cm

DATA

$A = 12$ cm², $l = 4.0$ cm, $w = ?$

BASIC EQUATION

$A = lw$

WORKING EQUATION

$w = \dfrac{A}{l}$

SUBSTITUTION

$w = \dfrac{12 \text{ cm}^2}{4.0 \text{ cm}} = 3.0$ cm

13. Find the effective value of an ac voltage whose maximum voltage is 2250 V.
14. Find the maximum current in an ac circuit with an effective value of 6.00 A.
15. Find the effective value of an ac voltage whose maximum voltage is 165 V.
16. Find the maximum current in an ac circuit with an effective value of 4.00 A.
17. Find the effective value of a current in an ac circuit that reaches a maximum of 17.0 A.
18. Find the effective value of an ac voltage whose maximum voltage is 1150 V.
19. What is the maximum current in a circuit in which an ac ammeter reads 8.50 A?
20. Find the maximum current in a circuit in which an ac ammeter reads 7.00 A.
21. A technician uses an oscilloscope to measure an effective voltage in an ac circuit. Find the effective value if the maximum voltage is 135 V.
22. A technician uses a cathode ray oscilloscope to measure current in an ac circuit that reaches a maximum of 125 A. What is the effective value of the current?
23. A maximum voltage of 34.0 V is developed by an ac generator that delivers a maximum current of 0.170 A to a circuit. (a) What is the effective voltage of the generator? (b) Find the effective current delivered to the circuit by the generator. (c) Find the resistance of the circuit.
24. A power plant generator develops a maximum voltage of $17\overline{0}$ V. What is the effective voltage?
25. A maximum current of 0.700 A flows through a 60.0-W light bulb in a circuit. (a) What is the effective current? (b) Find the resistance of the light bulb when it is lit.
26. If the average power dissipated by an electric light is $15\overline{0}$ W, what is the peak power?
27. If the average power dissipation by a hair dryer is $100\overline{0}$ W, what is the peak power?

19.2   ac Power

When the load has only resistance, power in ac circuits is found in the same way as in dc circuits:

$$P = I^2 R$$
$$P = VI \qquad \text{(using } V = IR\text{)}$$
$$P = \frac{V^2}{R} \qquad \text{(using } I = V/R\text{)}$$

EXAMPLE 1

What power is used in a resistance of 37.0 Ω if it has a current of 0.480 A flowing through it?

Data:

$$R = 37.0 \ \Omega$$
$$I = 0.480 \ A$$
$$P = ?$$

Basic Equation:

$$P = I^2 R$$

Working Equation: Same

Substitution:

$$P = (0.480 \ A)^2 (37.0 \ \Omega)$$
$$= 8.52 \ W$$

EXAMPLE 2

What power is used in a load of 12.0 Ω resistance if the voltage drop across it is 110 V?

Data:

$$R = 12.0 \ \Omega$$
$$V = 1\overline{1}0 \ V$$
$$P = ?$$

Basic Equation:

$$P = \frac{V^2}{R}$$

Working Equation: Same

Substitution:

$$P = \frac{(1\overline{1}0 \ V)^2}{12.0 \ \Omega}$$
$$= 1010 \ W \quad \text{or} \quad 1.01 \ kW$$

The preceding relationships are true only when e and i are in phase. Phase differences produced by capacitance and inductance in an ac circuit are due to reactance. Capacitance, inductance, and reactance will be studied later.

Note that, in the graphs comparing dc and ac power (Fig. 19.7), ac power varies but is always positive (+). The sign indicates only the direction of the current. Even so, p is positive in calculations because the product of $-e$ and $-i$ is positive: $p = (-e)(-i) = ei$.

Figure 19.7 dc and ac power compared

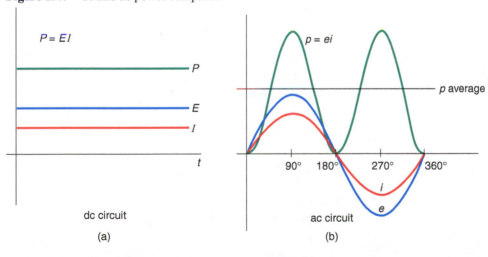

(a) dc circuit

(b) ac circuit

Transformers and Power

The uses of direct current in industry are somewhat limited. Primary applications are in charging storage batteries, electroplating, generating alternating current, electrolysis, electromagnets, and automobile ignition systems. However, ac can be changed to dc by a simple device called a **rectifier**.

Much more can be done with alternating current. From the kitchen toaster to the largest industrial motors, ac finds wide application. There are very practical reasons for this. The voltage of ac can be easily and efficiently changed in transformers to give almost any desired values. Actually, ac can be used for most purposes just as efficiently as dc. One advantage of ac is that it can be transmitted over long distances with very little heat loss. Heat lost in any electric device is found by the formula

$$\text{heat loss (power) } P = I^2 R$$

The energy wasted as heat can be reduced by making the current smaller. Transformers reduce the current by increasing the voltage because

$$P = \frac{E^2}{R}$$

The major advantage of ac over dc is that ac voltage can easily be changed to meet our needs.

EXAMPLE 3

A plant generates 50.0 kW (50,$\overline{0}$00 W) of power to be sent to a substation on a line with a resistance of 3.00 Ω. We know that some power will be lost as heat during the transmission. The power lost is $P_{\text{lost}} = I^2 R$.

(a) How much power is lost if the transmission is at 1150 V?
(b) What percent of the power generated is lost in transmission at 1150 V?
(c) How much power is lost if the transmission is at 11,500 V?
(d) What percent of the power generated is lost in transmission at 11,500 V?
(e) Compare the power losses at the two different transmission voltages.

At 1150 V:

(a)

$$P = VI$$
$$I = \frac{P}{V}$$
$$I = \frac{50,\overline{0}00 \text{ W}}{1150 \text{ V}}$$
$$= 43.5 \text{ A}$$
$$P_{\text{lost}} = I^2 R$$
$$P_{\text{lost}} = (43.5 \text{ A})^2 (3.00 \ \Omega)$$
$$= 5680 \text{ W}$$
$$= 5.68 \text{ kW}$$

(b)

$$\%_{\text{lost}} = \frac{\text{power lost}}{\text{power generated}} \times 100\%$$
$$\%_{\text{lost}} = \frac{5.68 \text{ kW}}{50.0 \text{ kW}} \times 100\% = 11.4\%$$

At 11,500 V:

(c)

$$P = VI$$
$$I = \frac{P}{V}$$
$$I = \frac{50,\overline{0}00 \text{ W}}{11,500 \text{ V}}$$
$$= 4.35 \text{ V}$$
$$P_{\text{lost}} = I^2 R$$
$$P_{\text{lost}} = (4.35 \text{ A})^2 (3.00 \ \Omega)$$
$$= 56.8 \text{ W}$$
$$= 0.0568 \text{ kW}$$

(d)

$$\%_{lost} = \frac{power\ lost}{power\ generated} \times 100\%$$

$$\%_{lost} = \frac{0.0568\ kW}{50.0\ kW} \times 100\% = 0.114\%$$

(e) This example shows that whereas 11.4% of the power is lost during transmission at 1150 V, only 0.114% is lost at 11,500 V; so by increasing the voltage, the current is correspondingly lowered and the power wasted in transmission is greatly reduced.

· · · · · · · · · · · · · · · · · ·

Changing Voltage with Transformers

The transformer is a device that is used to change the voltage to reduce the current and thereby lessen the power loss. A **transformer** consists of two coils of wire wrapped on an iron core (Fig. 19.8) and is used to change the voltage. When an alternating current passes through the **primary coil**, an induced current is produced in the **secondary coil** (Fig. 19.9). The magnitude of the voltage induced in the secondary coil depends on

1. The voltage applied to the primary coil.
2. The number of turns in the primary coil.
3. The number of turns in the secondary coil.
4. The power lost between primary and secondary coils.

Figure 19.9 An alternating current in the primary coil induces a current in the secondary coil.

Figure 19.8 Basic components of a transformer

Magnetic field induced in core by primary current

If we assume no power loss between the primary and secondary coils, we have

$$\frac{V_P}{V_S} = \frac{N_P}{N_S}$$

where V_P = primary voltage
V_S = secondary voltage
N_P = number of primary turns
N_S = number of secondary turns

EXAMPLE 4

A transformer on a neon sign has $10\overline{0}$ turns in its primary coil and $15,\overline{0}00$ turns in its secondary coil. If the voltage applied to the primary coil is $11\overline{0}$ V, what is the secondary voltage?

Data:

$$V_P = 11\overline{0}\ V$$
$$N_P = 10\overline{0}\ turns$$
$$N_S = 15,\overline{0}00\ turns$$
$$V_S = ?$$

Basic Equation:

$$\frac{V_P}{V_S} = \frac{N_P}{N_S}$$

Working Equation:

$$V_S = \frac{V_P N_S}{N_P}$$

Substitution:

$$V_S = \frac{(11\overline{0} \text{ V})(15,\overline{0}00 \text{ turns})}{10\overline{0} \text{ turns}}$$
$$= 16,500 \text{ V} \quad \text{or} \quad 16.5 \text{ kV}$$

· · · · · · · · · · · · · · · · · ·

Transformers used to raise or lower voltage are called step-up or step-down transformers. *Step-up transformers* are used when a high voltage is needed to operate X-ray tubes or neon signs and to transmit electric power over long distances. A **step-up transformer** increases the voltage by having more turns in the secondary coil than in the primary coil [Fig. 19.10(a)].

A **step-down transformer** lowers the voltage from high-voltage transmission lines to regular 110 V and 220 V for home and industrial use. Voltage is lowered in the step-down transformer because it has more turns in the primary coil than in the secondary coil [Fig. 19.10(b)].

Figure 19.10 Step-up and step-down transformers

Primary coil   Secondary coil

Step-up transformer

(a)

Primary coil   Secondary coil

Step-down transformer

(b)

Some common transformers are shown in Fig. 19.11.

Figure 19.11 Some large transformers can be reverse-connected to become either a step-up or a step-down transformer. Since transformers have no moving parts, they have a normal life expectancy of 20–25 years.

(a) A step-down electrical transformer near top of a utility pole

(b) Large transformers at an electrical transmission station

Auto transformers are used when a variable output voltage is needed. In this type of transformer, contact can be made across a variable number of the secondary coils using a brush contact. The output voltage is therefore variable from nearly zero to some maximum value. This type of transformer is often used to supply ac power to resistive heater elements to control the heating output.

Transformers do not create energy. In fact, some energy is lost during the change of voltage. Energy losses in transformers are of three types:

1. *Copper losses.* These result from the resistance of the copper wires in the coils and are unavoidable.
2. *Magnetic losses* (called *hysteresis losses*). Some energy is lost (turned into heat) by reversing the magnetism in the core.
3. *Eddy currents.* When a mass of metal (the core) is subjected to a changing magnetic field, *eddy currents* are set up in the metal that do no useful work, waste energy, and produce heat. These losses can be reduced by *laminating* the core. Instead of a solid block of metal for the core [Fig. 19.12(a)], thin sheets of metal with insulated surfaces are used [Fig. 19.12(b)], reducing these induced currents. These ac eddy currents cause the laminations to vibrate, producing the characteristic transformer "hum."

Figure 19.12 Laminating the core reduces the eddy current losses.

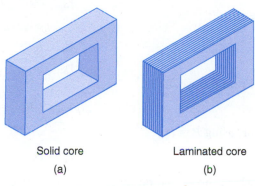

Solid core
(a)

Laminated core
(b)

When a transformer steps up the voltage applied to its primary coil, it reduces the current. Energy is conserved—we cannot get any more electrical energy out of a transformer than we put into it. The relationship between primary and secondary currents is

$$\frac{I_S}{I_P} = \frac{N_P}{N_S}$$

where I_S = current in secondary coil
  I_P = current in primary coil
  N_P = number of turns in primary
  N_S = number of turns in secondary

The primary current in a transformer is 10.0 A. If the primary coil has $55\overline{0}$ turns and the secondary has $250\overline{0}$ turns, what current flows in the secondary coil?

EXAMPLE 5

Data:

$$N_P = 55\overline{0} \text{ turns}$$
$$I_P = 10.0 \text{ A}$$
$$N_S = 250\overline{0} \text{ turns}$$
$$I_S = ?$$

Basic Equation:

$$\frac{I_S}{I_P} = \frac{N_P}{N_S}$$

Working Equation:

$$I_S = \frac{I_P N_P}{N_S}$$

Substitution:

$$I_S = \frac{(10.0 \text{ A})(55\overline{0} \text{ turns})}{250\overline{0} \text{ turns}}$$

$$= 2.20 \text{ A}$$

It follows from the last two shaded formulas that

$$\frac{V_P}{V_S} = \frac{N_P}{N_S} = \frac{I_S}{I_P}$$

so

$$\frac{V_P}{V_S} = \frac{I_S}{I_P}$$

From this we obtain

$$I_P V_P = I_S V_S$$

or, using $P = IV$,

$$P_P = P_S$$

which shows that power is conserved between the primary and secondary coils under these assumptions.

EXAMPLE 6

The power in the primary coil of a transformer is 375 W. If the current in the secondary coil is 11.4 A, what is the voltage in the secondary?

Data:

$$P_P = 375 \text{ W}$$
$$I_S = 11.4 \text{ A}$$
$$V_S = ?$$

Basic Equation:

$$P_P = P_S \quad \text{and} \quad P_S = V_S I_S$$

Working Equations:

$$P_P = V_S I_S \qquad (\textit{Note:} \text{ Substitute for } P_S.)$$

$$V_S = \frac{P_P}{I_S}$$

Substitution:

$$V_S = \frac{375 \text{ W}}{11.4 \text{ A}}$$

$$= 32.9 \text{ V}$$

Good transformers are more than 98% efficient. This is very important in power transmission. It is impractical to generate electricity at high voltage, but high voltage is desirable for transmission. Therefore, transformers are used to step up the voltage for transmission. High voltage is unsuitable, though, for consumer use, so transformers are used to reduce the voltage. A simplified diagram of a power distribution system is shown in Fig. 19.13.

Figure 19.13 Power distribution system

(a) Generating station 6600 V

(c) High-voltage power transmission lines 132,000 V

(e) Power pole step-down transformer 110 V 220 V

(b) Step-up transformer

(d) Substation step-down transformer 6000 V

PROBLEMS 19.2

1. A soldering iron is rated at $35\overline{0}$ W. If the current in the iron is 4.00 A, what is the resistance of the iron?
2. What power is developed by a device that draws 6.00 A with resistance 12.0 Ω?
3. Find the output power of a transformer with output voltage $50\overline{0}$ V and current 7.00 A.
4. A heater operates on a $11\overline{0}$-V line and is rated at $45\overline{0}$ W. What is the resistance of the element?
5. A heating element draws 6.00 A on a $22\overline{0}$-V line. What power is used in the element?
6. A 32.0-Ω resistance coil uses 375 W of power. What is the current in the coil?
7. What power is used by a heater that has a resistance of 12.0 Ω and draws a current of 7.00 A?
8. A heater operates on a $11\overline{0}$-V line and is rated at $75\overline{0}$ W. What is the resistance of the heater element?
9. A $11\overline{0}$-Ω resistance coil draws a current of 5.00 A. What power is used?
10. What power is used by a heater with resistance 19.5 Ω and that draws a current of 5.55 A?

11. $V_P = 30.0$ V
    $V_S = 45.0$ V
    $N_S = 15.0$ turns
    Find N_P.

12. $V_P = 25\overline{0}$ V
    $N_P = 73\overline{0}$ turns
    $N_S = 275$ turns
    Find V_S.

13. $I_P = 6.00$ A
    $I_S = 4.00$ A
    $V_P = 39.0$ V
    Find V_S.

14. A step-up transformer on a 115-V line provides a voltage of $230\overline{0}$ V. If the primary coil has 65.0 turns, how many turns does the secondary have?
15. A step-down transformer on a 115-V line provides a voltage of 11.5 V. If the secondary coil has 30.0 turns, how many turns does the primary have?
16. A transformer has 20.0 turns in the primary coil and $220\overline{0}$ turns in the secondary. If the primary voltage is 12.0 V, what is the secondary voltage?

SKETCH

12 cm² w

4.0 cm

DATA

$A = 12$ cm², $l = 4.0$ cm, $w = ?$

BASIC EQUATION

$A = lw$

WORKING EQUATION

$w = \dfrac{A}{l}$

SUBSTITUTION

$w = \dfrac{12 \text{ cm}^2}{4.0 \text{ cm}} = 3.0$ cm

17. If the current is 9.00 A in the primary coil in Problem 13, find the current in the secondary.
18. If the voltage in the secondary coil of a transformer is $11\overline{0}$ V and the current in it is 15.0 A, what power does it supply?
19. A neon sign has a transformer that changes electricity from $11\overline{0}$ V to $15,\overline{0}00$ V. (a) If the primary current is 8.00 A, find the current in the secondary coil. (b) Find the power in the primary coil.
20. A transformer has an output power of 990 W. If the current in the secondary coil is 0.45 A, what is its voltage?
21. The current in the secondary coil of a transformer is 5.00 A. Find the voltage in the secondary if the power is 775 W.
22. A transformer steps down $66\overline{0}0$ V to $12\overline{0}$ V. (a) If the secondary current is 14.0 A, what is the primary current? (b) Find the power in the primary coil.
23. The primary coil of a step-down transformer has $75\overline{0}0$ turns, and the secondary coil has 125 turns. The voltage across the primary is $72\overline{0}0$ V. (a) Find the voltage across the secondary. (b) If the current in the secondary coil is 36.0 A, find the primary current.
24. A step-up transformer has $30\overline{0}0$ turns in the secondary coil and $20\overline{0}$ turns in the primary coil. The primary is supplied with alternating current with an effective voltage of 90.0 V. (a) Find the voltage in the secondary coil. (b) If the current in the secondary coil is 2.00 A, what current flows in the primary coil? (c) What power is developed in the primary coil? (d) What power is developed in the secondary coil?
25. A mechanic is testing a transformer that increases voltage from $11\overline{0}$ V to $30\overline{0}0$ V. If the primary coil has $15\overline{0}$ turns, how many turns are in the secondary coil?
26. A construction site generator has a transformer with 175 turns in the primary coil and $75\overline{0}$ turns in the secondary coil. If it has a current of 20.0 A in the primary coil, what is the current in the secondary coil?

PHYSICS ON THE JOB

Power Distribution System Operators

The web of the nation's electrical grid is a massive, complicated, and vital component of the U.S. economy. Too much current traveling through high-voltage wires can overload systems and cause massive blackouts without skilled operators navigating, regulating, and monitoring the flow of electricity as it travels between stations and substations. National power distribution system operators are responsible for monitoring and switching electrical currents to and from regions involved in blackouts or natural disasters. Regional power distribution system operators must be able to adjust electrical currents away from local transformer failures to prevent electrical reflection in transmission lines that can lead to multiple power failures and larger power outages.

Figure 19.14   Power distribution system operator

Courtesy of fotolia © dgdimension

Power distribution system operators typically have a minimum of an associate's degree, yet on-the-job training is the most important part of the technical training. Regular professional development is necessary to keep current with modern equipment, especially in the nuclear power industry. In addition, mathematics and physics skills, particularly in the area of electricity and magnetism, are vital to the position (Fig. 19.14).

19.3 Inductance

Electronic circuitry in televisions, radios, computers, and electronic instruments has many components other than resistors. These components include capacitors, inductors, diodes, and transistors. With these components weak signals can be amplified, noise can be reduced, and signals at certain frequencies can be detected while signals from other frequencies can be rejected (that is, a circuit can be "tuned in" to a frequency). The analysis of the behavior of circuits with these components can become very complex. As a start toward understanding these circuits, we discuss inductors and capacitors. The operation of diodes and transistors will be discussed only briefly. We begin with a discussion of inductors.

Inductance measures the tendency of a coil of wire to resist a change in the current because the magnetism produced by one part of the coil acts to oppose the change of current in other parts of the coil. Inductance is the property of an electric circuit in which a varying current produces a varying magnetic field that induces voltage in the same circuit or in a nearby circuit. An **inductor** is a circuit component, such as a coil of wire, in which an induced emf opposes any change in the current (Fig. 19.15). The emf is induced in the coil itself as the magnetic field of the coil changes.

The unit of inductance, L, is the henry (H), named after **Joseph Henry**. A coil has an inductance of 1 H if an emf of 1 V is induced when the current changes at the rate of 1 A/s. We can express the henry as an ohm second:

$$1 \text{ H} = 1 \ \Omega \text{ s}$$

The henry is a large unit. A more practical unit is the millihenry (mH), which is one one-thousandth of a henry.

Inductance can be illustrated by connecting a coil with a large number of turns and a lamp in series. When connected to a dc source, the lamp burns brightly [Fig. 19.16(a)]. However, when this circuit is connected to an ac power source of the same voltage, the lamp is dimmer because of the inductance of the coil [Fig. 19.16(b)]. The circuit symbol for inductance is shown in Fig. 19.17.

Figure 19.15 Induced emf opposing current change in an ac circuit

Joseph Henry (1797–1898),

physicist, was born in Albany, New York. He constructed the first electromagnetic motor, discovered electric induction, and demonstrated the oscillatory nature of electric discharges. He also introduced a system of weather forecasting and investigated the propagation of light and sound waves. The unit of induction, the henry, is named after him.

Figure 19.16 A coil in an ac circuit produces inductance.

(a)   (b)

Figure 19.17 Circuit diagram symbol for inductance

Inductive Reactance

The opposition to ac current flow in an inductor is called **inductive reactance** and is measured in ohms. This is usually represented by X_L. The inductive reactance of a coil is directly proportional to frequency and is found by the following:

$$X_L = 2\pi f L$$

where X_L = inductive reactance
            f = frequency of the ac voltage, expressed in hertz (cycles per second), such as $6\overline{0}$ Hz or $6\overline{0}$/s
            L = inductance, in henries

If inductance is given in mH, a conversion must be made to H (henries).

The current in a circuit that has only an ac voltage source and an inductor is given by

$$I = \frac{E}{X_L}$$

where I = current
  E = voltage
  X_L = inductive reactance

EXAMPLE

A coil with inductance 1.00 mH is connected to a 60.0-kHz ac power source of $11\overline{0}$ V. What is the current in the circuit?

Sketch:

$E = 11\overline{0}$ V
$f = 60.0$ kHz
$L = 1.00$ mH

Data:

$$E = 11\overline{0} \text{ V}$$
$$L = 1.00 \text{ mH} = 1.00 \times 10^{-3} \text{ H}$$
$$f = 60.0 \text{ kHz} = 60.0 \times 10^3 \text{ Hz} = 6.00 \times 10^4/\text{s}$$
$$I = ?$$

Basic Equations:

$$X_L = 2\pi f L \quad \text{and} \quad I = \frac{E}{X_L}$$

Working Equations: Same

Substitutions:

$$X_L = 2\pi (6.00 \times 10^4/\text{s})(1.00 \times 10^{-3} \text{ H})$$

$$= 377 \frac{\text{H}}{\text{s}}$$

$$= 377 \frac{\text{H}}{\text{s}}\left(\frac{1 \ \Omega \ \text{s}}{1 \ \text{H}}\right) \qquad \text{(note conversion factor)}$$

$$= 377 \ \Omega$$

$$I = \frac{E}{X_L}$$

$$I = \frac{11\overline{0} \text{ V}}{377 \ \Omega}$$

$$= 0.292 \frac{\text{V}}{\Omega}\left(\frac{\text{A} \ \Omega}{\text{V}}\right) \qquad \text{(note conversion factor)}$$

$$= 0.292 \text{ A}$$

In an inductor, the current lags behind the voltage by one-fourth of a cycle (Fig. 19.18). For example, in a 60-Hz ac circuit, the frequency is 60 cycles/s; that is, the time for one complete cycle is 1/60 s. Thus, the maximum voltage in a 60-Hz circuit occurs

$$\frac{1}{4} \times \frac{1}{60} \text{ s} = \frac{1}{240} \text{ s}$$

before the maximum current. The current lag is usually measured in degrees. One fourth of a cycle is 90°.

Figure 19.18   In an inductive circuit, the current lags behind the voltage by one fourth of a cycle.

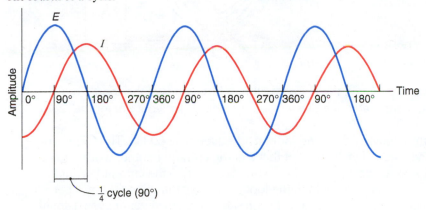

PROBLEMS 19.3

Find the inductive reactance (in ohms) of each inductance at the given frequency.

1. $L = 3.00$ mH, $f = 60.0$ Hz
2. $L = 20.0$ mH, $f = 75.0$ Hz
3. $L = 70.0$ mH, $f = 10.0$ kHz
4. $L = 8.00$ mH, $f = 8.00$ kHz
5. What is the inductive reactance (in ohms) of a 425-μH inductance at a frequency of 15.0 MHz?
6. Find the inductive reactance (in ohms) of a 655-μH inductance at a frequency of 125 MHz.

Find the current (in amperes) in each inductive circuit.

7. $L = 30.0$ mH, $f = 125$ Hz, $E = 14.0$ V
8. $L = 1.00$ mH, $f = 125$ kHz, $E = 145$ V
9. $L = 5.00$ mH, $f = 2.00$ kHz, $E = 50.0$ V
10. $L = 30.0$ mH, $f = 7.00$ MHz, $E = 75.0$ V
11. Find the current (in amperes) in an inductive circuit where $L = 72.0$ μH, $f = 2.00$ MHz, and $E = 105$ V.
12. Find the current (in amperes) in an inductive circuit where $L = 525$ μH, $f = 25.0$ MHz, and $E = 65.0$ V.

SKETCH

12 cm² $\quad w$

4.0 cm

DATA
$A = 12$ cm², $l = 4.0$ cm, $w = ?$

BASIC EQUATION
$A = lw$

WORKING EQUATION
$w = \dfrac{A}{l}$

SUBSTITUTION
$w = \dfrac{12 \text{ cm}^2}{4.0 \text{ cm}} = 3.0$ cm

PHYSICS CONNECTIONS

Induction and the Ground Fault Interrupter

Ground fault interrupt (GFI) outlets (Fig. 19.19) are required by most building codes for electric outlets around sinks or other devices where water can cause electric shocks. The purpose of a GFI is to detect minor losses of current in a circuit. If a working hairdryer were to fall into a sink or a bathtub, the water would cause the hairdryer to short out and send electric charge into the water and metal pipes. In such a

Figure 19.19   (a) A standard ground fault interrupt outlet (b) The basic components of a ground fault interrupt outlet

(a)

(b)   Detector coil

situation, the GFI outlet would automatically shut off the current within milliseconds of the short circuit. If someone touched the hairdryer while in the water, the person would be electrocuted if not for the rapid circuit-breaking ability of the GFI.

All ac electric appliances have current moving back and forth at a frequency of 60 Hz. The GFI is designed to measure the changes in the induced magnetic field of that electric current. In other words, the current going into the appliance establishes a magnetic field in the iron loop, while the current leaving the appliance creates an equal and opposite magnetic field in the loop, reversing direction 60 times per second. These two fields, in a normal operating situation, cancel each other out. However, if even a small amount of current does not return through the GFI outlet, the magnetic fields do not cancel out. In that situation, the magnetic field in the loop creates an electric current in the detection coil. Any current passing through the detection coil causes the breaker to trip, which quickly shuts off the circuit and prevents serious injury.

19.4   Inductance and Resistance in Series

Figure 19.20   ac circuit with resistance and inductance

In addition to inductance, most ac circuits have resistance in the form of lights or other resistors (Fig. 19.20). The current lags behind the voltage by any amount of time greater than zero and as large as one fourth of a cycle.

Impedance is a measure of the total opposition to current flow in an ac circuit resulting from the effect of both the resistance and the inductive reactance on the circuit. Ohm's law in an ac circuit can be written as

$$I = \frac{E}{Z}$$

where I = current
E = voltage
Z = impedance

The impedance of a series circuit containing a resistance and an inductance is

$$Z = \sqrt{R^2 + X_L^2}$$
$$Z = \sqrt{R^2 + (2\pi f L)^2}$$

where Z = impedance
R = resistance
X_L = inductive reactance
f = frequency
L = inductance

The impedance can be represented as a vector and as the hypotenuse of the right triangle shown in Fig. 19.21. The resistance is always drawn as a vector pointing in the positive x-direction. The inductive reactance is drawn as a vector pointing in the positive y-direction. The angle ϕ shown between the resistance and impedance vectors is the **phase angle** and equals the amount by which the current lags behind the voltage. The phase angle is given by

$$\tan\phi = \frac{X_L}{R}$$

Figure 19.21 Graphic representation of impedance

A lamp of resistance 40.0 Ω is connected in series with an inductance of 95.0 mH. This circuit is connected to a 115-V, 60.0-Hz power supply. (a) What is the current in the circuit? (b) What is the phase angle?

EXAMPLE

Sketch:

(a) Data:

$E = 115$ V

$f = 60.0$ Hz $= 60.0/\text{s}$

$R = 40.0$ Ω

$L = 95.0$ mH $= 95.0 \times 10^{-3}$ H $= 0.0950$ H

$Z = ?$

$I = ?$

Basic Equations:

$$Z = \sqrt{R^2 + (2\pi fL)^2} \quad \text{and} \quad I = \frac{E}{Z}$$

Working Equations: Same

Substitutions: First calculate the impedance.

$$Z = \sqrt{(40.0\ \Omega)^2 + [2\pi(60.0/\text{s})(0.0950\ \text{H})]^2}$$
$$= \sqrt{16\overline{0}0\ \Omega^2 + (35.8\ \text{H/s})^2}$$
$$= \sqrt{16\overline{0}0\ \Omega^2 + \left[\left(35.8\ \frac{\text{H}}{\text{s}}\right)\left(\frac{1\ \Omega\ \text{s}}{1\ \text{H}}\right)\right]^2} \quad \text{(note conversion factor)}$$
$$= \sqrt{16\overline{0}0\ \Omega^2 + 1280\ \Omega^2}$$
$$= \sqrt{2880\ \Omega^2}$$
$$= 53.7\ \Omega$$

$$I = \frac{E}{Z}$$
$$I = \frac{115\ \text{V}}{53.7\ \Omega}\left(\frac{1\ \text{A}\ \Omega}{1\ \text{V}}\right)$$
$$= 2.14\ \text{A}$$

Figure 19.22

(b) To find the phase angle ϕ, first find X_L. Then, construct the vector right triangle as in Fig. 19.22 to find ϕ.

Data:

$$f = 60.0/s$$
$$L = 0.0950 \text{ H}$$
$$X_L = ?$$
$$\phi = ?$$

Basic Equations:

$$X_L = 2\pi f L \quad \text{and} \quad \tan\phi = \frac{X_L}{R}$$

Working Equations: Same

Substitutions:

$$X_L = 2\pi(60.0/\text{s})\,(0.0950 \text{ H})$$

$$= 35.8 \,\frac{\text{H}}{\text{s}}$$

$$= 35.8 \,\frac{\text{H}}{\text{s}}\left(\frac{1\,\Omega\,\text{s}}{1\,\text{H}}\right)$$

$$= 35.8 \,\Omega \qquad [\textbf{Note:} \text{ This value can also be taken directly from the second term under the radical in the Substitution for Z in part (a).}]$$

From Fig. 19.22 we have

$$\tan\phi = \frac{35.8 \,\Omega}{40.0 \,\Omega} = 0.895$$

$$\phi = 41.8°$$

· · · · · · · · · · · · · · · ·

PROBLEMS 19.4

1. For a circuit with $R = 20\overline{0}\ \Omega$, $L = 10.0$ mH, and $f = 1.25$ kHz: (a) find the impedance (in ohms), (b) find the phase angle, and (c) find the current if the voltage is 45.0 V.

2. For a circuit with $R = 12.0\ \Omega$, $L = 1.00$ mH, and $f = 90\overline{0}$ Hz: (a) find the impedance (in ohms), (b) find the phase angle, and (c) find the current if the voltage is 10.0 V.

3. For a circuit with $R = 1.00$ kΩ, $L = 50.0$ mH, and $f = 10.0$ kHz: (a) find the impedance (in ohms), (b) find the phase angle, and (c) find the current if the voltage is 15.0 V.

4. For a circuit with resistance 2.00 kΩ, inductance 70.0 mH, and frequency 5.00 kHz: (a) find the impedance (in ohms), (b) find the phase angle, and (c) find the current if the voltage is 12.0 V.

5. For a circuit with resistance $30\overline{0}\ \Omega$, inductance 2.00 mH, and frequency 3.00 kHz: (a) find the impedance (in ohms), (b) find the phase angle, and (c) find the current if the voltage is 6.00 V.

19.5 Capacitance

An important component of many ac circuits is the capacitor. A **capacitor** is used to store an electric charge, and it consists of two conductors that are usually parallel plates separated by a thin insulator. The plates are often made of a metal foil rolled to a convenient size and inserted in a cylinder. Capacitors are represented in circuit diagrams as shown in Fig. 19.23.

Figure 19.23   Circuit diagram symbol for a capacitor

When a capacitor is connected to a battery, electrons flow from the negative terminal to one capacitor plate as shown in Fig. 19.24(b). When the capacitor is removed from the battery, the charges remain on the capacitor. If the capacitor is then connected to a resistor [Fig. 19.24(c)], electrons will flow through the circuit until the capacitor has lost its charge.

Figure 19.24

(a) Charged capacitor

(b) Capacitor and battery in a circuit

(c) An electron current flows when the capacitor is connected to a resistance

A capacitor will block a direct current from flowing in a circuit once the capacitor is charged. A low-frequency ac voltage in a capacitive circuit (Fig. 19.25) will cause only a small current to flow because of this blocking nature of a capacitor. A high-frequency ac voltage source will cause a larger current to flow in the capacitive circuit of Fig. 19.25 because a current is required to quickly change the polarity of the capacitor voltage. Capacitors can therefore be used to tune the frequency response of circuits, allowing the blocking of low-frequency electric signals and tuning the resonance frequency of circuits as described in Section 19.8.

Figure 19.25   Capacitor in an ac circuit

Capacitance is the ratio of the charge on either plate of a capacitor to the potential difference between the plates. That is,

$$C = \frac{Q}{V}$$

where C = capacitance of a capacitor
   Q = amount of charge on either plate of the capacitor
   V = potential difference between the two plates of the capacitor

The unit of capacitance is the *farad* (F), named after **Michael Faraday**. A more practical unit is the microfarad (μF or 10^{-6} F). **Capacitive reactance** is a measure of the opposition to ac current flow by a capacitor. The effect of a capacitor on a circuit is inversely proportional to frequency and given by

$$X_C = \frac{1}{2\pi f C}$$

where X_C = capacitive reactance (ohms)
   f = frequency
   C = capacitance (farads)

$$1 \text{ F} = 1 \text{ s}/\Omega$$

Michael Faraday (1791–1867),

chemist and experimental physicist, was born in England. He introduced the concept of capacitance, discovered electromagnetic induction, and made numerous other contributions to electricity and magnetism. The unit of capacitance, the farad, is named after him.

In a circuit that contains only capacitors, the current *leads* the voltage by 90° (one-fourth cycle) as shown in Fig. 19.26.

Figure 19.26   Current leads the voltage by one-fourth cycle.

EXAMPLE

Find the capacitive reactance of a 10.0-μF capacitor in a circuit of frequency 1.00 kHz.

Data:

$$C = 10.0 \ \mu\text{F} = 10.0 \times 10^{-6} \ \text{F} = 1.00 \times 10^{-5} \ \text{F}$$
$$f = 1.00 \ \text{kHz} = 1.00 \times 10^{3}\text{/s}$$
$$X_C = ?$$

Basic Equation:

$$X_C = \frac{1}{2\pi f C}$$

Working Equation: Same

Substitution:

$$X_C = \frac{1}{2\pi(1.00 \times 10^{3}\text{/s})(1.00 \times 10^{-5} \ \text{F})}$$

$$= \frac{1}{\left(0.0628 \ \dfrac{\text{F}}{\text{s}}\right)\left(\dfrac{1 \ \text{s}}{1 \ \text{F} \ \Omega}\right)} \qquad \text{(note conversion factor)}$$

$$= 15.9 \ \Omega$$

PROBLEMS 19.5

Find the capacitive reactance (in ohms) in each ac circuit.

1. $C = 20.0 \ \mu\text{F}, f = 1.00 \ \text{kHz}$
2. $C = 7.00 \ \text{mF}, f = 10\overline{0} \ \text{Hz}$
3. $C = 0.600 \ \mu\text{F}, f = 0.100 \ \text{kHz}$
4. $C = 30.0 \ \text{mF}, f = 2.50 \ \text{MHz}$
5. $C = 0.800 \ \mu\text{F}, f = 0.250 \ \text{MHz}$
6. Find the capacitive reactance of a 15.0-μF capacitor in a circuit of frequency 60.0 Hz.
7. Find the capacitive reactance of a 45.0-μF capacitor in a circuit of frequency 60.0 kHz.
8. Find the capacitive reactance of a 6.00-mF capacitor in a circuit of frequency 10$\overline{0}$ Hz.
9. Find the capacitive reactance of a 33$\overline{0}$-μF capacitor in a circuit of frequency 30$\overline{0}$ Hz.
10. Find the capacitive reactance of a 222-μF capacitor in a circuit of frequency 12$\overline{0}$ Hz.

19.6 Capacitance and Resistance in Series

The combined effect of capacitance and resistance in series is measured by the impedance, Z, of the circuit.

$$Z = \sqrt{R^2 + X_C^2}$$

$$Z = \sqrt{R^2 + \left(\frac{1}{2\pi f C}\right)^2}$$

where Z = impedance
   R = resistance
   X_C = capacitive reactance
   f = frequency
   C = capacitance

The current is given by Ohm's law:

$$I = \frac{E}{Z}$$

where I = current
   E = voltage
   Z = impedance

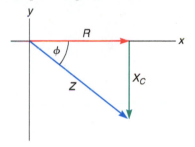

Figure 19.27 Determination of the phase angle in an ac circuit

The phase angle can be found by drawing the resistance as a vector in the positive x-direction and the capacitive impedance as a vector in the negative y-direction as shown in Fig. 19.27. The phase angle gives the amount by which the voltage lags behind the current.

$$\tan \phi = \frac{X_C}{R}$$

EXAMPLE

What current will flow in a 60.0-Hz ac circuit that includes a $11\overline{0}$-V source, a capacitor of 90.0 μF, and a 16.0-Ω resistance in series? Also find the phase angle.

Sketch:

$R = 16.0\ \Omega$
$C = 90.0\ \mu F$
$E = 10\overline{0}\ V$
$f = 60.0\ Hz$

Data:

$E = 11\overline{0}\ V$
$f = 60.0\ Hz = 60.0/s$
$R = 16.0\ \Omega$
$C = 90.0\ \mu F = 90.0 \times 10^{-6}\ F = 9.00 \times 10^{-5}\ F$
$Z = ?$
$I = ?$

Basic Equations:

$$Z = \sqrt{R^2 + \left(\frac{1}{2\pi fC}\right)^2} \quad \text{and} \quad I = \frac{E}{Z}$$

Working Equations: Same

Substitutions:

First, find Z.

$$Z = \sqrt{(16.0\ \Omega)^2 + \left(\frac{1}{2\pi(60.0/\text{s})(9.00 \times 10^{-5}\ \text{F})}\right)^2}$$

$$= \sqrt{256\ \Omega^2 + \left(\frac{1}{0.0339\ \text{F/s}}\right)^2}$$

$$= \sqrt{256\ \Omega^2 + \left(29.5\ \frac{\cancel{s}}{\cancel{F}} \times \frac{1\ \Omega\ \cancel{F}}{1\ \cancel{s}}\right)^2}$$

$$= \sqrt{256\ \Omega^2 + 87\overline{0}\ \Omega^2}$$

$$= 33.6\ \Omega$$

Then use Z to find I.

$$I = \frac{E}{Z}$$

$$I = \frac{11\overline{0}\ \text{V}}{33.6\ \Omega}$$

$$= 3.27\ \frac{\cancel{V}}{\cancel{\Omega}}\left(\frac{1\ \cancel{\Omega}\ \text{A}}{1\ \cancel{V}}\right)$$

$$= 3.27\ \text{A}$$

Then find the phase angle.

$$\tan\phi = \frac{X_C}{R}$$

$$\tan\phi = \frac{29.5\ \cancel{\Omega}}{16.0\ \cancel{\Omega}} = 1.84 \qquad (X_C \text{ from above})$$

$$\phi = 61.5°$$

. . . .

PROBLEMS 19.6

1. For an ac circuit with $R = 1.00\ \text{k}\Omega$, $C = 1.00\ \mu\text{F}$, $E = 10\overline{0}\ \text{V}$, and $f = 10\overline{0}\ \text{Hz}$:
   (a) find the impedance (in ohms), (b) find the phase angle, and (c) find the current.

2. For an ac circuit with $R = 375\ \Omega$, $C = 5.00\ \mu\text{F}$, $E = 20.0\ \text{V}$, and $f = 1.00\ \text{kHz}$:
   (a) find the impedance (in ohms), (b) find the phase angle, and (c) find the current.

3. For an ac circuit with $R = 4.80\ \text{k}\Omega$, $C = 45.0\ \mu\text{F}$, $E = 15.0\ \text{V}$, and $f = 1.75\ \text{kHz}$:
   (a) find the impedance (in ohms), (b) find the phase angle, and (c) find the current.

4. For an ac circuit with resistance 145 mΩ, capacitance 10.0 μF, frequency 72.5 kHz, and $E = 7.00$ mV: (a) find the impedance (in ohms), (b) find the phase angle, and (c) find the current.
5. For an ac circuit with resistance 10.0 mΩ, capacitance 5.00 μF, frequency 10.0 kHz, and $E = 15.0$ mV: (a) find the impedance (in ohms), (b) find the phase angle, and (c) find the current.

19.7 Capacitance, Inductance, and Resistance in Series

Many circuits that are important in the design of electronic equipment contain all three types of circuit elements discussed in this chapter. The impedance of a circuit containing resistance, capacitance, and inductance in series can be found from the equation

$$Z = \sqrt{R^2 + (X_L - X_C)^2}$$

where
Z = impedance
R = resistance
X_L = inductive reactance
X_C = capacitive reactance

The vector diagram for this type of circuit is shown in Fig. 19.28. The phase angle is given by

$$\tan\phi = \frac{X_L - X_C}{R}$$

Figure 19.28 Vector diagrams for inductive reactance and capacitive reactance

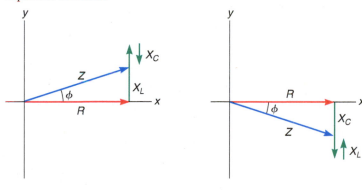

(a) Inductive reactance   (b) Capacitive reactance

In a circuit containing R, L, and C components, the circuit is *inductive* if $X_L > X_C$ and the current lags behind the voltage. A circuit is *capacitive* if $X_C > X_L$, in which case the voltage lags behind the current. If $X_C = X_L$, the circuit is *resistive*; the voltage and current are *in phase*. If the circuit power is to be maximized, the voltage and current must be kept in phase. The current in this type of circuit is given by

$$I = \frac{E}{Z} = \frac{E}{\sqrt{R^2 + \left(2\pi f L - \dfrac{1}{2\pi f C}\right)^2}}$$

EXAMPLE

An ac circuit contains a $10\overline{0}$-Ω resistance, a 10.0-μF capacitor, and a 10.0-mH inductance in series with a 25.0-V, $20\overline{0}$-Hz source. Find the impedance and the current.

Sketch:

Data:

$$R = 10\overline{0}\ \Omega$$
$$C = 10.0\ \mu\text{F} = 10.0 \times 10^{-6}\ \text{F} = 1.00 \times 10^{-5}\ \text{F}$$
$$L = 10.0\ \text{mH} = 10.0 \times 10^{-3}\ \text{H} = 1.00 \times 10^{-2}\ \text{H}$$
$$E = 25.0\ \text{V}$$
$$f = 20\overline{0}\ \text{Hz} = 20\overline{0}/\text{s}$$
$$Z = ?$$
$$I = ?$$

Basic Equations:

$$X_L = 2\pi f L$$
$$X_C = \frac{1}{2\pi f C}$$
$$Z = \sqrt{R^2 + (X_L - X_C)^2}$$
$$I = \frac{E}{Z}$$

Working Equations: Same

Substitutions:

First find X_L.

$$X_L = 2\pi f L$$
$$X_L = 2\pi (20\overline{0}/\text{s})(1.00 \times 10^{-2}\ \text{H})$$
$$= 12.6\ \frac{\text{H}}{\text{s}} \left(\frac{1\ \Omega\ \text{s}}{1\ \text{H}} \right)$$
$$= 12.6\ \Omega$$

Then find X_C.

$$X_C = \frac{1}{2\pi f C}$$
$$X_C = \frac{1}{2\pi (20\overline{0}/\text{s})(1.00 \times 10^{-5}\ \text{F})}$$
$$= 79.6\ \Omega$$

$$\boxed{\frac{1}{\dfrac{\text{F}}{\text{s}}} = 1 \div \frac{\text{F}}{\text{s}} = 1 \times \frac{\text{s}}{\text{F}} = \frac{\text{s}}{\text{F}} \times \frac{1\ \Omega\ \text{F}}{1\ \text{s}} = \Omega}$$

Then find the impedance.

$$Z = \sqrt{R^2 + (X_L - X_C)^2}$$
$$Z = \sqrt{(10\overline{0}\ \Omega)^2 + (12.6\ \Omega - 79.6\ \Omega)^2}$$

$$= \sqrt{(10\overline{0}\ \Omega)^2 + (-67.0\ \Omega)^2}$$
$$= \sqrt{1.00 \times 10^4\ \Omega^2 + 4490\ \Omega^2}$$
$$= 12\overline{0}\ \Omega$$

Then find the current.

$$I = \frac{E}{Z}$$

$$I = \frac{25.0\ \text{V}}{12\overline{0}\ \Omega}$$

$$= 0.208\ \frac{\text{V}}{\Omega}\left(\frac{1\ \text{A}\ \Omega}{1\ \text{V}}\right)$$

$$= 0.208\ \text{A}$$

PROBLEMS 19.7

Find the impedance and current in each ac circuit.

1. $R = 25.0\ \Omega$, $L = 50.0$ mH, $C = 50.0\ \mu\text{F}$, $f = 60.0$ Hz, $E = 5.00$ V
2. $R = 225\ \Omega$, $L = 10.0$ mH, $C = 0.200\ \mu\text{F}$, $f = 1.00$ kHz, $E = 15.0$ V
3. $R = 1.00$ kΩ, $L = 10.0$ mH, $C = 30.0$ mF, $f = 10.0$ kHz, $E = 15.0$ V
4. $R = 1.00$ kΩ, $L = 0.700$ H, $C = 30.0\ \mu\text{F}$, $f = 60.0$ Hz, $E = 8.00$ V
5. A circuit contains a 15$\overline{0}$-Ω resistance, a 35.0-μF capacitor, and a 0.600-H inductance in series with a 6.00-V, 12$\overline{0}$-Hz source. Find the impedance and the current.
6. A circuit contains a 225-Ω resistance, a 5.00-μF capacitor, and a 0.550-H inductance in series with a 7.50-V, 60.0-Hz source. Find the impedance and the current.
7. A circuit contains a 175-Ω resistance, a 4.50-μF capacitor, and a 0.735-H inductance in series with a 5.00-V, 10$\overline{0}$-Hz source. Find the impedance and the current.
8. A circuit contains a 575-Ω resistance, a 10$\overline{0}$-μF capacitor, and a 0.400-H inductance in series with a 10$\overline{0}$-V, 60$\overline{0}$-Hz source. Find the impedance and the current.
9. A circuit contains a 45$\overline{0}$-Ω resistance, a 35.0-μF capacitor, and a 45.0-mH inductance in series with a 25.0-V, 1.00-kHz source. Find the impedance and the current.
10. A circuit contains a 375-Ω resistance, a 50$\overline{0}$-μF capacitor, and a 0.500-H inductance in series with a 55.0-V, 50$\overline{0}$-Hz source. Find the impedance and the current.

SKETCH

12 cm² | w

4.0 cm

DATA
$A = 12$ cm², $l = 4.0$ cm, $w = ?$

BASIC EQUATION
$A = lw$

WORKING EQUATION
$w = \dfrac{A}{l}$

SUBSTITUTION
$w = \dfrac{12\ \text{cm}^2}{4.0\ \text{cm}} = 3.0\ \text{cm}$

19.8 Resonance

The current in a circuit containing resistance, capacitance, and inductance is given by the equation

$$I = \frac{E}{\sqrt{R^2 + (X_L - X_C)^2}}$$

When the inductive reactance equals the capacitive reactance, they nullify each other, and the current is given by

$$I = \frac{E}{R}$$

Figure 19.29   In this variable air–dielectric capacitor, rotating the movable plates between the fixed plates changes the overlap area and thus the capacitance. Such capacitors were common in tuning circuits in older radios.

Courtesy of Frank LaBua\Pearson Education/PH College

which is its maximum possible value. When this condition exists, the circuit is in **resonance** with the applied voltage. To have resonance, it is essential for the circuit to have both capacitance and inductance.

Resonant circuits are used in radios and televisions. The frequency of a certain station is tuned in when a resonant circuit (antenna circuit) is adjusted to that frequency (Fig. 19.29). This is accomplished by changing the capacitance until the capacitive reactance equals the inductive reactance. The applied voltage is the radio signal picked up by the antenna. A variable capacitor has one set of plates (usually aluminum) mounted on a rotating shaft. The plates have air between them, and the capacitance varies with the amount of overlap of the plates as they are rotated between each other.

The resonant frequency occurs when $X_L = X_C$; power transfer is maximized at resonance as discussed in the preceding section. We find this frequency as follows. We start with

$$X_L = X_C$$

$$2\pi fL = \frac{1}{2\pi fC} \qquad \text{(By substitution)}$$

$$f^2 = \frac{1}{4\pi^2 LC} \qquad \text{(Solve for } f^2.)$$

$$f = \frac{1}{\sqrt{4\pi^2 LC}} \qquad \text{(Take the square root of both sides.)}$$

$$f = \frac{1}{2\pi\sqrt{LC}}$$

The circuit can be adjusted to any frequency by varying the capacitance or the inductance.

EXAMPLE

Find the resonant frequency of a circuit containing a 5.00-nF capacitor in series with a 2.60-μH inductor.

Sketch:

$$C = 5.00 \text{ nF}$$

$$L = 2.60 \text{ }\mu\text{H}$$

Data:

$$C = 5.00 \text{ nF} = 5.00 \times 10^{-9} \text{ F}$$
$$L = 2.60 \text{ }\mu\text{H} = 2.60 \times 10^{-6} \text{ H}$$
$$f = ?$$

Basic Equation:

$$f = \frac{1}{2\pi\sqrt{LC}}$$

Working Equation: Same

Substitution:

$$f = \frac{1}{2\pi\sqrt{(2.60 \times 10^{-6} \text{ H})(5.00 \times 10^{-9} \text{ F})}}$$

$$= \frac{1}{2\pi\sqrt{1.30 \times 10^{-14} \text{ H F}\left(\dfrac{1 \,\Omega\, \text{s}}{1 \text{ H}}\right)\left(\dfrac{1 \text{ s}}{1 \text{ F } \Omega}\right)}}$$

$$= \frac{1}{7.16 \times 10^{-7} \text{ s}}$$

$$= 1.40 \times 10^{6} \,\frac{\text{cycles}}{\text{s}} \quad \text{or} \quad 14\overline{0}0 \,\frac{\text{kilocycles}}{\text{s}} \quad \text{or} \quad 14\overline{0}0 \text{ kHz}$$

This frequency is in the AM radio band.

....

PROBLEMS 19.8

Find the resonant frequency in each ac circuit.

1. $L = 1.00\ \mu\text{H}$ and $C = 4.00\ \mu\text{F}$
2. $L = 2.00\ \mu\text{H}$ and $C = 35.0\ \mu\text{F}$
3. $L = 2.50\ \mu\text{H}$ and $C = 7.00\ \mu\text{F}$
4. $L = 2.65\ \mu\text{H}$ and $C = 35.0\ \mu\text{F}$
5. $L = 42.5\ \mu\text{H}$ and $C = 40.0\ \mu\text{F}$
6. Find the resonant frequency of a circuit containing a 25.0-μF capacitor in series with a 75.0-μH inductor.
7. Find the resonant frequency of a circuit containing a 33.0-μF capacitor in series with a 43.5-μH inductor.
8. Find the resonant frequency of a circuit containing a 10.0-μF capacitor in series with a 37.5-μH inductor.
9. Find the resonant frequency of a circuit containing an 8.00-μF capacitor in series with a 10$\overline{0}$-μH inductor.
10. Find the resonant frequency of a circuit containing a 3.75-μF capacitor in series with a 30$\overline{0}$-μH inductor.

19.9   Rectification and Amplification

It is often necessary to change ac into dc to provide dc for charging batteries or to power the integrated circuits (ICs) of computers and other electronic units. This process of changing ac to dc is called **rectification**. A device that accomplishes this is called a *diode*. Early diodes were constructed as vacuum tubes. Modern diodes are made of a semiconductor material and are usually less than 5 mm long (Fig. 19.30).

A **diode** allows current to flow through it in only one direction and not the other. It is similar to a turnstile that revolves in only one direction. People can pass the turnstile in one direction but are blocked when they attempt to pass in the opposite direction (Fig. 19.31).

SKETCH

12 cm²   w

4.0 cm

DATA

$A = 12 \text{ cm}^2, l = 4.0 \text{ cm}, w = ?$

BASIC EQUATION

$A = lw$

WORKING EQUATION

$w = \dfrac{A}{l}$

SUBSTITUTION

$w = \dfrac{12 \text{ cm}^2}{4.0 \text{ cm}} = 3.0 \text{ cm}$

Figure 19.30   Semiconductor diodes

Courtesy of fotolia © a185ok

Thomas Edison (1847–1931),

inventor, was born in Milan, Ohio. He held patents to over 1000 inventions, which included the phonograph, the carbon transmitter that made Alexander Graham Bell's telephone practical, an incandescent lamp, a storage battery, and a system for widespread distribution of electric power from central generating stations. His only true scientific discovery in 1883, called "the Edison effect," for which he could see no commercial application, led to the vacuum tube, which became the basis for the radio. He spent his life inventing many of the devices that shaped the modern world.

Thomas Edison found that when a wire filament was heated near a metal plate, an electron charge would begin to flow from the filament across space to the plate. This is called the *Edison effect* and was the beginning of the electronics industry. The electron emitter filament is called the *cathode*. The plate is called the *anode*. The entire device is called a *diode*.

A rectifier changes alternating current to direct current by allowing it to pass in only one direction (Fig. 19.32). Additional circuit devices can be added to the rectifier to smooth out the direct current

Figure 19.31

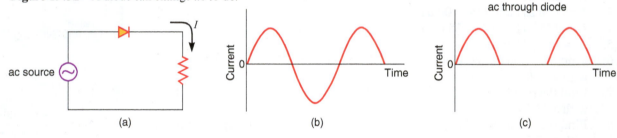

(a) A one-way turnstile allowing people to move through in only one direction is like a diode.

(b) A diode allows electrons to flow in only one direction.

so that it appears as shown in Fig. 19.33. Rectifiers are used in automobiles to change the alternating current produced by the alternator into direct current.

Figure 19.32 A diode can change ac to dc.

Figure 19.33 Rectified ac current

It is often necessary to increase the strength of an electronic signal. This is called **amplification**. Radios, stereos, and many other instruments contain one or more amplifier circuits. Early amplifiers utilized vacuum tubes together with other components. The transistor, developed by **John Bardeen**, has replaced the vacuum tube in most circuitry because of its smaller size and lesser power consumption. A *transistor* amplifier is typically composed of one or more transistors in addition to capacitors, resistors, and possibly inductors to provide the amplifier with the desired gain (amplification), frequency response, and power output. Amplifiers composed of individual transistors, resistors, and capacitors have been replaced for many applications by *integrated circuits* (ICs), which are only slightly larger in size than some transistors. An IC may contain millions of tiny transistors, diodes, resistors, and capacitors on a small chip of silicon less than 1 cm square. In addition to amplifying signals, ICs have been designed to serve as memory or logic units in computers or other applications. These ICs can be programmed to perform arithmetic operations, as in a calculator, or to perform control operations, as in most appliances.

John Bardeen (1908–1991),

physicist, was born in Madison, Wisconsin. He developed the transistor in 1947, for which he shared the Nobel Prize in Physics. He shared a second Nobel Prize in Physics in 1972 for the theory of superconductivity, which provided a breakthrough in electromagnetic design. Bardeen became the first person to win the Nobel Prize in Physics twice.

19.10 Commercial Generator Power Output

The power output of the generator is the product of voltage and current. The ac generator converts mechanical energy to electric energy by performing three functions:

1. *Production of voltage:* electric pressure, which pushes the current through the loads
2. *Production of power current:* current converted into heat, light, and mechanical power
3. *Production of magnetizing current:* current transferred back and forth for magnetizing purposes in the generation of electric power, called *reactive kVA* (kilovolt-amperes)

Apparent Power and Reactive kVA

If the current and voltage are not in phase, the product of effective values of alternating current and voltage is the **apparent power** instead of the actual power. Apparent power is measured in kVA (kilovolt-amperes). **Actual power** is a measure of the actual power available to be converted into other forms of energy; it is the product of apparent power and the **power factor**:

$$\text{power factor} = \frac{\text{actual power}}{\text{apparent power}}$$

where the actual power is measured in kW, the apparent power is measured in kVA and is called *reactive kVA*, and the power factor is a unitless ratio less than 1. Note that 1 VA = 1 W.

Mathematically, the power factor is equal to the cosine of the angle by which the current lags behind (or in rare cases leads) the voltage. The power factor is really a correction factor that must be applied to determine the actual power produced. The situation is very similar to finding the amount of work done when a force and the motion are not in the same direction.

Find the actual power produced by a generating system that produces 13,600 kVA with a power factor of 0.900.

EXAMPLE

Data:

$$\begin{aligned} \text{apparent power} &= 13{,}600 \text{ kVA} \\ \text{power factor} &= 0.900 \\ \text{actual power} &= ? \end{aligned}$$

Basic Equation:

$$\text{power factor} = \frac{\text{actual power}}{\text{apparent power}}$$

Working Equation:

$$\text{actual power} = (\text{apparent power})(\text{power factor})$$

Substitution:

$$\begin{aligned} \text{actual power} &= (13{,}600 \text{ kVA})(0.900) \\ &= 12{,}200 \text{ kVA} \\ &= 12{,}200 \text{ kW} \end{aligned}$$

SKETCH

12 cm^2 | w

4.0 cm

DATA

$A = 12 \text{ cm}^2, l = 4.0 \text{ cm}, w = ?$

BASIC EQUATION

$A = lw$

WORKING EQUATION

$w = \dfrac{A}{l}$

SUBSTITUTION

$w = \dfrac{12 \text{ cm}^2}{4.0 \text{ cm}} = 3.0 \text{ cm}$

PROBLEMS 19.10

1. Find the actual power produced by a generating station that produces 12,600 kVA with a power factor of 0.850.
2. A generating station operates with a power factor of 0.910. What actual power is available on the transmission lines if the apparent power is 12,800 kVA?
3. Find the apparent power produced by a generating station whose actual power is $12\overline{0},000$ kW and whose power factor is 0.900.
4. Find the apparent power produced by a generating station whose actual power is $1,9\overline{0}0,000$ kW and whose power factor is 0.800.
5. A generating station operates with a power factor of 0.880. What actual power is available on the transmission lines if the apparent power is 11,500 kVA?
6. Find the apparent power produced by a generating station whose actual power is 2,350,000 kW and whose power factor is 0.850.
7. Find the actual power produced by a generating station that produces 23,800 kVA with a power factor of 0.810.
8. A generating station operates with a power factor of 0.840. What actual power is available on the transmission lines if the apparent power is 13,500 kVA?
9. Find the apparent power produced by a generating station whose actual power is $35\overline{0},000$ kW and whose power factor is 0.860.
10. Find the apparent power produced by a generating station whose actual power is 1,250,000 kW and whose power factor is 0.820.
11. Find the power factor of a generating station whose actual power is 55,800 kW and whose apparent power is 63,400 kVA.
12. Find the power factor of a generating station whose apparent power is 645,000 kVA and whose actual power is 587,000 kW.

Glossary

Actual Power A measure of the actual power available to be converted into other forms of energy. (p. 553)

Alternating Current A current that flows in one direction in a conductor, changes direction, and then flows in the other direction. (p. 524)

Amplification The process of increasing the strength of an electronic signal. (p. 552)

Apparent Power The product of the effective values of alternating current and voltage. (p. 553)

Capacitance The ratio of the charge on either plate of a capacitor to the potential difference between the plates. (p. 543)

Capacitor A circuit component consisting of two parallel plates separated by a thin insulator used to build up and store charge. (p. 543)

Capacitive Reactance A measure of the opposition to ac current flow by a capacitor. (p. 543)

Diode A device that allows current to flow through it in only one direction. (p. 551)

Effective Value The number of amperes of alternating current that produces the same amount of heat in a resistance as an equal number of amperes of a steady direct current. (p. 526)

Impedance A measure of the total opposition to current flow in an ac circuit resulting from the effect of both the resistance and the inductive reactance on the circuit. (p. 540)

Inductance A measure of the tendency of a coil of wire to resist a change in the current because the magnetism produced by one part of the coil acts to oppose the change of current in other parts of the coil. (p. 537)

Inductive Reactance A measure of the opposition to ac current flow in an inductor. (p. 537)

Inductor A circuit component, such as a coil, in which an induced emf opposes any current change in the circuit. (p. 537)

Instantaneous Current The current at any instant of time. (p. 525)

Instantaneous Voltage The voltage at any instant of time. (p. 525)

Power Factor The ratio of the actual power to the apparent power. (p. 553)

Primary Coil The coil of a transformer that carries an alternating current and induces a current in the secondary coil. (p. 531)

Rectification The process of changing ac to dc. (p. 551)

Rectifier A device that changes ac to dc. (p. 529)

Resonance A condition in a circuit when the inductive reactance equals the capacitive reactance and they nullify each other. The current that flows in the circuit is then at its maximum value. (p. 550)

Secondary Coil The coil of a transformer in which a current is induced by the current in the primary coil. (p. 531)

Step-Down Transformer A transformer used to lower voltage; it has more turns in the primary coil. (p. 532)

Step-Up Transformer A transformer used to increase voltage; it has more turns in the secondary coil. (p. 532)

Transformer A device composed of two coils (primary and secondary) and a magnetic core. Used to step up or step down a voltage. (p. 531)

Formulas

19.1 $i = I_{max} \sin \theta$

$e = E_{max} \sin \theta$

$I = 0.707 I_{max}$ or $I_{max} = 1.41 I$

$E = 0.707 E_{max}$ or $E_{max} = 1.41 E$

19.2 $P = I^2 R = VI = \dfrac{V^2}{R}$

$\dfrac{V_P}{V_S} = \dfrac{N_P}{N_S}$

$\dfrac{I_S}{I_P} = \dfrac{N_P}{N_S}$

$I_P V_P = I_S V_S$

19.3 $X_L = 2\pi f L$

$I = \dfrac{E}{X_L}$

19.4 $I = \dfrac{E}{Z}$

$Z = \sqrt{R^2 + X_L^2}$

$Z = \sqrt{R^2 + (2\pi f L)^2}$

$\tan \phi = \dfrac{X_L}{R}$

19.5 $C = \dfrac{Q}{V}$

$X_C = \dfrac{1}{2\pi f C}$

19.6 $Z = \sqrt{R^2 + X_C^2}$

$Z = \sqrt{R^2 + \left(\dfrac{1}{2\pi f C}\right)^2}$

$I = \dfrac{E}{Z}$

$\tan\phi = \dfrac{X_C}{R}$

19.7 $Z = \sqrt{R^2 + (X_L - X_C)^2}$

$\tan\phi = \dfrac{X_L - X_C}{R}$

$I = \dfrac{E}{Z} = \dfrac{E}{\sqrt{R^2 + \left(2\pi f L - \dfrac{1}{2\pi f C}\right)^2}}$

19.8 $I = \dfrac{E}{\sqrt{R^2 + (X_L - X_C)^2}}$

$f = \dfrac{1}{2\pi\sqrt{LC}}$

19.10   power factor $= \dfrac{\text{actual power}}{\text{apparent power}}$

Review Questions

1. Which of the following describes alternating current electricity?
   (a) It can be produced by rotating a loop of wire through a magnetic field.
   (b) It flows in one direction for a period of time and then reverses direction.
   (c) It goes through one cycle when it flows in one direction and then reverses direction.
   (d) All of the above.
2. The voltage, e, and the current, i, in an alternating current circuit are in phase when
   (a) the peak values of both e and i occur at different times.
   (b) the peak values of e and i occur at the same time but their zero values do not.
   (c) the peak values and the zero values of e and i occur simultaneously.
   (d) none of the above.
3. Which of the following affect the voltage induced in the secondary coil of a transformer?
   (a) The current through the primary coil
   (b) The resistance of the primary coil
   (c) The number of turns in the primary coil
   (d) None of the above
4. Which of the following contribute to the energy loss in a transformer?
   (a) Resistance of the copper wires
   (b) Reversing the magnetic field in the core

(c)   Induced currents in the core
(d)   The emf in the outside circuit
(e)   All of the above
5.   Explain the difference between maximum current and effective current.
6.   Explain the difference between maximum voltage and instantaneous voltage.
7.   Explain how power in an ac circuit is related to voltage and current.
8.   Explain how power in an ac circuit is related to voltage and resistance.
9.   If the number of turns in the secondary coil of a transformer is doubled, how does the output voltage change?
10.   The unit of inductance is the _____.
11.   Discuss the importance of inductive reactance.
12.   How does the inductive reactance depend on frequency?
13.   Does the current lead or lag the voltage in an inductive circuit?
14.   Describe how energy is stored in a capacitor. How can the stored energy be used?
15.   Does the current lead or lag the voltage in a capacitive circuit?
16.   How does the reactance of a capacitor depend on frequency?
17.   Discuss the condition that leads to resonance.
18.   What is the function of a diode in a circuit?
19.   Explain the difference between amplification and rectification.
20.   Is the phase angle always constant in a circuit containing resistive, capacitive, and inductive elements?

Review Problems

1.   What is the maximum voltage in a circuit when the instantaneous value of the voltage is 95.4 V at $\theta = 62°$?
2.   If the maximum ac voltage on a line is 185 V, what is the instantaneous voltage at $\theta = 41°$?
3.   If the maximum ac voltage on a line is 175 V, what is the instantaneous voltage at $\theta = 23°$?
4.   What is the effective value of an ac voltage whose maximum voltage is 135 V?
5.   What is the maximum current in a circuit with a current rated at 6.35 A?
6.   What power is developed by a device that draws 6.87 A and has a resistance of 15.4 Ω?
7.   A heating element draws 4.50 A on a $11\overline{0}$-V line. What power is used in the element?
8.   What power is used by a heater with resistance 22.3 Ω and that draws a current of 7.65 A?
9.   A step-up transformer on a 115-V line provides a voltage of 2050 V. (a) If the primary coil has 75.0 turns, how many turns does the secondary coil have? (b) If there is a current of 4.55 A in the primary coil, what is the current in the secondary? (c) Find the power in the primary coil.
10.   An inductance of 48.0 mH is connected in series with a lamp with resistance 23.0 Ω. This circuit is connected to a 115-V, 60.0-Hz power supply. (a) What is the current in the circuit? (b) What is the phase angle? (c) What is the voltage drop across the inductance?
11.   A lamp with resistance 47.5 Ω is connected in series with an inductance of 43.2 mH. This circuit is connected to a 115-V, 60.0-Hz power supply. (a) What is the current in the circuit? (b) What is the phase angle? (c) What is the voltage drop across the resistance?
12.   (a) What current will flow in a 60.0-Hz ac series circuit that includes a $11\overline{0}$-V source, a 19.5-Ω resistor, and a 57.4-μF capacitor? (b) What is the phase angle?

SKETCH

| 12 cm² | w |

4.0 cm

DATA
$A = 12$ cm², $l = 4.0$ cm, $w = ?$

BASIC EQUATION
$A = lw$

WORKING EQUATION
$w = \dfrac{A}{l}$

SUBSTITUTION
$w = \dfrac{12 \text{ cm}^2}{4.0 \text{ cm}} = 3.0 \text{ cm}$

(c) What is the voltage across the resistor? (d) What is the voltage across the capacitor?

13. A 21.6-Ω resistor and a 38.5-μF capacitor are connected in series with a 60.0-Hz ac source with a voltage of $11\overline{0}$ V. (a) What is the current in the circuit? (b) What is the voltage across the capacitor? (c) What is the phase angle?

14. A circuit contains a 175-Ω resistance, a 25.0-μF capacitor, and a 62.0-mH inductance in series with a $11\overline{0}$-V, 60.0-Hz source. Find the impedance and the current.

15. A circuit contains a 115-Ω resistance, a 35.0-μF capacitor, and a 65.0-mH inductance in series with a $11\overline{0}$-V, 60.0-Hz source. Find the impedance and the current.

16. (a) Find the resonant frequency of a circuit containing a 7.50-μF capacitor in series with a 3.70-μH inductance, a 633-Ω resistor, and a $11\overline{0}$-V, 60.0-Hz source. (b) Find the impedance and the current.

17. (a) Find the resonant frequency of a circuit containing a 4.70-μF capacitor in series with a 4.50-μH inductance, a 25.0-Ω resistor, and a $11\overline{0}$-V, 60.0-Hz source. (b) Find the impedance and the current.

18. Find the apparent power produced by a generating station whose actual power is 2,9$\overline{0}$0,000 kW and whose power factor is 0.850.

APPLIED CONCEPTS

1. A microwave oven is designed to draw 11.8 A of current when connected to a $11\overline{0}$-V ac circuit. (a) What is the average power of the microwave? (b) What is the maximum current drawn by the microwave? (c) Provide a rationale for the current's value 180° after the beginning of a cycle.

2. Before converting alternating current to direct current for an electronic video game, a transformer first must reduce the voltage from $12\overline{0}$ V to 9.00 V. (a) If the primary coil has 307 turns, how many turns are in the secondary coil? (b) What is the power of each coil in the transformer? (c) Explain the relationship between the power for the primary coil and the power for the secondary coil.

3. A neighborhood requires 8.50×10^4 W of power from the local substation. (a) What is the current in the wire if the electricity is delivered at 22$\overline{0}$ V? (b) What is the current in the wire if the electricity is delivered at 60$\overline{0}$0 V? (c) If the resistance of the power line is 0.250 Ω, how much electric power will be converted into heat throughout the 22$\overline{0}$-V and 60$\overline{0}$0-V power lines? (d) Which power line voltage is the more efficient for transmitting electricity?

4. A 65.5-V, 60.0-Hz ac generator is connected to a radio circuit. (a) If the inductor coil has an inductive reactance of 125 Ω, find the inductance of the coil. (b) What is the effective current in the circuit? (c) Find the power of the circuit.

5. An AM radio tuner circuit has an inductance of 275 mH. (a) What is the capacitance of the variable capacitor when tuned to 88$\overline{0}$ kHz? (b) What is the capacitance when tuned to 1010 kHz? (c) Are the capacitor plates larger or smaller when the radio is tuned to a higher-frequency station?

CHAPTER 20

LIGHT

Objectives

The major goals of this chapter are to enable you to:

1. Describe the nature of light.
2. Solve problems involving the speed of light.
3. Contrast the wave and particle characteristics of light.
4. Apply principles of photometry to technical problems.

For centuries scientists have sought to explain the nature of light—why it is reflected and why it is refracted. Even the speed of light has been the subject of study since the 17th century.

Light seems to exhibit some characteristics of both particles and waves. The study of the measurement of light is photometry. We begin by examining the nature of light.

20.1 Nature of Light

Light may be defined as radiant energy that can be seen by the human eye. The search for an explanation for the nature of light has been going on for many centuries. A number of famous scientists, including Isaac Newton, Christiaan Huygens, Albert Einstein, and Louis de Broglie, made major contributions to the current theory of light and its interaction with matter.

Many of the foundations for modern scientific theories were developed in the 17th century. Two conflicting theories for the nature of light were proposed. The experimental observations that had to be explained were the following:

1. *Straight-line propagation of light.* An application of this property of light is found in survey work, in which sight lines are commonly used (Fig. 20.1).
2. ***Reflection***, *or turning back, of all or a part of a beam of light at the boundary between two different media* (Fig. 20.2). Examples are the reflection of light by a mirror and by the surface of still water.

Figure 20.2 Reflection of light by the surface of water

Figure 20.1 Surveyor using the straight-line travel characteristic of light

Courtesy of fotolia © CandyBox Images

Courtesy of fotolia © gijssmulders

3. ***Refraction***, *or bending, of light as it passes through the boundary between two media, such as air and water* (Fig. 20.3). This bending of light makes objects under water appear to be closer to the surface and farther away from the observer than they really are when viewed from above the surface. It also makes a straight object partially submerged in water appear bent at the surface.

Figure 20.3 Refraction of light as it passes from one medium two another

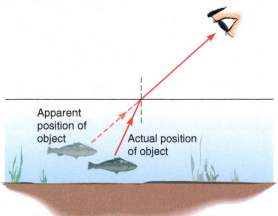

The two conflicting theories referred to above are the wave theory and the particle theory of light. **Christiaan Huygens** proposed the **wave theory** of light, according to which light consists of waves traveling out from light sources like water waves traveling out from the point at which a stone is dropped into still water. The waves continue long after the stone drops to the bottom and thus the succeeding waves cannot

TRY THIS ACTIVITY

Aquarium Tricks

An aquarium is a fun place to observe the refraction of light. The next time you are at an aquarium or have the opportunity to look into a small fish tank, observe an object from above the water line and note its apparent location (Fig. 20.4). Then, attempt to locate the same object by looking from the side of the aquarium. Note the difference in the apparent location of the object. Move your head from side to side and notice how the apparent position changes.

Figure 20.4   Note the apparent positions of the same light-colored stingray when viewed looking down into the aquarium from above the water line and when viewed from the side of the aquarium.

be caused by the stone's continuing activity. Huygens developed a geometric model as shown in Fig. 20.5 to explain how a wave front advances and reasoned the following important concept named after him:

    Huygens's principle: *Each point on a wave front can be regarded as a new source of small wavelets, which form succeeding waves that spread out uniformly in the forward direction at the same speed.*

    The new wave front is the result of all the wavelets. A similar type of wave behavior is sound propagation out from a source of sound. All three types of waves mentioned here—light, water, and sound—travel in straight lines. Waves also reflect and refract when they encounter new media.

    Isaac Newton proposed the **particle theory** of light as an alternative explanation of the experimental observations. Newton thought light was made up of streams of particles. These particles of light, which Newton referred to as *corpuscles*, he felt behaved in a manner similar to particles of matter, which travel in a straight line if the net force acting on them is zero. Particles of matter also rebound or reflect off surfaces. An example of this is a rubber ball bouncing off a wall. A change in the velocity of a particle could produce a change in direction. Newton's particle theory persisted until the early 19th century, when diffraction and interference of light were observed. Since these properties could only be explained by the wave theory, the particle theory fell out of favor.

    A new theory for the nature of light emerged as physicists began to consider light as an oscillating disturbance of an electric field and a corresponding magnetic field.

Christiaan Huygens (1629–1695),

physicist and astronomer, was born in The Netherlands. He made the first pendulum clock, proposed the wave theory of light, discovered the polarization of light, and discovered the ring and fourth satellite of Saturn in 1655.

Figure 20.5   Huygens's principle is used to determine the future position of both (a) an advancing spherical wave front *CD* and (b) a plane wave front *CD* when *AB* is given.

(a) Spherical wave front   (b) Plane wave front

It was discovered that an **electromagnetic wave** consists of two perpendicular transverse waves with one component of the wave being a vibrating electric field and the other being a corresponding vibrating magnetic field; the electromagnetic wave moves in a direction perpendicular to both electric and magnetic field components as shown in Fig. 20.6. All such waves travel at the same speed in a vacuum (3.00×10^8 m/s or 186,000 mi/h) but differ in their frequencies and wavelengths. Beginning in the late 19th century, experiments confirmed the **electromagnetic theory** that describes the interrelationship among electric fields, magnetic fields, electric charge, and electric current. Note that as the frequency increases to the right in Fig. 20.7, the wavelength decreases. Electromagnetic waves differ from other transverse and longitudinal waves in that they do not need a medium such as air, water, or a solid through which to travel. As a result, radio waves, visible light, gamma rays, and X rays travel through space at the speed of light. Electromagnetic waves are produced by accelerating electric charges, which create an electric field that in turn creates a corresponding magnetic field.

The electromagnetic theory, however, is still not a complete explanation of the nature of light. Early in the 20th century, the **photoelectric effect** was discovered: the emission (or giving off) of electrons by a substance when struck by electromagnetic radiation. **Albert Einstein** received a Nobel Prize for his work on the photoelectric effect and in the emerging field of theoretical physics. The energy of the emitted

Albert Einstein (1879–1955),

theoretical physicist, was born in Germany. His theories of relativity revolutionized modern thought. He won the Nobel Prize in Physics for his work on the photoelectric effect in 1921.

Figure 20.6   The electric and magnetic field components of an electromagnetic wave are perpendicular to each other as well as to the direction of travel of the electromagnetic wave.

Figure 20.7   Electromagnetic spectrum

Guglielmo Marconi (1874–1937), *physicist and inventor, was born in Italy. He experimented on ways to convert electromagnetic waves into electricity. His first successful experiments in wireless telegraphy were made in Bologna in 1895. He formed the Marconi Telegraph Company in London in 1899. In 1898 he transmitted signals across the English Channel and across the Atlantic Ocean in 1901. He later developed short-wave radio equipment and established a worldwide radio telegraph network for the British government. He shared the 1909 Nobel Prize in Physics.*

electron is related to the energy (frequency) of the incoming light in the same way that the energy of a billiard ball is related to the energy of the cue ball that strikes it. The particle theory of light explains the photoelectric effect, while the wave theory cannot. The electromagnetic wave theory fails to fully explain the nature of these emissions of electrons. **Guglielmo Marconi** used Einstein's theories as the basis for experiments on ways to convert electromagnetic waves into electricity.

Later work by **Max Planck** and Albert Einstein developed the idea that light was energy radiated at the speed of light in the form of wave packets of energy, which were called **photons**. Further work showed that in some circumstances light acted like a stream of particles. Based on the work of Planck and Einstein, the **quantum theory** was developed. This theory states that energy, including electromagnetic radiation, is not absorbed or radiated continuously but is radiated in multiples of definite and discontinuous units.

Light therefore appears to have at least a dual character, having properties of both waves and particles (Fig. 20.8). This dual character may be shown by considering how energy may be transported from one point to another.

The first method is to transport particles of matter that carry energy with them. Examples of this method are the conduction of electrons in a wire, the shooting of a bullet, the sending of natural gas through pipelines, and the transport of gasoline.

The second method is the propagation of a wave disturbance through the medium between two points. Sound and water waves are examples of this method of energy transport.

Light is unusual in that it appears to combine characteristics of each method. When light is traveling through a medium, it appears to behave like a wave, with the following characteristics: (1) reflection by the surface of the medium, (2) refraction when passing from one medium to another, (3) interference when two waves are superimposed, and (4) diffraction (bending) when the waves pass the corners of an obstacle. When light interacts with matter, such as when it is absorbed or emitted, it behaves as if it were a massless particle.

Max Planck (1858–1947), *theoretical physicist, was born in Germany. His work on the law of thermodynamics and black body radiation led him to question some principles of classical Newtonian physics and introduce in 1900 aspects of what would become the quantum theory. He received the Nobel Prize in Physics in 1918 for his work in quantum theory.*

Figure 20.8   Light has properties of both particles and waves.

TRY THIS ACTIVITY

Flickering Lights

Light-emitting diodes (LEDs) are often used to demonstrate the cyclical motion of electrons in an alternating current circuit. Find an extension cord with an LED indicator light in the receptacle end of the cord. Typically the LED is a yellow, green, or red light. After the extension cord is plugged into a 110-V ac outlet, the LED should appear as a constant light. Take the end of the extension cord and spin the light and 2 ft of cord around in a circle. Observe the light patterns and explain why the light from the LED no longer appears constant.

20.2 The Speed of Light

One of the most important measured quantities in physics is the **speed of light**, the speed at which light and other forms of electromagnetic radiation travel. At first, the travel of light was assumed to be instantaneous. **Galileo** probably first suggested that time was required for light to travel from place to place. **Ole Roemer** made the first estimate of the speed of light from his study of the time of eclipse of one of the moons of the planet Jupiter as viewed from different places of the orbit of the earth in 1675. Since then, laboratory methods for measuring the speed of light have been developed, most notably by **Albert Michelson**. His measurements were made by using rotating mirrors to reflect a beam of light to a mirror on a mountain 22 miles away and back.

Figure 20.9 shows the mirror arrangement used by Michelson to measure the speed of light. Light enters the eyepiece of the telescope only when the octagonal mirror is at rest or when it is spinning at exactly the right speed. By measuring the time for $\frac{1}{8}$ of a rotation of the mirror and the distance to the stationary mirror miles away, Michelson could calculate the speed of light with great accuracy. He also developed an instrument called an *interferometer*, with which in 1920 he made the first accurate measurement of a star's diameter.

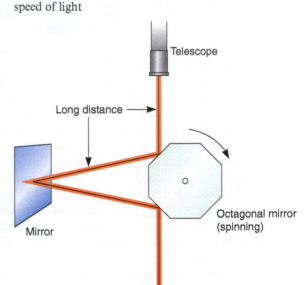

Figure 20.9   Michelson's method for determining the speed of light

Modern laboratory methods have been used to measure accurately the speed of light, which is now defined as 299,792,458 m/s. This is usually rounded to 3.00×10^8 m/s or 186,000 mi/s.

As stated earlier, light is one form of a class of radiation called *electromagnetic radiation*, which also includes radio and television waves, infrared, gamma rays, and X rays. A chart of the entire electromagnetic spectrum is shown in Fig. 20.7.

The distance traveled by any form of electromagnetic radiation can be found by substituting the speed of light c into the equation $s = vt$ as follows:

$$s = ct$$

where s = distance
       c = speed of light, 3.00×10^8 m/s or 186,000 mi/s
       t = time

EXAMPLE

Find the distance (in mi) traveled by an X ray in 0.100 s.

Data:

$$c = 186{,}000 \text{ mi/s}$$
$$t = 0.100 \text{ s}$$
$$s = ?$$

Basic Equation:

$$s = ct$$

Working Equation: Same

Substitution:

$$s = (186{,}000 \text{ mi/s})(0.100 \text{ s})$$
$$= 18{,}600 \text{ mi}$$

· · · · · · · · · · · · · · · · ·

Very large distances, such as those between stars, cannot be conveniently expressed in common distance units. Astronomers therefore use the unit light-year to measure such distances. A **light-year** is the distance traveled by light in one earth year, so 1 light-year equals 9.45×10^{15} m or 5.87×10^{12} mi.

PROBLEMS 20.2

1. Find the distance (in metres) traveled by a radio wave in 5.00 s.
2. Find the distance (in metres) traveled by a light wave in 6.40 s.
3. A television signal is sent to a communications satellite that is $20{,}\overline{0}00$ mi above a relay station. How long does it take for the signal to reach the satellite?
4. How long does it take for a radio signal from the earth to reach an astronaut on the moon? The distance from the earth to the moon is 2.40×10^5 mi.
5. The sun is 9.30×10^7 mi from the earth. How long does it take light to travel from the sun to the earth?
6. A radar wave is bounced off an airplane and returns to the radar receiver in 2.50×10^{-5} s. How far (in km) is the airplane from the radar receiver?
7. How long does it take for a radio wave to travel $30\overline{0}0$ mi across the United States?
8. How long does it take for a flash of light to travel $10\overline{0}$ m?
9. How long does it take for a police radar beam to travel to a truck and back if the truck is 115 m from the radar unit?
10. How far away (in km) is an airplane if the radar wave returns to the scanning radar unit in 1.24×10^{-3} s?
11. An auto mechanic uses a strobe light to time a classic car engine. If the light is held 0.20 m from the flywheel and the mechanic's eye is 0.75 m from the strobe, how long does it take the light from the strobe to reach the mechanic's eye?
12. A construction company uses GPS technology to monitor the location of certain equipment. If the satellite is 18,250 mi above the earth, how long does it take a signal to travel to the satellite and back?
13. (a) How long does it take for light to reach the earth from Mars when the separation of the two planets is at its smallest? The earth's orbital radius is 143 million kilometres. The orbital radius of Mars is 218 million kilometres. (b) How long does it take when the separation is at its maximum? Assume the planetary orbits are circular. Also make the (nonphysical) assumption that the sun is transparent to the transmission of light between the planets.
14. If it takes 4.31 years for light to reach the earth from Alpha Centauri, the closest star to the earth other than the sun, what is the distance (in miles) to the next-nearest neighbor (Barnard's Star), which is 25% farther away?
15. How long does it take light to reach the earth from Jupiter when the separation of the two planets is (a) at its smallest and (b) at its largest? The earth's orbital radius is 143 million kilometres. The orbital radius of Jupiter is 725 million kilometres. Assume the planetary orbits are circular. Also make the (nonphysical) assumption that the sun is transparent to the transmission of light between the planets.

SKETCH

12 cm² w

4.0 cm

DATA

$A = 12$ cm², $l = 4.0$ cm, $w = ?$

BASIC EQUATION

$A = lw$

WORKING EQUATION

$w = \dfrac{A}{l}$

SUBSTITUTION

$w = \dfrac{12 \text{ cm}^2}{4.0 \text{ cm}} = 3.0 \text{ cm}$

16. Preparing for reentry, astronauts use radar to determine the distance back to the earth. What is their altitude if it takes 0.330 s for the radar wave to travel to the earth and return?

17. The distance to the moon can be calculated by reflecting a ray of light off a mirror left by astronauts. The light travels to the mirror and back in 2.56 s. Find the distance to the moon.

18. Light from the sun travels 1.50×10^8 km to reach the earth. How long does its journey take in minutes?

PHYSICS ON THE JOB

Surveyor

Surveyors are responsible for making accurate measurements for construction projects, cartography (map-making), real estate, and other engineering projects. They must have a solid working knowledge of geography, geology, physics, and mathematics fields such as geometry and trigonometry. Surveyors work with Global Positioning Systems (GPS), levels, transits, theodolites, and laser range finders to precisely measure angles, elevations, and distances.

In addition to having a clear working knowledge of the surveying techniques and the science behind the tools used to gather surveying data, the surveyor must also be able to manage and lead a team of surveying technicians when out in the field.

A bachelor's degree is typically needed to become a surveyor. The advanced mathematical concepts and coursework in a degree program are necessary to receive accreditation by the Board for Engineering and Technology (ABET). It should be noted that surveying technicians, often found on job sites to assist surveyors in gathering measurements and/or using that data to create maps, typically need a high school diploma or an associate's degree to become a technician (Fig. 20.10).

Figure 20.10 Surveyor

Courtesy of fotolia © Kadmy

20.3 Light as a Wave

Light and the other forms of electromagnetic radiation are composed of oscillations in the electric and magnetic fields that exist in space. These oscillations are set up by the rapid movement of charged particles such as electrons in radio antennas and electrons in a hot object such as a light bulb filament. A wave is characterized by its **wavelength**, the distance between two successive corresponding points on the wave (Fig. 20.11). This distance is denoted by the Greek lowercase letter lambda, λ. The wavelength of visible light ranges from about 4.0×10^{-7} m to 7.6×10^{-7} m. The human eye perceives light in the visible spectrum as one or more colors depending upon the frequency or wavelength of the light hitting the retina of the eye. The longest visible wavelengths ($\lambda = {\sim}7.5 \times 10^{-7}$ m), which are also the lowest frequencies, are perceived as red. The shortest visible wavelengths ($\lambda = {\sim}4.0 \times 10^{-7}$ m), which are the highest frequencies, are perceived as blue. The wavelengths of other electromagnetic radiations are given in Fig. 20.7.

Another characteristic of waves is the frequency, f. **Frequency** is the number of vibrations or cycles per second of a wave. Frequency can be measured by counting the number of wavelengths that pass a stationary point in 1 s. The measurement unit of frequency (cycles/s) is named the hertz (Hz) after Heinrich Hertz, a leader in the study

Figure 20.11   The wavelength of a repeating wave is the distance between two successive corresponding points.

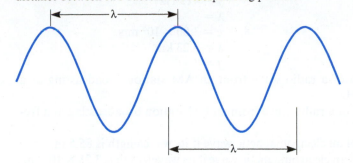

of electromagnetic theory. (1 Hz = 1 cycle per second = 1/s.) Since a "cycle" has no units, it does not appear in the hertz unit.

The following basic relationship exists for all electromagnetic waves:

$$c = \lambda f$$

where c = speed of light, 3.00×10^8 m/s or 186,000 mi/s
λ = wavelength
f = frequency

Find the frequency of a light wave with a wavelength of 5.00×10^{-7} m.   ◀   **EXAMPLE**

Data:

$$\lambda = 5.00 \times 10^{-7}\ \text{m}$$
$$c = 3.00 \times 10^8\ \text{m/s}$$
$$f = ?$$

Basic Equation:

$$c = \lambda f$$

Working Equation:

$$f = \frac{c}{\lambda}$$

Substitution:

$$f = \frac{3.00 \times 10^8\ \text{m/s}}{5.00 \times 10^{-7}\ \text{m}}$$
$$= 6.00 \times 10^{14}\ \text{Hz}\quad \text{(or cycles/s)}$$

PROBLEMS 20.3

1. $c = 3.00 \times 10^8$ m/s
  $\lambda = 4.55 \times 10^{-5}$ m
  $f = ?$

2. $c = 3.00 \times 10^8$ m/s
  $\lambda = 9.70 \times 10^{-10}$ m
  $f = ?$

3. $c = 3.00 \times 10^8$ m/s
  $f = 9.70 \times 10^{11}$ Hz
  $\lambda = ?$

4. $c = 3.00 \times 10^8$ m/s
  $f = 24.2$ MHz
  $\lambda = ?$

SKETCH

$$12 \text{ cm}^2 \quad w$$
$$4.0 \text{ cm}$$

DATA

$A = 12 \text{ cm}^2, l = 4.0 \text{ cm}, w = ?$

BASIC EQUATION

$A = lw$

WORKING EQUATION

$w = \dfrac{A}{l}$

SUBSTITUTION

$w = \dfrac{12 \text{ cm}^2}{4.0 \text{ cm}} = 3.0 \text{ cm}$

5. $c = 3.00 \times 10^8$ m/s
    $f = 45.6$ MHz
    $\lambda = ?$

6. $c = 3.00 \times 10^8$ m/s
    $f = 415$ Hz
    $\lambda = ?$

7. $c = 3.00 \times 10^8$ m/s
    $\lambda = 6.59 \times 10^{12}$ m
    $f = ?$

8. $c = 3.00 \times 10^8$ m/s
    $\lambda = 9.23$ km
    $f = ?$

9. Find the wavelength of a radio wave from an AM station broadcasting at a frequency of $14\overline{0}0$ kHz.

10. Find the wavelength of a radio wave from an FM station broadcasting at a frequency of $10\overline{0}$ MHz.

11. Find the frequency of an electromagnetic wave if its wavelength is 85.5 m.

12. Find the frequency of an electromagnetic wave if its wavelength is 3.25×10^{-8} m.

13. Find the frequency of blue light.

14. Find the frequency of red light.

15. Find the frequency of yellow light if its wavelength is midway between those of red and blue light.

16. An AM radio station broadcasts a signal with a wavelength of 237 m. Find its frequency in kHz.

20.4 Light as a Particle

As mentioned earlier, light sometimes behaves as if it were a particle. These particles are called *photons*, and each carries a portion of the energy of the wave. This energy is given by

$$E = hf$$

where $f = $ frequency
$\quad\quad h = 6.626 \times 10^{-34}$ J s   (*Planck's constant*)
$\quad\quad E = $ energy

When Planck introduced his concept of discrete values of energy proportional to the vibration frequencies of electrons, it was initially regarded by many as only a theoretical aid. However, this concept actually triggered the beginning of quantum physics. The only unknown in Planck's equation above was h, which was evaluated to make the mathematical expression fit the experimental data. This value, $h = 6.626 \times 10^{-34}$ J s, is known as **Planck's constant**.

EXAMPLE

What is the energy of a photon of electromagnetic radiation with frequency 5.00×10^{12} Hz?

Data:

$$h = 6.626 \times 10^{-34} \text{ J s}$$
$$f = 5.00 \times 10^{12} \text{ Hz} = 5.00 \times 10^{12}/\text{s}$$
$$E = ?$$

Basic Equation:

$$E = hf$$

Working Equation: Same

Substitution:

$$E = (6.626 \times 10^{-34} \text{ J s})(5.00 \times 10^{12}/\text{s})$$
$$= 3.31 \times 10^{-21} \text{ J}$$

PROBLEMS 20.4

1. What is the energy of a photon of electromagnetic radiation with frequency 8.95×10^{10} Hz?
2. What is the frequency of a photon of electromagnetic radiation with energy 3.96×10^{-22} J?
3. What is the energy of a photon of electromagnetic radiation with frequency 4.55×10^{8} Hz?
4. Find the frequency of electromagnetic radiation with energy 2.00×10^{-24} J.
5. Find the frequency of electromagnetic radiation with energy 5.50×10^{-26} J.
6. Find the energy of a photon of electromagnetic radiation with frequency 2.50×10^{12} Hz.
7. Find the frequency of electromagnetic radiation with energy 3.65×10^{-23} J.
8. Find the energy of a photon of electromagnetic radiation with frequency 9.20×10^{16} Hz.
9. Find the energy of a red photon.
10. Find the energy of a blue photon.
11. Find the energy of a yellow photon if the wavelength of its electromagnetic radiation is midway between those of red and blue photons.
12. An AM radio signal has a frequency of $65\overline{0}$ kHz. What is the energy of a photon of that electromagnetic radiation?
13. An AM radio station in a nearby town broadcasts a signal with 9.37×10^{-28} J of energy. Find the frequency of this signal in kHz.

SKETCH

```
┌──────────┐
│  12 cm²  │  w
└──────────┘
    4.0 cm
```

DATA

$A = 12 \text{ cm}^2, l = 4.0 \text{ cm}, w = ?$

BASIC EQUATION

$A = lw$

WORKING EQUATION

$w = \dfrac{A}{l}$

SUBSTITUTION

$w = \dfrac{12 \text{ cm}^2}{4.0 \text{ cm}} = 3.0 \text{ cm}$

PHYSICS CONNECTIONS

Light Amplification by Stimulated Emission of Radiation

Light amplification by stimulated emission of radiation is more commonly known by its acronym, LASER. A **laser** is a light source that produces a narrow beam of light with high intensity and is a good example of how light behaves both as a particle and as a wave (Fig. 20.12). A burst of energy from the laser's flash tube causes an increase in the energy level of some of the electrons in the tube. Immediately following the burst of energy, the energy level of the electrons goes back to its original state. When an electron goes to a lower energy level, it emits a particle of light called a photon. When the emitted photon strikes another

Figure 20.12   (a) A green argon laser beam, generated by a laser in the upper right corner, bounces off various reflective surfaces in a laser laboratory. (b) The inside of a ruby laser

Courtesy of Clive Streeter © Dorling Kindersley

(a)

(b)

atom, it causes a similar event for that atom as well. As this process repeats, more and more photons are generated, causing what is known as "stimulated emission of radiation." The photons located inside the flash tube begin to increase their energy as they travel back and forth and repeatedly reflect off the mirrors on each side of the tube. The increase in the photon energy is called "light amplification." As the energy in the tube increases, a small amount of the coherent laser light escapes out of one of the mirrors that is partially transparent. Coherent light is light in which the crests and troughs of the waves line up and are in synchronization with each other.

The first laser was constructed by T. H. Maiman in 1960. The type of laser described above is the typical ruby laser. Some other common types of lasers are the red helium–neon laser ($\lambda = 633$ nm), the green argon laser ($\lambda = 477$ nm, 488 nm, and 515 nm), and the infrared carbon dioxide laser ($\lambda = 10,600$ nm).

The laser has become an essential tool in virtually all areas of science and technology. In the medical field, lasers are used to destroy cancerous tumors, correct vision defects, and seal capillaries to prevent excessive bleeding during surgery. In manufacturing, lasers are used to measure or align parts for fabrication, drill accurate holes, and weld metals together. The telecommunications and electronics industries use lasers to transmit digital signals in fiber optic cables; create holograms; read bar-codes; and record or read data from CDs, DVDs, and CD-ROMs. Geologists, astronomers, and physicists use lasers to measure great distances or detect minor changes in position.

20.5 Photometry

Recall that light is produced along with other forms of radiation when substances are heated, like our greatest source of natural light, the sun. Light may also be produced by electrons bombarding molecules in a gas such as neon or chemically, as by fireflies. Our most common source of artificial light, however, is the incandescent lamp.

Incandescent lamps (like light bulbs) produce light by the heating of a material (the filament) and the giving off of a wide range of radiation in addition to visible light. Objects that produce light are called *luminous*. On the other hand, objects like the moon, which are not producers of light but only reflect light from another source, are called *illuminated*. When light strikes the surface of most objects, some light is reflected, some is transmitted, and some is absorbed.

The study of the measurement of light is called **photometry**. Two important measurable quantities in photometry are the luminous intensity, I, of a light source, and the illumination, E, of a surface.

Luminous intensity is a measure of the brightness of a light source. The unit for luminous intensity, I, is the candle or *candela* (cd). The early use of certain candles for standards of intensity led to the name of the unit. We now use a platinum source at a certain temperature as the standard for comparison. Another unit, the *lumen*, ℓm, is often used for the measurement of the intensity of a source. One candle produces 4π lumens (ℓm):

$$1 \text{ candle} = 4\pi \ \ell\text{m}$$

To determine the intensity rating in lumens of a 40-W bulb rated at 35 cd, use the following conversion:

$$35 \text{ cd} \times \frac{4\pi \ \ell\text{m}}{1 \text{ cd}} = 440 \ \ell\text{m}$$

Thus, a 40-W light bulb rated at 35 cd has an intensity rating of 440 ℓm.

Illumination is the amount of luminous intensity per unit area. The illumination on a surface may be varied by either changing the luminous intensity of the source (using a brighter bulb) or changing the position of the source (moving it closer to or farther from the surface to be illuminated). Of course, the illumination is also less if the surface illuminated is slanted and not directly facing the light source.

The amount of illumination on a surface varies inversely with the square of the distance from the source. For example, if the distance of the illuminated surface from the source is doubled, the illumination is reduced to one fourth of its former intensity. This can be illustrated by considering a point source of light at the center of concentric (having the same center) spheres (Fig. 20.13). The solid angle is measured in units called *steradians.*

If the source radiates light uniformly in all directions, the light is uniformly distributed over a spherical surface centered at the source. Since the surface area of a sphere is $4\pi r^2$, the illumination, E, at the surface is given by

$$E = \frac{I}{4\pi r^2}$$

where E = illumination
        I = luminous intensity of the source (in lumens)
        r = distance between the source and the illuminated surface
The unit of illumination, E, is the *lux:*

$$1 \text{ lux} = 1 \text{ } \ell m/m^2$$

We assume the surface being illuminated is perpendicular to the source.

Note that the inverse-square-law behavior of light is similar to the two inverse-square laws studied earlier, those of gravity and of the electric force between charges described by Coulomb's law.

Figure 20.13   The amount of illumination on a surface varies inversely with the square of the distance from the source.

1 cm
2 cm
3 cm

Find the illumination E on a surface located 2.00 m from a source with an intensity of $40\overline{0}$ ℓm.

EXAMPLE 1

Data:

$$I = 40\overline{0} \text{ } \ell m$$
$$r = 2.00 \text{ m}$$
$$E = ?$$

Basic Equation:

$$E = \frac{I}{4\pi r^2}$$

Working Equation: Same

Substitution:

$$E = \frac{40\overline{0} \text{ } \ell m}{4\pi (2.00 \text{ m})^2}$$

$$= 7.96 \text{ } \ell m/m^2 \quad \text{or} \quad 7.96 \text{ lux}$$

The unit used for illumination is the lux ($\ell m/m^2$) in the metric system as shown above and the foot-candle ($\ell m/ft^2$) in the U.S system:

$$1 \text{ ft-candle} = 1 \text{ } \ell m/ft^2$$

EXAMPLE 2

Find the illumination 4.00 ft from a source with an intensity of $60\overline{0}$ ℓm.

Data:

$$I = 60\overline{0} \text{ ℓm}$$
$$r = 4.00 \text{ ft}$$
$$E = ?$$

Basic Equation:

$$E = \frac{I}{4\pi r^2}$$

Working Equation: Same

Substitution:

$$E = \frac{60\overline{0} \text{ ℓm}}{4\pi(4.00 \text{ ft})^2}$$

$$E = 2.98 \frac{\text{ℓm}}{\text{ft}^2} \times \frac{1 \text{ ft-candle}}{1 \text{ ℓm/ft}^2}$$

$$= 2.98 \text{ ft-candles}$$

$$\frac{\text{ℓm}}{\text{ft}^2} \times \frac{1 \text{ ft-candle}}{1 \text{ ℓm/ft}^2} = \frac{\cancel{\text{ℓm}}}{\cancel{\text{ft}^2}} \times 1 \text{ ft-candle} \times \frac{\cancel{\text{ft}^2}}{\cancel{\text{ℓm}}}$$

$$= \text{ft-candle}$$

In photography, photoelectric cells are used in light meters or exposure meters to measure illumination for taking photographs. The electricity produced is proportional to the illumination and is directly calibrated on the instrument scale. The units of measurement of such meters, however, are not standardized, and the scale may be arbitrarily selected.

SKETCH

12 cm² | w

4.0 cm

DATA
$A = 12 \text{ cm}^2, l = 4.0 \text{ cm}, w = ?$

BASIC EQUATION
$A = lw$

WORKING EQUATION
$w = \dfrac{A}{l}$

SUBSTITUTION
$w = \dfrac{12 \text{ cm}^2}{4.0 \text{ cm}} = 3.0 \text{ cm}$

PROBLEMS 20.5

1. $I = 48.0$ cd
    $I = \underline{}$ ℓm
2. $I = 342$ cd
    $I = \underline{}$ ℓm
3. $I = 765$ ℓm
    $I = \underline{}$ cd
4. $I = 432$ ℓm
    $I = \underline{}$ cd
5. $I = 75.0$ cd
    $I = \underline{}$ ℓm
6. $I = 65\overline{0}$ ℓm
    $I = \underline{}$ cd
7. $I = 90\overline{0}$ ℓm
    $r = 7.00$ ft
    $E = ?$
8. $I = 741$ ℓm
    $r = 6.50$ m
    $E = ?$
9. $I = 893$ ℓm
    $r = 3.25$ ft
    $E = ?$
10. $E = 4.32$ lux
    $r = 9.00$ m
    $I = ?$
11. $E = 10.5$ ft-candles
    $r = 6.00$ ft
    $I = ?$

12. Find the intensity of a light source that produces an illumination of 5.50 ft-candles at 9.85 ft from the source.
13. Find the intensity of a light source that produces an illumination of 2.39 lux at 4.50 m from the source.
14. Find the intensity of a light source that produces an illumination of 5.28 lux at 6.50 m from the source.
15. If an observer triples her distance from a light source: (a) Does the illumination at that point increase or decrease? (b) In what proportion does the illumination increase or decrease?

16. If the illuminated surface is slanted at an angle of 35.0°, what part of the full-front illumination is lost?
17. Find the illumination on a surface by three light sources, each with intensity 150 ℓm, located at distances of 2.00 m, 2.70 m, and 2.98 m from the surface, respectively.
18. Find the intensity of two identical light sources located 1.40 m and 1.96 m, respectively, from a point where the illumination is 3.54 ℓm/m^2.
19. Find the intensity of two identical light sources located 0.880 m and 1.12 m from a point where the illumination is 5.86 ℓm/m^2.
20. A desk is 3.35 m below an 1850-ℓm incandescent lamp. What is the illumination on the desktop?

Glossary

Electromagnetic Theory Describes the interrelationship among electric fields, magnetic fields, electric charge, and electric current. (p. 562)

Electromagnetic Wave A wave consisting of two perpendicular transverse waves with one component of the wave being a vibrating electric field and the other component being a corresponding vibrating magnetic field; the electromagnetic wave moves in a direction perpendicular to both electric and magnetic field components. (p. 562)

Frequency The number of complete vibrations or cycles per second of a wave. (p. 566)

Huygens's Principle Each point on a wave front can be regarded as a new source of small wavelets, which form succeeding waves that spread out uniformly in the forward direction at the same speed. (p. 561)

Illumination The luminous intensity per unit area. (p. 570)

Laser A light source that produces a narrow beam with high intensity; an acronym for "light amplification by stimulated emission of radiation." (p. 569)

Light Radiant energy that can be seen by the human eye. (p. 559)

Light-Year The distance that light travels in one earth year: 9.45×10^{15} m or 5.87×10^{12} mi. (p. 565)

Luminous Intensity A measure of the brightness of a light source. (p. 570)

Particle Theory Theory that light consists of streams of particles. (p. 561)

Photoelectric Effect The emission of electrons by a surface when struck by electromagnetic radiation. (p. 562)

Photometry The study of the measurement of light. (p. 570)

Photons Wave packets of energy that carry light and other forms of electromagnetic radiation. (p. 563)

Planck's Constant A fundamental constant of quantum theory (6.626×10^{-34} J s). (p. 568)

Quantum Theory Theory initiated by Planck and Einstein that energy, including electromagnetic radiation, is radiated or absorbed in multiples of certain units of energy. (p. 563)

Reflection The turning back of all or part of a beam of light at the boundary between two different media. (p. 560)

Refraction The bending of light as it passes through the boundary between two media, such as air and water. (p. 560)

Speed of Light The speed at which light and other forms of electromagnetic radiation travel; equal to 3.00×10^8 m/s in a vacuum. (p. 564)

Wave Theory Theory that light consists of waves traveling out from light sources, like water waves traveling out from the point at which a stone is dropped into water. (p. 560)

Wavelength The distance between two successive corresponding points on a wave. (p. 566)

Formulas

20.2 $s = ct$

20.3 $c = \lambda f$

20.4 $E = hf$

20.5 $E = \dfrac{I}{4\pi r^2}$

Review Questions

1. Which of the following are examples of electromagnetic radiation?
   (a) Gamma rays   (b) Sound waves   (c) Radio waves
   (d) Water waves   (e) Visible light
2. The particle theory of light explains
   (a) diffraction of light around a sharp edge.
   (b) refraction of light at a boundary.
   (c) the photoelectric effect.
   (d) none of the above.
3. A light-year equals
   (a) the time it takes light to travel from the sun to the earth.
   (b) the distance from the sun to the earth.
   (c) the distance to the nearest star other than the sun.
   (d) the distance light travels in one earth year.
4. Light behaves
   (a) as a massive particle.
   (b) always as a wave.
   (c) sometimes as a wave, sometimes as a particle.
   (d) as none of the above.
5. Does the wavelength of light depend on its frequency? Explain.
6. How does the energy of a photon of light depend on its frequency?
7. How does the intensity of illumination depend on the distance from a source radiating uniformly in all directions?
8. In your own words, explain how the speed of light has been measured.
9. Does light always travel at the same speed? Explain.
10. What name is given to the entire range of waves that are similar to visible light?
11. Who proposed the particle theory of light?
12. Who developed the wave packet theory of light?
13. Who made the first estimate of the speed of light?
14. How was the first estimate of the speed of light made?
15. What are the units of luminous intensity?
16. In your own words, explain luminous intensity.

Review Problems

1. Find the distance (in metres) traveled by a radio wave in 21.5 h.
2. A radar wave that is bounced off an airplane returns to the radar receiver in 3.78×10^{-5} s. How far (in miles) is the airplane from the radar receiver?
3. How long does it take for a police radar beam to travel to a car and back if the car is 0.245 mi from the radar unit?

4.   How long does it take for a pulse of laser light to return to a police speed detector after bouncing off a speeding car 0.274 mi away?

5.   How long does it take for a radio signal to travel from the earth to a communications satellite 22,500 mi above the surface of the earth?

6.   Find the wavelength of a radio wave from an AM station broadcasting at a frequency of 1230 kHz.

7.   Find the frequency of a radio wave if its wavelength is 46.5 m.

8.   Find the frequency of a light wave if its wavelength is 5.415×10^{-8} m.

9.   What is the energy of a photon with frequency 1.45×10^{11} Hz?

10.   What is the frequency of a photon with energy of 4.75×10^{-23} J?

11.   What is the energy of a photon with frequency 8.25×10^{-15} Hz?

12.   Find the intensity of the light source necessary to produce an illumination of 3.75 ft-candles at 6.75 ft from the source.

13.   Find the intensity of the light source necessary to produce an illumination of 4.86 lux at 9.25 m from the source.

14.   What is the intensity of the light source required to produce the illumination of Problem 13 if the distance from the light source is doubled?

15.   What are the maximum and minimum transit times for light traveling from Jupiter to Mars? The orbital radii are 215 million kilometres for Mars and 725 million kilometres for Jupiter. Assume the planetary orbits are circular. Also make the (nonphysical) assumption that the sun is transparent to the transmission of light between the planets.

16.   Find the intensity of two identical light sources located 0.454 m and 0.538 m, respectively, from a point where the illumination is 8.46 ℓm/m^2.

17.   Find the illumination on a surface by three light sources, each with intensity 125 ℓm, located at 1.85 m, 1.92 m, and 2.43 m from the surface, respectively.

SKETCH

| 12 cm^2 | w |

4.0 cm

DATA
$A = 12$ cm^2, $l = 4.0$ cm, $w = ?$

BASIC EQUATION
$A = lw$

WORKING EQUATION
$w = \dfrac{A}{l}$

SUBSTITUTION
$w = \dfrac{12 \text{ cm}^2}{4.0 \text{ cm}} = 3.0 \text{ cm}$

A P P L I E D   C O N C E P T S

1.   The distance between New York City and London is 3470 mi. (a) If a radio wave from New York City is transmitted directly across the ocean, how long will it take to reach a receiver in London? (b) In fact, due to the curvature of the earth, radio waves cannot be transmitted directly across such large distances on the earth. Instead, a signal is typically transmitted to a communications satellite located 2.20×10^4 mi above the surface of the earth. If the satellite is located midway between the two cities, how long will the radio wave take to reach London? (Disregard the effect that the curvature of the earth has on the calculations.)

2.   (a) When the Apollo astronauts landed on the moon, it took the radio signal 1.28 s to reach Mission Control on the earth. How far away were the astronauts from the earth? (b) Since the sun is 1.50×10^{11} m from the earth, how much time does it take for light to travel from the sun to the earth? (c) Light from our next-closest star, Alpha Centauri, takes 4.31 years to reach the earth. How far away from the earth is Alpha Centauri?

3.   The range of electromagnetic wave frequencies on the FM radio band is 88.0 MHz to 108 MHz. (a) What is the range of wavelengths for the FM radio band? (b) What is the range of photon energy for the FM radio band? (c) Explain the relationship between frequency and photon energy.

4.   The individual rods on rooftop antennas are designed to be one quarter of a wavelength for each television frequency. What is the range of rod lengths needed for television Channels 2 through 6 if their frequencies are between 54.0 MHz and 88.0 MHz?

5.   An illumination of 180 lux on student desks and other work areas is the standard that architects use when designing lighting systems for schools. (a) If a ceiling is 2.50 m above a student work area, what intensity light source must be installed? (b) If the ceiling were twice its original height, what light intensity would be needed to meet the standard requirement?

CHAPTER 21

REFLECTION AND REFRACTION

Objectives

The major goals of this chapter are to enable you to:

1. Describe the laws of reflection.
2. Locate and describe images formed by plane, convex, and concave mirrors.
3. Apply the mirror formula to image formation.
4. Describe the law of refraction.
5. Describe total internal reflection.
6. Locate and describe images formed by converging and diverging lenses.

The nature of light may still be somewhat of a mystery. However, its characteristics have been the subject of intensive study for hundreds of years. Light may be transmitted, reflected, or absorbed by a medium.

Anyone wearing glasses can appreciate the refraction of light as it bends upon passing from one medium to another. The index of refraction is a tool that the scientist uses to describe the ability of certain substances to bend light as it passes through them.

Our examination of the behavior of light begins with the study of images and reflection.

21.1 Mirrors and Images

Although the nature of light is complex, we know very well how light behaves. Every day we experience and unconsciously use our knowledge of what light does and depend on the ways it works. **Reflection** is the turning or turning back of all or a part of a beam of light as it strikes a surface. Unlike sound, light does not require a medium (some kind of matter) to travel through and may be transmitted through empty space. When light does strike a medium, the light may be reflected, absorbed, transmitted, or undergo a combination of the three.

Mirrors show how light may be reflected. Any dark cloth shows how light may be absorbed. Window glass illustrates how light may be transmitted through a medium.

A medium can be classified according to how well light is transmitted through it:

1. *Transparent:* almost all light passes through.
    Examples: window glass, clear water
2. *Translucent:* some but not all light passes through.
    Examples: murky water, light fog, skylight panels for farm buildings, stained glass
3. *Opaque:* almost all light is reflected or absorbed.
    Examples: wood, metal, plaster

These classifications are relative because for any medium, some light is reflected from the surface and some passes into or through it.

In studying reflection we observe what happens when light is turned back from a surface. The beam of a flashlight directed at a mirror shows several things about reflection. First, upon striking the surface of the glass, some of the light is reflected in all directions. This is called *scattering*. If there were no scattering, no light would reach our eyes and we would be unable to observe the beam at all. However, only a very small part of the beam of light is scattered. Rough or uneven surfaces produce more scattering than do smooth ones. This scattering of light by uneven surfaces is called **diffusion**. Diffused lighting has many applications at home and in industry where bright glare is not desirable.

Nearly complete reflection (with very little scattering) is called **regular** (or specular) **reflection**. Regular reflection occurs when parallel or nearly parallel rays of light (such as sunlight and spotlight beams) remain parallel after being reflected from a surface (Fig. 21.1). Note that the incoming rays are referred to as *incident rays*.

A flashlight beam on a mirror in a darkened room also shows something else about light striking a regular reflecting surface: The reflected rays of light leave the surface at the same angle at which the incident (incoming) rays strike the surface (Fig. 21.2). Expressed another way, the angles measured from the normal (the perpendicular) to the reflecting surface are equal. These angles are shown in Fig. 21.3.

Figure 21.1   Regular reflection (reflected rays are parallel)

Incident rays   Reflected rays

Figure 21.2   On a regular surface, the reflected rays leave at the same angle as the incident rays.

Figure 21.3   The angle of incidence is equal to the angle of reflection.

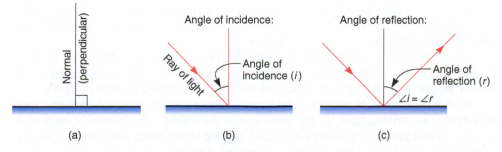

(a)   (b)   (c)

The same principle applies to curved surfaces (Fig. 21.4). This behavior of light rays is defined by the following **first law of reflection**:

FIRST LAW OF REFLECTION
The angle of incidence, *i*, is equal to the angle of reflection, *r*; that is,

$$\angle i = \angle r$$

Figure 21.4 Likewise, for curved surfaces, $\angle i = \angle r$.

Further observation of the light beam readily shows the following **second law of reflection**:

SECOND LAW OF REFLECTION
The incident ray, the reflected ray, and the normal (perpendicular) to the surface all lie in the same plane.

These laws of reflection apply not only to light, but to all kinds of waves.

We look next at how images are formed by three widely used kinds of mirrors: plane, concave, and convex. **Plane mirrors** are flat. **Concave mirrors** are curved away from the observer [like the inside of a bowl; Fig. 21.5(a)] and **convex mirrors** are curved toward the observer [like ball-shaped Christmas tree ornaments; Fig. 21.5(b)].

Figure 21.5 Spherical mirrors

(a) Concave mirror   (b) Convex mirror

Mirrors of glass or any highly reflecting surface have countless practical applications, from rear-view mirrors in automobiles to mirrors to watch for shoplifting in stores (Fig. 21.6).

For convenience we will use spherical mirrors (reflecting surfaces that are sections of spheres), although parabolic mirrors (reflecting surfaces in the shape of a parabola; Fig. 21.7) have wider practical use.

We consider next how images are formed by plane, concave, and convex mirrors. Images formed by mirrors may be **real images** (images formed by rays of light) or **virtual images** (images that only appear to the eye to be formed by rays of light).

Real images made by a single mirror are always inverted (upside down) and may be larger than, smaller than, or the same size as the object. They can be shown on a screen. Virtual images are always erect and may be larger than, smaller than, or the same size as the object. They cannot be shown on a screen.

Figure 21.6(a) Convex mirrors are used in traffic areas at sharp corners to guide drivers to avoid running into a person or another vehicle.

Figure 21.6(b) Convex security mirrors are used to watch for shoplifting because they reflect images from over wide areas.

Figure 21.7 Parabolic mirror

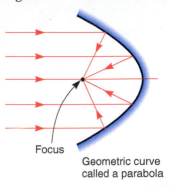

Focus

Geometric curve called a parabola

21.2 Images Formed by Plane Mirrors

Plane mirror images are always erect and virtual and appear as far behind the mirror as the distance the object is in front of the mirror. Note that plane mirrors also reverse right and left, so the right hand held in front of a plane mirror appears in the mirror to be the left hand.

Look at your image in a mirror. Then look at something on the mirror's surface like a speck of dirt. Notice that you have to adjust your eyes and refocus from looking at the image of your face to looking at the mirror surface. It should be clear that the image is not on the surface of the mirror but behind it.

We use light-ray diagrams to illustrate how our eyes see images in the various kinds of mirrors. We do this by representing rays of light and lines of sight with straight lines (Fig. 21.8). This method is used to construct diagrams and locate the images formed. Simply view the object from two or more separate places and construct the light-ray lines.

Figure 21.8

(a) Plane mirror image

(b) Location of an image in a plane mirror

21.3 Images Formed by Concave Mirrors

Find a shiny tablespoon and look at your image in it (Fig. 21.9). Now turn it over and look again. The images are very different; one is erect and the other, inverted (upside down). We use ray diagrams to show why this happens. As we shall see, the kind of image produced depends on the location of the object with respect to the mirror.

Figure 21.9   Reflected images as seen on opposite sides of a large spoon

(a)   (b)

We need to define some terms before we discuss how images are formed. Figure 21.10(a) shows a spherical mirror with the key terms identified. The *center of curvature*, *C*, is the center of the sphere that forms a part of the spherical mirror. The *vertex*, *V*, is the center of the mirror (sometimes called its optical center). The *principal axis* is the line *CV* drawn through the center of curvature and the vertex. The *principal focus*, *F*, is the point on the principal axis through which all rays parallel to the principal axis converge in a concave mirror as shown in Fig. 21.10(b) or from which they diverge in a convex mirror. Note that any ray through the center of curvature is along a radius of a spherical mirror and is reflected straight back because any such ray is perpendicular to the surface of the mirror at the point the ray strikes the mirror. The **focal length** is the distance between the principal focus of a mirror (or lens) and its vertex. The *aperture* is that very small portion of the sphere that forms the actual mirror.

Figure 21.10   Spherical mirror

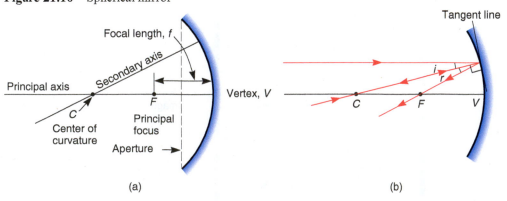

(a)   (b)

For mirrors with small apertures, the focal length, f, is one half the radius of curvature, R. That is, $f = \dfrac{R}{2}$.

If the object is placed at the focal point, no image will be formed because the rays of light will be reflected parallel to the principal axis (Fig. 21.11). The location of the reflected ray may be found by using the laws of reflection (Fig. 21.12). The angle of incidence is equal to the angle of reflection.

Figure 21.11 No image is formed if the object is located at the focal point.

Figure 21.12

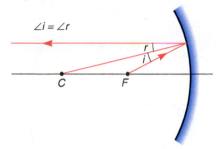

Now consider the more common case, in which the object is beyond the center of curvature [for example, looking into a tablespoon as in Fig. 21.9, whose mirror diagram is shown in Fig. 21.13(a)]. Note that again we use the fact that a ray parallel to the principal axis is reflected through the focal point and a ray through the focal point is reflected parallel to the principal axis. Then, where the two rays intersect, a point on the image is formed. (In this case, it is the flame of the candle. Note that the candle base image lies on the principal axis because the object base is also on that line.) The same method is used for the case where the candle base extends below the principal axis [Fig. 21.13(b)].

Figure 21.13 Images formed in spherical mirrors

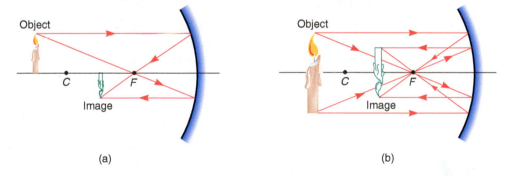

Now apply these principles to the diagrams of other images formed by concave mirrors (Fig. 21.14). Decide whether the image is real or virtual; erect or inverted; larger, smaller, or the same size; and where it is located. Note that the only time a virtual image is produced is when the object is between the focal point and the mirror [Fig. 21.14(f)]. The construction of the diagram is the same as the other cases except that the light rays are converging (coming together) and must be extended behind the mirror (where the image appears to be) forming the virtual image.

Figure 21.14 Formation of images in concave mirrors by objects that are located in various positions relative to the center of curvature

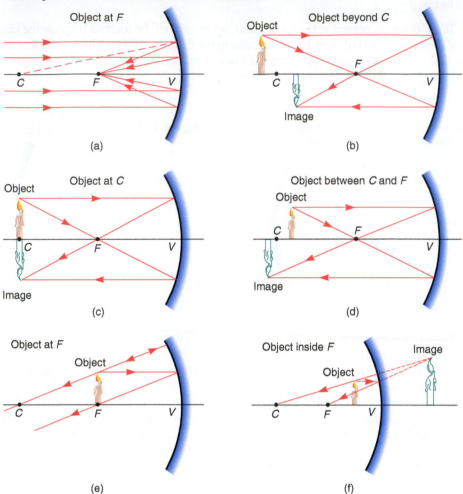

21.4 Images Formed by Convex Mirrors

By looking into the back side of our tablespoon in Fig. 21.9, we see an erect, virtual, smaller image. Use the mirror diagram shown in Fig. 21.15 to see how such an image is formed. Curved surface mirrors are used in some telescopes, spotlights, and automobile headlights. However, because spherical mirrors produce clear images over only a very small portion of their surfaces, the surfaces used commercially are another

Figure 21.15 Formation of images in convex mirrors

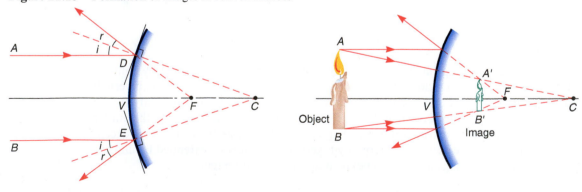

Note that $\angle i = \angle r$ for convex mirrors also.

(a)   (b)

geometric shape, that of a parabola, or parabolic, as mentioned before. For apertures wider than about 10°, spherical mirrors produce fuzzy images because all parallel rays are not reflected through the focal point. This is called *spherical aberration*.

Two common applications of convex mirrors are as security devices in convenience stores and on some automobile rearview mirrors because of the wide-angle view they produce.

21.5   The Mirror Formula

As we might expect from the previous cases, the focal length, the distance from the object to the mirror, and the distance from the image to the mirror are all related (Fig. 21.16). This relationship can be expressed as the *mirror formula*:

Figure 21.16   The mirror formula is expressed in terms of f, s_o, and s_i.

$$\frac{1}{f} = \frac{1}{s_o} + \frac{1}{s_i}$$

where f = focal length of mirror
      s_o = distance of object
         from mirror
      s_i = distance of image
         from mirror

Therefore, if two of the three distances f, s_o, and s_i are known, the third can be found.

A second formula shows the magnification of the mirror and how the height of the object and the height of the image depend on the object distance and the image distance:

$$M = \frac{h_i}{h_o} = \frac{-s_i}{s_o}$$

where M = magnification
     h_i = image height
     h_o = object height
     s_i = image distance
     s_o = object distance

In using *both* of the preceding formulas for concave and convex mirrors, remember that the distance to a virtual image is always negative; similarly, the focal length of a convex mirror is also negative. An inverted image has a negative magnification and an erect image has a positive magnification.

An object 10.0 cm in front of a convex mirror forms an image 5.00 cm behind the mirror. What is the focal length of the mirror?

EXAMPLE

Sketch:

Data:

$$s_o = 10.0 \text{ cm}$$

$$s_i = -5.00 \text{ cm}$$ *Note:* The image is virtual (appears behind the mirror) so s_i is given a (–) sign to show this. [Won't f also be (–)?]

$$f = ?$$

Basic Equation:

$$\frac{1}{f} = \frac{1}{s_o} + \frac{1}{s_i}$$

Working Equation: Same

Substitution:

$$\frac{1}{f} = \frac{1}{10.0 \text{ cm}} + \frac{1}{-5.00 \text{ cm}} = \frac{1}{10.0 \text{ cm}} - \frac{1}{5.00 \text{ cm}}$$

$$f = -10.0 \text{ cm}$$

The key entry sequence on a scientific calculator for this calculation is

10 $\boxed{x^{-1}}$ $\boxed{-}$ 5 $\boxed{x^{-1}}$ $\boxed{=}$ $\boxed{x^{-1}}$ $\boxed{=}$

$$\boxed{-10}$$

Remember that f and s_i may be negative only when forming virtual images and/or using convex mirrors.

PROBLEMS 21.5

Use the formulas $\dfrac{1}{f} = \dfrac{1}{s_o} + \dfrac{1}{s_i}$ and $M = \dfrac{h_i}{h_o} = \dfrac{-s_i}{s_o}$ for Problems 1–8.

1. Given $s_o = 1.65$ cm and $s_i = 6.00$ cm, find f.
2. Given $f = 15.0$ cm and $s_i = 3.00$ cm, find s_o.
3. Given $s_i = 14.5$ cm and $f = 10.0$ cm, find s_o.
4. Given $s_i = -10.0$ cm and $f = -5.00$ cm, find s_o.
5. Given $s_o = 7.35$ cm and $s_i = 17.0$ cm, find f.
6. Given $h_i = 2.75$ cm, $h_o = 4.50$ cm, and $s_i = 6.00$ cm, find s_o.
7. Given $h_o = 12.0$ cm, $s_i = 13.0$ cm, and $s_o = 25.0$ cm, find h_i.
8. Given $h_i = 3.50$ cm, $h_o = 2.50$ cm, and $s_i = 15.5$ cm, find s_o.
9. If an object is 2.50 m tall and 8.60 m from a large mirror with an image formed 3.75 m from the mirror, find the height of the image.
10. An object 30.0 cm tall is located 10.5 cm from a concave mirror with focal length 16.0 cm. (a) Where is the image located? (b) How high is it?
11. An object and its image in a concave mirror are the same height, yet inverted, when the object is 20.0 cm from the mirror. What is the focal length of the mirror?
12. An object 12.6 cm in front of a convex mirror forms an image 6.00 cm behind the mirror. What is the focal length of the mirror?

PROBLEM SOLVING

SKETCH

$\boxed{12 \text{ cm}^2} \; w$

4.0 cm

DATA

$A = 12 \text{ cm}^2, l = 4.0 \text{ cm}, w = ?$

BASIC EQUATION

$A = lw$

WORKING EQUATION

$w = \dfrac{A}{l}$

SUBSTITUTION

$w = \dfrac{12 \text{ cm}^2}{4.0 \text{ cm}} = 3.0 \text{ cm}$

13. What is the height of an image in a truck mirror when the object is 1.20 m tall and is standing 7.60 m from the mirror and the image is formed 1.50 m from the mirror?

14. A lift truck has a rearview mirror that is 0.76 m from the eye of its driver. How tall is the image of a 1.40-m-tall loaded pallet that is 6.20 m from the mirror?

15. Find the focal length of a convex mirror that forms an image 3.55 cm behind the mirror of an object 24.5 cm in front of the mirror.

16. Find the focal length of a mirror that forms an image 5.66 cm behind a mirror of an object 34.4 cm in front of the mirror.

17. Find the focal length of a mirror that forms an image 2.30 m behind a mirror of an object 6.50 m in front of the mirror.

18. An image of a statue appears to be 11.5 cm behind a convex mirror with focal length 13.5 cm. Find the distance from the statue to the mirror.

19. (a) What is the height of a figurine 7.33 cm in front of a concave mirror that produces an image −2.75 cm high? The image appears to be 5.03 cm in front of the mirror. (b) Find the focal length of the mirror. (c) What distance would an image appear to be from the mirror with double the focal length?

21.6   The Law of Refraction

Does light travel in a straight line? Most people's first answer would be, "Yes." But does it always? See Fig. 21.17. Why does a straw appear to bend at the surface when placed in water as in Fig. 21.17(b)? Place a straw in a glass of water and observe it from a variety of angles.

Figure 21.17   Refraction is the bending of light as it passes from one medium to another.

Courtesy of Dave King © Dorling Kindersley, Courtesy of The Science Museum, London

(a)   (b)

The answers to our questions may be found in the study of another property of light—refraction. **Refraction** is the bending of light as it passes at an angle from one medium to another of different optical density. **Optical density** is a property of a transparent material that is a measure of the speed of light through the given material. For example, water is optically denser than air and the speed of light in water is less than the speed of light in air. This change of speed of light when it passes from one medium to another produces refraction. The wave shown in Fig. 21.18 illustrates how this occurs. Note that when passing from one medium to another perpendicular to the

Figure 21.18 The speed of light is different in different media.

Figure 21.19 The wave bends when all parts of the wave do not strike the glass at the same time. It also bends when leaving the glass.

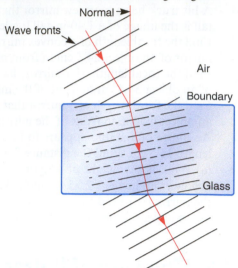

surface (called the *interface*), the wave is not bent, although the speed of the wave is slowed. When the light emerges back into the original medium, the wave returns to its original speed.

When the wave passes through at an angle, the entire wave front does not all strike the surface at the same time. The first part of the wave to strike the glass is slowed before the part striking later—thus the bending of the wave (Fig. 21.19). Draw your own diagram to show whether a fish in a pond, as viewed from the bank, is actually nearer the surface or the bottom of the pond than it appears to be (see Fig. 20.3).

Another way to understand refraction is to compare it to a child's wagon rolling along a sloping sidewalk. The front wheel rolls off the edge and goes into the grass. The interaction of the wheel with the rougher surface, the grass, causes the wagon to change direction and to be pulled toward the grass. The wheel that hits the grass slows down first while the opposite wheel is still rolling on the sidewalk. The wagon pivots, and its path is bent toward the grass.

A mirage is a good example of refraction. It is not a figment of a person's imagination. It is actually formed by light and can be photographed (Fig. 21.20). Light travels faster through the very hot and less dense air near a warm surface such as the ground on a hot day than through the cooler air above it. As a result, the image appears to the observer to shimmer as if it had been reflected from the surface of water. It is not reflected, however, but refracted, forming an image.

As we might expect, since the speed of light increases when a light wave leaves a denser medium to enter the air, bending also occurs. In this case, however, instead of the light bending toward the perpendicular or normal, the light is bent away from the normal when passing from the denser medium to the less dense one (Fig. 21.19). This illustrates the following **law of refraction**:

LAW OF REFRACTION

When a beam of light passes at an angle from a medium of lower optical density to a denser medium, the light is bent *toward* the normal. When a beam of light passes at an angle from a medium of greater optical density to one less dense, the light is bent *away from* the normal.

Figure 21.20 A mirage is actually formed by light and can be photographed.

Courtesy of gettyimages © Jeremy Woodhouse

Figure 21.21 The angles of incidence and refraction are measured from the normal.

Note: The angles of incidence and refraction are measured from the *normal* as in Fig. 21.21.

Willebord Snell developed a formula to determine the relative optical density of a material, called the **index of refraction**. This index is a constant for a particular material and is independent of the angle at which the light strikes. **Snell's law** may be expressed in two ways:

1. The index of refraction, *n*, equals the sine of the angle of incidence divided by the sine of the angle of refraction, where the incident ray travels through a vacuum or air:

$$n = \frac{\sin i}{\sin r}$$

2. The index of refraction, *n*, for a given substance is the ratio of the speed of light in a vacuum (nearly the same as in air) to the speed of light in the substance:

$$n = \frac{\text{speed of light in vacuum}}{\text{speed of light in substance}}$$

Willebord Snell (1580–1626), mathematician, was born in The Netherlands. He discovered the law of refraction known as Snell's law and extensively developed the use of triangulation in surveying.

The angle of incidence of light passing from air to water is 61.0°. The angle of refraction is 41.0°. What is the index of refraction of the water?

EXAMPLE 1

Data:

$$i = 61.0°$$
$$r = 41.0°$$
$$n = ?$$

Basic Equation:

$$n = \frac{\sin i}{\sin r}$$

Working Equation: Same

Substitution:

$$n = \frac{\sin 61.0°}{\sin 41.0°}$$

$$= 1.33$$

. . . .

EXAMPLE 2

The index of refraction of water is 1.33. What is the speed of light in water?

Data:

$$n = 1.33$$
$$c = 3.00 \times 10^8 \text{ m/s}$$
$$v_{\text{water}} = ?$$

Basic Equation:

$$n = \frac{\text{speed of light in vacuum}}{\text{speed of light in substance}}$$

Working Equation:

$$\text{speed of light in water} = \frac{\text{speed of light in vacuum}}{n}$$

Substitution:

$$\text{speed of light in water} = \frac{3.00 \times 10^8 \text{ m/s}}{1.33}$$

$$= 2.26 \times 10^8 \text{ m/s}$$

. . . .

The index of refraction for some common substances is given in Table 21.1.

Table 21.1   Indices of Refraction for Various Substances

Substance	Index of Refraction
Air, dry (STP)	1.00029
Alcohol, ethyl	1.360
Benzene	1.501
Carbon dioxide (STP)	1.00045
Carbon disulfide	1.625
Carbon tetrachloride	1.459
Diamond	2.417
Glass, crown flint	1.575
Lucite	1.50
Quartz, fused	1.45845
Water, distilled	1.333
Water vapor (STP)	1.00025
Zircon	1.92

Note: STP indicates standard temperature (0°C) and pressure (101.32 kPa).

TRY THIS ACTIVITY

Twinkling Stars

Look at an object at the bottom of a pool, tank, or shallow area of clear water. If the water is still, the object should appear quite clear; however, if there are ripples in the water, the light from the object becomes irregularly refracted by the shape of the ripples.

This is similar to the explanation of why stars appear to twinkle. Stars really do not twinkle, but they appear to do so when different densities of air get in the way of the light as it comes toward an observer's eye. Instead of a constant beam of light striking the eye, the irregularities in the atmospheric density cause the light to be refracted from time to time, resulting in various intensities of light striking the eye.

TRY THIS ACTIVITY

Apparent Depth

In the previous chapter, a suggested activity involved looking at the apparent location of an object from both above and below the water line (Fig. 21.22). The difference in the apparent location was due to refraction. In this activity, pay particular attention to the location and magnification of the object.

The next time you are at an aquarium, look at something that is protruding out of the water and look at the object from a location where you can see it from above and below the water line. Since water refracts light when the light strikes it at an angle, the submerged portion of the object will appear to be in a different location from each of the two perspectives. Enjoy the wildlife at the aquarium, but also enjoy the physics of optics!

Figure 21.22

(a) A sign and its base as seen from above the water line

(b) The sign as seen from above the water line and the base as seen from below the water line. Notice the apparent location and size of the sign's base.

21.7 Total Internal Reflection

If the angle of refraction is 90° or greater, a beam of light does not leave the medium but is reflected inside it. This is called **total internal reflection** (Fig. 21.23). Total internal reflection occurs when the angle of incidence is greater than the critical angle.

Figure 21.23 (a) A point light source S is shown in glass. For all angles of incidence less than the critical angle i_c, the light ray is partially reflected but mostly refracted and leaves the glass as shown at b, c, and d. When the angle of incidence equals the critical angle i_c, the refracted ray points along the glass–air surface as shown at e. For all angles of incidence greater than the critical angle i_c, total internal reflection of the light ray occurs as shown at f, g, and h. (b) Here, the first two beams of light have angles of incidence less than the critical angle [similar to b, c, and d in part (a)] and are mostly refracted upward into the second medium. The third beam of light has an angle of incidence greater than the critical angle [similar to f, g, and h in part (a)] and is totally reflected downward and stays within the first medium.

(a)

Courtesy of Ken Kay/Fundamental Photographs

(b)

The **critical angle** is the smallest angle of incidence at which all light striking the surface is totally internally reflected. It may be expressed by the formula

$$\sin i_c = \frac{1}{n}$$

where $i_c =$ critical angle of incidence
$n =$ index of refraction of denser medium

Note: The incident ray is traveling through a vacuum or air.

In Fig. 21.24, the angle of incidence from our perspective looking through the side of the tank is greater than the critical angle. Note that the fish in the water are reflected above in the surface boundary.

Figure 21.24

TRY THIS ACTIVITY

Water Fiber Optics

A 2-L plastic bottle and a pen laser are needed for this activity. Make a small, pencil-sized hole in the lower wall of an empty plastic bottle. Placing a finger over the hole, fill the bottle with water. Shine the laser pen through the bottle so the laser light emerges from the hole where your finger is located. Remove your finger from the hole. Have someone stand in front of the hole and observe what is happening to the laser light as the water and the light emerge from the hole. Explain, using internal reflection principles, why the laser light behaves as it does when it emerges from the bottle.

What is the critical angle of incidence for water that has an index of refraction of 1.33?

Data:

$$n = 1.33$$
$$i_c = ?$$

Basic Equation:

$$\sin i_c = \frac{1}{n}$$

Working Equation: Same

Substitution:

$$\sin i_c = \frac{1}{1.33} = 0.752$$
$$i_c = 49°$$

Where there is total internal reflection, no light enters the air; it is totally reflected within the glass (see Figs. 21.25 and 21.28). The property of having a very small critical angle gives a diamond its brilliance due to multiple internal reflections occurring before the light passes out through the top of the diamond. An example of the practical application of this principle is fiber optics (Fig. 21.25). Light may be transferred inside flexible glass or plastic fibers, which are transparent but keep nearly all the light inside because the light is reflected along the inside surface of the fibers.

Figure 21.25 (a) Light travels inside an optical or glass fiber like a light pipe in which the light is always incident at an angle greater than the critical angle. Thus, no light escapes the optical fiber by refraction. Fiber optics is essential to light-wave communications systems. (b) Total internal reflection within the tiny fibers of this fiber optic cable makes it possible to transmit light along complex paths with minimal loss.

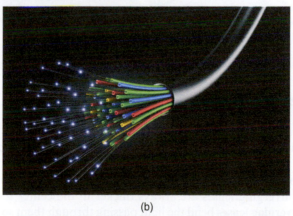

Glass fiber

(a)

(b)

Courtesy of fotolia © zentilia

PHYSICS CONNECTIONS

Fiber Optic Cables

Most transmission of information travels as electric impulses through electric and telephone lines and fiber optic cables. Electric signals travel relatively slowly, cause wires to heat up, and need transformers to boost the voltage of signals traveling over long distances. Electric signals and wires are being replaced with light signals traveling through flexible, low-cost strands of glass. Because light travels through glass optical fibers, there is no electrical resistance to weaken the signal, and the signal travels at the speed of light, which is much faster than the speed of conventional electric signals. Such advances in fiber optics communications are revolutionizing the way we communicate.

Light traveling in the same medium travels in a straight line, whereas fiber optic cables can transmit a signal while twisting and turning because of total internal reflection. The angle at which the light strikes the cladding of the fiber is always greater than the critical angle of the cladding and the core. The low critical angle allows the light to continually reflect and travel great distances without needing to be reamplified. In order for the cable to maintain a low critical angle, the glass must contain no imperfections or bubbles that would cause the light to be directed out or backward through the cable (see Fig. 21.26).

Fiber optic cables are used in telecommunications, computer networks, and medicine. A few strands of glass fiber can carry thousands of separate digital telephone conversations by slightly altering the frequency of the light for each phone conversation. A digital signal transmitted at one frequency cannot be confused by a signal carried at another frequency. Many computer networks and internal components in computers use fiber optic cables to carry data. By eliminating electric wiring, the fiber optic cable helps to reduce the temperature inside computers and servers. Finally, physicians use fiber optic bundles to perform minimally invasive procedures. A tool called an endoscope, composed of a bundle of fiber optic cables, transmits light into a patient's body while another bundle of fibers on the endoscope functions as a digital camera. The camera picks up the image and sends it back through the fiber optic cable to a monitor in the operating room.

Figure 21.26   (a) The red laser light entering the fiber optic cable is totally internally reflected, which results in the light emerging at the end of the cable. (b) An endoscope is a bundle of fiber optic cables used in many minimally invasive surgeries.

Courtesy of Helen King/Corbis

(a)   (b)

21.8   Types of Lenses

Many technical applications, ranging from apparatus to test the nature of liquids to microscopes to eyeglasses, use the principles of refraction. We now consider the use of refraction in applications using lenses. Lenses may be converging or diverging. **Converging lenses** bend the light passing through them to some point beyond the lens. Converging lenses are always thicker in the center than on the edges [Fig. 21.27(a)]. **Diverging lenses** bend the light passing through them so as to spread the light. Diverging lenses are thicker on the edges than at the center [Fig. 21.27(b)].

Figure 21.27

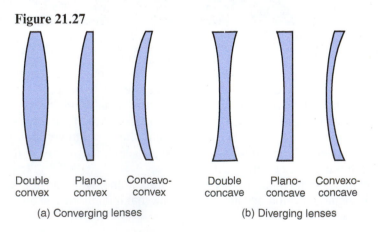

Double convex   Plano-convex   Concavo-convex   Double concave   Plano-concave   Convexo-concave

(a) Converging lenses   (b) Diverging lenses

Understanding how light is bent in lenses may be made easier by observing the bending of light passing through a prism (Fig. 21.28). Recall and apply the law of refraction, noting that light is bent toward the normal when passing into the glass and away from the normal when passing from the glass back to the air.

Figure 21.28 Bending of light passing through a triangular prism. Note that as light passes from air to glass, it is refracted toward the normal. As light passes from glass to air, it is refracted away from the normal. Note that the blue beam is totally internally reflected within the prism before being refracted as it passes from glass to air.

Courtesy of Richard Megna/Fundamental Photographs

21.9 Images Formed by Converging Lenses

As with mirrors, we use light-ray diagrams to show how light can be bent with lenses. Every lens has a focal length—that distance from the lens center to the point where parallel beams directed through the lens come together if converging or *appear* to come together if diverging (Fig. 21.29).

Figure 21.29

Courtesy of Richard Megna/Fundamental Photographs

Courtesy of Richard Megna/Fundamental Photographs

Optical center
Principal axis
F
Focus
Focal length
(a) Converging lens

Virtual focus
Optical center
F'
Focal length
(b) Diverging lens

The location of the focus depends upon the curvature of the lens and the index of refraction of the lens material. Rays of light passing through the optical center of the lens are not striking the lens at an angle, and so they pass straight through the lens (Fig. 21.30).

Now apply these principles to the diagrams shown in Fig. 21.31 of other images formed by converging lenses, depending on the object location, and decide whether the image is real or virtual or no image formed at all; and whether the image is erect or inverted; larger, smaller, or the same size; and where it is located.

Figure 21.30   Lens diagram for the formation of an image using a convex lens

Figure 21.31   Images formed by converging lenses

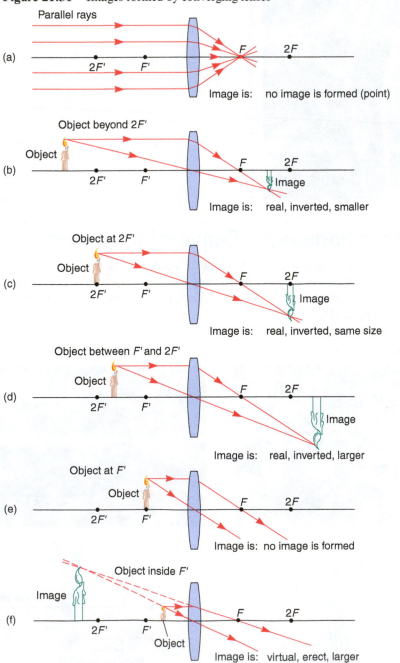

LASIK Surgery

Laser-Assisted Stromal In-situ Keratomileusis (LASIK) is a common procedure that alters the shape of the cornea. By altering the curvature of the cornea, the lens of the eye, surgeons make light refract correctly onto the retina. This procedure is meant to correct the vision of a patient, eliminating the need for glasses or contact lenses.

The eye must be mechanically immobilized for the procedure to be conducted. An eye tracking system follows the position of the eye throughout the procedure with tremendous precision. When immobilized, the outer layer of the cornea is cut to create a flap that exposes the middle section of the cornea, known as the stroma. An ultraviolet, 193-nanometer wavelength excimer laser then vaporizes up to a 10-micrometer-thick layer of the stroma. The pulse of laser light lasts between 10 and 20 nanoseconds to burn the corneal tissue using less than 1 millijoule (mJ) of energy. It generates little heat, therefore defining the vaporization photo ablation area to a narrow diameter of only 0.25 microns—half the diameter of a typical human hair! The laser is moved accordingly to create a corneal curvature that will reveal a clear image for the patient, eliminating the need for glasses or contact lenses (Fig. 21.32)!

Figure 21.32   Artist's rendition of a laser directed at the cornea

Courtesy of fotolia © hakandogu

21.10 Images Formed by Diverging Lenses

Virtual images are the only images produced by diverging lenses (Fig. 21.33). The same formulas that apply to mirrors also apply to converging and diverging lenses.

$$\frac{1}{f} = \frac{1}{s_o} + \frac{1}{s_i}$$

where f = focal length
s_o = object distance from lens center
s_i = image distance from lens center

Figure 21.33   Images formed by diverging lenses

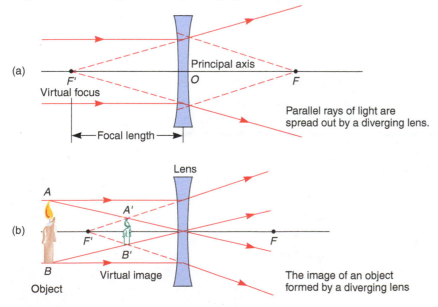

Therefore, if two of the three distances, f, s_o, and s_i are known, the third can be found. Also, the magnification of a lens is in the same form as for a mirror. An inverted image has a negative magnification and an erect image has a positive magnification.

$$M = \frac{h_i}{h_o} = \frac{-s_i}{s_o}$$

where M = magnification
h_i = height of image
h_o = height of object
s_i = image distance from lens center
s_o = object distance from lens center

Remember, when the image is virtual, s_i is negative (–), and for diverging lenses, both s_i and f are negative.

EXAMPLE

An object 3.00 cm tall is placed 24.0 cm from a converging lens. A real image is formed 8.00 cm from the lens.

(a)   What is the focal length of the lens?
(b)   What is the size of the image?

(a) Data:

$$s_o = 24.0 \text{ cm}$$
$$s_i = 8.00 \text{ cm}$$
$$f = ?$$

Basic Equation:

$$\frac{1}{f} = \frac{1}{s_o} + \frac{1}{s_i}$$

Working Equation: Same

Substitution:

$$\frac{1}{f} = \frac{1}{24.0 \text{ cm}} + \frac{1}{8.00 \text{ cm}}$$
$$f = 6.00 \text{ cm}$$

The key entry sequence on a scientific calculator for this calculation is

24 $\boxed{x^{-1}}$ $\boxed{+}$ 8 $\boxed{x^{-1}}$ $\boxed{=}$ $\boxed{x^{-1}}$ $\boxed{=}$

$$\boxed{6}$$

(b) Data:

$$s_o = 24.0 \text{ cm}$$
$$s_i = 8.00 \text{ cm}$$
$$h_o = 3.00 \text{ cm}$$
$$h_i = ?$$

Basic Equation:

$$\frac{h_i}{h_o} = \frac{-s_i}{s_o}$$

Working Equation:

$$h_i = \frac{-s_i h_o}{s_o}$$

Substitution:

$$h_i = \frac{-(8.00 \text{ cm})(3.00 \text{ cm})}{24.0 \text{ cm}}$$

$$= -1.00 \text{ cm}$$

· · · · · · · · · · · · · · · · · ·

PROBLEMS 21.10

1. Find the index of refraction of a medium for which the angle of incidence of a light beam is 31.5° and angle of refraction is 25.6°.
2. If the index of refraction of a medium is 2.40 and the angle of incidence is 14.6°, what is the angle of refraction?
3. If the index of refraction of a liquid is 1.50, find the speed of light in that liquid.
4. The angle of incidence of light passing from air to a liquid is 38.0°. The angle of refraction is 24.5°. What is the index of refraction of the liquid?
5. If the critical angle of a liquid is 42.4°, find the index of refraction for that liquid.
6. If the index of refraction of a substance is 2.45, find its critical angle of incidence.
7. A converging lens has a focal length of 15.0 cm. If it is placed 48.0 cm from an object, how far from the lens will the image be formed?
8. An object 2.50 cm tall is placed 20.0 cm from a converging lens. A real image is formed 9.00 cm from the lens. (a) What is the focal length of the lens? (b) What is the size of the image?
9. The focal length of a lens is 5.00 cm. How far from the lens must the object be to produce an image 1.50 cm from the lens?
10. If the distance from the lens in your eye to the retina is 19.0 mm, what is the focal length of the lens when reading a sign 40.0 cm from the lens?
11. An object 5.00 cm tall is placed 15.0 cm from a converging lens, and a real image is formed 7.50 cm from the lens. (a) What is the focal length of the lens? (b) What is the size of the image?
12. An object 4.50 cm tall is placed 18.0 cm from a converging lens with a focal length of 26.0 cm. (a) What is the location of the image? (b) What is its size?
13. What are the size and location of an image produced by a converging lens with a focal length of 19.5 cm of an object 5.76 cm from the lens and 1.45 cm high?
14. What are the size and location of an image produced by a convex lens with a focal length of 14.5 cm of an object 10.5 cm from the lens and 2.35 cm high?
15. What is the focal length of a convex lens that produces an inverted image twice as large as the object at a distance of 13.3 cm from the lens?

SKETCH

12 cm² | w

4.0 cm

DATA

$A = 12 \text{ cm}^2, l = 4.0 \text{ cm}, w = ?$

BASIC EQUATION

$A = lw$

WORKING EQUATION

$w = \dfrac{A}{l}$

SUBSTITUTION

$w = \dfrac{12 \text{ cm}^2}{4.0 \text{ cm}} = 3.0 \text{ cm}$

Glossary

Concave Mirror A mirror with a surface that curves away from an observer. (p. 578)

Converging Lens A lens that bends the light passing through it to some point beyond the lens. Converging lenses are thicker in the center. (p. 592)

Convex Mirror A mirror with a surface that curves inward toward an observer. (p. 578)

Critical Angle The smallest angle of incidence at which all light striking a surface is totally internally reflected. (p. 590)

Diffusion Scattering of light by an uneven surface. (p. 577)

Diverging Lens A lens that bends the light passing through it so as to spread the light. Diverging lenses are thicker at the edges than at the center. (p. 592)

First Law of Reflection The angle of incidence equals the angle of reflection. (p. 578)

Focal Length The distance between the principal focus of a mirror or lens and its vertex. (p. 580)

Index of Refraction A measure of the optical density of a material; equal to the ratio of the speed of light in a vacuum to the speed of light in the material. (p. 587)

Law of Refraction When a beam of light passes at an angle from a medium of lower optical density to a denser medium, the light is bent toward the normal. When a beam passes from a medium of greater optical density to one less dense, the light is bent away from the normal. (p. 586)

Opaque Absorbing or reflecting almost all light. (p. 577)

Optical Density A property of a transparent material that is a measure of the speed of light through the given material. (p. 585)

Plane Mirror A mirror with a flat surface. (p. 578)

Real Image An image formed by rays of light. (p. 578)

Reflection The turning or turning back of all or part of a beam of light as it strikes a surface. (p. 576)

Refraction The bending of light as it passes at an angle from one medium to another of different optical density. (p. 585)

Regular Reflection Reflection of light with very little scattering. (p. 577)

Second Law of Reflection The incident ray, the reflected ray, and the normal (perpendicular) to the reflecting surface all lie in the same plane. (p. 578)

Snell's Law The index of refraction equals the sine of the angle of incidence divided by the sine of the angle of refraction. (p. 587)

Total Internal Reflection A condition such that light striking a surface does not pass through the surface but is completely reflected inside it. (p. 589)

Translucent Allowing some but not all light to pass through. (p. 577)

Transparent Allowing almost all light to pass through so that objects or images can be seen clearly. (p. 577)

Virtual Image An image that only appears to the eye to be formed by rays of light. (p. 578)

Formulas

21.5 $\dfrac{1}{f} = \dfrac{1}{s_o} + \dfrac{1}{s_i}$

$M = \dfrac{h_i}{h_o} = \dfrac{-s_i}{s_o}$

21.6 $n = \dfrac{\sin i}{\sin r}$

$n = \dfrac{\text{speed of light in vacuum}}{\text{speed of light in substance}}$

21.7 $\sin i_c = \dfrac{1}{n}$

21.10 $\dfrac{1}{f} = \dfrac{1}{s_o} + \dfrac{1}{s_i}$

$M = \dfrac{h_i}{h_o} = \dfrac{-s_i}{s_o}$

Review Questions

1. Stained glass is an example of
    (a) a transparent material.
    (b) a translucent material.
    (c) an opaque material.
    (d) none of the above.
2. A virtual image may be
    (a) larger than the object.
    (b) smaller than the object.
    (c) erect.
    (d) all of the above.
    (e) none of the above.
3. A real image may be
    (a) erect.
    (b) shown on a screen.
    (c) formed by a plane mirror.
    (d) none of the above.
4. Explain the difference between diffusion and regular reflection.
5. In your own words, explain the first law of reflection.
6. In your own words, explain the second law of reflection.
7. Describe the type of images formed by plane mirrors.
8. Explain the difference between real and virtual images.
9. Explain the difference between a concave and a convex mirror.
10. Explain the effect of spherical aberration.
11. For a mirror of given focal length, how does the image distance change if the object distance is decreased?
12. For a given object distance from a mirror, how does the image distance change if the focal length is increased?
13. The index of refraction depends on
    (a) the focal length.
    (b) the speed of light.
    (c) the image distance.
    (d) none of the above.
14. Snell's law involves
    (a) the lens equation.
    (b) the index of refraction.
    (c) the focal length.
    (d) none of the above.
15. Explain the difference between converging and diverging lenses.
16. Give several examples of total internal reflection.
17. In your own words, explain the law of refraction.
18. How does the speed of light in a high-index-of-refraction material compare to the speed of light in a vacuum?
19. In your own words, explain why light waves are refracted at a boundary between two materials.
20. What types of images are formed by diverging lenses?
21. What types of images are formed by converging lenses?

22. How do water waves affect the escape of light from below the surface of the water?
23. Explain why a fish under water appears to be at a different depth below the surface than it actually is. Does it appear deeper or shallower?
24. Does light always travel in a straight line? Explain.
25. Explain how total internal reflection allows light in a glass fiber to be guided along the fiber.
26. Under what conditions will a converging lens form a virtual image?
27. Under what conditions will a converging lens form a real image that is the same size as the object?
28. Under what conditions will a diverging lens form a virtual image that is smaller than the object?

Review Problems

SKETCH

12 cm² w

4.0 cm

DATA
$A = 12 \text{ cm}^2, l = 4.0 \text{ cm}, w = ?$

BASIC EQUATION
$A = lw$

WORKING EQUATION
$w = \dfrac{A}{l}$

SUBSTITUTION
$w = \dfrac{12 \text{ cm}^2}{4.0 \text{ cm}} = 3.0 \text{ cm}$

1. Using $\dfrac{1}{f} = \dfrac{1}{s_o} + \dfrac{1}{s_i}$, $s_o = 3.50$ cm, and $s_i = 7.25$ cm, find f.

2. Using $\dfrac{1}{f} = \dfrac{1}{s_o} + \dfrac{1}{s_i}$, $s_o = 8.50$ cm, and $f = 25.0$ cm, find s_i.

3. Using $M = \dfrac{h_i}{h_o} = \dfrac{-s_i}{s_o}$, $h_o = 6.50$ cm, $s_i = 7.50$ cm, and $s_o = 14.0$ cm, find h_i.

4. If an object is 3.75 m tall and 7.35 m from a large mirror with an image formed 4.35 m from the mirror, what is the height of the image?

5. An object 43.0 cm tall is located 23.4 cm from a concave mirror with focal length 21.4 cm. (a) Where is the image located? (b) How high is the image?

6. An object and its image in a concave mirror are the same height, but inverted, when the object is 45.3 cm from the mirror. What is the focal length of the mirror?

7. The angle of incidence of light passing from air to a liquid is 41.0°. The angle of refraction is 29.0°. Find the index of refraction of the liquid.

8. If the index of refraction of a liquid is 1.44, find the speed of light in that liquid.

9. If the critical angle of a liquid is 45.6°, find the index of refraction for that liquid.

10. If the index of refraction of a substance is 1.50, find its critical angle of incidence.

11. A converging lens has a focal length of 12.0 cm. If it is placed 36.0 cm from an object, how far from the lens will the image be formed?

12. An object 4.50 cm tall is placed 20.0 cm from a converging lens. A real image is formed 12.0 cm from the lens. (a) What is the focal length of the lens? (b) What is the size of the image?

13. The focal length of a lens is 4.00 cm. How far from the lens must the object be to produce an image 7.20 cm from the lens?

14. What is the focal length of a convex lens that produces an image three times as large as the object at a distance of 25.0 cm from the lens?

15. What is the focal length of a mirror that forms an image 3.44 m behind a convex mirror of an object 5.33 m in front of the mirror?

16. What are the size and location of an image produced by a convex lens with a focal length of 21.0 cm of an object 11.5 cm from the lens and 3.25 cm high?

17. What is the speed of light passing through a diamond? See Table 21.1.

18. Find the critical angle of incidence for Lucite. See Table 21.1.

19. Find the focal length of a concave mirror with an object 39.3 cm in front of it that projects an image 17.8 cm in front of the mirror.
20. (a) Find the height and location of an image produced by a concave mirror with focal length 8.70 cm and an object that is 13.2 cm tall and 19.3 cm from the mirror. (b) Find the height of the image produced by the mirror if the object is twice as far from the mirror.

APPLIED CONCEPTS

1. Tamera uses a concave mirror when applying makeup. (a) The mirror has a radius of curvature of 38.0 cm. What is the focal length of the mirror? (b) Tamera's face is located 12.5 cm from the mirror. Where will her image appear? (c) Will the image be upright or inverted? (d) How long will her face appear to be in the mirror if her face is actually 25.0 cm long?

2. A convex security mirror has a radius of curvature of −1.50 m. (a) A shoplifter picks up a 0.255-m-high purse 11.5 m from the mirror. How far behind the mirror is the image of the purse? (b) Is the image real or virtual? (c) How high is the image of the purse? (d) Are convex security mirrors better for viewing detailed images or large fields of view?

3. A fish tank made of crown glass is full of fresh water. (a) What is the angle of refraction when the light travels from the air and enters the crown glass at an incident angle of 30.0°? (b) Determine the angle of refraction when the light leaves the crown glass and enters the water. (c) If the aquarium is full of saltwater, will the angle of refraction be greater or less than the angle of refraction for fresh water?

4. Diamonds are cut to take advantage of internal reflections of light. The angles of a diamond are cut so that all of the light that enters it reflects out the top. The light or brilliance that emerges from the top of a diamond is one of the qualities that determines its monetary value. (a) What is the critical angle for light traveling from a diamond to air? (b) Would diamond jewelers want to create angles smaller or larger than the critical angle? (c) Zircon is an inexpensive substitute for diamond. Zircon's index of refraction is 1.92. What is its critical angle? (d) If a zircon has the same cut as a diamond, which will have a greater brilliance by internally reflecting more light to the top of the stone?

5. A photographer uses a 60.0-mm lens. (a) How far away should the lens be from the film if the object is located 9.20 m from the lens? (b) What will be the height of the image if the object is 2.00 m tall?

CHAPTER 22

COLOR

Objectives

The major goals of this chapter are to enable you to:

1. Describe how the colors of the visible spectrum are formed through dispersion of light.
2. Describe color as a property of light and how it is related to its frequency or its wavelength.
3. Describe the result of mixing colors.
4. Describe diffraction as a property of light.
5. Describe interference as a property of light.
6. Describe polarization of light and some of its applications.

Understanding why a beam of sunlight forms a spectrum when directed through a glass prism, why the sky is blue, why sunsets are red, why clouds are white, why the ocean is blue, why we see rainbows, how colors of paint are mixed, how Polaroid sunglasses work, and how the LCD screen in a calculator works are basic questions many people commonly ask. Our examination of these basic questions forms the basis of the development and discussion of basic color properties.

22.1 The Color of Light

Let us observe a narrow beam of sunlight that is directed into and passes through a glass prism in a dark room as in Fig. 22.1. Note the band of colors with one color shade gradually blending into another. This band of colors is called the **visible spectrum**. This spreading of white light into the full spectrum is called **dispersion**. Dispersion was described by Newton, who observed six colors: red, orange, yellow, green, blue, and violet. Sunlight is an example of white light. The *color* of the light is related to its wavelength or its frequency. Red light has lower frequency and longer wavelength, whereas at the opposite end of the spectrum violet light has higher frequency and shorter wavelength. *Polychromatic light* is light consisting of several colors. *Monochromatic light* consists of only one color.

As you can see from the dispersion of light, the refraction of the red light is less than that of the others and the refraction of the violet light is the greatest (Fig. 22.2). The shorter wavelengths are refracted more than the longer wavelengths. Thus, **color** is a property of the light that reaches our eyes and is determined by its wavelength or its frequency. Light waves with wavelengths outside the visible spectrum as shown in Fig. 20.7 are not visible to the human eye. Light waves within the visible spectrum have wavelengths in the range of 4.0×10^{-7} m (for violet) to 7.5×10^{-7} m (for red). Some photographic films are sensitive to ultraviolet (UV) light and others, to infrared (IR) light.

Figure 22.1 A narrow beam of sunlight directed into a clear glass prism in a dark room is dispersed into the visible spectrum.

Courtesy of fotolia © Peter Hermes Furian

Figure 22.2 White light is dispersed by a prism into the visible spectrum on a white screen.

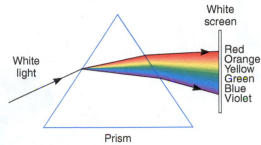

Most of the objects we see reflect rather than emit light. A shirt is called red if it reflects only red light and absorbs all other colors [Fig. 22.3(a)]. A shirt is called blue if it reflects only blue light and absorbs all other colors [Fig. 22.3(b)]. The color of an opaque object depends on the color of the light it reflects.

Figure 22.3 (a) A red shirt reflects only red light and absorbs light of all other colors. (b) A blue shirt reflects only blue light and absorbs light of all other colors.

Courtesy of fotolia © eurobanks

Courtesy of fotolia © pathdoc

(a) (b)

A white object reflects all the colors (light rays of various wavelengths) it receives. A black object absorbs all the colors it receives. A shirt that is blue in sunlight appears black in red light because there is no blue light present for it to reflect, and it absorbs all the other colors. Similarly, this is why a red shirt appears black in blue light.

Figure 22.4 Red glass transmits only red light and absorbs light of all other colors. Blue glass transmits only blue light and absorbs light of all other colors.

Courtesy of fotolia © Anton Petrus

Regular window glass transmits all colors and is often called colorless. Red glass absorbs all colors except red, which it transmits. Blue glass absorbs all colors except blue, which it transmits (Fig. 22.4). The color of a transparent object depends on the color of the light it transmits.

Mixing Colors of Light

Just as polychromatic light, such as sunlight (sometimes called white light), can be dispersed into its separate colors, the separate colors can be combined to form polychromatic light. White light can also be produced by projecting overlapping red, green, and blue light onto a white screen as shown in Fig. 22.5. For this reason, red, green, and blue are called the **primary colors**. This is an additive mixture of the three colors of light.

Figure 22.5 When the three primary colors overlap (light mixed by addition), white light is produced.

Courtesy of Richard Ward © Dorling Kindersley

Note in Fig. 22.5 the colors cyan, magenta, and yellow are produced when only two of the primary colors overlap; that is,

$$\text{blue} + \text{green} = \text{cyan}$$
$$\text{red} + \text{blue} = \text{magenta}$$
$$\text{red} + \text{green} = \text{yellow}$$

The color cyan is often called the opposite of red, magenta is the opposite of green, and yellow is the opposite of blue. Thus,

$$\text{cyan} + \text{red} = \text{white} = (\text{blue} + \text{green}) + \text{red}$$
$$\text{magenta} + \text{green} = \text{white} = (\text{red} + \text{blue}) + \text{green}$$
$$\text{yellow} + \text{blue} = \text{white} = (\text{red} + \text{green}) + \text{blue}$$

When any two colors are combined to form white, they are called **complementary colors**. Thus, cyan and red, magenta and green, and yellow and blue are complementary colors.

Colors of light are commonly mixed on theatrical stages to produce many different color effects (Fig. 22.6).

Figure 22.6   Bright spotlights at this theatre show multiple colors using overlapping colored lights.

Courtesy of fotolia © imaginando

Figure 22.7   Primary pigments. The results of the subtractive combination of cyan, magenta, and yellow filters are shown. Note that the combination of any two primary pigments gives the complement of the third by subtraction, namely, red, green, and blue. The subtractive combination of all three primary pigments transmits no light.

Mixing Pigments

Mixing pigments in paints and dyes is completely different from mixing lighting. Artists and painters know that mixing red, green, and blue paint results in a murky dark brown—not white! Pigments are actually small particles that absorb, rather than reflect, given colors. When paint pigments are mixed, each one subtracts, or absorbs, certain colors from white light with the resulting reflected color depending on the light waves *not* absorbed. The **primary pigments** are the complements of the three primary colors, namely, cyan (the complement of red), magenta (the complement of green), and yellow (the complement of blue) (see Fig. 22.7). Comparing additive mixtures and the corresponding subtractive mixtures, we have the following:

Additive Mixture	*Corresponding Subtractive Mixture*
cyan + red = white	red = white − cyan
magenta + green = white	green = white − magenta
yellow + blue = white	blue = white − yellow

That is, pigments that produce red absorb its complement, cyan, and reflect red. In other words, taking cyan away from white produces red. Pigments that produce green absorb its complement, magenta, and reflect green. Or taking magenta away from white produces green. Pigments that produce blue absorb its complement, yellow, and reflect blue. Or taking yellow away from white produces blue.

When the three primary pigments are mixed in appropriate proportions, a large range of colors can be produced.

Color Printing

All color photographs printed in books, magazines, and newspapers use only four colors of ink: cyan, magenta, and yellow for the colors and black for the shadow areas and definition. Four negatives are made using different filters so that each negative has the given amount of its color corresponding to the original color photograph. A red filter is used for making the cyan-printer negative, a green filter is used for making the magenta-printer negative, and a blue filter is used for making the yellow-printer negative. Each filter makes the color to be printed by its corresponding plate photograph as black and the other two colors photograph as white. The negative for the black plate is made by partial exposures through each of the three color filters so that black and shadows will print where all three colors are present.

A metal printing plate is made from each of these four negatives. Each metal printing plate is treated to hold the printer's ink in areas that have been exposed to light. Each of the four plates is printed in the same position, one after the other, with great precision on white paper. (Recall seeing a color photograph in a newspaper or magazine where either the plates or the paper was not within the proper precision and the colors were out of synchronization.) The lightness and the darkness of each color are controlled by the size of the very small printing dots. Use a magnifying glass to examine the very small printing dots in any color photograph in this text, a magazine, or a newspaper. Note that not only do the dots provide the basic colors, but also the overlapping dots of these basic colors give the appearance of many other colors. Figure 22.8 illustrates how only four colors produce color photographs and illustrations with many vibrant colors.

Figure 22.8 Four-color printing process. The colors (a) cyan, (b) magenta, and (c) yellow printed separately. (d) Only these three colors combined. (e) The fourth (black) color. (f) The combined, finished result. Note the importance of black for the background, shadows, and definition.

(a)   (b)   (c)

(d)   (e)   (f)

Courtesy of fotolia © tacna

Inkjet printers deposit simultaneous combinations of cyan, magenta, yellow, and black inks on white paper, which results in a color photograph or illustration using the same four-color principle.

Color digital cameras use filters and record as pixels the three primary colors red, blue, and green of different intensities to create the variety of colors. A pixel is a tiny unit or dot that makes up the image on a television screen, computer monitor, or similar display. The greater the number of pixels per inch, the greater is the resolution, that is, the better the clarity of the detail that can be distinguished in the image on the screen.

Why the Sky Is Blue

When sunlight passes through the atmosphere, the light is *scattered* in all directions by air molecules and other small particles in the atmosphere. The amount of scattering depends on the wavelength of the light. Air molecules and particles much smaller than the wavelengths of the various colors of light are less of an obstruction to long wavelengths than to short wavelengths. Thus, the longer wavelengths of red, orange, yellow, and green light are scattered much less than the shorter wavelengths of blue and violet light. Since our eyes are not very sensitive to violet light, the scattered blue light in the lower atmosphere is the dominant color (Fig. 22.9).

Figure 22.9   The sky is blue because sunlight is scattered as it passes through the atmosphere. We see the most scattered, shorter wavelengths of blue light, which makes the sky appear blue.

A deep blue sky is most likely on a clear, low-humidity day. When the atmosphere contains water molecules, dust, and other particles larger than the oxygen and nitrogen molecules, the longer wavelength light (red, orange, yellow) is also significantly scattered, which turns the sky into a gray-white haze. This haze is caused by very small particles that scatter and absorb light before it reaches our eyes. As the number of particles increase, more light is absorbed and scattered, which results in less clarity, color, and visual range. Factories and car and truck engines emit pollution in the form of many billions of particles per second. Although these particles are very small, other particles adhere to them and scatter the longer-wavelength light, which produces gray-white or brown hazes (Fig. 22.10). A rainstorm often cleanses the lower atmosphere of the polluted particles so that we can enjoy the freshened air and see a clear blue sky.

Why Sunsets Are Red

At sunset (and sunrise), sunlight travels through significantly more of the atmosphere than at other times of the day (Fig. 22.11). This results in (1) even more scattering of the shorter wavelengths of violet, blue, and green, which are taken out before reaching us, and (2) some orange but primarily red light—the light of longest wavelength and least scattered—most readily passing through the atmosphere to reach our eyes for a red sunrise or sunset (Fig. 22.12).

Figure 22.10   (a) Small particles in the air due to pollution scatter longer-wavelength sunlight that produces a gray-white haze. Here a smog alert day in the Los Angeles Basin near Hollywood at sunrise is triggered as the downtown freeways generate hazardous air. (b) Heavy air pollution particles actually absorb and block out more of the sunlight to produce a brown haze. Here clouds of pollution belch smoke from factory chimneys in one of the worst-polluted cities when this photo was taken. The smoke and exhaust from high-sulfur coal block out the generally polluted sky.

Courtesy of fotolia © trekandshoot

(a)

Courtesy of Dick Durrance II/National Geographic/Getty Images

(b)

Figure 22.11   A ray of sunlight must travel through more of the atmosphere to a given point on the earth's surface at sunset than at any other time during the day. At point *A*, the shorter wavelengths of violet, blue, and some green have been scattered so that the light that reaches us tends to be the longer wavelengths of some orange and mostly red. At point *D*, a ray of sunlight travels most directly through the least amount of atmosphere with the least amount of scattering.

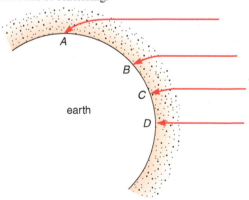

Figure 22.12   A sunrise over the Washington, DC Jefferson Memorial displays a mostly red and orange sky.

Why Clouds and Snow Are White

Clouds and snow consist of many different-sized water droplets. Small droplets tend to scatter the short-wavelength (blue) light, whereas the medium droplets tend to scatter the medium-wavelength (green) light, and the large droplets tend to scatter the long-wavelength (red) light. The overall result is that all of the colors are scattered and combined so that we see a general white reflection (Fig. 22.13).

Dark clouds are mostly composed of large droplets, which absorb much of the light passing through them. As the size of these droplets increases even more, they fall as rain.

Figure 22.13   Clouds consist of all sizes of water droplets. Small droplets scatter the short-wavelength (blue) light, the medium-size droplets scatter the medium-wavelength (green) light, and the large droplets scatter the long-wavelength (red) light. The overall result is that all of the colors are scattered and we see a general white reflection.

Figure 22.14   Water reflects the blue color of the sky, reflects and scatters the short-wavelength (blue and green) light, and absorbs the long-wavelength (orange and red) light. This results in blue to blue-green water.

Courtesy of fotolia © sattriani

Why the Ocean Is Blue

The ocean and most bodies of water appear blue because (1) the water surface acts like a mirror and reflects the blue color of the sky and (2) the water tends to reflect and scatter the short-wavelength light (yellow, green, blue, and violet) and absorb the long-wavelength light (orange and red) as well as infrared, which helps increase its temperature (Fig. 22.14). From Table 14.2, we see that the thermal conductivity of water is relatively low, which is why the water near the surface in large bodies of water is much warmer in the summer, due to the infrared absorption of heat, and much colder in the winter than the deeper water.

The spray above water waves breaking up is white because, like clouds, it contains many small water droplets, which scatter all colors of sunlight. Bodies of water that take on other colors usually contain other particles, such as mud, minerals, or pollution.

Why We See Rainbows

A **rainbow** is a spectrum of light formed when sunlight strikes raindrops, refracts into them, reflects within them, and then refracts out of them. Figure 22.15 shows the path

Figure 22.15

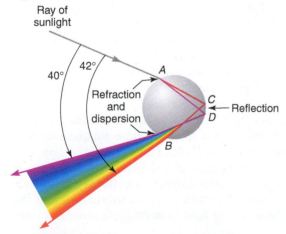

(a) The path of a ray of sunlight being refracted and dispersed, reflected, and then refracted and dispersed again as it passes through one raindrop of a rainbow. The angle between the ray entering the raindrop and the ray to the person's eyes must be 42° for the red color and 40° for violet as shown.

(b) This illustration shows a ray of sunlight being refracted and reflected as it enters and passes through a raindrop.

of a ray of sunlight in passing through a single raindrop. As the ray enters at point A, it is refracted and dispersed with the red end of the spectrum to point C and the violet end of the spectrum to point D, where the spectrum is then reflected to the raindrop surface near B, where the spectrum is again refracted upon leaving the water drop. For a person to see a rainbow, (1) the person must be between the sun and the raindrops and (2) the angle between the sun to the raindrops and back to the person's eyes must be between 40° and 42°. A rainbow is shown in Fig. 22.16.

Figure 22.16(a)   A rainbow

Courtesy of fotolia© makam1969

(a)

Figure 22.16(b)   A rainbow can form anywhere there are water molecules and the angle between the sun and the viewer is between 40° and 42°.

(b)

TRY THIS ACTIVITY

Making Rainbows

Rainbows don't just occur before or after a rain shower. With a sunny day and a garden hose, anyone can create a rainbow. Figure 22.15 refers to the formation of a rainbow as light enters a water droplet. Use it as a guide to make your own rainbow. Set a garden hose nozzle to a mist setting and position yourself so the sun is behind you and the mist is released up and in front of you. Adjust the position of the mist and the sun until you create your own section of a rainbow!

22.2 Diffraction of Light

As we saw in Section 16.1, **diffraction** is a property of a wave that describes its ability to bend around obstacles in its path. We used water waves to illustrate this property and noted that diffraction is commonly observed only when the obstacle or opening is nearly the same size as the wavelength. The diffraction of sound waves at doorways and other openings makes it possible to "hear around corners."

When a narrow beam of light passes by a sharp edge, around a fine wire, or through a narrow slit or a pinhole, the beam tends to spread or flare out and a distinct shadow is formed (Fig. 22.17). This distinct shadow is called a *diffraction pattern*. For best diffraction effects, monochromatic light or light of a single wavelength from a point source should be used. We often do not notice diffraction patterns because our most common sources of light are not monochromatic or point sources, so that the light coming from different points of the source or light from multiple sources washes out any diffraction pattern.

Huygens's principle (Section 20.1) also explains what happens when a wave front hits an obstacle and is partially obstructed, such as by a sharp edge, around a fine wire, or through a narrow opening. It shows that waves bend, or diffract, behind the obstacle.

Figure 22.17   Diffraction patterns resulting from a point source precisely centered behind (a) a dime, (b) a razor blade, and (c) a single slit

(a)

(b)

(c)

Since diffraction only occurs for waves and not for particles, it verifies the wave nature of light. Reminder: Light is an electromagnetic wave that exhibits both wave- and particle-like behavior, as discussed in Chapter 20.

22.3 Interference

We saw the results of the interference of water waves in a ripple tank in Fig. 16.9 in Section 16.1. We saw in Section 16.3 that when sound waves interfere with each other, beats are produced. Now let's study the interference of light waves. In 1801, **Thomas Young** first performed his famous *double-slit experiment*. Light from a single source (Young used the sun) falls on a screen that contains two closely spaced narrow slits, S_1 and S_2, as shown in a schematic diagram of his experiment in Fig. 22.18. If light consisted of particles, two bright lines would be expected to appear on a viewing screen placed behind the screen with the two slits. However, Young found not two, but a series of bright lines. Young explained his result as a *wave-interference* phenomenon.

Young's double-slit experiment was conducted using sunlight as shown in Fig. 22.19. The narrow slits, S_1 and S_2, are about 1 mm apart on a black piece of paper. A series of dark and bright narrow bands is seen on the viewing screen similar

Thomas Young (1773–1829),

English physicist and physician, is often acknowledged for his famous double-slit experiment and his work on light and interference, which established the wave theory of light. He also contributed to better understanding in the areas of work and energy, the elastic properties of materials, fluids, hemodynamics (the study of the forces involved in the circulation of blood), color, and egyptology, in which he provided the key to the deciphering of Egyptian hieroglyphics.

Figure 22.19   The light wave fronts in Young's experiment

Figure 22.18   Basic schematic of Young's double-slit experiment. Young found not two, but a series of bright lines behind the screen with the two slits.

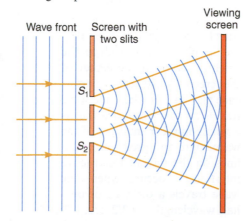

Figure 22.20   An interference pattern produced by a double-slit experiment with red laser light

Courtesy of Pearson Education

to the one in Fig. 22.20. We have used red laser light to produce a dramatic series of red and dark bands to illustrate the constructive and destructive interference patterns.

As we saw in Section 16.1, when two waves of a similar type pass through the same medium, a new wave is created by the *superposition of waves*. This new wave is the algebraic sum of the separate displacements of the individual waves as in Fig. 16.6. *Constructive interference* occurs when the waves add together to form a larger displace-ment [more intense light (red bands) in Fig. 22.20]. *Destructive interference* occurs when the waves add together to form a smaller displacement [less intense light (dark bands) in Fig. 22.20]. The series of bright and dark bands in Fig. 22.20 results from the different wave paths. Bright areas occur when the waves from both slits arrive in phase, whereas dark areas occur when the overlapping waves arrive out of phase (see Fig. 22.21). If the waves from the two slits travel the same distance, they are in phase and produce the brightest color at the center of the screen. Constructive interfer-ence also occurs when one wave travels an extra distance that is a whole-number multiple of the wavelength and produces bright lines on the screen. Destructive interference occurs when one wave travels a distance of one-half wavelength (or 3/2, 5/2, 7/2,

Figure 22.21   The bright and dark bands of light on the viewing screen result from the constructive and destructive interference, respectively, of the light waves after they pass through the double slits.

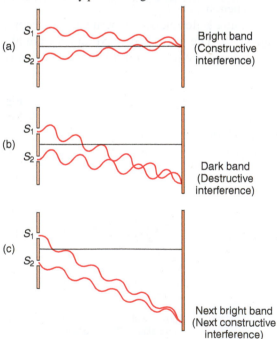

etc.) more than the other and produces dark lines on the screen. In general, constructive interference occurs for a given wavelength of light when

$$d \sin \theta = n\lambda$$

where d = distance between slits
θ = angle the light rays make with the horizontal
n = 1, 2, 3, . . . (called the *order* of the interference fringes)
λ = given wavelength of light

Destructive interference occurs for a given wavelength of light when

$$d \sin \theta = (n + \tfrac{1}{2}\lambda)$$

The first-order fringes ($n = 1$) occur on either side of the central bright spot, the second-order fringes occur next, and so forth. The intensity is greatest for $n = 1$ and decreases for higher orders.

22.4 Polarization of Light

Light waves are usually emitted in all directions with many orientations. Some are vertical, some are horizontal, and some are at other angles. An ordinary incandescent light bulb and the sun emit light waves in multiple orientations. **Polarized light** has its waves restricted to a single plane that is perpendicular to the direction of the wave motion. Ordinary light may become polarized using a Polaroid filter (Polaroid materials were invented by Edward Land in 1929). Figure 22.22 shows a simple model to illustrate polarized light. Figure 22.22(a) shows a vibrating rope attached to a wall through two vertically aligned pieces of picket fence. Note that only a vertical transverse wave can pass through the vertical slats of the first fence and then continue

Figure 22.22   A uniform transverse wave generated by a rope (a) passes through uniformly when the filters are aligned but (b) cannot pass through the second filter when they are not aligned.

(a)

(b)

Figure 22.23   Light waves generated by a candle in a dark room (a) pass through both filters as polarized light when the filters are aligned but (b) cannot pass through the second filter when they are not aligned.

(a)

(b)

through the vertical slats of the second. In Fig. 22.22(b) note that the vertical transverse wave that passes through the vertical slats of the first fence is *stopped* by the horizontal slats of the second fence and cannot pass through.

Similarly, if you look at a candle in an otherwise dark room through two vertically aligned Polaroid filters, you will see the polarized light from the candle as in Fig. 22.23(a). The two filters will allow only the vertically aligned light waves to pass through from the candle to your eyes. In Fig. 22.23(b), the second filter is horizontally aligned. Note that the vertically aligned light waves now pass through only the first filter and not through the second filter, so that you cannot see the candle. In this arrangement, the first filter is called the *polarizer* and the second is called the *analyzer*. Polaroid sunglasses block out horizontally vibrating light waves and roughly 50% of the light. When two pairs of sunglasses overlap at right angles, no light waves pass through (similar to what is demonstrated in Fig. 22.24). Whenever unwanted glare occurs, polarized sunglasses are very useful, especially for driving, sailing, or walking on a beach.

Figure 22.24 shows a red flower behind two Polaroid filters. In Fig. 22.24(a) the filters are aligned vertically and one can see the partially shaded red flower through the filters. In Fig. 22.24(b) the second filter is rotated at a 45° angle. Note that one can still see the red flower, but the red flower is more shaded. In Fig. 22.24(c) the second filter is rotated so that the filters are at right angles and no light passes through.

Much of the light that is reflected from nonmetallic surfaces is polarized. Good examples are the glare off smooth water or the glass of a windshield. The reflected light waves from the sun tend to be the horizontal-component waves parallel to the reflecting surface as shown in Fig. 22.25, whereas the vertical-component waves tend to be refracted through the surface into the water or glass. Because most reflected nonmetallic surfaces are horizontal, Polaroid sunglasses are aligned so that the polarized slits are vertical. Figure 22.26(a) shows a photograph of an automobile with glare from its windows. Figure 22.26(b) shows a photograph of the same automobile through a polarized filter; note the absence of glare, so one can more easily see into the automobile.

Electronic devices that display information using a liquid crystal display (LCD) use polarized filters to form the black segments that form the numbers and letters (see Fig. 22.27).

Figure 22.24   When a red flower is behind two overlapping Polaroid filters that (a) are aligned, one can easily see the red flower; (b) are 45° out of alignment, one can see the partially shaded red flower; and (c) are at right angles, one cannot see the red flower.

(a)

(b)

(c)

Figure 22.25   Much of the sunlight reflected from water is polarized. The horizontal-component waves reflect parallel to the water's surface, whereas the vertical-component waves tend to be refracted through the surface into the water.

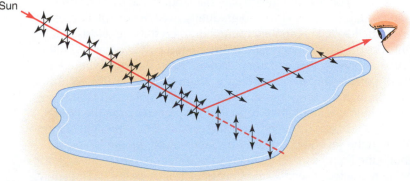

Figure 22.26 (a) An automobile with glare from its windows (b) The same automobile seen through a polarized filter; note that there is no glare, so one can see into the automobile.

(a)   (b)

Figure 22.27 The segments that form letters and numbers appearing in an LCD are formed using rotating polarized filters.

Courtesy of fotolia © MyLightscapes.com

PHYSICS CONNECTIONS

Polarized Calculators

Polarized filters are often found in sunglasses, but they are also used in devices such as stress analyzers, flat-screen computer displays, digital clocks, and calculators. Polarized filters play a part in displaying numbers on calculator screens.

There are three main components in calculators with liquid crystal display screens. A reflective mirror is placed at the rear of the display to reflect the light coming into the display back out to the user. A liquid crystal is located in front of the mirror. Liquid crystals have the ability to gradually twist the orientation of the light 90° from its original orientation. When a small voltage is applied to the liquid crystal, it untwists and maintains the original orientation of the light. The third component of the display is a pair of polarized filters. The polarized filters are oriented perpendicular to one another and are placed on each side of the liquid crystal. The inner polarized filter is aligned horizontally and the outer filter is aligned vertically (Fig. 22.28).

When light enters the display and passes through the first filter, only the vertically polarized light is permitted to pass through the filter and reach the liquid crystal. If there is no voltage placed across the crystal, the light is forced to alter its orientation by 90° as it passes through the liquid crystal. Since the light's orientation is now horizontal, the light passes through the second polarized filter, reflects off the mirror, passes through the second filter again, moves through the liquid crystal, changes its orientation, passes through the initial filter, and is directed to the user. The light emitted is seen as the gray background of the LCD display.

If the calculator is programmed to apply a voltage across the liquid crystal, the crystal will untwist and prevent the light from passing through the second filter and reflecting off the mirror. A black segment would appear to the calculator user because the horizontally polarized filter absorbs the vertically oriented light. Therefore, when a calculator displays a number, voltage must be placed across the liquid crystal to create that black segment.

Figure 22.28   (a) The path that light takes as it enters the liquid crystal display of a calculator (b) A standard liquid crystal display found on many calculators

PHYSICS CONNECTIONS

Televisions

For decades, the cathode ray tube was the primary technology for creating pictures on television screens. As illustrated in Fig. 17.12, the cathode ray tube utilizes electromagnetic fields to transmit electrons from one side of a television screen to the other side of the screen. The electrons excite phosphoric atoms on the screen, which produce specific colors that collectively form recognizable images.

Today, liquid crystal displays (LCDs) and plasma televisions have replaced conventional cathode ray tube televisions as the popular choice among consumers. Both LCD and plasma televisions produce significantly clearer, better-defined pictures. In addition, the size of the screen can be significantly larger because plasma and LCD televisions are dramatically thinner than conventional televisions.

The inner workings of LCD and plasma televisions differ from one another. LCD televisions create specific colors by energizing each pixel's liquid crystal. The electrical current sent through the polarizing filter of the liquid crystal determines the specific color that will be produced in that pixel by blocking out unwanted frequencies of light. This produces the desired color in the specific location on the screen. The screen consists of millions of pixels, which collectively form the picture on the screen.

Plasma televisions use a plasma gas inside tiny cells to stimulate photons that ultimately create visible light on the television screen. When electricity is sent into one of the millions of tiny cells within the television, the plasma causes ultraviolet light to be emitted within the cell. Each cell is coated with a red, blue, or green substance that causes the cell to emit that color on the screen because the ultraviolet light is invisible to the eye. Similar to the LCD television, the millions of cells create a clear and consistent picture.

Glossary

Color A property of the light that reaches our eyes and is determined by its wavelength or its frequency. (p. 603)

Complementary Colors Two colors that, when combined, form white; for example, cyan and red, magenta and green, and yellow and blue. (p. 604)

Diffraction A property of a wave that describes its ability to bend around obstacles in its path. (p. 610)

Dispersion The spreading of white light into the full spectrum. (p. 602)

Polarized Light Light waves restricted to a single plane that is perpendicular to the direction of the wave motion. (p. 613)

Primary Colors The colors red, green, and blue; an additive mixture of the three colors of light resulting in white. (p. 604)

Primary Pigments The complements of the three primary colors; namely, cyan (the complement of red), magenta (the complement of green), and yellow (the complement of blue). (p. 605)

Rainbow A spectrum of light formed when sunlight strikes raindrops refracts into them, reflects within them, and then refracts out of them. (p. 609)

Visible Spectrum The colors resulting from the dispersion of white light through a glass prism: red, orange, yellow, green, blue, and violet. (p. 602)

Review Questions

1. Name the colors of the visible spectrum.
2. What property of light determines its color?
3. Name the light waves whose wavelengths are (a) slightly longer than visible light and (b) slightly shorter than visible light.
4. What is the apparent color of a green dress in a closed room with only a red light? Explain.
5. What would the U.S. flag look like through a piece of (a) red glass and (b) blue glass?
6. What are the primary colors?
7. What is a complementary color? Name the complement of each primary color.
8. How are the primary pigments related to the primary colors?
9. What are the four colors used to print color photographs in books, magazines, and newspapers?
10. Why is the sky blue?
11. Why is a sunset red?
12. Why are clouds white?
13. Why is the ocean blue?
14. True or false: For you to see a rainbow, the raindrops must be between the sun and you.
15. Name the property of a wave that describes its ability to bend around obstacles in its path.
16. When a narrow beam of light passes by a sharp edge, around a fine wire, or through a narrow slit or a pinhole, the beam tends to spread or flare out. The distinct shadow formed is called a(n) _____.
17. Name the person who first performed the famous double-slit experiment.
18. When waves add together to form a larger displacement or more intense light, this is called _____.
19. When waves add together to form a smaller displacement or less intense light, this is called _____.
20. How do Polaroid sunglasses reduce the glare of bright sunlight?

CHAPTER 23

SURVEY OF MODERN PHYSICS

Objectives

The major goals of this chapter are to enable you to:

1. Describe the basis of quantum theory.
2. Describe the development of the current model of the atom.
3. Describe the structure and properties of the atomic nucleus.

4. Analyze problems of radioactive decay.
5. Describe nuclear fission and fusion.
6. Describe principles of detection and measurement of radioactivity.

To this point we have studied classical or Newtonian physics. Physics is also concerned with the building blocks of all matter, such as atoms, subatomic particles, and the forces between them.

Ancient Greek philosophers thought that all matter could be reduced to four basic elements: air, earth, fire, and water. Modern physics is much more complex and probes deeply into the nature of matter.

Quantum theory, the atom, radioactivity, nuclear fission, and fusion are all topics being studied by physicists today. While an in-depth study of these topics is beyond the scope of this text, some familiarity with the concepts is essential to those in all technical fields.

23.1 Quantum Theory

In the 1860s, **James Clerk Maxwell** developed a theory relating magnetic fields and electric fields in the transmission of energy across empty space, which became known as the electromagnetic wave theory. Maxwell's theory led to a complete description

James C. Maxwell (1831–1879),

physicist, was born in Scotland. His early work led to the mathematical development of the theory of electricity and magnetism. His greatest work was his theory of electromagnetic radiation and the basic equations of electromagnetism, which established him as the leading theoretical physicist of the 19th century.

of electricity and magnetism and was the basis for the development of radio, television, and countless electronic devices. It was also the starting point in the study of the nature of the atom.

Later experiments by Heinrich Hertz confirmed Maxwell's predictions and confirmed that light was a form of wave—almost. Only two problems remained: the wave theory of light could not explain why hot objects change color when heated and why ultraviolet light could discharge electrically charged metal plates. Solving these two problems led to the development of the quantum theory. Its confirmation forever changed our view of the physical world.

Particles Behave Like Waves and Waves Behave Like Particles

Incandescent objects (like a glowing light bulb filament) emit light because of the vibrations of charged particles inside their atoms. The spectrum of color produced by the light depends on the heat of the object and the frequency range of the light. Max Planck explained this with the then revolutionary idea that particles can have only certain energies, which are whole-number multiples of a constant now known as Planck's constant. Planck proposed that atoms could only emit radiation when their vibration energy changed in multiples of this constant.

In addition, certain metals emit electrons when they are exposed to light. This is called the *photoelectric effect* and could not be explained by simple wave theory. The light produced an effect like a stream of particles. Einstein explained this effect by suggesting that light consists of packets of energy, called *photons*, that act like particles. He thought that, although photons have no mass, they travel at the speed of light, have energy related to their frequency, and have momentum. Later experiments confirmed his theory.

Louis Victor de Broglie (1892–1987),

physicist, was born in France. In 1923, while a graduate student at the University of Paris, he theorized that since light could be seen to behave under some conditions as particles and other times as waves, atomic particles have both particle and wave properties. This led to his pioneering work on the wave nature of the electron.

Later, **Louis Victor de Broglie** in 1923 suggested that particles have wave properties. This revolutionary idea was largely ignored until Einstein supported it; de Broglie won a Nobel Prize for it in 1929. Two wave properties exhibited by particles are diffraction and interference. These can easily be reproduced in the laboratory.

Most physicists today believe that the particle and wave aspects of light are complementary. That is, they must be viewed together to have a complete picture of the true nature of light. This complementary nature in exhibiting properties of both waves and particles is the essence of the quantum theory.

23.2 The Atom

In Section 12.1 we saw that an **atom** is the smallest particle of an element that can exist in a stable or independent state. The three most important particles of the atom are (1) the **proton**, a particle with a positive charge; (2) the **electron**, a particle with a negative charge; and (3) the **neutron**, a particle that does not have an electric charge. The **nucleus** is the center part of an atom, made up of protons and neutrons, while the electrons surround the nucleus. Historically, how was this current concept of the atom developed?

By 1900, most scientists agreed atoms exist. In 1890, **J. J. Thomson** discovered small, negatively charged particles, which he called electrons. Between 1909 and 1911, Ernest Rutherford discovered that atoms have a relatively more massive, positively charged center, or nucleus. Another 30 years was required for agreement to be reached on how these parts of the atom were arranged.

J. J. Thomson (1856–1940),

physicist, was born in England. His experiments in 1897 led to the discovery of the electron, which he found to be 2000 times smaller in mass than the lightest known atomic particle, the hydrogen ion. He received the Nobel Prize in Physics in 1906.

The Rutherford Model

Thomson originally thought that electrons were embedded in a spherical volume of positive charge like raisins in a muffin. However, Rutherford, with Hans Geiger and Ernst Marsden, was able to show that the atom was mostly space with most of the mass concentrated in a relatively small region he called the nucleus. He thought electrons orbited the nucleus the way planets orbit the sun, which is the most popular oversimplification of atomic structure still found today. The flaw in Rutherford's model is that if electrons were like planets, they would quickly lose energy and collapse into the nucleus, which does not happen.

Scientists sought a model of the atom that could explain the mystery of atomic spectra. An emission spectrum is a spectrum of electromagnetic radiation emitted by a luminous source. Many substances emit a characteristic color when heated. For example, applying a high voltage to gas atoms, as in a neon light, causes them to emit their characteristic color. The light can be studied with a diffraction grating or prism. A *diffraction grating* is a glass or polished metal surface containing many closely spaced, very fine parallel lines in the surface, which separate the colors of light by interference and produce a light spectrum. In contrast to an emission spectrum, as a gas cools, the opposite process occurs and an absorption spectrum is produced. An absorption spectrum is a spectrum of electromagnetic radiation absorbed by matter when radiation of all frequencies is passed through it. Based on the above concepts, atomic spectra were used to determine the structure of the atom.

Niels Bohr developed the next recognized model of the atom, using Einstein's idea that since emission spectra contain only specific wavelengths, an electron can emit or absorb only specific amounts of energy; that is, the energy is quantized. Bohr knew his model was not complete and continued to search for a better model. The Bohr model did, however, provide an explanation for some of the chemical properties of elements. The foundation of much of our knowledge of chemical reactions and bonding is based on the idea that each atom has a unique electron arrangement.

In 1926, **Erwin Schroedinger** used de Broglie's idea of matter behaving like waves to create a quantum model of the atom based on waves. **Werner Heisenberg** and others developed the theory into a complete description of the atom. The theory does not provide a simple picture like the planetary model. The wave–particle nature of matter means it is impossible to know both the momentum and the position of an electron at the same time. The quantum model of the atom predicts only the probability that an electron is at a specific location. This region in which there is a high probability of finding an electron is called an electron cloud. Though difficult to visualize, quantum mechanics has been very successful in predicting many details of the structure of atoms.

Niels Bohr (1885–1962), Danish physicist, developed an early model of atomic structure in which electrons travel around the nucleus in given stable orbits determined by quantum conditions.

Erwin Schroedinger (1887–1961), physicist, was born in Vienna. His study of the wave behavior of matter within quantum mechanics led to the understanding of matter at the subatomic level. He also made significant contributions to molecular biology.

Werner Heisenberg (1901–1976), theoretical physicist, was born in Germany. He developed a method of expressing quantum mechanics mathematically and formulated his revolutionary uncertainty principle—increasing the accuracy of measurement of one observable quantity increases the uncertainty with which a related quantity may be known.

23.3   Atomic Structure and Atomic Spectra

Niels Bohr suggested the energy of an electron is quantized, that is, restricted only to certain fixed values called *energy levels*. Each of these energy levels was shown by Bohr to be given by the equation

$$E = -\frac{kZ^2}{n^2}$$

where E = energy of electron
   k = a constant (2.179×10^{-18} J or 13.595 eV)
   Z = atomic number (the number of positive charges in the nucleus)
   n = integer that characterizes the energy level ($n = 1, 2, 3, 4, \ldots$), also called the quantum number

An *electron volt* (eV) is the energy acquired by an electron as a result of moving through a potential difference of 1 V. One eV is equal to 1.602×10^{-19} J. So one MeV (one million electron volts) is equal to 1.602×10^{-13} J. The distance from an electron in an atom to the nucleus increases as the integer n increases. (The formula is $r_n = n^2 r_1$ for the hydrogen atom.) In the lowest-energy level or lowest orbit, its energy is minimum. This lowest-energy level ($n = 1$) is called the stable or **ground state** of the electron. The higher-energy levels ($n = 2, 3, 4, \ldots$) are called **excited states** of the electron.

The **Bohr model** was an early model of atomic structure in which electrons travel around the nucleus in a number of discrete stable energy levels determined by quantum conditions. Of the several early proposed models of atomic structure, Bohr's was the first successful model for the atomic structure of the hydrogen atom (Fig. 23.1). The Bohr model is clearly limited in scope, but it was the forerunner to the modern quantum theory.

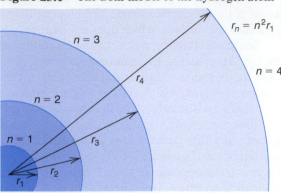

Figure 23.1 The Bohr model of the hydrogen atom

EXAMPLE 1

Find the energy of an electron in the $n = 2$ energy level in a hydrogen atom.

Data:

$$n = 2$$
$$Z = 1$$
$$E = ?$$

Basic Equation:

$$E = -\frac{kZ^2}{n^2}$$

Working Equation: Same

Substitution:

$$E = -\frac{(13.595 \text{ eV})(1)^2}{2^2}$$
$$= -3.3988 \text{ eV}$$

All objects emit radiation with intensity that is proportional to the fourth power of the Kelvin temperature (T_K^4). At normal temperatures, we are not aware of this electromagnetic radiation because of its low intensity. At higher temperatures, sufficient infrared radiation exists so that we feel the heat if we are close to the object. At still higher temperatures (approximately 1000 K), an object, such as molten steel, an electric toaster element, or an electric stove burner, glows. The filament of a light bulb glows at temperatures above 2000 K. This radiation is due to the rapid vibrations of the atoms and molecules.

Rarefied (low density and low pressure) gases can also be excited to emit light by intense heating in a tube. The radiation from excited gases emits light of only certain wavelengths. When the light is analyzed through the slit of a spectroscope, a line spectrum is observed instead of a continuous color spectrum.

Neon lights work on the principle that certain colors of light are given off by different atoms when the electrons in the atoms undergo a transition from the ground state to an excited state due to an electric current passing through the neon tube containing the gas at low pressure. As the electrons return to their ground state, they give off their excess energy in the form of electromagnetic radiation. A neon tube containing hydrogen gives off the color blue. Other atoms give off other colors, leading to the rich colors available through this type of light tube.

Fluorescence is a property of certain substances in which radiation is absorbed at one frequency and then reemitted at a lower frequency. When radiation is absorbed by a particle (an atom or a molecule), the energy of the electrons is increased to a higher energy level. This results in the particle being in an abnormal, excited state. Some absorbed energy is lost by collisions with other particles so that the particle cannot maintain its energy level in its excited state. The electrons give up their remaining extra energy and the particle returns to its ground state in a series of two or more steps and emits radiation in the form of visible light. The image on a television screen is the result of fluorescence due to electron bombardment of the screen.

A *fluorescent lamp* is a source of artificial light that consists of a long, sealed glass tube with a small amount of mercury and an electrode at each end. The inside of the tube is coated with a mixture of fluorescent powders. An electric current through the tube vaporizes the mercury and causes it to emit ultraviolet radiation. That radiation is absorbed by the fluorescent coating, which then emits visible light. See Fig. 23.2.

Figure 23.2 Fluorescent lamp. An electric current through the glass tube vaporizes the mercury and causes it to emit ultraviolet light. That radiation is absorbed by the fluorescent coating, which then emits visible light.

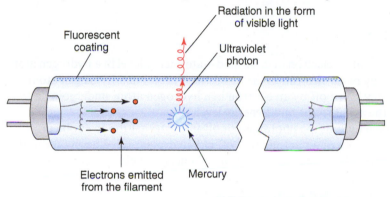

The wavelength of a photon emitted during the transition between energy levels can be determined using concepts developed in earlier chapters: $E = hf$ gives the energy of a photon and $c = \lambda f$ relates the wavelength and frequency of light.

EXAMPLE 2

What is the wavelength of a photon given off by a hydrogen atom undergoing a transition from the $n = 2$ to the $n = 1$ energy level?

Data: For $n = 2$ energy level, $E = -3.3988$ eV (from Example 1)

For $n = 1$ energy level,

$$n = 1$$
$$k = 13.595 \text{ eV}$$
$$Z = 1$$

Basic Equations:

$$E = -\frac{kZ^2}{n^2}$$
$$E = hf$$
$$c = \lambda f$$

Working Equations:

$$E_1 = -\frac{kZ^2}{n^2} \qquad E_{\text{transition}} = E_2 - E_1 \qquad \lambda = \frac{c}{f} = \frac{c}{E/h} = \frac{ch}{E}$$

Substitutions:

$$E_1 = -\frac{(13.595 \text{ eV})(1)^2}{1^2}$$

$$= -13.595 \text{ eV}$$

Therefore,

$$E_{\text{transition}} = -3.3988 \text{ eV} - (-13.595 \text{ eV})$$

$$= 10.196 \text{ eV}$$

So,

$$\lambda = \frac{(3.00 \times 10^8 \text{ m/s})(6.626 \times 10^{-34} \text{ J s})}{10.196 \text{ eV} \times \dfrac{1.602 \times 10^{-19} \text{ J}}{1 \text{ eV}}}$$

$$= 1.22 \times 10^{-7} \text{ m}$$

· · · · · · · · · · · · · · · ·

PROBLEMS 23.3

1. Find the energy of an electron in its ground-state energy level in a hydrogen atom.
2. Find the energy of an electron in the $n = 3$ energy level in a hydrogen atom.
3. Find the energy given off by an electron in a hydrogen atom undergoing a transition from the $n = 3$ to the $n = 1$ energy level.
4. Find the wavelength of the photon given off in the transition in Problem 3.
5. Find the wavelength of a photon given off in a hydrogen-filled neon tube undergoing a transition from the $n = 3$ to the $n = 2$ energy level.

23.4 Quantum Mechanics and Atomic Properties

The agreement between the Bohr model and the energy levels of the hydrogen atom does not account for the properties of the more complex atoms, that is, those with more than one electron. The quantum model of the atom better fits the more complex atoms. **Quantum mechanics** is a theory that unifies the wave–particle dual nature of electromagnetic radiation. The quantum model is based on the theory that matter, just like light, behaves sometimes as a particle and sometimes as a wave. The wave theory of the atom can be used in a simple way to understand the Bohr model structure of an atom. In the Bohr model, the electron can have only specific discrete energy levels; other energy levels are not possible. When an electron in an atom behaves as a wave, its distance from the nucleus or its energy level must consist of an integral number of wavelengths; otherwise, destructive interference occurs (Fig. 23.3).

When an electron behaves as a wave, its exact location cannot be determined; instead, the probability can be calculated, using wave equations, of where it is likely to be found. This probability describes for any position in space how likely the electron is to be found at that point at any given time (Fig. 23.4). A thin spherical shell region of space (or electron cloud) surrounding the atomic nucleus in which an electron is likely to be found is its energy level. In the ground state of the hydrogen atom, the electron spends most of its time in a spherical shell region, or energy level, centered about 0.5 Å (angstrom) from the nucleus as shown in Fig. 23.5. *Note:* 1 Å = 1 × 10⁻¹⁰ m.

The shell regions, or energy levels, are characterized by a number, n, which is called the *principal quantum number*, where $n = 1, 2, 3, 4, \ldots$. Just as in the Bohr model, the shells extend farther from the nucleus as n increases. As the shells extend farther from the nucleus, there are different shell shapes that become possible for each principal quantum number. Each of these shapes is described by a *secondary quantum number* or letter ($s, p, d, f, \ldots$) as shown in Fig. 23.6.

SKETCH

| 12 cm² | w |

4.0 cm

DATA
$A = 12 \text{ cm}^2, l = 4.0 \text{ cm}, w = ?$

BASIC EQUATION
$A = lw$

WORKING EQUATION
$w = \dfrac{A}{l}$

SUBSTITUTION
$w = \dfrac{12 \text{ cm}^2}{4.0 \text{ cm}} = 3.0 \text{ cm}$

Figure 23.3 A shell must contain an integral number of wavelengths of the electron.

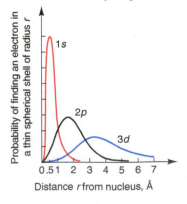

Figure 23.4 The probability of finding the electron at a distance r from the nucleus for the three lowest shells of a hydrogen atom

Figure 23.5   Electron density for an electron in the ground state of a hydrogen atom

Figure 23.6   Shapes for (a) an *s* subshell, (b) a *p* subshell, (c) a *d* subshell, and (d) an *f* subshell

(a)   (b)   (c)   (d)

Each possible shape represents an energy level, or shell, which can contain up to two electrons. In the ground state (lowest energy state) of an atom, the electrons fill up the lowest energy levels. The types of energy levels, or shells, for each principal quantum number are shown in Table 23.1. Table 23.2 summarizes the number of electrons in each shell for the atomic elements with up to 21 protons in the nucleus and therefore up to 21 electrons in energy levels. Note in Table 23.2 that the $n = 4$ shell starts filling up (see potassium) before the $n = 3$ shell is completely filled. This is because the 4s shell has lower energy than the 3d shells.

Table 23.1   Maximum Number of Electrons in Atomic Shells or Energy Levels

n Value	Types of Energy Levels	Maximum Number of Electrons
1	1s	2
2	2s, 2p	8
3	3s, 3p, 3d	18
4	4s, 4p, 4d, 4f	32
5	5s, 5p, 5d, 5f, 5g	50

Table 23.2   Table of the First 21 Elements

Element	Atomic No.	Number of Protons	Number of Electrons	Electrons in $n = 1$ Shell	Electrons in $n = 2$ Shell	Electrons in $n = 3$ Shell	Electrons in $n = 4$ Shell
Hydrogen	1	1	1	1			
Helium	2	2	2	2			
Lithium	3	3	3	2	1		
Beryllium	4	4	4	2	2		
Boron	5	5	5	2	3		
Carbon	6	6	6	2	4		
Nitrogen	7	7	7	2	5		
Oxygen	8	8	8	2	6		
Fluorine	9	9	9	2	7		
Neon	10	10	10	2	8		
Sodium	11	11	11	2	8	1	
Magnesium	12	12	12	2	8	2	
Aluminum	13	13	13	2	8	3	
Silicon	14	14	14	2	8	4	
Phosphorus	15	15	15	2	8	5	
Sulfur	16	16	16	2	8	6	
Chlorine	17	17	17	2	8	7	
Argon	18	18	18	2	8	8	
Potassium	19	19	19	2	8	8	1
Calcium	20	20	20	2	8	8	2
Scandium	21	21	21	2	8	9	2

The theory of electron shells provides a basis for understanding the properties of different atoms, based on how many electrons are in the outer shell and how much energy it takes to add or remove an electron from that shell during chemical reactions between atoms. Each atom in all matter attempts to have its outer shell complete and does this by attempting to borrow, lend, or share electrons with other atoms surrounding it.

An atom is classified as a metal if it tends to lend electrons. Copper is an example of a metal. If an atom tends to borrow electrons, it is called a nonmetal. Atoms are classified as inert when they have complete outer shells and therefore tend not to borrow or share electrons. Helium and neon are examples of inert atoms. They both have completely filled outer shells (see Table 23.2). The fewer electrons an atom tends to borrow, lend, or share, the more reactive it tends to be in chemical reactions.

PROBLEMS 23.4

1. Which atom has the $n = 2$ energy level filled?
2. Which atom has the $n = 1$ energy level filled?
3. Which atom is one electron short of having the $n = 2$ energy level filled?
4. Which atom has just one electron in the $n = 3$ energy level?

23.5 The Nucleus—Structure and Properties

Henri Becquerel discovered radioactivity in 1896 when he was working with compounds containing uranium and found that his photographic plates were partially exposed when these uranium compounds were anywhere near his plates. The plates were exposed by penetrating rays that went through the plate coverings. This phenomenon was called radiation. Many scientists then began to study radiation, and one of the first results was an understanding of the composition of the atomic nucleus.

The nucleus of an atom contains over 99.9% of the mass of an atom, yet it occupies less than one trillionth of 1% of the volume of an atom. The fundamental particles that make up the nucleus of an atom are called *nucleons*. There are two types of nucleons: *protons*, which carry a positive charge, and *neutrons,* which are electrically neutral (Fig. 23.7). Protons have a mass of

$$m_p = 1.6726 \times 10^{-27} \text{ kg}$$

while neutrons have a mass of

$$m_n = 1.6749 \times 10^{-27} \text{ kg}$$

Both are much more massive than the third fundamental particle of an atom, the negatively charged electron, which has a mass of

$$m_e = 9.1094 \times 10^{-31} \text{ kg}$$

Negatively charged electrons are attracted to the nucleus by the positively charged protons (Fig. 23.8). The *size* of the positive charge of a proton is the same as the *size* of the negative charge of an electron. An electrically neutral atom contains an equal number of protons and electrons.

The simplest atom, hydrogen, contains one proton in its nucleus. Other atoms contain two or more protons together with some number of neutrons in their nuclei. The role of neutrons in the nucleus of these more complex atoms is to bind the positively charged protons together and prevent the nucleus from flying apart under the repulsive electric force between the protons. In addition to the repulsive **Coulomb force** (electric force) between protons, there is a nuclear force, referred to as the **strong force**, which is a very short-range attractive force among all nucleons (neutrons and protons) independent of their charge. The neutrons provide additional strong force to overcome Coulomb repulsion. Two neutrons in a nucleus with two protons provide the additional attractive strong force that holds the nucleus together. For heavier atoms it is necessary to have more neutrons than protons to hold the nucleus together (Fig. 23.9). For example, there are 82 protons and 126 neutrons in an atom of

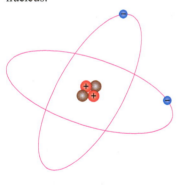

Figure 23.7   The nucleus is composed of neutrons and protons. Electrons surround the nucleus.

Proton   Electron

Figure 23.8   The attractive strong force operates at short distances.

Figure 23.9 Number of neutrons versus number of protons for stable nuclei (dots)

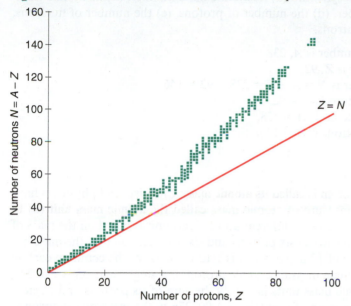

common lead. Atoms heavier than lead cannot be completely stabilized even by the addition of extra neutrons.

The number of protons in a nucleus is called its **atomic number**, Z. The total number of nucleons (protons and neutrons) in a nucleus is called the **atomic mass number**, A. A **nuclide** is a specific type of atom characterized by its nuclear properties, such as its number of neutrons and protons and the energy state of its nucleus. The **neutron number**, N, is the number of neutrons in an atomic nucleus of a particular nuclide. Also, $N = A - Z$, the difference between the atomic mass number and the atomic number. N can be different for nuclei with the same atomic number. Each type of nuclide is specified using a symbol of the form

$$^A_Z X$$

where X = chemical symbol for the given element
A = atomic mass number
Z = atomic number

For example, $^{14}_6$C refers to a carbon nucleus containing 6 protons with an atomic mass number of 14. This atom thus contains 8 neutrons. Other carbon atoms exist that contain 5, 6, 7, 9, or 10 neutrons. All carbon atoms contain 6 protons. (Otherwise, they wouldn't be carbon!) Nuclei that contain the same number of protons but a different number of neutrons are called **isotopes**. See Fig. 23.10 for an illustration of the isotopes of hydrogen.

Figure 23.10 Three isotopes of hydrogen. Each has a single electron and a single proton but different numbers of neutrons.

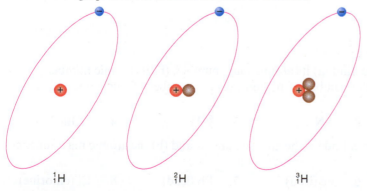

$^1_1$H $^2_1$H $^3_1$H

EXAMPLE

For the uranium isotope $^{238}_{92}U$, find (a) its atomic mass number, (b) its atomic number, (c) its neutron number, (d) the number of protons, (e) the number of nucleons, and (f) the number of neutrons.

(a)   The atomic mass number is A, 238.
(b)   The atomic number is Z, 92.
(c)   The neutron number is $N = A - Z = 238 - 92 = 146$.
(d)   The number of protons is $Z = 92$.
(e)   The number of nucleons is $A = 238$.
(f)   The number of neutrons is $N = 146$.

· · · · · · · · · · · · · · · · ·

The mass of a single atom is called its **atomic mass**. Chemists and physicists have developed a unit of the measure of atomic mass called the **atomic mass unit** (u) to more easily compare the masses of different atoms. This unit is based on the mass of the common carbon atom with its six protons and six neutrons, so that the mass has been given the exact value of 12 u. Thus, 1 u is 1/12 the mass of the common carbon atom and so the approximate mass of a single proton or a single neutron is 1 u. The mass of an atom in atomic mass units is simply the sum of its protons and its neutrons. More precisely, the mass of a proton is 1.007276 u, the mass of a neutron is 1.008665 u, the mass of an electron is 0.00054858 u, and the mass of a neutral hydrogen atom is 1.007825 u.

Periodic Table

Dmitri Mendeleev (1834–1907),

chemist, was born in Russia. He developed the Periodic Table of the elements and predicted the existence of several elements that were subsequently discovered.

In 1889, years before the atomic theory of the atom was developed, **Dmitri Mendeleev** proposed a table containing the elements arranged in a periodic manner that showed various trends in the properties of atoms. This **Periodic Table** (see Table 21 of Appendix C) contains all of the atomic elements arranged according to their atomic numbers, which can be used to predict their chemical properties. The existence of eight columns in his table and the trends in the properties of the atoms were later understood in terms of the atomic structure described in the last section.

The horizontal rows of the Periodic Table are called periods or rows. There are seven periods, each of which starts with an atom having one electron in its outer shell and ends with an atom with a complete outer shell structure (an inert element). The first three rows consist of 2, 8, and 8 elements. Rows 4 and 5 are longer rows consisting of 18 elements, and row 6 has 32 elements. Most row 7 elements are radioactive and do not occur in nature. The number of elements in each row can be understood in terms of the atomic theory described earlier and the number of electrons in each energy level, or shell (see Table 23.2).

Metals are found on the left side of the table, with the most active metals in the lower left corner. Nonmetals are found on the right side. The noble or inert gases are on the far right. The acid-forming properties increase toward the right. Base-forming properties increase toward the left. Other properties, including the atomic weight and the atomic size, are given in a common form in the Periodic Table.

PROBLEMS 23.5

For each given isotope, find (a) its atomic mass number, (b) its atomic number, (c) its neutron number, (d) the number of protons, (e) the number of nucleons, and (f) the number of neutrons.

1. $^{12}_{6}C$ 2. $^{13}_{7}N$ 3. $^{48}_{22}Ti$ 4. $^{141}_{58}Ba$

From the Periodic Table, find (a) the atomic number and (b) the atomic mass for each element.

5. Na (sodium)   6. Fe (iron)   7. Pb (lead)   8. Cl (chlorine)

9.   The nucleus of a certain element contains 6 protons and 8 neutrons. Write its chemical symbol.
10. The nucleus of a certain element contains 92 protons and 142 neutrons. Write its chemical symbol.

Use the Periodic Table to determine the following:

11. Which of the elements Br and Ca would be expected to be more acidic?
12. Which of the elements Br and Ca would be expected to be a stronger base?
13. Which of the elements Cl, Ar, I, Ca, and K is inert?
14. Which of the elements C, N, O, and Fe is a metal?

23.6 Nuclear Mass and Binding Energy

When the total mass of a nucleus is compared with the sum of the individual masses of the protons and neutrons when unbound, or broken apart, the nuclear mass is found to be smaller. How can this be? Early in the 20th century, Einstein stated as a physical law that mass and energy are equivalent forms of "matter." According to this **mass–energy equivalence**, mass can be converted into energy as follows:

$$E = \Delta m c^2$$

where E = energy
          Δm = change in mass
          c = speed of light, 3.00×10^8 m/s

   When the total mass in a reaction is decreased by some amount, energy is given off in the form of radiation or kinetic energy. The difference in mass, converted using the equation above, is called the **binding energy of the nucleus**. This energy represents the total energy required to break a nucleus into its separate nucleons. For instance, if the mass of a $^4_2$He nucleus were exactly equal to the mass of two protons and two neutrons, the nucleus would break apart without any additional energy. To be stable, the mass of the nucleus must be less than the mass of its component nucleons. You may think of the binding energy as the energy a nucleus lacks relative to the total mass of its separate nucleon components. The binding energies of nuclei are approximately 10^6 times greater than the binding energies of the electrons in atoms and therefore are much more significant. *Note:* In this chapter we often use more than three significant digits because of the precision of atomic and nuclear measurements.

Find the binding energy of a $^4_2$He nucleus. The mass of the neutral $^4_2$He atom is 6.6463×10^{-27} kg.

EXAMPLE 1

Data: The mass of two individual neutrons, two individual protons, and two individual electrons is

$$m = 2m_p + 2m_n + 2m_e$$
$$= 2(1.6726 \times 10^{-27} \text{ kg}) + 2(1.6749 \times 10^{-27} \text{ kg}) + 2(9.1094 \times 10^{-31} \text{ kg})$$
$$= 6.6968 \times 10^{-27} \text{ kg}$$

Since the mass of a neutral $^4_2$He atom $= 6.6463 \times 10^{-27}$ kg,

$$\Delta m = 6.6968 \times 10^{-27} \text{ kg} - 6.6463 \times 10^{-27} \text{ kg}$$
$$= 0.0505 \times 10^{-27} \text{ kg} = 5.05 \times 10^{-29} \text{ kg}$$

Basic Equation:

$$E = \Delta m c^2$$

Working Equation: Same

Substitution:

$$E = (5.05 \times 10^{-29} \text{ kg})(3.00 \times 10^8 \text{ m/s})^2$$
$$= 4.55 \times 10^{-12} \text{ kg m}^2/\text{s}^2 \qquad\qquad (1 \text{ J} = 1 \text{ kg m}^2/\text{s}^2)$$
$$= 4.55 \times 10^{-12} \text{ J}$$

The conversion factor relating atomic mass units and metric mass is

$$1 \text{ u} = 1.6605 \times 10^{-27} \text{ kg}$$

The mass of atomic and nuclear particles is also expressed in million electron volts (MeV). Recall that an electron volt is the energy that one electron would gain in passing through an electric potential difference of 1 V. One MeV is given by

$$1 \text{ MeV} = 1.602 \times 10^{-13} \text{ J}$$

Thus, the binding energy of a ^4_2He nucleus from Example 1 in MeV is

$$4.55 \times 10^{-12} \text{ J} \times \frac{1 \text{ MeV}}{1.602 \times 10^{-13} \text{ J}} = 28.4 \text{ MeV}$$

EXAMPLE 2

Express the MeV unit in terms of its equivalent mass in kilograms and atomic mass units using the equation $E = \Delta mc^2$.

Data:

$$E = 1 \text{ MeV} = 1.602 \times 10^{-13} \text{ J}$$
$$c = 2.998 \times 10^8 \text{ m/s} \qquad\qquad (\textbf{\textit{Note:}} \text{ We need to use four}$$
$$\Delta m = ? \qquad\qquad\qquad\qquad \text{significant digits here.})$$

Basic Equation:

$$E = \Delta mc^2$$

Working Equation:

$$\Delta m = \frac{E}{c^2}$$

Substitution:

$$\Delta m = \frac{1.602 \times 10^{-13} \text{ J}}{(2.998 \times 10^8 \text{ m/s})^2} \times \frac{1 \text{ kg m}^2/\text{s}^2}{1 \text{ J}}$$
$$= 1.782 \times 10^{-30} \text{ kg}$$

Next, convert kilograms to atomic mass units:

$$1.782 \times 10^{-30} \text{ kg} \times \frac{1 \text{ u}}{1.6605 \times 10^{-27} \text{ kg}} = 1.073 \times 10^{-3} \text{ u}$$

Thus, 1 MeV (1.602×10^{-13} J) has an equivalent mass of 1.782×10^{-30} kg or 1.073×10^{-3} u.

The masses of some atomic particles are listed in the following table.

Particle	Mass	
	kg	u
Electron	9.1094×10^{-31}	0.00054858
Proton	1.67262×10^{-27}	1.007276
Neutron	1.67493×10^{-27}	1.008665
$^1_1$H atom	1.67353×10^{-27}	1.007825

The **average binding energy per nucleon** in a nucleus is the total binding energy of the nucleus divided by the total number of nucleons, A. Figure 23.11 gives the average binding energy per nucleon in MeV for stable nuclei. For $^4_2$He the average binding energy is its total binding energy in MeV, which we found following Example 1, divided by its total number of nucleons; that is, 28.4 MeV/4 = 7.1 MeV. Binding energies are largest for the number of nucleons, A, between about 30 and 60. This allows nuclei below 30 and above 60 to undergo reactions or decay, which produce substantial amounts of energy.

Figure 23.11 Average binding energy per nucleon for a given mass number A for stable nuclei

Estimate the average binding energy for $^{238}_{92}$U from Fig. 23.11.

EXAMPLE 3

Note that $^{238}_{92}$U has 238 nucleons as indicated by its atomic mass number, the upper number. Place a ruler or straight edge vertically at 238 on the horizontal axis. Note its point of intersection with the graph. Then, read the number on the vertical axis, 7.6 MeV, corresponding to this point of intersection. Thus, the average binding energy for $^{238}_{92}$U is approximately 7.6 MeV.

PROBLEMS 23.6

1. Find the mass in kilograms of the $^{232}_{92}$U atom if its mass in atomic mass units is 232.037131 u.
2. Find the mass in kilograms of the $^{228}_{90}$Th atom if its mass in atomic mass units is 228.028716 u.

SKETCH

12 cm² | w

4.0 cm

DATA

$A = 12 \text{ cm}^2, l = 4.0 \text{ cm}, w = ?$

BASIC EQUATION

$A = lw$

WORKING EQUATION

$w = \dfrac{A}{l}$

SUBSTITUTION

$w = \dfrac{12 \text{ cm}^2}{4.0 \text{ cm}} = 3.0 \text{ cm}$

3. Find the binding energy of a $^{14}_{7}N$ nucleus in MeV if the mass of the neutral N atom is 14.00307 u.

4. Find the binding energy of a $^{232}_{92}U$ nucleus in MeV if the mass of the neutral U atom is 232.037131 u.

5. Find the binding energy of a $^{228}_{90}Th$ nucleus in MeV if the mass of the neutral Th atom is 228.028716 u.

Estimate the average binding energy per nucleon for each of the following nuclei from Fig. 23.11.

6. $^{48}_{22}Ti$　　　　　7. $^{141}_{58}Ba$　　　　　8. $^{12}_{6}C$

23.7 Radioactive Decay

The size of an atomic nucleus is limited by the fact that neutrons are unstable. On average, an isolated neutron will decay into a proton and an electron in about 12 min. In the presence of protons, the neutron is more stable (Fig. 23.12). In many atomic nuclei, neutrons are stable for many billions of years. For heavy atoms, the large number of neutrons required to hold the protons together leads to a situation where not enough protons exist to prevent one or more of the neutrons from decaying into a proton and an electron. Such unstable nuclei are called *radioactive*. **Radioactive decay** occurs when an unstable nucleus of an atom is transformed into a new element through the spontaneous disintegration of its atomic nuclei. Examples of radioactive elements are uranium, plutonium, radium, thorium, and their products. Often, one radioactive isotope decays into another isotope that is also radioactive, which decays into yet another radioactive isotope. A *radioactive series* is a series of isotopes of various elements that is successively transformed into lighter elements before the series of elements reaches a stable state, usually lead.

All elements heavier than bismuth exhibit radioactive decay. The radioactive elements exhibit three types of decay in which one or more radioactive rays are emitted from the nucleus: alpha (α), beta (β), and gamma (γ) rays. An **α ray** consists of alpha particles, each having two protons and two neutrons, and is positively charged; that is, it curves in the direction that known positive charges curve in a magnetic field. The helium nucleus is identical to an α particle. As we discuss later, the α particle (helium nucleus) is the most stable nuclear combination and can therefore be easily formed in a nuclear reaction. A **β ray** consists of a stream of beta particles (electrons) that are emitted from neutrons in a nucleus as the neutrons decay into protons and electrons. A β ray is negatively charged and curves in the opposite direction of an α ray in a magnetic field. A **γ ray** is composed of photons of electromagnetic radiation, which have no mass. A γ ray is similar to light but has much higher energy; it is not affected by a magnetic field and is uncharged. (See Fig. 23.13.)

Figure 23.12　(a) A neutron by itself is unstable; the lone neutron decays into a proton and an electron. (b) A neutron with a proton is stable.

Unstable　　　　Stable

(a)　　　　　(b)

Figure 23.13　Alpha and beta rays curve in opposite directions in a magnetic field; gamma rays are undeflected.

Alpha Decay

A nucleus that emits an α particle becomes different because it has lost two protons and two neutrons (Fig. 23.14). When radium 226 ($^{226}_{88}$Ra) gives off an α particle, it becomes a nucleus with $A = 226 - 4 = 222$ and $Z = 88 - 2 = 86$. This nucleus is that of the element radon ($^{222}_{86}$Rn). This nuclear reaction is written in the form

$$^{226}_{88}\text{Ra} \longrightarrow {}^{222}_{86}\text{Rn} + {}^{4}_{2}\text{He}$$

Alpha decay occurs because there are not enough neutrons in the nucleus to keep it stable. The total energy released in radioactive decay is called the **disintegration energy**, Q, and is given by

$$Q = (M_p - M_d - m_\alpha)c^2$$

where Q = disintegration energy
      M_p = mass of the parent nucleus
      M_d = mass of the daughter nucleus
      m_α = mass of the α particle
      c = speed of light

The emission of α particles changes the *parent* nucleus into a different nucleus, called a *daughter* nucleus. This disintegration energy is in the form of kinetic energy of (1) the α particle as it moves away from the nucleus and (2) the recoiling nucleus that moves away in the opposite direction.

Beta Decay

A nucleus can also decay by the emission of a β particle (an electron). How can a nucleus made up of only protons and neutrons emit an electron? One of the neutrons in the nucleus can break up into a proton and an electron and give off the electron in a manner similar to the decay of free neutrons. In β decay, a nucleus thus changes its charge, and the atom will have to pick up an additional electron in its charge clouds to remain neutral. An element involved in this process is thereby changed into a different element with one more electron. An example of β decay is the changing of carbon into nitrogen (Fig. 23.15):

$$^{14}_{6}\text{C} \longrightarrow {}^{14}_{7}\text{N} + e^-$$

The radioactivity of an isotope is commonly measured by the decay rate or the half-life (Fig. 23.16). The **decay rate** is the probability per unit time that a decay of radioactive isotopes will occur. The decay of radioactive isotopes is a completely random process. That is, it cannot be predicted exactly when a given nucleus will decay. For a large number of nuclei in a sample, the **half-life** is the length of time required

Figure 23.14 Uranium 238 is unstable and decays by giving off an α particle.

$$^{238}_{92}\text{U} \longrightarrow {}^{234}_{90}\text{Th} + {}^{4}_{2}\text{He}$$

Figure 23.15 Carbon 14 can decay into a nitrogen 14 atom by giving off a β particle (an electron).

$$^{14}_{6}\text{C} \longrightarrow {}^{14}_{7}\text{N} + e^-$$

Figure 23.16 An exponential decrease of radioactive nuclei is observed.

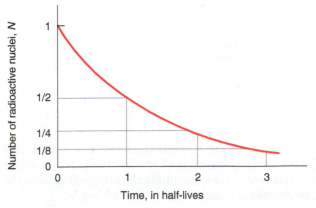

for one half of the original amount of the radioactive atoms in the sample to decay (Fig. 23.16). This means that after the first half-life, one half of the original amount remains. After the second half-life, one half of one half, or one fourth, of the original amount remains. After the third half-life, one half of one fourth, or one eighth, of the original amount remains. This process continues.

EXAMPLE 1

Draw a graph through four half-lives of the radioactive decay of 80.0 g of radium 226, which has a half-life of 1620 years (see Fig. 23.17).

Note that the amount of radium is halved as we move down the vertical axis in Fig. 23.17, whereas the number of years doubles as we move across the horizontal axis. That is, one half of the original amount (80.0 g) of radium 226 decays in the first 1620 years (leaving 40.0 g). During the second half-life, one half of the remaining 40.0 g decays in the next 1620 years (leaving 20.0 g). During the third half-life, one half of the remaining 20.0 g decays in the next 1620 years (leaving 10.0 g). During the fourth half-life, one half of the remaining 10.0 g decays in the next 1620 years (leaving 5.0 g).

Figure 23.17   Radioactive decay of 80.0 g of radium 226 through four half-lives

Radioactive decay can be expressed by the equation

$$N = N_0\, e^{-\lambda t}$$

where N = number of remaining radioactive atoms
  N_0 = original number of radioactive atoms
  e = the natural number ($e = 2.718$ approximately)
  λ = decay constant
  t = elapsed time

The half-life $T_{1/2}$ is related to the decay constant by the equation

$$T_{1/2} = \frac{0.693}{\lambda}$$

where $T_{1/2}$ = half-life
  λ = decay constant

Note: The half-life formula involves an exponential equation with the natural number e, which is beyond the basic mathematics used in this book. However, to gain a basic sense of the results of this formula, exponential calculations involving the natural number e can easily be performed on calculators with an e^x key.

Find the half-life of $^{13}_{7}N$ if its decay constant is 1.16×10^{-3} decay/s.

EXAMPLE 2

Data:

$$\lambda = 1.16 \times 10^{-3}/s$$
$$T_{1/2} = ?$$

Basic Equation:

$$T_{1/2} = \frac{0.693}{\lambda}$$

Working Equation: Same

Substitution:

$$T_{1/2} = \frac{0.693}{1.16 \times 10^{-3}/s}$$
$$= 597\ s$$

That is, one half of the original amount will decay in 597 s.

Find the remaining amount of radioactive radium after 2450 years if the original amount was 3.54×10^{23} atoms. The half-life of radium is 1620 years.

EXAMPLE 3

Data:

$$N_0 = 3.54 \times 10^{23}$$
$$\lambda = \frac{0.693}{T_{1/2}} = \frac{0.693}{1620\ yr} = 4.28 \times 10^{-4}/yr$$
$$t = 2450\ yr$$
$$N = ?$$

Basic Equation:

$$N = N_0 e^{-\lambda t}$$

Working Equation: Same

Substitution:

$$N = (3.54 \times 10^{23})e^{-(4.28 \times 10^{-4}/yr)(2450\ yr)}$$
$$= (3.54 \times 10^{23})e^{-1.0486}$$
$$= 1.24 \times 10^{23}\ atoms$$

Note: $\frac{1.24 \times 10^{23}}{3.54 \times 10^{23}} = 35.0\%$ of the radioactive atoms remain after 2450 years.

Half-lives of atoms range from a small fraction of a second to billions of years. For short times, radioactive half-lives can be measured by waiting until one half of the atoms decay. For longer times, the rate at which a sample decays can be measured by using one of several types of radiation detectors.

Radiocarbon Dating

Radiocarbon dating is a method used to obtain age estimates of organic materials. The method is used in archaeology, geology, and other branches of science to find the age of wood, charcoal, bone and antler, peat and organic-bearing sediments, and other organic materials. Radioactive carbon 14 is produced when nitrogen 14 is bombarded by cosmic rays in the atmosphere, drifts down to earth, and is absorbed from the air by plants. Animals eat the plants and take the carbon into their bodies. Humans take the carbon into their bodies by eating both plants and animals. When a living organism dies, it stops absorbing the carbon; the carbon already in the organism begins to disintegrate. Scientists then can measure how much carbon 14 has disintegrated and how much remains. Carbon 14 decays at a slow, steady rate (half-life 5730 years) and reverts to nitrogen 14. The method involves counting the number of β radiations per minute per gram of material. Modern carbon 14 emits approximately 15 β radiations per gram of material; carbon material that is 5730 years old will emit only half that amount. Radiocarbon dating of organic materials older than 50,000 years is not reliable because the amount of carbon 14 remaining is so small that the fossil cannot be dated reliably. The accuracy of radiocarbon dating for some materials less than 50,000 years old has been verified through comparison with ancient historical records.

PHYSICS CONNECTIONS

Food Irradiation

The widespread threat of food-borne diseases and contamination of food from bacteria and insects as well as food's relatively short shelf-life have resulted in significant economic losses and health problems. In the 1950s, the U.S. Army began irradiating food to prolong its shelf-life for soldiers. In the 1960s, the U.S. Food and Drug Administration (FDA) approved the irradiation of wheat flour to limit the growth of mold. Since then, the FDA has tested and approved the irradiation of fruits, vegetables, spices, pork, chicken, and red meat to increase shelf-life and reduce or eliminate insects, parasites, and bacteria in food (Fig. 23.18).

Figure 23.18   (a) This symbol must be placed on all food that is exposed to food irradiation. (b) Food that has not been exposed and that has been exposed to irradiation, respectively

(a)   (b)

   Most food is irradiated using γ radiation from cobalt 60 and cesium 137 sources. The γ radiation produced from the radioactive sources breaks up the DNA molecules in the microbes, insects, and bacteria. The exposure to the radiation either kills the organisms or prevents them from reproducing. Insects are quickly killed upon breakup of their DNA, whereas bacteria may take longer to destroy. If fruits and vegetables are exposed to radiation, their DNA is destroyed, which delays unwanted sprouting and/or ripening.

   Upon seeing the irradiated food symbol, many people will not purchase the item. When food has been irradiated, it means that the food was exposed to radiation to destroy unwanted bacteria, insects, and microbes, but the food itself does not become radioactive. Irradiating food to increase shelf-life reduces the cost of rushed preparation and shipping costs. Such benefits would be realized not only by individuals in the United States, but also by individuals in third-world countries, where a great deal of food is prevented from being imported because it spoils before arriving at the ports. A few studies have indicated that irradiation reduces some essential vitamins and other nutrient levels that were present before the irradiation process. However, other studies have shown that such reductions are also found when food is canned or cooked.

Irradiation, although relatively new for the majority of our population, has been used for years by the military and on NASA space missions. A crop can be preserved for a significant amount of time as a result of irradiation.

PROBLEMS 23.7

1. Find the half-life of a radioactive sample if its decay constant is 1.72×10^4 decays/s.
2. Find the half-life of a radioactive sample if its decay constant is 8.25×10^{-6} decay/s.
3. Find the remaining quantity of $^{124}_{55}Cs$ from an original sample of 50.0 g after 4.00 min. Its half-life is 30.8 s.
4. Find the remaining quantity of radon 222 from an original sample of 75.0 g after 10.0 days. Its half-life is 3.82 days.
5. Find the percent of a sample of $^{124}_{55}Cs$ that will decay in the next 10.0 s. Its half-life is 30.8 s.
6. Find the percent of a sample of $^{238}_{92}U$ that will decay in the next 975 years. Its half-life is 4.47×10^9 years.
7. Find the remaining quantity of uranium 238 atoms from an original sample of 5.50×10^{20} atoms after 2.45 billion years. Its half-life is 4.50 billion years.
8. Find the remaining quantity of $^{14}_{6}C$ from an original sample of 3.75×10^{21} atoms after 1000 years. Its half-life is 5730 years.
9. Find the percent of a $^{14}_{6}C$ sample that will decay in the next 3000 years. Its half-life is 5730 years.
10. Find the percent of a radioactive sample of half-life 2.35 s that will decay in the next second.

SKETCH

$12\ cm^2$ | w

4.0 cm

DATA

$A = 12\ cm^2, l = 4.0\ cm, w = ?$

BASIC EQUATION

$A = lw$

WORKING EQUATION

$w = \dfrac{A}{l}$

SUBSTITUTION

$w = \dfrac{12\ cm^2}{4.0\ cm} = 3.0\ cm$

23.8   Nuclear Reactions—Fission and Fusion

Nuclear reactions take place when a nucleus is struck by a neutron, a γ ray, or another nucleus, causing an interaction. Many nuclear reactions have been viewed since Ernest Rutherford's 1919 observation of a reaction in which nitrogen, when bombarded by α particles, emitted a proton and was transformed into an isotope of oxygen. One of the most significant findings occurred during an experiment using neutron bombardment of existing elements in an attempt to produce new elements. Scientists, while attempting to form heavier elements using neutron bombardment of uranium, the heaviest known element at that time, made an amazing discovery, which has had a great impact on the world. This discovery led to the development of nuclear bombs and nuclear reactors used to produce heat energy for electric power, propulsion of nuclear submarines, various industrial processes, and other applications. They found that the bombardment of uranium with neutrons produces lighter nuclei, each nearly half the size of uranium, along with a number of neutrons (Fig. 23.19). The reaction can be written

$$n + {}^{235}_{92}U \longrightarrow {}^{236}_{92}U \longrightarrow {}^{141}_{56}Ba + {}^{92}_{36}Kr + 3n$$

Figure 23.19 Fission of $^{235}_{92}$U after neutron capture

$$n + ^{235}_{92}U \longrightarrow ^{236}_{92}U \longrightarrow \qquad\qquad ^{141}_{56}Ba + ^{92}_{36}Kr + 3n$$

This nuclear reaction, in which an atomic nucleus splits into fragments with the release of energy, is known as **nuclear fission**. A substantial amount of energy is released in this process because the mass of the uranium nucleus plus the mass of the bombarding neutron is larger than the combined mass of the fission fragments produced by the reaction. The total energy release per fission is approximately 200 MeV. On the nuclear scale, this is an extremely large amount of energy per atom!

The neutrons released in each nuclear fission reaction can be used to create further reactions. The process is called a **chain reaction**. **Enrico Fermi** and his associates succeeded in producing the first self-sustaining chain reaction at the University of Chicago in 1942. Chain reactions are used in nuclear reactors to produce electric power, to produce intense neutron sources for research and medical use, and in atomic bombs. **Robert J. Oppenheimer** led the research that resulted in the construction of the first atomic bomb. Many of the end products of fission reactions are radioactive for long periods of time. These radioactive wastes must be disposed of with extreme care.

Another process for releasing energy is **nuclear fusion**, a nuclear reaction in which light nuclei interact to form heavier nuclei with the release of energy. Nuclear fusion can occur because the total mass of the reaction products can be less than that of the initial nuclei and particles. Very high temperatures are required for nuclear fusion to occur. An example of a sequence of fusion reactions is

$$\begin{aligned}
^{1}_{1}H + ^{1}_{1}H &\longrightarrow ^{2}_{1}H + e^{+} &&(0.42 \text{ MeV}) \\
^{1}_{1}H + ^{2}_{1}H &\longrightarrow ^{3}_{2}He + \gamma &&(5.49 \text{ MeV}) \\
^{3}_{2}He + ^{3}_{2}He &\longrightarrow ^{4}_{2}He + ^{1}_{1}H + ^{1}_{1}H &&(12.86 \text{ MeV})
\end{aligned}$$

This reaction sequence is constantly going on in many stars, including our sun. Fusion is the source of energy in stars. The sun's energy is believed to be produced largely by this sequence of reactions, which starts with protons and produces helium nuclei (α particles) with the release of substantial energy. Nuclear fusion has been used in hydrogen bombs. Fusion is also a subject of considerable research as a "clean" energy source because the end products of the reaction are not radioactive.

23.9 Detection and Measurement of Radiation

Since we cannot see, touch, or feel nuclear radiation, it is necessary to have instruments available to detect its presence. The most common detector of radiation is the Geiger counter (Fig. 23.20), named after **Hans Wilhelm Geiger**. This instrument contains a cylindrical metal tube filled with gas. A wire runs along the center of the metal tube. A high voltage (approximately 1000 V) is maintained across the wire and cylinder. When a charged particle enters the tube and ionizes some of the gas, the voltage on the tube causes the gas to "break down" and conduct an electric pulse between the wire and the cylinder. This electric pulse is detected by electronics that count the number of pulses. A loudspeaker is often hooked up to the electronics, allowing a "click" to be heard each time a charged particle enters the tube.

Figure 23.20 Diagram of a Geiger counter

Since radiation can cause severe damage to living organisms and materials, it is necessary to be able to measure the amount, or *dose*, of radiation. There are two common units of *source activity*: the **curie** (Ci), which is defined as

$$1 \text{ Ci} = 3.70 \times 10^{10} \text{ disintegrations/s}$$

and the **becquerel** (Bq), which is the SI unit and defined as

$$1 \text{ Bq} = 1 \text{ disintegration/s}$$

The **source activity** is the strength of a source of radiation that can be specified at a given time according to the equation

$$A = \lambda N = \lambda N_0 e^{-\lambda t} = A_0 e^{-\lambda t}$$

where
A = source activity
N = remaining quantity of radioactive atoms
N_0 = original quantity
λ = decay constant
t = time
A_0 = original source activity

Find the source activity of a $^{222}_{88}\text{Ra}$ (radium) source 6.54 days after it was originally certified to have an activity of 0.356 Ci. Its half-life is 3.82 days.

EXAMPLE 1

Data:

$$t = 6.54 \text{ days}$$
$$A_0 = 0.356 \text{ Ci}$$
$$T_{1/2} = 3.82 \text{ days}$$
$$\lambda = \frac{0.693}{T_{1/2}} = \frac{0.693}{3.82 \text{ days}} = 0.181/\text{day}$$
$$A = ?$$

Basic Equation:

$$A = A_0 e^{-\lambda t}$$

Working Equation: Same

Substitution:

$$A = (0.356\ \text{Ci})e^{-(0.181/\text{day})(6.54\ \text{day})}$$
$$= (0.356\ \text{Ci})e^{-1.184}$$
$$= 0.109\ \text{Ci}$$

· · · · · · · · · · · · · · · · ·

PROBLEMS 23.9

1. Find the source activity of a 1.24-Ci sample of $^{13}_{7}\text{N}$ (nitrogen) 20.0 min after certification. Its half-life is 10.0 min.
2. Find the source activity of a 2.64-Ci sample of $^{14}_{6}\text{C}$ (carbon) 4000 years after certification. Its half-life is 5370 yr.
3. Find the source activity of a 0.476-Ci sample of $^{24}_{11}\text{Na}$ (sodium) 36.5 h after certification. Its half-life is 14.95 h.
4. Find the source activity of a 3.98-Ci sample of $^{11}_{6}\text{C}$ (carbon) 10.3 h after certification. Its half-life is 20.4 min.
5. Find the source activity of a 10.0-Ci sample of $^{226}_{88}\text{Ra}$ (radium) 125 yr after certification. Its half-life is 1590 yr.
6. Find the source activity of a 4.00-Ci sample of $^{131}_{53}\text{I}$ (iodine) 6.00 h after certification. Its half-life is 8.06 days.
7. Find the source activity of a 75.0-μCi sample of $^{214}_{84}\text{Po}$ (polonium) 5.00 μs after certification. Its half-life is 1.50×10^{-4} s.

23.10 Radiation Penetrating Power

The three types of radiation rays (α, β, and γ) can be stopped by matter. Some are stopped more easily than others (Fig. 23.21). Alpha rays, which are massive and carry a double positive charge, are stopped quite easily during "collisions" with atoms in the material. These collisions occur when the positively charged α particle comes close enough to an atom to feel a Coulomb force from some of the electrons or protons. The α particle gives up its energy to the material through these "collisions." A few pieces of paper or a few centimetres of air are sufficient to stop an α particle. As soon as an α particle slows enough to pick up two electrons while passing through matter, it becomes a harmless helium atom.

Beta rays, which are electrons, can easily be stopped by a thin sheet of metal, such as aluminum. They lose their energy due to many collisions with electrons in atoms in

Figure 23.21 Penetrating power of different forms of nuclear radiation

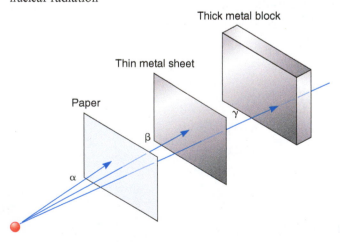

the material they are passing through. They become electrons in the material, indistinguishable from others.

Gamma rays have the greatest penetrating power because they have no charge and therefore cannot be electrically attracted or repelled by electrons or protons. The γ ray, which is an energetic photon, is slowed only through a direct hit of an electron or a nucleus. A dense material, such as a heavy metal, is the best absorber of γ radiation.

As radiation passes through matter and loses its energy, it can do much damage. Many materials become weakened and brittle upon exposure to substantial levels of radiation. This major engineering problem had to be solved in order to build safe nuclear reactors.

Radiation damage to living organisms can be very severe. As radiation is absorbed in material, atoms become ionized as electrons are captured or emitted. If these atoms are basic in bonding molecules together, the molecule may break apart. The functioning of the living cell may be altered substantially if a large number of key molecules are damaged. If the damage occurs to the DNA molecule, which controls the growth and replication of the cell, the cell may be severely damaged and die. Obviously, if a large number of cells die, the organism itself will become sick (radiation sickness) or die. If the damaged cell replicates itself rapidly with faulty DNA, cancer may result. To prevent this type of damage during normal medical X-ray procedures, it is common for patients to wear lead aprons around tissue not being imaged by the X ray. This is also why X-ray technicians step behind a lead shield when X rays are being taken.

Radiation can be useful for medical diagnostics (X rays and radioactive tracers) and radiation therapy. Cancer cells can be destroyed by the localized application of radiation therapy to a cancerous tumor.

PHYSICS ON THE JOB

Radiologic Technologist

Radiology is the science of X-ray imaging and other high-energy radiation as it pertains to the diagnosis and treatment of diseases. Radiologic technologists are hands-on medical personnel who are primarily responsible for operating various high-energy, radiological related scanners and imaging equipment.

Radiologic technologists typically receive, prepare, and administer imaging solutions and position patients in testing equipment. A doctor who specializes in radiology, called a radiologist, then analyzes and interprets the images produced by the equipment. Although radiologic technologists are not qualified to make medical decisions, magnetic resonance imaging (MRI) and radiologic technologists are acutely aware and trained to operate the high-energy, radioactive equipment that will yield accurate and clear images for diagnosing an issue with a patient.

An associate's degree is a minimum requirement to become a radiologic technologist. Specific coursework is needed in chemistry and physics to understand the functionality of the equipment, while coursework in anatomy, physiology, and biology is required to understand the human body undergoing the scan.

Radiologic technologists are typically educated on the basics of various medical imaging technologies. Individuals studying to become radiologic technologists are often trained to use X-ray equipment, computed tomography (CT) scans, nuclear medicines, positron emission tomography (PET) scans, and MRI scans. Other common equipment used by radiological technologists include ultrasound and mammography equipment (Fig. 23.22).

Figure 23.22   Radiologic technologist operating a CT scan

Courtesy of fotolia © Tyler Olson

Glossary

Alpha Ray A ray consisting of alpha particles, each having two protons and two neutrons, and positively charged; that is, it curves in the direction that known positive charges curve in a magnetic field. (p. 632)

Atom The smallest particle of an element that can exist in a stable or independent state. (p. 620)

Atomic Mass The mass of a single atom, usually expressed in atomic mass units, u. (p. 628)

Atomic Mass Number The total number of nucleons (protons and neutrons) in a nucleus, A. (p. 627)

Atomic Mass Unit (u) A unit of measure of atomic mass based on the mass of the common carbon atom, which has six protons and six neutrons, so the mass of the carbon 12 atom has been given the exact value of 12 u. Thus, the approximate mass of a single proton or a single neutron is 1 u, and the mass of an atom in atomic mass units is simply the sum of the number of its protons and its neutrons. (p. 628)

Atomic Number The number of protons in a nucleus, Z. (p. 627)

Average Binding Energy per Nucleon The total binding energy of the nucleus divided by the total number of nucleons. (p. 631)

Becquerel (Bq) Unit of source activity; 1 Bq = 1 disintegration/s. (p. 639)

Beta Ray A ray consisting of a stream of beta particles (electrons), which are emitted from neutrons in a nucleus as they decay into protons and electrons. This ray is negatively charged and curves in the opposite direction of an alpha ray in a magnetic field. (p. 632)

Binding Energy of the Nucleus The total energy required to break a nucleus apart into separate nucleons. (p. 629)

Bohr Model An early model of atomic structure in which electrons travel around the nucleus in a number of discrete stable energy levels determined by quantum conditions. (p. 622)

Chain Reaction The process of using the neutrons released in each nuclear fission reaction to create further reactions. (p. 638)

Coulomb Force An electric, repulsive force between protons. (p. 626)

Curie (Ci) Unit of source activity; 1 Ci = 3.70×10^{10} disintegrations/s. (p. 639)

Decay Rate The probability per unit time that a decay of radioactive isotopes will occur. (p. 633)

Disintegration Energy The total energy released in radioactive decay in the form of kinetic energy. (p. 633)

Electron A fundamental particle of an atom; negatively charged. (p. 620)

Excited State A high-energy level for an electron in an atom. (p. 621)

Fluorescence A property of certain substances in which radiation is absorbed at one frequency and then reemitted at a lower frequency. (p. 623)

Gamma Ray A ray composed of photons of electromagnetic radiation, which have no mass. (p. 632)

Ground State The lowest energy level for an electron in an atom. (p. 621)

Half-Life The length of time required for one half of the original amount of the radioactive atoms in a sample to decay. (p. 633)

Isotopes Nuclei that contain the same number of protons but a different number of neutrons. (p. 627)

Mass–Energy Equivalence A physical law stating that mass and energy are equivalent forms of matter; $E = \Delta mc^2$; stated by Einstein. (p. 629)

Neutron A fundamental particle in the nucleus of an atom; neutrally charged. (p. 620)

Neutron Number The number of neutrons in an atomic nucleus of a particular nuclide, N. Also, $N = A - Z$, the difference between the atomic mass number and the atomic number. N can be different for nuclei with the same atomic number. (p. 627)

Nuclear Fission A nuclear reaction in which an atomic nucleus splits into fragments with the release of energy. (p. 638)

Nuclear Fusion A nuclear reaction in which light nuclei interact to form heavier nuclei with the release of energy. (p. 638)

Nucleus The center part of the atom, made up of protons and neutrons. (p. 620)

Nuclide A specific type of atom characterized by its nuclear properties, such as the number of neutrons and protons and the energy state of its nucleus. (p. 627)

Periodic Table A table that contains all of the atomic elements arranged according to their atomic numbers and which can be used to predict their chemical properties. (p. 628)

Proton A fundamental particle in the nucleus of the atom; positively charged. (p. 620)

Quantum Mechanics A theory that unifies the wave–particle dual nature of electromagnetic radiation. The quantum model is based on the idea that matter, just like light, behaves sometimes as a particle and sometimes as a wave. (p. 624)

Radioactive Decay A type of nuclear decay that occurs when an unstable atom is transformed into a new element through the spontaneous disintegration of its nucleus. (p. 632)

Radiocarbon Dating A method used to obtain age estimates of organic materials using carbon 14 decay. (p. 636)

Source Activity The strength of a source of radiation that can be specified at a given time. (p. 639)

Strong Force An attractive force among all nucleons (neutrons and protons) independent of their charge. (p. 626)

Formulas

23.3 $E = -\dfrac{kZ^2}{n^2}$

23.6 $E = \Delta mc^2$

23.7 $Q = (M_p - M_d - m_\alpha)c^2$

$N = N_0 e^{-\lambda t}$

$T_{1/2} = \dfrac{0.693}{\lambda}$

23.9 $A = \lambda N = \lambda N_0 e^{-\lambda t} = A_0 e^{-\lambda t}$

Review Questions

1. Which of the following are nuclear particles?
   - (a) neutrons
   - (b) protons
   - (c) nucleons
   - (d) atoms
   - (e) all of the above

2. Einstein's equivalence principle relates to
   - (a) weight and time.
   - (b) space and gravity.
   - (c) mass and energy.
   - (d) all of the above.
   - (e) none of the above.

3. The amount of radioactive material remaining after a period of time is related to the
   - (a) atomic mass.
   - (b) pressure.
   - (c) half-life.
   - (d) volume.
   - (e) none of the above.

4. Explain the difference between the ground state and the excited states of electrons in atoms.
5. Describe the Bohr model of the atom.
6. Describe the similarities of protons and neutrons. Describe the differences.
7. Describe the differences between the electric force and the strong force.
8. If the strong force suddenly became much weaker in a nucleus while the electric force remained unchanged, what might happen to the nucleus?
9. Explain the principle of mass–energy equivalence in your own words.
10. What is the difference among the following atoms: $^{11}_{6}C$, $^{12}_{6}C$, and $^{13}_{6}C$? What are the similarities?
11. Explain the term *electron volt* in your own words.
12. Describe the importance of the neutron in atomic nuclei.
13. Describe an α ray in your own words.
14. Describe a γ ray in your own words.
15. Describe a β ray in your own words.
16. What important discovery was made by Enrico Fermi?
17. Explain a self-sustaining chain reaction.
18. Describe nuclear fusion.
19. Describe nuclear fission.
20. What fraction of a radioactive sample has not decayed after four half-lives have elapsed?
21. What damage can be caused to living organisms by radiation?
22. What medical uses does radiation have?

Review Problems

SKETCH

12 cm^2 | w

4.0 cm

DATA

$A = 12 \text{ cm}^2, l = 4.0 \text{ cm}, w = ?$

BASIC EQUATION

$A = lw$

WORKING EQUATION

$w = \dfrac{A}{l}$

SUBSTITUTION

$w = \dfrac{12 \text{ cm}^2}{4.0 \text{ cm}} = 3.0 \text{ cm}$

1. Find the energy of the electron in the $n = 5$ energy level of a hydrogen atom.
2. Find the energy of the photon given off when the electron in Problem 1 transitions down to the $n = 3$ energy level.
3. List four inert atoms from the Periodic Table.
4. List four metals from the Periodic Table.
5. Find the mass in kilograms of the $^{15}_{8}O$ atom. Its mass in atomic mass units is 15.003065 u.
6. Find the mass in kilograms of the $^{19}_{9}F$ atom. Its mass in atomic mass units is 18.998404 u.
7. Find the mass in kilograms of the $^{166}_{68}Er$ atom. Its mass in atomic mass units is 165.930292 u.
8. Find the binding energy in MeV of a $^{86}_{38}Sr$ atom. The mass of the neutral Sr atom is 85.909266 u.
9. Estimate the average binding energy for $^{102}_{44}Ru$ from Fig. 23.11.
10. Estimate the average binding energy for $^{153}_{63}Eu$ from Fig. 23.11.
11. Estimate the average binding energy for $^{187}_{75}Re$ from Fig. 23.11.
12. Find the remaining quantity of osmium 191 atoms from an original sample of 8.25×10^{13} atoms after 54 days. Its half-life is 15.4 days.
13. Find the remaining quantity of iodine 131 atoms from an original sample of 8.33×10^{18} atoms after 34.4 days. Its half-life is 8.04 days.
14. Find the percent of a strontium 88 sample that will decay in the next 2.40 yr. Its half-life is 29.1 yr.
15. Find the percent of an osmium 191 sample that will decay in the next 2.00 h. Its half-life is 15.4 days.
16. Find the source activity of a 2.43-Ci sample of osmium 191 43.3 days after certification. Its half-life is 15.4 days.
17. Find the source activity of a 3.79-Ci sample of nitrogen 13 43.0 min after certification. Its half-life is 10.0 min.

18. Find the source activity of a 9.41-Ci sample of carbon 11 95.4 min after certification. Its half-life is 20.4 min.

19. Find the source activity of a 6.75-Ci sample of $^{238}_{92}U$ (uranium) one billion years after certification. Its half-life is 4.5×10^9 yr.

20. Find the source activity of a 50.0-μCi sample of $^{214}_{82}Pb$ (lead) 1.00 h after certification. Its half-life is 26.8 min.

APPLIED CONCEPTS

1. To produce the blue neon light used in many storefront signs, an electron from a hydrogen atom must undergo a transition from one energy level to another. (a) What is the transition energy (eV) if the electron changes its energy level from $n = 4$ to $n = 2$? (b) What is the frequency for the photon that is released? (c) What is the photon's velocity? (d) What is the wavelength for the released photon?

2. A photon is released from a hydrogen atom when an electron moves from a higher energy level to a lower energy level. At what level does an electron begin its journey to the level $n = 1$ if the emitted photon has a wavelength of 1.22×10^{-7} m?

3. The binding energy for a ^4_2He nucleus is 28.40 MeV. (a) What is the binding energy (in J)? (b) How many protons, neutrons, and electrons are in the ^4_2He atom? (c) Find the total mass of the various subatomic particles in the ^4_2He atom. (d) What is the mass of the neutral ^4_2He atom?

4. A source of radon 222 has 2.45×10^{23} atoms remaining after 10.0 days. (a) If its half-life is 3.82 days, what is its decay rate? (b) Find the original number of atoms before the decay.

5. Find the source activity of a 0.875-Ci sample of $^{24}_{11}Na$ (sodium) 32.4 h before certification if its half-life is 14.95 h.

CHAPTER 24

SPECIAL AND GENERAL RELATIVITY

Objectives

The major goals of this chapter are to enable you to:

1. Discuss the theories, successes, and milestones that made Albert Einstein an influential scientist.
2. Explain that the speed of light is constant, but the rate at which time passes can vary according to motion.
3. Find the amount of energy contained in a particular amount of mass.
4. Describe the theory of a four-dimensional space-time universe.
5. Express Einstein's theory that acceleration and gravity are equivalent.

At the age of 16, Albert Einstein dreamed of what it would be like to ride on a beam of light. His enthusiasm for science, specifically the topics of light, space, and time, led him to develop several theories that would make him the most famous scientist of all time. In this chapter we will explore the life of Einstein, the warping of space-time, and the formula that changed the world, $E = mc^2$.

24.1 Albert Einstein

Born in Ulm, Germany, in 1879, **Albert Einstein** spent most of his younger years reading and engaging in science experiments, while rebelling against the traditional regimentation found in many German schools at the time. He was given a compass at the age of five by his father and was intrigued by how the compass needle would move even though nothing was physically touching it. The physical mystery surrounding the compass fascinated Einstein and set him on a course that revolutionized science. However, his path was not easy and included being expelled, dropping out of school, and failing academic entrance exams. Einstein was eventually able to enroll in the Zurich Polytechnic Institute, where he was largely left to explore the wonders of science on his own.

After graduation, Einstein was unable to find a full-time job as an assistant in a laboratory. Instead, he obtained a job at the Swiss Patent Office. During this period he did some of his greatest work, including his 1905 paper about the **special theory of relativity**, which is based on the following basic assumptions: (1) *the laws of physics are the same in both moving and nonmoving frames of reference* and (2) *the speed of light is constant no matter what the speed of the observer or the source of light.* The special theory boosted Einstein's physics career and caused many universities to take notice of the previously unknown patent clerk.

Ten years later, Einstein produced a paper that extended the special theory to accelerated frames of reference. The **general theory of relativity** is based on the assumption that *gravity and acceleration are equivalent and that light has mass and its path can be warped by gravity.* In 1921, Einstein won the Nobel Prize in Physics, although not for relativity, but for his work on the photoelectric effect. He began to focus on a *theory that would link what came to be understood as the five fundamental forces: gravitational, electric, magnetic, strong, and weak forces.* The work that Einstein began is continued today by physicists around the world and is now called the **grand unified theory**.

During the turbulent years leading up to World War II, Germany was not a safe place for the world's most famous Jewish scientist. In the 1930s, Einstein decided to leave Germany and live in the United States. He relocated to Princeton, New Jersey, where he continued to lecture and work on his unified theory. His famous formula $E = mc^2$ was the scientific principle behind the atomic bomb, and he urged the United States to develop such a weapon because he feared that Germany might develop one first. After the war he opposed its use.

Albert Einstein died in Princeton in 1955. Throughout his life, he addressed many major scientific matters and was also a strong moral voice on social and political issues. Einstein showed great compassion for humankind and was known throughout the world as the genius professor who attempted, as Einstein said, "to read the mind of God" (Fig. 24.1).

Figure 24.1   Albert Einstein at work in his study at Princeton University in 1943

Photo courtesy of the National Archives and Records Administration

24.2   Special Theory of Relativity

Let's say you are reading this textbook while sitting on the ground. Are you moving? You would probably say no because when you look around, nothing seems to be changing its position. However, you suddenly hear a noise and see a man driving a car down the street. You see that the car is changing position, but how do you know that the car is moving and you are not?

The driver of the car looks out the car window and sees you moving closer and then away from him after passing you. The driver can adopt the perspective that you are moving while he is at rest. In fact, he can see everything outside the car as moving and himself as stationary. This is a simplistic example of **relative motion**, which is the concept that motion can be described differently depending upon the observer's perspective. In order for a motion to be fully described, it must be described in reference to something or someone's perspective. Both the driver and you see the same relative motion in two different ways. You see the car moving toward you, while the driver can argue that you are moving toward him. The difference can be expressed through the notion of *relativity*: the same event can be observed differently depending on the perspective of the observer.

As a result of Einstein's thoughts on motion and light, he concluded that the laws of physics must be the same for all nonaccelerating objects. This is called Einstein's **first postulate of special relativity**: *the laws of physics are the same for both moving and nonmoving frames of reference.* If an individual in a moving train drops a ball while the train is moving at a constant velocity, it will appear to that person that the ball accelerates directly down toward the floor of the train. A ball also will fall directly toward the ground if a person standing on the side of the tracks drops it. In other words, Einstein believed that the laws of physics are the same for objects at rest as they are for objects moving at a constant velocity.

Einstein also proposed the **second postulate of special relativity**: *the speed of light is constant regardless of the speed of the observer or the light source.*

$$c = \text{speed of light} = 3.00 \times 10^8 \text{ m/s} = 1.86 \times 10^5 \text{ mi/s}$$

Einstein created a "thought experiment" to demonstrate the implications of the postulate that the speed of light is constant. In his thought experiment, two telephone poles are struck simultaneously by lightning at the instant a train moving close to the speed of light is passing the midpoint of the two poles. From the perspective of a person standing on the side of the tracks, the lightning strikes both poles at exactly the same time [Fig. 24.2(a)]. A person on the train sees the lightning first strike pole B in front of the train and then strike pole A behind the train an instant later [Fig. 24.2(b)] because, as the train moves closer to pole B, the light from pole B takes less time than the light from pole A to reach the person on the train [Fig. 24.2(c)]. Einstein's thought experiment showed, in theory, that two people can see the same event happen at different points in time. The distance and time may vary, but the speed of light will always be constant. By changing your state of motion, as was the case with the individual on the train relative to the person on the side of the tracks, the temporary relationship of events can be altered.

PHYSICS CONNECTIONS

Albert Einstein: "The Person of the Century"

"Why is it that nobody understands me, yet everybody likes me?" This quote from Einstein, taken from an interview with *The New York Times* in 1944, is reflective of Einstein's fame and genius. Although Albert Einstein is recognized around the world as a genius, very few people understand what his genius meant for the scientific world. Despite his superior intellect, he was not able to understand why so many people, who did not understand his scientific work, adored him.

Einstein became world famous after two British scientists experimentally proved Einstein's theory that the path of light could be warped due to gravity. This occurred in 1919, just after the end of World War I, when the world needed something to rally around and get excited about. His revolutionary scientific discovery; his "crazy professor" appearance; and his outspoken views on life, human nature, and politics carried his fame to nonscientists throughout the world.

Today, more than a half-century after his death, Albert Einstein's name is still synonymous with "genius." Einstein's unique quirkiness, funny facial expressions, and thoughtful quotes can now be seen on coffee mugs, T-shirts, and posters, and in books and magazines. In addition, numerous fictional and documentary films have been made about his life and scientific work. *Time Magazine* recognized Albert Einstein's impact on the world by awarding him the title, "Person of the Century."

Figure 24.2 (a) Light travels the same distance from each telephone pole to the observer on the ground, who sees the two lightning strikes at the same time. (b) A person on the moving train experiences the same event differently. (c) As the train moves closer to pole B, the light from pole B reaches the person on the train sooner than the light from pole A because pole B is now a bit closer and the light from it takes less time to reach this observer.

$E = mc^2$

Albert Einstein's development of the equation $E = mc^2$ became one of the most important scientific achievements of the 20th century. This equation defines the equivalence between energy and mass.

$$E = mc^2$$

where E = energy
m = mass
$c = 3.00 \times 10^8$ m/s $= 1.86 \times 10^5$ mi/s (the speed of light)

A very little mass is equivalent to a tremendous amount of energy. However, converting mass to 100% energy is virtually impossible. If you take your pencil eraser and convert it into energy by burning it, some of the eraser's mass would be converted into energy in the form of heat, light, and sound. Yet some quantity of the eraser would remain as burnt ash, which still has mass and therefore energy. Nuclear reactions, as discussed in Chapter 23, are the best methods scientists have for converting mass to pure energy.

How much energy is contained in 1.00 g of matter?

EXAMPLE

Data:

$m = 1.00$ g $= 1.00 \times 10^{-3}$ kg
$c = 3.00 \times 10^8$ m/s
$E = ?$

Basic Equation:

$$E = mc^2$$

Working Equation: Same

Substitution:

$$E = (1.00 \times 10^{-3} \text{ kg}) (3.00 \times 10^8 \text{ m/s})^2$$
$$= 9.00 \times 10^{13} \text{ J}$$

$$\boxed{\text{kg (m/s)}^2 = (\text{kg m/s}^2)\, \text{m} = \text{N m} = \text{J}}$$

This very large amount of energy, produced from only 1.00 g of mass, can light a 60-W light bulb for 47,600 yr!

· · · · · · · · · · · · · · · · · ·

PROBLEMS 24.2

SKETCH

12 cm² | w

4.0 cm

DATA
$A = 12 \text{ cm}^2, l = 4.0 \text{ cm}, w = ?$

BASIC EQUATION
$A = lw$

WORKING EQUATION
$w = \dfrac{A}{l}$

SUBSTITUTION
$w = \dfrac{12 \text{ cm}^2}{4.0 \text{ cm}} = 3.0 \text{ cm}$

1. If the tip of a pencil has a mass of 2.30 g, how much energy can be produced if all the mass is converted to energy?
2. If a textbook has a mass of 1.30 kg, how much energy can be released if the textbook is converted to energy?
3. How much mass is needed to create $60\overline{0}$ J of energy?
4. How much mass is needed to create 67.0 J of energy?

24.3 Space-Time

Scientists and mathematicians use coordinates to locate objects in our physical world. A pilot defines her location by specifying her longitude, latitude, and altitude. Similarly, mathematicians and scientists use coordinates in relation to the x-, y-, and z-axes to define a position in three dimensions (Fig. 24.3).

Albert Einstein agreed that objects exist in three spatial dimensions, but he felt that three spatial dimensions alone could not accurately locate an object at any

Figure 24.3   Spatial location can be determined on a three-axis coordinate system.

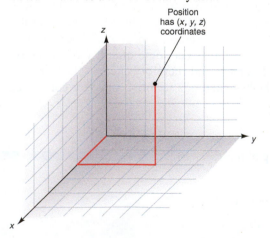

particular instant. He felt that a time dimension needed to be taken into consideration when defining the dimensions of the physical universe. His argument was that anything could be located in a particular spatial location, but it could also change its position over time. **Space-time** allows an object's position in the universe to be defined using the three spatial dimensions (x, y, z) and one time dimension.

Space-time plays an important role in Einstein's theory of general relativity. Einstein's theory describes the universe as a place where massive objects, such as the sun, planets, and stars, warp the four dimensions of space-time. The warping of space-time influences the motion of an object traveling through the universe (Fig. 24.4). Black holes warp space to such an extent that mass and light cannot escape.

Figure 24.4   Illustration of light drawn into a black hole

Photo courtesy of Getty Images, Inc.—Photodisc/Royalty Free

24.4 General Theory of Relativity

Although the special theory of relativity focuses on the relationship between motion and time, Einstein's **general theory of relativity** states that gravity is equivalent to acceleration and that the path of light is warped by gravity.

Einstein designed several thought experiments to illustrate general relativity. In one thought experiment, he pictured a person in a spaceship accelerating at 9.80 m/s^2. In such a situation, the spaceship applies a force on the astronaut equal to the force of gravity on the earth. As a result, Einstein concluded, gravity and acceleration are equivalent. Einstein continued in his thought experiment. If a beam of light comes through a window as the spaceship accelerates, the light will strike the opposite wall just a bit lower than the height at which it entered. As a result of the accelerating spaceship, the light appears to bend in the accelerating/gravitational field. Finally, since gravity only alters the path of objects with mass, Einstein concluded that light must have mass (Fig. 24.5).

In 1919, British scientists traveled to South America to observe a solar eclipse and experimentally prove Einstein's theory of relativity. On the night before the solar eclipse, the scientists observed the exact positions of several stars. The next day, when the moon eclipsed the sun, the scientists observed the same stars in the darkened sky. However, the stars appeared to have slightly changed their positions in the sky. The light from the stars curved around the warp in space-time created by the massive sun and moon. The apparent shift in the stars' positions experimentally proved Einstein's theory of general relativity and made him a world-renowned celebrity (Fig. 24.6).

Figure 24.5 (a) If the spaceship accelerates at 9.80 m/s², the force on the person inside will feel the same as the pull of gravity on earth, showing that acceleration is the same as gravity. (b) Although this depiction is exaggerated, the light would strike the wall lower than where it entered, showing that gravity bends light.

Figure 24.6 Massive objects warp space, causing light from distant stars to bend, altering their apparent position in the sky.

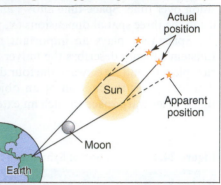

Glossary

$E = mc^2$ The equation that illustrates and defines the equivalence between energy and mass. (p. 649)

First Postulate of Special Relativity The laws of physics are the same for both moving and nonmoving frames of reference. (p. 648)

General Theory of Relativity Extends the special theory of relativity to accelerated frames of reference with the assumption that gravity and acceleration are equivalent and that light has mass and its path can be warped by gravity. (p. 647)

Grand Unified Theory A theory linking the five fundamental forces: gravitational, electric, magnetic, strong, and weak forces. (p. 647)

Relative Motion The concept that motion can be described differently depending upon the observer's perspective. (p. 647)

Second Postulate of Special Relativity The speed of light is constant regardless of the speed of the observer or of the light source. (p. 648)

Space-Time An object's position in the universe can be pinpointed using three spatial dimensions and one time dimension. (p. 651)

Special Theory of Relativity The laws of physics are the same in moving and nonmoving frames of reference and the speed of light is constant no matter what the speed of the observer or light source. (p. 647)

Formula

24.2 $E = mc^2$

Review Questions

1. What field or fields of physics intrigued Einstein at a very young age?
   (a) Momentum
   (b) Thermodynamics
   (c) Electricity and magnetism
   (d) Sound and waves
2. Which of the following did Albert Einstein not complete?
   (a) General theory of relativity
   (b) Special theory of relativity
   (c) Principle of the warping of space-time
   (d) Grand unified Theory
3. Describe Einstein's first postulate of special relativity.
4. Explain a situation where you experienced the first postulate of special relativity.
5. If you are riding a bike at 10.0 m/s and throw a ball forward with a velocity of 4.00 m/s, will the ball travel 4.00 m/s relative to the ground, or will it travel at 10.0 m/s plus another 4.00 m/s relative to the ground? Explain your reasoning.
6. Describe Einstein's second postulate of special relativity.
7. According to Einstein's second postulate, if you are moving 10.0 m/s on a bicycle with a headlight on, will the light from the bicycle have a velocity of 3.00×10^8 m/s or will it have a velocity of 3.00×10^8 m/s plus 10.0 m/s? Explain your reasoning.
8. What does traveling close to the speed of light do to the dimension of time?
9. While you are sitting and reading this question, are you traveling through the space dimensions or the time dimension?
10. Explain what $E = mc^2$ represents.
11. What are the four dimensions of our space-time universe? Why is time an important dimension?
12. What does Einstein's general theory of relativity state about the relationship between gravity and acceleration?
13. What happens to light and other electromagnetic waves around massive objects such as the sun?
14. Explain how the solar eclipse of 1919 proved Einstein's theory that light has mass.

Review Problems

1. A train is moving at a speed of 65.0 mi/h. The ticket collector is walking 2.00 mi/h toward the front of the train. How fast is the ticket collector moving from the point of view of a person on the train?
2. How fast is the ticket collector in Problem 1 moving relative to an observer on the side of the tracks?
3. The ticket collector in Problem 1 turns around and walks toward the rear of the train at 2.00 mi/h. How fast is he walking relative to an observer on the side of the tracks?
4. Convert the mass of one electron ($m = 9.10 \times 10^{-31}$ kg) to energy.
5. Convert the mass of one proton ($m = 1.67 \times 10^{-27}$ kg) to energy.
6. A particular task requires 9.80 J of energy. Using $E = mc^2$, determine how much mass is needed to accomplish this task.

MATHEMATICS REVIEW

A.1 Signed Numbers

Signed numbers have many applications in the study of physics. The rules for working with signed numbers follow.

Adding Signed Numbers

To add two positive numbers, add their absolute values.* A positive sign may or may not be placed before the result. It is usually omitted.

Add: ◄ **EXAMPLE 1**

(a)　+ 4
　　 + 7
　　────
　　+ 11　or　11

(b)　$(+3) + (+5) = +8$　or　8

To add two negative numbers, add their absolute values and place a negative sign before the result.

Add: ◄ **EXAMPLE 2**

(a)　−2
　　 −5
　　────
　　−7

(b)　$(-6) + (-7) = -13$
(c)　$(-8) + (-4) = -12$

To add a negative number and a positive number, find the difference of their absolute values. Place the sign of the number having the larger absolute value before the result.

Add: ◄ **EXAMPLE 3**

(a)　+4
　　 −6
　　────
　　−2

(b)　−2
　　 +8
　　────
　　+6

(c)　−8
　　 +3
　　────
　　−5

(d)　+9
　　 −4
　　────
　　+5

(e)　$(+7) + (-2) = +5$
(g)　$(-3) + (+10) = +7$

(f)　$(-9) + (+6) = -3$
(h)　$(+4) + (-12) = -8$

───────────

*The absolute value of a number is its nonnegative value. For example, the absolute value of −6 is 6; the absolute value of +10 is 10; and the absolute value of 0 is 0.

To add three or more signed numbers:

1. Add the positive numbers.
2. Add the negative numbers.
3. Add the sums from steps 1 and 2 according to the rules for addition of signed numbers.

EXAMPLE 4

Add: $(-2) + 4 + (-6) + 10 + (-7)$.

Step 1: $+4$ Step 2: -2 Step 3: -15
$\ +10$ $\ -6$ $\ +14$
$\ \overline{+14}$ $\ -7$ $\ \overline{-1}$
$\ \overline{-15}$

Therefore, $(-2) + 4 + (-6) + 10 + (-7) = -1$.

Subtracting Signed Numbers

To subtract two signed numbers, change the sign of the *number being subtracted* and *add* according to the rules for addition.

EXAMPLE 5

Subtract:

(a) Subtract: $+3$ ↔   Add: $+3$ To subtract, change the sign of the
$+7$ -7 number being subtracted, $+7$, and
$\overline{-4}$ $\overline{-4}$ add.

(b) Subtract: -9 ↔   Add: -9 To subtract, change the sign of the
-6 $+6$ number being subtracted, -6, and
$\overline{-3}$ $\overline{-3}$ add.

(c) Subtract: $+8$ ↔   Add: $+8$
-4 $+4$
$\overline{+12}$ $\overline{+12}$

(d) Subtract: -6 ↔   Add: -6
$+8$ -8
$\overline{-14}$ $\overline{-14}$

(e) $(+6) - (+8) = (+6) + (-8) = -2$ To subtract, change the sign of the number being subtracted, $+8$, and add.

(f) $(-3) - (-5) = (-3) + (+5) = +2$
(g) $(+10) - (-3) = (+10) + (+3) = +13$
(h) $(-5) - (+2) = (-5) + (-2) = -7$

When more than two signed numbers are involved in subtraction, change the sign of *each* number being subtracted and add the resulting signed numbers.

EXAMPLE 6

Subtract: $(-2) - (+4) - (-1) - (-3) - (+5)$.
$= (-2) + (-4) + (+1) + (+3) + (-5)$

Step 1: +1   Step 2: −2   Step 3: +4
         +3   −4   −11
         ---   −5   ---
         +4   ---   −7
                            −11

Therefore, $(-2) - (+4) - (-1) - (-3) - (+5) = -7$.

. . . .

When combinations of addition and subtraction of signed numbers occur in the same problem, change *only* the sign of each number being subtracted. Then add the resulting signed numbers.

EXAMPLE 7

Find the result:

$(-2) + (-4) - (-3) - (+6) + (+1) - (+2) + (-7) - (-5)$
$= (-2) + (-4) + (+3) + (-6) + (+1) + (-2) + (-7) + (+5)$

Step 1:   +3   Step 2:   −2   Step 3:   +9
          +1   −4   −21
          +5   −6   ---
          ---   −2   −12
          +9   −7

                              −21

Therefore, $(-2) + (-4) - (-3) - (+6) + (+1) - (+2) + (-7) - (-5) = -12$.

. . . .

Multiplying Signed Numbers

To multiply two signed numbers:

1. If the signs of the numbers are both positive or both negative, find the product of their absolute values. This product is always positive.
2. If the signs of the numbers are unlike, find the product of their absolute values and place a negative sign before the result.

EXAMPLE 8

Multiply:

(a)   +3   (b)   −5   (c)   −6   (d)   +2
       +4   −8   +7   −3
       ---   ---   ---   ---
       +12   +40   −42   −6

(e) $(+3)(+5) = +15$ (f) $(-7)(-8) = +56$
(g) $(-1)(+6) = -6$ (h) $(+4)(-2) = -8$

. . . .

To multiply more than two signed numbers, first multiply the absolute values of the numbers. If there is an odd number of negative factors, place a negative sign before the result. If there is an even number of negative factors, the product is positive. *Note:* An *even* number is a number divisible by 2.

EXAMPLE 9

Multiply:

(a) $(+5)(-6)(+2)(-1) = +60$
(b) $(-3)(-3)(+4)(-5) = -180$

. . . .

Dividing Signed Numbers

The rules for dividing signed numbers are similar to those for multiplying signed numbers.

To divide two signed numbers:

1. If the signs of the numbers are both positive or both negative, divide their absolute values. This quotient is always positive.
2. If the two numbers have different signs, divide their absolute values and place a negative sign before the quotient.

Note: Division by 0 is undefined.

EXAMPLE 10

Divide:

(a) $\dfrac{+10}{+2} = +5$ (b) $\dfrac{-18}{-3} = +6$ (c) $\dfrac{+20}{-4} = -5$ (d) $\dfrac{-24}{+2} = -12$

Problems A.1

Perform the indicated operations.

1. $(-5) + (-6)$
2. $(+1) + (-10)$
3. $(-3) + (+8)$
4. $(+5) + (+7)$
5. $(-5) + (+3)$
6. $0 + (-3)$
7. $(-7) - (-3)$
8. $(+2) - (-9)$
9. $(-4) - (+2)$
10. $(+4) - (+7)$
11. $0 - (+3)$
12. $0 - (-2)$
13. $(-9)(-2)$
14. $(+4)(+6)$
15. $(-7)(+3)$
16. $(+5)(-8)$
17. $(+6)(0)$
18. $(0)(-4)$
19. $\dfrac{+36}{+12}$
20. $\dfrac{-9}{-3}$
21. $\dfrac{+16}{-2}$
22. $\dfrac{-15}{+3}$
23. $\dfrac{0}{+6}$
24. $\dfrac{4}{0}$
25. $(+2) + (-1) + (+10)$
26. $(-7) + (+2) + (+9) + (-8)$
27. $(-9) + (-3) + (+3) + (-8) + (+4)$
28. $(+8) + (-2) + (-6) + (+7) + (-6) + (+9)$
29. $(-4) - (+5) - (-4)$
30. $(+3) - (-5) - (-6) - (+5)$
31. $(-7) - (-4) - (+6) - (+4) - (-5)$
32. $(-8) - (+7) - (+3) - (-7) - (-8) - (-2)$
33. $(+5) + (-2) - (+7)$
34. $(-3) - (-8) - (+3) + (-9)$
35. $(-2) - (+1) - (-10) + (+12) + (-9)$
36. $(-1) - (-11) + (+2) - (-10) + (+8)$
37. $(+3)(-5)(+3)$
38. $(-1)(+2)(+2)(-1)$
39. $(+2)(-4)(-6)(-3)(+2)$
40. $(-1)(+3)(-2)(-4)(+5)(-1)$

A.2 Powers of 10

The ability to work quickly and accurately with powers of 10 is important in scientific and technical fields.

> When multiplying two powers of 10, add the exponents. That is,
>
> $$10^a \times 10^b = 10^{a+b}$$

EXAMPLE 1

Multiply:

(a) $(10^6)(10^3) = 10^{6+3} = 10^9$ (b) $(10^4)(10^2) = 10^{4+2} = 10^6$
(c) $(10^1)(10^{-3}) = 10^{1+(-3)} = 10^{-2}$ (d) $(10^{-2})(10^{-5}) = 10^{[-2+(-5)]} = 10^{-7}$

When dividing two powers of 10, subtract the exponents as follows:
$$10^a \div 10^b = 10^{a-b}$$

Divide: ◀ **EXAMPLE 2**

(a) $\dfrac{10^7}{10^4} = 10^{7-4} = 10^3$ (b) $\dfrac{10^3}{10^5} = 10^{3-5} = 10^{-2}$

(c) $\dfrac{10^{-2}}{10^{+3}} = 10^{(-2)-(+3)} = 10^{-5}$ (d) $\dfrac{10^4}{10^{-2}} = 10^{4-(-2)} = 10^6$

· · · · · · · · · · · · · · · · ·

To raise a power of 10 to a power, multiply the exponents as follows:
$$(10^a)^b = 10^{ab}$$

Find each power: ◀ **EXAMPLE 3**

(a) $(10^2)^3 = 10^{(2)(3)} = 10^6$ (b) $(10^{-3})^2 = 10^{(-3)(2)} = 10^{-6}$
(c) $(10^4)^{-5} = 10^{(4)(-5)} = 10^{-20}$ (d) $(10^{-3})^{-4} = 10^{(-3)(-4)} = 10^{12}$

· · · · · · · · · · · · · · · · ·

Next, we will show that $10^0 = 1$. To do this, we need to use the substitution principle, which states that

$$\text{if} \quad a = b \quad \text{and} \quad a = c, \quad \text{then} \quad b = c.$$

First,

$$\dfrac{10^n}{10^n} = 10^{n-n} \qquad \text{To divide powers, subtract the exponents.}$$
$$= 10^0$$

Second,

$$\dfrac{10^n}{10^n} = 1 \qquad \text{Any number other than zero divided by itself equals 1.}$$

That is, since

$$\dfrac{10^n}{10^n} = 10^0 \quad \text{and} \quad \dfrac{10^n}{10^n} = 1$$

then $10^0 = 1$.

We also will use the fact that $\dfrac{1}{10^a} = 10^{-a}$. To show this, we write

$$\dfrac{1}{10^a} = \dfrac{10^0}{10^a} \qquad (1 = 10^0)$$
$$= 10^{0-a} \qquad \text{To divide powers, subtract the exponents.}$$
$$= 10^{-a}$$

We also need to show that $\frac{1}{10^{-a}} = 10^a$. We write

$$\frac{1}{10^{-a}} = \frac{10^0}{10^{-a}}$$
$$= 10^{0-(-a)}$$
$$= 10^a$$

In summary,

$$10^0 = 1 \qquad \frac{1}{10^a} = 10^{-a} \qquad \frac{1}{10^{-a}} = 10^a$$

Actually, any number (except zero) raised to the zero power equals 1.

Problems A.2
Do as indicated. Express the results using positive exponents.

1. $(10^5)(10^3)$
2. $10^6 \div 10^2$
3. $(10^2)^4$

4. $(10^{-2})(10^{-3})$
5. $\dfrac{10^3}{10^6}$
6. $(10^{-3})^3$

7. $10^5 \div 10^{-2}$
8. $(10^{-2})^{-3}$
9. $(10^4)(10^{-1})$

10. $\dfrac{10^0}{10^{-4}}$
11. $(10^0)(10^{-4})$
12. $\dfrac{10^{-4}}{10^{-3}}$

13. $(10^0)^{-2}$
14. 10^{-3}
15. $\dfrac{1}{10^{-5}}$

16. $\dfrac{(10^4)(10^{-2})}{(10^6)(10^3)}$
17. $\dfrac{(10^{-2})(10^{-3})}{(10^3)^2}$
18. $\dfrac{(10^2)^4}{(10^{-3})^2}$

19. $\left(\dfrac{1}{10^3}\right)^2$
20. $\left(\dfrac{10^2}{10^{-3}}\right)^2$
21. $\left(\dfrac{10 \cdot 10^2}{10^{-1}}\right)^2$

22. $\left(\dfrac{1}{10^{-3}}\right)^2$
23. $\dfrac{(10^4)(10^{-2})}{10^{-8}}$
24. $\dfrac{(10^4)(10^6)}{(10^0)(10^{-2})(10^3)}$

A.3 Solving Linear Equations
An equation is a mathematical sentence stating that two quantities are equal. To solve an equation means to find the number or numbers that can replace the variable in the equation to make the equation a true statement. The value we find that makes the equation a true statement is called the *root* or solution of the equation. When the root of an equation is found, we say that we have *solved* the equation.

If $a = b$, then $a + c = b + c$ or $a - c = b - c$. (If the same quantity is added to or subtracted from both sides of an equation, the resulting equation is equivalent to the original equation.)

To solve an equation using this rule, think first of undoing what has been done to the variable.

Solve: $x - 5 = -9$.   **EXAMPLE 1**

$$x - 5 = -9$$
$$x - 5 + 5 = -9 + 5 \qquad \text{Undo the subtraction by adding 5 to both sides.}$$
$$x = -4$$

. . . .

Solve $x + 4 = 29$.   **EXAMPLE 2**

$$x + 4 = 29$$
$$x + 4 - 4 = 29 - 4 \qquad \text{Undo the addition by subtracting 4 from both sides.}$$
$$x = 25$$

. . . .

> If $a = b$, then $ac = bc$ or $a/c = b/c$ with $c \neq 0$. (If both sides of an equation are multiplied or divided by the same nonzero quantity, the resulting equation is equivalent to the original equation.)

Solve: $3x = 18$.   **EXAMPLE 3**

$$3x = 18$$
$$\frac{3x}{3} = \frac{18}{3} \qquad \text{Undo the multiplication by dividing both sides by 3.}$$
$$x = 6$$

. . . .

Solve: $x/4 = 9$.   **EXAMPLE 4**

$$\frac{x}{4} = 9$$
$$4\left(\frac{x}{4}\right) = 4 \cdot 9 \qquad \text{Undo the division by multiplying both sides by 4.}$$
$$x = 36$$

. . . .

When more than one operation is indicated on the variable in an equation, undo the additions and subtractions first, then undo the multiplications and divisions.

Solve: $3x + 5 = 17$.   **EXAMPLE 5**

$$3x + 5 = 17$$
$$3x + 5 - 5 = 17 - 5 \qquad \text{Subtract 5 from both sides.}$$
$$3x = 12$$
$$\frac{3x}{3} = \frac{12}{3} \qquad \text{Divide both sides by 3.}$$
$$x = 4$$

. . . .

EXAMPLE 6

Solve: $2x - 7 = 10$.

$$2x - 7 = 10$$
$$2x - 7 + 7 = 10 + 7 \qquad \text{Add 7 to both sides.}$$
$$2x = 17$$
$$\frac{2x}{2} = \frac{17}{2} \qquad \text{Divide both sides by 2.}$$
$$x = \frac{17}{2} = 8.5$$

EXAMPLE 7

Solve: $(x/5) - 10 = 22$.

$$\frac{x}{5} - 10 = 22$$
$$\frac{x}{5} - 10 + 10 = 22 + 10 \qquad \text{Add 10 to both sides.}$$
$$\frac{x}{5} = 32$$
$$5\left(\frac{x}{5}\right) = 5(32) \qquad \text{Multiply both sides by 5.}$$
$$x = 160$$

To solve an equation with variables on both sides:

1. Add or subtract either variable term from both sides of the equation.
2. Add or subtract from both sides of the equation the constant term that now appears on the same side of the equation with the variable. Then solve.

EXAMPLE 8

Solve: $3x + 6 = 7x - 2$.

$$3x + 6 = 7x - 2$$
$$3x + 6 - 3x = 7x - 2 - 3x \qquad \text{Subtract } 3x \text{ from both sides.}$$
$$6 = 4x - 2$$
$$6 + 2 = 4x - 2 + 2 \qquad \text{Add 2 to both sides.}$$
$$8 = 4x$$
$$\frac{8}{4} = \frac{4x}{4} \qquad \text{Divide both sides by 4.}$$
$$2 = x$$

EXAMPLE 9

Solve: $4x - 2 = -5x + 10$.

$$4x - 2 = -5x + 10$$
$$4x - 2 + 5x = -5x + 10 + 5x \qquad \text{Add } 5x \text{ to both sides.}$$
$$9x - 2 = 10$$
$$9x - 2 + 2 = 10 + 2 \qquad \text{Add 2 to both sides.}$$
$$9x = 12$$

$$\frac{9x}{9} = \frac{12}{9}$$ Divide both sides by 9.

$$x = \frac{4}{3}$$

· · · · · · · · · · · · · · · ·

To solve equations containing parentheses, first remove the parentheses and then proceed as before. The rules for removing parentheses follow:

1. If the parentheses are preceded by a plus (+) sign, they may be removed without changing any signs.
    Examples:

    $$2 + (3 - 5) = 2 + 3 - 5$$
    $$3 + (x + 4) = 3 + x + 4$$
    $$5x + (-6x + 9) = 5x - 6x + 9$$

2. If the parentheses are preceded by a minus (−) sign, the parentheses may be removed if *all* the signs of the numbers (or letters) within the parentheses are changed.
    Examples:

    $$2 - (3 - 5) = 2 - 3 + 5$$
    $$5 - (x - 7) = 5 - x + 7$$
    $$7x - (-4x - 11) = 7x + 4x + 11$$

3. If the parentheses are preceded by a number, the parentheses may be removed if each of the terms inside the parentheses is multiplied by that (signed) number.
    Examples:

    $$2(x + 4) = 2x + 8$$
    $$-3(x - 5) = -3x + 15$$
    $$2 - 4(3x - 5) = 2 - 12x + 20$$

Solve: $3(x - 4) = 15$. ◄ **EXAMPLE 10**

$$3(x - 4) = 15$$
$$3x - 12 = 15$$ Remove parentheses.
$$3x - 12 + 12 = 15 + 12$$ Add 12 to both sides.
$$3x = 27$$
$$\frac{3x}{3} = \frac{27}{3}$$ Divide both sides by 3.
$$x = 9$$

· · · · · · · · · · · · · · · ·

Solve: $2x - (3x + 15) = 4x - 1$. ◄ **EXAMPLE 11**

$$2x - (3x + 15) = 4x - 1$$
$$2x - 3x - 15 = 4x - 1$$ Remove parentheses.
$$-x - 15 = 4x - 1$$ Combine like terms.
$$-x - 15 + x = 4x - 1 + x$$ Add x to both sides.
$$-15 = 5x - 1$$

$$-15 + 1 = 5x - 1 + 1 \qquad \text{Add 1 to both sides.}$$
$$-14 = 5x$$
$$\frac{-14}{5} = \frac{5x}{5} \qquad \text{Divide both sides by 5.}$$
$$-2.8 = x$$

Problems A.3

Solve each equation.

1. $3x = 4$
2. $\dfrac{y}{2} = 10$
3. $x - 5 = 12$
4. $x + 1 = 9$
5. $2x + 10 = 10$
6. $4x = 28$
7. $2x - 2 = 33$
8. $4 = \dfrac{x}{10}$
9. $172 - 43x = 43$
10. $9x + 7 = 4$
11. $6y - 24 = 0$
12. $3y + 15 = 75$
13. $15 = \dfrac{105}{y}$
14. $6x = x - 15$
15. $2 = \dfrac{50}{2y}$
16. $9y = 67.5$
17. $8x - 4 = 36$
18. $10 = \dfrac{136}{4x}$
19. $2x + 22 = 75$
20. $9x + 10 = x - 26$
21. $4x + 9 = 7x - 18$
22. $2x - 4 = 3x + 7$
23. $-2x + 5 = 3x - 10$
24. $5x + 3 = 2x - 18$
25. $3x + 5 = 5x - 11$
26. $-5x + 12 = 12x - 5$
27. $13x + 2 = 20x - 5$
28. $5x + 3 = -9x - 39$
29. $-4x + 2 = -10x - 20$
30. $9x + 3 = 6x + 8$
31. $3x + (2x - 7) = 8$
32. $11 - (x + 12) = 100$
33. $7x - (13 - 2x) = 5$
34. $20(7x - 2) = 240$
35. $-3x + 5(x - 6) = 12$
36. $3(x + 117) = 201$
37. $5(2x - 1) = 8(x + 3)$
38. $3(x + 4) = 8 - 3(x - 2)$
39. $-2(3x - 2) = 3x - 2(5x + 1)$
40. $\dfrac{x}{5} - 2\left(\dfrac{2x}{5} + 1\right) = 28$

A.4 Solving Quadratic Equations

A **quadratic equation** in one variable is an equation that can be written in the form

$$ax^2 + bx + c = 0 \qquad \text{(where } a \neq 0\text{)}$$

EXAMPLE 1

Solve: $x^2 = 16$.

To solve a quadratic equation of this type, take the square root of both sides of the equation.

$$x^2 = 16$$
$$x = \pm 4 \qquad \text{Take the square root of both sides.}$$

In general, solve equations of the form $ax^2 = b$, where $a \neq 0$, as follows:

$$ax^2 = b$$
$$x^2 = \frac{b}{a} \qquad \text{Divide both sides by } a.$$
$$x = \pm\sqrt{\frac{b}{a}} \qquad \text{Take the square root of both sides.}$$

Solve: $2x^2 - 18 = 0$.

EXAMPLE 2

$$2x^2 - 18 = 0$$

$\qquad 2x^2 = 18 \qquad$ Add 18 to both sides.

$\qquad x^2 = 9 \qquad$ Divide both sides by 2.

$\qquad x = \pm 3 \qquad$ Take the square root of both sides.

. . . .

Solve: $5y^2 = 100$.

EXAMPLE 3

$$5y^2 = 100$$

$\qquad y^2 = 20 \qquad$ Divide both sides by 5.

$\qquad y = \pm \sqrt{20} \qquad$ Take the square root of both sides.

$\qquad y = \pm 4.47$

. . . .

The solutions of the general quadratic equation

$$ax^2 + bx + c = 0 \quad \text{(where } a \neq 0\text{)}$$

are given by the formula (called the *quadratic formula*)

$$x = \frac{-b \pm \sqrt{b^2 - 4ac}}{2a}$$

where $a =$ coefficient of the x^2 term

$\qquad b =$ coefficient of the x term

$\qquad c =$ constant term

The symbol ($\pm$) is used to combine two expressions or equations into one. For example, $a \pm 2$ means $a + 2$ or $a - 2$. Similarly,

$$x = \frac{-b \pm \sqrt{b^2 - 4ac}}{2a}$$

means

$$x = \frac{-b + \sqrt{b^2 - 4ac}}{2a} \quad \text{or} \quad x = \frac{-b - \sqrt{b^2 - 4ac}}{2a}$$

In the equation $4x^2 - 3x - 7 = 0$, identify a, b, and c.

EXAMPLE 4

$$a = 4, \quad b = -3, \quad \text{and} \quad c = -7$$

. . . .

Solve $x^2 + 2x - 8 = 0$ using the quadratic formula.

EXAMPLE 5

First, $a = 1$, $b = 2$, and $c = -8$. Then

$$x = \frac{-b \pm \sqrt{b^2 - 4ac}}{2a}$$

$$x = \frac{-2 \pm \sqrt{(2)^2 - 4(1)(-8)}}{2(1)}$$

$$= \frac{-2 \pm \sqrt{4 - (-32)}}{2}$$

$$= \frac{-2 \pm \sqrt{36}}{2}$$

$$= \frac{-2 \pm 6}{2}$$

$$= \frac{-2 + 6}{2} \quad \text{or} \quad \frac{-2 - 6}{2}$$

$$= \frac{4}{2} \quad \text{or} \quad \frac{-8}{2}$$

$$= 2 \quad \text{or} \quad -4 \qquad \text{The solutions are 2 and } -4.$$

If the number under the radical sign is not a perfect square, find the square root of the number by using a calculator and proceed as before.

EXAMPLE 6

Solve $4x^2 - 7x = 32$ using the quadratic formula.

Before identifying a, b, and c, the equation must be set equal to zero. That is,

$$4x^2 - 7x - 32 = 0$$

First, $a = 4$, $b = -7$, and $c = -32$. Then

$$x = \frac{-b \pm \sqrt{b^2 - 4ac}}{2a}$$

$$= \frac{-(-7) \pm \sqrt{(-7)^2 - 4(4)(-32)}}{2(4)}$$

$$= \frac{7 \pm \sqrt{49 - (-512)}}{8}$$

$$= \frac{7 \pm \sqrt{561}}{8}$$

$$= \frac{7 \pm 23.7}{8} \qquad (\sqrt{561} = 23.7)$$

$$= \frac{7 + 23.7}{8} \quad \text{or} \quad \frac{7 - 23.7}{8}$$

$$= 3.84 \quad \text{or} \quad -2.09$$

The approximate solutions are 3.84 and −2.09.

Problems A.4

Solve each equation.

1. $x^2 = 36$
2. $y^2 = 100$
3. $2x^2 = 98$
4. $5x^2 = 0.05$
5. $3x^2 - 27 = 0$
6. $2y^2 - 15 = 17$
7. $10x^2 + 4.9 = 11.3$
8. $2(32)(48 - 15) = v^2 - 27^2$
9. $2(107) = 9.8t^2$
10. $65 = \pi r^2$
11. $2.50 = \pi r^2$
12. $24^2 = a^2 + 16^2$

Find the values of a, b, and c, in each quadratic equation.

13. $3x^2 + x - 5 = 0$
14. $-2x^2 + 7x + 4 = 0$
15. $6x^2 + 8x + 2 = 0$
16. $5x^2 - 2x - 15 = 0$
17. $9x^2 + 6x = 4$
18. $6x^2 = x + 9$
19. $5x^2 + 6x = 0$
20. $7x^2 - 45 = 0$
21. $9x^2 = 64$
22. $16x^2 = 49$

Solve each quadratic equation using the quadratic formula.

23. $x^2 - 10x + 21 = 0$ 24. $2x^2 + 13x + 15 = 0$ 25. $6x^2 + 7x = 20$
26. $15x^2 = 4x + 4$ 27. $6x^2 - 2x = 19$ 28. $4x^2 = 28x - 49$
29. $18x^2 - 15x = 26$ 30. $48x^2 + 9 = 50x$
31. $16.5x^2 + 8.3x - 14.7 = 0$ 32. $125x^2 - 167x + 36 = 0$

A.5   Right-Triangle Trigonometry

A **right triangle** is a triangle with one right angle (90°), two acute angles (less than 90°), two legs, and a hypotenuse (the side opposite the right angle) (Fig. A.1).

When it is necessary to label a triangle, the vertices are often labeled using capital letters and the sides opposite the vertices are labeled using the corresponding lower-case letters (Fig. A.2).

The side opposite angle A is a.
The side adjacent to angle A is b.
The side opposite angle B is b.
The side adjacent to angle B is a.
The side opposite angle C is c and is called the *hypotenuse*.

If we consider a certain acute angle of a right triangle, the two legs can be identified as the side opposite or the side adjacent to the acute angle.

The side opposite angle A is the same as the side adjacent to angle B.
The side adjacent to angle A is the same as the side opposite angle B.
The side opposite angle B is the same as the side adjacent to angle A.
The side adjacent to angle B is the same as the side opposite angle A.

A **ratio** is a comparison of two quantities by division. In a right triangle (Fig. A.3), there are three very important ratios:

$\dfrac{\text{side opposite angle } A}{\text{hypotenuse}}$ is called **sine** A (abbreviated sin A)

$\dfrac{\text{side adjacent to angle } A}{\text{hypotenuse}}$ is called **cosine** A (abbreviated cos A)

$\dfrac{\text{side opposite angle } A}{\text{side adjacent to angle } A}$ is called **tangent** A (abbreviated tan A)

Figure A.1   Parts of a right triangle

Figure A.2   Labeling a right triangle

Figure A.3

$$\sin A = \frac{\text{side opposite angle } A}{\text{hypotenuse}}$$

$$\cos A = \frac{\text{side adjacent to angle } A}{\text{hypotenuse}}$$

$$\tan A = \frac{\text{side opposite angle } A}{\text{side adjacent to angle } A}$$

The ratios are defined similarly for angle B:

$$\sin B = \frac{\text{side opposite angle } B}{\text{hypotenuse}}$$

$$\cos B = \frac{\text{side adjacent to angle } B}{\text{hypotenuse}}$$

$$\tan B = \frac{\text{side opposite angle } B}{\text{side adjacent to angle } B}$$

EXAMPLE 1

Figure A.4

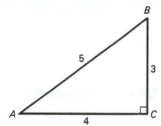

Find the three trigonometric ratios of angle A in Fig. A.4.

$$\sin A = \frac{\text{side opposite angle } A}{\text{hypotenuse}} = \frac{3}{5} = 0.60$$

$$\cos A = \frac{\text{side adjacent to angle } A}{\text{hypotenuse}} = \frac{4}{5} = 0.80$$

$$\tan A = \frac{\text{side opposite angle } A}{\text{side adjacent to angle } A} = \frac{3}{4} = 0.75$$

EXAMPLE 2

Find the three trigonometric ratios of angle B in Fig. A.4.

$$\sin B = \frac{\text{side opposite angle } B}{\text{hypotenuse}} = \frac{4}{5} = 0.80$$

$$\cos B = \frac{\text{side adjacent to angle } B}{\text{hypotenuse}} = \frac{3}{5} = 0.60$$

$$\tan B = \frac{\text{side opposite angle } B}{\text{side adjacent to angle } B} = \frac{4}{3} = 1.33$$

Note: Every acute angle has three trigonometric ratios associated with it.

In this book we assume that you will be using a calculator. When calculations involve a trigonometric ratio, we will use the following generally accepted practice for significant digits:

Angle Expressed to Nearest:	Length of Side Contains:
1°	Two significant digits
0.1°	Three significant digits
0.01°	Four significant digits

A useful and time-saving fact about right triangles is that *the sum of the two acute angles of any right triangle is always* 90°. That is,

$$A + B = 90°$$

Why is this true? We know that the sum of the three interior angles of any triangle is 180°. A right triangle must contain a right angle, whose measure is 90°. This leaves 90° to be divided between the two acute angles. Therefore, if one acute angle is known, the other acute angle may be found by subtracting the known angle from 90°. That is,

$$A = 90° - B$$
$$B = 90° - A$$

EXAMPLE 3

Find angle B and side a in the right triangle in Fig. A.5.
To find angle B, we use

$$B = 90° - A = 90° - 30.0° = 60.0°$$

To find side a, we use a trigonometric ratio. Note that we are looking for the *side opposite* angle A and that the *hypotenuse* is given. The trigonometric ratio having these two quantities is sine.

$$\sin A = \frac{\text{side opposite angle } A}{\text{hypotenuse}}$$

$$\sin 30.0° = \frac{a}{20.0 \text{ m}}$$

$$(\sin 30.0°)(20.0 \text{ m}) = \left(\frac{a}{20.0 \text{ m}}\right)(20.0 \text{ m}) \qquad \text{Multiply both sides by 20.0 m.}$$

$$10.0 \text{ m} = a$$

. . . .

EXAMPLE 4

Find angle A, angle B, and side a in the right triangle in Fig. A.6.

First, find angle A. The *side adjacent* to angle A and the *hypotenuse* are given. Therefore, we use cos A to find angle A because cos A uses these two quantities:

Figure A.6

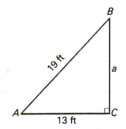

$$\cos A = \frac{\text{side adjacent to } A}{\text{hypotenuse}}$$

$$\cos A = \frac{13 \text{ ft}}{19 \text{ ft}} = 0.684$$

Using a calculator as in Section D.3 of Appendix D, we find that $A = 47°$.

To find angle B, we use

$$B = 90° - A = 90° - 47° = 43°$$

To find side a, we use sin A because the *hypotenuse* is given and side a is the *side opposite* angle A:

$$\sin A = \frac{\text{side opposite angle } A}{\text{hypotenuse}}$$

$$\sin 47° = \frac{a}{19 \text{ ft}}$$

$$(\sin 47°)(19 \text{ ft}) = \left(\frac{a}{19 \text{ ft}}\right)(19 \text{ ft}) \qquad \text{Multiply both sides by 19 ft.}$$

$$14 \text{ ft} = a$$

. . . .

EXAMPLE 5

Find angle A, angle B, and the hypotenuse in the right triangle in Fig. A.7.

To find angle A, use tan A:

$$\tan A = \frac{\text{side opposite angle } A}{\text{side adjacent to angle } A}$$

$$\tan A = \frac{12.00 \text{ km}}{19.00 \text{ km}} = 0.6316$$

$$A = 32.28°$$

Figure A.7

To find angle B,

$$B = 90° - A = 90° - 32.28° = 57.72°$$

To find the hypotenuse, use sin A:

$$\sin A = \frac{\text{side opposite angle } A}{\text{hypotenuse}}$$

$$\sin 32.28° = \frac{12.00 \text{ km}}{c}$$

$$(\sin 32.28°)(c) = \left(\frac{12.00 \text{ km}}{c}\right)(c) \qquad \text{Multiply both sides by } c.$$

$$(\sin 32.28°)(c) = 12.00 \text{ km}$$

$$\frac{c(\sin 32.28°)}{\sin 32.28°} = \frac{12.00 \text{ km}}{\sin 32.28°} \qquad \text{Divide both sides by } \sin 32.28°.$$

$$c = \frac{12.00 \text{ km}}{\sin 32.28°}$$

$$= 22.47 \text{ km}$$

Figure A.8

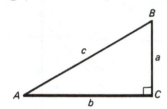

When the two legs of a right triangle are given, the hypotenuse can be found without using trigonometric ratios. From geometry, *the sum of the squares of the legs of a right triangle is equal to the square of the hypotenuse* (**Pythagorean theorem**; see Fig. A.8):

$$a^2 + b^2 = c^2$$

or, taking the square root of each side of the equation,

$$c = \sqrt{a^2 + b^2}$$

Also, if one leg and the hypotenuse are given, the other leg can be found by using

$$a = \sqrt{c^2 - b^2}$$
$$b = \sqrt{c^2 - a^2}$$

EXAMPLE 6

Figure A.9

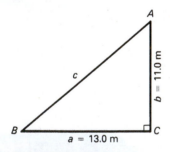

Find the hypotenuse of the right triangle in Fig. A.9.

$$c = \sqrt{a^2 + b^2}$$
$$c = \sqrt{(13.0 \text{ m})^2 + (11.0 \text{ m})^2}$$
$$= \sqrt{169 \text{ m}^2 + 121 \text{ m}^2}$$
$$= \sqrt{290 \text{ m}^2}$$
$$= 17.0 \text{ m}$$

EXAMPLE 7

Figure A.10

Find side b in the right triangle in Fig. A.10.

$$b = \sqrt{c^2 - a^2}$$
$$b = \sqrt{(12.2 \text{ km})^2 - (7.30 \text{ km})^2}$$
$$= 9.77 \text{ km}$$

Problems A.5

Figure A.11

Use right triangle ABC in Fig. A.11 to fill in each blank.

1. The side opposite angle A is _____.
2. The side opposite angle B is _____.
3. The hypotenuse is _____.
4. The side adjacent to angle A is _____.
5. The side adjacent to angle B is _____.
6. The angle opposite side a is _____.
7. The angle opposite side b is _____.
8. The angle opposite side c is _____.
9. The angle adjacent to side a is _____.
10. The angle adjacent to side b is _____.

Use a calculator to find each trigonometric ratio rounded to four significant digits.

11. $\sin 71°$
12. $\cos 40°$
13. $\tan 61°$
14. $\tan 41.2°$
15. $\cos 11.5°$
16. $\sin 79.4°$
17. $\cos 49.63°$
18. $\tan 53.45°$
19. $\tan 17.04°$
20. $\cos 34°$
21. $\sin 27.5°$
22. $\cos 58.72°$

Find each angle rounded to the nearest whole degree.

23. $\sin A = 0.2678$
24. $\cos B = 0.1046$
25. $\tan A = 0.9237$
26. $\sin B = 0.9253$
27. $\cos B = 0.6742$
28. $\tan A = 1.351$

Find each angle rounded to the nearest tenth of a degree.

29. $\sin B = 0.5963$
30. $\cos A = 0.9406$
31. $\tan B = 1.053$
32. $\sin A = 0.9083$
33. $\cos A = 0.8660$
34. $\tan B = 0.9433$

Find each angle rounded to the nearest hundredth of a degree.

35. $\sin A = 0.3792$
36. $\cos B = 0.06341$
37. $\tan B = 0.3010$
38. $\sin A = 0.4540$
39. $\cos B = 0.8141$
40. $\tan A = 2.369$

Solve each triangle (find the missing angles and sides) using trigonometric ratios.

41.

42.

43.

44.

45.

46.

47.

48.

49.

50.

Find the missing side in each right triangle using the Pythagorean theorem.

51.

52.

53.

54.

55.

56.

57.

58.

A.6 Law of Sines and Law of Cosines

For those classes and individuals with general triangle trigonometry skills, this section is written as a review. This text is written so that no more than right-triangle trigonometry needs to be used, but those with the prerequisite skills may prefer to use the more advanced trigonometry techniques.

An **oblique**, or **general**, triangle is a triangle that contains no right angles. We shall use the standard notation of labeling the vertices of a triangle by the capital letters A, B, and C and using the small letters a, b, and c as the labels for the sides opposite angles A, B, and C, respectively (see Fig. A.12).

Solving a triangle means finding all those sides and angles that are not given or known. To solve a triangle, we need three parts (including at least one side). Solving any oblique triangle falls into one of four cases where the following parts of a triangle are known:

1. Two sides and an angle opposite one of them (SSA).
2. Two angles and a side opposite one of them (AAS).
3. Two sides and the included angle (SAS).
4. Three sides (SSS).

One law that we use to solve triangles is called the **law of sines**. In words, for any triangle the ratio of any side to the sine of the opposite angle is a constant. The formula for the law of sines is as follows:

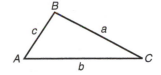

Figure A.12   Oblique, or general, triangle ABC.

LAW OF SINES

$$\frac{a}{\sin A} = \frac{b}{\sin B} = \frac{c}{\sin C}$$

In order to use the law of sines, we must know either of the following:

1. Two sides and an angle opposite one of them (SSA).
2. Two angles and a side opposite one of them (AAS). *Note:* Knowing two angles and any side is sufficient because knowing two angles, we can easily find the third.

You must select the proportion that contains three parts that are known and the unknown part.

When calculations with measurements involve a trigonometric function, we shall use the following rule for significant digits:

Angle Expressed to Nearest	Length of Side Contains
1°	Two significant digits
0.1°	Three significant digits
0.01°	Four significant digits

The following relationship is very helpful as a check when solving a general triangle: The longest side of any triangle is opposite the largest angle and the shortest side is opposite the smallest angle.

If $A = 65.0°$, $a = 20.0\,\text{m}$, and $b = 15.0\,\text{m}$, solve the triangle.

EXAMPLE 1

First, draw a triangle as in Fig. A.13 and find angle B by using the law of sines:

$$\frac{a}{\sin A} = \frac{b}{\sin B}$$

$$\frac{20.0\,\text{m}}{\sin 65.0°} = \frac{15.0\,\text{m}}{\sin B}$$

$$\sin B = \frac{(15.0\,\text{m})(\sin\ 65.0°)}{20.0\,\text{m}} = 0.6797$$

$$B = 42.8°$$

Figure A.13

$b = 15.0$ m $a = 20.0$ m

65.0°

c

This angle may be found using a calculator as follows:

$$15 \quad \boxed{\times} \quad \boxed{\sin} \quad 65 \quad \boxed{)} \quad \boxed{\div} \quad 20 \quad \boxed{=} \quad \boxed{\sin^{-1}} \quad \boxed{\text{ANS}} \quad \boxed{=}$$

| 42.82261353 |

So $B = 42.8°$ rounded to the nearest tenth of a degree.

To find C, use the fact that the sum of the angles of any triangle is 180°. Therefore,

$$C = 180° - 65.0° - 42.8°$$
$$= 72.2°$$

Finally, find c using the law of sines.

$$\frac{a}{\sin\ A} = \frac{c}{\sin\ C}$$

$$\frac{20.0\,\text{m}}{\sin\ 65.0°} = \frac{c}{\sin\ 72.2°}$$

$$c = \frac{(20.0\,\text{m})(\sin\ 72.2°)}{\sin\ 65.0°} = 21.0\,\text{m} \qquad \text{(Rounded to three significant digits)}$$

This side may be found using a calculator as follows:

$$20 \quad \boxed{\times} \quad \boxed{\sin} \quad 72.2 \quad \boxed{)} \quad \boxed{\div} \quad \boxed{\sin} \quad 65 \quad \boxed{=}$$

| 21.01117096 |

That is, side $c = 21.0\,\text{m}$ rounded to three significant digits.
The solution is $B = 42.8°$, $C = 72.2°$, and $c = 21.0\,\text{m}$.

· · · · · · · · · · · · · · · · ·

EXAMPLE 2

If $C = 25°$, $c = 59\,\text{ft}$, and $B = 108°$, solve the triangle.
First, draw a triangle as in Fig. A.14 and find b:

Figure A.14

B

$c = 59$ ft   108°

a

A b 25° C

$$\frac{c}{\sin\ C} = \frac{b}{\sin\ B}$$

$$\frac{59\,\text{ft}}{\sin\ 25°} = \frac{b}{\sin\ 108°}$$

$$b = \frac{(59\,\text{ft})(\sin\ 108°)}{\sin\ 25°} = 130\,\text{ft} \qquad \text{(Rounded to two significant digits)}$$

59 [×] [sin] 108 [ )] [÷] [sin] 25 [=]

```
132.7730946
```

$A = 180° - 25° - 108° = 47°$

Find a:

$$\frac{a}{\sin A} = \frac{c}{\sin C}$$

$$\frac{a}{\sin 47°} = \frac{59\text{ ft}}{\sin 25°}$$

$$a = \frac{(59\text{ ft})(\sin 47°)}{\sin 25°} = 1\overline{0}0\text{ ft}$$

The solution is $A = 47°$, $a = 1\overline{0}0$ ft, and $b = 130$ ft.

· · · · · · · · · · · · · · · · ·

The Ambiguous Case

The solution of a triangle when two sides and an angle opposite one of the sides (SSA) are given requires special care. There may be one, two, or no triangles formed from the given data. By construction and discussion, let's study the possibilities.

Construct a triangle given that $A = 35°$, $b = 10$, and $a = 7$.   ◄   **EXAMPLE 3**

As you can see from Fig. A.15, two triangles that satisfy the given information can be drawn: triangles ACB and ACB'. Note that in one triangle angle B is acute and in the other triangle angle B is obtuse.

Figure A.15

· · · · · · · · · · · · · · · · ·

Construct a triangle given that $A = 45°$, $b = 10$, and $a = 5$.   ◄   **EXAMPLE 4**

As you can see from Fig. A.16, no triangle can be drawn that satisfies the given information. Side a is simply not long enough to reach the side opposite angle C.

Figure A.16

· · · · · · · · · · · · · · · · ·

EXAMPLE 5

Construct a triangle given that $A = 60°$, $b = 6$, and $a = 10$.

As you can see from Fig. A.17, only one triangle that satisfies the given information can be drawn. Side a is too long for two solutions.

Figure A.17

In summary, let's list the possible cases when two sides and an angle opposite one of the sides are given. Assume that *acute* angle A and adjacent side b are given. As a result of $h = b \sin A$, h is also determined. Depending on the length of the opposite side, a, we have the four cases shown in Fig. A.18.

Figure A.18   Possible triangles when two sides and an acute angle opposite one of the sides are given.

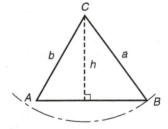

(a) When $h < a < b$, there are two possible triangles. In words, when the side opposite the given *acute* angle is less than the known adjacent side but greater than the altitude, there are two possible triangles.

(b) When $h < b < a$, there is only one possible triangle. In words, when the side opposite the given *acute* angle is greater than the known adjacent side, there is only one possible triangle.

(c) When $a = h$, there is one possible (right) triangle. In words, when the side opposite the given *acute* angle equals the length of the altitude, there is only one possible (right) triangle.

(d) When $a < h$, there is no possible triangle. In words, when the side opposite the given *acute* angle is less than the length of the altitude, there is no possible triangle.

If angle *A* is *obtuse*, we have two possible cases (Fig. A.19).

Figure A.19   Possible triangles when two sides and an obtuse angle opposite one of the sides are given.

(a) When *a* > *b*, there is one possible triangle. In words, when the side opposite the given *obtuse* angle is greater than the known adjacent side, there is only one possible triangle.

(b) When *a* ≤ *b*, there is no possible triangle. In words, when the side opposite the given *obtuse* angle is less than or equal to the known adjacent side, there is no possible triangle.

Note: If the given parts are not angle *A*, side opposite *a*, and side adjacent *b* as in our preceding discussions, then you must substitute the given angle and sides accordingly. This is why it is so important to understand the general word description corresponding to each case.

If $A = 26°$, $a = 25\,\text{cm}$, and $b = 41\,\text{cm}$, solve the triangle.

EXAMPLE 6

First, find *h*:

$$h = b \sin A = (41\,\text{cm})(\sin 26°) = 18\,\text{cm}$$

Since $h < a < b$, there are two solutions. First, let's find *B* in triangle *ACB* in Fig. A.20:

$$\frac{a}{\sin A} = \frac{b}{\sin B}$$

$$\frac{25\,\text{cm}}{\sin 26°} = \frac{41\,\text{cm}}{\sin B}$$

$$\sin B = \frac{(41\,\text{cm})(\sin 26°)}{25\,\text{cm}} = 0.7189$$

$$B = 46°$$

Figure A.20

45.96610103

$$C = 180° - 26° - 46° - 46° = 108°$$

Find *c*:

$$\frac{c}{\sin C} = \frac{a}{\sin A}$$

$$\frac{c}{\sin 108°} = \frac{25\,\text{cm}}{\sin 26°}$$

$$c = \frac{(25\,\text{cm})(\sin 108°)}{\sin 26°} = 54\,\text{cm}$$

Therefore, the first solution is $B = 46°$, $C = 108°$, and $c = 54\,\text{cm}$.

The second solution occurs when B is obtuse, as in triangle ACB'. That is, find the obtuse angle whose sine is 0.7189.

$$B' = 180° - 46° = 134°$$

Then $C = 180° - 26° - 134° = 2\overline{0}°$.

For c,

$$\frac{c}{\sin C} = \frac{a}{\sin A}$$

$$\frac{c}{\sin 2\overline{0}°} = \frac{25\,\text{cm}}{\sin 26°}$$

$$c = \frac{(25\,\text{cm})(\sin 2\overline{0}°)}{\sin 26°} = 2\overline{0}\,\text{cm}$$

The second solution is $B' = 134°$, $C = 2\overline{0}°$, and $c = 2\overline{0}$ cm.

. . . .

EXAMPLE 7

If $A = 62.0°$, $a = 415$ m, and $b = 855$ m, solve the triangle.

First, find h:

$$h = b \sin A$$

$$h = (855\,\text{m})(\sin 62.0°)$$

$$= 755\,\text{m}$$

Since $a < h$, there is no possible solution. What would happen if you applied the law of sines anyway? You would obtain

$$\frac{a}{\sin A} = \frac{b}{\sin B}$$

$$\frac{415\,\text{m}}{\sin 62.0°} = \frac{855\,\text{m}}{\sin B}$$

$$\sin B = \frac{(855\,\text{m})(\sin 62.0°)}{415\,\text{m}} = 1.819 \qquad \text{(Tilt!)}$$

Note: $\sin B = 1.819$ is impossible because $-1 \leq \sin B \leq 1$.

. . . .

In summary:

1. ***Given two angles and one side (AAS):*** There is only one possible triangle.
2. ***Given two sides and an angle opposite one of them (SSA):*** There are three possibilities. If the side opposite the given angle is
   (a) greater than the known adjacent side, there is only one possible triangle.
   (b) less than the known adjacent side but greater than the altitude, there are two possible triangles.
   (c) less than the altitude, there is no possible triangle.

Since solving a general triangle requires several operations, errors are often introduced. The following points may be helpful in avoiding some of these errors:

1. Always choose a given value over a calculated value when doing calculations.
2. Always check your results to see that the largest angle is opposite the largest side and the smallest angle is opposite the smallest side.
3. Avoid finding the largest angle by the law of sines whenever possible because it is often not clear whether the resulting angle is acute or obtuse.

Law of Cosines

When the law of sines cannot be used, we use the **law of cosines**. In words, the square of any side of a triangle is equal to the sum of the squares of the other two sides minus twice the product of these two sides and the cosine of their included angle (see Fig. A.21). By formula, the law is stated as follows.

Figure A.21

> ### LAW OF COSINES
> $$a^2 = b^2 + c^2 - 2bc \cos A$$
> $$b^2 = a^2 + c^2 - 2ac \cos B$$
> $$c^2 = a^2 + b^2 - 2ab \cos C$$

There are two cases when the law of sines does not apply and we use the law of cosines to solve triangles:

1. Two sides and the included angle are known (SAS).
2. All three sides are known (SSS).

Do you see that when the law of cosines is used, there is no possibility of an ambiguous case? If not, draw a few triangles for each of these two cases (SAS and SSS) to convince yourself intuitively.

If $a = 112$ m, $b = 135$ m, and $C = 104.3°$, solve the triangle. ◄

EXAMPLE 8

First, draw a triangle as in Fig. A.22 and find c by using the law of cosines:

Figure A.22

$$c^2 = a^2 + b^2 - 2ab \cos C$$
$$c^2 = (112 \text{ m})^2 + (135 \text{ m})^2 - 2(112 \text{ m})(135 \text{ m})(\cos 104.3°)$$
$$c = 196 \text{ m}$$

This side may be found using a calculator as follows:

So $c = 196$ m rounded to three significant digits.

To find A, use the law of sines since it requires less computation:

$$\frac{a}{\sin A} = \frac{c}{\sin C}$$
$$\frac{112 \text{ m}}{\sin A} = \frac{196 \text{ m}}{\sin 104.3°}$$
$$\sin A = \frac{(112 \text{ m})(\sin 104.3°)}{196 \text{ m}} = 0.5537$$
$$A = 33.6°$$
$$B = 180° - 104.3° - 33.6° = 42.1°$$

The solution is $A = 33.6°$, $B = 42.1°$, and $c = 196$ m.

EXAMPLE 9

If $a = 375.0$ ft, $b = 282.0$ ft, and $c = 114.0$ ft, solve the triangle.

First, draw a triangle as in Fig. A.23 and find A by using the law of cosines:

Figure A.23

$$a^2 = b^2 + c^2 - 2bc \cos A$$

$$(375.0 \text{ ft})^2 = (282.0 \text{ ft})^2 + (114.0 \text{ ft})^2 - 2(282.0 \text{ ft})(114.0 \text{ ft}) \cos A$$

$$\cos A = \frac{(375.0 \text{ ft})^2 - (282.0 \text{ ft})^2 - (114.0 \text{ ft})^2}{-2(282.0 \text{ ft})(114.0 \text{ ft})}$$

$$\cos A = -0.7482$$

$$A = 138.43° \quad \text{(Rounded to the nearest hundredth of a degree)}$$

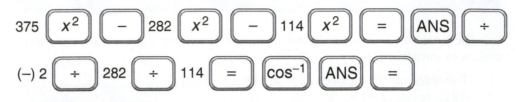

$$138.432994$$

Next, to find B, use the law of sines:

$$\frac{a}{\sin A} = \frac{b}{\sin B}$$

$$\frac{375.0 \text{ ft}}{\sin 138.43°} = \frac{282.0 \text{ ft}}{\sin B}$$

$$\sin B = \frac{(282.0 \text{ ft})(\sin 138.43°)}{375.0 \text{ ft}} = 0.4990$$

$$B = 29.93°$$

$$C = 180° - 138.43° - 29.93° = 11.64°$$

The solution is $A = 138.43°$, $B = 29.93°$, and $C = 11.64°$.

· · · · · · · · · · · · · · · · ·

Problems A.6

Figure A.24

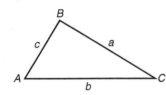

Solve each triangle using the labels as shown in Fig. A.24.[*]

Express the lengths of sides to three significant digits and the angles to the nearest tenth of a degree.

1. $A = 69.0°$, $a = 25.0$ m, $b = 16.5$ m
2. $C = 57.5°$, $c = 166$ mi, $b = 151$ mi
3. $B = 61.4°$, $b = 124$ cm, $c = 112$ cm
4. $A = 19.5°$, $a = 487$ km, $c = 365$ km
5. $B = 75.3°$, $A = 57.1°$, $b = 257$ ft
6. $C = 59.6°$, $B = 43.9°$, $b = 4760$ m
7. $A = 115.0°$, $a = 5870$ m, $b = 4850$ m
8. $A = 16.4°$, $a = 205$ ft, $b = 187$ ft

Express the lengths of sides to four significant digits and the angles to the nearest hundredth of a degree.

9. $C = 72.58°$, $b = 28.63$ cm, $c = 42.19$ cm
10. $A = 58.95°$, $a = 3874$ m, $c = 2644$ m
11. $B = 28.76°$, $C = 19.30°$, $c = 39,750$ mi
12. $A = 35.09°$, $B = 48.64°$, $a = 8.362$ km

---

[*]Because of differences in rounding, your answers may differ slightly from the answers in the text if you choose to solve for the parts of a triangle in an order different from that chosen by the authors.

Express the lengths of sides to two significant digits and the angles to the nearest degree.

13. $A = 25°$, $a = 5\bar{0}$ cm, $b = 4\bar{0}$ cm   14. $B = 42°$, $b = 5.3$ km, $c = 4.6$ km
15. $C = 8°$, $c = 16$ m, $a = 12$ m   16. $A = 105°$, $a = 460$ mi, $c = 380$ mi

For each general triangle, (a) determine the number of solutions and (b) solve the triangle, if possible.

Express the lengths of sides to three significant digits and the angles to the nearest tenth of a degree.

17. $A = 37.0°$, $a = 21.5$ cm, $b = 16.4$ cm 18. $B = 55.0°$, $b = 182$ m, $c = 203$ m
19. $C = 26.5°$, $c = 42.7$ km, $a = 47.2$ km 20. $B = 40.4°$, $b = 81.4$ m, $c = 144$ m
21. $A = 71.5°$, $a = 3.45$ m, $c = 3.50$ m   22. $C = 17.2°$, $c = 2.20$ m, $b = 2.00$ m
23. $B = 105.0°$, $b = 16.5$ mi, $a = 12.0$ mi 24. $A = 98.8°$, $a = 707$ ft, $b = 585$ ft

Express the lengths of sides to two significant digits and the angles to the nearest degree.

25. $C = 18°$, $c = 24$ mi, $a = 45$ mi   26. $B = 36°$, $b = 75$ cm, $a = 95$ cm
27. $C = 6\bar{0}°$, $c = 150$ m, $b = 180$ m   28. $A = 3\bar{0}°$, $a = 4800$ ft, $c = 3600$ ft
29. $B = 8°$, $b = 450$ m, $c = 850$ m   30. $B = 45°$, $c = 2.5$ m, $b = 3.2$ m

Express the lengths of sides to four significant digits and the angles to the nearest hundredth of a degree.

31. $B = 41.50°$, $b = 14.25$ km, $a = 18.50$ km
32. $A = 15.75°$, $a = 642.5$ m, $c = 592.7$ m
33. $C = 63.85°$, $c = 29.50$ cm, $b = 38.75$ cm
34. $B = 50.00°$, $b = 41,250$ km, $c = 45,650$ km
35. $C = 8.75°$, $c = 89.30$ m, $a = 61.93$ m
36. $A = 31.50°$, $a = 375.0$ mm, $b = 405.5$ mm

Express the lengths of sides to three significant digits and the angles to the nearest tenth of a degree.

37. $A = 60.0°$, $b = 19.5$ m, $c = 25.0$ m   38. $B = 19.5°$, $a = 21.5$ ft, $c = 12.5$ ft
39. $C = 109.0°$, $a = 14\bar{0}$ km, $b = 215$ km 40. $A = 94.7°$, $c = 875$ yd, $b = 185$ yd
41. $a = 19.2$ m, $b = 21.3$ m, $c = 27.2$ m
42. $a = 125$ km, $b = 195$ km, $c = 145$ km
43. $a = 4.25$ ft, $b = 7.75$ ft, $c = 5.50$ ft
44. $a = 3590$ m, $b = 7950$ m, $c = 4650$ m

Express the lengths of sides to two significant digits and the angles to the nearest degree.

45. $A = 45°$, $b = 51$ m, $c = 39$ m
46. $B = 6\bar{0}°$, $a = 160$ cm, $c = 230$ cm
47. $a = 7\bar{0}00$ m, $b = 5600$ m, $c = 4800$ m
48. $a = 5.8$ cm, $b = 5.8$ cm, $c = 9.6$ cm
49. $C = 135°$, $a = 36$ ft, $b = 48$ ft
50. $A = 5°$, $b = 19$ m, $c = 25$ m

Express the lengths of sides to four significant digits and the angles to the nearest hundredth of a degree.

51. $B = 19.25°$, $a = 4815$ m, $c = 1925$ m
52. $C = 75.00°$, $a = 37,550$ mi, $b = 45,250$ mi
53. $C = 108.75°$, $a = 405.0$ mm, $b = 325.0$ mm
54. $A = 111.05°$, $b = 1976$ ft, $c = 325\bar{0}$ ft
55. $a = 207.5$ km, $b = 105.6$ km, $c = 141.5$ km
56. $a = 19.45$ m, $b = 36.50$ m, $c = 25.60$ m

PROBLEM-SOLVING STRATEGY: DIMENSIONAL AND UNIT ANALYSIS

B.1 Dimensional Analysis

In physics we rarely perform mathematical operations using only numbers without an associated unit. Physics describes in mathematical terms the motion and energy of all matter throughout the universe. It is a science of measurement. So we measure time, distance, length, width, volume, forces, work, temperature, current, and many other physical quantities.

Many of the things we measure are a combination of other measured quantities. For instance, velocity is distance traveled during a period of time in a specific direction. Mathematically, it is distance divided by time. Force equals mass times acceleration. In these examples distance, time, mass, and acceleration are called **dimensions**. They are how we define velocity and force, and they never change. What does change, however, are the **units** by which each dimension is measured. For example, velocity may have units of metres per second (m/s), kilometres per hour (km/h), feet per second (ft/s), or miles per hour (mi/h). The units may change, but the dimensions never change.

In problem solving in physics we must always be aware of the dimensions we are working with. If we are solving a problem that asks for velocity and our answer comes out in metres, we immediately know we have made an error in solving the problem. If the problem asks for velocity, we need to make sure that our answer is in units of velocity.

Dimensional analysis is an extremely useful procedure that makes use of the fact that dimensions can be treated as algebraic quantities. For example, quantities to be added or subtracted must be of the same dimensions. You can add two distances, but you cannot add distance and time.

In addition, quantities on both sides of an equation must have the same dimensions. Time is not equal to distance. A great deal of effort in problem solving can be saved and accuracy can be greatly increased by using dimensional analysis and taking the time to study the dimensions involved in the problem before starting to manipulate the numbers.

The fact that an equation is dimensionally correct is a valuable check, and the technique can actually help solve the problem. For example, how long will it take a motorcycle traveling at a speed of 20.0 metres per second to go $12\overline{0}0$ metres? Speed is given in dimensions of distance over time. The distance traveled has the dimension of distance. Multiplying these numbers does not yield the dimension of time:

$$\frac{\text{distance}}{\text{time}} \times \text{distance} = \frac{\text{distance}^2}{\text{time}}$$

To arrive at an answer that will have the dimension of time, you must divide the distance by the speed of the motorcycle, so that

$$\text{distance} \div \frac{\text{distance}}{\text{time}} = \text{distance} \times \frac{\text{time}}{\text{distance}} = \text{time}$$

Substituting the given data gives

$$12\overline{0}0 \, \text{m} \times \frac{1 \, \text{s}}{20 \, \text{m}} = 60.0 \, \text{s}$$

This simple example could be solved by many students without dimensional analysis; however, in more complex calculations it is an extremely valuable tool.

B.2 Unit Analysis and Conversion Factors

While we now know that dimensions never change, the units of measured quantities do vary, and great care must be taken in problem solving to ensure that we are using the correct units of measurement in our calculations.

Problem solving in physics usually requires more thought than simply substituting numbers into formulas. For instance, we may know the volume in cubic feet of a patio we want to pour with concrete, but concrete is sold in cubic yards. Therefore our data in cubic feet must be converted to cubic yards.

Suppose we have 513 ft^3 to pour. The units of our given data are ft^3. We want to convert this measurement to cubic yards (yd^3). We will use a conversion factor to accomplish this. A conversion factor is an expression used to change from one unit or set of units to another. To do this, we use two mathematical principles: (1) Any number or quantity can be multiplied or divided by 1 without changing the value of the original quantity, and (2) any fraction equals 1 when its numerator and denominator are equal. For example, $\frac{3}{3} = 1$, $\frac{10 \, \text{m}}{10 \, \text{m}} = 1$, and $\frac{30 \, \text{kg}}{30 \, \text{kg}} = 1$, and since $1 \, \text{kg} = 1000 \, \text{g}$, $\frac{1 \, \text{kg}}{1000 \, \text{g}} = 1$. Similarly, $\frac{1000 \, \text{g}}{1 \, \text{kg}} = 1$ because in each case the numerator equals the denominator.

So in our example, to find the number of yd^3 needed, we will multiply our given data by a factor of one (1). Since $1 \, \text{yd}^3 = 27 \, \text{ft}^3$, we create the conversion factors $\frac{1 \, \text{yd}^3}{27 \, \text{ft}^3} = 1$ and $\frac{27 \, \text{ft}^3}{1 \, \text{yd}^3} = 1$. We use unit analysis to determine that the units we want to end up with are yd^3, so we choose $\frac{27 \, \text{ft}^3}{1 \, \text{yd}^3} = 1$ and set up the problem as follows:

$$513 \, \text{ft}^3 \times \frac{1 \, \text{yd}^3}{27 \, \text{ft}^3} = ?$$

so that the units ft^3 cancel, and our answer is 19.0 yd^3.

Change 320 cm to metres.

EXAMPLE 1

Possible conversion factors are $\frac{1 \, \text{m}}{100 \, \text{cm}}$ and $\frac{100 \, \text{cm}}{1 \, \text{m}}$. We choose the conversion factor with centimetres in the denominator so that the cm units cancel each other:

$$320 \, \text{cm} \times \frac{1 \, \text{m}}{100 \, \text{cm}} = 3.20 \, \text{m}$$

EXAMPLE 2

Change 0.0200 km to centimetres.
    First, change km to metres and then metres to centimetres.

$$0.0200 \text{ km} \times \frac{1000 \text{ m}}{\text{km}} = 20.0 \text{ m}$$

$$20.0 \text{ m} \times \frac{100 \text{ cm}}{1.0 \text{ m}} = 2000 \text{ cm}$$

Or

$$0.0200 \text{ km} \times \frac{100 \text{ m}}{\text{km}} \times \frac{100 \text{ cm}}{\text{m}} = 2000 \text{ cm}$$

Unit Conversions

These principles equally apply to unit conversions.

EXAMPLE 3

Change 4.00 km/h to m/s.

$$\frac{4.00 \text{ km}}{\text{h}} \times \frac{\text{h}}{3600 \text{ s}} \times \frac{1000 \text{ m}}{\text{km}} = 1.11 \frac{\text{m}}{\text{s}}$$

EXAMPLE 4

Change an acceleration of 10.5 m/s^2 to km/h^2.

$$10.5 \frac{\text{m}}{\text{s}^2} \times \frac{1 \text{ km}}{1000 \text{ m}} \times \left(\frac{3600 \text{ s}}{1 \text{ h}}\right)^2 = 1.36 \times 10^5 \frac{\text{km}}{\text{h}^2}$$

Sample problems with solutions.

1. Change 40.0 km to m.

$$40.0 \text{ km} \times \frac{1000 \text{ m}}{1 \text{ km}} = 40{,}\bar{0}00 \text{ m}$$

2. Change $35\bar{0}$ cm to in. (see Appendix C, Table 2).

$$35\bar{0} \text{ cm} \times \frac{0.394 \text{ in}}{1 \text{ cm}} = 138 \text{ in.}$$

3. Change 56.0 m/s to km/h.

$$56.0 \frac{\text{m}}{\text{s}} \times \frac{3600 \text{ s}}{1 \text{ h}} \times \frac{1 \text{ km}}{1000 \text{ m}} = 202 \frac{\text{km}}{\text{h}}$$

4. Change $16\bar{0}$ acres to ft^2.
   Use the conversion table for area (Appendix C, Table 3).

$$1 \text{ acre} = 43{,}560 \text{ ft}^2$$

$$16\bar{0} \text{ acres} \times \frac{43{,}560 \text{ ft}^2}{1 \text{ acre}} = 6{,}970{,}000 \text{ ft}^2$$

Problems B.2

1. Change 20.0 km to m.
2. Change 2460 m to km.
3. Change 38.0 min to s.
4. Change $40\overline{0}0$ s to min.
5. Change $72\overline{0}0$ s to h.
6. Change 3.00 h to s.
7. Change 88.0 ft/s to mi/h.
8. Change 30.0 mi/h to ft/s.
9. Change 80.0 m/s to km/h.
10. Change 2.00 km/h to m/s.
11. Change 5.00 yd^2 to ft^2.
12. Change 50.0 ft^2 to yd^2.
13. Change 54.0 ft^3 to yd^3.
14. Change 10.0 yd^3 to ft^3.
15. Change 0.0300 km to cm.
16. Change $50\overline{0}0$ cm to km.
17. Change 3.00 km/h to m/s.
18. Change 75.0 m/s to km/h.
19. Change 65.0 mi/h to ft/s.
20. Change $10\overline{0}$ ft/s to mi/h.
21. Change 65.3 m/s^2 to km/h^2.
22. Change 45.3 km/h^2 to m/s^2.
23. Change 24.2 m/s^2 to ft/s^2.

TABLES

Table 1   U.S. Weights and Measures

Units of Length	Units of Volume	Units of Weight
Standard unit—inch (in. or ″) 12 inches = 1 foot (ft or ′) 3 feet = 1 yard (yd) $5\frac{1}{2}$ yards or $16\frac{1}{2}$ feet = 1 rod (rd) 5280 feet = 1 mile (mi)	*Liquid* 16 ounces (fl oz) = 1 pint (pt) 2 pints = 1 quart (qt) 4 quarts = 1 gallon (gal) *Dry* 2 pints (pt) = 1 quart (qt) 8 quarts = 1 peck (pk) 4 pecks = 1 bushel (bu)	Standard unit—pound (lb) 16 ounces (oz) = 1 pound 2000 pounds = 1 ton (T)

Table 2   Conversion Table for Length

	cm	m	km	in.	ft	mi
1 centimetre =	1	10^{-2}	10^{-5}	0.394	3.28×10^{-2}	6.21×10^{-6}
1 metre =	100	1	10^{-3}	39.4	3.28	6.21×10^{-4}
1 kilometre =	10^5	1000	1	3.94×10^4	3280	0.621
1 inch =	2.54	2.54×10^{-2}	2.54×10^{-5}	1	8.33×10^{-2}	1.58×10^{-5}
1 foot =	30.5	0.305	3.05×10^{-4}	12	1	1.89×10^{-4}
1 mile =	1.61×10^5	1610	1.61	6.34×10^4	5280	1

Table 3   Conversion Table for Area

Metric	U.S.
$1\,m^2 = 10{,}000\ cm^2$ $\quad = 1{,}000{,}000\ mm^2$ $1\,cm^2 = 100\ mm^2$ $\quad = 0.0001\ m^2$ $1\,km^2 = 1{,}000{,}000\ m^2$	$1\ ft^2 = 144\ in^2$ $1\ yd^2 = 9\ ft^2$ $1\ rd^2 = 30.25\ yd^2$ $1\ acre = 160\ rd^2$ $\quad = 4840\ yd^2$ $\quad = 43{,}560\ ft^2$ $1\ mi^2 = 640\ acres$

	m^2	cm^2	ft^2	in^2
1 square metre =	1	10^4	10.8	1550
1 square centimetre =	10^{-4}	1	1.08×10^{-3}	0.155
1 square foot =	9.29×10^{-2}	929	1	144
1 square inch =	6.45×10^{-4}	6.45	6.94×10^{-3}	1

1 circular mil = $5.07 \times 10^{-6}\ cm^2 = 7.85 \times 10^{-7}\ in^2$

1 hectare = $10{,}000\ m^2 = 2.47\ acres$

Table 4   Conversion Table for Volume

	Metric	U.S.
	$1\text{ m}^3 = 10^6\text{ cm}^3$ $1\text{ cm}^3 = 10^{-6}\text{ m}^3$ $= 10^3\text{ mm}^3$	$1\text{ ft}^3 = 1728\text{ in}^3$ $1\text{ yd}^3 = 27\text{ ft}^3$

	m^3	cm^3	L	ft^3	in^3
$1\text{ m}^3 =$	1	10^6	1000	35.3	6.10×10^4
$1\text{ cm}^3 =$	10^{-6}	1	1.00×10^{-3}	3.53×10^{-5}	6.10×10^{-2}
$1\text{ litre} =$	1.00×10^{-3}	1000	1	3.53×10^{-2}	61.0
$1\text{ ft}^3 =$	2.83×10^{-2}	2.83×10^4	28.3	1	1728
$1\text{ in}^3 =$	1.64×10^{-5}	16.4	1.64×10^{-2}	5.79×10^{-4}	1

1 U.S. fluid gallon = 4 U.S. fluid quarts = 8 U.S. pints = 128 U.S. fluid ounces = 231 in^3 = 0.134 ft^3

1 L = 1000 cm^3 = 1.06 qt   1 fl oz = 29.5 cm^3 1 ft^3 = 7.47 gal = 28.3 L

Table 5   Conversion Table for Mass

	g	kg	slug	oz	lb	ton
1 gram =	1	0.001	6.85×10^{-5}	3.53×10^{-2}	2.21×10^{-3}	1.10×10^{-6}
1 kilogram =	1000	1	6.85×10^{-2}	35.3	2.21	1.10×10^{-3}
1 slug =	1.46×10^4	14.6	1	515	32.2	1.61×10^{-2}
1 ounce =	28.4	2.84×10^{-2}	1.94×10^{-3}	1	6.25×10^{-2}	3.13×10^{-5}
1 pound =	454	0.454	3.11×10^{-2}	16	1	5.00×10^{-4}
1 ton =	9.07×10^5	907	62.2	3.2×10^4	2000	1

1 metric ton = 1000 kg = 2205 lb

Quantities in the shaded areas are not mass units. When we write, for example, 1 kg "=" 2.21 lb, this means that a kilogram is a mass that weighs 2.21 pounds under standard conditions of gravity ($g = 9.80\text{ m/s}^2 = 32.2\text{ ft/s}^2$).

Table 6   Conversion Table for Density

	$slug/ft^3$	kg/m^3	g/cm^3	lb/ft^3	lb/in^3
1 slug per ft^3 =	1	515	0.515	32.2	1.86×10^{-2}
1 kilogram per m^3 =	1.94×10^{-3}	1	0.001	6.24×10^{-2}	3.61×10^{-5}
1 gram per cm^3 =	1.94	1000	1	62.4	3.61×10^{-2}
1 pound per ft^3 =	3.11×10^{-2}	16.0	1.60×10^{-2}	1	5.79×10^{-4}
1 pound per in^3 =	53.7	2.77×10^4	27.7	1728	1

Quantities in the shaded areas are weight densities and, as such, are dimensionally different from mass densities.

Note that

$$D_w = D_m g$$

where D_w = weight density

          D_m = mass density

            $g = 9.80\text{ m/s}^2 = 32.2\text{ ft/s}^2$

Table 7 **Conversion Table for Time**

	yr	day	h	min	s
1 year =	1	365	8.77×10^3	5.26×10^5	3.16×10^7
1 day =	2.74×10^{-3}	1	24	1440	8.64×10^4
1 hour =	1.14×10^{-4}	4.17×10^{-2}	1	60	3600
1 minute =	1.90×10^{-6}	6.94×10^{-4}	1.67×10^{-2}	1	60
1 second =	3.17×10^{-8}	1.16×10^{-5}	2.78×10^{-4}	1.67×10^{-2}	1

Table 8 **Conversion Table for Speed**

	ft/s	km/h	m/s	mi/h	cm/s
1 foot per second =	1	1.10	0.305	0.682	30.5
1 kilometre per hour =	0.911	1	0.278	0.621	27.8
1 metre per second =	3.28	3.60	1	2.24	100
1 mile per hour =	1.47	1.61	0.447	1	44.7
1 centimetre per second =	3.28×10^{-2}	3.60×10^{-2}	0.01	2.24×10^{-2}	1
1 mi/min = 88.0 ft/s = 60.0 mi/h					

Table 9 **Conversion Table for Force**

	N	lb	oz
1 newton =	1	0.225	3.60
1 pound =	4.45	1	16
1 ounce =	0.278	0.0625	1

Table 10 **Conversion Table for Power**

	Btu/h	ft lb/s	hp	kW	W
1 British thermal unit per hour =	1	0.216	3.93×10^{-4}	2.93×10^{-4}	0.293
1 foot pound per second =	4.63	1	1.82×10^{-3}	1.36×10^{-3}	1.36
1 horsepower =	2550	550	1	0.746	746
1 kilowatt =	3410	738	1.34	1	1000
1 watt =	3.41	0.738	1.34×10^{-3}	0.001	1

Table 11 Conversion Table for Pressure

	atm	Inches of Water	mm Hg	N/m^2 (Pa)	lb/in^2	lb/ft^2
1 atmosphere =	1	407	$76\bar{0}$	1.01×10^5	14.7	2120
1 inch of water[a] at 4°C =	2.46×10^{-3}	1	1.87	249	3.61×10^{-2}	5.20
1 millimetre of mercury[a] at 0°C =	1.32×10^{-3}	0.535	1	133	1.93×10^{-2}	2.79
1 newton per metre² (pascal) =	9.87×10^{-6}	4.02×10^{-3}	7.50×10^{-3}	1	1.45×10^{-4}	2.09×10^{-2}
1 pound per in² =	6.81×10^{-2}	27.7	51.7	6.90×10^3	1	144
1 pound per ft² =	4.73×10^{-4}	0.192	0.359	47.9	6.94×10^{-3}	1

[a]Where the acceleration of gravity has the standard value, $g = 9.80$ m/s² $= 32.2$ ft/s².

Table 12 Mass and Weight Density

Substance	Mass Density (kg/m^3)	Weight Density (lb/ft^3)
Solids		
Aluminum	2,700	169
Brass	8,700	540
Concrete	2,300	140
Copper	8,890	555
Cork	240	15
Ice	917	57
Iron	7,800	490
Lead	11,300	708
Wood, white pine	420	26
Liquids		
Alcohol	790	49.4
Gasoline	680	42.0
Mercury	13,600	846
Oil	870	54.2
Seawater	1,025	64.0
Water	$1,0\bar{0}0$[a]	62.4

	At 0°C and 1 atm Pressure	At 32°F and 1 atm Pressure
Gases[b]		
Air	1.29	0.081
Ammonia	0.760	0.047
Carbon dioxide	1.96	0.123
Carbon monoxide	1.25	0.078
Helium	0.178	0.011
Hydrogen	0.0899	0.0056
Nitrogen	1.25	0.078
Oxygen	1.43	0.089
Propane	2.02	0.126

[a]Metric weight density of water $= 98\bar{0}0$ N/m³.
[b]The density of a gas is found by pumping the gas into a container, measuring its volume and mass or weight, and then using the appropriate density formula.

Table 13   Specific Gravity of Certain Liquids[a]

Liquid	Specific Gravity
Benzene	0.90
Ethyl alcohol	0.79
Gasoline	0.68
Kerosene	0.82
Mercury	13.6
Seawater	1.025
Sulfuric acid	1.84
Turpentine	0.87
Water	1.000

[a]At room temperature (20°C or 68°F)

Table 14   Conversion Table for Energy, Work, and Heat

	Btu	ft lb	J	cal	kWh
1 British thermal unit =	1	778	1060	252	2.93×10^{-4}
1 foot pound =	1.29×10^{-3}	1	1.36	0.324	3.77×10^{-7}
1 joule =	9.48×10^{-4}	0.738	1	0.239	2.78×10^{-7}
1 calorie =	3.97×10^{-3}	3.09	4.19	1	1.16×10^{-6}
1 kilowatt-hour =	3410	2.66×10^6	3.60×10^6	8.60×10^5	1

Table 15   Heat Constants

	Melting Point (°C)	Boiling Point (°C)	Specific Heat cal/g °C or kcal/kg °C or Btu/lb °F	J/kg °C	Heat of Fusion cal/g or kcal/kg	J/kg	Heat of Vaporization cal/g or kcal/kg	J/kg
Alcohol, ethyl	−117	78.5	0.58	2400	24.9	1.04×10^5	204	8.54×10^5
Aluminum	660	2057	0.22	920	76.8	3.21×10^5		
Brass	840		0.092	390				
Copper	1083	2330	0.092	390	49.0	2.05×10^5		
Glass			0.21	880				
Ice	0		0.51	2100	$8\overline{0}$	3.35×10^5		
Iron (steel)	1540	3000	0.115	481	7.89	3.30×10^4		
Lead	327	1620	0.031	130	5.86	2.45×10^4		
Mercury	−38.9	357	0.033	140	2.82	1.18×10^4	65.0	2.72×10^5
Silver	961	1950	0.056	230	26.0	1.09×10^5		
Steam			0.48	$200\overline{0}$				
Water (liquid)	0	$10\overline{0}$	1.00	4190	$8\overline{0}$	3.35×10^5	$54\overline{0}$	2.26×10^6
Zinc	419	907	0.092	390	23.0	9.63×10^4		

Table 16   Coefficient of Linear Expansion

Material	α (metric)	α (U.S.)
Aluminum	$2.3 \times 10^{-5}/C°$	$1.3 \times 10^{-5}/F°$
Brass	$1.9 \times 10^{-5}/C°$	$1.0 \times 10^{-5}/F°$
Concrete	$1.1 \times 10^{-5}/C°$	$6.0 \times 10^{-6}/F°$
Copper	$1.7 \times 10^{-5}/C°$	$9.5 \times 10^{-6}/F°$
Glass	$9.0 \times 10^{-6}/C°$	$5.1 \times 10^{-6}/F°$
Pyrex	$3.0 \times 10^{-6}/C°$	$1.7 \times 10^{-6}/F°$
Steel	$1.3 \times 10^{-5}/C°$	$6.5 \times 10^{-6}/F°$
Zinc	$2.6 \times 10^{-5}/C°$	$1.5 \times 10^{-5}/F°$

Table 17   Coefficient of Volume Expansion

Liquid	β (metric)	β (U.S.)
Acetone	$1.49 \times 10^{-3}/C°$	$8.28 \times 10^{-4}/F°$
Alcohol, ethyl	$1.12 \times 10^{-3}/C°$	$6.62 \times 10^{-4}/F°$
Carbon tetrachloride	$1.24 \times 10^{-3}/C°$	$6.89 \times 10^{-4}/F°$
Mercury	$1.8 \times 10^{-4}/C°$	$1.0 \times 10^{-4}/F°$
Petroleum	$9.6 \times 10^{-4}/C°$	$5.33 \times 10^{-4}/F°$
Turpentine	$9.7 \times 10^{-4}/C°$	$5.39 \times 10^{-4}/F°$
Water	$2.1 \times 10^{-4}/C°$	$1.17 \times 10^{-4}/F°$

Table 18   Charge

$$\text{Charge on one electron} = 1.60 \times 10^{-19}\,\text{coulomb}$$
$$1\ \text{coulomb} = 6.25 \times 10^{18}\ \text{electrons of charge}$$
$$1\ \text{ampere-hour} = 3600\ \text{C}$$

Table 19   Relationships of Metric SI Base and Derived Units

This chart shows graphically how the 17 SI-derived units with special names are derived from the base and supplementary units. It was provided by the National Institute of Standards and Technology.

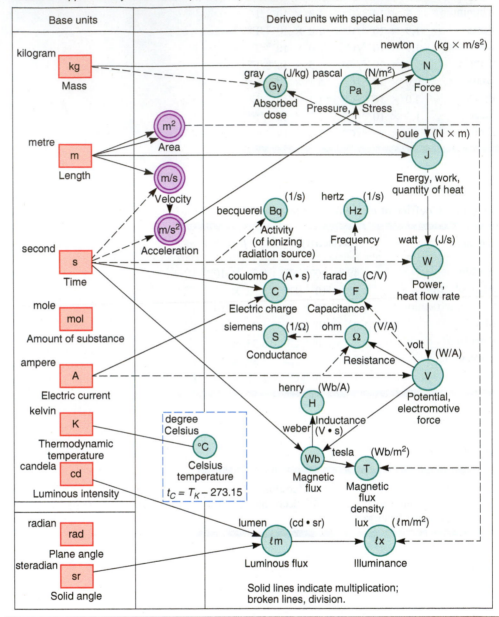

Table 20   Electric Symbols

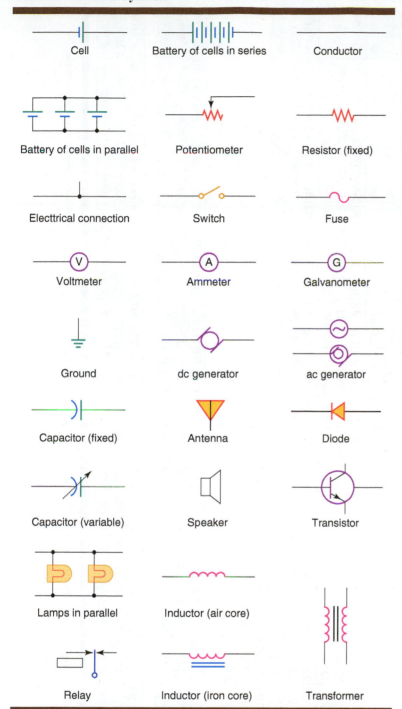

Cell	Battery of cells in series	Conductor
Battery of cells in parallel	Potentiometer	Resistor (fixed)
Electtrical connection	Switch	Fuse
Voltmeter	Ammeter	Galvanometer
Ground	dc generator	ac generator
Capacitor (fixed)	Antenna	Diode
Capacitor (variable)	Speaker	Transistor
Lamps in parallel	Inductor (air core)	
Relay	Inductor (iron core)	Transformer

Table 21 Periodic Table

Group I	Group II	Transition Elements										Group III	Group IV	Group V	Group VI	Group VII	Group 0
H 1 1.0079 $1s^1$																	He 2 4.00260 $1s^2$
Li 3 6.94 $2s^1$	Be 4 9.01218 $2s^2$											B 5 10.81 $2p^1$	C 6 12.011 $2p^2$	N 7 14.0067 $2p^3$	O 8 15.9994 $2p^4$	F 9 18.9984 $2p^5$	Ne 10 20.18 $2p^6$
Na 11 22.9898 $3s^1$	Mg 12 24.305 $3s^2$											Al 13 26.9815 $3p^1$	Si 14 28.0855 $3p^2$	P 15 30.974 $3p^3$	S 16 32.06 $3p^4$	Cl 17 35.453 $3p^5$	Ar 18 39.948 $3p^6$
K 19 39.0983 $4s^1$	Ca 20 40.08 $4s^2$	Sc 21 44.9559 $3d^1 4s^2$	Ti 22 47.9 $3d^2 4s^2$	V 23 50.9415 $3d^3 4s^2$	Cr 24 51.996 $3d^5 4s^1$	Mn 25 54.938 $3d^5 4s^2$	Fe 26 55.847 $3d^6 4s^2$	Co 27 58.9332 $3d^7 4s^2$	Ni 28 58.7 $3d^8 4s^2$	Cu 29 63.546 $3d^{10} 4s^1$	Zn 30 65.39 $3d^{10} 4s^2$	Ga 31 69.73 $4p^1$	Ge 32 72.6 $4p^2$	As 33 74.9216 $4p^3$	Se 34 78.96 $4p^4$	Br 35 79.904 $4p^5$	Kr 36 83.80 $4p^6$
Rb 37 85.47 $5s^1$	Sr 38 87.62 $5s^2$	Y 39 88.9059 $4d^1 5s^2$	Zr 40 91.22 $4d^2 5s^2$	Nb 41 92.9064 $4d^4 5s^1$	Mo 42 95.94 $4d^5 5s^1$	Tc 43 (98) $4d^5 5s^2$	Ru 44 101.07 $4d^7 5s^1$	Rh 45 102.906 $4d^8 5s^1$	Pd 46 106.4 $4d^{10} 5s^0$	Ag 47 107.868 $4d^{10} 5s^1$	Cd 48 112.41 $4d^{10} 5s^2$	In 49 114.82 $5p^1$	Sn 50 118.7 $5p^2$	Sb 51 121.75 $5p^3$	Te 52 127.60 $5p^4$	I 53 126.90 $5p^5$	Xe 54 131.3 $5p^6$
Cs 55 132.905 $6s^1$	Ba 56 137.33 $6s^2$	57–71†	Hf 72 178.49 $5d^2 6s^2$	Ta 73 180.95 $5d^3 6s^2$	W 74 183.85 $5d^4 6s^2$	Re 75 186.207 $5d^5 6s^2$	Os 76 190.2 $5d^6 6s^2$	Ir 77 192.22 $5d^7 6s^2$	Pt 78 195.08 $5d^9 6s^1$	Au 79 196.97 $5d^{10} 6s^1$	Hg 80 200.59 $5d^{10} 6s^2$	Tl 81 204.38 $6p^1$	Pb 82 207.2 $6p^2$	Bi 83 208.980 $6p^3$	Po 84 (209) $6p^4$	At 85 (210) $6p^5$	Rn 86 (222) $6p^6$
Fr 87 (223) $7s^1$	Ra 88 226.025 $7s^2$	89–103‡	Rf 104 (261) $6d^2 7s^2$	Ha 105 (262) $6d^3 7s^2$	106 (263)	107 (262)	108 (265)	109 (266)									

Legend:
Symbol — Cl 17 — Atomic number
Atomic mass* — 35.453
$3p^5$ — Electron configuration

†**Lanthanide series**

La 57 138.906 $5d^1 6s^2$	Ce 58 140.12 $5d^1 4f^1 6s^2$	Pr 59 140.908 $4f^3 6s^2$	Nd 60 144.24 $4f^4 6s^2$	Pm 61 (145) $4f^5 6s^2$	Sm 62 150.4 $4f^6 6s^2$	Eu 63 151.96 $4f^7 6s^2$	Gd 64 157.25 $5d^1 4f^7 6s^2$	Tb 65 158.925 $5d^1 4f^8 6s^2$	Dy 66 162.50 $4f^{10} 6s^2$	Ho 67 164.930 $4f^{11} 6s^2$	Er 68 167.26 $4f^{12} 6s^2$	Tm 69 168.934 $4f^{13} 6s^2$	Yb 70 173.04 $4f^{14} 6s^2$	Lu 71 174.967 $5d^1 4f^{14} 6s^2$

‡**Actinide series**

Ac 89 (227) $6d^1 7s^2$	Th 90 232.038 $6d^2 7s^2$	Pa 91 231.036 $5f^2 6d^1 7s^2$	U 92 238.029 $5f^3 6d^1 7s^2$	Np 93 237.048 $5f^4 6d^1 7s^2$	Pu 94 (244) $5f^6 6d^0 7s^2$	Am 95 (243) $5f^7 6d^0 7s^2$	Cm 96 (247) $5f^7 6d^1 7s^2$	Bk 97 (247) $5f^9 6d^0 7s^2$	Cf 98 (251) $5f^{10} 6d^0 7s^2$	Es 99 (252) $5f^{11} 6d^0 7s^2$	Fm 100 (257) $5f^{12} 6d^0 7s^2$	Md 101 (258) $5f^{13} 6d^0 7s^2$	No 102 (259) $5f^{14} 6d^0 7s^2$	Lr 103 (260) $6d^1 7s^2$

*Atomic mass values averaged over isotopes in percentages in which they occur on the earth's surface. For many unstable elements, the mass number of the most stable known isotope is given in parentheses.

Table 22   The Greek Alphabet

Capital	Lowercase	Name
A	α	alpha
B	β	beta
Γ	γ	gamma
Δ	δ	delta
E	ε	epsilon
Z	ζ	zeta
H	η	eta
Θ	θ	theta
I	ι	iota
K	κ	kappa
Λ	λ	lambda
M	μ	mu
N	ν	nu
Ξ	ξ	xi
O	o	omicron
Π	π	pi
P	ρ	rho
Σ	σ	sigma
T	τ	tau
Υ	υ	upsilon
Φ	ϕ	phi
X	χ	chi
Ψ	ψ	psi
Ω	ω	omega

SCIENTIFIC CALCULATOR

There are several kinds and brands of calculators. Some are very simple to use; others do more difficult calculations. We demonstrate various operations on common basic scientific calculators that use algebraic logic, which follows the steps commonly used in mathematics. Yours may differ. If so, consult your manual.

To demonstrate how to use a calculator, we show what buttons are pushed and the order in which they are pushed. We assume that you know how to add, subtract, multiply, and divide on the calculator.

D.1 Scientific Notation

Numbers expressed in scientific notation can be entered into many calculators. The results may then also be given in scientific notation.

EXAMPLE 1

Multiply $(6.5 \times 10^8)(1.4 \times 10^{-15})$ and write the result in scientific notation.

6.5 [EE] 8 [×] 1.4 [EE] [(−)] 15 [=]

$$9.1 \times 10^{-7}$$

The result is 9.1×10^{-7}.

. . . .

EXAMPLE 2

Divide $\dfrac{3.24 \times 10^{-5}}{7.2 \times 10^{-12}}$ and write the result in scientific notation.

3.24 [EE] [(−)] 5 [÷] 7.2 [EE] [(−)] 12 [=]

$$4.5 \times 10^6$$

The result is 4.5×10^6.

. . . .

EXAMPLE 3

Find the value of $\dfrac{(-6.3 \times 10^4)(-5.07 \times 10^{-9})(8.11 \times 10^{-6})}{(5.63 \times 10^{12})(-1.84 \times 10^7)}$ and write the result in scientific notation rounded to three significant digits.

The result rounded to three significant digits is -2.50×10^{-29}.

D.2 Squares and Square Roots

Find the value of $(46.8)^2$. ◄ **EXAMPLE 1**

46.8 $\boxed{x^2}$ $\boxed{=}$

$\boxed{2190.24}$

The result is 2190.24.

Find the value of $(6.3 \times 10^{-18})^2$. ◄ **EXAMPLE 2**

6.3 $\boxed{\text{EE}}$ $\boxed{(-)}$ 18 $\boxed{x^2}$ $\boxed{=}$

$\boxed{3.969 \times 10^{-35}}$

The result is 3.969×10^{-35}.

Find the value of $\sqrt{158.65}$ and round to four significant digits. ◄ **EXAMPLE 3**

 $\boxed{=}$

$\boxed{12.59563416}$

The result rounded to four significant digits is 12.60.

Find the value of $\sqrt{6.95 \times 10^{-15}}$ and round to three significant digits. ◄ **EXAMPLE 4**

 6.95   15 $\boxed{=}$

$\boxed{8.336666 \times 10^{-8}}$

The result rounded to three significant digits is 8.34×10^{-8}.

EXAMPLE 5

Find the value of $\sqrt{15.7^2 + 27.6^2}$ and round to three significant digits.

$$\boxed{\sqrt{}}\ \boxed{15.7}\ \boxed{x^2}\ \boxed{+}\ \boxed{27.6}\ \boxed{x^2}\ \boxed{=}$$

$$\boxed{31.75295262}$$

The result is 31.8 rounded to three significant digits.

Note: You may need to use parentheses with some calculators.

EXAMPLE 6

Find the value of $\dfrac{14}{\sqrt{5}} - \sqrt{\dfrac{15}{8}}$ and round the result to three significant digits.

$$14\ \boxed{\div}\ \boxed{\sqrt{}}\ 5\ \boxed{)}^{*}\ \boxed{-}\ \boxed{\sqrt{}}\ 15\ \boxed{\div}\ 8\ \boxed{)}^{*}\ \boxed{=}$$

$$\boxed{4.891683943}$$

The result is 4.89 rounded to three significant digits.

EXAMPLE 7

Find the value of $\sqrt{\left(\dfrac{16}{1.3}\right)^2 + \left[\dfrac{1}{2\pi(60)(6 \times 10^{-5})}\right]^2}$ rounded to three significant digits.

$$\boxed{\sqrt{}}\ \boxed{(}\ 16\ \boxed{\div}\ 1.3\ \boxed{)}\ \boxed{x^2}\ \boxed{+}\ \boxed{(}\ \boxed{(}\ 2$$

$$\boxed{\times}\ \boxed{\pi}\ \boxed{\times}\ 60\ \boxed{\times}\ 6\ \boxed{\text{EE}}\ \boxed{(-)}\ 5\ \boxed{)}\ \boxed{x^{-1}}$$

$$\boxed{)}\ \boxed{x^2}\ \boxed{)}\ \boxed{=}$$

$$\boxed{45.89092973}$$

The result is 45.9 rounded to three significant digits.

Note: You will need to insert parentheses to clarify the order of operations. You may need to supply other parentheses.

D.3 Trigonometric Operations

Calculators must have sine, cosine, and tangent buttons for use in this book.

Note: Make certain that your calculator is in the degree mode (not in the radian mode) for this section and when working problems in degrees.

---

*Note: You may need to insert a right parenthesis to clarify the order of operations. The square root key sometimes includes the left parenthesis. If not, you may need to supply it.

Find sin 26° rounded to four significant digits. ◄ **EXAMPLE 1**

0.438371147

That is, sin 26° = 0.4384 rounded to four significant digits.

. . . .

Find cos 36.75° rounded to four significant digits. ◄ **EXAMPLE 2**

0.801253813

That is, cos 36.75° = 0.8013 rounded to four significant digits.

. . . .

Find tan 70.6° rounded to four significant digits. ◄ **EXAMPLE 3**

| tan | 70.6 | = |

2.839653913

That is, tan 70.6° = 2.840 rounded to four significant digits.

. . . .

Since we use right triangles almost exclusively, we show how to find the angle of a right triangle when the value of a given trigonometric ratio is known. That is, we will find angle A when $0° \leq A \leq 90°$.

Given sin A = 0.4321, find angle A to the nearest tenth of a degree. ◄ **EXAMPLE 4**

25.60090542

Note: Make certain your calculator is in the degree mode. Thus, A = 25.6° to the nearest tenth of a degree.

. . . .

Given cos B = 0.6046, find angle B to the nearest tenth of a degree. ◄ **EXAMPLE 5**

52.79993633

Thus, B = 52.8° to the nearest tenth of a degree.

. . . .

EXAMPLE 6

Given tan $A = 2.584$, find angle A to the nearest tenth of a degree.

$\boxed{\tan^{-1}}$ 2.584 $\boxed{=}$

$\boxed{68.8437168}$

Thus, $A = 68.8°$ to the nearest tenth of a degree.

· · · · · · · · · · · · · · · · ·

Trigonometric functions often occur in expressions that must be evaluated.

EXAMPLE 7

Given $a = (\tan 54°)(25.6 \text{ m})$, find a rounded to three significant digits.

$\boxed{\tan}$ 54 $\boxed{)}$ $\boxed{\times}$ 25.6 $\boxed{=}$

$\boxed{35.23537716}$

Thus, $a = 35.2$ m rounded to three significant digits.

· · · · · · · · · · · · · · · · ·

EXAMPLE 8

Given $b = \dfrac{452 \text{ m}}{\cos 37.5°}$, find b rounded to three significant digits.

452 $\boxed{\div}$ $\boxed{\cos}$ 37.5 $\boxed{=}$

$\boxed{569.7335311}$

Thus, $b = 57\bar{0}$ m rounded to three significant digits.

· · · · · · · · · · · · · · · · ·

D.4 Finding a Power

To raise a number to a power, use the button as follows.

EXAMPLE 1

Find the value of 4^5.

4 $\boxed{\wedge}$ or $\boxed{y^x}$ 5 $\boxed{=}$

$\boxed{1024}$

That is, $4^5 = 1024$.

· · · · · · · · · · · · · · · · ·

Find the value of 1.5^{-4} rounded to three significant digits. ◀

EXAMPLE 2

1.5 [∧] or [y^x] [(−)] 4 [=]

| 0.197530864 |

That is, $1.5^{-4} = 0.198$ rounded to three significant digits.

. . . .

Problems D.4
Do as indicated and round each result to three significant digits.

1. $(6.43 \times 10^8)(5.16 \times 10^{10})$
2. $(4.16 \times 10^{-5})(3.45 \times 10^{-7})$
3. $(1.456 \times 10^{12})(-4.69 \times 10^{-18})$
4. $(-5.93 \times 10^9)(7.055 \times 10^{-12})$
5. $(7.45 \times 10^8) \div (8.92 \times 10^{18})$
6. $(1.38 \times 10^{-6}) \div (4.324 \times 10^6)$

7. $\dfrac{-6.19 \times 10^{12}}{7.755 \times 10^{-8}}$
8. $\dfrac{1.685 \times 10^{10}}{1.42 \times 10^{24}}$

9. $\dfrac{(5.26 \times 10^{-8})(8.45 \times 10^6)}{(-6.142 \times 10^9)(1.056 \times 10^{-12})}$

10. $\dfrac{(-2.35 \times 10^{-9})(1.25 \times 10^{11})(4.65 \times 10^{17})}{(8.75 \times 10^{23})(-5.95 \times 10^{-6})}$

11. $(68.4)^2$
12. $(3180)^2$
13. $\sqrt{46,500}$

14. $\sqrt{0.000634}$
15. $(1.45 \times 10^5)^2$
16. $(1.095 \times 10^{-18})^2$

17. $\sqrt{4.63 \times 10^{18}}$
18. $\sqrt{9.49 \times 10^{-15}}$

19. $\sqrt{(4.68)^2 + (9.63)^2}$
20. $\sqrt{(18.4)^2 - (6.5)^2}$

21. $18\sqrt{3} + \left(\dfrac{28.1}{19}\right)^2$
22. $\dfrac{8}{\sqrt{2}} + \sqrt{\dfrac{58}{14.5}}$

23. $25^2 - \sqrt{\dfrac{29.8}{0.0256}}$
24. $\dfrac{18.3}{6\sqrt{5}} - \left(\dfrac{225}{147}\right)^2$

25. $(12.6^2 + 21.5^2)^2 + (34.2^2 - 26.4^2)^2$
26. $\sqrt{21.4^2 + 18.7^2} + \sqrt{31.5^2 - 16.3^2}$

27. $\dfrac{91.4 - 48.6}{91.4 - 15.9}$
28. $\dfrac{14.7 + 9.6}{45.7 + 68.2}$

29. $\sqrt{\left(\dfrac{80.5}{25.6}\right)^2 + \left[\dfrac{1}{2\pi(60)(1.5 \times 10^{-7})}\right]^2}$

30. $\sqrt{\left(\dfrac{175}{36.5}\right)^2 + \left[\dfrac{1}{2\pi(60)(8.5 \times 10^{-10})}\right]^2}$

31. $\dfrac{(17.2)(11.6) + (8)(17.6) - (6)(16)}{(5)(15) + (8.5)(15) + (10)(26.5)}$

32. $\dfrac{(18.8)(5.5) + (7.75)(16.5) - (9.25)(13.85)}{(6.25)(12.5) + (4.75)(16.5) + (11.5)(14.1)}$

33. $\sin 13°$
34. $\cos 22°$
35. $\tan 52.3°$
36. $\tan 31.25°$
37. $\cos 59.36°$
38. $\sin 84.55°$
39. $\sin 48°$
40. $\cos 48°$
41. $\tan 75°$
42. $\sin 8°$
43. $\sin 8.7°$
44. $\cos 35°$

Find each angle rounded to the nearest tenth of a degree.

45.	$\sin A = 0.6527$	46.	$\cos B = 0.2577$	47.	$\tan A = 0.4568$
48.	$\sin B = 0.4658$	49.	$\cos A = 0.5563$	50.	$\tan B = 1.496$
51.	$\sin B = 0.1465$	52.	$\cos A = 0.4968$	53.	$\tan B = 1.987$
54.	$\sin A = 0.2965$	55.	$\cos B = 0.3974$	56.	$\tan A = 0.8885$

Find each angle to the nearest tenth of a degree between 0° and 90° and each side to three significant digits.

57. $b = (\sin 58.2°)(296 \text{ m})$

58. $a = (\cos 25.2°)(54.5 \text{ m})$

59. $c = \dfrac{37.5 \text{ m}}{\cos 65.2°}$

60. $b = \dfrac{59.7 \text{ m}}{\tan 41.2°}$

61. $\tan A = \dfrac{512 \text{ km}}{376 \text{ km}}$

62. $\cos B = \dfrac{75.2 \text{ m}}{89.5 \text{ m}}$

63. $a = (\cos 19.5°)(15.7 \text{ cm})$

64. $c = \dfrac{235 \text{ km}}{\sin 65.2°}$

65. $b = \dfrac{36.7 \text{ m}}{\tan 59.2°}$

66. $a = (\tan 5.7°)(135 \text{ m})$

Find the value of each power and round each result to three significant digits.

67.	12^4	68.	1.8^3	69.	0.46^5
70.	9^{-3}	71.	14^{-5}	72.	0.65^{-4}

GLOSSARY

A

Absolute Pressure The actual air pressure given by the gauge reading plus the normal atmospheric pressure.

Absolute Zero The lowest possible temperature.

Acceleration Change in velocity per unit time.

Acceleration Due to Gravity The acceleration of a freely falling object. On the earth's surface the acceleration due to gravity is 9.80 m/s^2 (metric) or 32.2 ft/s^2 (U.S.).

Accuracy The number of digits, called significant digits, in a measurement, which indicates the number of units that we are reasonably sure of having counted. The greater the number of significant digits, the better is the accuracy.

Actual Mechanical Advantage The ratio of the resistance force to the actual effort force.

Actual Power A measure of the actual power available to be converted into other forms of energy.

Adhesion The force of attraction between different or unlike molecules.

Alpha Ray A ray consisting of alpha particles, each having two protons and two neutrons, and positively charged; that is, it curves in the direction that known positive charges curve in a magnetic field.

Alternating Current A current that flows in one direction in a conductor, changes direction, and then flows in the other direction.

Ammeter An instrument that measures the current flowing in a circuit.

Ampère's Rule To find the direction of a magnetic field near a current and a straight wire, hold the wire in your right hand with your thumb extended in the direction of the current. Your fingers circle the wire in the direction of the flux lines.

Amplification The process of increasing the strength of an electronic signal.

Amplitude The maximum displacement of any part of a wave or a vibration from its equilibrium, or rest, position.

Angular Acceleration The rate of change of angular velocity (change in angular velocity/time).

Angular Displacement The angle through which any point on a rotating body moves.

Angular Momentum For a rotating body about a fixed axis, the angular momentum is the product of the moment of inertia and the angular velocity of the body.

Angular Velocity The rate of angular displacement (angular displacement/time).

Apparent Power The product of the effective values of alternating current and voltage.

Approximate Number A number that has been determined by some measurement or estimation process.

Archimedes' Principle Any object placed in a fluid apparently loses weight equal to the weight of the displaced fluid.

Area The number of square units contained in a figure.

Armature The rotating coil or electromagnet in a generator.

Astronomy The branch of science that studies everything that takes place outside the earth's atmosphere.

Atmospheric Pressure The pressure caused by the weight of the air in the atmosphere.

Atom The smallest particle of an element that can exist in a stable or independent state.

Atomic Mass The mass of a single atom, usually expressed in atomic mass units, u.

Atomic Mass Number The total number of nucleons (protons and neutrons) in a nucleus, A.

Atomic Mass Unit (u) A unit of measure of atomic mass based on the mass of the common carbon atom, which has six protons and six neutrons, so the mass of the carbon 12 atom has been given the exact value of 12 u. Thus, the approximate mass of a single proton or a single neutron is 1 u, and the mass of an atom in atomic mass units is simply the sum of the number of its protons and its neutrons.

Atomic Number The number of protons in a nucleus, Z.

Average Binding Energy per Nucleon The total binding energy of the nucleus divided by the total number of nucleons.

B

Becquerel (Bq) Unit of source activity; 1 Bq = 1 disintegration/s.

Bending Consists of both tension and compression stresses. It occurs when a force is placed on a beam, causing it to sag.

Bernoulli's Principle For the horizontal flow of a fluid through a tube, the sum of the pressure and energy of motion (kinetic energy) per unit volume of the fluid is constant.

Beta Ray A ray consisting of a stream of beta particles (electrons), which are emitted from neutrons in a nucleus as they decay into protons and electrons. This ray is negatively charged and curves in the opposite direction of an alpha ray in a magnetic field.

Binding Energy of the Nucleus The total energy required to break a nucleus apart into separate nucleons.

Biology The branch of science that studies living organisms.

Bohr Model An early model of atomic structure in which electrons travel around the nucleus in a number of discrete stable energy levels determined by quantum conditions.

Boyle's Law If the temperature of a gas is constant, the volume is inversely proportional to the absolute pressure, $V/V' = P'/P$.

Brinell Method Common industrial method used to measure the hardness of a metal.

Btu   (British thermal unit) The amount of heat (energy) necessary to raise the temperature of 1 lb of water 1°F.

Buoyant Force   The upward force exerted on a submerged or partially submerged object.

C

Calorie   The amount of heat necessary to raise the temperature of 1 g of water 1°C.

Capacitance   The ratio of the charge on either plate of a capacitor to the potential difference between the plates.

Capacitive Reactance   A measure of the opposition to ac current flow by a capacitor.

Capacitor   A circuit component consisting of two parallel plates separated by a thin insulator used to build up and store charge.

Capillary Action   The behavior of liquids that causes the liquid level in very small-diameter tubes to be different than in larger-diameter tubes. This behavior is due both to adhesion of the liquid molecules with the tube and to the surface tension of the liquid.

Celsius Scale   The metric temperature scale on which ice melts at 0° and water boils at 100°.

Center of Gravity   The point of any body at which all of its weight can be considered to be concentrated.

Centripetal Force   The force acting on a body in circular motion that causes it to move in a circular path. This force is exerted toward the center of the circle.

Chain Reaction   The process of using the neutrons released in each nuclear fission reaction to create further reactions.

Change of Phase   (sometimes called *change of state*) A change in a substance from one form of matter (solid, liquid, or gas) to another.

Charles's Law   If the pressure on a gas is constant, the volume is directly proportional to its Kelvin or Rankine temperature, $V/T = V'/T'$.

Chemistry   The branch of science that studies the composition, structure, properties, and reactions of matter.

Coefficient of Friction   The ratio between the frictional force and the normal force of an object. The number represents how rough or smooth two surfaces are when moving across one another.

Coefficient of Linear Expansion   A constant that indicates the amount by which a solid expands or contracts when its temperature is changed 1 degree.

Cohesion   The force of attraction between like molecules that holds the closely packed molecules of a solid together.

Color   A property of the light that reaches our eyes and is determined by its wavelength or its frequency.

Commutator   A device in an ac generator that produces a direct current. Composed of a split ring that replaces the slip rings in an ac generator and produces a direct current in the circuit connected to the split ring of the generator.

Compass   A small magnetic needle that is free to rotate on a bearing.

Complementary Colors   Two colors that, when combined, form white; for example, cyan and red, magenta and green, and yellow and blue.

Component Vector   When two or more vectors are added, each of the vectors is called a component of the resultant, or sum, vector.

Compound   A substance containing two or more elements.

Compound Machine   A combination of simple machines. Its total mechanical advantage is the product of the mechanical advantage of each simple machine.

Compression   A stress caused by two forces acting directly toward each other. This stress tends to cause objects to become shorter and thicker.

Concave Mirror   A mirror with a surface that curves away from an observer.

Concurrent Forces   Two or more forces applied to, or acting at, the same point.

Condensation   The change of phase from gas or vapor to a liquid.

Conduction   A form of heat transfer from a warmer part of a substance to a cooler part as a result of molecular collisions, which cause the slower-moving molecules to move faster. A transfer of charge from one place to another.

Conductor   A material through which an electron charge is readily transferred.

Constructive Interference   The superposition of waves to form a larger disturbance (wave) in a medium; occurs when two crests or troughs of superimposed waves meet.

Convection   A form of heat transfer by the movement of warm molecules from one region of a gas or a liquid to another.

Converging Lens   A lens that bends the light passing through it to some point beyond the lens. Converging lenses are thicker in the center.

Conversion Factor   An expression used to convert from one set of units to another; often expressed as a fraction whose numerator and denominator are equal to each other, although in different units.

Convex Mirror   A mirror with a surface that curves inward toward an observer.

Coulomb Force   An electric, repulsive force between protons.

Coulomb's Law   The force between two point charges is directly proportional to the product of their magnitudes and inversely proportional to the square of the distance between them.

Critical Angle   The smallest angle of incidence at which all light striking a surface is totally internally reflected.

Curie   (Ci) Unit of source activity; $1 \text{ Ci} = 3.70 \times 10^{10}$ disintegrations/s.

Current   The flow of charge that passes through a conductor.

Curvilinear Motion   Motion along a curved path.

D

Decay Rate   The probability per unit time that a decay of radioactive isotopes will occur.

Deceleration   An acceleration that indicates an object is slowing down.

Degree   An angular unit of measure; defined as 1/360 of one complete revolution.

Destructive Interference   The superposition of waves to form a smaller disturbance (wave) in a medium.

Dew Point   The temperature at which air becomes saturated with water vapor and condensation occurs.

Diffraction   The property of a wave that describes its ability to bend around obstacles in its path.

Diffusion   The process by which molecules of a gas mix with the molecules of a solid, a liquid, or another gas. Scattering of light by an uneven surface.

Diode   A device that allows current to flow through it in only one direction.

Direct Current   Current that flows in one direction.

Disintegration Energy   The total energy released in radioactive decay in the form of kinetic energy.

Dispersion   The spreading of white light into the full spectrum.

Displacement   The net change in position of an object, or the direct distance and direction it moves; a vector. Also, the distance of an object is simple harmonic motion from its equilibrium, or rest, position.

Diverging Lens   A lens that bends the light passing through it so as to spread the light. Diverging lenses are thicker at the edges than at the center.

Doppler Effect   The variation of the frequency heard when a source of sound and the ear are moving relative to each other.

Dry Cell   A voltage-generating cell that consists of a chemical paste (electrolyte) and two electrodes of unlike materials, one of which reacts chemically with the electrolyte.

Ductility   A property of a metal that enables it to be drawn through a die to produce a wire.

E

$E = mc^2$ The equation that illustrates and defines the equivalence between mass and energy forms of matter.

Effective Value   The number of amperes of alternating current that produce the same amount of heat in a resistance as an equal number of amperes of a steady direct current.

Efficiency   The ratio of the work output to the work input of a machine.

Effort   The force applied to a machine.

Effort Arm   The distance from the effort force to the fulcrum of a lever.

Elastic Collision   A collision in which two objects return to their original shape without being permanently deformed.

Elastic Limit   The point beyond which a deformed object cannot return to its original shape.

Elasticity   A measure of a deformed object's ability to return to its original size and shape once the deforming force is removed.

Electric Circuit   A conducting loop in which electrons carrying electric energy may be transferred from a suitable source to do useful work and returned to the source.

Electric Field   An electric field exists where an electric force acts on a charge brought into the area.

Electrolyte   An acid solution that produces large numbers of free electrons at the negative pole of a cell.

Electromagnet   A combination of a solenoid and a magnetic material, such as iron, in the core of the solenoid. When a current is passed through the solenoid, the magnetic fields of the atoms in the magnetic material line up to produce a strong magnetic field.

Electromagnetic Spectrum   The entire range of electromagnetic waves classified according to frequency.

Electromagnetic Theory   Describes the interrelationship among electric fields, magnetic fields, electric charge, and electric current.

Electromagnetic Wave   A wave consisting of two perpendicular transverse waves with one component of the wave being a vibrating electric field and the other component being a corresponding vibrating magnetic field; the electromagnetic wave moves in a direction perpendicular to both electric and magnetic field components.

Electron   A fundamental particle of an atom; negatively charged.

Element   A substance that cannot be separated into simpler substances.

emf   The potential difference across a source.

Energy   The ability to do work. There are many forms of energy, such as mechanical, electrical, thermal, fluid, chemical, atomic, and sound. Work delivered to an electric component or appliance (power × time).

Equilibrant Force   The force that, when applied at the same point as the resultant force, produces equilibrium.

Equilibrium   An object is in equilibrium when the net force acting on it is zero. A body that is in equilibrium is either at rest or moving at a constant velocity.

Equivalent Resistance   The single resistance that can replace a series and/or parallel combination of resistances in a circuit and provide the same current flow and voltage drop.

Evaporation   The process by which high-energy molecules of a liquid continually leave its surface.

Exact Number   A number that has been determined as a result of counting or by some definition, such as $1\,h = 60\,min$.

Excited State   A high-energy level for an electron in an atom.

Expansion   Property of a gas in which the rapid random movement of its molecules causes the gas to completely occupy the volume of its container.

Experimental Physicist   A physicist who performs experiments to develop and confirm physical theories.

F

Fahrenheit Scale   The U.S. temperature scale on which ice melts at 32° and water boils at 212°.

First Condition of Equilibrium   The sum of all parallel forces on a body in equilibrium must be zero.

First Law of Reflection   The angle of incidence equals the angle of reflection.

First Postulate of Special Relativity   The laws of physics are the same for both moving and nonmoving frames of reference.

Fixed Pulley   A pulley that is fastened to a fixed object.

Flow Rate   The volume of fluid flowing past a given point in a pipe per unit time.

Fluid   A substance that takes the shape of its container; either a liquid or a gas.

Fluorescence   A property of certain substances in which radiation is absorbed at one frequency and then reemitted at a lower frequency.

Flux Lines   Lines indicating the direction of the magnetic field near a magnetic pole.

Focal Length   The distance between the principal focus of a mirror or lens and its vertex.

Force A push or a pull that tends to change the motion of an object or prevent an object from changing motion. Force is a vector quantity with both magnitude and direction.

Formula An equation, usually expressed in letters (called *variables*) and numbers.

Freezing The change of phase from liquid to solid; also called *solidification*.

Frequency The number of complete vibrations or cycles per second of a wave.

Friction A force that resists the relative motion of two objects in contact caused by the irregularities of two surfaces sliding or rolling across each other.

Fulcrum A pivot about which a lever is free to turn.

Fusion The change of phase from solid to liquid; also called *melting*.

G

Gamma Ray A ray composed of photons of electromagnetic radiation, which have no mass.

Gas A substance that takes the shape of its container and has the same volume as its container.

Gauge Pressure The amount of air pressure excluding the normal atmospheric pressure.

Gear Train A series of gears that transfers rotational motion from one gear to another.

General Theory of Relativity Extends the special theory of relativity to accelerated frames of reference with the assumption that gravity and acceleration are equivalent and that light has mass and its path can be warped by gravity.

Generator An apparatus consisting of a coil of wire rotating in a magnetic field. A current is induced in the coil, converting mechanical energy into electric energy.

Geology The branch of science that studies the origin, history, and structure of the earth.

Grand Unified Theory A theory linking the five fundamental forces: gravitational, electric, magnetic, strong, and weak forces.

Gravitational Field The area around a massive body in which an object experiences a gravitational force. The more massive and closer an object is to that body, the stronger is the gravitational field.

Gravitational Potential Energy The energy determined by the position of an object relative to a particular reference level.

Ground State The lowest energy level for an electron in an atom.

H

Half-Life The length of time required for one half of the original amount of the radioactive atoms in a sample to decay.

Hardness A measure of the internal resistance of the molecules of a solid being forced farther apart or closer together.

Heat A form of internal kinetic and potential energy contained in an object associated with the motion of its atoms or molecules and that may be transferred from an object at a higher temperature to one at a lower temperature.

Heat of Fusion The heat required to melt 1 g or 1 kg or 1 lb of a substance.

Heat of Vaporization The amount of heat required to vaporize 1 g or 1 kg or 1 lb of a liquid.

Heat Pump A device that warms or cools by transferring heat. It contains a vapor (refrigerant) that is easily condensed to a liquid when under pressure. It produces heat during compression and cooling during vaporization.

Hooke's Law A principle of elasticity in solids: The ratio of the force applied to an object to its change in length (resulting in its being stretched or compressed by the applied force) is constant as long as the elastic limit has not been exceeded.

Huygens's Principle Each point on a wave front can be regarded as a new source of small wavelets, which form succeeding waves that spread out uniformly in the forward direction at the same speed.

Hydraulic Principle (Pascal's Principle) The pressure applied to a confined liquid is transmitted without measurable loss throughout the entire liquid to all inner surfaces of the container.

Hydrometer A sealed glass tube weighted at one end so that it floats vertically in a liquid.

Hydrostatic Pressure The pressure a liquid at rest exerts on a submerged object.

Hypothesis A scientifically based prediction that needs testing to verify its validity.

I

Ideal Mechanical Advantage The ratio of the resistance force to the effort force (assuming 100% efficiency and disregarding friction).

Illumination The luminous intensity per unit area.

Impedance A measure of the total opposition to current flow in an ac circuit resulting from the effect of both the resistance and the inductive reactance on the circuit.

Impulse The product of the force exerted and the time interval during which the force acts on the object. Impulse equals the change in momentum of an object in response to the exerted force.

Impulse–Momentum Theorem If the mass of an object is constant, then a change in its velocity results in a change of its momentum. That is, $F\Delta t = \Delta p = m\Delta v = mv_f - mv_i$.

Inclined Plane A plane surface set at an angle from the horizontal used to raise objects that are too heavy to lift vertically.

Index of Refraction A measure of the optical density of a material; equal to the ratio of the speed of light in a vacuum to the speed of light in the material.

Induced Current A current produced in a circuit by motion of the circuit through the flux lines of a magnetic field.

Induced Magnetism Magnetism produced in a magnetic material such as iron when the material is placed in a magnetic field, such as that produced in the core of a current-carrying solenoid.

Inductance A measure of the tendency of a coil of wire to resist a change in the current because the magnetism produced by one part of the coil acts to oppose the change of current in other parts of the coil.

Induction A method of charging one object by bringing a charged object near to, but not touching, it.

Induction Motor An ac motor with an electromagnetic current induced by the moving magnetic field of the ac current.

Inductive Reactance A measure of the opposition to ac current flow in an inductor.

Inductor A circuit component, such as a coil, in which an induced emf opposes any current change in the circuit.

Inelastic Collision A collision in which two objects couple together.

Inertia The property of a body that causes it to remain at rest if it is at rest or to continue moving with a constant velocity unless an unbalanced force acts upon it.

Instantaneous Current The current at any instant of time.

Instantaneous Voltage The voltage at any instant of time.

Insulator A substance that does not allow electric current to flow readily through it.

Intensity The energy transferred by sound per unit time through unit area.

Interference The effect of two intersecting waves resulting in a loss of displacement in certain areas and an increase in displacement in others.

Internal Potential Energy The energy determined by the nature or condition of a substance.

Internal Resistance The resistance within a cell that opposes movement of the electrons.

Isotopes Nuclei that contain the same number of protons but a different number of neutrons.

K

Kelvin Scale The metric absolute temperature scale on which absolute zero is 0 K and the units are the same as on the Celsius scale.

Kilocalorie The amount of heat necessary to raise the temperature of 1 kg of water 1°C.

Kilogram The basic metric unit of mass.

Kinetic Energy The energy due to the mass and the velocity of a moving object.

L

Laser A light source that produces a narrow beam with high intensity; an acronym for "light amplification by stimulated emission of radiation."

Lateral Surface Area The area of all the lateral (side) faces of a geometric solid.

Law The highest level of certainty for an explanation of physical occurrences. A law is often accompanied by a formula.

Law of Acceleration The total force acting on a body is equal to the mass of the body times its acceleration (Newton's second law).

Law of Action and Reaction For every force applied by object A to object B (action), there is an equal but opposite force exerted by object B on object A (reaction) (Newton's third law).

Law of Conservation of Angular Momentum The angular momentum of a system remains unchanged unless an outside torque acts on it.

Law of Conservation of Mechanical Energy The sum of the kinetic energy and the potential energy in a system is constant if no resistant forces do work.

Law of Conservation of Momentum When no outside forces are acting on a system of moving objects, the total momentum of the system remains constant.

Law of Inertia A body that is in motion continues in motion with the same velocity (at constant speed and in a straight line) and a body at rest continues at rest unless an unbalanced (outside) force acts upon it (Newton's first law).

Law of Refraction When a beam of light passes at an angle from a medium of lower optical density to a denser medium, the light is bent toward the normal. When a beam passes from a medium of greater optical density to one less dense, the light is bent away from the normal.

Law of Simple Machines Resistance force × resistance distance = effort force × effort distance.

Lever A rigid bar free to turn on a pivot called a fulcrum.

Light Radiant energy that can be seen by the human eye.

Light-Year The distance that light travels in one earth year: 9.45×10^{15} m or 5.87×10^{12} mi.

Liquid A substance that takes the shape of its container and has a definite volume.

Load The object in a circuit that converts electric energy into other forms of energy or work.

Longitudinal Wave A disturbance in a medium in which the motion of the particles is along the direction of the wave travel.

Loudness The strength of the sensation of sound to an observer.

Luminous Intensity A measure of the brightness of a light source.

M

Machine An object or system that is used to transfer energy from one place to another and allows work to be done that could not otherwise be done or could not be done as easily.

Magnetic Property of metals or other materials that can attract iron or steel.

Magnetic Field A field of force near a magnetic pole or a current that can be detected using a magnet.

Malleability A property of a metal that enables it to be hammered and rolled into a sheet.

Mass A measure of the inertia of a body. A measure of the quantity of material making up an object.

Mass Density The mass per unit volume of a substance.

Mass–Energy Equivalence A physical law stating that mass and energy are equivalent forms of matter; $E = \Delta mc^2$; stated by Einstein.

Matter Anything that occupies space and has mass.

Mechanical Advantage The ratio of the resistance force to the effort force.

Mechanical Equivalent of Heat The relationship between heat and mechanical work.

Melting The change of phase from solid to liquid; also called *fusion*.

Meniscus The crescent-shaped surface of a liquid column in a tube.

Method of Mixtures When two substances at different temperatures are mixed together, heat flows from the warmer body to the cooler body until they reach the same temperature. Part of the heat lost by the warmer body is transferred to the cooler body and to surrounding objects. If the two substances are well insulated from surrounding objects, the heat lost by the warmer body is equal to the heat gained by the cooler body.

Metre   The basic metric unit of length.

Molecule   The smallest particle of a substance that exists in a stable and independent state.

Moment of Inertia   Rotational inertia; the property of a rotating body that causes it to continue to turn until a torque causes it to change its rotational motion.

Momentum   A measure of the amount of inertia and motion an object has or the difficulty in bringing a moving object to rest. Momentum equals the mass times the velocity of an object.

Motion   A change of position.

Motor   A device that is composed of an armature and a stator. When a current is passed through the armature, the armature rotates in the magnetic field of the stator and converts electric energy to mechanical energy.

Movable Pulley   A pulley that is fastened to the object to be moved.

Multimeter   An instrument used to measure current flow, voltage drop, and resistance.

N

Natural Frequency   The frequency at which an object vibrates when struck by another object, such as a rubber hammer.

Neutron   A fundamental particle in the nucleus of an atom; neutrally charged.

Neutron Number   The number of neutrons in an atomic nucleus of a particular nuclide, N. Also, $N = A - Z$, the difference between the atomic mass number and the atomic number. N can be different for nuclei with the same atomic number.

Newton's Law of Universal Gravitation   All objects that have mass are attracted to one another by a gravitational force.

Normal Force   Force perpendicular to the contact surface.

Nuclear Fission   A nuclear reaction in which an atomic nucleus splits into fragments with the release of energy.

Nuclear Fusion   A nuclear reaction in which light nuclei interact to form heavier nuclei with the release of energy.

Nucleus   The center part of an atom made up of protons and neutrons.

Nuclide   A specific type of atom characterized by its nuclear properties, such as the number of neutrons and protons and the energy state of its nucleus.

Number Plane   A plane determined by the horizontal line called the x-axis and a vertical line called the y-axis intersecting at right angles at a point called the origin. These two lines divide the number plane into four quadrants. The x-axis contains positive numbers to the right of the origin and negative numbers to the left of the origin. The y-axis contains positive numbers above the origin and negative numbers below the origin.

O

Ohm's Law   When a voltage is applied across a resistance in an electric circuit, the current equals the voltage drop across the resistance divided by the resistance, $I = V/R$.

Ohmmeter   An instrument that measures the resistance of a circuit component.

Opaque   Absorbing or reflecting almost all light.

Optical Density   A property of a transparent material that is a measure of the speed of light through the given material.

Orbit   The path taken by an object during its revolution around another object, such as the path of the moon or a satellite about the earth or of a planet about the sun.

P

Parallel Circuit   An electric circuit with more than one path for the current to flow. The current is divided among the branches of the circuit.

Particle Theory   Theory that light consists of streams of particles.

Pendulum   An object suspended so that it swings freely back and forth about a pivot.

Period   The time required for a single wave to pass a given point or the time required for one complete vibration of an object in simple harmonic motion.

Periodic Table   A table that contains all of the atomic elements arranged according to their atomic numbers and that can be used to predict their chemical properties.

Photoelectric Effect   The emission of electrons by a surface when struck by electromagnetic radiation.

Photometry   The study of the measurement of light.

Photons   Wave packets of energy that carry light and other forms of electromagnetic radiation.

Physicist   A person who is an expert in or who studies physics.

Physics   The branch of science that describes the motion and energy of all matter throughout the universe.

Pitch   The distance a screw advances in one revolution of the screw. Also the distance between two successive threads. Also the effect of the frequency of sound waves on the ear.

Planck's Constant   A fundamental constant of quantum theory $(6.626 \times 10^{-34} \text{ J s})$.

Plane Mirror   A mirror with a flat surface.

Platform Balance   An instrument consisting of two platforms connected by a horizontal rod that balances on a knife edge. The pull of gravity on objects placed on the two platforms is compared.

Polarized Light   Light waves restricted to a single plane that is perpendicular to the direction of the wave motion.

Potential Energy   The stored energy of a body due to its internal characteristics or its position.

Power   The rate of doing work (work divided by time). Energy per unit time consumed in a circuit.

Power Factor   The ratio of the actual power to the apparent power.

Precision   Refers to the smallest unit with which a measurement is made, that is, the position of the last significant digit.

Pressure   The force applied per unit area.

Primary Cell   A cell that cannot be recharged.

Primary Coil   The coil of a transformer that carries an alternating current and induces a current in the secondary coil.

Primary Colors   The colors red, green, and blue; an additive mixture of the three colors of light resulting in white.

Primary Pigments The complements of the three primary colors, namely, cyan (the complement of red), magenta (the complement of green), and yellow (the complement of blue).

Principle A rule or fundamental assumption that has been proven in the laboratory.

Problem-Solving Method An orderly procedure that aids in understanding and solving problems.

Projectile A launched object that travels through the air but has no capacity to propel itself.

Projectile Motion The motion of a projectile as it travels through the air influenced only by its initial velocity and gravitational acceleration.

Propagation Velocity The velocity of energy transfer of a wave, given by the distance traveled by the wave in one period divided by the period.

Proton A fundamental particle in the nucleus of the atom; positively charged.

Pulley A grooved wheel that turns readily on an axle and is supported in a frame.

Pulse Nonrepeated disturbance that carries energy through a medium or through space.

Q

Quantum Mechanics A theory that unifies the wave–particle dual nature of electromagnetic radiation. The quantum model is based on the idea that matter, just like light, behaves sometimes as a particle and sometimes as a wave.

Quantum Theory Theory initiated by Planck and Einstein that energy, including electromagnetic radiation, is radiated or absorbed in multiples of certain units of energy.

R

Radian An angular unit of measurement. Defined as that angle with its vertex at the center of a circle whose sides cut off an arc on the circle equal to its radius. Equal to approximately 57.3°.

Radiation A form of heat transfer through energy being radiated or transmitted in the forms of rays, waves, or particles.

Radioactive Decay A type of nuclear decay that occurs when an unstable atom is transformed into a new element through the spontaneous disintegration of its nucleus.

Radiocarbon Dating A method used to obtain age estimates of organic materials using carbon 14 decay.

Rainbow A spectrum of light formed when sunlight strikes raindrops, refracts into them, reflects within them, and then refracts out of them.

Range The horizontal distance that a projectile will travel before striking the ground.

Rankine Scale The U.S. absolute temperature scale on which absolute zero is 0° R and the degree units are the same as on the Fahrenheit scale.

Real Image An image formed by rays of light.

Recharging The passing of an electric current through a secondary cell to restore the original chemicals.

Rectification The process of changing ac to dc.

Rectifier A device that changes ac to dc.

Rectilinear Motion Motion in a straight line.

Reflection The turning back of all or part of a beam of light at the boundary between two different media.

Refraction The bending of light as it passes at an angle from one medium to another of different optical density.

Regular Reflection Reflection of light with very little scattering.

Relative Humidity Ratio of the actual amount of vapor in the atmosphere to the amount of vapor required to reach 100% of saturation at the existing temperature.

Relative Motion The concept that motion can be described differently depending upon the observer's perspective.

Resistance The force overcome by a machine; the opposition to current flow.

Resistance Arm The distance from the resistance force to the fulcrum of a lever.

Resistivity The resistance per unit length of a material with uniform cross section.

Resonance A sympathetic vibration of an object caused by the transfer of energy from another object vibrating at the natural frequency of vibration of the first object. A condition in a circuit when the inductive reactance equals the capacitive reactance and they nullify each other. The current that flows in the circuit is then at its maximum value.

Resultant Force The sum of the forces applied at the same point; the single force that has the same effect as the two or more forces acting together.

Resultant Vector The sum of two or more vectors.

Revolution A unit of measurement in rotational motion; one complete rotation of a body.

Rotational Motion Spinning motion of a body.

Rotor The rotating coil in a generator.

S

Scalar A physical quantity that can be completely described by a number (called its magnitude) and a unit.

Science A system of knowledge that is concerned with establishing accurate conclusions about the behavior of everything in the universe.

Scientific Method An orderly procedure used by scientists in collecting, organizing, and analyzing new information that refutes or supports a scientific hypothesis.

Scientific Notation A form in which a number can be written as a product of a number between 1 and 10 and a power of 10. The general form is $M \times 10^n$, where M is a number between 1 and 10 and n is the exponent or power of 10.

Screw An inclined plane wrapped around a cylinder.

Second The basic unit of time.

Second Condition of Equilibrium The sum of the clockwise torques on a body in equilibrium must be equal to the sum of the counterclockwise torques about any point.

Second Law of Reflection The incident ray, the reflected ray, and the normal (perpendicular) to the reflecting surface all lie in the same plane.

Second Postulate of Special Relativity The speed of light is constant regardless of the speed of the observer or of the light source.

Secondary Cell A rechargeable type of cell.

Secondary Coil The coil of a transformer in which a current is induced by the current in the primary coil.

Semiconductors A small number of materials that fall between conductors and insulators in their ability to conduct electric current.

Series Circuit An electric circuit with only one path for the current to flow. The current in a series circuit is the same throughout.

Shearing A stress caused by two forces applied in parallel, opposite directions.

SI (Système International d'Unités) The international modern metric system of units of measurement.

Significant Digits The number of digits in a measurement, which indicates the number of units we are reasonably sure of having counted.

Simple Harmonic Motion A type of linear motion of an object in which the acceleration is directly proportional to its displacement from its equilibrium position and the motion is always directed to the equilibrium position.

Simple Machine Any one of six mechanical devices in which an applied force results in useful work. The six simple machines are the lever, the wheel and axle, the pulley, the inclined plane, the screw, and the wedge.

Snell's Law The index of refraction equals the sine of the angle of incidence divided by the sine of the angle of refraction.

Solenoid A coil of tightly wrapped wire. Commonly used to create a strong magnetic field by passing current through the wire.

Solid A substance that has a definite shape and a definite volume.

Solidification The change of phase from liquid to solid; also called *freezing*.

Sound Those waves transmitted through a medium with frequencies capable of being detected by the human ear.

Source The object that supplies electric energy for the flow of electric charge (electrons) in a circuit.

Source Activity The strength of a source of radiation that can be specified at a given time.

Space-Time An object's position in the universe can be pinpointed using three spatial dimensions and one time dimension.

Special Theory of Relativity The laws of physics are the same in moving and nonmoving frames of reference and the speed of light is constant no matter what the speed of the observer or the light source.

Specific Gravity The ratio of the density of any material to the density of water.

Specific Heat The amount of heat necessary to change the temperature of 1 kg of a substance 1°C in the metric system or 1 lb of a substance 1°F in the U.S. system.

Speed The distance traveled per unit of time. A scalar described by a number and a unit.

Speed of Light The speed at which light and other forms of electromagnetic radiation travel: 3.00×10^8 m/s in a vacuum.

Speed of Sound The speed at which sound waves travel in a medium: 331 m/s in dry air at 1 atm pressure and 0°C.

Spring Balance An instrument containing a spring, which stretches in proportion to the force applied to it, and a pointer attached to the spring with a calibrated scale read directly in given units.

Standard Position A vector is in standard position when its initial point is at the origin of the number plane. The vector is expressed in terms of its length and its angle, measured counterclockwise from the positive *x*-axis to the vector.

Standard Temperature and Pressure (STP) A commonly used reference in gas laws. Standard temperature is the freezing point of water. Standard pressure is equivalent to atmospheric pressure.

Standards of Measure A set of units of measurement for length, weight, and other quantities defined in such a way as to be useful to a large number of people.

Standing Waves A special case of superposition of two waves when no energy propagation occurs along the wave. The wave displacements are constant and remain fixed in location.

Statics The study of objects that are in equilibrium.

Stator The field magnets in a generator.

Step-Down Transformer A transformer used to lower voltage; it has more turns in the primary coil.

Step-Up Transformer A transformer used to increase voltage; it has more turns in the secondary coil.

Strain The deformation of an object due to an applied force.

Streamline Flow The smooth flow of a fluid through a tube.

Stress The ratio of an outside applied distorting force to the area over which the force acts.

Strong Force An attractive force among all nucleons (neutrons and protons) independent of their charge.

Superconductor A material that continuously conducts electric current without resistance when cooled to typically very low temperatures, often near absolute zero.

Superposition of Waves The algebraic sum of the separate displacements of two or more individual waves passing through a medium.

Surface Tension The ability of the surface of a liquid to act like a thin, flexible film.

Synchronous Motor An ac motor whose speed of rotation is constant and is directly proportional to the frequency of its ac power supply.

T

Technology The field that uses scientific knowledge to develop material products or processes that satisfy human needs and desires.

Temperature A measure of the hotness or coldness of an object.

Tensile Strength A measure of a solid's resistance to being pulled apart.

Tension A stress caused by two forces acting directly opposite each other. This stress tends to cause objects to become longer and thinner.

Terminal Speed The speed attained by a freely falling body when the air resistance equals its weight and no further acceleration occurs.

Theoretical Physicist A physicist who predominantly uses previous theories and mathematical models to form new theories in physics.

Theory A scientifically accepted principle that attempts to explain natural occurrences.

Thermal Conductivity   The ability of a material to transfer heat by conduction.

Torque   A measure of the tendency to produce change in rotational motion; equal to the applied force times the length of the torque arm.

Torsion   A stress related to a twisting motion. This type of stress severely compromises the strength of most materials.

Total Internal Reflection   A condition such that light striking a surface does not pass through the surface but is completely reflected inside it.

Total Surface Area   The total area of all the surfaces of a geometric solid, that is, the lateral surface area plus the area of the bases.

Transformer   A device composed of two coils (primary and secondary) and a magnetic core; used to step up or step down a voltage.

Translucent   Allowing some but not all light to pass through.

Transparent   Allowing almost all light to pass through so that objects or images can be seen clearly.

Transverse Wave   A disturbance in a medium in which the motion of the particles is perpendicular to the direction of the wave motion.

Turbulent Flow   The erratic, unpredictable flow of a fluid resulting from excessive speed of the flow or sudden changes in direction or size of the tube.

U

Universal Motor   A motor that can be run on either ac or dc power.

V

Vaporization   The change of phase from liquid to a gas or vapor.

Variable   A symbol, usually a letter, used to represent some unknown number or quantity.

Vector   A physical quantity that requires both magnitude (size) and direction to be completely described.

Velocity   The rate of motion in a particular direction. The time rate of change of an object's displacement. Velocity is a vector that gives the direction of travel and the distance traveled per unit of time.

Virtual Image   An image that only appears to the eye to be formed by rays of light.

Viscosity   The internal friction of a fluid caused by molecular attraction, which makes it resist a tendency to flow.

Visible Spectrum   The colors resulting from the dispersion of white light through a glass prism: red, orange, yellow, green, blue, and violet.

Volatility   A measure of a liquid's ability to vaporize. The more volatile the liquid, the greater is its rate of evaporation.

Voltage Drop   The potential difference across a load in a circuit.

Voltmeter   An instrument that measures the difference in potential (voltage drop) between two points in a circuit.

Volume   The number of cubic units contained in a figure.

W

Wave   A disturbance that moves through a medium or through space.

Wave Theory   Theory that light consists of waves traveling out from light sources, like water waves traveling out from the point at which a stone is dropped into water.

Wavelength   The distance between two successive corresponding points on a wave.

Wedge   An inclined plane in which the plane is moved instead of the resistance.

Weight   A measure of the gravitational force or pull exerted on an object by the earth or by another large body.

Weight Density   The weight per unit volume of a substance.

Wheel-and-Axle   A large wheel attached to an axle so that both turn together.

Work   The product of the force in the direction of motion and the magnitude of the displacement.

X

x-component   The horizontal component of a vector that lies along the x-axis.

Y

y-component   The vertical component of a vector that lies along the y-axis.

ANSWERS TO ODD-NUMBERED PROBLEMS AND TO ALL CHAPTER REVIEW QUESTIONS AND PROBLEMS

Chapter 0 Review Questions Pages 8–9
1. d   **2.** b   **3.** c   **4.** b   **5.** d
6. Archimedes conducted and documented his physical theories.
7. Science is a system of knowledge, while technology uses that knowledge to develop material products or processes that satisfy human needs and desires.
8. Behavior of light—fiber optics; electricity—light bulb.
9. The scientific method is used to discover facts about the natural world. The problem-solving method uses the scientific method to create something useful.
10. Knowledgeable of the physical world. It allows one to understand and answer new questions about everyday occurrences. It is also important in many technical fields.

Chapter 1

1.2 Pages 14–15
1. kilo   **3.** hecto   **5.** milli   **7.** mega   **9.** h   **11.** m   **13.** M   **15.** c
17. 135 mm   **19.** 28 kL   **21.** 49 cg   **23.** 75 hm   **25.** 24 metres
27. 59 grams   **29.** 27 millimetres   **31.** 45 dekametres   **33.** 26 megametres
35. metre   **37.** litre and cubic metre   **39.** second

1.3 Page 17
1. 3.26×10^2 **3.** 2.65×10^3 **5.** 8.264×10^2 **7.** 4.13×10^{-3} **9.** 6.43×10^0
11. 6.5×10^{-5} **13.** 5.4×10^5 **15.** 7.5×10^{-6} **17.** 5×10^{-8} **19.** 7.32×10^{17}
21. 86,200   **23.** 0.000631   **25.** 0.768   **27.** 777,000,000   **29.** 69.3
31. 96,100   **33.** 1.4   **35.** 0.0000084   **37.** 700,000,000,000   **39.** 0.00000072
41. 4,500,000,000,000   **43.** 0.000000000055

1.4 Pages 21–22
1. 1 metre   **3.** 1 kilometre   **5.** 1 kilometre   **7.** cm   **9.** m   **11.** mm
13. km   **15.** mm   **17.** m   **19.** km   **21.** cm   **23.** km   **25.** km   **27.** mm
29. 1000   **31.** 100   **33.** 0.1   **35.** 100   **37.** 10   **39.** 0.25 km   **41.** 178,000 m
43. 8.3 m   **45.** 3750 mm   **47.** Answers vary   **49.** (a) 11,000 yd   (b) 33,000 ft
51. 2.80 in.

1.5 Pages 29–31

1. 40 cm^2 **3.** 39 in^2 **5.** 22 in^2 **7.** 72 in^3 **9.** 40 cm^3 **11.** 1 litre
13. 1 cubic centimetre **15.** 1 square kilometre **17.** L **19.** m^2
21. m^3 **23.** ha **25.** mL **27.** m^3 **29.** L **31.** L **33.** m^2
35. L **37.** ha **39.** m^3 **41.** m^2 **43.** L **45.** 1000 **47.** 0.1
49. 0.01 **51.** 1 **53.** 1,000,000 **55.** 0.001 **57.** 10,000 **59.** 100
61. 100 **63.** 7.5 L **65.** 275,000 mm^3 **67.** 275 mL **69.** 1000 L
71. 7500 cm^3 **73.** 50 cm^2 **75.** 50,000 cm^2 **77.** 400 km^2 **79.** 45 ft^2
81. 13,935 cm^2 **83.** 7.91 in^2 **85.** 156,816 in^2 **87.** 513 ft^3
89. 30.1 yd^3 **91.** 13,824 in^3 **93.** 623.2 cm^3 **95.** (a) 36 cm^2 (b) 76 cm^2
97. 40 mL

1.6 Pages 34–35

1. 1 gram **3.** 1 kilogram **5.** 1 kilogram **7.** kg **9.** kg
11. metric ton **13.** g **15.** mg **17.** kg **19.** g **21.** kg **23.** g
25. g **27.** kg **29.** metric ton **31.** kg **33.** 1000 **35.** 100
37. 0.1 **39.** 1000 **41.** 0.001 **43.** 575,000 mg **45.** 0.65 g
47. 30,000,000 mg **49.** 0.4 mg **51.** 750 g **53.** 15,575 N **55.** 890 N
57. 450 lb **59.** 7.5 lb **61.** 36 oz **63.** 418.3 N **65.** second, s
67. newton, N **69.** 1 millisecond **71.** 1 ms **73.** 3.45×10^{-4} s
75. 15,915 s **77.** 4×10^9 ns

1.7 Page 37

1. 3 **3.** 4 **5.** 2 **7.** 3 **9.** 5 **11.** 3 **13.** 5 **15.** 4 **17.** 3
19. 2 **21.** 1 **23.** 4 **25.** 2 **27.** 3 **29.** 4

1.8 Page 39

1. 1 ft **3.** 1 m **5.** 0.0001 in. **7.** 10 km **9.** 0.01 m **11.** 0.00001 in.
13. 0.01 m **15.** 1 kg **17.** 0.0001 in. **19.** 1000 N **21.** 1 N **23.** $\overline{0}$,01 m^2
25. 100 kg **27.** 0.000001 kg or 1×10^{-6} kg **29.** 10,000 kg or 1×10^4 kg
31. (a) 15.7 in. (b) 0.018 in. **33.** (a) 16.01 cm (b) 0.734 cm
35. (a) 0.0350 s (b) 0.00040 s **37.** (a) 27,0$\overline{0}$0 L (b) 4.75 L
39. (a) All have one significant digit. (b) 50 N **41.** (a) 0.05 in. (b) 16.4 in.
43. (a) 0.65 m (b) 27.5 m **45.** (a) 0.00005 g (b) 0.75 g
47. (a) 3 N (b) 45,000 N **49.** (a) 20 kg (b) 40$\overline{0}$,000 kg

1.9 Pages 44–45

1. 14,200 ft **3.** 83.3 cm **5.** 7$\overline{0}$,000 N **7.** 802 m or 80,200 cm **9.** 18 s
11. 500 kg **13.** 41.0 g **15.** 3200 km **17.** 900,000 kg **19.** 0.40 m or 4$\overline{0}$ cm
21. 4900 m^2 **23.** 1,4$\overline{0}$0,000 km^2 or 1.40×10^6 km^2 **25.** 737.7 m^2
27. 5560 cm^3 **29.** 2.91×10^7 in^3 **31.** 3$\overline{0}$ ft **33.** 3.06 cm **35.** 75 km/h
37. 1$\overline{0}$00 mi/h **39.** 1100 ft lb/s **41.** 370 mi/h **43.** 43.2 m
45. 4530 kg m/s^2 **47.** 10,300 m^3 **49.** 6100 m^2 **51.** 28,800 ft

Chapter 1 Review Questions Page 46

1. c **2.** b **3.** b **4.** c **5.** a
6. (1) Pieces made separately may not fit together.
  (2) Workers could not communicate directions to each other.
  (3) Workers could not tell each other how much material to buy.
7. It is based on the decimal system.
8. (1) The distance to the moon. (2) The thickness of aluminum foil. **9.** Yes
10. The surface that would be seen by cutting a geometric solid with a thin plate.

11. Yes   **12.** Hectare   **13.** Litre
14. (1) Medicines   (2) Perfumes   (3) Wine
15. Mass measures the quantity of matter. Weight measures the gravitational pull on an object.
16. Newton (N)   **17.** Millionth
18. Because nearly all measurements are approximate numbers rather than exact numbers.
19. No   **20.** Yes

Chapter 1 Review Problems Pages 46–47

1. k   **2.** m   **3.** μ **4.** M   **5.** 45 mg   **6.** 138 cm   **7.** 1 L   **8.** 1 kg
9. 1 m^3 **10.** 0.25   **11.** 0.85   **12.** 5400   **13.** 550,000   **14.** 25,000
15. 75,000   **16.** 27,500   **17.** 0.035   **18.** 150,000   **19.** 500
20. 68.2   **21.** 11.0   **22.** 98.4   **23.** 968   **24.** 216   **25.** 212   **26.** 71.2
27. 4 h 20 min   **28.** 3   **29.** 4   **30.** 2   **31.** 3   **32.** 0.1 ft   **33.** 0.0001 s
34. 1000 mi   **35.** 100,000 N
36. (a) 12.00 m   (b) 0.008 m   (c) 0.150 m   (d) 2600 m
37. (a) 18,050 L   (b) 0.75 L   (c) 0.75 L   (d) 18,050 L   **38.** 0.125 s
39. 63,000 N   **40.** 1,800,000 cm^3 **41.** 150 m^2 **42.** 9.73 kg m/s^2
43. 9.90 m^2 **44.** $7\overline{0}0$ cm^3

Chapter 2

2.1 Pages 51–52

1. $s = vt$ **3.** $m = \dfrac{w}{g}$ **5.** $R = \dfrac{E}{I}$ **7.** $g = \dfrac{E_p}{mh}$ **9.** $h = \dfrac{v^2}{2g}$ **11.** $W = Pt$

13. $t = \dfrac{W}{P}$ **15.** $m = \dfrac{2E_k}{v^2}$ **17.** $s = \dfrac{W}{F}$ **19.** $I = \dfrac{V - E}{-r}$ or $I = \dfrac{E - V}{r}$

21. $P = \dfrac{\pi}{2R}$ **23.** $C = \dfrac{5F - 160}{9}$ or $C = \dfrac{5}{9}(F - 32)$ **25.** $f = \dfrac{1}{2\pi CX_C}$

27. $R_3 = R_T - R_1 - R_2 - R_4$ **29.** $I_P = \dfrac{I_S N_S}{N_P}$ **31.** $v_i = 2v_{\text{avg}} - v_f$

33. $s = \dfrac{v^2 - v_i^2 + 2as_i}{2a}$ **35.** $R = \dfrac{QJ}{I^2 t}$ **37.** $r = \sqrt{\dfrac{A}{\pi}}$ **39.** $d = \sqrt{\dfrac{kL}{R}}$

41. $I = \pm\sqrt{\dfrac{QJ}{Rt}}$

2.2 Pages 54–55

1. (a) $A = bh$ (b) 162 cm^2 **3.** (a) $b = \dfrac{A}{h}$ (b) 7.50 cm

5. (a) $c = P - a - b$ (b) 6.0 cm   **7.** (a) $r = \dfrac{C}{2\pi}$ (b) 10.9 yd

9. (a) $b = \dfrac{P - 2a}{2}$ or $b = \dfrac{P}{2} - a$ (b) 33.2 km

11. (a) $h = \dfrac{V}{\pi r^2}$ (b) 6.11 m   **13.** (a) $B = \dfrac{V}{h}$ (b) 154 m^2

15. (a) $b = \sqrt{A}$ (b) 21.6 in.   **17.** (a) $C = 2\pi r$ (b) 121.6 m

19. (a) $B = \dfrac{3V}{h}$ (b) 122.4 ft^2

2.3 Pages 60–62

1. 25,900 cm³   **3.** 284 cm³   **5.** 102.1 cm²   **7.** 10,100 ft³   **9.** 864 ft³
11. 12.0 cm²   **13.** 1.58 cm²   **15.** 36.0 m   **17.** 52.8 cm³   **19.** 9.39 m
21. 137 m   **23.** 65.5 ft   **25.** $6\bar{0}$ panels   **27.** 24.1 m³   **29.** 4.44 yd³
31. 266 in³

Chapter 2 Review Questions Page 63

1. c   **2.** b   **3.** a
4. (1) To find the volume of liquid storage tanks
    (2) To determine the amount of concrete needed for a driveway
5. As a shorthand way to designate different measured quantities of the same type
6. Most mistakes are made in problem solving by missing needed information or misinterpreting the information given.
7. Making a sketch helps to visualize what is happening in the problem.
8. The basic equation
9. The working equation is found by solving the basic equation for the unknown quantity.
10. Carrying the units through a problem shows whether the answer is the kind expected.
11. Making an estimate of the correct answer shows whether the solution is reasonable.

Chapter 2 Review Problems Pages 63–64

1. (a) $m = \dfrac{F}{a}$ (b) $a = \dfrac{F}{m}$ **2.** $h = \dfrac{v^2}{2g}$ **3.** $v_f = \dfrac{2s}{t} - v_i$ **4.** $v = \sqrt{\dfrac{2\,E_k}{m}}$

5. 18 ft   **6.** 12.0 m   **7.** 2.19 m   **8.** 122 cm²   **9.** 12.0 cm

10. 42.3 mm   **11.** 606 cm²   **12.** 6.0 cm   **13.** 6.27 m   **14.** 14.4 m

15. 430 cm³   **16.** 4680 m²

Chapter 2 Applied Concepts Page 64

1. 2.62 ¢/yd²   **2.** 2.13 ft³/s   **3.** 2.91   **4.** 16
5. (a) 6.73 m³   (b) Not safe; the spool's mass is 52,200 kg.

Chapter 3

3.1 Page 69

1. 2.0   **3.** 2.8   **5.** 6.3

7.

9.

11.

13. $1\frac{1}{4}$ **15.** $2\frac{5}{8}$ **17.** $\frac{15}{16}$

19.

21.

23.

3.2 Pages 75–76

1.
x-comp: −4
y-comp: −4

3.
x-comp: +5
y-comp: +4

5. *x*-comp: −4
y-comp: −8

7.
x-comp: −7
y-comp: +4

9.
x-comp: +3
y-comp: +9

11.

$A = 2\bar{0}$ m at 25°

13.

$C = 25$ m at 245°

15.
$E = 15$ m at 105°

17.

$G = 3\bar{0}$ m at 405°

	x-*component*	y-*component*
19.	9.96 m	8.97 m
21.	−18.2 km	−45.1 km
23.	97.4 km	−14.4 km
25.	38.2 m	7.09 m
27.	6.17 km	−7.35 km
29.	−5.88 m	28.9 m

3.3 Pages 85–88

1. 61 km at 55° north of east
3. 1300 mi at 1° west of south
5. 36 km at 5° east of north
7. 38 km at 25° north of west
9. 1500 mi at 71° north of east
11. 120 km at 72° south of east
13. 47 mi at 49° north of east

15.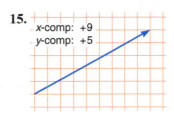
x-comp: +9
y-comp: +5

17.
x-comp: −10
y-comp: +10

19.
x-comp: +4
y-comp: +12

21.
x-comp: +8
y-comp: −4

23.
x-comp: −6.5
y-comp: 0

25.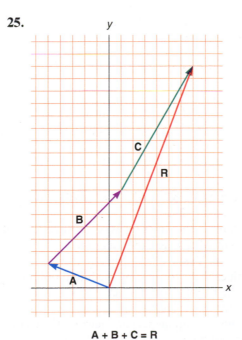

A + B + C = R

Vector	x-component	y-component
A	−5	2
B	6	6
C	6	10
R	7	18

A + B + C = R

27.

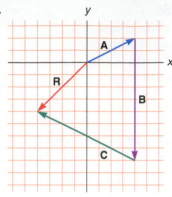

A + B + C = R

Vector	x-component	y-component
A	4	2
B	0	−10
C	−8	4
R	−4	−4

A + B + C = R

29.

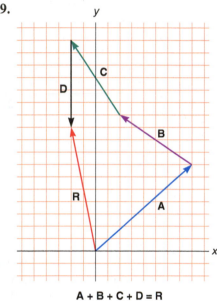

A + B + C + D = R

Vector	x-component	y-component
A	8	7
B	−6	4
C	−4	6
D	0	−7
R	−2	10

A + B + C + D = R

31. 10.0 m at 36.9°   **33.** 26.2 mi at 315.0°   **35.** 9.70 m at 98.3°
37. 53.3 m at 291.5°   **39.** 22.3 mi at 48.8°   **41.** 10.6 m at 155.7°
43. 18$\overline{0}$ m   **45.** 307 km at 48.4° north of west
47. 155 km at 58.4° south of west

Chapter 3 Review Questions Page 90

1. d   **2.** a   **3.** a
4. Two intersecting lines (axes) determine a plane (the number plane). The origin is the point where the axes intersect.
5. Yes   **6.** Yes
7. Graph each vector placing the initial point of each at the endpoint of the previous one; the resultant is the vector from the beginning point of the first vector to the endpoint of the last vector.
8. Add the x-components and then add the y-components. These are the components of the resultant.

9. No   **10.** Counterclockwise
11. In the third quadrant, the angle measure must be between 180° and 270°.
12. Complete the right triangle with the legs being the *x*- and *y*-components of the vector. Then find the lengths of *x* and *y* and determine their signs.
13. Complete the right triangle with the legs being the *x*- and *y*-axes, respectively.

Chapter 3 Review Problems Pages 90–91

1. $R_x = 11.3$ cm; $R_y = 6.50$ cm   **2.** $R_x = 5.00$ cm; $R_y = 8.66$ cm
3. $R_x = 17.3$ cm; $R_y = 10.0$ cm   **4.** $R_x = -4.50$ cm; $R_y = -7.79$ cm
5. $R_x = 6.89$ cm; $R_y = 5.79$ cm   **6.** $R_x = 10.3$ cm; $R_y = -14.7$ cm
7. 3.00 km north   **8.** 3.00 km south   **9.** 284 km at 39.3° north of west
10. 318 km at 53.4° north of east   **11.** 14.3   **12.** 11.2   **13.** 8.25
14. 5.00   **15.** $R_x = 6.00$; $R_y = -2.00$ **16.** $R_x = 9.00$; $R_y = 9.00$
17. $R_x = -6.00$; $R_y = 2.00$ **18.** $R_x = -9.00$; $R_y = -9.00$

19.

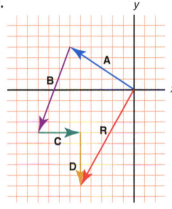

$$A + B + C + D = R$$

Vector	x-component	y-component
A	−6	4
B	−3	−8
C	4	0
D	0	−5
R	−5	−9

20.

$$A + B + C + D = R$$

Vector	x-component	y-component
A	−2	6
B	−5	0
C	−8	0
D	2	−4
R	−3	2

21. 389 km at 63.6° south of west   **22.** 344 km at 25.3° south of east

Chapter 3 Applied Concepts Page 92
1. 417 ft; 184 ft   **2.** 1090 m   **3.** 13.6 m   **4.** 2.80 mi   **5.** 656 km

Chapter 4

4.1 Pages 100–101
1. $5\overline{0}$ mi/h   **3.** 21.6 m/s   **5.** 68.3 ft/s   **7.** 320 km
9. (a) 81 ft/s   (b) 25 m/s   (c) 89 km/h   **11.** 275 km   **13.** 4.42 h
15. $8\overline{0}$ km/h, east   **17.** 125 mi/h, south   **19.** 61.1 km/h at $3\overline{0}$° south of east

21. (a) 53.7 mi/h   (b) 25.5 mi/h, west   **23.** $37\overline{0}$ km/h at 90.0°
25. 239 km/h at 190.8°   **27.** 128 mi/h at 239.4°   **29.** 227 km/h at 162.4°

4.2 Pages 104–105
1. 15 m/s^2 **3.** $1\overline{0}$ ft/s^2 **5.** 3.2 m/s^2 **7.** 6.25 m/s^2
9. 0.206 m/s^2 **11.** 67.5 mi/h   **13.** $72\overline{0}$ km/h
15. 0.534 m/s^2 **17.** 16.7 s   **19.** 0.714 m/s^2
21. 4.00 m/s^2 **23.** −8.33 m/s^2 **25.** −16.7 m/s^2

4.3 Pages 112–114
1. 5.05 m/s   **3.** 127.8 m   **5.** 2.13 ft/s^2 **7.** 9.00 mi/h   **9.** 2.19 m
11. $55\overline{0}$ ft   **13.** 1.1 m/s^2 **15.** −16.4 m/s^2 **17.** −1.6 m/s^2
19. (a) 23.5 m/s   (b) 28.2 m   **21.** (a) 3190 m   (b) 25.5 s   (c) 51.0 s
23. (a) 1.02 s   (b) 5.10 m   (c) 24.3 m/s   (d) 3.50 s
25. (a) 44.0 ft/s   (b) 4.21 s   (c) 135 ft/s   **27.** 2.06 s
29. 116 km/h   **31.** 9.05 s

4.4 Page 119
1. 62.5 m   **3.** 118 m   **5.** 124 m   **7.** 381 m   **9.** 288 ft; the ball will bounce.
11. 0.442 m

Chapter 4 Review Questions Page 120
1. c   **2.** c   **3.** a   **4.** b   **5.** d
6. Gravity accelerates all objects at the same rate.
7. Horizontal motion is independent of the vertical pull of gravity.
8. Velocity has magnitude and direction, whereas speed has only magnitude.
9. No; anything speeding up or slowing down has a changing velocity.
10. Vectors are necessary to determine the direction of velocity, acceleration, and other vector quantities. Examples: navigating airplanes, ships, and spacecraft.
11. (a) An automobile speeding up or slowing down;   (b) anything being dropped to the ground;   (c) a bullet being fired.
12. Acceleration is change in velocity; deceleration is a special case of acceleration where the object is slowing; acceleration need not be uniform and so may be subject to averaging.
13. 9.80 m/s^2 (metric) and 32.2 ft/s^2 (U.S.)

Chapter 4 Review Problems Page 121
1. 25.5 mi   **2.** 5.45 h   **3.** 226 km/h at 107.8°   **4.** 33.4 km/h at 337.1°
5. 4.00 ft/s^2 **6.** 6.9 s   **7.** 15.0 km/h   **8.** 3.90 m/s   **9.** $3\overline{0}$ h
10. 13 m/s   **11.** 4.00 m/s   **12.** 2.6 m/s^2 **13.** 2.52 m/s^2 **14.** 60.4 m/s
15. (a) 205 m/s   (b) 20.9 s   (c) 41.8 s   **16.** (a) 37.0 m/s   (b) 64.6 m
17. 8.55 m   **18.** Since $s_x = 52.8$ ft, the arrow will miss the bull's eye.

Chapter 4 Applied Concepts Page 122
1. (a) 3.76 m/s   (b) 2.01 m/s   **2.** 16.6 mi/h, 18.8° downriver   **3.** 633 ft
4. 4.71 m/s   **5.** 25.2 m/s

Chapter 5

5.2 Pages 127–128
1. 30.0 N   **3.** 744 lb   **5.** 252 lb   **7.** $23\overline{0}$ N   **9.** 40.0 m/s^2 **11.** 11.7 m/s^2
13. 0.518 m/s^2 **15.** 1.39 ft/s^2 **17.** 5250 N   **19.** 1320 lb   **21.** 2440 N

23. 1.69 m/s^2 **25.** 5740 N **27.** 6.00 kg **29.** 12.8 m/s^2
31. (a) 7.37 kg (b) It is halved (9.50 m/s^2).   (c) It is doubled (38.0 m/s^2).
33. (a) 9880 N in the opposite direction of the truck.   (b) Double the force would be required to stop the truck in half the distance.

5.3 Page 131

1. 294 N **3.** 322 lb **5.** 1.73 kg **7.** 1220 kg **9.** $6.8 \times 10^{11} \text{ kg}$
11. $11,300 \text{ N}$ **13.** 85.4 slugs **15.** 765 kg **17.** (a) $13,200 \text{ N}$ (b) $22\overline{0}0 \text{ N}$
19. (a) 65.0 kg (b) 106 N **21.** Answers vary.   **23.** Answers vary.
25. (a) 3.57 slugs (b) 303 lb **27.** Answers vary.   **29.** 102 kg

5.4 Pages 134–135

1. 380 N **3.** 1100 N **5.** 0.080 **7.** $25,000 \text{ lb}$ **9.** 1600 N **11.** $26\overline{0}0 \text{ N}$
13. 0.350 **15.** (a) -3680 N (the $-$ sign indicates the frictional force acts in opposite direction to that of the truck)   (b) -4.91 m/s^2 (c) 91.6 m

5.5 Pages 137–138

1. 3.0 N; right   **3.** 15.0 N; left   **5.** 4 N; left   **7.** 4.00 ft/s^2 **9.** 0.509 m/s^2
11. 0.504 m/s^2 **13.** $12\overline{0} \text{ N}$

Chapter 5 Review Questions Pages 140–141

1. d   **2.** b   **3.** b   **4.** c   **5.** a
6. (a) A bridge being supported over a river   (b) Isometric exercises   (c) A magnet on a refrigerator
7. (a) No   (b) No
8. A car hit from behind is forced into the car ahead of it.
9. A body in motion continues in motion and a body at rest continues at rest unless an outside force acts on it.
10. Acceleration is change of velocity.
11. Only if they have the same mass
12. Yes; $3.00 \text{ lb} = 13.35 \text{ N}$
13. More difficult; everything would slide.
14. Weight is a measure of gravitational pull. The moon, having less mass than the earth, exerts less gravitational pull.
15. For every force applied by object A to B, there is an equal force applied by B to A in the opposite direction.
16. For every action, there is an equal and opposite reaction.

Chapter 5 Review Problems Page 141

1. 3.00 m/s^2 **2.** 75.0 kg **3.** 19 ft/s^2 **4.** $19,000 \text{ N}$ **5.** 64.0 N **6.** 0.17
7. 3800 N **8.** 11 N **9.** 127 N **10.** 12 lb **11.** $20\overline{0} \text{ N}$ **12.** 896 N
13. 0.273 **14.** 150 N **15.** 1.20 m/s^2 **16.** 3680 N **17.** 41.3 kg
18. 1.22 kg

Chapter 5 Applied Concepts Page 142

1. (a) 1.55 m/s^2 (b) 39.4 m/s (c) 68.2 m/s
2. (a) 12.1 m/s^2 (b) $1.36 \times 10^5 \text{ m}$ or 136 km
3. (a) $12\overline{0} \text{ lb}$ (b) 137 lb (c) $12\overline{0} \text{ lb}$ (d) 99.6 lb
4. (a) 714 N (b) -2.94 m/s^2 (c) 13.7 s **5.** $-1.23 \times 10^4 \text{ N}$

Chapter 6

6.1 Pages 153–154
1. 80.0 kg m/s   **3.** 765 slug ft/s   **5.** 9.5×10^8 kg m/s   **7.** 6.89×10^8 kg m/s
9. (a) 12,700 slug ft/s   (b) 159 ft/s
11. (a) 55,200 kg m/s   (b) $17\overline{0}$ km/h   **13.** 7.67 m/s
15. (a) 14.0 m/s   (b) 1750 kg m/s
17. (a) 0.00287 s   (b) $12,\overline{0}00$ N   (c) 34.4 kg m/s   (d) 34.5 kg m/s
19. (a) 8250 N   (b) 52.8 m   **21.** 5830 N   **23.** (a) 2.57 lb s   (b) 367 lb

6.2 Pages 156–158
1. 8.58 m/s, right   **3.** 1.75 m/s, right   **5.** 0.964 m/s, right
7. 1.71 m/s, north   **9.** 3.75 m/s, right   **11.** 13.1 m/s
13. (a) 3.95 m/s   (b) 10.4 m/s

6.3 Pages 160–161
1. 854 kg km/h   **3.** 9.43×10^5 kg km/h
5. (a) 2.46×10^4 kg m/s   (b) 1.72×10^4 kg m/s   (c) 24.6 m/s   (d) 17.2 m/s
7. (a) 8730 kg m/s   (b) 7330 kg m/s   (c) 9.19 m/s   (d) 7.72 m/s

Chapter 6 Review Questions Page 162
1. b   **2.** d
3. The slow-moving truck has a large mass and a small velocity, whereas the rifle bullet has a small mass and a large velocity; in both cases, the product of mass and velocity is large.
4. They are the same.
5. The longer the bat (applied force) is on the ball, the greater is the impulse.
6. Total momentum in a system remains constant.
7. The momentum of the escaping gas molecules is equal to the momentum of the rocket.
8. Elastic   **9.** Inelastic   **10.** They are equal.

Chapter 6 Review Problems Pages 162–163
1. 1.23×10^5 slug ft/s   **2.** 0.204 m/s   **3.** $15,\overline{0}00$ N s   **4.** 8.5 kg m/s
5. 0.556 m/s   **6.** (a) 12 m/s   (b) 1800 kg m/s
7. (a) 4.62×10^{-4} s   (b) 106,000 N   (c) 49.0 kg m/s   (d) 48.8 kg m/s
8. (a) 70.4 m   (b) 4.74 s   **9.** 0.521 m/s, right   **10.** 1.50 m/s, east
11. 8.12 m/s, right   **12.** 1.42×10^5 kg km/h at 48.0°
13. (a) 0.210 kg m/s   (b) 0.158 kg m/s   (c) 0.600 m/s   (d) 0.451 m/s

Chapter 6 Applied Concepts Page 163
1. (a) 8.46 slug ft/s   (b) The outgoing velocity is less, thereby reducing the change in momentum and the impulse.   **2.** (a) $p_{adult} = -1680$ N s; $p_{child} = 842$ N s
(b) $F_{adult} = 2980$ N; $F_{child} = 3240$ N   **3.** (a) 3690 N   (b) 1010 N
(c) Bungee cords increase the time, thereby decreasing the force of the impulse.
4. (a) 6.68 ft/s   (b) It is easier to step out of a heavier canoe. The canoe has a greater inertia and does not move backward with as large a velocity.
5. (a) -26.5 m/s $= -95.5$ km/h (the negative sign represents west)   (b) Yes, the Jeep was speeding.

Chapter 7

7.1 Pages 169–170
1. $3\overline{0}$ N (right)   **3.** (a) $40\overline{0}$ N   (b) $5\overline{0}$ N   **5.** 1640 N at 55.6°   **7.** $18\overline{0}0$ N at 33.7°
9. 2730 lb at 140.4°   **11.** 4620 N at 123.2°   **13.** 190 N at 126.3° from $\mathbf{F}_1$

7.2 Pages 178–180

1. $10\overline{0}$ N **3.** $26\overline{0}$ N **5.** $69\overline{0}0$ N **7.** $57\overline{0}$ N **9.** Yes **11.** 322 N
13. $F_1 = 70.7$ N; $F_2 = 70.7$ N **15.** $F_1 = 577$ lb; $F_2 = 289$ lb
17. $F_1 = 433$ lb; $F_2 = 50\overline{0}$ lb **19.** $T_1 = T_2 = 731$ lb **21.** $T_1 = 565$ lb; $T_2 = 613$ lb
23. $C = 30\overline{0}0$ lb; $T = 26\overline{0}0$ lb **25.** 5540 N **27.** $T = 2330$ lb; $C = 3690$ lb

7.3 Page 184

1. 96.0 lb ft **3.** 2.00 m **5.** 187 N **7.** 1.60×10^3 N **9.** 0.357 m
11. 159 lb **13.** $40\overline{0}$ N **15.** 217 N **17.** (a) 25 lb (b) It is halved.
19. 133 N **21.** 171 N

7.4 Pages 190–191

1. $10\overline{0}$ lb **3.** $50\overline{0}$ N **5.** $45\overline{0}$ N **7.** $40\overline{0}$ N **9.** 551 N; 331 N
11. 2.77×10^5 N; 1.07×10^4 N **13.** 2.67 m **15.** 872 N; 948 N
17. 34.2 kg; 41.8 kg

7.5 Pages 193–195

1. 22.6 **3.** $F_1 = 99$ lb; $F_2 = 88.0$ lb **5.** 909 N; 591 N
7. 9.90×10^4 N; 8.82×10^4 N **9.** 117 lb; 51 lb **11.** 8.99×10^4 N; 1.01×10^5 N
13. 197 N; 118 N **15.** 2390 N; 943 N **17.** 1020 N; 775 N
19. 1525 N (up) 4.54 m from A

Chapter 7 Review Questions Page 197

1. b **2.** b **3.** a **4.** c **5.** c **6.** a **7.** b
8. b **9.** No (e.g., bridge)
10. They are equal.
11. Equilibrium is the condition of a body where the net force acting on it is zero.
12. Toward the center of the earth.
13. They are in equilibrium.
14. A diagram showing how forces act on a body
15. No, only when the pedals are parallel to the ground.
16. Even if the vector sum of opposing forces is zero, they must also be positioned so there is no rotation in the system.
17. Choose a point through which a force acts to eliminate a variable.
18. (a) Stacking bricks (b) Riding a bicycle (c) Lifting any object (d) Hitting a baseball (e) Leaning into the wind
19. No; only if the object is of uniform composition and shape
20. The support closer to the bricks

Chapter 7 Review Problems Pages 198–199

1. 569 N (left) **2.** (a) $50\overline{0}$ lb (b) $5\overline{0}$ lb **3.** 3700 N at 156°
4. 8450 lb at 334.5° **5.** 94,600 N at 80.3° **6.** 1610 N at 114.2° from F_1
7. 470 N **8.** 6.0 tons **9.** -525 **10.** 1080 N at 33.7°
11. $F_1 = 645$ N; $F_2 = 1520$ N **12.** $F_1 = 5240$ lb; $F_2 = 2210$ lb
13. $T = 1790$ N; $C = 1370$ N **14.** $T_1 = 348$ lb; $T_2 = 426$ lb; $T_3 = 475$ lb
15. $T_1 = 2900$ N; $T_2 = 1900$ N; $T_3 = 2200$ N
16. $T_1 = 6250$ N; $T_2 = 6250$ N; $C = 6250$ N **17.** 26.5 N m
18. 27.0 lb **19.** 880 N **20.** 160 lb **21.** 100.0 kg
22. 51.0 cm **23.** $560\overline{0}$ kg **24.** End closest to truck, 62,300 N; other end, 47,700 N
25. 343 N; 483 N **26.** 2580 N **27.** 8750 N

Chapter 7 Applied Concepts Page 200

1. (a) $230\overline{0}$ N m (b) 1150 N m (c) Continually push perpendicular to the door
2. (a) 18.6° (b) 58.3 lb (c) 79.5 lb (d) The angle between the ropes increased so that Sean and Greg not only pull up but also horizontally.

3. (a) 2.28 N m more torque.   (b) 18.0 N less force   **4.** (a) $T = 45.5$ N m, which will not support the torque.   (b) Reduce the angle between the wall and the bracket.
5. (a) $F_{fulcrum} = 2210$ N; $F_{bracket} = -1420$ N (negative sign means that the bracket pulls down)   (b) Between the bracket and the fulcrum

Chapter 8

8.1 Pages 206–208
1. 34.3 N m   **3.** 917 kJ   **5.** 24.4°   **7.** 16,200 J   **9.** 4410 J   **11.** 34,000 ft lb
13. 24.1 kJ   **15.** 163,000 ft lb   **17.** 6.2 MJ   **19.** 28,900 J
21. (a) 903 J   (b) −903 J   **23.** 10,700 J

8.2 Pages 213–214
1. 18.9 N m/s or 18.9 W   **3.** 0.533 s   **5.** 12.4 W   **7.** (a) 1.68 hp   (b) 219 N
9. 59.4 s   **11.** 1.49 kW   **13.** 6.17 kW   **15.** 1.94 kW
17. (a) 6.9 kW   (b) 9.2 hp   (c) 9.3 kW   **19.** Four times
21. (a) 110 passengers   (b) 22 kW   **23.** 1.15 kW   **25.** 0.350 kW

8.3 Pages 217–218
1. 2460 N m   **3.** 217 J   **5.** (a) 80.7 ft/s   (b) 3.09×10^6 ft lb
7. 14.2 ft/s   **9.** (a) 294 kJ   (b) 392 kJ   **11.** 342 m/s   **13.** 220 kJ
15. (a) 2650 kW   (b) 3550 hp   (c) 1.59×10^8 J or 159 MJ
17. 4   **19.** (a) 1960 J   (b) 1960 J   (c) 2160 N   (d) 1.31×10^5 N
21. (a) 3150 J   (b) 3150 J   (c) The increase in gravitational potential energy is equal to the work done in raising the painter. The work done by the painter is the source of the increase in energy.

8.4 Pages 224–225
1. 7.00 m/s   **3.** 72.7 m/s   **5.** 87.3 ft   **7.** 49.5 m/s   **9.** 8.40 m/s
11. $19\overline{0}$ ft/s
13.

t	s	v	KE	h	PE	Total
0.000	0.00	0.00	0	300.0	11,760	11,760
1.000	4.90	9.80	190	295.1	11,570	11,760
2.000	19.6	19.60	770	280.4	10,990	11,760
3.000	44.1	29.40	1,730	255.9	10,030	11,760
4.000	78.4	39.20	3,070	221.6	8690	11,760
5.000	122.5	49.00	$4,8\overline{0}0$	177.5	6960	11,760
6.000	176.4	58.80	6,910	123.6	4850	11,760
7.000	240.1	68.60	9,410	59.9	2350	11,760
7.800	300.0	76.68	11,760	0.00	0.00	11,760

Chapter 8 Review Questions Page 228
1. c   **2.** a   **3.** a   **4.** d   **5.** c   **6.** No   **7.** No
8. $J = \left(\dfrac{\text{kg m}}{\text{s}^2}\right)$ (m)   **9.** No   **10.** No   **11.** Yes
12. By measuring the force applied, the distance traveled, and the time taken
13. It possesses the ability to do work (e.g., turn a wheel) in falling to the lower level.

14. (a) Elevator counterweights (b) Roller coasters
15. Yes; $KE = \frac{1}{2}mv^2$ 16. At its lowest point 17. At its highest point
18. No 19. Yes 20. The bolt has accelerated to a higher velocity.

Chapter 8 Review Problems Page 229

1. 3.6×10^6 J 2. 0 3. 10.2 m 4. 9.80×10^5 J
5. (a) 27.2 W (b) 0.0365 hp 6. 0.102 m 7. 2.00 ft
8. 1.41 m/s 9. 1.40 m/s 10. 303,000 ft lb 11. 2350 J
12. $30\overline{0}0$ J 13. 1670 J 14. (a) $30\overline{0}0$ J (b) 1880 J 15. 31.3 m/s

Chapter 8 Applied Concepts Pages 229–230

1. (a) 4.49 W (b) The mass of the water, the distance the substance is carried, and the time to move the water 2. (a) 2.09×10^5 J (b) A, 34.3 m/s; B, 28.0 m/s; C, 24.3 m/s; D, 19.8 m/s (c) The higher the elevation, the less is the velocity
3. (a) 4.80×10^7 J (same as the jet's original kinetic energy) (b) 4.17×10^5 N
(c) The force needed to stop the jet is less because the time is more.
4. (a) 5.27×10^9 W (this is enough energy to power all of Paraguay)
(b) 1.05×10^{10} W (c) The higher the dam, the more potential energy changes to kinetic energy. 5. (a) 1.56×10^5 J (b) 1.56×10^5 J (c) 3.57×10^5 N
(d) When the wrecking ball is at its lowest point, all of its energy has been converted to kinetic energy, and thus it strikes the wall with the greatest velocity.

Chapter 9

9.1 Pages 237–239

1. (a) 40.8 rad (b) 2340° 3. (a) 12.5 rev (b) 4500° 5. 154 rpm
7. 8.38 rad/s 9. 354 rev/s 11. 6450 rpm 13. 14.2 rad/s
15. $66\overline{0}$ rpm 17. (a) 6.67 rev/s (b) $40\overline{0}$ rpm (c) 41.9 rad/s
19. (a) 0.0571 s (b) 87.5 rev 21. 8.40 rad 23. 0.131 m 25. 1.33 m/s
27. (a) 68.6 rad/s (b) 12,300 rad (c) 157 cm (d) 3080 m (e) 17.2 m/s
29. (a) 50.5 rad/s (b) 1520 rad (c) 502 m (d) 502 m 31. 188 in./s or 15.7 ft/s
33. (a) 1680 km/h (b) 1450 km/h (c) 1190 km/h (d) 838 km/h
35. 1.33 rad/s² 37. (a) -3.60 rad/s² (b) 432 m/s (c) 18.3 rev

9.3 Pages 244–245

1. 4350 N 3. 79.4 slugs 5. 5.53 m/s 7. 1.92 m 9. 3.60×10^3 lb
11. 28.8 m 13. 5420 N 15. 19 N 17. $13\overline{0}0$ N 19. $20\overline{0}0$ N
21. 34,400 N

9.4 Page 248

1. 7260 3. 18.7 5. 1860 N m 7. 343 hp 9. 4.95/s
11. 4600 W 13. (a) 299 hp (b) 599 hp 15. 2.65 N m 17. 37 kJ/min
19. 251 kW 21. 6.67/s 23. 2800 W 25. 7.17 N m 27. 118 W

9.6 Pages 256–259

1. 52 rpm 3. 24 teeth 5. 207 rpm 7. 42.5 rpm 9. 75 teeth 11. 63 teeth
13. 144 rpm 15. 60 teeth 17. 9 teeth 19. Counterclockwise 21. Clockwise
23. Clockwise 25. Clockwise 27. Counterclockwise 29. 1050 rpm
31. 576 rpm 33. 1480 rpm 35. 40 teeth 37. 20 teeth
39. Gear B is reversed in all problems.

9.7 Page 261

1. 2250   **3.** 3600   **5.** 147   **7.** 62.5 rpm   **9.** 22.5 in.   **11.** Clockwise
13. Counterclockwise   **15.** Counterclockwise

Chapter 9 Review Questions Pages 263–264

1. d   **2.** a   **3.** b
4. Curvilinear motion is motion along a curved path; rotational motion occurs when the body itself is spinning.
5. Radians and revolutions
6. A radian is an angle with its vertex at the center of a circle whose sides cut off an arc on the circle equal to its radius.
7. Angular displacement is the angle through which any point on a rotating body moves. It can be measured in radians or revolutions.
8. Linear velocity = angular velocity × radius
9. They are alike except for the substitution of θ for s, ω for v, and α for a.
10. Law of conservation of angular momentum   **11.** Yes
12. No; it tends to cause a body to continue in a straight line (tangent to the curve).
13. Number of teeth of driver times number of revolutions of driver = number of teeth of driven gear times number of revolutions of driven gear
14. An idler changes direction of rotation of the driver gear.
15. Opposite   **16.** Since the gears are connected, they both rotate together.
17. Diameters are used in similar equations rather than teeth.
18. The small pulley
19. The belt is one continuous piece of material.

Chapter 9 Review Problems Page 264

1. (a) 26π or 81.7 rad   (b) 4680°   **2.** 1.7 rad/s   **3.** 0.885 m/s   **4.** 7.50 N
5. 2.75 m/s   **6.** 5.67 ft lb/s   **7.** 6.00 rad/s   **8.** 6.00 rpm   **9.** 20 teeth
10. 57.5 rpm   **11.** Yes   **12.** 105 rpm   **13.** 2.00 cm   **14.** Counterclockwise
15. 720 rpm   **16.** 32 teeth

Chapter 9 Applied Concepts Page 265

1. (a) 1.78 rad/s = 17.0 rpm   (b) 2.51 rad/s = 24.0 rpm   (c) No, the mass of the person is unrelated to the rotational speed.   **2.** (a) 240 Nm   (b) 405 kg m^2
(c) Increase the radius of the waterwheel or make the waterwheel a ring and not a solid disk. More mass farther away from the axis increases the rotational inertia.
3. (a) B, 8.00 g's; D, 2.00 g's   (b) B is not safe. The B loop should have a larger radius to reduce the centripetal force. D is safe.   **4.** (a) 86.8 ft/s = 59.2 mi/h
(b) 123 ft/s = 83.9 mi/h   (c) As the radius increases, the centripetal force is reduced and the maximum velocity increases.   **5.** (a) 2.26×10^4 lb ft/s = 41.1 hp
(b) 43.2 lb ft   (c) The lower the gear, the lower is the speed and the greater is the torque. More torque is required to move up steep hills.

Chapter 10

10.2 Pages 273–274

1. 14.8   **3.** 36.3   **5.** 52.4   **7.** 2.39   **9.** 48.8   **11.** 1.55   **13.** 2.27
15. (a) 1100 lb   (b) 4.00   **17.** (a) 1.50 ft   (b) 4.00

10.3 Pages 275–276
1. 14.0   **3.** 271   **5.** 524   **7.** 48.8   **9.** 20.4   **11.** 181 N
13. (a) 438 lb   (b) 6.00   **15.** (a) 2.99 cm   (b) 3.34

10.4 Pages 279–280
1. 1   **3.** 3
5. 6   **7.** 2
15. 2
17. (a) 388 N   (b) 82.0 m
19. No

9.

4 : 1

11.

13.

21.   Mechanical Advantage (MA)

Number of Pulleys	1	2	3	4	5	6	7	8
Fixed	1	1	2	2	3	3	4	4
Movable	0	1	1	2	2	3	3	4
Fixed		0	1	1	2	2	3	3
Movable		1	1	2	2	3	3	4

10.5 Pages 283–284
1. 12.8   **3.** 21.2   **5.** 36.3   **7.** 4.62   **9.** 5.61
11. (a) 4.00   (b) 182 N   (c) No   **13.** 3.00   **15.** (a) 1.33 m   (b) 4.33
17. 1790 N

10.6 Page 286
1. 2.30   **3.** 14.2   **5.** 2.28   **7.** 29.3   **9.** 35.2
11. (a) 4.84 lb   (b) 754 N   **13.** (a) 37.7   (b) 566 lb   (c) 188 lb

10.8 Pages 288–289
1. 15.0   **3.** 40.0   **5.** 75.0 lb   **7.** 125 N   **9.** 1.44×10^5 N

10.9 Page 292
1. 80%   **3.** 13   **5.** 80%   **7.** 75%   **9.** 18

Chapter 10 Review Questions Pages 294–295
1. d   **2.** a   **3.** b   **4.** d   **5.** b   **6.** b
7. (a) Bicycle   (b) Auto transmission   (c) High-speed drill   **8.** Resistance force
9. Effort force × effort distance = resistance force × resistance distance

10. Mechanical advantage   11. Efficiency   12. No

13. The fulcrum   14. $\text{MA} = \dfrac{\text{effort arm length}}{\text{resistance arm length}}$ 15. First class

16. $F_R \times s_R = F_E \times s_E$
17. The opposite end of the resistance force with the effort force in between
18. Resistance force × resistance radius = effort force × effort radius
19. No; it depends on the radii.
20. A fixed pulley does not move. It is suspended by its center axle. A movable pulley is free to move and is suspended by the strand around the groove.
21. No   22. $\text{MA} = \dfrac{\text{length of plane}}{\text{height of plane}}$
23. The distance a screw advances into the wood in one revolution
24. It is greater because the handle of the jackscrew can be longer than the radius of the screwdriver.

Chapter 10 Review Problems Pages 295–296
1. 2.00   2. 73.0%   3. 3.82   4. (a) 540 N   (b) 360 N
5. (a) 6.54 cm   (b) 2.60   6. 12   7. 675 N   8. (a) 292 N   (b) 3.00
9. 2.30 ft   10. 16,400 N   11. 1.08 mm   12. 151
13. (a) 0.112   (b) 0.106   (c) 16.4 N
14. (a) The MA is doubled.   (b) 1.57 cm   15. 1.74   16. 81.4 cm
17. 5   18. 4   19. (a) 64.0   (b) 4800 N   20. 6.0

Chapter 10 Applied Concepts Pages 296–297
1. (a) 2.62×10^{31} m   (b) $\text{MA}_{\text{lever}} = 6.82 \times 10^{22}$ This was clearly a theoretical statement by Archimedes.   2. (a) 0.120   (b) 2.54 m   (c) 4.58 m/s   (d) Increases the speed of the resistance end. The tip of the fishing pole needs to be moving at a high speed to cast the lure.   3. (a) 7.16   (b) 3.27 lb   (c) The MA would increase.   4. (a) 0.118   (b) 6.42 lb   (c) 10.4 in.   (d) An MA less than 1 is advantageous for speed. The bike must be in a high gear and moving quickly.
5. (a) 3.29   (b) 3.53   (c) Increase the length of the lever arm and/or the radius of the wheel.

Chapter 11

11.1 Page 302
1. (a) 826 N   (b) 0.497 N   3. 1.23×10^{22} N   5. 5.70×10^{16} N
7. 3.37×10^5 N   9. 3.61×10^{-51} N

11.3 Page 307
1. 1020 m/s   3. 4.79×10^4 m/s   5. 9630 m/s   7. 7.59×10^6 s or 0.241 yr
9. 9.32×10^8 s or 29.6 yr

Chapter 11 Review Questions Page 308
1. d   2. b   3. a   4. c   5. a
6. The earth's mass gives it a greater gravitational force.
7. The increased radius would cause you to weigh one fourth of your original weight.
8. The increased mass would cause you to weigh twice your original weight.
9. The satellite's horizontal velocity enables it to continually miss the earth as it falls.
10. With no horizontal velocity, the apple struck the ground. The moon's horizontal velocity enables it to maintain its orbit.
11. The force at the perigee would be greater than that at the apogee.
12. No, the mass in the period equation is the mass of the object being orbited.

Chapter 11 Review Problems Page 309

1. 9.60×10^{-11} N   **2.** 2.82×10^{-7} N   **3.** 636 N; 143 lb   **4.** 1610 N; 362 lb
5. 49.2 N; 11.1 lb   **6.** 8.37×10^5 s or 9.69 days   **7.** 6.70×10^6 s or 77.5 days
8. 1.90×10^7 s or $22\overline{0}$ days   **9.** 832 N   **10.** 745 N

Chapter 11 Applied Concepts Page 309

1. (a) 3.70 m/s^2 (b) 315 N; 833 N   (c) The astronaut's muscles would weaken and have a difficult time supporting the returning astronaut on the earth.
2. (a) 1.57×10^7 m   (b) 2.46 earth radii   **3.** (a) $r = 4.22 \times 10^7$ m from the center of the earth; $r = 3.58 \times 10^7$ m from the surface of the earth   (b) 3.07×10^3 ms = 1.11×10^4 km/h   **4.** (a) 1.61×10^3 m/s $= 5.80 \times 10^3$ km/h   (b) 7370 s or 2.05 h
(c) Increasing the altitude would decrease the velocity and increase the period.
5. (a) 565 N   (b) The astronauts only feel weightless because they are in continuous free fall.

Chapter 12

12.2 Pages 324–325

1. 2.50 m $\times$ 0.80 m: 71 kPa; 2.50 m $\times$ 0.45 m: 130 kPa; 0.80 m $\times$ 0.45 m: 390 kPa
3. (a) 46.2 in.   (b) 20.3 lb   **5.** $20\overline{0}$ N/m   **7.** (a) 0.0250 cm   (b) 1.56×10^6 N
9. 3.57×10^5 lb   **11.** 5.97×10^7 Pa or 59.7 MPa

13. (a)   (b) 180 N   (c) 9.0 cm   (d) 25 N/cm

15. 16.8 cm; 12.6 cm   **17.** 489 N The breaking point is dependent on the diameter of the wire and not on its length.   **19.** 0.827 J

12.5 Pages 336–337

1. 2870 kg/m^3 **3.** 1750 lb   **5.** 5600 cm^3 **7.** 1210 lb/ft^3 **9.** 2710 kg/m^3
11. 684 kg/m^3 **13.** 3.5 m^3 **15.** 58.8 lb/ft^3 **17.** 1.49 m^3 **19.** 2820 kg/m^3
21. (a) $100\overline{0}$ L   (b) 1500 L   (c) 73.5 L   **23.** 850 kg   **25.** 0.917   **27.** 7.8
29. 0.68   **31.** Floats   **33.** Floats   **35.** 5.8×10^{15} **37.** 2330 kg/m^3

Chapter 12 Review Questions Pages 339–340

1. a, c, and e   **2.** All   **3.** d   **4.** d   **5.** d   **6.** b   **7.** d   **8.** e
9. Mass density refers to the mass per unit volume; weight density refers to weight per unit volume.

10. Yes; no.

11. Capillary action refers to the effect of surface tension of liquids that causes the level of liquid in a small-diameter tube to be higher or lower than that of the liquid in a large-diameter tube due to the adhesive force between the liquid and the tube.

12. Adhesion refers to the attractive force between different molecules; cohesion refers to the attractive force between similar molecules.

13. The surface tension of water allows the base of many insects' legs to be supported by the surface of water, allowing them to "walk" on the surface of a pond.

14. 1800   **15.** Pressure

16. Stress is directly proportional to strain as long as the elastic limit is not exceeded.

17. kPa

18. The specific gravity of an object can be found by dividing the density of the object by the density of water. The density can be found by determining the mass of the object and dividing by its volume.

19. Mass density   **20.** Viscosity   **21.** Elastic limit   **22.** Solid, liquid, gas

23. An atom consists of one nucleus and its surrounding electrons; a molecule consists of two or more atoms.

24. A proton has a positive charge; a neutron has a neutral charge.

25. Tension, compression, shear, bending, and twisting

26. The hydrometer measures the density of the battery electrolyte. The density is related to the amount of sulfuric acid in the electrolyte and therefore the charge on the battery. Temperature does affect the measurement.

Chapter 12 Review Problems Page 340

1. 165 N   **2.** 24.6 cm   **3.** 1.63×10^6 lb/in.; 0.0317 in.
4. 311 MPa   **5.** 2020 lb/ft^3 **6.** 1.3 N   **7.** 8950 kg/m^3
8. 83.4 lb/ft^3 **9.** 55.8 cm^3 **10.** 21.9 m^3 **11.** 2160 kg
12. 59,300 lb   **13.** 18.9 lb/ft^3 **14.** 54.8 lb/ft^3 **15.** 0.117 ft^3
16. 0.694   **17.** 77.8 lb   **18.** 9.26 kg   **19.** 2960 kg/m^3 **20.** Sink

Chapter 12 Applied Concepts Page 341

1. (a) 3.08×10^5 N/m^2 (b) 1.27×10^5 N/m^2 (c) High-speed driving and the high-pressure tire together increase the heat energy in the tire. This could eventually break down the tire, resulting in a blowout.   **2.** (a) A shearing strain causes the warping of the lines.   (b) An automobile exerts a tremendous amount of force on the pavement to begin its acceleration. On the open road, the vehicle exerts much less force on the pavement to continue its motion.   **3.** 3.72 ft^3
4. (a) 7.26×10^4 kg   (b) 6.3×10^4 kg   (c) 4.9×10^4 kg   **5.** (a) 178 lb/in.
(b) 201 lb   (c) At some point the spring will not be able to compress any farther.

Chapter 13

13.1 Pages 349–350

1. 21.7 lb/in^2 **3.** 16.5 lb/ft^3 **5.** 55.2 lb/in^2 **7.** 245 kPa   **9.** 9.69 m
11. 757 kg/m^3 **13.** 57.8 lb/in^2 **15.** 233 lb/in^2 **17.** 2.22×10^8 N
19. $49\overline{0}$ kPa   **21.** 728 kPa

13.2 Pages 353–354
1. 60.0 lb   **3.** 22.0   **5.** 24.0   **7.** 48 N   **9.** 23,000 N   **11.** (a) 146 N
(b) 11.6 N/cm^2 (c) 36.0   **13.** (a) 4800 N   (b) 9.55 kPa   (c) 9.55 kPa
(d) 64.0   (e) The lift will exert twice the force on the large piston; the MA and the pressure will double.   (f) The lift will exert four times the force on the large piston; the MA and the pressure will increase by a factor of 4.   **15.** Increased by a factor of 4   **17.** 2610 N   **19.** 1570 cm^2

13.3 Pages 357–358
1. 118 lb/in^2 **3.** 293 kPa   **5.** (a) 303.96 kPa   (b) 202.64 kPa   (c) 88.2 lb/in^2
(d) 506.6 kPa   (e) 33.77 kPa   (f) 25.33 kPa   **7.** 586 kPa   **9.** 30.3 lb/in^2
11. 306 kPa   **13.** 1174 kPa   **15.** 401 kPa

13.4 Pages 360–361
1. 13.0 lb   **3.** 3.9 N   **5.** 75.5 lb   **7.** 3.20 N   **9.** 105 lb   **11.** 431 N
13. (a) 3.06×10^4 ft^3 (b) 795 tons   **15.** 7.65 m^3 **17.** 440 N

13.5 Pages 365–366
1. 358 L/min   **3.** (a) 9.6 cm   (b) 0.34 m/s   **5.** (a) 34.1 ft/min   (b) 136 ft/min
7. 16.7 cm   **9.** 49,500 L/min

Chapter 13 Review Questions Pages 367–368
1. c   **2.** b   **3.** b   **4.** c   **5.** a   **6.** kPa
7. Pressure is the force applied per unit area.
8. $F_s = \frac{1}{2} A h D_w$ or one half the force exerted by the water on the bottom of the tank.
9. External pressure applied to a fluid is transmitted to all inner surfaces of the liquid's container.
10. A ship floats because it is lighter than an equal volume of displaced water.
11. The spinning baseball creates a different air velocity on one side than on the other. This creates a higher pressure on one side, which causes the ball to "curve."
12. The top side of the wing is curved more than the bottom, so the velocity of air rushing past the top must be larger than that going past the bottom. This creates a low-pressure area at the top of the wing and creates lift.
13. In streamline flow, all particles of the fluid passing a given point follow the same path. In turbulent flow, the particles passing a given point may follow different paths from that point.
14. A balloon filled with a gas with a lower density than air, such as helium, rises.
15. Absolute pressure is the actual air pressure given by the gauge reading plus normal atmospheric pressure. Gauge pressure is the pressure measured relative to atmospheric pressure. Gauge pressure is used in an automobile tire gauge.
16. The pressures are identical. The forces are different.
17. This force depends on the horizontal surface area and the average pressure on that surface. The average pressure depends on the density and the height of the liquid.
18. The pressure at the bottom is greater than at the top.
19. Smaller. The total force exerted on the brake pads must be larger than the force applied to the brake pedal.
20. Yes, but only if the fluid to be drawn through the straw is in an airtight sealed container.

Chapter 13 Review Problems Page 368
1. 31.8 kPa   **2.** 455 ft   **3.** 30.3 m   **4.** 9300 lb   **5.** 196 kPa
6. 9.60×10^7 N   **7.** 7.01×10^8 N   **8.** 2.80×10^4 N   **9.** 2610 kPa
10. 192 lb/in^2 **11.** 9410 kPa   **12.** (a) 77.2 lb   (b) 8.42   **13.** 35.5 N
14. (a) 103 N   (b) 111 N/cm^2 (c) 42.2   **15.** 303 kPa   **16.** 554 kPa
17. 299 lb/in^2 **18.** 3.3 N   **19.** 6.30 N   **20.** 2.4 N   **21.** (a) 26,000 ft^3

(b) 654 tons   **22.** 188 L/min   **23.** (a) 3610 L/min   (b) 90,300 L
24. (a) 49.3 N/m^3 (b) Alcohol

Chapter 13 Applied Concepts Page 369

1. (a) 372 m^2 (b) 2.76×10^6 N   (c) The bands should be spaced closer together because of the increased pressure at the bottom of the tank.   **2.** (a) 44.6 lb/in^2
(b) 429 N on each piston or brake drum   **3.** (a) 0.703 m^3 (b) 16,000 N
(c) 7060 N   (d) 8900 N   **4.** (a) 50.5 mi/h   (b) The space between the buildings acts as a Venturi tube by limiting the area that the air can pass through.
5. (a) 201 in./s = 16.8 ft/s   (b) Limit the number of holes or make the holes smaller. Limiting the area for the water to come out causes it to travel faster.

Chapter 14

14.1 Pages 373–374

1. 25°C   **3.** 125°C   **5.** 293°F   **7.** −12.2°C   **9.** 203°F   **11.** 298 K
13. 133°C   **15.** 223 K   **17.** 5727°C   **19.** −38.9°C   **21.** 29°C
23. −321°F   **25.** 12$\overline{0}$0   **27.** 941   **29.** 1175   **31.** 59$\overline{0}$

14.2 Page 376

1. 23 cal   **3.** 1.21×10^6 ft lb   **5.** 3.21×10^6 J or 3.21 MJ   **7.** 4450 Btu
9. 2.62 MJ   **11.** 4.82×10^7 J   **13.** 42.2 MJ   **15.** (a) 2.77×10^5 J/s
(b) 277 kW   **17.** 2.10×10^8 J

14.3 Pages 381–382

1. 0.021 ft^2 °F h/Btu   **3.** 0.45 ft^2 °F h/Btu   **5.** 1.7 ft^2 °F h/Btu
7. 8.1×10^6 Btu   **9.** 11 J   **11.** 1400 Btu   **13.** 14 MJ   **15.** 6800 J

14.4 Page 384

1. 173 Btu   **3.** 38$\overline{0}$0 Btu   **5.** 2.1×10^5 J or 210 kJ   **7.** 280 cal
9. 3.01×10^4 J or 30.1 kJ   **11.** 64.0 kcal   **13.** 0.259 kg   **15.** 29,000 kcal
17. 1.89×10^8 J or 189 MJ   **19.** 8.07×10^7 J or 80.7 MJ   **21.** 252°C
23. 5770 J   **25.** (a) 10.1°C   (b) 22°C   (c) Water is the better coolant because its temperature increase is only about half that of alcohol in absorbing the same amount of heat.

14.5 Pages 386–387

1. 428°F   **3.** 0.051 cal/g °C   **5.** 95°F   **7.** 81°C   **9.** 0.104 cal/g °C
11. 286°C   **13.** 0.259 kg

14.6 Pages 392–393

1. 0.30 ft   **3.** 0.10 m   **5.** 200.10 m   **7.** 0.54 ft   **9.** 0.752 in.
11. 1.8 cm^2 **13.** 88.97 cm^2 **15.** 60.10 cm^3 **17.** 4.9 mm
19. 1.08×10^{-5}/°C   **21.** 15 cm^3 **23.** 0.3397 m^3

14.7 Pages 395–396

1. 653 L   **3.** 12.2 m^3 **5.** 3754 ft^3 **7.** 0.58 cm^3 **9.** $416.16   **11.** 23°C
13. 12.5 L

14.8 Pages 406–407

1. 1100 cal   **3.** 10,700 Btu   **5.** 26,000 cal   **7.** 6.70×10^6 J or 6.70 MJ
9. 3.39×10^6 J or 3.39 MJ   **11.** 6070 Btu   **13.** 11,650 Btu
15. 5$\overline{0}$00 kcal   **17.** 3.12×10^6 J or 3.12 MJ   **19.** 467,000 cal   **21.** 3090 kcal

Chapter 14 Review Questions Pages 409–410

1. a, b, and d   **2.** b and d   **3.** a, b, and d   **4.** a, b, and d   **5.** d   **6.** c

7. The total heat lost by warm objects is gained by cold objects.

8. 778 ft lb of work is equivalent to 1 Btu.

9. The Rankine scale.   **10.** The Kelvin scale.

11. Each Celsius degree is 1.8 times the Fahrenheit degree. The freezing point of water is 0°C and 32°F. The boiling point of water is 100°C and 212°F.

12. Heat is an amount of energy. Temperature is a measure of the average kinetic energy of atoms and molecules in matter.

13. Burning fuel in a furnace. Burning fuel in an engine. Conversion of heat into electricity in a steam-driven generator.

14. Heat generated in a solid by a drill bit. Heat generated by rubbing two objects together. Heat generated in automobile brakes. All are due to friction.

15. Light clothing should be worn. It absorbs less heat.

16. It increases. The metal block increases in size, as does the hole.

17. (a) 4°C.   (b) Most other liquids have their highest density at the temperature at which the change of state to a solid occurs.   (c) Water below 4°C is less dense than water at 4°C. Above 4°C, water expands normally as its temperature increases.

18. Ten kilograms of ice at 0°C. The difference is the heat of fusion, which must be added to the ice to turn it to water.

19. The large amount of heat released when steam changes to water can cause severe burns.

20. Because water expands as it solidifies

21. Conduction is the transfer of kinetic energy from atoms or molecules in a warmer location through nearby atoms or molecules to atoms or molecules in a colder section of the material. Convection is the transfer of kinetic energy from one region to another via the motion of warmer atoms or molecules from one region to another. Radiation is the transfer of energy through the emission and absorption of electromagnetic radiation.

22. Coolants boil at a higher temperature at high pressure. Therefore, the engine coolant can be at a higher temperature and transfer more heat from the engine to the atmosphere because of the greater temperature difference.

23. Heat is extracted from the outside air in winter to vaporize the refrigerant. The heat is given up by the condensing fluid to warm the inside air. In summer, heat is extracted from the inside air to vaporize the refrigerant. The condensing fluid then gives up this heat to the outside air.

Chapter 14 Review Problems Page 410

1. 71°C   **2.** 297 K   **3.** 9230°F   **4.** 335°C   **5.** 10.3 cal   **6.** 1.56 kcal

7. 3.38×10^5 ft lb   **8.** 1.4×10^5 Btu   **9.** 2.4×10^4 J or 24 kJ

10. 4.18×10^4 Btu   **11.** 1.5×10^4 kcal   **12.** 1.22×10^4 kcal

13. 1300°F   **14.** 0.353 cal/g °C   **15.** 85°C   **16.** 12.524 m

17. 7.51 mm; 44.3 mm^2 **18.** 0.0295 cm^3 **19.** 2.57 L   **20.** 123 ft^3

21. 11,600 kcal   **22.** 12$\overline{0}$0 Btu   **23.** 3440 kcal   **24.** 4.80×10^5 J

Chapter 14 Applied Concepts Page 411

1. 1.25×10^4 J   **2.** 4.52×10^5 kcal   **3.** (a) 8.4×10^6 kcal   (b) 8.9×10^6 kcal
(c) 1.8×10^5 kcal   **4.** (a) Steel will contract less.   (b) 0.0161 ft = 0.193 in.
5. (a) 2.43 gal   (b) $0.14

Chapter 15

15.1 Pages 414–415
1. 288 K **3.** 44°C **5.** 532°R **7.** 9$\overline{0}$°F **9.** 258 cm^3 **11.** −16°F
13. 38$\overline{0}$ m^3 **15.** 1430 L **17.** −59°C **19.** 30.5 L **21.** 107 K
23. −111°C **25.** 154 L **27.** 1920 L **29.** 277 L

15.2 Pages 418–419
1. 265 cm^3 **3.** 75.0 kPa **5.** 2.08 kg/m^3 **7.** 0.180 lb/ft^3 **9.** 90.4 kPa
11. 3.34 kg/m^3 **13.** 801 kPa **15.** 19.0 ft^3 **17.** 162 psi **19.** 2.48 m^3
21. (a) 37.5 in^3 (b) 25.0 in^3 (c) 15$\overline{0}$ in^3 **23.** 4.37 kg/m^3 **25.** 1.16 L
27. 0.352 kg/m^3 **29.** (a) Lower (b) 12.4 kPa

15.3 Pages 420–421
1. 1270 in^3 **3.** 506 m^3 **5.** −39°C **7.** 143°C **9.** 22$\overline{0}$0 psi **11.** 399 kPa
13. (a) 70$\overline{0}$ kPa (b) 891°C (c) 3.59 m^3 (d) 0.563 m^3 **15.** 12.3 L

Chapter 15 Review Questions Pages 422–423
1. c **2.** a **3.** c **4.** c **5.** b **6.** b
7. Standard pressure is 101.32 kPa or 14.7 lb/in^2 and standard temperature is 0°C or 32°F.
8. A temperature increase tends to cause a volume increase. A pressure increase tends to cause a volume decrease. If both temperature and pressure are increased, the volume change is given by the combined Charles's and Boyle's laws.
9. The temperature will increase.
10. Heating a gas increases the kinetic energy of the gas molecules. This causes an increase in pressure and volume.
11. When a gas is compressed, work must be done on it to decrease its volume. This work is transferred into increased kinetic energy of the molecules of the gas, causing a temperature increase.
12. When the volume is increased, the number of gas molecules striking the surface of the container per unit area decreases. Thus the pressure exerted by the gas molecules on the container decreases.

Chapter 15 Review Problems Page 423
1. 14.9 ft^3 **2.** 4.40 m^3 **3.** 131°F **4.** 30.9°C **5.** 34.1°C
6. 138 kPa **7.** 23.9 ft^3 **8.** 3.68 kg/m^3 **9.** 478 in^3
10. 19.9°C **11.** (a) 2180 psi (b) 18°F **12.** 304 kPa **13.** 135 kPa
14. 822 L **15.** 3.07 kg/m^3

Chapter 15 Applied Concepts Page 424
1. (a) The temperature of the helium begins to decrease, and the volume of the balloon decreases. (b) 1.66 ft^3 (c) The volume would increase to 1.97 ft^3. (d) Overinflate the balloon in winter; underinflate the balloon in summer.
2. (a) 39 lb/in^2 (b) 26 lb/in^2 **3.** (a) 1.88 × 10^{-4} m^3 (b) 138 kPa (c) 148 kPa
4. (a) 6.68 m^3 (b) 1980 m^3 **5.** (a) 98$\overline{0}$ L (b) No, the balloon should not be fully inflated. If it were, the balloon would probably burst on the way up.

Chapter 16

16.2 Page 435

1. 2.00 ms   **3.** $80\overline{0}$ m/s   **5.** 1.93 m/s   **7.** 0.813 m   **9.** 12.0 Hz
11. 1.2×10^9 Hz   **13.** 2.00×10^8 Hz   **15.** 5.21×10^{-3} m   **17.** 2.06×10^{10} m/s

16.4 Pages 441–442

1. 337 m/s   **3.** 317 m/s   **5.** 5.00 s   **7.** 2640 m   **9.** 622 Hz   **11.** 507 Hz
13. 653 Hz   **15.** 0.325 s   **17.** (a) 695 Hz   (b) 522 Hz   **19.** 0.0870 s

16.6 Page 446

1. 55.9 cm   **3.** 2.24 s   **5.** 4.13 in.   **7.** 1.36 s   **9.** It is $\sqrt{2}$ times the original period.   **11.** 1.74 s   **13.** 1.0 s

Chapter 16 Review Questions Pages 448–449

1. a, b, and c   **2.** c   **3.** a   **4.** b   **5.** b   **6.** a   **7.** c and d
8. Interference is the result of two or more waves traveling through the same region at the same time. Diffraction is the bending of a single wave passing near an obstacle with an opening nearly the same size as the wavelength.
9. The addition of two or more waves to form a larger wave is constructive interference. The formation of a smaller wave is destructive interference.
10. Sound would not be heard if some obstacle came between you and the stereo speakers and there were no nearby reflecting surfaces.
11. Waves passing through a break in a seawall   **12.** It increases.
13. A wave is a periodic disturbance. A pulse is a one-time disturbance.
14. A sharp explosion creates a sound pulse.
15. The speed of sound increases. The higher kinetic energy of air molecules at high temperature leads to a higher velocity of sound through the air.
16. A seismograph detects slight vibrations of the earth's surface by detecting the relative motion of the earth's surface and a massive object.
17. The speed of sound is higher in water than in air. The higher density of water provides a higher speed of sound.
18. Motion toward an oncoming sound wave increases the frequency at which the maximum-pressure regions in the sound wave strike an observer, therefore producing a higher-frequency sound. Motion away from an oncoming sound wave decreases the frequency at which the maximum-pressure regions strike an observer, therefore producing a lower-frequency sound.
19. Sympathetic vibrations occur when an object vibrates at its natural resonance frequency in response to the vibration of a nearby object at some other frequency. Forced vibration occurs when an object vibrates at the same frequency as another nearby vibrating object.
20. Resonance occurs when an object vibrates at its natural resonance frequency in response to the vibration at the same frequency of a nearby object.
21. The light from these stars is shifted toward the red as a result of their motion away from the earth.
22. Amplitude is maximum displacement.
23. Period is the time required for one full vibration. Frequency is the number of complete vibrations per unit of time.
24. No; the period of a pendulum is independent of its mass.

Chapter 16 Review Problems Page 449

1. 2.82×10^{-6} s or $2.82\,\mu$s   **2.** 3.13 Hz   **3.** (a) 6.67×10^{14} Hz   (b) 1.50×10^{-15} s
4. 5.87 m/s   **5.** 1.24 m   **6.** 65 Hz   **7.** 4.58 m   **8.** 383 m/s   **9.** $31\overline{0}$ m/s
10. 0.967 s   **11.** 29 s   **12.** 602 Hz   **13.** (a) 5360 Hz   (b) $46\overline{0}0$ Hz
14. 368 Hz   **15.** 5.59×10^{14} Hz   **16.** 1.35 s   **17.** 4.80 in.

Chapter 16 Applied Concepts Page 450

1. (a) 0.248 m   (b) No   (c) 0.0411 m   **2.** (a) One wavelength   (b) 560 ft/s
(c) 1400 ft   **3.** (a) AM: 182 to 545 m; FM: 2.78 to 3.41 m   (b) FM wavelengths
are closer to the sizes of openings of tunnels and underpasses. As stated in the text,
"Wave diffraction is commonly observed only when the opening is nearly the same
size as the wavelength."   **4.** (a) 1.16×10^5 W   (b) 2.50 W/m^2 (c) Twenty-
five percent of the initial intensity   **5.** (a) 29 m/s   (b) 706 Hz

Chapter 17

17.3 Page 457

1. 922 N   **3.** 1.3 cm   **5.** 1.5×10^{-8} C   **7.** (a) -0.93 N   (b) $+0.707$ N
(c) -2.08 N

17.4 Page 460

1. 1.50×10^4 N/C   **3.** 0.500 N/C   **5.** 1.78×10^{-3} C   **7.** 0.676 N

17.5 Pages 464–465

1. $1.07\ \Omega$ **3.** $0.0165\ \Omega$/ft   **5.** $3.95\ \Omega$ **7.** $0.684\ \Omega$ **9.** 0.0131 cm^2

17.6 Pages 467–468

1. 4.79 A   **3.** $22\overline{0}$ V   **5.** $153\ \Omega$ **7.** $17.6\ \Omega$ **9.** 1.26 A   **11.** (a) 0.067 A
(b) $3\overline{0}$ V   (c) 0.13 A   **13.** $18.8\ \Omega$ **15.** 1.85 A

17.7 Pages 471–472

1. $13.50\ \Omega$ **3.** $60.0\ \Omega$ **5.** 0.750 A   **7.** 378 V   **9.** $23.0\ \Omega$
11. $10.8\ \Omega$, $14.2\ \Omega$, $19.0\ \Omega$ **13.** $2.78\ \Omega$, 40.0 V, $3.40\ \Omega$

17.8 Page 478

1. (a) $4.28\ \Omega$ (b) 11.7 A   (c) 4.55 A   (d) 7.14 A   **3.** (a) $10\overline{0}\ \Omega$ (b) 1.67 A
(c) 0.250 A

17.9 Pages 482–483

1. (a) R_2, R_3 (b) $3.00\ \Omega$ **3.** 8.89 A   **5.** (a) 2.23 A   (b) 6.68 A   **7.** $21.24\ \Omega$
9. 45 V   **11.** 5.41 A   **13.** $10.0\ \Omega$ **15.** $17.3\ \Omega$ **17.** 30.0 V   **19.** 6.93 V
21. $13.3\ \Omega$ **23.** 60.2 V   **25.** 0.370 A

17.10 Pages 484–486

1.

	V	I	R
Batt.	12.0 V	9.00 A	1.33 Ω
R_1	12.0 V	6.00 A	2.00 Ω
R_2	12.0 V	3.00 A	4.00 Ω

3.

	V	I	R
Batt.	36.0 V	6.00 A	6.00 Ω
R_1	36.0 V	2.00 A	18.0 Ω
R_2	36.0 V	3.00 A	12.0 Ω
R_3	36.0 V	1.00 A	36.0 Ω

5.

	V	I	R
Batt.	50.0 V	5.00 A	10.0 Ω
R_1	25.0 V	2.00 A	12.5 Ω
R_2	25.0 V	2.00 A	12.5 Ω
R_3	10.0 V	3.00 A	3.33 Ω
R_4	40.0 V	3.00 A	13.3 Ω

7.

	V	I	R
Batt.	36.0 V	9.00 A	4.00 Ω
R_1	12.0 V	6.00 A	2.00 Ω
R_2	12.0 V	3.00 A	4.00 Ω
R_3	24.0 V	6.00 A	4.00 Ω
R_4	24.0 V	3.00 A	8.00 Ω

9.

	V	I	R
Batt.	80.0 V	12.0 A	6.67 Ω
R_1	18.0 V	4.00 A	4.50 Ω
R_2	18.0 V	2.00 A	9.00 Ω
R_3	18.0 V	6.00 A	3.00 Ω
R_4	48.0 V	12.0 A	4.00 Ω
R_5	8.00 V	4.00 A	2.00 Ω
R_6	8.00 V	8.00 A	1.00 Ω
R_7	6.00 V	12.0 A	0.500 Ω

11.

	V	I	R
Batt.	65.0 V	5.00 A	13.0 Ω
R_1	10.0 V	0.500 A	20.0 Ω
R_2	10.0 V	1.00 A	10.0 Ω
R_3	10.0 V	2.50 A	4.00 Ω
R_4	10.0 V	1.00 A	10.0 Ω
R_5	25.0 V	5.00 A	5.00 Ω
R_6	30.0 V	5.00 A	6.00 Ω

17.12 Pages 490–491

1. 1.49 V   **3.** 1.33 A   **5.** (a) 3.75 A   (b) 9.00 V   (c) 0.0200 Ω   **7.** 0.160 A
9. 0.120 A

17.13 Pages 494–495

1. 957 W   **3.** 0.682 A   **5.** 6.82 A   **7.** $0.56   **9.** Yes   **11.** $0.026
13. $0.046   **15.** (a) Microwave oven, two fluorescent bulbs, two light bulbs
(b) Projection TV, personal computer, two light bulbs   **17.** 2.36 W
19. 60.0 W   **21.** $90\overline{0}$ J   **23.** (a) $24\overline{0}$ W   (b) 50.4 kWh   (c) $5.54
25. (a) 9.58 mA   (b) 1.10 W   (c) $0.07   **27.** 6150 J

Chapter 17 Review Questions Pages 498–499

1. b   **2.** c   **3.** c   **4.** b   **5.** e   **6.** a and e
7. a   **8.** c   **9.** b   **10.** c   **11.** c   **12.** b and c
13. Materials can become charged when a charged object is brought nearby, inducing a polarization (separation) of charge on the material. If one side of the material is touched by another object, the charge at one side of the material can then be "drained" off, leaving the charge at the other side of the material.
14. Protons, electrons, and neutrons   **15.** Protons and neutrons
16. Electrons are located in charge clouds surrounding the nucleus.
17. Positive and negative. Protons carry a positive charge; electrons carry a negative charge.
18. A charged object is brought into contact with the electroscope, thereby providing some of that charge to the electroscope.

19. A charged object is brought near the conducting ball on an electroscope, causing a polarization (separation) of charge on the electroscope. The conducting ball is touched by a "ground," allowing one type of charge to leave the electroscope and go to ground, leaving the other charge behind.

20. Coulomb's law states that the force between two charges is directly proportional to the product of the magnitude of the charges and inversely proportional to the square of the separation between the two charges.

21. An electric field at a point represents the magnitude and direction of the force that would be exerted on a single unit of charge if placed at that point.

22. Lightning is the discharge of built-up static charge on a portion of a cloud.

23. Current   24. (a) Ampere   (b) Volt   (c) Ohm

25. It decreases the resistance by a factor of 4.

26. The voltage drop across a segment of a circuit equals the product of the current through that segment and the resistance of that segment of the circuit.

27. Current flows sequentially through each portion of a series circuit. The current is divided among different segments of a parallel circuit.

28. The equivalent resistance in a series circuit is the sum of the resistances in the circuit. In a parallel circuit, the equivalent resistance is given by the reciprocal of the sum of the reciprocals of all resistances.

29. The highest range

30. Water flow is split and flows in parallel through different segments of a water distribution system. In a similar manner, current is divided and flows in parallel through resistors connected in parallel.

31. The current decreases by a factor of 2.   32. The current increases by a factor of 2.

33. The resistance increases by a factor of 2.

34. Electrical charges move from regions of higher potential to regions of lower potential. Chemical reactions in batteries raise charges to higher potential energy. These charges can flow through a circuit and do work on circuit elements (create heat, light, or motion).

35. Chemical energy is transferred into electrical potential energy in the dry cell. Charges flow through the two lamps, giving up their energy in the form of light and heat.

36. Primary cells are not rechargeable, whereas secondary cells are.

37. An electric current flows in the reverse direction through a secondary cell, causing the normal chemical reaction to proceed in reverse, thus restoring charge in the battery.

38. The electrolyte causes a chemical reaction at the plates and conducts current between the plates.

39. An electrolyte causes a chemical reaction at the plates that releases energy to force electrical charge to move through the battery and the outside circuit.

40. It decreases the voltage through the outside circuit.

41. Watt   42. Power is given by the product of the voltage and the current.

43. We pay for our energy use. Energy consumed is the total work done. Power is the instantaneous use of electricity; energy is the power multiplied by the duration of time the power is used.

44. Power is the square of the voltage divided by the resistance.

45. The power increases by a factor of 2.

46. The power increases by a factor of 4.   47. The power decreases by a factor of 4.

48. The cost increases by a factor of 2.

Chapter 17 Review Problems Pages 499–501

1. 8.10×10^4 N   2. 9.13×10^{-6} C   3. 0.0671 m
4. 2.00×10^5 N/C   5. 3.33×10^6 C   6. 1.06 N   7. 1.17 Ω
8. 3.47 Ω 9. 1.11 Ω 10. 145 ft   11. 0.0105 cm^2
12. 7.47 A   13. 25.1 Ω 14. 0.491 A   15. 19.60 Ω
16. 0.612 A   17. 48.5 V   18. 16.2 Ω 19. 3.5 Ω

20. 2.45 Ω **21.** 44.9 A **22.** 19.5 A **23.** 25.5 A
24. 5.00 Ω **25.** 10.0 A **26.** 5.95 A; 4.05 A **27.** 4.42 Ω
28. 17.1 A **29.** 58.5 V **30.** 2.83 A **31.** 12.3 V

32.

	V	I	R
Batt.	35.0 V	4.70 A	7.45 Ω
R_1	5.00 V	2.75 A	1.82 Ω
R_2	5.00 V	1.95 A	2.56 Ω
R_3	13.2 V	4.70 A	2.80 Ω
R_4	7.50 V	0.97 A	7.73 Ω
R_5	7.50 V	3.73 A	2.01 Ω
R_6	9.30 V	4.70 A	1.98 Ω

33. 1.43 V **34.** 12.2 V **35.** (a) 3.95 A (b) 6.00 V (c) 0.0165 Ω
36. 0.528 A **37.** 6.67 Ω **38.** 39.3 W **39.** 1.36 A **40.** $1.01
41. 240 h **42.** 0.91 A

Chapter 17 Applied Concepts Page 501

1. (a) -8.23×10^{-8} N (b) 3.63×10^{-47} N (c) We would need an extremely large amount of charge on our bodies to feel that attraction. The earth has so much mass that we are noticeably attracted to the earth via gravity.
2. (a) 3.06×10^{-4} N (b) 3.06×10^{-4} N (c) 8.38×10^{-9} C (d) As the distance between the particles decreases, the force increases, which results in a stronger electric field in the sawdust. **3.** (a) 2.20×10^{-4} A (b) 1.10 A
4. (a) $I_{microwave} = 8.33$ A; $I_{bulb} = 0.333$ A; $I_{computer} = 4.58$ A
(b) $R_{microwave} = 14.4$ Ω; $R_{bulb} = 36\overline{0}$ Ω; $R_{computer} = 26.2$ Ω **5.** 3.32×10^{-4} m

Chapter 18

18.2 Pages 510–511

1. 1.20×10^{-5} T **3.** 57.5 A **5.** 1.71×10^{-6} T **7.** 0.0196 T **9.** 0.249 A
11. 0.0126 T **13.** 0.239 A

Chapter 18 Review Questions Page 522

1. d **2.** b **3.** b **4.** Tesla
5. A tightly wound solenoid with many turns per unit length carrying a large current produces a strong magnetic field. A magnetic core such as iron significantly increases the magnetic field.
6. Use Ampère's rule to find the direction of the flux line from any single turn in the solenoid. The magnetic field direction of the solenoid is in this direction.
7. The current-carrying coil produces a magnetic field that causes the magnetic domains in the magnetic material to align in the direction of the coil's field. This produces a stronger induced field in the core.
8. A moving magnet induces a current in the generator's coil.
9. The commutator is a split ring that allows the current produced by a generator always to flow in the same direction.
10. An induced magnetic field in the motor's electromagnet is repelled by the permanent magnet, causing the rotor to spin. The commutator in a dc motor allows the current through the electromagnet to change polarity, causing the rotor to continue spinning.

11. A synchronous motor rotates at a frequency that depends on the number of coils and the frequency of the ac power source. The motor has a number of poles along the stator, which cause the rotor to spin at a fixed frequency.
12. A universal motor can operate on either dc or ac. The induction motor can operate only on ac.
13. The stator is a static magnet. The armature is an electromagnet that is free to rotate.
14. An electromagnet produces a strong magnetic field when a current is run through a solenoid. This magnetic field in turn induces a stronger magnetic field in a magnetic core.
15. The magnetic field increases by a factor of 2.
16. The field does not change as long as the length of the solenoid is much greater than the original diameter.
17. The magnetic field increases by a factor of 4.
18. The flux lines can be found by placing a small compass or magnetic filings near the magnet.
19. A spinning armature in the field of a stator crosses the magnetic flux lines of the stator, thereby inducing a current to flow in the armature coil. As the armature rotates and is reversed in the field, the direction of the current in the coil is reversed.

Chapter 18 Review Problems Page 522
1. 1.08×10^{-6} T **2.** 4.90×10^{-6} T **3.** 41.6 A **4.** 0.0253 T
5. 12.4 A **6.** 15.4°

Chapter 18 Applied Concepts Page 523
1. (a) 1.98×10^{-5} T (b) 3.17 m **2.** (a) 7.77×10^{-6} T (b) The wire's magnetic field is 14.9% of the strength of the earth's magnetic field. **3.** The magnetic field of the inner cable cancels out the opposite magnetic field orientation of the outer braid. **4.** (a) For both solenoids, N is on top and S is on the bottom. (b) "b" The electron is coming out of the page. **5.** (a) 0.525 Ω (b) 8.57 A (c) 98.0 loops (d) 0.0896 m (e) 1.05×10^{-3} T

Chapter 19

19.1 Pages 527–528
1. 47.1 V **3.** 117 V **5.** 9.19 A **7.** 83.2 V **9.** 5.66 A **11.** 70.3°
13. 1590 V **15.** 117 V **17.** 12.0 A **19.** 12.0 A **21.** 95.4 V
23. (a) 24.0 V (b) 0.120 A (c) $20\overline{0}$ Ω
25. (a) 0.495 A (b) 242 Ω **27.** $200\overline{0}$ W

19.2 Pages 535–536
1. 21.9 Ω **3.** $350\overline{0}$ W or 3.50 kW **5.** 1320 W or 1.32 kW **7.** 588 W
9. 2750 W or 2.75 kW **11.** 10.0 turns **13.** 58.5 V **15.** $30\overline{0}$ turns **17.** 6.00 A
19. (a) 0.0587 A or 58.7 mA (b) $88\overline{0}$ W
21. 155 V **23.** (a) $12\overline{0}$ V (b) 0.600 A **25.** 4090 V

19.3 Page 539
1. 1.13 Ω **3.** $440\overline{0}$ Ω **5.** 40.1 kΩ **7.** 0.594 A **9.** 0.796 A **11.** 0.116 A

19.4 Page 542
1. (a) 215 Ω (b) 21.4° (c) 0.209 A **3.** (a) $330\overline{0}$ Ω (b) 72.3° (c) 4.55 mA
5. (a) 302 Ω (b) 7.2° (c) 19.9 mA

19.5 Page 544
1. 7.96 Ω   **3.** 2650 Ω or 2.65 kΩ   **5.** 0.796 Ω   **7.** 58.9 Ω   **9.** 1.61 Ω

19.6 Pages 546–547
1. (a) 1880 Ω   (b) 57.9°   (c) 53.2 mA   **3.** (a) 48$\overline{0}$0 Ω   (b) 0.0241°
(c) 3.13 mA   **5.** (a) 3.18 Ω   (b) 89.8°   (c) 4.72 mA

19.7 Page 549
1. 42.4 Ω; 0.118 A   **3.** 1180 Ω; 12.7 mA   **5.** 44$\overline{0}$ Ω; 13.6 mA
7. 206 Ω; 24.3 mA   **9.** 529 Ω; 47.3 mA

19.8 Page 551
1. 79.6 kHz   **3.** 38.0 kHz   **5.** 3.86 kHz   **7.** 4.20 kHz   **9.** 5.63 kHz

19.10 Page 554
1. 10,700 kW   **3.** 133,000 kVA   **5.** 10,100 kW   **7.** 19,300 kW
9. 407,000 kVA   **11.** 0.880

Chapter 19 Review Questions Pages 556–557
1. d   **2.** c   **3.** a and e   **4.** a, b, and c
5. The maximum current is the maximum instantaneous current. The effective value of an alternating current is the number of amperes that produce the same amount of heat in a resistance as an equal number of amperes of a steady direct current.
6. The maximum voltage is the maximum instantaneous voltage. The instantaneous voltage is the voltage at any instant.
7. Power is the product of the effective values of the voltage and the current.
8. Power is the square of the effective value of the voltage divided by the resistance.
9. The output voltage doubles.   **10.** Henry
11. Inductive reactance allows the analysis of circuits containing inductors.
12. The inductive reactance is directly proportional to the frequency.
13. The current lags the voltage in an inductive circuit.
14. Energy is stored in the form of potential energy associated with a sheet of positive charge on one side of the capacitor and a sheet of negative charge on the other side.
15. The current leads the voltage in a capacitive circuit.
16. The frequency and reactance of a capacitor are inversely proportional.
17. Resonance occurs when the inductive reactance equals the capacitive reactance. The current is then given by its maximum value.
18. A diode allows current to flow in one direction but not in the reverse direction.
19. Amplification produces an increase in the value of a voltage or current in a circuit. Rectification produces a current or voltage in only one direction.
20. No; the phase angle depends on the frequency.

Chapter 19 Review Problems Pages 557–558
1. 110 V   **2.** 120 V   **3.** 68 V   **4.** 95.4 V   **5.** 8.95 A   **6.** 727 W
7. 495 W   **8.** 1310 W   **9.** (a) 1340 turns   (b) 0.255 A   (c) 523 W
10. (a) 3.92 A   (b) 38.2°   (c) 71.0 V   **11.** (a) 2.29 A   (b) 18.9°   (c) 109 V
12. (a) 2.20 A   (b) 67.1°   (c) 42.9 V   (d) 102 V   **13.** (a) 1.52 A   (b) 32.8 V
(c) 72.6°   **14.** 194 Ω; 0.567 A   **15.** 126 Ω; 0.873 A   **16.** (a) 30,200 Hz
(b) 633 Ω, 0.174 A   **17.** (a) 3.46×10^5 Hz   (b) 565 Ω, 0.195 A
18. 3.41×10^6 kVA

Chapter 19 Applied Concepts Page 558

1. (a) $13\overline{0}0$ W   (b) 16.6 A   (c) $i = 0$ A because the current is reversing at this instant.
2. (a) 23.0 turns   (b) $P_p = 114$ W; $P_s = 114$ W
(c) The power is conserved in a transformer.   **3.** (a) 386 A   (b) 14.2 A
(c) $P_{L\,220\text{v}} = 37,200$ W; $P_{L\,6000\text{v}} = 50.4$ W   (d) The $60\overline{0}0$-V power line is better because it only loses 50.4 W of power.
4. (a) 0.332 H   (b) 0.524 A   (c) 34.3 W   **5.** (a) 1.19×10^{-13} F
(b) 9.04×10^{-14} F   (c) Tuning to a high frequency lowers the capacitance because the plates are smaller.

Chapter 20

20.2 Pages 565–566
1. 1.50×10^9 m   **3.** 0.108 s   **5.** $50\overline{0}$ s   **7.** 16.1 ms   **9.** 7.67×10^{-7} s
11. 3.2×10^{-9} s   **13.** (a) $25\overline{0}$ s   (b) $12\overline{0}0$ s   **15.** (a) 1940 s   (b) 2890 s
17. 3.84×10^8 m

20.3 Pages 567–568
1. 6.59×10^{12} Hz   **3.** 3.09×10^{-4} m   **5.** 6.58 m   **7.** 4.55×10^{-5} Hz
9. 214 m   **11.** 3.51 MHz   **13.** 7.5×10^{14} Hz   **15.** 5.2×10^{14} Hz

20.4 Page 569
1. 5.93×10^{-23} J   **3.** 3.01×10^{-25} J   **5.** 83.0 MHz   **7.** 5.51×10^{10} Hz
9. 2.7×10^{-19} J   **11.** 3.4×10^{-19} J   **13.** 1410 kHz

20.5 Page 572
1. 603 ℓm   **3.** 60.9 cd   **5.** 942 ℓm   **7.** 1.46 ft-cd
9. 6.73 ft-cd   **11.** 4750 ℓm   **13.** 608 ℓm   **15.** (a) Decrease
(b) $\frac{1}{9}$ **17.** 5.96 lux   **19.** 35.3 ℓm

Chapter 20 Review Questions Page 574
1. a, c, and e   **2.** c   **3.** d   **4.** c
5. Yes; the wavelength varies inversely with the frequency.
6. The energy is directly proportional to the frequency.
7. The intensity falls off as 1 divided by the square of the distance.
8. The speed of light is measured by determining the time light takes to travel a measured distance and using the relationship $v = s/t$.
9. It always travels at the same speed in a vacuum. It has a slower speed in any other medium.
10. Electromagnetic radiation   **11.** Max Planck
12. Albert Einstein and Max Planck   **13.** Olaus Roemer
14. By measuring the time difference for the start of the eclipse of the moons of Jupiter as viewed from different parts of earth's orbit
15. Candela and lumen
16. Luminous intensity is the brightness of a light source.

Chapter 20 Review Problems Pages 574–575
1. 2.32×10^{13} m   **2.** 3.52 mi   **3.** 2.63×10^{-6} s or 2.63 μs
4. 1.47×10^{-6} s or 1.47 μs   **5.** 0.121 s   **6.** 244 m
7. 6.45×10^6 Hz or 6.45 MHz   **8.** 5.54×10^{15} Hz   **9.** 9.61×10^{-23} J
10. 7.17×10^{10} Hz   **11.** 5.47×10^{-18} J   **12.** 2150 ℓm   **13.** 5230 ℓm
14. 20,900 ℓm   **15.** 3130 s; $17\overline{0}0$ s   **16.** 12.8 ℓm   **17.** 7.29 lux

Chapter 20 Applied Concepts Page 575

1. (a) 0.0187 s   (b) 0.238 s   **2.** (a) 3.84×10^5 km or 239,000 mi
(b) $50\overline{0}$ s or 8.33 min   (c) 4.08×10^{16} m or 2.53×10^{13} mi
3. (a) 3.41 to 2.78 m   (b) 5.83×10^{-26} to 7.16×10^{-26} J   (c) The higher the frequency, the greater is the energy.   **4.** Channel 2: $\frac{1}{4}\lambda = 1.39$ m to Channel 6: $\frac{1}{4}\lambda = 0.852$ m   **5.** (a) 14,100 ℓm   (b) 56,500 ℓm (4 times the original)

Chapter 21

21.5 Pages 584–585

1. 1.29 cm   **3.** 32.2 cm   **5.** 5.13 cm   **7.** −6.24 cm   **9.** −1.09 m
11. 10.0 cm   **13.** 0.237 m   **15.** −4.15 cm   **17.** −3.56 m   **19.** (a) 4.01 cm
(b) 2.98 cm   (c) 31.9 cm

21.10 Page 597

1. 1.21   **3.** 2.00×10^8 m/s   **5.** 1.48   **7.** 21.8 cm   **9.** −2.14 cm
11. (a) 5.00 cm   (b) −2.50 cm   **13.** $s_i = -8.17$ cm, $h_i = 2.06$ cm   **15.** 4.43 cm

Chapter 21 Review Questions Pages 599–600

1. a   **2.** d   **3.** a and b
4. Parallel light rays reflected off a rough surface may scatter in many different directions (diffusion), whereas parallel light rays reflect off a smooth surface as parallel rays.
5. At a smooth surface, the incident angle of a light ray is the same as the angle of the reflected ray.
6. At a smooth surface, the normal to the surface and both the incident and reflected rays lie in the same plane.
7. They are virtual, erect, and lie as far behind the mirror as the object is in front of the mirror.
8. A real image can be shown on a screen. A virtual image cannot.
9. Concave mirrors curve away from the observer; convex mirrors curve out toward the observer.
10. For large apertures, not all parallel rays are reflected through the focal point. This produces a fuzzy or aberrant image.
11. The image distance is increased.
12. The image distance is increased.
13. b   **14.** b
15. Converging lenses convert parallel rays into converging rays. Diverging lenses convert parallel rays into diverging rays.
16. Light propagating in an optical fiber, light reflected into a swimming pool from an underwater light
17. Light passing at an angle through an interface from a medium of low optical density to a medium of higher optical density is bent toward the normal to the interface.
18. The speed of light is lower in the high-index material.
19. A wave passing at an angle from one medium to another with a different wave velocity will be bent toward or away from the normal depending on whether the speed is higher or lower in the new material, respectively.
20. Virtual   **21.** Real or virtual
22. They roughen the surface and scatter the light.
23. It appears shallower because of the refraction of light at the surface.
24. Light travels in a straight line unless it is reflected or refracted.
25. Light traveling at an angle greater than the critical angle is reflected from one side of the fiber to another as it travels down the length of the fiber.
26. When the object is closer to the lens than the focal length

27. When the object is located a distance from the lens equal to the focal length
28. When the object is located outside the focal point

Chapter 21 Review Problems Pages 600–601

1. 2.36 cm   **2.** −12.9 cm   **3.** −3.48 cm   **4.** −2.22 m
5. (a) $25\overline{0}$ cm from the mirror   (b) −459 cm   **6.** 22.7 cm   **7.** 1.35
8. 2.08×10^8 m/s   **9.** 1.40   **10.** 41.8°   **11.** 18.0 cm
12. (a) −7.50 cm   (b) 2.70 cm   **13.** 9.00 cm   **14.** 6.25 cm   **15.** −9.70 m
16. −25.4 cm, 7.18 cm high   **17.** 1.24×10^8 m/s   **18.** 41.8°   **19.** 12.3 cm
20. (a) $s_i = 15.8$ cm; $h_i = -10.8$ cm   (b) −3.83 cm

Chapter 21 Applied Concepts Page 601

1. (a) 19.0 cm   (b) $s_i = -36.5$ cm (the minus sign represents the opposite side of the mirror)   (c) Upright   (d) 73.0 cm   **2.** (a) −0.704 m   (b) Virtual
(c) 0.0156 m   (d) Better for increasing the field of view   **3.** (a) 18.5°   (b) 22.1°
(c) The angle will be smaller because light refracts toward the normal line when entering a denser medium.   **4.** (a) 24.44°   (b) Larger than 24.44°   (c) 31.4°
(d) Diamond; a zircon will allow light to escape when angles are less than 31.4°.
5. (a) 60.4 mm   (b) −13.1 mm

Chapter 22

Chapter 22 Review Questions Page 618

1. Red, orange, yellow, green, blue, and violet
2. The color of light is determined by its wavelength or its frequency.
3. (a) Infrared   (b) Ultraviolet
4. Black. The cloth in the dress absorbs all colors except green and reflects only green. The red light is absorbed and no light is reflected.
5. (a) Red stars on a black field, rest of the flag would be red.   (b) Blue field with no stars, rest of the flag would have blue and black stripes.
6. Red, green, and blue
7. Two colors that combine to form white are called complementary colors. The complement of red is cyan, the complement of green is magenta, and the complement of blue is yellow.
8. The primary pigments are the complements of the three primary colors.
9. Cyan, magenta, and yellow for the colors and black for the shadow areas and definition
10. The sky is blue because sunlight is scattered as it passes through the atmosphere. The amount of scattering depends on the wavelength of the light. The longer wavelengths of red, orange, and yellow are scattered much less than the shorter wavelengths of blue and violet. Because our eyes are not very sensitive to violet light, the scattered blue light in the atmosphere dominates our vision so that we see a blue sky.
11. At sunset, sunlight travels through significantly more of the atmosphere than at other times of the day. This results in even more scattering of the shorter wavelengths of violet, blue, and green, which are taken out before reaching us, with some orange but primarily red, the light of longest wavelength and least scattered, most readily passing through the atmosphere to reach our eyes for a red sunset.
12. Clouds consist of many different-sized water droplets. The small droplets tend to scatter the short-wavelength (blue) light, the medium droplets tend to scatter the medium-wavelength (green) light, and the large droplets tend to scatter the long-wavelength (red) light. The overall result is that all of the colors are scattered and combined so that we see a general white reflection.

13. The ocean is blue because the water surface acts like a mirror and reflects the blue color of the sky and the water tends to reflect and scatter the short-wavelength yellow, green, blue, and violet light and absorb the long-wavelength orange and red light.
14. False   15. Diffraction   16. Diffraction pattern
17. Thomas Young   18. Constructive interference   19. Destructive interference
20. Polarized sunglasses restrict the light waves passing through to a single plane rather than normal sunlight, which is emitted in all directions with many orientations.

Chapter 23

23.3 Page 624
1. -13.595 eV   3. 12.084 eV   5. 6.57×10^{-7} m

23.4 Page 626
1. Neon   3. Fluorine

23.5 Pages 628–629
1. (a) 12   (b) 6   (c) 6   (d) 6   (e) 12   (f) 6   3. (a) 48   (b) 22
(c) 26   (d) 22   (e) 48   (f) 26   5. (a) 11   (b) 22.9898 u
7. (a) 82   (b) 207.2 u   9. ^{14}C   11. Bromine   13. Argon

23.6 Pages 631–632
1. 3.8530×10^{-25} kg   3. 6680 MeV   5. 1750 MeV   7. 8.5 MeV

23.7 Page 637
1. 4.03×10^{-5} s   3. 0.226 g   5. 79.9%   7. 3.77×10^{20} atoms   9. 30.4%

23.9 Page 640
1. 0.310 Ci   3. 0.0875 Ci   5. 9.47 Ci   7. 73.3 µCi

Chapter 23 Review Questions Pages 643–644
1. a, b, and c   2. c   3. c
4. The ground state is the lowest energy level for the electron in the atom. The excited states are the higher energy levels, which are unstable. An electron in an excited state will in time decay through lower-energy-level excited states to the ground state. The ground state has the lowest energy level.
5. In the Bohr theory of the atom, the energy levels of the electrons are restricted to certain values; that is, the energy is quantized. The energy levels in a hydrogen atom are given by the equation $E = -kZ^2/n^2$.
6. Protons and neutrons are both nucleons, which exert the attractive strong force on other nucleons. They have similar masses, although the neutron is slightly more massive. The proton is a stable particle as an individual nucleon. The neutron, however, is unstable by itself. The proton has a positive charge, equal in magnitude to that of the electron. The neutron is uncharged.
7. The electric force can be either attractive or repulsive and is exerted between charged particles. The strong force is always attractive and is exerted between nucleons, whether charged or not. The strong force is a very short-range force. The electric force is exerted over larger distances.
8. The nucleus would expand in size due to the repulsive electric force between protons. The nucleus might even break apart.
9. Mass and energy are equivalent forms. Mass can be changed into energy under the proper conditions, and vice versa.

10. All three atoms have six protons in the nucleus and six electrons in the orbital shells. They all exhibit the chemical properties of carbon. They each have a different number of neutrons in the nucleus and therefore have a different mass.

11. An electron volt is the energy that is gained or lost by an electron in passing through a potential difference of 1 volt.

12. The neutron is the "glue" that binds the positively charged protons together in the nucleus. Without neutrons, the positively charged protons would be repelled by their similar electric charge.

13. An α ray is composed of particles that have a double positive charge and are composed of two protons and two neutrons. The α particles are identical to the nucleus of a helium atom.

14. A γ ray is composed of very energetic photons of uncharged electromagnetic radiation.

15. A β ray is composed of negatively charged particles (electrons).

16. Enrico Fermi discovered that the neutron bombardment of uranium can result in the formation of lighter nuclei that are approximately one half the mass of uranium.

17. A self-sustaining chain reaction is a nuclear reaction in which a sufficient number of neutrons are produced to cause subsequent nuclear reactions to continue at a fixed rate.

18. In nuclear fusion reactions, nuclei bombard each other, producing heavier nuclei. The original nuclei "fuse" together.

19. In nuclear fission reactions, nuclei are bombarded by nuclear particles such as neutrons or α particles, causing the original nuclei to split apart.

20. 6.25%

21. Molecules in plant or animal cells can be damaged as a result of changes in the chemicals caused by nuclear reactions produced by the radiation. Some of these damaged molecules may be in the genetic material, which might cause cells produced by the division of this cell to be defective.

22. Radiation is used for diagnostic purposes by the injection or ingestion of radioactive tracer materials that may allow the identification of cancerous or defective cells. It can also be used to treat cancer by bombardment of the cancer cells, therefore destroying the cancerous material.

Chapter 23 Review Problems Pages 644–645

1. -0.5438 eV 2. 0.967 eV 3. He, Ne, Ar, Kr, Xe, Rn
4. Cu, Ag, Au, Fe, Ti, Cr, Mn, etc. 5. 2.4913×10^{-26} kg
6. 3.1547×10^{-26} kg 7. 2.7553×10^{-25} kg 8. 749 MeV
9. 8.7 MeV 10. 8.3 MeV 11. 8.1 MeV 12. 7.26×10^{12} atoms
13. 4.29×10^{17} atoms 14. 5.6% 15. 0.37% 16. 0.346 Ci
17. 0.193 Ci 18. 0.367 Ci 19. 5.8 Ci 20. 10.6 µCi

Chapter 23 Applied Concepts Page 645

1. (a) 2.5491 eV (b) 6.163×10^{14} Hz (c) 3.00×10^8 m/s (d) 4.87×10^{-7} m
2. $n = 2$ 3. (a) 4.550×10^{-12} J (b) Two protons, two neutrons, two electrons
(c) 6.6968×10^{-27} kg (d) 6.64×10^{-27} kg 4. (a) 0.181/day
(b) 1.50×10^{24} atoms 5. 0.393 Ci

Chapter 24

24.2 Page 650
1. 2.07×10^{14} J 3. 6.67×10^{-15} kg

Chapter 24 Review Questions Page 653

1. c   **2.** d

3. Physics is the same for moving and nonmoving objects.

4. If you were in a moving car and flipped a coin, the coin would flip into the air and fall back into your hand the same as would happen if you flipped a coin while standing on the side of the road.

5. It would travel at 14 m/s because the ball was already traveling at 10 m/s while attached to the bike.

6. The speed of light is not relative, but is constant.

7. The light would travel at only 3.00×10^8 m/s.

8. Time passes more slowly than for someone not traveling close to the speed of light.

9. Relative to the ground (considering I am not walking and reading), just the time dimension.

10. That energy and mass are equivalent, just in different forms. Mass has energy and energy has mass.

11. Spatially we exist in three dimensions (x, y, z), yet we also move through the dimension of time. Space and time are needed to locate a particular occurrence in the universe.

12. Gravity and acceleration are the same.

13. Light can be warped around massive objects.

14. The light curved around the warping of space-time that was created by the mass of the sun and the moon.

Chapter 24 Review Problems Page 653

1. 2.00 mi/h   **2.** 67.0 mi/h   **3.** 63.0 mi/h   **4.** 8.19×10^{-14} J   **5.** 1.50×10^{-10} J

6. 1.09×10^{-16} kg

Appendix A

A.1 Page 658

1. -11 **3.** 5   **5.** -2 **7.** -4 **9.** -6 **11.** -3 **13.** 18   **15.** -21

17. 0   **19.** 3   **21.** -8 **23.** 0

25. 11   **27.** -13 **29.** -5 **31.** -8 **33.** -4 **35.** 10

37. -45 **39.** -288

A.2 Page 660

1. 10^8 **3.** 10^8 **5.** $\dfrac{1}{10^3}$ **7.** 10^7 **9.** 10^3 **11.** $\dfrac{1}{10^4}$

13. 1   **15.** 10^5 **17.** $\dfrac{1}{10^{11}}$ **19.** $\dfrac{1}{10^6}$ **21.** 10^8 **23.** 10^{10}

A.3 Page 664

1. $\dfrac{4}{3}$ **3.** 17   **5.** 0   **7.** 17.5   **9.** 3   **11.** 4

13. 7   **15.** 12.5   **17.** 5   **19.** 26.5   **21.** 9   **23.** 3

25. 8   **27.** 1   **29.** $-\dfrac{11}{3}$ or $-3\dfrac{2}{3}$ **31.** 3   **33.** 2

35. 21   **37.** $\dfrac{29}{2}$ or $14\dfrac{1}{2}$ **39.** -6

A.4 Pages 666–667

1. ± 6 **3.** ± 7 **5.** ± 3 **7.** ± 0.8 **9.** ± 4.67

11. ± 0.892 **13.** $a = 3; b = 1; c = -5$ **15.** $a = 6; b = 8; c = 2$

17. $a = 9; b = 6; c = -4$ **19.** $a = 5; b = 6; c = 0$ **21.** $a = 9; b = 0; c = -64$

23. 3; 7   **25.** $\dfrac{4}{3}$; $-\dfrac{5}{2}$ **27.** 1.95; -1.62 **29.** 1.69; -0.855

31. 0.725; -1.23

A.5 Pages 671–672

1. a **3.** c **5.** a **7.** B **9.** B **11.** 0.9455
13. 1.804 **15.** 0.9799 **17.** 0.6477 **19.** 0.3065
21. 0.4617 **23.** 16° **25.** 43° **27.** 48° **29.** 36.6°
31. 46.5° **33.** 30.0° **35.** 22.28° **37.** 16.75° **39.** 35.50°
41. $B = 65.0°$; $a = 8.45$ m; $b = 18.1$ m **43.** $A = 47.7°$; $B = 42.3°$; $a = 12.4$ km
45. $A = 24.4°$; $B = 65.6°$; $c = 24.2$ mi **47.** $B = 70°$; $b = 24$ m; $c = 25$ m
49. $A = 49.35°$; $a = 17.98$ cm; $b = 15.44$ cm **51.** $b = 8.49$ cm
53. $c = 21.6$ mi **55.** $a = 10.2$ ft **57.** $c = 24.8$ cm

A.6 Pages 680–681

1. $B = 38.0°$, $C = 73.0°$, $c = 25.6$ m **3.** $C = 52.5°$, $A = 66.1°$, $a = 129$ cm
5. $C = 47.6°$, $a = 223$ ft, $c = 196$ ft **7.** $B = 48.5°$, $C = 16.5°$, $c = 1840$ m
9. $A = 67.07°$, $B = 40.35°$, $a = 40.72$ cm **11.** $A = 131.94°$, $a = 89,460$ mi,
$b = 57,870$ mi **13.** $B = 20°$, $C = 135°$, $c = 84$ cm **15.** $A = 6°$, $B = 166°$,
$b = 28$ m **17.** $B = 27.3°$, $C = 115.7°$, $c = 32.2$ cm **19.** $A = 29.6°$,
$B = 123.9°$, $b = 79.4$ km; or $A = 150.4°$, $B = 3.1°$, $b = 5.18$ km **21.** $B = 34.3°$,
$C = 74.2°$, $b = 2.05$ m; or $B = 2.7°$, $C = 105.8°$, $b = 0.171$ m **23.** $A = 44.6°$,
$C = 30.4°$, $c = 8.64$ mi **25.** $A = 35°$, $B = 127°$, $b = 62$ mi; or $A = 145°$, $B = 17°$,
$b = 23$ mi **27.** No triangle **29.** $A = 157°$, $C = 15°$, $a = 1300$ m; or $A = 7°$, $C = 165°$,
$a = 390$ m **31.** $A = 59.34°$, $C = 79.16°$, $c = 21.12$ km; or $A = 120.66°$,
$C = 17.84°$, $c = 6.588$ km **33.** No triangle **35.** $A = 6.06°$, $B = 165.19°$,
$b = 150.1$ m **37.** $B = 47.8°$, $C = 72.2°$, $a = 22.8$ m **39.** $A = 27.0°$,
$B = 44.0°$, $c = 292$ km **41.** $A = 44.6°$, $B = 51.2°$, $C = 84.2°$ **43.** $A = 32.1°$,
$B = 104.6°$, $C = 43.3°$ **45.** $B = 85°$, $C = 50°$, $a = 36$ m **47.** $A = 84°$,
$B = 53°$, $C = 43°$ **49.** $A = 19°$, $B = 26°$, $c = 78$ ft **51.** $A = 148.80°$,
$C = 11.95°$, $b = 3064$ m **53.** $A = 40.11°$, $B = 31.14°$, $c = 595.2$ mm
55. $A = 113.43°$, $B = 27.84°$, $C = 38.73°$

Appendix B

B.2 Page 685

1. 20,000 m **3.** 2280 s **5.** 2.00 h **7.** 60.0 mi/h
9. 288 km/h **11.** 45.0 ft^2 **13.** 2.00 yd^3 **15.** 3000 cm **17.** 0.830 m/s
19. 95.3 m/s **21.** 8.46×10^5 km/h^2 **23.** 79.4 ft/s^2

Appendix D

D.4 Pages 701–702

1. 3.32×10^{19} **3.** -6.83×10^{-6} **5.** 8.35×10^{-11} **7.** -7.98×10^{19}
9. -68.5 **11.** 4680 **13.** 216 **15.** 2.10×10^{10} **17.** 2.15×10^9
19. 10.7 **21.** 33.4 **23.** 591 **25.** 609,000 **27.** 0.567
29. 1.77×10^4 **31.** 0.523 **33.** 0.225 **35.** 1.29 **37.** 0.510
39. 0.743 **41.** 3.73 **43.** 0.151 **45.** 40.7° **47.** 24.6° **49.** 56.2°
51. 8.4° **53.** 63.3° **55.** 66.6° **57.** 252 m **59.** 89.4 m **61.** 53.7°
63. 14.8 cm **65.** 21.9 m **67.** 20,700 **69.** 0.0206 **71.** 1.86×10^{-6}

INDEX

FORMULAS FROM GEOMETRY

Plane Figures

In the following, a, b, c, d, and h are lengths of sides and altitudes, respectively.

		Perimeter	*Area*
Rectangle		$P = 2(a + b)$	$A = ab$
Square	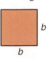	$P = 4b$	$A = b^2$
Parallelogram	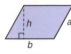	$P = 2(a + b)$	$A = bh$
Rhombus		$P = 4b$	$A = bh$
Trapezoid	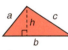	$P = a + b + c + d$	$A = \left(\dfrac{a + b}{2}\right)h$
Triangle	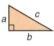	$P = a + b + c$	$A = \dfrac{1}{2}bh$
Right triangle		$c^2 = a^2 + b^2$ or $c = \sqrt{a^2 + b^2}$	

		Circumference	*Area*
Circle		$C = \pi d$	$A = \pi r^2 \quad d = 2r$
		$C = 2\pi r$	$A = \dfrac{\pi d^2}{4}$

Geometric Solids

In the following, B, r, and h are the area of base, length of radius, and height, respectively.

		Volume	*Lateral Surface Area*
Prism		$V = Bh$	
Cylinder		$V = \pi r^2 h$ $V = \dfrac{\pi d^2 h}{4}$	$A = 2\pi rh$
Pyramid		$V = \dfrac{1}{3}Bh$	
Cone		$V = \dfrac{1}{3}\pi r^2 h$	$A = \pi rs$, s is the slant height.
Sphere		$V = \dfrac{4}{3}\pi r^3$ $V = \dfrac{\pi}{6}d^3$	$A = 4\pi r^2$

RED RIVER COLLEGE
OF APPLIED ARTS, SCIENCE AND TECHNOLOGY

School of Hospitality And Culinary Arts

Recipe Name: Ranch Dressing

Recipe No.:

Yield & portion size: 500mL

Prep time:

Ingredients	Quantity	Method
Mayonnaise	180ml	Whisk all ingredients together
Sour Cream	180ml	
Buttermilk	60ml	
White Wine Vinegar	15ml	
Lemon Juice	10ml	
Worcestershire	10ml	
Chopped Parsley	15ml	
Fresh Dill	30ml	
Garlic, minced	10ml	
Chopped Green onion	1	
Mustard, prepared	5ml	
Celery seed	5ml	
Salt	5ml	
Black Pepper	2.5ml	

Other Instructions

00004332PR-13   001 Princess St PrismPOS
002 RRC - PRINCESS ST CAMPUS STORE PR REG 2
D-ANNEK   09/07/2022 12:03

Qty Description   Tax Sale Price
    SKU   Cat # / ISBN

          -1 10001 - Buyback Pay Out PR
                         N N ($84.25)

Sub Total: ($84.25)

Total: ($84.25)

                               Tenders:
    DEBIT CARD   ($84.25)

Change Due: $0.00
                     Original Rcpt Number:
                     _____

                       Customer Name:
                     _____

             RRC ID #:_____

                       Customer Phone:
                     _____

                     Reason for Return:

                     Buy Back
